Higher Mathematics for Physics and Engineering

Hiroyuki Shima · Tsuneyoshi Nakayama

Higher Mathematics for Physics and Engineering

Springer

Dr. Hiroyuki Shima, Assistant Professor
Department of Applied Physics
Hokkaido University
Sapporo 060-8628, Japan
shima@eng.hokudai.ac.jp

Dr. Tsuneyoshi Nakayama, Professor
Toyota Physical and Chemical Research Institute
Aichi 480-1192, Japan
Riken-nakayama@mosk.tytlabs.co.jp

ISBN 978-3-540-87863-6 e-ISBN 978-3-540-87864-3
DOI 10.1007/b138494
Springer Heidelberg Dordrecht London New York

Library of Congress Control Number: 2009940406

Cover design: eStudio Calamar Steinen

Printed on acid-free paper

Springer is part of Springer Science+Business Media (www.springer.com)

To our friends and colleagues

Preface

Owing to the rapid advances in the physical sciences and engineering, the demand for higher-level mathematics is increasing yearly. This book is designed for advanced undergraduates and graduate students who are interested in the mathematical aspects of their own fields of study. The reader is assumed to have a knowledge of undergraduate-level calculus and linear algebra.

There are any number of books available on mathematics for physics and engineering but they all fall into one of two categories: the one emphasizes mathematical rigor and the exposition of definitions or theorems, whereas the other is concerned primarily with applying mathematics to practical problems. We believe that neither of these approaches alone is particularly helpful to physicists and engineers who want to understand the mathematical background of the subjects with which they are concerned. This book is different in that it provides a short path to higher mathematics via a combination of these approaches. A sizable portion of this book is devoted to theorems and definitions with their proofs, and we are convinced that the study of these proofs, which range from trivial to difficult, is useful for a grasp of the general idea of mathematical logic. Moreover, several problems have been included at the end of each section, and complete solutions for all of them are presented in the greatest possible detail. We firmly believe that ours is a better pedagogical approach than that found in typical textbooks, where there are many well-polished problems but no solutions.

This book is essentially self-contained and assumes only standard undergraduate preparation such as elementary calculus and linear algebra. The first half of the book covers the following three topics: real analysis, functional analysis, and complex analysis, along with the preliminaries and four appendixes. Part I focuses on sequences and series of real numbers of real functions, with detailed explanations of their convergence properties. We also emphasize the concepts of Cauchy sequences and the Cauchy criterion that determine the convergence of infinite real sequences. Part II deals with the theory of the Hilbert space, which is the most important class of infinite vector spaces. The completeness property of Hilbert spaces allows one to develop various types of complex orthonormal polynomials, as described in the middle of Part II. An introduction to the Lebesgue integration theory, a subject of ever-increasing importance in physics, is also presented. Part III describes the theory of complex-valued functions of one complex variable. All relevant elements including analytic functions, singularity, residue, continuation, and conformal mapping are described in a self-contained manner. A thorough understanding of the fundamentals treated is important in order to proceed to more advanced branches of mathematical physics.

In the second half of the volume, the following three specific topics are discussed: Fourier analysis, differential equations, and tensor analysis. These three are the most important subjects in both engineering and the physical sciences, but their rigorous mathematical structures have hardly been covered in ordinary textbooks. We know that mathematical rigor is often unnecessary for practical use. However, the blind usage of mathematical methods as a tool may lead to a lack of understanding of the symbiotic relationship between mathematics and the physical sciences. We believe that readers who study the mathematical structures underlying these three subjects in detail will acquire a better understanding of the theoretical backgrounds associated with their own fields. Part IV describes the theory of Fourier series, the Fourier transform, and the Laplace transform, with a special emphasis on the proofs of their convergence properties. A more contemporary subject, the wavelet transform, is also described toward the end of Part IV. Part V deals with ordinary and partial differential equations. The existence theorem and stability theory for solutions, which serve as the underlying basis for differential equations, are described with rigorous proofs. Part VI is devoted to the calculus of tensors in terms of both Cartesian and non-Cartesian coordinates, along with the essentials of differential geometry. An alternative tensor theory expressed in terms of abstract vector spaces is developed toward the end of Part VI.

The authors hope and trust that this book will serve as an introductory guide for the mathematical aspects of the important topics in the physical sciences and engineering.

Sapporo,
November 2009

Hiroyuki Shima
Tsuneyoshi Nakayama

Contents

Part IV Fourier Analysis

1

Preliminaries

This chapter provides the basic notation, terminology, and abbreviations that we will be using, particularly in real analysis, and is designed to serve a reference rather than as a systematic exposition.

1.1 Basic Notions of a Set

1.1.1 Set and Element

A **set** is a collection of **elements** (or **points**) that are definite and separate objects. If a is an element of a set S, we write

$$a \in S.$$

Otherwise, we write

$$a \notin S$$

to indicate that a does not belong to S. If a set contains no elements, it is called an **empty set** and is designated by $\emptyset$.

A set may be defined by listing its elements or by providing a rule that determines which elements belong to it. For example, we write

$$X = \{x_1, x_2, x_3, \cdots, x_n\}$$

to indicate that X is a set with n elements: $x_1, x_2, \cdots x_n$. When a set contains a finite (infinite) number of elements, it is called a **finite** (**infinite set**).

A set X is said to be a **subset** of Y if every element in X is also an element in Y. This relationship is expressed as

$$X \subseteq Y.$$

When $X \subseteq Y$ and $Y \subseteq X$, the two sets have the same elements and are said to be **equal**, which is expressed by

$$X = Y.$$

But when $X \subseteq Y$ and $X \neq Y$, then X is called a **proper subset** of Y, and we use the more specific expression

$$X \subset Y.$$

The **intersection** of two sets X and Y, denoted by

$$X \cap Y,$$

consists of elements that are contained in both X and Y. The **union**

$$X \cup Y$$

consists of all the elements contained in either X or Y, including those contained in both X and Y. When the two sets X and Y have no element in common (i.e., when $X \cap Y = \emptyset$), X and Y are said to be **disjoint**.

For two sets A and B, we define their **difference** by the set

$$\{x : \ x \in A, \ x \notin B\}$$

and denote it by $A \backslash B$ (see Fig. 1.1). In particular, if A contains all the sets under discussion, we say that A is the **universal set** and $A \backslash B$ is called the **complementary set** or **complement** of A.

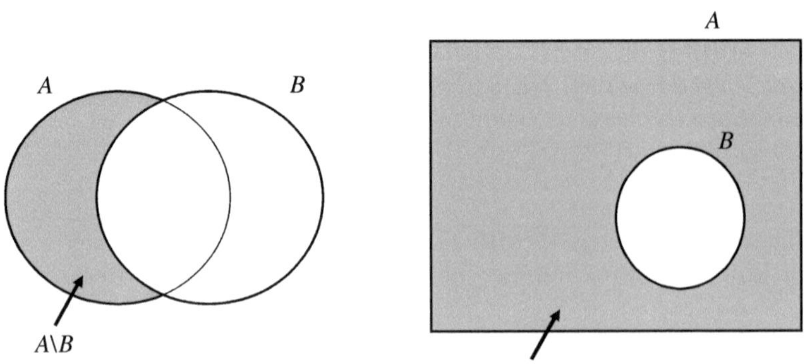

Fig. 1.1. *Left*: The difference of two sets A and B. *Right*: The complementary set or complement of B in A

1.1.2 Number Sets

Our abbreviations for fundamental number systems are given by

N : The set of all positive integers not including zero.
Z : The set of all integers.
Q : The set of all rational numbers.
R : The set of all real numbers.
C : The set of all complex numbers.

The symbol $\mathbf{R}^n$ denotes an n-dimensional **Euclidean space** (see Sects. 4.1.3 and 19.2.3). Points in $\mathbf{R}^n$ are denoted by bold face, say, $\boldsymbol{x}$; the coordinates of $\boldsymbol{x}$ are denoted by the ordered n-tuple $(x_1, x_2, \cdots, x_n)$, where $x_i \in \mathbf{R}$. We also use the **extended real number** defined by

$$\overline{\boldsymbol{R}} = \boldsymbol{R} \cup \{-\infty, \infty\}.$$

1.1.3 Bounds

The precise terminology for bounds of real number sets follow. Meanwhile we assume S to be a set of real numbers.

♠ **Bounds of a set:**

1. A real number b such that $x \leq b$ for all $x \in S$ is called an **upper bound** of S.

2. A real number a such that $x \geq a$ for all $x \in S$ is called a **lower bound** of S.

Figure 1.2 illustrates the point. We say that a set S is **bounded above** or **bounded below** if it has an upper bound or a lower bound, respectively. In particular, when a set S is bounded above and below simultaneously, it is a **bounded set**. If a set S is not bounded, then it is said to be an **unbounded set**.

It follows from these definitions that if b is an upper bound of S, any number greater than b will also be an upper bound of S. Thus it makes sense to seek the *smallest* among such upper bounds. This is also the case for a lower bound of S if it is bounded below. In fact, the two extrema bounds, the smallest and the largest, are referred to by specific names as follows:

♠ **Least upper bound:**

An element $b \in \boldsymbol{R}$ is called the **least upper bound** (abbreviated by **l.u.b.**) or **supremum**, of S if

(i) b is an upper bound of S, and
(ii) there is no upper bound of S that is smaller than b.

♠ **Greatest lower bound:**

An element $a \in \boldsymbol{R}$ is called the **greatest lower bound** (abbreviated by **g.l.b.**) or **infimum**, if
(i) a is a lower bound of S, and
(ii) there is no lower bound of S that is greater than a.

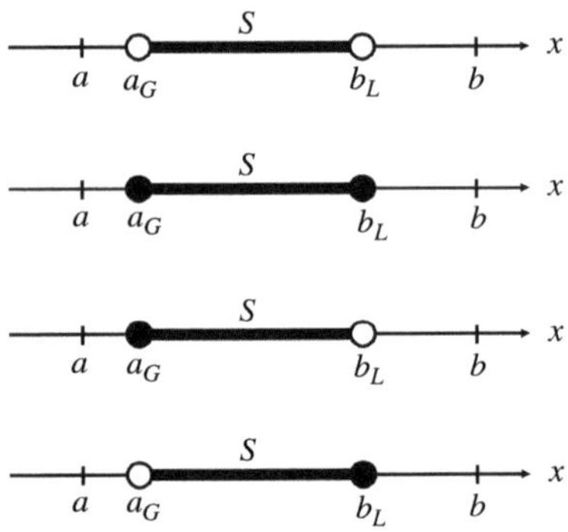

Fig. 1.2. In all the figures, the points a and b are lower and upper bounds of S, respectively. In particular, the point a_G is the greatest lower bound, and the b_L is the least upper bound

In symbols, the supremum and infimum of S are denote, respectively, by

$$\sup S \quad \text{and} \quad \inf S.$$

We must emphasize the fact that the supremum and infimum of the set S may or may not belong to S. For instance, the set $S = \{x : x < 1\}$ has the supremum 1, which it does not belong to S. Nevertheless, particularly when S is finite, we have

$$\sup S = \max S \quad \text{and} \quad \inf S = \min S,$$

where $\max S$ and $\min S$ denote the **maximum** and **minimum** of S, respectively, both of which belong to S.

1.1.4 Interval

When a set of real numbers is bounded above or below (or both), it is referred to as an **interval**; there are several classes of intervals as listed below.

♠ **Intervals:** Given a real variable x, the set of all values of x such that

1. $a \leq x \leq b$ is a **closed interval**, denoted by $[a, b]$.
2. $a < x < b$ is a **bounded open interval**, denoted by (a, b).
3. $a < x$ and $x < b$ are **unbounded open intervals**, denoted by (a, ∞) and $(-\infty, b)$, respectively.

Sets of points $\{x\}$ such that

$$a \leq x < b, \; a < x \leq b, \; a \leq x, \; x \leq b$$

may be referred to as **semiclosed intervals**; see Sect. 1.1.5 for more rigorous definitions. Every interval I_1 contained in another interval I_2 is a **subinterval** of I_2.

1.1.5 Neighborhood and Contact Point

The following is a preliminary definition that will be significant in the discussions on continuity and convergence properties of sets and functions.

♠ **Neighborhoods:**

Let $x \in \boldsymbol{R}$. A set $V \subseteq \boldsymbol{R}$ is called a **neighborhood** of x if there is a number $\varepsilon > 0$ such that

$$(x - \varepsilon, x + \varepsilon) \subseteq V.$$

In line with the idea of neighborhoods, we introduce the following important concept (see Fig. 1.3):

♠ **Contact points:**

Assume a point $x \in \boldsymbol{R}$ and a set $S \subseteq \boldsymbol{R}$. Then x is called a **contact point** of S if and only if *every* neighborhood of x contains at least one point of S.

Remark. A contact point of S may or may not belong to S. In contrast, a point $x \in S$ is necessarily a contact point of S.

Obviously, every point of S is a contact point of S. In particular, when S is a single-element set given by $S = \{x_0\}$ with $x_0 \in \boldsymbol{R}$, then x_0 is a contact point of S since every neighborhood of x_0 contains x_0 itself. The collection of all contact points of a set S is called the **closure** of S and is denoted by $[S]$.

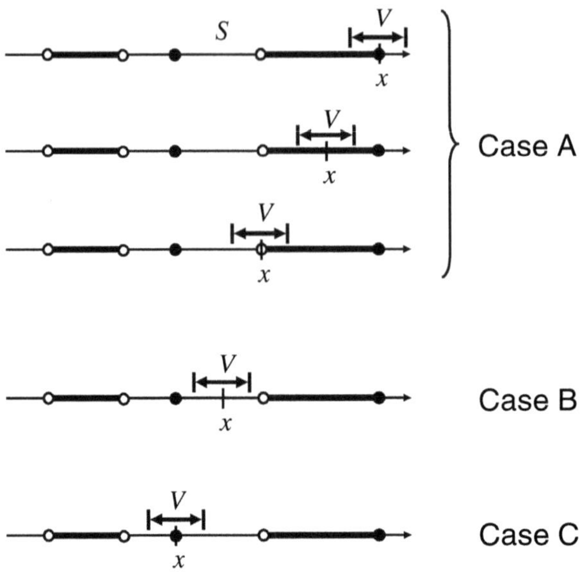

Fig. 1.3. Case A: x is a limit point (and thus a contact point) of S. Case B: x is *not* a contact point of S. Case C: x is an isolated point (and thus a contact point) of S

Contact points can be classified as follows (see again Fig. 1.3):

♠ **Limit points:**
A contact point $x \in \boldsymbol{R}$ is called a **limit point** of the set $S \subseteq \boldsymbol{R}$ if and only if every neighborhood V of x contains a point of S *different from* x.

♠ **Isolated points:**
A contact point x is called an **isolated point** of S if and only if x has a neighborhood V in which x is the only point belonging to S.

In plain words, a limit point x is a point such that every interval $(x-\varepsilon,\ x+\varepsilon)$ contains an infinite number of points, regardless of the smallness of ε. A limit point may be referred to as a **cluster point** or **accumulation point**, depending on the context. The symbol $\hat{S}$ is commonly used to denote the set of limit points of S.

Examples **1.** If S is the set of rational numbers in the interval $[0,1]$, then every point of $[0,1]$, rational or not, is a limit point of S.

2. The integer set $\boldsymbol{Z}$ has no limit point; it has an infinite number of isolated points.

3. The origin is the limit point of the set $\{1/m : m \in \boldsymbol{N}\}$.

Remark. From the definition, a limit point of a set need not belong to the set. For instance, $x = 1$ is the limit point of the set $S = \{x : x \in \boldsymbol{R},\ x > 1\}$, but it does not belong to S. In contrast, an isolated point of S must lie in S.

Limit points are further divided into two classes. A limit point x of a set S is called an **interior point** of S if and only if x has a neighborhood $V \subseteq S$. Otherwise, it is called a **boundary point** of S. Figure 1.4 is a schematic illustration of the difference between interior and boundary points.

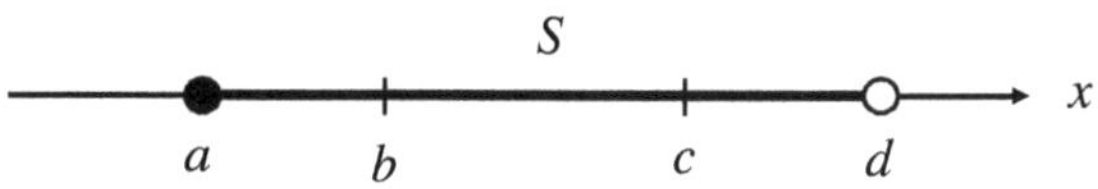

Fig. 1.4. All four points are limit points of S. Among them, b and c are interior points, whereas a and d are boundary points

1.1.6 Closed and Open Sets

Closed and open sets are defined in terms of the concepts of contact points and closure. Recall that a closure of S, denoted by $[S]$, is a set of all contact points of S, which is a union of the two sets: all limit and all isolated points of S.

♠ **Closed sets:**
A set $S \subseteq \boldsymbol{R}$ is **closed** if $[S] = S$, i.e., if S coincides with its own closure.

♠ **Open sets:**
A set $S \subseteq \boldsymbol{R}$ is **open** if S consists entirely of its interior points and has no boundary points.

It follows intuitively that a set $S \subseteq \boldsymbol{R}$ is open if and only if its **complementary set** is closed. The proof is given in Exercise **4** in this chapter. Note that the condition $[S] \neq S$ is inconclusive as to whether S is open or not.

Examples **1.** Every single-element set $S = \{x_0\}$ with $x_0 \in \boldsymbol{R}$ is closed since $[S] = S$.

2. Every set consisting of a finite number of points is closed.
3. For any real number x, the set $\boldsymbol{R}\backslash\{x\}$ is open since $\{x\}$ is closed.
4. The intervals $[a, b]$, $[a, \infty)$, and $(-\infty, b]$ are all closed, which is proven by considering their closures.
5. The interval $[a, b)$ is neither closed nor open. In fact, it is not closed since it excludes its boundary point b and it is not open since it contains its boundary point a.

Exercises

1. Give the supremum and infimum of each of the following sets:
(1) $S = \{x : 0 \le x \le 5\}$.
(2) $S = \{x : \mathrm{x} \in \boldsymbol{Q} \text{ and } \mathrm{x}^2 < 2\}$.
(3) $S = \{x : x = 3 + \frac{1}{n^2}, n \in \mathbf{N}\}$.

Solution: (1) $\sup S = 5$, $\inf S = 0$. (2) $\sup S = \sqrt{2}$, $\inf S = 0$. (3) $\sup S = 4$, $\inf S = 3$. ♣

2. Suppose S to be any of the intervals: $(a, b), [a, b), (a, b]$, or $[a, b]$. Show that

$$\sup S = b, \quad \inf S = a.$$

Solution: Take $S = (a, b)$. Since $x \le b$ for all $x \in S$, b serves as one of upper bounds of S. We show that b is surely the *least* upper bound. To see this, we first assume that u is another upper bound of S such that $u < b$; then $a < u < (u + b)/2 < b$. This implies that

$$\frac{u+b}{2} \in S \quad \text{and} \quad u < \frac{u+b}{2},$$

which contradicts the assumption that u is an upper bound of S. Hence, $u \ge b$; i.e., any upper bound other than b must be larger than b. We thus conclude that $b = \sup S$. The proof is similar for the other three cases. ♣

3. Show that the set of integers has no limit point, i.e., $\hat{\boldsymbol{Z}} = \emptyset$.

Solution: Take any $x \in \boldsymbol{Z}$, and let $\varepsilon = \min\{|n - x| : n \in \boldsymbol{Z}\}$. The interval $(x-\varepsilon,\ x+\varepsilon)$ contains no integers other than x; hence, $x \notin \hat{\boldsymbol{Z}}$. Since this is the case for any $x \in \boldsymbol{Z}$, we conclude that $\boldsymbol{Z}$ is totally composed of isolated points. ♣

4. Show that a set $S \subseteq \boldsymbol{R}$ is open if and only if its complementary set $\boldsymbol{R}\backslash S$ is closed.

Solution: If S is open, then every point $x \in S$ has a neighborhood contained in S. Therefore no point $x \in S$ can be a contact point of $\boldsymbol{R}\backslash S$. In other words, if x is a contact point of $\boldsymbol{R}\backslash S$, then $x \in \boldsymbol{R}\backslash S$, i.e., $\boldsymbol{R}\backslash S$ is closed.

Conversely, if $\boldsymbol{R}\backslash S$ is closed, then any point $x \in S$ must have a neighborhood contained in S, since otherwise every neighborhood of x would contain points of $\boldsymbol{R}\backslash S$, i.e., x would be a contact point of $\boldsymbol{R}\backslash S$ not in $\boldsymbol{R}\backslash S$. Therefore S is open. ♣

1.2 Conditional Statements

Phrases such as *if... then...*, and ... *if and only if* ... are frequently used to connect **simple statements** that can be described as either **true** or **false**. For the sake of typographical convenience, there are conventional logical symbols for representing such phrases.

Suppose P and Q are two different statements. The compound statements

$$\text{if } P \text{ then } Q$$

and

$$P \text{ implies } Q$$

mean that if P is true then Q is true. This is written symbolically as

$$P \Rightarrow Q. \tag{1.1}$$

We say that

$$P \text{ is a } \textbf{sufficient condition} \text{ for } Q$$

or

$$Q \text{ is a } \textbf{necessary condition} \text{ for } P.$$

In the above context, P stands for the hypothesis or assumption, and Q is the conclusion.

Remark. To prove the implication (1.1) in actual problems, it suffices to exclude the possibility that P is true and Q is false. This may be done in one of three ways.

1. Assume that P is true and prove that Q is true (**direct proof**).
2. Assume that Q is false and prove that P is false (**contrapositive proof**).
3. Assume that P is true and Q is false, and then prove that this leads to a contradiction (**proof by contradiction**).

When P implies Q and Q implies P, we abbreviate this to

$$P \iff Q,$$

and we say that

$$P \text{ is } \textbf{equivalent} \text{ to } Q$$

or, more commonly,

$$P \textbf{ if and only if } Q.$$

This also means that P is a **necessary and sufficient condition** for Q.

Examples Observe that

$$x = 1 \Rightarrow x^2 = 1 \text{ and } x = -1 \Rightarrow x^2 = 1.$$

Conversely, we see that

$$x^2 = 1 \Rightarrow x = -1 \text{ or } 1.$$

Therefore, we conclude that

$$x^2 = 1 \iff x \in \{-1, 1\}.$$

1.3 Order of Magnitude

1.3.1 Symbols O, o, and $\sim$

We use the notations O, o, and $\sim$ to express orders of magnitude. To explain their use, we consider the behavior of functions $f(x)$ and $g(x)$ in a neighborhood of a point x_0.

1. We write
$$f(x) = O(g(x)), \quad x \to x_0$$
if there exists a positive constant A such that
$$|f(x)| \le A|g(x)|$$
for all values of x in some neighborhood of x_0.

2. We write
$$f(x) = o(g(x)), \quad x \to x_0$$
if
$$\lim_{x\to x_0} \left| \frac{f(x)}{g(x)} \right| = 0.$$

3. We write
$$f(x) \sim g(x), \quad x \to x_0$$
if
$$\lim_{x\to x_0} \frac{f(x)}{g(x)} = 1.$$

In addition to the formal definitions above, we summarize the actual meaning of these symbols:

1. $f(x) = O(g(x))$ means that $f(x)$ does *not* grow *faster* than $g(x)$ as $x \to x_0$.
2. $f(x) = o(g(x))$ means that $f(x)$ grows more *slowly* than $g(x)$ as $x \to x_0$.
3. $f(x) \sim g(x)$ means that $f(x)$ and $g(x)$ grow *at the same rate* as $x \to x_0$.

We occasionally employ the symbols

$$f(x) = O(1) \text{ as } x \to x_0.$$

This simply means that $f(x)$ is bounded on the order of 1. The symbol

$$f(x) = o(1) \text{ as } x \to x_0$$

means that $f(x)$ approaches zero as $x \to x_0$.

Examples The relations **1–3** below hold for $x \to \infty$.

1. $$\frac{1}{1+x^2} = O\left(\frac{1}{x^2}\right), \quad \frac{1}{1+x^2} = o\left(\frac{1}{x}\right), \quad \frac{1}{1+x^2} \sim \frac{1}{x^2}.$$

2. $$\frac{1}{1+x^2} = \frac{1}{x^2} + O\left(\frac{1}{x^4}\right), \quad \frac{1}{1+x^2} = \frac{1}{x^2} + o\left(\frac{1}{x^2}\right).$$

3. $$\sqrt{x^2+1} = x + O\left(\frac{1}{x}\right), \quad \sqrt{x^2+1} = x + o(1), \quad \sqrt{x^2+1} \sim x.$$

The following hold for $x \to 0$:

4. $$\sin x = O(1), \quad \sin x \sim x, \quad \cos x = 1 + O\left(x^2\right).$$

1.3.2 Asymptotic Behavior

Asymptotic behavior of $f(x)$ as $x \to a$ can be quantified by using the powers of $(x-a)$ as comparison functions. As an example, suppose that a function $f(x)$ satisfies the relation

$$f(x) = O\left((x-a)^p\right) \text{ for } x \to a \tag{1.2}$$

for some real number $p = p_0$. Then, the relation (1.2) clearly holds for all p for $p \le p_0$, and it may or may not hold for some p if $p \ge p_0$. Thus we can define the **supremum** of such p's that satisfy (1.2), and denote it by q, i.e.,

$$q = \sup\{p \mid f(x) = O\left((x-a)^p\right)\}. \tag{1.3}$$

In this case, we say that f **vanishes** at $x = a$ to **order** q. The quantity q defined by (1.3) is useful for describing the asymptotic behavior of $f(x)$ in the vicinity of $x = a$.

Remark. Note that (1.3) itself does not imply that

$$f(x) = O\left((x-a)^q\right), \quad x \to a.$$

For instance, the function $f(x) = \log x$ defined within the interval $(0,1)$ yields $q = 0$, since for $x \to 0$,

$$\log x \begin{cases} = O\left(x^p\right) & p < 0, \\ \neq O\left(x^p\right) & p > 0. \end{cases}$$

But it is obvious that $\log x \neq O(1)$.

1.4 Values of Indeterminate Forms

1.4.1 l'Hôpital's Rule

A function $f(x)$ of the form $u(x)/v(x)$ is not defined for $x = a$ if $f(a)$ takes the form $0/0$. Still, if the limit $\lim_{x \to a} f(x)$ exists, then it is often desirable to define $f(a) \equiv \lim_{x \to a} f(x)$. In such a case, the value of the limit can be evaluated by using the following theorem:

♠ **l'Hôpital's rule:**
Let $u(a) = v(a) = 0$. If there exists a neighborhood of $x = a$ such that
(i) $v(x) \neq 0$ except for $x = a$, and
(ii) $u'(x)$ and $v'(x)$ exist and do not vanish simultaneously, then,

$$\lim_{x \to a} \frac{u(x)}{v(x)} = \lim_{x \to a} \frac{u'(x)}{v'(x)}$$

whenever the limit on the right exists.

For the proof of the theorem, see Exercise **3** in Sect. 8.1.

Remark. If $u'(x)/v'(x)$ is itself an indeterminate form, the above method may be applied to $u'(x)/v'(x)$ in turn, so that

$$\lim_{x \to a} \frac{u(x)}{v(x)} = \lim_{x \to a} \frac{u'(x)}{v'(x)} = \lim_{x \to a} \frac{u''(x)}{v''(x)}.$$

If necessary, this process may be continued.

1.4.2 Several Examples

In the following, we show several examples of indeterminate forms other than the form of 0/0 previously discussed. Often functions $f(x)$ of the forms $u(x)v(x)$, $[u(x)]^{v(x)}$, and $u(x) - v(x)$ can be reduced to the form $p(x)/q(x)$ with the aid of the following relations:

$$
\begin{aligned}
u(x)v(x) &= \frac{u(x)}{1/v(x)} = \frac{v(x)}{1/u(x)}, \\
[u(x)]^{v(x)} &= e^{g(x)}, \quad \text{where} \quad g(x) = \frac{\log u(x)}{1/v(x)} = \frac{\log v(x)}{1/u(x)}, \\
u(x) - v(x) &= \frac{\frac{1}{v(x)} - \frac{1}{u(x)}}{\frac{1}{u(x)}\frac{1}{v(x)}} = \log h(x), \quad \text{where} \quad h(x) = \frac{e^{u(x)}}{e^{v(x)}}.
\end{aligned}
$$

After the reduction, the l'Hôpital method given in Sect. 1.4.1 becomes applicable.

Part I

Real Analysis

2

Real Sequences and Series

Abstract In this chapter, we deal with the fundamental properties of sequences and series of real numbers. We place particular emphasis on the concept of "convergence," a thorough understanding of which is important for the study of the various branches of mathematical physics that we are concerned with subsequent chapters.

2.1 Sequences of Real Numbers

2.1.1 Convergence of a Sequence

This section describes the fundamental definitions and ideas associated with **sequences of real numbers** (called **real sequences**). We must emphasize that the sequence

$$(x_n : n \in \boldsymbol{N})$$

is not the same as the set

$$\{x_n : n \in \boldsymbol{N}\}.$$

In fact, the former is the *ordered* list of x_n, some of which may be repeated, whereas the latter is merely the defining *range* of x_n. For instance, the constant sequence $x_n = 1$ is denoted by $(1, 1, 1, \cdots)$, whereas the set $\{1\}$ contains only one element.

We start with a precise definition of the convergence of a real sequence, which is an initial and crucial step for various branches of mathematics.

♠ **Convergence of a real sequence:**

A real sequence (x_n) is said to be **convergent** if there exists a real number x with the following property: For every $\varepsilon > 0$, there is an integer N such that

$$n \geq N \;\Rightarrow\; |x_n - x| < \varepsilon. \tag{2.1}$$

We must emphasize that the magnitude of ε is arbitrary. No matter how small an ε we choose, it must always be possible to find a number N that will increase as ε decreases.

Remark. In the language of **neighborhoods**, the above definition is stated as follows: *The sequence* (x_n) *converges to* x *if every neighborhood of* x *contains all but a finite number of elements of the sequence.*

When (x_n) is convergent, the number x specified in this definition is called a **limit** of the sequence (x_n), and we say that x_n converges to x. This is expressed symbolically by writing

$$\lim_{n\to\infty} x_n = x,$$

or simply by

$$x_n \to x.$$

If (x_n) is not convergent, it is called **divergent**.

Remark. The limit x may or may not belong to (x_n); this situation is similar to the case of the **limit point** of a set of real numbers discussed in Sect. 1.1.5.

An example in which $x = \lim x_n$ but $x \neq x_n$ for any n is given below.

Examples Suppose that a sequence (x_n) consisting of rational numbers is defined by

$$(x_n) = (3.1,\ 3.14,\ 3.142,\ \cdots,\ x_n,\ \cdots),$$

where $x_n \in \boldsymbol{Q}$ is a rational number to n decimal places close to π. Since the difference $|x_n - \pi|$ is less than 10^{-n}, it is possible to find an N for any $\varepsilon > 0$ such that

$$n \geq N \quad \Rightarrow \quad |x_n - \pi| < \varepsilon.$$

This means that

$$\lim_{n\to\infty} x_n = \pi.$$

However, as the limit, π, is an irrational number it is not in $\boldsymbol{Q}$.

Remark. The above example indicates that only a restricted class of convergent sequences has a limit in the same sequence.

2.1.2 Bounded Sequences

In the remainder of this section, we present several fundamental concepts associated with real sequences. We start with the boundedness properties of sequences.

♠ **Bounded sequences:**
A real sequence (x_n) is said to be **bounded** if there is a positive number M such that

$$|x_n| \leq M \text{ for all } n \in \boldsymbol{N}.$$

The following is an important relation between convergence and boundedness of a real sequence:

♠ **Theorem:**
If a sequence is convergent, then it is bounded.

Proof Suppose that $x_n \to x$. If we choose $\varepsilon = 1$ in (2.1), there exists an integer N such that

$$|x_n - x| < 1 \text{ for all } n \geq N.$$

Since $|x_n| - |x| \leq |x_n - x|$, it follows that

$$|x_n| < 1 + |x| \text{ for all } n \geq N.$$

Setting $M = \max\{|x_1|, |x_2|, \cdots, |x_{N-1}|, 1 + |x|\}$ yields

$$|x_n| < M \text{ for all } n \in \boldsymbol{N},$$

which means that (x_n) is bounded. ♣

Remark. Observe that the converse of the theorem is false. In fact, the sequence

$$(\,1, -1, 1, -1, \cdots, (-1)^n, \cdots\,)$$

is divergent, although it is bounded.

2.1.3 Monotonic Sequences

Another important concept in connection with real sequences is monotonicity, defined as follows:

♠ **Monotonic sequences:**
A sequence (x_n) is said to be

1. **increasing** (or monotonically increasing) if $x_{n+1} \geq x_n$ for all $n \in \boldsymbol{N}$,
2. **strictly increasing** if $x_{n+1} > x_n$ for all $n \in \boldsymbol{N}$,

3. **decreasing** (or monotonically decreasing) if $x_{n+1} \leq x_n$ for all $n \in \boldsymbol{N}$, and

4. **strictly decreasing** if $x_{n+1} < x_n$ for all $n \in \boldsymbol{N}$.

These four kinds of sequences are collectively known as **monotonic sequences**. Note that a sequence (x_n) is increasing if and only if $(-x_n)$ is decreasing. Thus, the properties of monotonic sequences can be fully investigated by restricting ourselves solely to increasing (or decreasing) sequences.

Once a sequence assumes monotonic properties, its convergence is determined only by its boundedness, as stated below.

♠ **Theorem:**

A monotonic sequence is convergent if and only if it is bounded. More specifically,

(i) If (x_n) is increasing and bounded above, then its limit is given by

$$\lim_{n\to\infty} x_n = \sup x_n.$$

(ii) If (x_n) is decreasing and bounded below, then

$$\lim_{n\to\infty} x_n = \inf x_n.$$

Proof If (x_n) is convergent, then it must be bounded as proven earlier (see Sect. 2.1.2). Now we consider the converses for cases (i) and (ii).

(i) Assume (x_n) is increasing and bounded. The set $S = \{x_n\}$ will then have the supremum denoted by $\sup S = x$. By the definition of the supremum, for arbitrary small $\varepsilon > 0$ there is an $x_N \in S$ such that

$$x_N > x - \varepsilon. \tag{2.2}$$

Since x_n is increasing, we obtain

$$x_n \geq x_N \quad \text{for all} \quad n \geq N. \tag{2.3}$$

Moreover, since x is the supremum of S, we have

$$x \geq x_n \quad \text{for all} \quad n \in \boldsymbol{N}. \tag{2.4}$$

From (2.2), (2.3), and (2.4), we arrive at

$$|x_n - x| = x - x_n \leq x - x_N < \varepsilon \quad \text{for all} \quad n \geq N,$$

which gives us the desired conclusion, i.e,

$$\lim_{n\to\infty} x_n = x = \sup S.$$

(ii) If (x_n) is decreasing and bounded, then $(-x_n)$ is increasing and bounded. Hence, from (i), we have

$$\lim_{n\to\infty} (-x_n) = \sup(-S).$$

Since $\sup(-S) = -\inf S$, it follows that

$$\lim_{n\to\infty} x_n = \inf S. \quad \clubsuit$$

2.1.4 Limit Superior and Limit Inferior

We close this section by introducing two specific limits an any bounded sequence. Let (x_n) be a bounded sequence and define two sequences (y_n) and (z_n) as follows:

$$\begin{aligned} y_n &= \sup\{x_k : \ k \geq n\}, \\ z_n &= \inf\{x_k : \ k \geq n\}. \end{aligned} \tag{2.5}$$

Note that y_n and z_n differ, respectively, from $\sup\{x_n\}$ and $\inf\{x_n\}$. It follows from (2.5) that

$$y_1 = \sup\{x_k : \ k \geq 1\} \geq y_2 = \sup\{x_k : \ k \geq 2\} \geq y_3 \cdots,$$

which means that the sequence (y_n) is monotonically decreasing and bounded below by $\inf x_n$. Thus in view of the theorem in Sect. 2.1.3, the sequence (y_n) must be convergent. The limit of (y_n) is called the **limit superior** or the **upper limit** of (x_n) and is denoted by

$$\limsup_{n\to\infty} x_n \quad (\text{or } \overline{\lim}\, x_n).$$

Likewise, since (z_n) is increasing and bounded above by $\sup x_n$, it possesses the limit known as the **limit inferior** or **lower limit** of x_n denoted by

$$\liminf_{n\to\infty} x_n \quad (\text{or } \underline{\lim}\, x_n).$$

In terms of the two specific limits, we can say that a bounded sequence (x_n) converges if and only if

$$\lim_{n\to\infty} x_n = \limsup_{n\to\infty} x_n = \liminf_{n\to\infty} x_n.$$

(A proof will be given in Exercise **4** in Sect. 2.1.4.) Note that by definition, it readily follows that

$$\begin{aligned} \limsup_{n\to\infty} x_n &\geq \liminf_{n\to\infty} x_n, \\ \limsup_{n\to\infty}(-x_n) &= -\liminf_{n\to\infty} x_n. \end{aligned}$$

Examples **1.** $x_n = (-1)^n \;\Rightarrow\; \limsup_{n\to\infty} x_n = 1,\; \liminf_{n\to\infty} x_n = -1.$

2. $x_n = (-1)^n + \dfrac{1}{n} \;\Rightarrow\; \limsup_{n\to\infty} x_n = 1,\; \liminf_{n\to\infty} x_n = -1.$

3. $x_{2n} = 1 + \dfrac{(-1)^n}{n},\; x_{2n-1} = \dfrac{(-1)^n}{n},\; \Rightarrow\; \limsup_{n\to\infty} x_n = 1,\; \liminf_{n\to\infty} x_n = 0.$

4. $(x_n) = (2, 0, -2, 2, 0, -2, \cdots) \;\Rightarrow\; \limsup_{n\to\infty} x_n = 2,\; \liminf_{n\to\infty} x_n = -2.$

The four cases noted above are illustrated schematically in Fig. 2.1. All the sequences (x_n) are not convergent and thus the limit $\lim_{n\to\infty} x_n$ does not exist. This fact clarifies the crucial difference between $\lim_{n\to\infty} x_n$ and $\limsup_{n\to\infty} x_n$ (or $\liminf_{n\to\infty} x_n$).

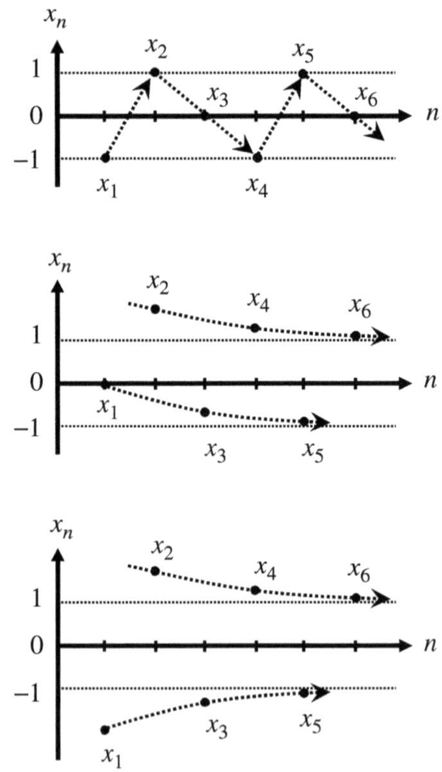

Fig. 2.1. All the sequences of $\{x_n\}$ in the figures do not converge, but they all possess $\limsup_{n\to\infty} x_n = 1$ and $\liminf_{n\to\infty} x_n = -1$

The limit superior of x_n has the following features and similar features are found for the limit inferior.

♠ **Theorem:**

1. For any small $\varepsilon > 0$, we can find an N such that
$$n > N \;\Rightarrow\; x_n < \limsup_{n\to\infty} x_n + \varepsilon.$$
2. For any small $\varepsilon > 0$, there are an infinite number of terms of x_n such that
$$\limsup_{n\to\infty} x_n - \varepsilon < x_n.$$

Proof **1.** Recall that $\limsup_{n\to\infty} x_n = \lim_{n\to\infty} y_n$, where y_n is defined in (2.5). For any $\varepsilon > 0$, there is an integer N such that
$$n > N \;\Rightarrow\; \limsup_{n\to\infty} x_n - \varepsilon < y_n < \limsup_{n\to\infty} x_n + \varepsilon.$$
Since $y_n \geq x_n$ for all n, we have
$$n > N \;\Rightarrow\; x_n < \limsup_{n\to\infty} x_n + \varepsilon. \quad ♣$$

2. Suppose that there is an integer m such that
$$n > m \;\Rightarrow\; \limsup_{n\to\infty} x_n - \varepsilon \geq x_n.$$
Then for all $k \geq n > m$, we have
$$x_k \leq \limsup_{n\to\infty} x_n - \varepsilon,$$
which means that
$$y_n \leq \limsup_{n\to\infty} x_n - \varepsilon \text{ for all } n > m.$$
In the limit of $n \to \infty$, we find a contradiction such that
$$\limsup_{n\to\infty} x_n \leq \limsup_{n\to\infty} x_n - \varepsilon.$$
This completes the proof. ♣

Exercises

1. Prove that if the sequence (x_n) is convergent, then its limit is unique.

Solution: Let $x = \lim x_n$ and $y = \lim x_n$ with the assumption $x \neq y$. Then we can find a neighborhood V_1 of x and a neighborhood V_2 of y such that $V_1 \cap V_2 = \emptyset$. For example, take $V_1 = (x - \varepsilon, x + \varepsilon)$ and $V_2 = (y - \varepsilon, y + \varepsilon)$, where $\varepsilon = |x - y|/2$. Since $x_n \to x$, all but a finite number of terms of the sequence lie in V_1. Similarly, since $y_n \to y$, all but a finite number of its terms also lie in V_2. However, these results contradict the fact that $V_1 \cap V_2 = \emptyset$, which means that the limit of a sequence should be unique. ♣

2. If $x_n \to x \neq 0$, then there is a positive number A and an integer N such that $n > N \Rightarrow |x_n| > A$. Prove it.

Solution: Let $\varepsilon = |x|/2$, which is a positive number. Hence, there is an integer N such that $n > N \Rightarrow |x_n - x| < \varepsilon \Rightarrow ||x_n| - |x|| < \varepsilon$. Consequently, $|x| - \varepsilon < |x_n| < |x| + \varepsilon$ for all $n \geq N$. From the left-hand inequality, we see that $|x_n| > |x|/2$, and we can take $M = |x|/2$ to complete the proof. ♣

3. Prove that the sequence $x_n = [1 + (1/n)]^n$ is convergent.

Solution: The proof is completed by observing that the sequence is monotonically increasing and bounded. To see this, we use the **binomial theorem**, which gives

$$\begin{aligned} x_n &= \sum_{k=0}^{n} {}_nC_{n-k} \frac{1}{n^k} \\ &= 1 + 1 + \frac{1}{2!}\left(1 - \frac{1}{n}\right) + \frac{1}{3!}\left(1 - \frac{1}{n}\right)\left(1 - \frac{2}{n}\right) + \cdots \\ &\quad + \frac{1}{n!}\left(1 - \frac{1}{n}\right)\left(1 - \frac{2}{n}\right)\cdots\left(1 - \frac{n-1}{n}\right). \end{aligned}$$

Likewise we have

$$\begin{aligned} x_{n+1} &= 1 + 1 + \frac{1}{2!}\left(1 - \frac{1}{n+1}\right) + \frac{1}{3!}\left(1 - \frac{1}{n+1}\right)\left(1 - \frac{2}{n+1}\right) + \cdots \\ &\quad + \frac{1}{(n+1)!}\left(1 - \frac{1}{n+1}\right)\left(1 - \frac{2}{n+1}\right)\cdots\left(1 - \frac{n}{n+1}\right). \end{aligned}$$

Comparing these expressions for x_n and x_{n+1}, we see that every term in x_n is no more than corresponding term in x_{n+1}. In addition, x_{n+1} has an extra positive term. We thus conclude that $x_{n+1} \geq x_n$ for all $n \in \boldsymbol{N}$, which means that the sequence (x_n) is monotonically increasing.

We next prove boundedness. For every $n \in \boldsymbol{N}$, we have $x_n < \sum_{k=0}^{n}(1/k!)$. Using the inequality $2^{n-1} \leq n!$ for $n \geq 1$ (which can be easily seen by taking the logarithm of both sides), we obtain

$$x_n < 1 + \sum_{k=1}^{n} \frac{1}{2^{k-1}} = 1 + \frac{1-(1/2)^n}{1-(1/2)} < 3.$$

Thus (x_n) is bounded above by 3. Thus, view of the theorem in Sect. 2.1.3, the sequence is convergent. ♣

4. Denote $\bar{x} = \limsup x_n$ and $\underline{x} = \liminf x_n$. Prove that a sequence (x_n) converges to x if and only if $x = \bar{x} = \underline{x}$.

Solution: In view of the theorem in Sect. 2.1.4, it follows that $(-\infty, \bar{x}+\varepsilon)$ contains all but a finite number of terms of (x_n). The same property applied to $(-x_n)$ implies that $(\underline{x}-\varepsilon, \infty)$ contains all but a finite number of such terms. If $x = \bar{x} = \underline{x}$, then $(x - \varepsilon, x+\varepsilon)$ contains all but a finite number of terms of (x_n). This is the assertion that $x_n \to x$.

Now suppose that $x_n \to x$. For any $\varepsilon > 0$, there is an integer N such that $n > N \Rightarrow x_n < x+\varepsilon \Rightarrow y_n \le x+\varepsilon$, where $y_n = \sup\{x_k : k \ge n\}$, as was introduced in (2.5). Hence, $\bar{x} \le x+\varepsilon$. Since $\varepsilon > 0$ is arbitrary, we obtain $\bar{x} \le x$. Working with the sequence $(-x_n)$, whose limit is $-x$, following same procedure, we get $\underline{x} \ge x$. Since $\underline{x} \le \bar{x}$, we conclude that $x = \bar{x} = \underline{x}$. ♣

2.2 Cauchy Criterion for Real Sequences

2.2.1 Cauchy Sequence

To test the convergence of a general (nonmonotonic) real sequence, we have thus far only the original definition given in Sect. 2.1.1 to rely on; in that case we must first have a candidate for the limit of the sequence in question before we can examine its convergence. Needless to say, it is more convenient if we can determine the convergence property of a sequence without having to guess its limit. This is achieved by applying the so-called **Cauchy criterion**, which plays a central role in developing the fundamentals of real analysis.

To begin with, we present a preliminary notion for subsequent discussions.

♠ **Cauchy sequence:**

The sequence (x_n) is called a **Cauchy sequence** (or **fundamental sequence**) if for every positive number ε, there is a positive integer N such that

$$m, n > N \;\Rightarrow\; |x_n - x_m| < \varepsilon. \tag{2.6}$$

This means that in every Cauchy sequence, the terms can be as close to one another as we like. This feature of Cauchy sequences is expected to hold for any convergent sequence, since the terms of a convergent sequence have to approach each other as they approach a common limit. This conjecture is ensured in part by the following theorem.

♠ **Theorem:**
If a sequence (x_n) is convergent, then it is a Cauchy sequence.

Proof Suppose $\lim x_n = x$ and ε is any positive number. From hypothesis, there exists a positive integer N such that

$$n > N \;\Rightarrow\; |x_n - x| < \frac{\varepsilon}{2}.$$

Now if we take $m, n \geq N$, then

$$|x_n - x| < \frac{\varepsilon}{2} \text{ and } |x_m - x| < \frac{\varepsilon}{2}.$$

It thus follows that

$$|x_n - x_m| \leq |x_m - x| + |x_n - x| < \varepsilon,$$

which means that (x_n) is a Cauchy sequence. ♣

This theorem naturally gives rise to a question as to whether converse true. In other words, we would like to know whether all Cauchy sequences are convergent or not. The answer is exactly what the Cauchy criterion states, as we prove in the next subsection.

2.2.2 Cauchy Criterion

The following is one of the fundamental theorems of real sequences.

♠ **Cauchy criterion:**
A sequence of real numbers is convergent if and only if it is a Cauchy sequence.

Bear in mind that the validity of this criterion was partly proven by demonstrating the previous theorem (see Sect. 2.2.1). Hence, in order to complete the proof of the criterion, we need only prove that *every Cauchy sequence is convergent.* The following serves as a lemma for developing the proof.

♠ **Bolzano – Weierstrass theorem:**
Every infinite and bounded sequence of real numbers has at least one limit point in $\boldsymbol{R}$. (The proof is given in Appendix A.)

We are now ready to prove that *every Cauchy sequence is convergent.*

Proof (of the Cauchy criterion): Let (x_n) be a Cauchy sequence and $S = \{x_n : n \in \boldsymbol{N}\}$. We consider two cases in turn: **(i)** the set S is finite, and **(ii)** S is infinite.

(i) It follows from the hypothesis that given $\varepsilon > 0$, there is an integer N such that

$$m, n > N \quad \Rightarrow \quad |x_n - x_m| < \varepsilon. \tag{2.7}$$

Since S is finite, one of the terms of the sequence (x_n), say x, should be repeated infinitely often in order to satisfy (2.7). This implies the existence of an $m > N$ such that $x_m = x$. Hence, we have

$$n > N \quad \Rightarrow \quad |x_n - x| < \varepsilon,$$

which means that $x_n \to x$.

(ii) Next we consider the case that S is infinite. It can be shown that every Cauchy sequence is bounded (see Exercise **1**). Hence, in view of the Bolzano – Weierstrass theorem, the sequence (x_n) necessarily has a limit point x. We shall prove that $x_n \to x$. Given $\varepsilon > 0$, there is an integer N such that

$$m, n > N \quad \Rightarrow \quad |x_n - x_m| < \varepsilon.$$

From the definition of a limit point, we see that the interval $(x - \varepsilon, x + \varepsilon)$ contains an infinite number of terms of the sequence (x_n). Hence, there is an $m \geq N$ such that $x_m \in (x - \varepsilon, x + \varepsilon)$, i.e., such that $|x_n - x_m| < \varepsilon$. Now, if $n \geq N$, then

$$|x_n - x| \leq |x_n - x_m| + |x_m - x| < \varepsilon + \varepsilon = 2\varepsilon,$$

which proves $x_n \to x$.

The results for **(i)** and **(ii)** shown above indicate that *every Cauchy sequence (finite and infinite) is convergent.* Recall again that its converse, *every convergent sequence is a Cauchy sequence,* was proven earlier in Sect. 2.2.1. This completes the proof of the Cauchy criterion. ♣

Exercises

1. Show that every Cauchy sequence is bounded.

   Solution: Let (x_n) be a Cauchy sequence. Taking $\varepsilon = 1$, there is an integer N such that

   $$n > N \Rightarrow |x_n - x_N| < 1.$$

Since $|x_n| - |x_N| \le |x_n - x_N|$, we have

$$n > N \Rightarrow |x_n| < |x_N| + 1.$$

Thus $|x_n|$ is bounded by $\max\{|x_1|, |x_2|, \cdots, |x_{N-1}|, |x_N|+1\}$. ♣

2. Let $x_1 = 1$, $x_2 = 2$, and $x_n = (x_{n-1} + x_{n-2})/2$ for all $n \ge 3$. Show that (x_n) is a Cauchy sequence.

Solution: Since for $n \ge 3$, $x_n - x_{n-1} = -(x_{n-1} - x_{n-2})/2$, we use the induction on n to obtain $x_n - x_{n+1} = (-1)^n/2^{n-1}$ for all $n \in \boldsymbol{N}$. Hence, if $m > n$, then

$$\begin{aligned} |x_n - x_m| &\le |x_n - x_{n+1}| + |x_{n+1} - x_{n+2}| + \cdots + |x_{m-1} - x_m| \\ &= \sum_{k=n}^{m-1} \frac{1}{2^{k-1}} = \frac{1}{2^{n-1}} \sum_{k=0}^{m-n-1} \frac{1}{2^k} \\ &= \frac{1}{2^{n-1}} \frac{1-(1/2)^{m-n}}{1-(1/2)} < \frac{1}{2^{n-1}} \frac{1}{1-(1/2)} = \frac{1}{2^{n-2}}. \end{aligned}$$

Since $1/2^{n-2}$ decreases monotonically with n, it is possible to choose N for any $\varepsilon > 0$ such that $(1/2^{N-2}) < \varepsilon$. We thus conclude that

$$m > n \ge N \Rightarrow |x_n - x_m| < \left(\frac{1}{2}\right)^{n-2} < \left(\frac{1}{2}\right)^{N-2} < \varepsilon,$$

which means that (x_n) is a Cauchy sequence. ♣

3. Suppose that the two sequences (x_n) and (y_n) converge to a common limit c and consider their **shuffled sequence** (z_n) defined by

$$(z_1, z_2, z_3, z_4, \cdots) = (x_1, y_1, x_2, y_2, \cdots).$$

Show that the sequence (z_n) also converges to c.

Solution: Let ε be any positive number. Since $x_n \to c$ and $y_n \to c$, there are two positive integers N_1 and N_2 such that

$$n \ge N_1 \Rightarrow |x_n - c| < \varepsilon \text{ and } n \ge N_2 \Rightarrow |y_n - c| < \varepsilon.$$

Define $N = \max\{N_1, N_2\}$. Since $x_k = z_{2k-1}$ and $y_k = z_{2k}$ for all $k \in \boldsymbol{N}$, we have

$$k \ge N \Rightarrow |x_k - c| = |z_{2k-1} - c| < \varepsilon \text{ and } |y_k - c| = |z_{2k} - c| < \varepsilon.$$

Hence, $n \ge 2N-1 \Rightarrow |z_n - c| < \varepsilon$, which just means $\lim z_n = c$. ♣

4. Show that $\lim_{n\to\infty} (a^n/n^k) \to \infty$, where $a > 1$ and $k > 0$.

Solution: We consider three cases in turn: **(i)** $k = 1$, **(ii)** $k < 1$, and **(iii)** $k > 1$.

(i) Let $k = 1$. Then set $a = 1 + h$ to obtain

$$a^n = (1+h)^n = 1 + nh + \frac{n(n-1)}{2}h^2 + \cdots > \frac{n(n-1)}{2}h^2,$$

which results in

$$a^n/n = (1+h)^n/n > (n-1)h^n/2 \to \infty. \ (n \to \infty).$$

(ii) The case of $k < 1$ is trivial since $a^n/n^k > a^n/n$ for any $n > 1$.

(iii) If $k > 1$, then $a^{1/k} > 1$ since $a > 1$. Hence, it follows from the result of **(i)** that for any $M > 1$, we can find an n so that $n > M \ \Rightarrow \ a^{1/k}/n > M$. This means that

$$\frac{a^n}{n^k} = \left[\frac{\left(a^{1/k}\right)^n}{n}\right]^k > M^k > M,$$

which implies that $a^n/n^k \to \infty$. ♣

5. Let $x_n = a^n/n!$ with $a > 0$. Show that the sequence (x_n) converges to 0.

Solution: Let k be a positive integer such that $k > 2a$, and define $c = a^k/k!$. Then for any $a > 0$ and for any $n > k$, we have

$$\frac{a^n}{n!} = c\frac{a}{k+1}\cdot\frac{a}{k+2}\cdots\frac{a}{n} < \frac{c}{2^{n-k}} = \frac{c\cdot 2^k}{2^n} < \frac{c\cdot 2^k}{n}. \qquad (2.8)$$

Since (2.8) holds for a sufficiently large n ($> k$), it also holds for n satisfying $n > 2^k c/\varepsilon$, where ε is an arbitrarily small number. In the latter case, we have

$$\frac{a^n}{n!} < \frac{2^k c}{n} < \varepsilon,$$

which means that

$$\lim_{n\to\infty} x_n = \lim_{n\to\infty}\frac{a^n}{n!} = 0. \ ♣$$

2.3 Infinite Series of Real Numbers

2.3.1 Limits of Infinite Series

This section focuses on convergence properties of **infinite series**. The importance of this issue will become apparent, particularly in connection with certain branches of functional analysis such as Hilbert space theory and orthogonal polynomial expansions, where infinite series of numbers (or of functions) enter quite often (see Chaps. 4 and 5).

To begin with, we briefly review the basic properties of infinite series of real numbers. Assume an infinite sequence $(a_1, a_2, \cdots, a_n, \cdots)$ of real numbers. We can then form another infinite sequence $(A_1, A_2, \cdots, A_n, \cdots)$ with the definition

$$A_n = \sum_{k=1}^{n} a_k.$$

Here, A_n is called the nth **partial sum** of the sequence (a_n), and the corresponding infinite sequence (A_n) is called the **sequence of partial sums** of (a_n). The infinite sequence (A_n) may or may not be convergent, which depends on the features of (a_n).

Let us introduce an **infinite series** defined by

$$\sum_{k=1}^{\infty} a_k = a_1 + a_2 + \cdots . \tag{2.9}$$

The infinite series (2.9) is said to converge if and only if the sequence (A_n) converges to the limit denoted by A. In other words, the series (2.9) converges if and only if the **sequence of the remainder** $R_{n+1} = A - A_n$ converges to zero. When (A_n) is convergent, its limit A is called the **sum of the infinite series** of (2.9), and we may write

$$\sum_{k=1}^{\infty} a_k = \lim_{n\to\infty} \sum_{k=1}^{n} a_k = \lim_{n\to\infty} A_n = A.$$

Otherwise, the series (2.9) is said to diverge.

The limit of the sequence (A_n) is formally defined in line with Cauchy's procedure as shown below.

♠ **Limit of a sequence of partial sums:**

The sequence of partial sums (A_n) has a limit A if for any small $\varepsilon > 0$, there exists a number N such that

$$n > N \;\Rightarrow\; |A_n - A| < \varepsilon. \tag{2.10}$$

Examples **1.** The infinite series $\sum_{k=1}^{\infty}\left(\frac{1}{k} - \frac{1}{k+1}\right)$ converges to 1 because

$$A_n = \sum_{k=1}^{n}\left(\frac{1}{k} - \frac{1}{k+1}\right) = 1 - \frac{1}{n+1} \to 1 \quad (n \to \infty).$$

2. The series $\sum_{k=1}^{\infty}(-1)^k$ diverges because the sequence

$$A_n = \sum_{k=1}^{n}(-1)^k = \begin{cases} 0 & n \text{ (is even)}, \\ -1 & n \text{ (is odd)} \end{cases}$$

does not approache any limit.

3. The series $\sum_{k=1}^{\infty} 1 = 1+1+1+\cdots$ diverges since the sequence $A_n = \sum_{k=1}^{n} 1 = n$ increases without limit as $n \to \infty$.

2.3.2 Cauchy Criterion for Infinite Series

The following is a direct application of the Cauchy criterion to the sequence (A_n), which consists of the partial sum $A_n = \sum_{k=1}^{n} a_k$:

♠ **Cauchy criterion for infinite series:**

The sequence of partial sums (A_n) converges if and only if for any small $\varepsilon > 0$ there exists a number N such that

$$n, m > N \;\Rightarrow\; |A_n - A_m| < \varepsilon. \tag{2.11}$$

Similarly to the case of real sequences, the Cauchy criterion alluded to above provides a necessary and sufficient condition for convergence of the sequence (A_n). Moreover, from the definition, it also gives a necessary and sufficient condition for convergence of an infinite series $\sum_{k=1}^{\infty} a_k$. Below is an important theorem associated with the latter statement.

♠ **Theorem:**

If an infinite series $\sum_{k=1}^{\infty} a_k$ is convergent, then

$$\lim_{n\to\infty} a_n = 0.$$

Proof From hypothesis, we have

$$\lim_{n\to\infty} \sum_{k=1}^{n} a_k = \lim_{n\to\infty} A_n = A.$$

Hence,

$$\lim_{n\to\infty} a_n = \lim_{n\to\infty} (A_n - A_{n-1}) = A - A = 0. \quad ♣$$

According to the theorem above, $\lim a_n = 0$ is a necessary condition for the convergence of A_n. However, it is not a sufficient condition, as shown in the following example.

Examples Let $a_k = 1/\sqrt{k}$. Although $\lim_{k\to\infty} a_k = 0$, the corresponding infinite series $\sum a_k$ diverges, as seen from

$$\begin{aligned}\sum_{k=1}^{n} a_k &= 1 + \frac{1}{\sqrt{2}} + \cdots + \frac{1}{\sqrt{n}} \\ &\geq \frac{1}{\sqrt{n}} + \frac{1}{\sqrt{n}} + \cdots + \frac{1}{\sqrt{n}} = \frac{n}{\sqrt{n}} = \sqrt{n} \to \infty.\end{aligned}$$

Remark. The contraposition of the previous theorem serves as a **divergent test** of the infinite series in question; we can say that

$$\lim_{n\to\infty} a_n \neq 0 \ \Rightarrow \ \sum_{k=1}^{\infty} a_k \text{ is divergent.}$$

2.3.3 Absolute and Conditional Convergence

Assume an infinite series

$$\sum_{k=1}^{\infty} a_k, \tag{2.12}$$

and an associated auxiliary series

$$\sum_{k=1}^{\infty} |a_k|, \tag{2.13}$$

in the latter of which all terms are positive. If the series (2.13) converges, then the series (2.12) is said to **converge absolutely**. The necessary and sufficient condition for absolute convergence of (2.12) is obtained by replacing A_n in (2.11) by

$$B_n = |a_1| + |a_2| + \cdots + |a_n|.$$

If the series (2.13) diverges and the original series (2.12) converges, we say that the series (2.12) **converges conditionally**. These results are summarized by the statement below.

♠ **Absolute convergence:**
The infinite series $\sum a_k$ is absolutely convergent if $\sum |a_k|$ is convergent.

♠ **Conditional convergence:**
The infinite series $\sum a_k$ is conditionally convergent if $\sum a_k$ is convergent and $\sum |a_k|$ is divergent.

Examples The infinite series

$$\sum_{k=1}^{\infty} \frac{(-1)^{k+1}}{k} \tag{2.14}$$

converges conditionally, since it converges while its absolute-value series $\sum_{k=1}^{\infty} |(-1)^{k+1}/k| = \sum_{k=1}^{\infty}(1/k)$ diverges. See Exercises **1** and **2** in this section.

The following is an important theorem that we use many times in the remainder of this book.

♠ **Theorem:**
An infinite series converges if it converges absolutely.

Proof Suppose that the series (B_n) consisting of

$$B_n = \sum_{k=1}^{n} |a_k|$$

converges as $n \to \infty$. This means that for any $\varepsilon > 0$ a number N exists such that

$$n, m > N \quad \Rightarrow \quad |B_n - B_m| < \varepsilon. \tag{2.15}$$

Assuming $n \geq m$, we rewrite the left-hand inequality in (2.15) as

$$\begin{aligned} |B_n - B_m| &= |a_{m+1}| + |a_{m+2}| + \cdots + |a_n| \\ &\geq |a_{m+1} + a_{m+2} + \cdots + a_n| \\ &= |A_n - A_m|, \end{aligned} \tag{2.16}$$

where we used the law of inequalities for sums. Hence, it follows from (2.15) and (2.16) that

$$n, m > N \quad \Rightarrow \quad |A_n - A_m| < \varepsilon,$$

which means that the series $\sum a_k$ converges. ♣

The converse of the above theorem is not true. Below we present a well-known example of a convergent series that is not absolutely convergent.

2.3.4 Rearrangements

Observe that the conditionally convergent series (2.14) expressed by

$$1-\frac{1}{2}+\frac{1}{3}-\frac{1}{4}+\frac{1}{5}-\cdots, \tag{2.17}$$

may be rearranged in a number of ways, such as

$$1+\frac{1}{3}-\frac{1}{2}+\frac{1}{5}-\frac{1}{4}+\cdots \tag{2.18}$$

or

$$-\frac{1}{2}-\frac{1}{8}+\frac{1}{7}+1+\frac{1}{3}-\cdots \tag{2.19}$$

or in any other way in which the terms $1, -1/2, 1/3, -1/4, \cdots$ are added in a certain order. Series such as (2.18) and (2.19) are called **rearrangements** of the series (2.17).

Of importance is the fact that rearranging procedures may change the convergence property of a conditionally convergent series; in what way this happens depends on the nature of the original series, as we shall now see. Suppose a series $\sum a_n$ to be conditionally convergent. Then, the sum of its positive terms or that of its negative term goes, respectively to $+\infty$ or $-\infty$; otherwise the original series would diverge or converge absolutely. Let (b_n) and (c_n) be, respectively, the subsequences of positive and negative terms of (a_n). Since $\sum_{k=1}^{n} b_k$ is monotonically increasing with respect to n, there is a positive integer m_1 such that

$$\sum_{k=1}^{m_1} b_k \geq 1-c_1.$$

Here the right-hand side is positive since c_1 is negative. We rewrite it as

$$\sum_{k=1}^{m_1} b_k + c_1 \geq 1.$$

Similarly, there is an integer $m_2 > m_1$ such that

$$\sum_{k=1}^{m_2} b_k + c_2 \geq 1.$$

Continue on the same process for $m_3, m_4, \cdots, m_n$ and take the sum of each side to obtain

$$\sum_{k=1}^{m_n} b_k + \sum_{k=1}^{n} c_k \geq n. \tag{2.20}$$

Note that the left-hand side is a partial sum of the rearrangement of the sequence (a_k) that may, for instance, take the form of

$$(b_1, b_2, \cdots, b_{m_1}, c_1, b_{m_1+1}, b_{m_1+2}, \cdots, b_{m_2}, c_2, \cdots). \tag{2.21}$$

Clearly, the left-hand side of (2.20) diverges as $n \to \infty$, which means that the rearrangement (2.21) diverges. Therefore, the conditionally convergent series may become divergent through the rearranging procedure. In fact, the discussion above serves as part of the proof of the theorem below.

♠ **Riemann theorem:**
Given any conditionally convergent series and any $r \in \overline{\boldsymbol{R}} = \boldsymbol{R} \cap \infty$, there is a rearrangement of the series that converges to r.

Proof The case of $r = \infty$ was proved in the previous discussion. Now let $r \in \boldsymbol{R}$ and assume that (b_n) and (c_n) is the subsequence of positive and negative terms, respectively, in the same order in which they appear in (a_n). It is possible to obtain the smallest sum such that

$$s_1 = \sum_{k=1}^{m_1} b_k$$

exceeds r. Then, add the least number of negative terms c_k to obtain the largest sum. Such that

$$s_2 = \sum_{k=1}^{m_1} b_k + \sum_{k=1}^{n_1} c_k$$

is less than r. Proceeding in this fashion, we obtain a sequence $s_1, s_2, s_3, \cdots$ that converges to r, since

$$\lim_{n\to\infty} b_n = \lim_{n\to\infty} c_n = 0.$$

This result is the case for an arbitrary real number r. Hence, the proof is complete. ♣

Exercises

1. Determine the convergent property of the series

$$\sum_{k=1}^{\infty} \frac{1}{k}. \tag{2.22}$$

This is known as a **harmonic series**.

Solution: Let $A_n = \sum_{k=1}^{n}(1/k)$. We then have

$$A_{2n} - A_n = \frac{1}{n+1} + \frac{1}{n+2} + \cdots + \frac{1}{2n} \geq \frac{1}{2n} \times n = \frac{1}{2},$$

which implies that the sequence (A_n) is not a Cauchy sequence. Thus, view of the Cauchy criterion, the harmonic series (2.22) diverges. ♣

2. Determine the convergence of the series

$$\sum_{k=1}^{\infty} \frac{1}{k^p}. \tag{2.23}$$

This is called a **hyperharmonic series** (or **zeta function**) and is denoted by $\xi(p)$.

Solution: When $p \leq 1$, a partial sum A_{2^n} consisting of the first 2^n terms reads

$$\begin{aligned} A_{2^n} &= \left(1 + \frac{1}{2^p}\right) + \left(\frac{1}{3^p} + \frac{1}{4^p}\right) + \left(\frac{1}{5^p} + \cdots + \frac{1}{8^p}\right) + \cdots \\ &\quad + \left[\frac{1}{(2^{n-1}+1)^p} + \cdots + \frac{1}{(2^n)^p}\right] \\ &\geq \left(1 + \frac{1}{2}\right) + \left(\frac{1}{3} + \frac{1}{4}\right) + \left(\frac{1}{5} + \cdots + \frac{1}{8}\right) + \cdots \\ &\quad + \left[\frac{1}{(2^{n-1}+1)} + \cdots + \frac{1}{2^n}\right] \\ &\geq \frac{1}{2} + \frac{1}{4} \times 2 + \frac{1}{8} \times 4 + \cdots + \frac{1}{2^n} \times 2^{n-1} \quad = \frac{n}{2}. \end{aligned}$$

This means that the series (2.23) diverges for $p \leq 1$.

For $p > 1$, we have

$$\begin{aligned} A_{2^{n+1}-1} &= 1 + \left(\frac{1}{2^p} + \frac{1}{3^p}\right) + \left(\frac{1}{4^p} + \cdots + \frac{1}{7^p}\right) + \cdots \\ &\quad + \left[\frac{1}{(2^n)^p} + \cdots + \frac{1}{(2^{n+1}-1)^p}\right] \\ &< 1 + \frac{1}{2^p} \times 2 + \frac{1}{4^p} \times 4 + \cdots + \frac{1}{(2^n)^p} \times 2^n \\ &< \sum_{k=0}^{n} \left(\frac{1}{2^{p-1}}\right)^k = \frac{1-(1/2^{p-1})^{n+1}}{1-(1/2^{p-1})} < \frac{2^{p-1}}{2^{p-1}-1}. \end{aligned}$$

Hence, the monotonically increasing sequence $\{A_n\}$ is bounded above and is thus convergent. ♣

3. Determine the convergence of the series

$$\sum_{k=1}^{\infty} \frac{(-1)^{k+1}}{k}. \tag{2.24}$$

Solution: Let n be an even integer, say $n = 2m$. Then, it follows that

$$\begin{aligned} A_{2m} &= \sum_{k=1}^{2m} \frac{(-1)^{k+1}}{k} = \left(1 - \frac{1}{2}\right) + \left(\frac{1}{3} - \frac{1}{4}\right) + \cdots + \left(\frac{1}{2m-1} - \frac{1}{2m}\right) \\ &= \frac{1}{2 \cdot 1} + \frac{1}{4 \cdot 3} + \cdots + \frac{1}{2m(2m-1)}, \end{aligned}$$

which means that (A_{2m}) is increasing with m. In addition, we have

$$A_{2m} = 1 - \left(\frac{1}{2} - \frac{1}{3}\right) - \left(\frac{1}{4} - \frac{1}{5}\right) - \cdots - \frac{1}{2m} < 1,$$

which indicates that (A_{2m}) is bounded above. Hence, (A_{2m}) converges to a limit A. Further more, since $A_{2m+1} = A_{2m} + 1/(2m+1)$, the same discussion as above tells us that the sequence (A_{2m+1}) also converges to the common limit A. By applying the result from shuffled sequences (see Exercise **3** in Sect. 2.1.2), we find that $\lim A_n$ exists, so the series (2.24) converges. It is thus proven that the series converges conditionally. ♣

4. Suppose that the infinite series $\sum_k a_k$ and $\sum_k b_k$ are both convergent absolutely. Let $(a_i b_j)$ be an infinite sequence in which the terms $a_i b_j$ are arranged in an arbitrary order, say, as

$$(a_2 b_1, a_1 b_3, a_2 b_4, a_5 b_1, \cdots).$$

Show that the sequence of the partial sums of $(a_i b_j)$ converges absolutely regardless of the order of the terms $a_i b_j$.

Solution: Let m and n be the maximum values of i and j, respectively, that are involved in the partial sum $\sum_{(i,j)} a_i b_j$; here (i, j) denotes the possible combinations of i and j that are arranged in the same order as in thesequence $(a_i b_j)$. The partial sum is a portion of the product of the finite sums given by $\left(\sum_{i=1}^m a_i\right)\left(\sum_{j=1}^n b_j\right)$. Hence, we have

$$\left|\sum_{(i,j)} a_i b_j\right| = \sum_{(i,j)} |a_i b_j| = \left|\sum_{i}^{m} a_i\right| \cdot \left|\sum_{j}^{n} b_j\right| \leq \sum_{i=1}^{m} |a_i| \sum_{j=1}^{n} |b_j|. \tag{2.25}$$

From hypothesis, the left-hand side in (2.25) converges as $m, n \to \infty$. This means that the partial sum $\sum_{(i,j)} |a_i b_j|$ is bounded above. In addition, it is obviously increasing. Therefore, $\sum_{(i,j)} |a_i b_j|$ converges (i.e., $\sum_{(i,j)} a_i b_j$ converges absolutely) independently of the order of i and j in the sequence of $(a_i b_j)$. ♣

5. Show that rearrangements of absolutely convergent series always converge absolutely to the same limit.

Solution: Let $\sum_{k=1}^\infty a_k$ be absolutely convergent and assume that $\sum_{k=1}^\infty b_k$ is its rearrangement. Define $A_n = \sum_{k=1}^n |a_k|$, $A = \lim_{n\to\infty} A_n$, $B_n = \sum_{k=1}^n |b_k|$, and let $\varepsilon > 0$. By hypothesis, there is an integer N such that $|A - A_N| = |a_{N+1}| + |a_{N+2}| + \cdots < \frac{\varepsilon}{2}$. Now we choose the integer M so that all the terms $a_1, a_2, \cdots, a_N$ appear in the first M terms of the rearranged series, i.e., within the finite sequence $(b_1, b_2, \cdots, b_M)$. Hence, these terms do not contribute to the difference $B_m - A_N$, where $m \geq N$. Consequently, we obtain

$$\begin{aligned} m \geq N \Rightarrow |B_m - A_N| &\leq |a_{N+1}| + |a_{N+2}| + \cdots < \frac{\varepsilon}{2} \\ \Rightarrow |A - B_m| &\leq |A - A_N| + |A_N - B_m| < \varepsilon, \end{aligned}$$

which shows that $\lim_{n\to\infty} B_n = A$. ♣

2.4 Convergence Tests for Infinite Real Series

2.4.1 Limit Tests

This section covers the important tests for convergence of infinite series. In general, these tests provide sufficient, not necessary, conditions for convergence. This is in contrast to the Cauchy criterion, which provides a necessary and sufficient condition for convergence, though it is difficult to apply in practice. The first test to be shown is called the **limit test**, by which we can examine the absolute convergence of infinite series quite easily.

♠ **Limit test for convergence:**
If

$$\lim_{k\to\infty} k^p a_k \text{ exists for some } p > 1,$$

then $\sum_{k=1}^{\infty} a_k$ converges absolutely (and thus converges ordinary).

Proof By hypothesis, we set $\lim_{k\to\infty} k^p a_k = A$ for certain $p > 1$, which implies that

$$\lim_{k\to\infty} k^p |a_k| = |A|.$$

Hence, there exists an integer m such that

$$k > m \quad \Rightarrow \quad k^p|a_k| - |A| < 1,$$

or equivalently,

$$k > m \quad \Rightarrow \quad |a_k| < \frac{|A|+1}{k^p}. \tag{2.26}$$

We know that the series $\sum_{k=m}^{\infty} 1/k^p$ converges for any $p > 1$ (see Exercise **2** in Sect. 2.3). Thus it follows from (2.26) that the series $\sum_{k=m}^{\infty} |a_k|$ also converges, from which the desired conclusion follows at once. ♣

There is a counterpart of the limit test for convergence that determines *divergence* properties of series as follows.

♠ **Limit test for divergence:**
If

$$\lim_{k\to\infty} k a_k \neq 0,$$

then $\sum_{k=1}^{\infty} a_k$ diverges. The test fails if the limit equals zero.

Proof Suppose $\lim k a_k = A > 0$. Then there exists an integer m such that

$$k \geq m \quad \Rightarrow \quad k a_k > \frac{A}{2}.$$

Hence, by employing the result from harmonic series (see Exercise **1** in Sect. 2.3), we obtain

$$\sum_{k=m}^{\infty} a_k > \sum_{k=m}^{\infty} \frac{1}{k} = \infty,$$

from which the desired result follows. The same procedure can be applied to the case of $A < 0$, in which case the series $\sum_{k=1}^{\infty}(-a_k)$ may be treated by the procedure above. The proof is thus complete. ♣

Remark.

1. The test is valid even when A goes to infinity.

2. The divergence test described above is inconclusive when $\lim ka_k = 0$. To see why, consider the two series

$$\sum_{k=1}^{\infty} \frac{1}{k^2} \quad \text{and} \quad \sum_{k=2}^{\infty} \frac{1}{k \log k}.$$

The former converges and the latter diverges, but both yield $\lim ka_k = 0$.

2.4.2 Ratio Tests

The following provides another test for absolute convergence of infinite series that is sometimes easier to use than the previous one.

♠ **Ratio test:**
A series $\sum_{k=0}^{\infty} a_k$ converges absolutely (and thus converges ordinary) if

$$\limsup_{k\to\infty} \left| \frac{a_{k+1}}{a_k} \right| < 1 \tag{2.27}$$

and diverges if

$$\limsup_{k\to\infty} \left| \frac{a_{k+1}}{a_k} \right| > 1. \tag{2.28}$$

If the limit superior is 1, the test is inconclusive.

Remark. When $|a_{k+1}/a_k|$ converges, the limits superior used in (2.27) and (2.28) reduce to the ordinary limits.

Proof **(i)** Suppose that $\ell \equiv \limsup_{k\to\infty} \left| \frac{a_{k+1}}{a_k} \right| < 1$. Then, for any $r \in (\ell, 1)$, we can find the number m such that

$$k > m \quad \Rightarrow \quad \left| \frac{a_{k+1}}{a_k} \right| < r.$$

It follows that $\left| \frac{a_{m+1}}{a_m} \right| \times \left| \frac{a_{m+2}}{a_{m+1}} \right| \times \cdots \times \left| \frac{a_{m+p}}{a_{m+p+1}} \right| < r^p$ or equivalently, $|a_{m+p}| < r^p |a_m|$, which holds for any $p \in \boldsymbol{N}$. Hence, we have

$$\sum_{p=1}^{\infty} |a_{m+p}| = \sum_{k=m+1}^{\infty} |a_k| < \sum_{p=1}^{\infty} r^p |a_m| = \frac{r}{1-r}|a_m|.$$

The last term is a finite constant. Therefore, the series $\sum_{k=m+1}^{\infty} |a_k|$ remains finite and the series $\sum_{k=0}^{\infty} a_k$ converges absolutely.

(ii) Next we assume that

$$\limsup_{k\to\infty} \left| \frac{a_{k+1}}{a_k} \right| = \ell > 1.$$

Then there is an integer m such that

$$k > m \quad \Rightarrow \quad \left| \frac{a_{k+1}}{a_k} \right| > 1.$$

That is,

$$k > m \quad \Rightarrow \quad |a_k| > |a_m| > 0,$$

which means that

$$\lim_{k\to\infty} a_k \neq 0.$$

In view of the remark in Sect. 2.3.2, the series $\sum_{k=0}^{\infty} a_k$ diverges. ♣

2.4.3 Root Tests

We now give an alternative absolute-convergence test based on examining the kth root of $|a_k|$.

♠ **Root test:**
A series $\sum_{k=0}^{\infty} a_k$ converges absolutely (and ordinary) if

$$\limsup_{k\to\infty} \sqrt[k]{|a_k|} < 1$$

and diverges if

$$\limsup_{k\to\infty} \sqrt[k]{|a_k|} > 1.$$

If the limit superior is 1, the test fails and does not provide any information.

Proof Let $r = \limsup_{k\to\infty} \sqrt[k]{|a_k|}$. We first prove that the series converges absolutely if $r < 1$. We choose a positive number $c \in (r, 1)$. Then there is a positive integer N such that

$$k \geq N \quad \Rightarrow \quad \sqrt[k]{|a_k|} < c \quad \Rightarrow \quad |a_k| < c^k.$$

Since the geometric series $\sum c^k$ with $c < 1$ converges, $\sum |a_k|$ converges, so that $\sum a_k$ converges absolutely.

When $r > 1$, it follows from the definition of the limit superior (see Sect. 2.1.4) that there are an infinite number of terms of $\sqrt[k]{|a_k|}$ greater than 1. This implies that $\lim a_k \neq 0$, which means that the infinite series $\sum a_k$ diverges. ♣

Examples Assume the series

$$\sum_{k=0}^{\infty} a_k = 1 - \frac{1}{2} + \frac{1}{4^2} - \frac{1}{2^3} + \frac{1}{4^4} - \frac{1}{2^5} + \cdots . \tag{2.29}$$

Since

$$\sqrt[0]{|a_0|} = 1, \quad \sqrt[1]{|a_1|} = \frac{1}{2}, \quad \sqrt[2]{|a_2|} = \frac{1}{4}, \quad \sqrt[3]{|a_3|} = \frac{1}{2}, \quad \sqrt[4]{|a_4|} = \frac{1}{4}, \cdots ,$$

we have

$$\limsup_{k\to\infty} \sqrt[k]{a_k} = \frac{1}{2} < 1.$$

Thus the series (2.29) converges (absolutely and ordinary).

2.4.4 Alternating Series Test

All the convergence tests presented so far are tests for absolute convergence, which assumes ordinary convergence. Nonetheless, certain kinds of series can exhibit conditional convergence, i.e., ordinary convergence with absolute divergence, whose convergence properties cannot be addressed by the tests given thus far. Hence, the significance of the test described below, known as the **alternating series test**, is that it may be used to test the conditional convergence of some absolutely divergent series.

We say that (x_k) is an **alternating sequence** if the sign of x_k is different from that of x_{k+1} for every k. The resulting series $\sum x_k$ is called the **alternating series**, whose convergence properties are partly determined by the following theorem:

♠ **Alternating series test:**
An alternating series given by

$$a_1 - a_2 + a_3 - a_4 + \cdots = \sum_{k=1}^{\infty} (-1)^{k+1} a_k \quad \text{with } a_k > 0 \text{ for all } k$$

converges if

$$a_k > a_{k+1} \quad \text{and} \quad \lim_{k\to\infty} a_k = 0.$$

Proof First we show that the sequence of partial sums S_n converges. It follows that

$$A_{2n} = (a_1 - a_2) + (a_3 - a_4) + \cdots + (a_{2n-1} + a_{2n}).$$

Since $a_k - a_{k+1} > 0$ for all k, the sequence A_{2n} is increasing. It is also bounded above because

$$A_{2n} = a_1 - (a_2 - a_3) - (a_4 - a_5) - \cdots - (a_{2n-2} + a_{2n-1}) - a_{2n} \le a_1 \quad (2.30)$$

for all $n \in \boldsymbol{N}$. Thus, $\lim A_{2n}$ exists and we call it A. On the other hand, we have

$$|A_{2n+1} - A| = |A_{2n}a_{2n+1} - A| \le |A_{2n} - A| + |a_{2n+1}|.$$

In the limit as $n \to \infty$, the left-hand side vanishes so that we obtain $\lim A_{2n+1} = A$. Therefore, we conclude that $S_n \to S$ ♣

Exercises

1. Show that $\displaystyle\sum_{k=1}^{\infty} \frac{(k+1)^{1/2}}{(k^5+k^3-1)^{1/3}}$ converges.

Solution: Taking $p = 7/6 > 1$ into the limit test for convergence, we have

$$\lim_{k\to\infty} k^{7/6} a_k = \lim_{k\to\infty} \frac{(1+k^{-1})^{1/2}}{(1-k^{-2}+k^{-5})^{1/3}} = 1. \quad ♣$$

2. Show that $\displaystyle\sum_{k=1}^{\infty} (-1)^k \frac{\log k}{k^2}$ converges.

Solution: With use of the limit test for convergence by taking $p = 3/2$, we obtain

$$\lim_{k\to\infty} k^{3/2} a_k = \lim_{k\to\infty} (-1)^k \frac{\log k}{\sqrt{k}} = 0. \quad ♣$$

3. Show that $\displaystyle\sum_{k=1}^{\infty} \frac{k \log k}{1+k^2}$ diverges.

Solution: From the limit test for divergence, we have

$$\lim_{k\to\infty} k a_k = \lim_{k\to\infty} \frac{k^2 \log k}{1+k^2} = \infty. \quad ♣$$

4. Show that $\sum_{k=0}^{\infty} \frac{(k!)^2}{(2k)!}$ converges.

Solution: The ratio test yields

$$\left|\frac{a_{k+1}}{a_k}\right| = \frac{(2k)!}{(k!)^2}\frac{[(k+1)!]^2}{(2k+2)!} = \frac{(k+1)^2}{(2k+2)(2k+1)}$$

$$= \frac{(1+\frac{1}{k})^2}{(2+\frac{2}{k})(2+\frac{1}{k})} \to \frac{1}{4} < 1 \quad (k \to \infty). \clubsuit$$

5. Show that $\sum_{k=1}^{\infty} \left(1+\frac{1}{k}\right)^{-k^2}$ converges.

Solution: The root test yields

$$\left[\left(1+\frac{1}{k}\right)^{-k^2}\right]^{1/k} = \frac{1}{[1+(1/k)]^k} \to \frac{1}{e} < 1.$$

3

Real Functions

Abstract Infinite sequences and series of real functions are encountered frequently in mathematical physics. The convergence of such sequences and series does not generally preserve the nature of their constituents; e.g., a sequence of "continuous" functions can converge into a "discontinuous" function. In this chapter, we show that this is not true in cases of uniform convergence (Sect. 3.2.2), which is a special class of convergence that preserves the continuity, integrability, and differentiability of the constituent functions of sequences and series, as we explain in detail in Sects. 3.2.4–3.2.6.

3.1 Fundamental Properties

3.1.1 Limit of a Function

Having discussed the limits of sequences and series of real numbers, we now turn our attention to the limit of functions. Let A be a real number and $f(x)$ a real-valued function of a real variable $x \in \boldsymbol{R}$. A formal notation of the above function is given by the mapping relation $f : \boldsymbol{R} \to \boldsymbol{R}$. The statement "*the limit of* $f(x)$ *at* $x = a$ *is* A" means that the value of $f(x)$ can be set as close to A as desired by setting x sufficiently close to a. This is stated formally by the following definition.

♠ **The limit of a function:**

A function $f(x)$ is said to have the **limit** A as $x \to a$ if and only if for every $\varepsilon > 0$, there exists a number $\delta > 0$ such that

$$|x - a| < \delta \Rightarrow |f(x) - A| < \varepsilon. \tag{3.1}$$

The limit of $f(x)$ is written symbolically as

$$\lim_{x \to a} f(x) = A$$

or

$$f(x) \to A \text{ for } x \to a.$$

If the first inequality in (3.1) is replaced by $0 < x - a < \delta$ (or $0 < a - x < \delta$), we say that $f(x)$ approaches A as $x \to a$ from above (or below) and write

$$\lim_{x \to a+} f(x) = A \quad \left(\text{or} \quad \lim_{x \to a-} f(x) = A \right).$$

This is called the **right-hand** (or **left-hand**) **limit** of $f(x)$. The two together are known as **one-sided limits**.

A necessary and sufficient condition for the existence of $\lim_{x \to a} f(x)$ is shown below.

♠ **Theorem:**
The limit of $f(x)$ at $x = a$ exists if and only if

$$\lim_{x \to a+} f(x) = \lim_{x \to a-} f(x). \tag{3.2}$$

Proof If $\lim_{x \to a} f(x)$ exists and is equal to A, it readily follows that

$$\lim_{x \to a+} f(x) = \lim_{x \to a-} f(x) = A. \tag{3.3}$$

We now consider the converse. Assume that (3.2) holds. This obviously means that both one-sided limits exist at $x = a$. Hence, given $\varepsilon > 0$, we have $\delta_1 > 0$ and $\delta_2 > 0$ such that

$$0 < x - a < \delta_1 \quad \Rightarrow \quad |f(x) - A| < \varepsilon,$$
$$0 < a - x < \delta_2 \quad \Rightarrow \quad |f(x) - A| < \varepsilon.$$

Let $\delta = \min\{\delta_1, \delta_2\}$. If x satisfies $0 < |x - a| < \delta$, then either

$$0 < x - a < \delta \le \delta_1 \text{ or } 0 < a - x < \delta \le \delta_2.$$

In either case, we have $|f(x) - A| < \varepsilon$. That is, we have seen that for a given ε, there exists δ such that

$$0 < |x - a| < \delta \quad \Rightarrow \quad |f(x) - A| < \varepsilon.$$

Therefore we conclude that

$$\text{Equation (3.2) holds} \quad \Rightarrow \quad \lim_{x \to a} f(x) = A,$$

and the proof is complete. ♣

3.1.2 Continuity of a Function

In general, the value of $\lim_{x\to a} f(x)$ has nothing to do with the value (and the existence) of $f(a)$. For instance, the function given by

$$f(x) = \begin{cases} 2 - x^2 & x \neq 1, \\ 2 & x = 1 \end{cases}$$

gives

$$\lim_{x\to 1} f(x) = 0 \text{ and } f(1) = 2,$$

which are quantitatively different from one another. This mismatch occurring at $x = 1$ results in a lack of geographical continuity in the curve of $y = f(x)$, as depicted in Fig. 3.1. In mathematical language, continuity of the curve of $y = f(x)$ is accounted for by the following statement.

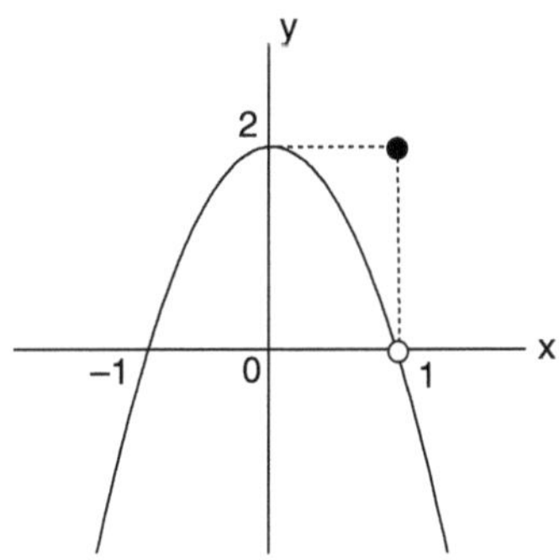

Fig. 3.1. A discontinuous function $y = f(x)$ at $x = 1$

♠ **Continuous functions:**

The function $f(x)$ is said to be **continuous** at $x = a$ if and only if for every $\varepsilon > 0$, there exists $\delta > 0$ such that

$$|x - a| < \delta \Rightarrow |f(x) - f(a)| < \varepsilon.$$

Remark. The definition noted above seems to be similar to the definition of the limit of $f(x)$ at $x = a$ (see Sect. 3.1.1). However, there is a crucial difference between them. When considering the limit of $f(x)$ at $x = a$, we are only interested in the behavior of $f(x)$ *in the vicinity of* the point a, not *just at* a. However, the continuity of $f(x)$ at $x = a$ requires the further condition that the value of $f(x)$ just at $x = a$ has to be defined. In symbols, we write

$$f(x) \text{ is continuous at } x = a \Rightarrow \lim_{x\to a-0} f(x) = \lim_{x\to a+0} f(x) = f(a).$$

We must emphasize that given a function $f(x)$ on a domain D, the limit of $f(x)$ is defined at limit points in D that may or may not lie in D. In contrast, the continuity of $f(x)$ is defined only at points contained in D. An illustrative example is given below.

Examples Assume a function given by

$$f(x) = x \text{ for all but } x = 1.$$

It has a limit at $x = 1$,

$$\lim_{x\to\infty} f(x) = 1,$$

but there is no way to examine its continuity because $x = 1$ is out of the defining domain.

When $f(x)$ is continuous, we can say that $f(x)$ belongs to the class of functions designated by the symbol C. Then, it follows that

$$f(x) \in C \text{ at } x = a \quad \Longleftrightarrow \quad \lim_{x\to a} f(x) = f(a).$$

If the symbol $x \to a$ appearing in the right-hand statement is replaced by $x \to a+$ (or $x \to a-$), $f(x)$ is said to be **continuous on the right** (or **left**) at $x = a$. We encounter the latter kind of a limit particularly when we consider the continuity of a function defined within a finite interval $[a, b]$; we say that

$$\begin{aligned} &f(x) \in C \text{ on } [a,b] \iff \\ &f(x) \in C \text{ on } (a,b) \quad \text{and} \quad \lim_{x\to a+} f(x) = f(a), \quad \lim_{x\to b-} f(x) = f(b). \end{aligned}$$

We also say that a function $f(x)$ on $[a, b]$ is **piecewise continuous** if

(i) $f(x)$ is continuous on $[a, b]$ except at a finite number of points $x_1, x_2, \cdots, x_n$;

(ii) at each of the points $x_1, x_2, \cdots, x_n$, there exist both the left-hand and right-hand limits of $f(x)$ defined by

$$f(x_k - 0) = \lim_{x=x_k-0} f(x), \quad f(x_k + 0) = \lim_{x=x_k+0} f(x).$$

3.1.3 Derivative of a Function

The following is a rigorous definition of the derivative of a real function.

♠ **Derivative of a function:**
If the limit

$$\lim_{x\to a} \frac{f(x) - f(a)}{x - a}$$

exists, it is called the **derivative** of $f(x)$ at $x = a$ and is denoted by $f'(a)$. The function $f(x)$ is said to be **differentiable** at $x = a$ if $f'(a)$ exists.

Similar to the case of one-sided limits, it is possible to define **one-sided derivatives** of real functions such as

$$f'(a+) = \lim_{x\to a+} \frac{f(x)-f(a)}{x-a},$$
$$f'(a-) = \lim_{x\to a-} \frac{f(x)-f(a)}{x-a}.$$

♠ **Theorem:**
If $f(x)$ is differentiable at $x = a$, then it is continuous at $x = a$. (The converse is not true.)

Proof Assume $x \neq a$. Then

$$f(x) - f(a) = \frac{f(x)-f(a)}{x-a}(x-a).$$

From hypothesis, each function $[f(x) - f(a)]/(x - a)$ as well as $x - a$ has the limit at $x = a$. Hence, we obtain

$$\lim_{x\to a}[f(x) - f(a)] = \lim_{x\to a} \frac{f(x)-f(a)}{x-a} \cdot \lim_{x\to a}(x-a) = f'(a) \times 0 = 0.$$

Therefore,

$$\lim_{x\to a} f(x) = f(a),$$

i.e., $f(x)$ is continuous at $x = a$. That the converse is false can be seen by considering $f(x) = |x|$; it is continuous at $x = 0$ but not differentiable. ♣

The term C^n functions is used to indicate that all the derivatives on the order of $\leq n$ exist; this is denoted by

$$f(x) \in C^n \quad \Longleftrightarrow \quad f^{(n)}(x) \in C.$$

Such an $f(x)$ is said to be a C^n function or to be of class C^n.

Examples **1.** $f(x) = \begin{cases} 0, & x < 0 \\ x, & x \geq 0 \end{cases}$

$$\Rightarrow \quad f(x) \in C^0(= C), \;\text{ but }\; f(x) \notin C^1 \;\text{ at }\; x = 0.$$

2. $f(x) = \begin{cases} 0 & x < 0 \\ x^2 & x \geq 0 \end{cases}$

$$\Rightarrow \quad f(x) \in C^1, \;\text{ but }\; f(x) \notin C^2 \;\text{ at }\; x = 0.$$

3. **Taylor series expansion** for functions $f \in C^n$ is given by

$$f(x) = \sum_{k\leq n} \frac{1}{k!} \frac{\partial f}{\partial x^k}\bigg|_{x=x_0} (x-x_0)^k + o\left(|x-x_0|^n\right).$$

3.1.4 Smooth Functions

We now introduce a new class of functions for which the derivative is continuous over the defining domain.

♠ **Smooth functions:**
The function $f(x)$ is said to be **smooth** for any $x \in [a, b]$ if $f'(x)$ exists and is continuous on $[a, b]$.

In geometrical language, the above statement means that the direction of the tangent changes continuously, without jumps, as it moves along the curve $y = f(x)$ (see Fig. 3.2). Thus, the graph of a smooth function is a smooth curve without any point at which the curve has two distinct tangents.

Similar to the case of piecewise continuity, the function $f(x)$ is said to be **piecewise smooth** on the interval $[a, b]$ if $f(x)$ and its derivatives are all piecewise continuous on $[a, b]$. The graph of a piecewise smooth function is either a continuous or a discontinuous curve; furthermore, it can have a finite number of points (called **corners**) at which the derivatives show jumps (see Fig. 3.2). Every piecewise smooth function $f(x)$ is bounded and has a bounded derivative everywhere, except at its corners and points of discontinuity; $f'(x)$ does not exist in the sense of continuity at any of these points.

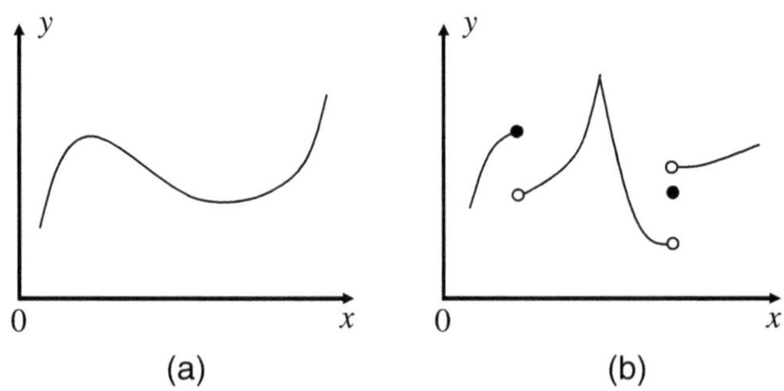

Fig. 3.2. (**a**) A continuous function $y = f(x)$. (**b**) A piecewise smooth function $y = f(x)$ having two discontinuous points and one corner

3.2 Sequences of Real Functions

3.2.1 Pointwise Convergence

In this section we focus on convergence properties of sequences consisting of real-valued *functions* of a real variable. Suppose that for each $n \in \boldsymbol{N}$, we have a function $f_n(x)$ defined on a domain $D \subseteq \boldsymbol{R}$. We then say that we have a sequence

$$(f_n(x) : \; n \in \boldsymbol{N})$$

of real-valued functions on D. If the sequence $(f_n(x))$ converges for every $x \in D$, the sequence of functions is said to **converge pointwise** on D, and the function defined by

$$f(x) = \lim_{n \to \infty} f_n(x)$$

is called the **pointwise limit** of $(f_n(x))$. The formal definition is given below.

♠ **Pointwise convergence:**

The sequence of functions (f_n) is said to **converge pointwise** to f on D if, given $\varepsilon > 0$, there is a natural number $N = N(\varepsilon, x)$ (which depends on ε and x) such that

$$n > N \;\; \Rightarrow \;\; |f_n(x) - f(x)| < \varepsilon.$$

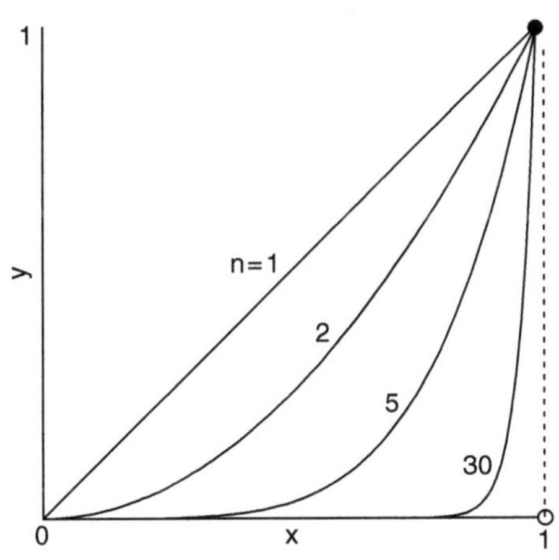

Fig. 3.3. Converging behavior of $f_n(x) = x^n$ given in (3.4)

Examples Assume a sequence (f_n) consisting of the function

$$f_n(x) = x^n \tag{3.4}$$

that is defined on a closed interval $[0, 1]$. It follows that the sequence converges pointwise to

$$f(x) = \lim_{n\to\infty} f_n(x) = \begin{cases} 0 \text{ for } 0 \le x < 1, \\ 1 \text{ at } x = 1. \end{cases} \tag{3.5}$$

See Fig. 3.3 for the converging behavior of $f_n(x)$ with increasing n.

The important point is the fact that under pointwise convergence, continuity of functions of $f_n(x)$ is not preserved. In fact, $f_n(x)$ given in (3.4) is continuous for each n over the whole interval $[0, 1]$, whereas the limit $f(x)$ given in (3.5) is discontinuous at $x = 1$. This indicates that interchanging the order of the limiting processes under pointwise convergence may produce different results, as expressed by

$$\lim_{x\to 1}\lim_{n\to\infty} f_n(x) \ne \lim_{n\to\infty}\lim_{x\to 1} f_n(x).$$

Similar phenomena might occur in connection with, integrability and differentiability of terms of functions $f_n(x)$. That is, under pointwise convergence, the limit of a sequence of integrable or differentiable functions may not be integrable or differentiable, respectively. Illustrative examples are given in Exercises **1** and **2** in Sect. 3.2.

3.2.2 Uniform Convergence

We know that if the sequence $(f_n(x))$ is pointwise convergent to $f(x)$ on $x \in D$, it is possible to choose $N(x)$ for any small ε such that

$$m > N(x) \quad \Rightarrow \quad |f_m(x) - f(x)| < \varepsilon. \tag{3.6}$$

In general, the least value of $N(x)$ that satisfies (3.6) will depend on x. But in certain cases, we can choose N *independent* of x such that $|f_m(x) - f(x)| < \varepsilon$ for all $m > N$ and *for all* x over the domain D. If this is true for any small ε, the sequence $(f_n(x))$ is said to **converge uniformly** to $f(x)$ on D. The formal definition is given below.

♠ **Uniform convergence:**

The sequence (f_n) of real functions on $D \subseteq \boldsymbol{R}$ **converges uniformly** to a function f on D if, given $\varepsilon > 0$, there is a positive integer $N = N(\varepsilon)$ (which depends on ε) such that

$$n > N \quad \Rightarrow \quad |f_n(x) - f(x)| < \varepsilon \text{ for all } x \in D.$$

Emphasis is placed on the fact that the integer $N = N(\varepsilon, x)$ in the pointwise convergence *depends* on x in general, whereas $N = N(\varepsilon)$ in the uniform convergence is *independent* of x. Under uniform convergence, therefore, by taking n large enough we can always force the graph of $y = f_n(x)$ into a band of width less than 2ε centered around the graph of $y = f(x)$ over the whole domain D (see Fig. 3.4).

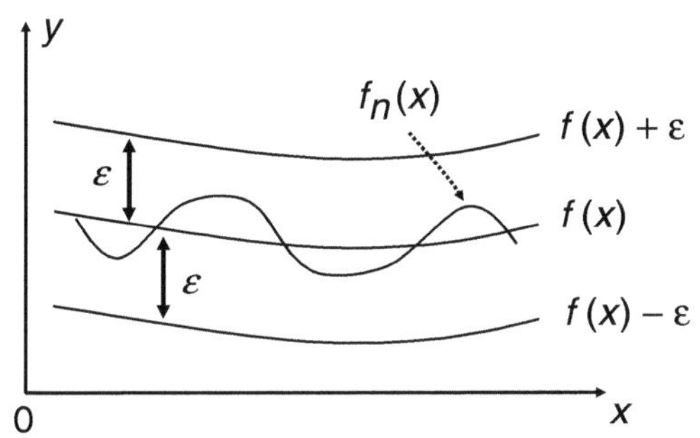

Fig. 3.4. A function $y = f_n(x)$ contained overall within a band of width less than 2ε

The definition of uniform convergence noted above is equivalent to the following statement.

♠ **Theorem:**
The sequence (f_n) of real functions on $D \subseteq \boldsymbol{R}$ converges uniformly to f on D if and only if

$$\sup_{x \in D} |f_n(x) - f(x)| = 0 \text{ as } n \to \infty.$$

3.2.3 Cauchy Criterion for Series of Functions

As in the case of real sequences, the Cauchy criterion is available for testing uniform convergence for sequences of functions.

♠ **Cauchy criterion for uniform convergence:**
The sequence of f_n defined on $D \subseteq \boldsymbol{R}$ converges uniformly to f on D if and only if, given $\varepsilon > 0$, there is a positive integer $N = N(\varepsilon)$ such that

$$m, n > N \;\Rightarrow\; |f_m(x) - f_n(x)| < \varepsilon \text{ for all } x \in D, \tag{3.7}$$

or equivalently,

$$m, n > N \;\Rightarrow\; \sup_{x \in D} |f_m(x) - f_n(x)| < \varepsilon.$$

Proof Suppose that $f_n(x)$ converges uniformly to $f(x)$ on D. Let $\varepsilon > 0$ and choose $N \in \boldsymbol{N}$ such that

$$n > N \quad \Rightarrow \quad |f_n(x) - f(x)| < \frac{\varepsilon}{2} \text{ for all } x \in D.$$

If $m, n \geq N$, we have

$$|f_n(x) - f_m(x)| \leq |f_n(x) - f(x)| + |f(x) - f_n(x)| < \varepsilon \text{ for all } x \in D.$$

This result implies that if $f_n(x)$ is uniformly convergent to $f(x)$ on D, there exists an N that satisfies (3.7) for any small ε.

Next we consider the converse. Suppose that (f_n) satisfies the criterion given by (3.7). Then, for each point of $x \in D$, $(f_n(x))$ forms a Cauchy sequence and thus converges pointwise to

$$f(x) = \lim_{n \to \infty} f_n(x) \text{ for all } x \in D.$$

We now show that this convergence is uniform. Let $n > N$ be fixed and take the limit $m \to \infty$ in (3.7) to obtain

$$n > N \quad \Rightarrow \quad |f_n(x) - f(x)| < \varepsilon \text{ for all } x \in D,$$

where N is independent of x, from which we conclude that the convergence of (f_n) to f is uniform. ♣

3.2.4 Continuity of the Limit Function

The most important feature of uniform convergence is that it overcomes some of the shortcomings of pointwise convergence demonstrated in Sect. 3.2.1; i.e., pointwise convergence does not preserve continuity, integrability, and differentiability of terms of the functions $f_n(x)$. We now examine the situation under uniform convergence, starting with the continuity of $f_n(x)$.

♠ **Theorem:**

If f_n converges uniformly to f on $D \subseteq \boldsymbol{R}$, then, if f_n is continuous at $c \in D$, so is f.

Remark. Note that the uniform convergence of f_n on D is a sufficient, but not a necessary, condition for f to be continuous. In fact, if f_n is not uniformly convergent on D, then its limit f may or may not be continuous at $c \in D$.

For the proof, it suffices to see that

$$\lim_{x \to c} f(x) = \lim_{x \to c} \lim_{n \to \infty} f_n(x) = \lim_{n \to \infty} \lim_{x \to c} f_n(x) = \lim_{n \to \infty} f_n(c) = f(c), \qquad (3.8)$$

which guarantees the continuity of the limit function $f(x)$ at $x = c$. In (3.8), we have used the interchangeability of limiting processes expressed by

$$\lim_{x\to c}\lim_{n\to\infty} f_n(x) = \lim_{n\to\infty}\lim_{x\to c} f_n(x),$$

which follows from the lemma below.

♠ **Lemma:**

Let c be a limit point of $D \subseteq \boldsymbol{R}$ and assume that f_n converges uniformly to f on $D\backslash\{c\}$. If

$$\lim_{x\to c} f_n(x) = \ell_n \tag{3.9}$$

exists for each n, then

(i) (ℓ_n) is convergent, and

(ii) $\lim_{x\to c} f(x)$ exists and coincides with $\lim_{n\to\infty}\ell_n$; i.e.,

$$\lim_{n\to\infty}\lim_{x\to c} f_n(x) = \lim_{x\to c}\lim_{n\to\infty} f_n(x). \tag{3.10}$$

Proof Let $\varepsilon > 0$. Since (f_n) converges uniformly on $D\backslash\{c\}$, it satisfies the Cauchy criterion; i.e., there is a positive integer N such that

$$m, n > N \;\Rightarrow\; |f_n(x) - f_m(x)| < \varepsilon \text{ for all } x \in D\backslash\{c\}. \tag{3.11}$$

Take the limit $x \to c$ in (3.11) to obtain

$$m, n > N \;\Rightarrow\; |\ell_n - \ell_m| < \varepsilon. \tag{3.12}$$

This implies that (ℓ_n) is a Cauchy sequence and thus convergent, which proves statement **(i)** above.

To prove **(ii)**, let

$$\ell = \lim_{n\to\infty} \ell_n.$$

Set $n = N$ and $m \to \infty$ in (3.9), (3.11), and (3.12) to set the following results:

$$\lim_{x\to c} f_N(x) = \ell_N, \tag{3.13}$$

$$|f_N(x) - f(x)| < \varepsilon \text{ for all } x \in D\backslash\{c\}, \tag{3.14}$$

and

$$|\ell_N - \ell| < \varepsilon. \tag{3.15}$$

In addition, the existence of (3.13) implies that there exists a $\delta > 0$ such that

$$|x - c| < \delta \text{ with } x \in D\backslash\{c\} \;\Rightarrow\; |f_N(x) - \ell_N| < \varepsilon. \tag{3.16}$$

Using (3.14), (3.15) and (3.16), we obtain

$$|x-c|<\delta \text{ with } x \in D\backslash\{c\}$$
$$\Rightarrow \quad |f(x)-\ell| \le |f(x)-f_N(x)| + |f_N(x)-\ell_N| + |\ell_N-\ell| < 3\varepsilon.$$

This means that

$$\lim_{x\to c} f(x) = \ell,$$

which is equivalent to the desired result of (3.10). ♣

Remark. The contraposition of the theorem tells us that if the limit function f is discontinuous, the convergence of f_n is not uniform. The example in Sect. 3.2.1 demonstrated such a sequence.

3.2.5 Integrability of the Limit Function

We know that the limit function $f(x)$ becomes continuous if the sequence $(f_n(x))$ of continuous functions is uniformly convergent. This immediately results in the following theorem.

♠ **Theorem:**

Suppose f_n be integrable on $[a,b]$ for each n. Then, if f_n converges uniformly to f on $[a,b]$, the limit function f is also integrable, so that

$$\int_a^b f(x)dx = \lim_{n\to\infty} \int_a^b f_n(x)dx, \tag{3.17}$$

or equivalently,

$$\int_a^b \lim_{n\to\infty} f_n(x)dx = \lim_{n\to\infty} \int_a^b f_n(x)dx.$$

Proof Since f_n for every n is integrable on $[a,b]$, it is continuous (piecewise, at least) on $[a,b]$. Thus $f(x)$ is also continuous (piecewise at least) on $[a,b]$ in view of the theorem given in Sect. 3.2.4, so that $f(x)$ is integrable on $[a,b]$. Furthermore, we observe that

$$\begin{aligned}\left|\int_a^b f_n(x)dx - \int_a^b f(x)dx\right| &\le \int_a^b |f_n(x)-f(x)|\, dx \\ &\le \int_a^b \sup_{x\in[a,b]} |f_n(x)-f(x)|\, dx \\ &\le (b-a) \sup_{x\in[a,b]} |f_n(x)-f(x)|\,.\end{aligned}$$

The uniform convergence of (f_n) ensures that

$$\sup_{x\in[a,b]} |f_n(x) - f(x)| \to 0 \text{ as } n \to \infty,$$

which immediately gives the desired result shown in (3.17). ♣

Remark.

1. Note again that uniform convergence is a *sufficient* but not a *necessary* condition for (3.17) to be valid, so (3.17) may be valid even in the absence of uniform convergence. For instance, the convergence of (f_n) with $f_n(x) = x^n$ on $[0, 1]$ is not uniform but we have

$$\int_0^1 f_n(x)dx = \int_0^1 x^n dx = \frac{1}{n+1} \to 0 = \int_0^1 f(x)dx.$$

2. The conditions on f_n stated in the theorem will be significantly relaxed when we take up the Lebesgue integral in Chap. 6.

3.2.6 Differentiability of the Limit Function

After the last two subsections, readers may expect that results for differentiability will be similar to those for continuity and integrability; i.e., they may be tempted to conclude that the differentiability of terms of functions $f_n(x)$ will be preserved if (f_n) converges uniformly to f. However, this is not the case. In fact, even if f_n converges uniformly to f on $[a, b]$ and f_n is differentiable at $c \in [a, b]$, it may occur that

$$\lim_{n\to\infty} f_n'(c) \neq f'(c).$$

Consider the following example:

Examples Suppose the sequence (f_n) is defined by

$$f_n(x) = \sqrt{x^2 + \frac{1}{n^2}}, \quad x \in [-1, 1]. \tag{3.18}$$

Clearly (3.18) is differentiable for each n, and the sequence (f_n) converges uniformly on $[-1, 1]$ to

$$f(x) = |x| \tag{3.19}$$

since

$$\begin{aligned} |f_n(x) - f(x)| &= \sqrt{x^2 + \frac{1}{n^2}} - \sqrt{x^2} \\ &= \frac{\frac{1}{n^2}}{\sqrt{x^2 + \frac{1}{n^2}} + \sqrt{x^2}} \leq \frac{1}{n} \to 0, \quad \text{for all } x \in [-1, 1]. \end{aligned}$$

However, the limit function f of (3.19) is not differentiable at $x = 0$. Hence, the desired result

$$\lim_{n\to\infty} f_n'(x) = f'(x) \tag{3.20}$$

breaks down at $x = 0$.

The following theorem provides sufficient conditions for (3.20) to be satisfied. The important point is that it requires the uniform convergence of the *derivatives* f_n', not of the functions f_n themselves.

♠ **Theorem:**
Suppose (f_n) to be a sequence of differentiable functions on $[a, b]$ that converge at a certain point $x_0 \in [a, b]$. If the sequence (f_n') is uniformly convergent on $[a, b]$, then

(i) (f_n) is also uniformly convergent on $[a, b]$ to f,

(ii) f is differentiable on $[a, b]$, and

(iii) $\lim_{n\to\infty} f_n'(x) = f'(x)$.

Proof Let $\varepsilon > 0$. From the convergence of $(f_n(x_0))$ and the uniform convergence of (f_n'), we conclude that there is an $N \in \boldsymbol{N}$ such that

$$m, n > N \;\Rightarrow\; |f_n'(x) - f_m'(x)| < \varepsilon \text{ for all } x \in [a, b] \tag{3.21}$$

and

$$m, n > N \;\Rightarrow\; |f_n(x_0) - f_m(x_0)| < \varepsilon. \tag{3.22}$$

Given any two points $x, t \in [a, b]$, it follows from the mean value theorem applied to $f_n - f_m$ that there is a point c between x and t such that

$$f_n(x) - f_m(x) - [f_n(t) - f_m(t)] = (x - t)\,[f_n'(c) - f_m'(c)]\,.$$

Using (3.21), we have

$$m, n > N \;\Rightarrow\; |f_n(x) - f_m(x) - [f_n(t) - f_m(t)]| < \varepsilon |x - t|. \tag{3.23}$$

From (3.22) and (3.23), it follows that

$$\begin{aligned} |f_n(x) - f_m(x)| &\le |f_n(x) - f_m(x) - [f_n(x_0) - f_m(x_0)]| + |f_n(x_0) - f_m(x_0)| \\ &< \varepsilon |x - x_0| + \varepsilon \\ &< \varepsilon(b - a + 1) \;= C\varepsilon, \quad \text{for all } x \in [a, b], \end{aligned}$$

Which means that (f_n) converges uniformly to some limit f. Hence, statement **(i)** has been proven.

Next we consider the proofs of **(ii)** and **(iii)**. For any fixed point $x \in [a,b]$, define

$$f_n(t) = \frac{f_n(t) - f_n(x)}{t - x}, \quad t \in [a,b] \backslash \{x\}$$

and

$$g(t) = \frac{f(t) - f(x)}{t - x}, \quad t \in [a,b] \backslash \{x\}.$$

Clearly, $f_n \to g$ as $n \to \infty$; furthermore, if $m, n \geq N$, the result of (3.23) tells us that

$$|f_n(t) - f_m(t)| < \varepsilon \text{ for all } t \in [a,b] \backslash \{x\}.$$

Thus in view of the Cauchy criterion, we see that f_n converges uniformly to g on $[a,b] \backslash \{x\}$. Now we observe that

$$\lim_{t \to x} f_n(t) = f_n'(x) \text{ for all } n \in \boldsymbol{N}. \tag{3.24}$$

Then, uniform convergence of f_n ensures taking the limit of $n \to \infty$ in (3.24) followed by interchanging the order of the limit processes, which yields

$$\lim_{n \to \infty} \lim_{t \to x} f_n(t) = \lim_{t \to x} g(t) = \lim_{t \to x} \frac{f(t) - f(x)}{t - x} = f'(x) = \lim_{n \to \infty} f_n'(x).$$

This proves that f is differentiable at x and that

$$f'(x) = \lim_{n \to \infty} f_n'(x). \quad ♣$$

Remark. That the uniform convergence of (f_n') is just sufficient, not necessary, is seen by considering the sequence

$$f_n(x) = \frac{x^{n+1}}{n+1}, \quad x \in (0,1).$$

This converges uniformly to 0, and its derivative $f_n'(x) = x^n$ also converges to 0. The conclusions **(i)**–**(iii)** given in the theorem above are thus all satisfied. But the convergence of (f_n') is not uniform.

Exercises

1. For the function

$$f_n(x) = nx(1 - x^2)^n, \quad x \in [0,1],$$

check that an interchange of the order of the limiting process $n \to \infty$ and integration gives different results.

Solution: The given function is integrable for each n so that

$$\int_0^1 f_n(x)dx = n\int_0^1 x(1-x^2)^n dx = \left[\frac{-n}{2(n+1)}(1-x^2)^{n+1}\right]_0^1$$
$$= \frac{n}{2(n+1)} \to \frac{1}{2}.$$

On the other hand, the limit given by

$$f(x) = \lim_{n\to\infty} f_n(x) = 0 \quad \text{for all } x \in [0,1]$$

yields $\int_0^1 f(x)dx = 0$. We thus conclude that

$$\lim_{n\to\infty}\int_0^1 f_n(x)dx \neq \int_0^1 \lim_{n\to\infty} f_n(x)dx;$$

i.e., interchanging the order of integration and limiting processes is not in general allowed under pointwise convergence. ♣

2. For $f_n(x)$ given by

$$f_n(x) = \begin{cases} -1 & x < -\frac{1}{n}, \\ \sin\left(\frac{n\pi x}{2}\right) & -\frac{1}{n} < x < \frac{1}{n}, \\ 1 & x > \frac{1}{n}, \end{cases}$$

check the continuity of its limit $f(x) = \lim_{n\to\infty} f_n(x)$ at $x = 0$.

Solution: $f_n(x)$ is differentiable for any $x \in \boldsymbol{R}$ for all n, and thus is continuous at $x = 0$ for all n. However, its limit,

$$f(x) = \begin{cases} -1 & x < 0, \\ 0 & x = 0, \\ 1 & x > 0 \end{cases}$$

is not continuous at $x = 0$. Hence, for the sequence of functions $\{f_n(x)\}$, the order of the limiting process $n \to \infty$ and the differentiation with respect to x is not interchangeable. ♣

3. Show that the sequence of functions $(f_n(x))$ defined by

$$f_n(x) = nxe^{-nx} \tag{3.25}$$

converges uniformly to $f(x) = 0$ on $x > 0$.

Solution: In view of the previous theorem, we show that

$$\sup\{f_n(x) : x \geq a\} = 0 \text{ as } n \to \infty,$$

where $a > 0$. To prove this, we consider the derivative

$$f_n'(x) = ne^{-nx}(1 - nx). \tag{3.26}$$

It follows from (3.26) that $x = 1/n$ is the only critical point of f_n. Now we choose a positive integer N such that $a > 1/N$. Then, the function f_n for each $n \geq N$ has no critical point on $x \geq a$, and is monotonically decreasing. Therefore, the maximum of $f_n(x)$ is attained at $x = a$ for any $n > N$, with the result that

$$\sup_{x \in [a,\infty)} f_n(x) = f_n(a) = nae^{-na} \to 0 \quad (n \to \infty).$$

This holds for any $a > 0$; hence, we conclude that f_n converges uniformly to 0 on $(0, \infty)$, i.e., on $x > 0$. ♣

Remark. Note that the range of uniform convergence of (3.25) is the open interval $(0, \infty)$, not the closed one $[0, \infty)$. Since in the latter case we have

$$\sup_{x \in [0,\infty)} f_n(x) = f_n\left(\frac{1}{n}\right) = \frac{1}{e} \not\to 0,$$

it is clear that (f_n) does not converge uniformly on $[0, \infty)$.

3.3 Series of Real Functions

3.3.1 Series of Functions

We close this chapter by considering convergence properties of series of real-valued functions. Assume a sequence (f_n) of functions defined on $D \subseteq \boldsymbol{R}$. By analogy with series of real numbers, we can define a series of functions by

$$S_n(x) = \sum_{k=1}^{n} f_k(x), \quad x \in D,$$

which gives a sequence $(S_n) = (S_1, S_2, \cdots)$.

As n increases, the sequence (S_n) may or may not converge to a finite value, depending on the feature of functions $f_k(x)$ as well as the point x in question. If the sequence converges for each point $x \in D$ (i.e., converges pointwise on D), then the limit of S_n is called the **sum of the infinite series of functions** $f_k(x)$ and is denoted by

$$S(x) = \lim_{n\to\infty} S_n(x) = \sum_{k=1}^{\infty} f_k(x), \quad x \in D.$$

It is obvious that the convergence of the series $S_n(x)$ implies the pointwise convergence $\lim_{n\to\infty} f_n(x) = 0$ on D. A series (S_n) that does not converge at a point $x \in D$ is said to diverge at that point.

Applied to series of functions, the Cauchy criterion for uniform convergence takes the following form:

♠ **Cauchy criterion for series of functions:**
The series S_n is uniformly convergent on D if and only if for every small $\varepsilon > 0$, there is a positive integer N such that

$$n > m > N$$
$$\Rightarrow |S_n(x) - S_m(x)| = \left| \sum_{k=m+1}^{n} f_k(x) \right| < \varepsilon \quad \text{for all } x \in D.$$

Set $n = m + 1$ in the above criterion to obtain

$$n > N \quad \Rightarrow \quad |f_n(x)| < \varepsilon \; \text{ for all } x \in D.$$

This results implies that the uniform convergence of $f_n(x) \to 0$ on D is a necessary condition for the convergence of $S_n(x)$ to be uniform on D. We will use this theorem when proving a more practical test for uniform convergence known as the Weierstrass M-test, which is presented in Sect. 3.3.3.

3.3.2 Properties of Uniformly Convergent Series of Functions

When a given series of functions $\sum f_k(x)$ is uniformly convergent, the properties of the sum $S(x)$ in terms of continuity, integrability, and differentiability can be easily inferred from the properties of the separate terms $f_k(x)$. In fact, applying the theorems given in Sects. 3.2.4–3.2.6 to the sequence (S_n) and using the linearity regarding the limiting process, integration, and differentiation, we obtain the parallel theorems shown below.

♠ **Continuity of the sum:**
Suppose $f_k(x)$ to be continuous for each k. If the sequence (S_n) of the series

$$S_n(x) = \sum_{k=1}^{n} f_k(x)$$

converges uniformly to $S(x)$, then $S(x)$ is also continuous, so that

$$\lim_{t\to x} S(t) = \lim_{t\to x} \sum_{k=1}^{\infty} f_k(t) = \sum_{k=1}^{\infty} \lim_{t\to x} f_k(t).$$

♠ **Integrability of the sum:**
Suppose f_k to be integrable on $[a, b]$ for all k. If (S_n) converges uniformly to S on $[a, b]$, we have

$$\int_a^b S(x)dx = \int_a^b \sum_{k=1}^{\infty} f_k(x)dx = \sum_{k=1}^{\infty} \int_a^b f_k(x)dx.$$

♠ **Differentiability of the sum:**
Let f_k be differentiable on $[a, b]$ for each k and suppose that (S_n) converges to S at some point $x_0 \in [a, b]$. If the series $\sum f_k'$ is uniformly convergent on $[a, b]$, then $S_n(x)$ is also uniformly convergent on $[a, b]$ and the sum $S(x)$ is differentiable on $[a, b]$, so that

$$\frac{d}{dx}S(x) = \frac{d}{dx}\left[\sum_{k=1}^{\infty} f_k(x)\right] = \sum_{k=1}^{\infty} \frac{df_k(x)}{dx} \quad \text{for all } x \in [a, b].$$

Observe that the second and third theorems provide a sufficient condition for performing term-by-term integration and differentiation, respectively, of an infinite series of functions. Without uniform convergence, such term-by-term calculations do not work.

3.3.3 Weierstrass M-test

The following is a very useful and simple test for the uniform convergence of a series of functions.

♠ **Weierstrass M test:** If there is a sequence of positive constants M_k for any x on the interval $[a, b]$ such that

$$|f_k(x)| \le M_k \tag{3.27}$$

and if the series

$$\sum_{k=0}^{\infty} M_k \tag{3.28}$$

converges, then the series of functions $\sum_{k=0}^{\infty} f_k(x)$ converges uniformly on $x \in [a, b]$.

Proof Since (3.28) converges, it follows from the Cauchy criterion that for any $\varepsilon > 0$ there exists a number N such that

$$n > m > N \Rightarrow \left| \sum_{k=0}^{n} M_k - \sum_{k=0}^{m} M_k \right| = \sum_{k=m}^{n} M_k < \varepsilon. \tag{3.29}$$

Furthermore, in view of the inequality rule for absolute values of sums and the relation (3.27), it follows that

$$\left| \sum_{k=m}^{n} f_k(x) \right| \le \sum_{k=m}^{n} |f_k(x)| \le \sum_{k=m}^{n} M_k \tag{3.30}$$

for all $x \in [a, b]$. Note that the left-hand term in (3.30) can be rewritten as

$$\left| \sum_{k=m}^{n} f_k(x) \right| = \left| \sum_{k=0}^{n} f_k(x) - \sum_{k=0}^{m} f_k(x) \right|. \tag{3.31}$$

From (3.29), (3.30), and (3.31), it follows that

$$n \ge m > N \Rightarrow \left| \sum_{k=0}^{n} f_k(x) - \sum_{k=0}^{m} f_k(x) \right| < \varepsilon \text{ for all } x \in [a, b],$$

which clearly indicates the uniform convergence of $\sum f_k(x)$ on $[a, b]$. ♣

Exercises

1. Determine the convergence of the series $\displaystyle\sum_{k=0}^{\infty} x^k$.

   Solution: It obviously converges to $1/(1-x)$ on the interval $[-a, a]$ if $0 < a < 1$. We show that this convergence is uniform on $[-a, a]$ for any $0 < a < 1$. A partial sum yields $S_n(x) = \sum_{k=0}^{n} x^k = (1 - x^n)/(1 - x)$, so that

   $$|S(x) - S_n(x)| = \frac{|x|^n}{|1 - x|} \le \frac{a^n}{1 - a} \quad \text{for } |x| \le a.$$

   Since $0 < a < 1$, the last term decreases monotonically with n; hence, for a given $\varepsilon > 0$, we can find an N such that $n > N \Rightarrow a^n/(1 - a) < \varepsilon$. Clearly the value of N does not depend on x. Therefore, we conclude that the infinite series $\sum x^k$ is uniformly convergent on $[-a, a]$ with $0 < a < 1$. ♣

2. Determine the convergence of the series $\sum_{k=0}^{\infty}(1-x)x^k$.

Solution: This converges to

$$S(x) = \begin{cases} 1, \text{ for } 0 < x < 1 \\ 0, \text{ at } x = 1 \end{cases}$$

but not uniformly. Actually, we have

$$|S(x) - S_n(x)| = \begin{cases} x^n & 0 < x < 1 \\ 0 & x = 1 \end{cases}$$

and if $\varepsilon = 1/4$, for instance, the inequality $x^n < 1/4$ $(0 < x < 1)$ is false for every fixed n because $x^n \to 1$ as $x \to 1$. ♣

3. Examine the uniform convergence of the series $\sum_{k=1}^{\infty} f_k(x)$, where **(i)** $f_k(x) = \frac{\cos kx}{k^2}$, **(ii)** $f_k(x) = \sin\left(\frac{x}{k^2}\right)$, and **(iii)** $f_k(x) = \frac{1}{k^2x^2}$.

Solution:

(i) The series converges uniformly for every real x. Check this by taking $M_k = 1/k^2$.

(ii) Let D be a subset of $\boldsymbol{R}$ bounded by c, i.e., $|x| \le c$ for all $x \in D$. Then we have

$$\left|\sin\left(\frac{x}{k^2}\right)\right| \le \frac{x}{k^2} \le \frac{c}{k^2} \quad \text{for all } x \in D.$$

Taking $M_k = c/k^2$ and noting that $\sum M_k$ is convergent, we conclude that $\sum f_k$ is uniformly convergent on any *bounded subset* of $\boldsymbol{R}$. Notably, however, this uniform convergence disappears when we extend the domain D to the whole $\boldsymbol{R}$. This is seen by noting that $f_k \to 0$ pointwise on $\boldsymbol{R}$, but

$$\sup_{x\in\boldsymbol{R}} |f_k(x)| \ge \left|\sin\left(\frac{k^2\pi/2}{k^2}\right)\right| = 1 \not\to 0,$$

which means that the convergence of (f_k) to 0 is not uniform on $\boldsymbol{R}$. In view of the theorem in Sect. 3.3.1, therefore, the series $\sum f_k$ fails to converge uniformly on $\boldsymbol{R}$.

(iii) The series $\sum_k 1/(k^2x^2)$ clearly converges pointwise on the open set $\boldsymbol{R}\backslash\{0\}$. Now let $c > 0$. For all $x \in \boldsymbol{R}$ such that $|x| > c$, we have $|f_k(x)| \le 1/(k^2c^2)$ for all k. Since $\sum_k 1/(k^2c^2)$ is convergent, the series $\sum f_k$ converges uniformly, by the M-test, on the closed set $\boldsymbol{R}\backslash(-c, c) = (-\infty, -c] \cup [c, \infty)$ for all $c > 0$. But, although $f_k \to 0$ pointwise on $\boldsymbol{R}\backslash\{0\}$, we have $\sup_{x \neq 0} |f_k(x)| \geq |f_k(1/k)| = 1 \not\to 0$. Hence, (f_k) does not converge uniformly to 0 on $\boldsymbol{R}\backslash\{0\}$, so the series $\sum f_k$ does not converge uniformly on $\boldsymbol{R}\backslash\{0\}$. ♣

3.4 Improper Integrals

3.4.1 Definitions

Suppose that a given function $f(x)$ is integrable on every open subinterval of (a, b). We try to perform the integration $\int_a^b f(x)dx$ under the following conditions:

1. $f(x)$ is unbounded in a neighborhood of $x = a$ or $x = b$.
2. The interval (a, b) itself is unbounded.

In Case **1**, we define a definite integral,

$$\int_a^b f(x)dx = \lim_{X \to b-0} \int_a^X f(x)dx,$$

if $f(x)$ is bounded and integrable on every finite interval (a, X) for $a < X < b$. Similarly, if $f(x)$ is bounded and integrable on every (X, b) for $a < X < b$, we can define

$$\int_a^b f(x)dx = \lim_{X \to a+0} \int_X^b f(x)dx.$$

These definite integrals are called **improper integrals**. Straightforward extensions of these results to Case **2** yields the other improper integrals:

$$\int_a^\infty f(x)dx = \lim_{X \to \infty} \int_a^X f(x)dx$$

and

$$\int_{-\infty}^b f(x)dx = \lim_{X \to \infty} \int_{-X}^b f(x)dx.$$

Examples **1.** The improper integral $\displaystyle\int_1^\infty \frac{dx}{x^2}$ has the value 1 since

$$\int_1^\infty \frac{dx}{x^2} = \lim_{A \to \infty} \int_1^A \frac{dx}{x^2}.$$

2. The improper integral $\displaystyle\int_0^4 \frac{dx}{\sqrt{x}}$ has the value 1 since

$$\int_0^4 \frac{dx}{\sqrt{x}} = \lim_{\varepsilon \to +0} \int_\varepsilon^4 \frac{dx}{\sqrt{x}} = \lim_{\varepsilon \to +0} \frac{2 - \sqrt{\varepsilon}}{2} = 1.$$

3.4.2 Convergence of an Improper Integral

An improper integral over $f(x)$ is said to **converge** if and only if the corresponding limit exists. Furthermore, it is said to **converge absolutely** if and only if the corresponding improper integral over $|f(x)|$ converges. (Keep in mind that absolute convergence implies convergence in the ordinary sense.) A convergent improper integral that does not converge absolutely is **conditionally convergent**.

An improper integral $\int_a^b f(x,y)dx$ **converges uniformly** on a set S of values of y if and only if the corresponding limit converges uniformly on S. A relevant theorem is given below.

♠ **Continuity theorem**

If $f(x,y)$ is a continuous function, then $\int_a^b f(x,y)dx$ is a continuous function of y in every open interval where the integral converges uniformly.

3.4.3 Principal Value Integral

Suppose that a bounded or unbounded open or closed interval, (a,b) or $[a,b]$, contains a discrete set of points $x = c_1, c_2, \cdots$, such that $f(x)$ is unbounded in a neighborhood of $x = c_i$ $(i = 1, 2, \cdots)$. Then, the integral $\int_a^b f(x)dx$ may be defined as a sum of improper integrals, introduced in the previous subsection; i.e.,

$$\int_a^b f(x)dx = \lim_{X_1 \to a+0} \int_{X_1}^c f(x)dx + \lim_{X_2 \to b-0} \int_c^{X_2} f(x)dx \quad (a < c < b), \tag{3.32}$$

$$\int_a^b f(x)dx = \lim_{X_1 \to c-0} \int_a^{X_1} f(x)dx + \lim_{X_2 \to c+0} \int_{X_2}^b f(x)dx \quad (a < c < b), \tag{3.33}$$

$$\int_{-\infty}^{\infty} f(x)dx = \lim_{X_1 \to \infty} \int_{-X_1}^c f(x)dx + \lim_{X_2 \to \infty} \int_c^{X_2} f(x)dx \tag{3.34}$$

if the limits exists.

Even though the integrals (3.32), (3.33) and (3.34) do not exist, the limits of integrals

$$\lim_{x \to \infty} \int_{-X}^{X} f(x)dx \quad \text{and} \quad \lim_{\delta \to 0} \left[\int_a^{c-\delta} f(x)dx + \int_{c+\delta}^b f(x)dx \right]$$

may exist. If any of these limits exist, the corresponding integral, (3.32), (3.33) or (3.34), is necessarily equal to its **principal value integral** (see Sect. 9.4.1).

3.4.4 Conditions for Convergence

In what follows we give the convergence criteria for improper integrals of the form

$$\int_a^{\infty} f(x)dx$$

and

$$\int_a^b f(x)dx = \lim_{X \to b-0} \int_a^X f(x)dx.$$

We assume that $f(x)$ is bounded and integrable on every bounded interval (a, X) that does not contain the upper limit of integration.

♠ **Cauchy's test** (= **necessary and sufficient conditions for convergence**):

The improper integral $\int_a^{\infty} f(x)dx$ converges if and only if for every positive real number ε, there exists a real number $M > a$ such that

$$X_2 > X_1 > M \;\Rightarrow\; \left| \int_{X_1}^{X_2} f(x)dx \right| < \varepsilon.$$

Similarly, $\int_a^{\infty} f(x)dx$ converges if and only if for every positive ε, there exists a positive $\delta < b - a$ such that

$$b - X_2 < b - X_1 < \delta \;\Rightarrow\; \left| \int_{X_1}^{X_2} f(x)dx \right| < \varepsilon.$$

Necessary and sufficient conditions for an improper integral to converge uniformly are stated below.

♠ **Weierstrass test**

The improper integral $\int_a^{\infty} f(x,y)dx$ [or $\int_a^b f(x,y)dx$] converges uniformly and absolutely on every set S of values of y such that $|f(x,y)| \le g(x)$ on the interval of integration, where $g(x)$ is a real comparison function whose integral $\int_a^{\infty} g(x)dx$ [or $\int_a^b g(x)dx$, respectively] converges.

Exercises

1. Show that the integral $\int_\pi^\infty \frac{\sin x}{x} dx$ converges.

Solution: We have

$$\int_\pi^A \frac{\sin x}{x} dx \leq \left[\frac{-\cos x}{x}\right]_\pi^A - \int_\pi^A \frac{\cos x}{x^2} dx,$$

so that

$$\left|\int_\pi^A \frac{\sin x}{x} dx\right| \leq \frac{1}{\pi} + \frac{1}{A} + \int_\pi^A \frac{dx}{x^2} = \left(\frac{1}{\pi} + \frac{1}{A}\right) + \left(\frac{1}{\pi} - \frac{1}{A}\right) = 2\pi.$$

This completes the proof. ♣

2. Show that $\int_1^\infty \left|\frac{\sin x}{x}\right| dx$ diverges.

Solution: It follows that

$$\begin{aligned}\int_{n\pi}^{(n+1)\pi} \left|\frac{\sin x}{x}\right| dx &= \int_0^\pi \frac{\sin x}{n\pi + x} dx > \frac{1}{(n+1)\pi} \int_0^\pi \sin x dx \\ &= \frac{2}{(n+1)\pi} > \frac{2}{\pi} \int_{n+1}^{n+2} \frac{dx}{x}.\end{aligned}$$

Hence, for $n > 1$ we have

$$\int_\pi^{n\pi} \left|\frac{\sin x}{x}\right| dx > \frac{2}{\pi} \int_2^{n+1} \frac{dx}{x} = \frac{1}{\pi} \log(n+1) \to \infty, \quad (n \to \infty).$$ ♣

3. Suppose that $f(x)$ is continuous within an interval $(a, b]$ and diverges at $x = a$ Prove that $\int_a^b f(x)dx$ converges if $(x-a)^p|f(x)|$ is bounded on the interval for $0 < p < 1$.

Solution: We assume that there is an appropriate positive number M such that

$$(x-a)^p|f(x)| < M \quad \text{for all } x \in (a, b].$$

Then we obtain

$$\begin{aligned}\int_{a+\varepsilon}^b |f(x)|dx &< M \int_{a+\varepsilon}^b \frac{dx}{(x-a)^p} = M\left[\frac{(x-p)^{1-p}}{1-p}\right]_{a+\varepsilon}^b \\ &= \frac{M}{1-p}\left[(b-a)^{1-p} - \varepsilon^{1-p}\right] < \frac{M}{1-p}(b-a)^{1-p} \\ &\quad (\text{since } 1-p > 0).\end{aligned} \tag{3.35}$$

Note that the integral on the left-hand side of (3.35) is monotonically increasing with decreasing ε, since $|f(x)| \geq 0$ over the integration interval. Yet it is bounded from above, as proved in (3.35). Hence, we conclude that the given integral is convergent (absolutely). ♣

4. Suppose that $f(x)$ is continuous within $[a, \infty)$ and that $x^p |f(x)|$ is bounded there for $p > 1$. Show that the integral $\int_a^b f(x)dx$ converges.

Solution: It follows from hypothesis that there is a positive number M such that

$$x^p |f(x)| < M \quad \text{for all } x \geq a.$$

Hence, we have for any $X > a$,

$$\int_a^X |f(x)|dx < M \int_a^X \frac{dx}{x^p} = \frac{-M}{p-1} \left[\frac{1}{x^{p-1}} \right]_a^X < \frac{M}{p-1} \frac{1}{a^{p-1}},$$

which completes the proof. ♣

Part II

Functional Analysis

4

Hilbert Spaces

Abstract A Hilbert space is an abstract vector space with the following two properties: the inner product property (Sect. 4.1.3), which determines the geometry of the vector space, and the completeness property (Sect. 4.1.6), which guarantees the self-consistency of the space. Most of the mathematical topics covered in this volume are based on Hilbert spaces. In particular, L^p spaces and l^p spaces (Sect. 4.3), which are specific classes of Hilbert spaces, are crucial for the formulation of the theories of orthonormal polynomials, Lebesgue integrals, Fourier analyses, and others, as we discuss in subsequent chapters.

4.1 Hilbert Spaces

4.1.1 Introduction

This section provides a framework for an understanding of **Hilbert spaces**. Plainly speaking, Hilbert spaces are the generalization of familiar finite-dimensional spaces to the infinite-dimensional case. In fact, the geometric structure of Hilbert spaces is very similar to that of ordinary Euclidean geometry. This analogy comes from the fact that the concept of **orthogonality** can be introduced in any Hilbert space so that the familiar Pythagorean theorem holds for elements involved in the space. Moreover, owing to its generality, a large number of problems in physics and engineering can be successfully treated with a geometric point of view in Hilbert spaces.

As we shall see later, Hilbert spaces are defined as a specific class of **vector spaces** endowed with the following two properties: **inner product** and **completeness**. The former property leads to a rich geometric structure and the latter enables us to describe an element in the space in terms of a set of **orthonormal bases**. These facts result in the possibility of establishing a wide variety of **complete orthonormal sets of functions** in Hilbert spaces; we discussed this point in detail in Sects. 5.1 and 5.2. For a better understanding of subsequent discussions, we provide all necessary definitions in this section, and then describe several important consequences relevant to an understanding of the nature of Hilbert spaces.

4.1.2 Abstract Vector Spaces

In order to make this text self-contained, we first give a brief summary of the definition of **vector spaces**. A more precise description of vector spaces and some related matters will be provided in Sect. 4.2.1.

♠ **Vector spaces:**

A vector space V is a collection of elements called **vectors**, which we denote by $\boldsymbol{x}, \boldsymbol{y}, \cdots$, that satisfy the following postulates:

1. There exists an operation $(+)$ on the vectors $\boldsymbol{x}$ and $\boldsymbol{y}$ such that $\boldsymbol{x}+\boldsymbol{y} = \boldsymbol{y}+\boldsymbol{x}$, where the resultant quantity $\boldsymbol{y}+\boldsymbol{x}$ also must be a vector.
2. There exists an **identity vector** (denoted by $\mathbf{0}$) that yields $\boldsymbol{x}+\mathbf{0} = \boldsymbol{x}$.
3. For every $\boldsymbol{x} \in V$, there exists a vector $\alpha\boldsymbol{x} \in V$ in which α is an arbitrary scalar (real and complex). In addition,

$$\alpha(\beta\boldsymbol{x}) = (\alpha\beta)\boldsymbol{x}, \qquad 1(\boldsymbol{x}) = \boldsymbol{x} \text{ for all } \boldsymbol{x},$$
$$\alpha(\boldsymbol{x}+\boldsymbol{y}) = \alpha\boldsymbol{x} + \alpha\boldsymbol{y}, \qquad (\alpha+\beta)\boldsymbol{x} = \alpha\boldsymbol{x} + \beta\boldsymbol{x}.$$

Emphasis is placed on the fact that vector spaces are not limited to a set of geometric *arrows* embedded in a **Euclidean space** (see Sects. 4.1.3 and 19.2.3); rather, they are general mathematical systems that have a specific algebraic structure. Several examples of such abstract vector spaces are given below.

Examples **1.** The set of all n-tuples of complex numbers denoted by

$$\boldsymbol{x} = (\xi_1, \xi_2, \cdots, \xi_n)$$

forms a vector space if the addition of vectors and the multiplication of a vector by a scalar are defined by

$$\begin{aligned} \boldsymbol{x}+\boldsymbol{y} &= (\xi_1, \xi_2, \cdots, \xi_n) + (\eta_1, \eta_2, \cdots, \eta_n) \\ &= (\xi_1+\eta_1, \xi_2+\eta_2, \cdots, \xi_n+\eta_n), \\ \alpha\boldsymbol{x} &= \alpha(\xi_1, \xi_2, \cdots, \xi_n) = (\alpha\xi_1, \alpha\xi_2, \cdots, \alpha\xi_n). \end{aligned}$$

2. The set of all complex numbers $\{z\}$ forms a **complex vector space** (see Sect. 4.2.1), where $z_1 + z_2$ and αz are interpreted as ordinary complex numerical addition and multiplication,

3. The set of all polynomials in a real variable x, constituting the set $\{1, x, x^2, x^3, \cdots\}$, with complex coefficients is a complex vector space if vector addition and scalar multiplication are the ordinary addition of two polynomials and the multiplication of a polynomial by a complex number, respectively.

4.1.3 Inner Product

The structure of a vector space is enormously enriched by introducing the concept of **inner product**, which enables us to define the length of a vector in a given vector space or the angle between the two vectors involved.

♠ **Inner product:**
An inner product is a scalar-valued function of the ordered pair of vectors $\boldsymbol{x}$ and $\boldsymbol{y}$ such that
1. $(\boldsymbol{x}, \boldsymbol{y}) = (\boldsymbol{y}, \boldsymbol{x})^*$.
2. $(\alpha \boldsymbol{x} + \beta \boldsymbol{y}, \boldsymbol{z}) = \alpha^*(\boldsymbol{x}, \boldsymbol{z}) + \beta^*(\boldsymbol{y}, \boldsymbol{z})$, where α and β are certain complex numbers.
3. $(\boldsymbol{x}, \boldsymbol{x}) \geq 0$ for any $\boldsymbol{x}$; $(\boldsymbol{x}, \boldsymbol{x}) = 0$ if and only if $\boldsymbol{x} = 0$.

Here, the asterisk $(*)$ indicates that one is to take the **complex conjugate.**

Remark. Vector spaces endowed with an inner product are called **inner product spaces**. In particular, a real inner product space is called a **Euclidean space** and a complex inner product space is called a **unitary space**.

The algebraic properties **1** and **2** are in principle the same as those governing the **scalar product** in ordinary vector algebra in a real vector space. The only property that is not obvious is that in a complex space, the inner product is not linear, but rather **conjugate linear** with respect to the first factor, i.e.,

$$(\alpha \boldsymbol{x}, \boldsymbol{y}) = \alpha^*(\boldsymbol{x}, \boldsymbol{y}).$$

Examples **1.** The simplest, but an important, example of an inner product space is the space, denoted by $\boldsymbol{C}$, that consists of a set of complex numbers $\{z_1, z_2, \cdots, z_n\}$. For two vectors $\boldsymbol{x} = (\xi_1, \xi_2, \cdots \xi_n)$ and $\boldsymbol{y} = (\eta_1, \eta_2, \cdots \eta_n)$ on $\boldsymbol{C}$, the inner product is defined by

$$(\boldsymbol{x}, \boldsymbol{y}) = \sum_{i=1}^{n} \xi_i^* \eta_i.$$

2. Suppose that $f(x)$ and $g(x)$ are polynomials in the complex vector space defined on the closed interval $x \in [0, 1]$. They then constitute an inner product space under the inner product defined by

$$(f, g) = \int_0^1 f(x)^* g(x) w(x) dx,$$

where $w(x)$ is a **weight function**. The weight function becomes important when defining the inner product of polynomials, which is treated in Chap. 5.

3. If $\boldsymbol{x} = [\xi_1, \xi_2, \xi_3, \xi_4]$ and $\boldsymbol{y} = [\eta_1, \eta_2, \eta_3, \eta_4]$ are column **four-vectors** having real-valued elements, then the quantity

$$(\boldsymbol{x}, \boldsymbol{y}) \equiv \xi_1\eta_1 + \xi_2\eta_2 + \xi_3\eta_3 - \xi_4\eta_4 \tag{4.1}$$

satisfies requirements **1** and **2** for an inner product, but not **3** since the quantity $(\boldsymbol{x}, \boldsymbol{x})$ is *not* positive-definite. Thus the entity (4.1) is not an inner product, but it plays an important role in the theory of special relativity.

For a complex vector space, the inner product is not symmetrical as it is in a real vector space. That is, $(\boldsymbol{x}, \boldsymbol{y}) \neq (\boldsymbol{y}, \boldsymbol{x})$ but rather $(\boldsymbol{x}, \boldsymbol{y}) = (\boldsymbol{y}, \boldsymbol{x})^*$. This implies that $(\boldsymbol{x}, \boldsymbol{x})$ is real for every $\boldsymbol{x}$, so we can define the *length* of the vector $\boldsymbol{x}$ by

$$||\boldsymbol{x}|| = (\boldsymbol{x}, \boldsymbol{x})^{1/2}.$$

Since $(\boldsymbol{x}, \boldsymbol{x}) \geq 0$, $||\boldsymbol{x}||$ is always nonnegative and real. The quantity $||\boldsymbol{x}||$ is referred to as the **norm** of the vector $\boldsymbol{x}$. Note also that

$$||\alpha\boldsymbol{x}|| = (\alpha\boldsymbol{x}, \alpha\boldsymbol{x})^{1/2} = [\alpha^*\alpha(\boldsymbol{x}, \boldsymbol{x})]^{1/2} = |\alpha| \cdot ||\boldsymbol{x}||.$$

Remark. Precisely speaking, the quantity $\|\boldsymbol{x}\|$ introduced above is a special kind of norm that is associated with an inner product; in fact, the norm was originally a more general concept that was independent of the inner product (see Sect. 4.2.2).

4.1.4 Geometry of Inner Product Spaces

Once a vector space is endowed with an inner product, several important theorems that can be easily interpreted in analogy with Euclidean geometry can be applied. The following three theorems characterize the geometric nature of inner product spaces ($\boldsymbol{x} \neq 0$ and $\boldsymbol{y} \neq 0$ are assumed; otherwise the theorems all become trivial).

♠ **Schwarz inequality:**
For any two elements $\boldsymbol{x}$ and $\boldsymbol{y}$ of an inner product space, we have

$$|(\boldsymbol{x}, \boldsymbol{y})| \leq \|\boldsymbol{x}\| \; \|\boldsymbol{y}\|. \tag{4.2}$$

The equality holds if and only if $\boldsymbol{x}$ and $\boldsymbol{y}$ are linearly independent.

Proof From the definition of the inner product, we have

$$0 \leq (\boldsymbol{x} + \alpha\boldsymbol{y}, \boldsymbol{x} + \alpha\boldsymbol{y}) = (\boldsymbol{x}, \boldsymbol{x}) + \alpha(\boldsymbol{x}, \boldsymbol{y}) + \alpha^*(\boldsymbol{y}, \boldsymbol{x}) + |\alpha|^2(\boldsymbol{y}, \boldsymbol{y}). \tag{4.3}$$

Now, set $\alpha = -(\boldsymbol{x},\boldsymbol{y})/(\boldsymbol{y},\boldsymbol{y})$ and multiply by $(\boldsymbol{y},\boldsymbol{y})$ to obtain

$$0 \le (\boldsymbol{x},\boldsymbol{x})(\boldsymbol{y},\boldsymbol{y}) - |(\boldsymbol{x},\boldsymbol{y})|^2,$$

which gives Schwarz's inequality.

Next, we prove the statement of the equality in (4.2). If $\boldsymbol{x}$ and $\boldsymbol{y}$ are linearly dependent, then $\boldsymbol{y} = \alpha\boldsymbol{x}$ for some complex number α so that we have

$$|(\boldsymbol{x},\boldsymbol{y})| = |(\boldsymbol{x},\alpha\boldsymbol{x})| = |\alpha|(\boldsymbol{x},\boldsymbol{x}) = |\alpha|\|\boldsymbol{x}\|\,\|\boldsymbol{x}\| = \|\boldsymbol{x}\|\,\|\alpha\boldsymbol{x}\| = \|\boldsymbol{x}\|\,\|\boldsymbol{y}\|.$$

The converse is also true; let $\boldsymbol{x}$ and $\boldsymbol{y}$ be vectors such that $|(\boldsymbol{x},\boldsymbol{y})| = \|\boldsymbol{x}\|\,\|\boldsymbol{y}\|$, or equivalently,

$$|(\boldsymbol{x},\boldsymbol{y})|^2 = (\boldsymbol{x},\boldsymbol{y})(\boldsymbol{y},\boldsymbol{x}) = (\boldsymbol{x},\boldsymbol{x})(\boldsymbol{y},\boldsymbol{y}) = \|\boldsymbol{x}\|^2\,\|\boldsymbol{y}\|^2. \tag{4.4}$$

Then we set

$$\begin{aligned}&\|(\boldsymbol{y},\boldsymbol{y})\boldsymbol{x} - (\boldsymbol{y},\boldsymbol{x})\boldsymbol{y}\|^2\\ &= \|\boldsymbol{y}\|^4\|\boldsymbol{x}\|^2 + |(\boldsymbol{y},\boldsymbol{x})|^2\,\|\boldsymbol{y}\|^2 - \|\boldsymbol{y}\|^2(\boldsymbol{y},\boldsymbol{x})(\boldsymbol{x},\boldsymbol{y}) - \|\boldsymbol{y}\|^2(\boldsymbol{y},\boldsymbol{x})^*(\boldsymbol{y},\boldsymbol{x})\\ &= 0,\end{aligned} \tag{4.5}$$

where the postulate (4.4) and the relation $(\boldsymbol{y},\boldsymbol{x})^* = (\boldsymbol{x},\boldsymbol{y})$ were used. The result (4.5) means that

$$(\boldsymbol{y},\boldsymbol{y})\boldsymbol{x} - (\boldsymbol{x},\boldsymbol{y})\boldsymbol{y} = \boldsymbol{0},$$

which clearly shows that $\boldsymbol{x}$ and $\boldsymbol{y}$ are linearly dependent, which completes the proof. ♣

♠ **Triangle inequality:**
For any two elements $\boldsymbol{x}$ and $\boldsymbol{y}$ of an inner product space, we have

$$\|\boldsymbol{x}+\boldsymbol{y}\| \le \|\boldsymbol{x}\| + \|\boldsymbol{y}\|.$$

Proof Setting $\alpha = 1$ in (4.3), we have

$$\begin{aligned}\|\boldsymbol{x}+\boldsymbol{y}\|^2 &= (\boldsymbol{x},\boldsymbol{x}) + (\boldsymbol{y},\boldsymbol{y}) + 2\mathrm{Re}(\boldsymbol{x},\boldsymbol{y})\\ &\le (\boldsymbol{x},\boldsymbol{x}) + (\boldsymbol{y},\boldsymbol{y}) + 2\,|(\boldsymbol{x},\boldsymbol{y})|\\ &\le \|\boldsymbol{x}\|^2 + \|\boldsymbol{y}\|^2 + 2\|\boldsymbol{x}\|\|\boldsymbol{y}\| \quad \text{(by Schwarz's inequality)}\\ &= (\|\boldsymbol{x}\| + \|\boldsymbol{y}\|)^2,\end{aligned}$$

which proves the desired inequality. ♣

♠ **Parallelogram law:**
For any two elements $\boldsymbol{x}$ and $\boldsymbol{y}$ of an inner product space, we have

$$\|\boldsymbol{x}+\boldsymbol{y}\|^2 + \|\boldsymbol{x}-\boldsymbol{y}\|^2 = 2\left(\|\boldsymbol{x}\|^2 + \|\boldsymbol{y}\|^2\right).$$

Proof We have

$$\begin{aligned}\|\boldsymbol{x}+\boldsymbol{y}\|^2 &= (\boldsymbol{x},\boldsymbol{x}) + (\boldsymbol{x},\boldsymbol{y}) + (\boldsymbol{y},\boldsymbol{x}) + (\boldsymbol{y},\boldsymbol{y}) \\ &= \|\boldsymbol{x}\|^2 + (\boldsymbol{x},\boldsymbol{y}) + (\boldsymbol{y},\boldsymbol{x}) + \|\boldsymbol{y}\|^2. \end{aligned} \tag{4.6}$$

Now replace $\boldsymbol{y}$ by $-\boldsymbol{y}$ to obtain

$$\|\boldsymbol{x}-\boldsymbol{y}\|^2 = \|\boldsymbol{x}\|^2 - (\boldsymbol{x},\boldsymbol{y}) - (\boldsymbol{y},\boldsymbol{x}) + \|\boldsymbol{y}\|^2. \tag{4.7}$$

By adding (4.6) and (4.7), we attain our objective. ♣

4.1.5 Orthogonality

One of the most important consequences of having the inner product is being able to define the **orthogonality** of vectors. The orthogonality allows us to establish a set of **orthonormal bases** that span the inner product space in question, thus yielding a useful way to analyze both the nature of the space itself and the relation between the constituents involved in that space.

♠ **Orthogonality:**
Two vectors $\boldsymbol{x}$ and $\boldsymbol{y}$ in an inner product space are called **orthogonal** if and only if $(\boldsymbol{x},\boldsymbol{y}) = 0$.

Notably, if $(\boldsymbol{x},\boldsymbol{y}) = 0$, then $(\boldsymbol{x},\boldsymbol{y}) = (\boldsymbol{y},\boldsymbol{x})^* = 0$ so that $(\boldsymbol{y},\boldsymbol{x}) = 0$ as well. Thus, the orthogonality is a symmetric relation, although the inner product is not symmetric. Note also that the **zero vector 0** is orthogonal to every vector in the inner product space.

A set of n vectors $\{\boldsymbol{x}_1, \boldsymbol{x}_2, \cdots \boldsymbol{x}_n\}$ is called **orthonormal** if $(\boldsymbol{x}_i, \boldsymbol{x}_j) = \delta_{ij}$ for all i and j, where δ_{ij} is the **Kronecker delta**. That is, the orthonormality of a set of vectors means that each vector is orthogonal to all the others in the set and is normalized to unit length.

It follows that any vector $\boldsymbol{x}$ may be normalized by dividing by its length to form the new vector $\boldsymbol{x}/||\boldsymbol{x}||$ with unit length. An example of an orthonormal set of vectors is the set of three unit vectors, $\{\boldsymbol{e}_i\}$ $(i = 1, 2, 3)$, for the three-dimensional **Cartesian space**.

The following theorem is important in various fields of mathematical physics.

♠ **Theorem:**

An orthonormal set is linearly independent.

(Proof of the theorem is given in Exercise **1**). Importantly, the above theorem suggests that any orthonormal set serves as a **basis** for an inner product space of interest (see Sect. 4.2.5). Below is another consequence of the orthonormal set of vectors; its proof is given in Exercise **2**.

♠ **Bessel inequality:**

If $\{\boldsymbol{x}_1, \boldsymbol{x}_2, \cdots, \boldsymbol{x}_n\}$ is a set of orthonormal vectors and $\boldsymbol{x}$ is any vector defined in the same inner product space, then

$$||\boldsymbol{x}||^2 \geq \sum_i |r_i|^2, \tag{4.8}$$

where $r_i = (\boldsymbol{x}_i, \boldsymbol{x})$. Furthermore, the vector $\boldsymbol{x}' = \boldsymbol{x} - \sum_i r_i \boldsymbol{x}_i$ is orthogonal to each $\boldsymbol{x}_j$.

4.1.6 Completeness of Vector Spaces

Having described features of inner product spaces, we turn now to another important concept relevant to the nature of Hilbert spaces, i.e., **completeness**. When a vector space is *finite* dimensional, the completeness of an orthonormal set involved in the space may be characterized by the fact that it is not contained in any larger orthonormal set. (This is intuitively understood by considering the **Cartesian basis** $\boldsymbol{e}_i$ $(i = 1, 2, 3)$ in a three-dimensional Euclidean space.) When considering an *infinite*-dimensional space, however, the completeness must be determined via the Cauchy criterion, which we discussed in Sect. 2.2. The following is a preliminary definition

♠ **Cauchy sequence of vectors:**

A sequence $\{\boldsymbol{x}_1, \boldsymbol{x}_2, \cdots\}$ of vectors is called a **Cauchy sequence** of vectors if for any positive $\varepsilon > 0$, there exists an appropriate number N such that $\|\boldsymbol{x}_m - \boldsymbol{x}_n\| < \varepsilon$ for all $m, n > N$.

In plain words, a sequence is a Cauchy sequence if the terms $\boldsymbol{x}_m$ and $\boldsymbol{x}_n$ in the sequence come closer and closer to each other as $m, n \to \infty$

♠ **Convergence of a sequence of vectors:**
A sequence $\{\boldsymbol{x}_1, \boldsymbol{x}_2, \cdots\}$ is said to be convergent if there exists an element $\boldsymbol{x}$ such that $\|\boldsymbol{x}_n - \boldsymbol{x}\| \to 0$.

♠ **Completeness of a vector space:**
If every Cauchy sequence in a space is convergent, we say that the space is **complete.**

Remark. Here the norm $\|\boldsymbol{x}\| = (\boldsymbol{x}, \boldsymbol{x})^{1/2}$ associated with an inner product is employed to define a Cauchy sequence, since we are focusing on inner product spaces. However, the concepts of Cauchy sequence and completeness both apply to more general vector spaces in which even a norm is unnecessary (see Sect. 4.1.6 for details).

4.1.7 Several Examples of Hilbert Spaces

Now we are ready to define Hilbert spaces.

♠ **Hilbert space:**
If an inner product space is complete, it is called a **Hilbert space.**

Examples **1.** Column-vector spaces with n real and complex components, denoted by $\boldsymbol{R}^n$ and $\boldsymbol{C}^n$, respectively, are finite-dimensional Hilbert spaces if endowed with an inner product $(\boldsymbol{x}, \boldsymbol{y}) \equiv \sum_{i=1}^n x_i^* y_i$. Completeness can be proved using the Bolzano–Weierstrass theorem (see Appendix A).

2. Assume an infinite-dimensional vector $\boldsymbol{x} = (x_1, x_2, \cdots)$, where x_i is a real or complex number satisfying the condition

$$\sum_{i=1}^{\infty} |x_i|^2 < \infty.$$

Then, vector spaces spanned by a set of vectors $\{\boldsymbol{x}\}$, called ℓ^2 **spaces** (see Sect. 4.3), are Hilbert spaces under the inner product

$$(\boldsymbol{x}, \boldsymbol{y}) = \sum_{i=1}^{\infty} x_i^* y_i.$$

Completeness will be proved in Sect. 4.3.1.

4. Assume a set of square-integrable functions $f(x)$ expressed by

$$\int_a^b |f(x)|^2 dx < \infty.$$

Then, the collection of all square-integrable functions, called the L^2 **space**, is a Hilbert space endowed with the inner product

$$(f, g) = \int_a^b f(x)^* g(x) dx. \tag{4.9}$$

Completeness will be proved given in Sect. 4.3.2.

5. Finally we show an example of an *incomplete* inner product space. Assume the following sequence of real-valued continuous functions $\{f_1(x), f_2(x), \cdots\}$, each of which is defined within the interval $[0, 1]$:

$$f_n(x) = \begin{cases} 1, & \text{for } 0 \le x \le \frac{1}{2}, \\ 1 - 2n\left(x - \frac{1}{2}\right) & \text{for } \frac{1}{2} \le x \le \frac{1}{2n} + \frac{1}{2}, \\ 0, & \text{for } \frac{1}{2n} + \frac{1}{2} \le x \le 1. \end{cases} \tag{4.10}$$

The graphs of $f_n(x)$ for $n = 1, 2, 3$ are given in Fig. 4.1. After some algebra, we obtain

$$\begin{aligned} \|f_n(x) - f_m(x)\| &= \left[\int_0^1 (f_n - f_m)^2 \, dx\right]^{1/2} \\ &= \left(1 - \frac{n}{m}\right)\sqrt{\frac{1}{6n}} \to 0 \quad \text{as } m, n \to \infty \quad (m > n). \end{aligned}$$

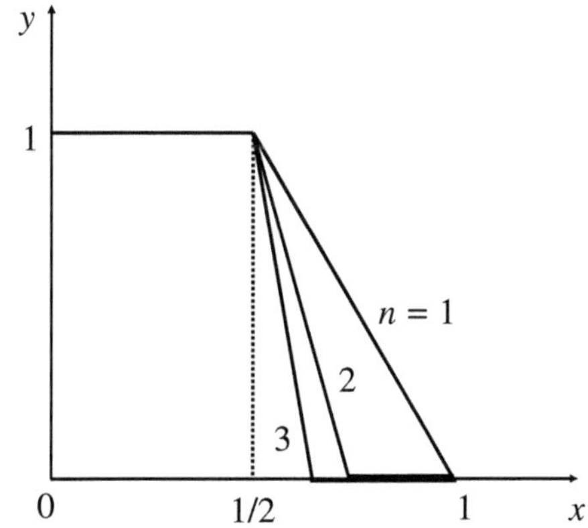

Fig. 4.1. The function $f_n(x)$ given in (4.10). The sequence $\{f_n(x)\}$ converges to a step function in the limit of $n \to \infty$

Thus, $\{f_n\}$ is a Cauchy sequence owing to the inner product given by (4.9). However, this sequence converges to the limit function

$$f(x) = \begin{cases} 1 \text{ if } 0 \le x \le \frac{1}{2}, \\ 0 \text{ if } \frac{1}{2} \le x \le 1, \end{cases}$$

which is not continuous and, hence, is not an element of the original inner product space. Consequently, the sequence is not complete, and thus is not a Hilbert space.

Exercises

1. Show that an orthonormal set is linearly independent.

Solution: Recall that a set of vectors $\{\boldsymbol{x}_1, \boldsymbol{x}_2, \cdots \boldsymbol{x}_n\}$ is said to be **linearly independent** if and only if

$$\sum_{i=1}^{n} \alpha_i \boldsymbol{x}_i = 0 \;\Rightarrow\; \alpha_i = 0 \text{ for all } i.$$

Now suppose that a set $\{\boldsymbol{x}_1, \boldsymbol{x}_2, \cdots, \boldsymbol{x}_n\}$ is orthonormal and satisfies the relation $\sum_i \alpha_i \boldsymbol{x}_i = 0$. Then, for any j, the orthonormal condition $(\boldsymbol{x}_i, \boldsymbol{x}_j) = \delta_{ij}$ results in

$$0 = \left(\boldsymbol{x}_j, \sum_{i=1}^{n} \alpha_i \boldsymbol{x}_i\right) = \sum_{i=1}^{n} \alpha_i (\boldsymbol{x}_j, \boldsymbol{x}_i) = \sum_i \alpha_i \delta_{ij} = \alpha_j.$$

Therefore, the set is linearly independent. ♣

2. Show the Bessel inequality for $\boldsymbol{x}$ given by 4.8 and the orthogonality of the vector $\boldsymbol{x}' = \boldsymbol{x} - \sum_i r_i \boldsymbol{x}_i$ to each $\boldsymbol{x}_j$.

Solution: We consider the inequality

$$\begin{aligned} 0 \le ||\boldsymbol{x}'||^2 = (\boldsymbol{x}', \boldsymbol{x}') &= \left(\boldsymbol{x} - \sum_{i=1}^{n} r_i \boldsymbol{x}_i,\ \boldsymbol{x} - \sum_{j=1}^{n} r_j \boldsymbol{x}_j\right) \\ &= (\boldsymbol{x}, \boldsymbol{x}) - \sum_{i=1}^{n} r_i^* (\boldsymbol{x}_i, \boldsymbol{x}) - \sum_{j=1}^{n} r_j (\boldsymbol{x}, \boldsymbol{x}_j) + \sum_{i,j=1}^{n} r_i^* r_j (\boldsymbol{x}_i, \boldsymbol{x}_j) \\ &= ||\boldsymbol{x}||^2 - \sum_{i=1}^{n} |r_i|^2 - \sum_{j=1}^{n} |r_j|^2 + \sum_{j=1}^{n} |r_j|^2 \\ &= ||\boldsymbol{x}||^2 - \sum_{i=1}^{n} |r_i|^2. \end{aligned}$$

Thus we have $||\boldsymbol{x}||^2 \geq \sum_i |r_i|^2$. The second part of the theorem is proven by

$$(\boldsymbol{x}', \boldsymbol{x}_j) = (\boldsymbol{x}, \boldsymbol{x}_j) - \sum_i r_i^*(\boldsymbol{x}_i, \boldsymbol{x}_j) = r_j^* - r_j^* = 0. \quad \clubsuit$$

4.2 Hierarchical Structure of Vector Spaces

4.2.1 Precise Definitions of Vector Spaces

In this section, we look at the hierarchical structure of abstract vector spaces. We will find that the Hilbert spaces that we have considered form a very limited, special class of general vector spaces under strict conditions. We begin with an exact definition of vector spaces.

♠ **Vector spaces:**

A vector space V is a set of elements $\boldsymbol{x}$ (called vectors) that satisfy the following sets of axioms:

1. V is a commutative group under addition:
- **(i)** $\boldsymbol{x} + \boldsymbol{y} = \boldsymbol{y} + \boldsymbol{x} \in V$ forany $\boldsymbol{x}, \boldsymbol{y} \in V$ (**closedness**).
- **(ii)** $\boldsymbol{x} + (\boldsymbol{y} + \boldsymbol{z}) = (\boldsymbol{x} + \boldsymbol{y}) + \boldsymbol{z}$ (**associativity**).
- **(iii)** There exists an addition identity, the zero vector $\boldsymbol{0}$, for every $\boldsymbol{x} \in V$ such that $\boldsymbol{x} + \boldsymbol{0} = \boldsymbol{x}$.
- **(iv)** There exists an additive inverse $-\boldsymbol{x}$ for every $\boldsymbol{x} \in V$ such that $\boldsymbol{x} + (-\boldsymbol{x}) = \boldsymbol{0}$.

2. V satisfies the following additional axioms with respect to a number field $\boldsymbol{F}$, whose elements α are called **scalars**:
- **(i)** V is closed under scalar multiplication:

$$\alpha\boldsymbol{x} \in V \quad \text{for arbitrary } \boldsymbol{x} \in V \text{ and } \alpha \in \boldsymbol{F}.$$

- **(ii)** Scalar multiplication is distributive with respect to elements of both V and $\boldsymbol{F}$:

$$\alpha(\boldsymbol{x} + \boldsymbol{y}) = \alpha\boldsymbol{x} + \alpha\boldsymbol{y}, \quad (\alpha + \beta)\boldsymbol{x} = \alpha\boldsymbol{x} + \beta\boldsymbol{x}.$$

- **(iii)** Scalar multiplication is associative: $\alpha(\beta\boldsymbol{x}) = \beta(\alpha\boldsymbol{x})$.
- **(iv)** Multiplication with the zero scalar $0 \in \boldsymbol{F}$ gives the zero vector such that $0\boldsymbol{x} = \boldsymbol{0} \in V$.
- **(v)** The unit scalar $1 \in \boldsymbol{F}$ has the property that $1\boldsymbol{x} = \boldsymbol{x}$.

In these definitions, $\boldsymbol{F}$ is either the set of real numbers, $\boldsymbol{R}$, or the set of complex numbers, $\boldsymbol{C}$. A vector space over $\boldsymbol{R}$ is called a **real vector space**. If $\boldsymbol{F} = \boldsymbol{C}$, then V is a **complex vector space**.

4.2.2 Metric Space

Once a vector space is endowed with the concept of a *distance* between the elements, say, $\boldsymbol{x} \in V$ and $\boldsymbol{y} \in V$, it is called a **metric space**.

♠ **Metric space:**

Assume a vector space V. A metric space is the pair (V, ρ) in which the function $\rho : V \times V \to \boldsymbol{R}$, called the **distance function**, is a single-valued, nonnegative, real function that satisfies:

1. $\rho(\boldsymbol{x}, \boldsymbol{y}) = 0$ if and only if $\boldsymbol{x} = \boldsymbol{y}$.
2. $\rho(\boldsymbol{x}, \boldsymbol{y}) = \rho(\boldsymbol{y}, \boldsymbol{x})$.
3. $\rho(\boldsymbol{x}, \boldsymbol{y}) \leq \rho(\boldsymbol{x}, \boldsymbol{z}) + \rho(\boldsymbol{z}, \boldsymbol{y})$ for any $\boldsymbol{z} \in V$.

Remark. Strictly speaking, the above is called a metric *vector* space as a subset of more general metric spaces. The latter consists of a pair (U, ρ), where U is a set of points (not necessarily vectors) and ρ is a distance function. If U is a vector space V, then (V, ρ) is called a **metric vector space**.

Examples **1.** If we set

$$\rho(\boldsymbol{x}, \boldsymbol{y}) = \begin{cases} 0 \text{ if } \boldsymbol{x} = \boldsymbol{y}, \\ 1 \text{ if } \boldsymbol{x} \neq \boldsymbol{y} \end{cases}$$

for arbitrary $\boldsymbol{x}, \boldsymbol{y} \in V$, we obtain a metric space.

2. The set of real numbers $\boldsymbol{R}$ with the distance function $\rho(x, y) = |x - y|$ forms a metric space.

3. The set of ordered n-tuples of real numbers $\boldsymbol{x} = (x_1, x_2, \cdots, x_n)$ with the distance function

$$\rho(\boldsymbol{x}, \boldsymbol{y}) = \left[\sum_{i=1}^{n} (x_i - y_i)^2 \right]^{1/2}$$

is a metric space. This is in fact the Euclidean n-space, denoted by $\boldsymbol{R}^n$.

4. Consider again the set of ordered n-tuples of real numbers $\boldsymbol{x} = (x_1, x_2, \cdots, x_n)$ with an alternative distance function:

$$\rho(\boldsymbol{x}, \boldsymbol{y}) = \max \left[|x_i - y_i| ; \ 1 \leq i \leq n \right].$$

This also serves as a metric space. The validity of Axioms **1**–**3** mentioned above is obvious.

Comparison between Examples **3** and **4** tells us that the same vector space V can be metrized in different ways. These two examples call attention to the importance of distinguishing a metric space (V, ρ) from the vector space V.

4.2.3 Normed Spaces

A metric space is said to be normed if for each element $\boldsymbol{x} \in V$ there is a corresponding nonnegative number $\|\boldsymbol{x}\|$, which is called the **norm of $\boldsymbol{x}$**.

♠ **Normed space:**

A metric space equipped with a **norm** is called a **normed space**. The norm is defined as a real-valued function (denoted by $\| \ \|$) on a vector space V, which satisfies

1. $\|\lambda \boldsymbol{x}\| = |\lambda| \|\boldsymbol{x}\|$ for all $\lambda \in \boldsymbol{F}$ and $\boldsymbol{x} \in X$.
2. $\|\boldsymbol{x} + \boldsymbol{y}\| \leq \|\boldsymbol{x}\| + \|\boldsymbol{y}\|$.
3. $\|\boldsymbol{x}\| = 0$ if and only if $\boldsymbol{x} = \boldsymbol{0}$.

Obviously, a normed space is a metric space under the definition of the distance $\rho(\boldsymbol{x}, \boldsymbol{y}) = \|\boldsymbol{x} - \boldsymbol{y}\|$.

Examples **1.** The space consisting of all n-tuples of real numbers: $\boldsymbol{x} = (x_1, x_2, \cdots, x_n)$ in which the norm is defined by

$$\|\boldsymbol{x}\| = \left(\sum_{i=1}^{n} x_i^2 \right)^{1/2}$$

is a normed space.

2. The space above can be normed by a more general form:

$$\|\boldsymbol{x}\| = \left(\sum_{i=1}^{n} x_i^p \right)^{1/p} \qquad (p \geq 1).$$

This norm is referred to as a **p-norm** of the vector $\boldsymbol{x}$.

3. We further obtain an alternative normed space if we set the norm of the vector $\boldsymbol{x} = (x_1, x_2, \cdots, x_n)$ equal to the max $\{|x_k|; \ 1 \leq k \leq n\}$.

4. The collection of all continuous functions defined on the closed interval $[a, b]$ in which

$$\|f(x)\| \equiv \max\{|f(x)| : \ x \in [a, b]\}$$

is a normed space.

5. The space consisting of all sequences $\boldsymbol{x} = (x_1, x_2, \cdots, x_n)$ of real numbers that satisfy the condition $\lim_{n\to\infty} x_n = 0$ is a normed space if we set

$$\|\boldsymbol{x}\| = \max\{|x_k| : \ 1 \leq n \leq \infty\}.$$

4.2.4 Subspaces of a Normed Space

A class of normed spaces involves the following two subclasses: one endowed with completeness and the other with the inner product. The normed spaces of the former class, i.e., a class of complete normed vector spaces, are called **Banach spaces**.

♠ **Banach space:**

If a normed space is complete, it is called a **Banach space**.

Here, the completeness of a space implies that every Cauchy sequence in the space is convergent. Refer to the arguments in Sect. 4.1.6 for details.

Remark. Every finite-dimensional normed space is a Banach space, since it is necessarily complete.

Examples **1.** Suppose that a set of infinite-dimensional vectors $\boldsymbol{x} = (x_1, x_2, \cdots, x_n, \cdots)$ satisfies the condition

$$\sum_{i=1}^{\infty} |x_i|^p < \infty, \quad (p \geq 1).$$

Then, this set is a Banach space, called an ℓ^p **space**, under the p-norm defined by

$$\|\boldsymbol{x}\|_p = \left(\sum_{i=1}^{\infty} |x_i|^p \right)^{1/p}. \tag{4.11}$$

The proof of its completeness is given in Sect. 4.3.1.

2. Assume a set of functions $f(x)$ expressed by

$$\int_a^b |f(x)|^p dx < \infty.$$

Then, this set constitutes a specific class of Banach spaces, called an L^p **spaces**, under the p-norm:

$$\|f\|_p = \left(\int_a^b |f(x)|^p dx \right)^{1/p}. \tag{4.12}$$

Completeness is proved in Sect. 4.3.2.

Now we focus on the counterpart, i.e., a noncompleted normed space endowed with an inner product known as a **pre-Hilbert space**.

♠ **Pre-Hilbert space:**

If a normed space is equipped with an inner product (not necessarily complete), then it is called a **pre-Hilbert space.**

Finally, we are at a point at which we can appreciate the definition of Hilbert spaces. They are defined as the intersection between Banach spaces and pre-Hilbert spaces as stated below

♠ **Hilbert space:**

A complete pre-Hilbert space, i.e., a complete normed space endowed with an inner product is called a **Hilbert space.**

Examples The ℓ^p spaces and L^p spaces with $p = 2$, known as the **ℓ^2 spaces** and **L^2 spaces**, are Hilbert spaces. The inner product of each space, respectively, is given by

$$(\boldsymbol{x}, \boldsymbol{y}) = \sum_{i=1}^{\infty} x_i y_i \text{ and } (f, g) = \int_a^b f^*(x) g(x) dx. \tag{4.13}$$

Remark. Clearly the quantities $(\boldsymbol{x}, \boldsymbol{x})^{1/2}$ and $(f, f)^{1/2}$, defined through the inner products (4.13), are special cases of the p-norm given by (4.11) and (4.12), respectively, with $p = 2$. In fact, for the ℓ^2 and L^2 spaces, the inner products are defined such that

$$(\boldsymbol{x}, \boldsymbol{x}) = \|\boldsymbol{x}\|^2 \text{ and } (f, f) = \|f\|^2.$$

However, for ℓ^p and L^p spaces with $p \neq 2$, we cannot introduce inner products as

$$(\boldsymbol{x}, \boldsymbol{x}) = (\|\boldsymbol{x}\|_p)^p \text{ and } (f, f) = (\|f\|_p)^p$$

because unless $p = 2$ the p-norm violates the parallelogram law. Accordingly, among the family of ℓ^p and L^p, only the spaces ℓ^2 and L^2 can be Hilbert spaces because they have an inner product.

4.2.5 Basis of a Vector Space: Revisited

For use in Sect. 4.2.6, we briefly review the definition of a **basis** in a *finite-dimensional* vector space and related matters.

♠ **Linearly independent vector:**

A *finite* set of vectors, say, $\boldsymbol{e}_1, \boldsymbol{e}_2, \cdots, \boldsymbol{e}_n$ is **linearly independent** if and only if

$$\sum_i c_i \boldsymbol{e}_{i=1}^n = 0 \iff c_i = 0 \text{ for all } i. \tag{4.14}$$

This definition applies to *infinite* sets of vectors $\boldsymbol{e}_1, \boldsymbol{e}_2, \cdots$ if the vector space under consideration admits a definition of convergence (see Sect. 4.2.6 for details).

♠ **Basis of a vector space:**

A **basis** of the vector space V is a set of linearly independent vectors $\{\boldsymbol{e}_i\}$ of V such that every vector $\boldsymbol{x}$ of V can be expressed as

$$\boldsymbol{x} = \sum_{i=1}^n \alpha_i \boldsymbol{e}_i. \tag{4.15}$$

Here, the numbers $\alpha_1, \alpha_2, \cdots, \alpha_n$ are **coordinates** of the vector $\boldsymbol{x}$ with respect to the basis, and they are uniquely determined owing to the linear independence property.

Therefore, every set of n linearly independent vectors is a basis in a finite-dimensional vector space spanned by n vectors. The number n is called the **dimension** of the vector space. Obviously, an infinite-dimensional vector space does not admit a finite basis, which is why it is called infinite-dimensional.

4.2.6 Orthogonal Bases in Hilbert Spaces

For any vector space (finite- or infinite-dimensional), a set of orthogonal vectors $\{\boldsymbol{x}_n\}$ is called an **orthogonal basis** if it is complete. Similarly, a complete orthogonal set of vectors is called an **orthonormal basis** if the norm $\|\boldsymbol{x}_n\| = 1$ for all n. It is convenient to use orthonormal bases in studying Hilbert spaces, since any vector in the space can be decomposed into a linear combination of orthonormal bases. However, when we choose some basis for an infinite-dimensional space, some care must be taken to examine its completeness property; i.e., an infinite sum of vectors in a vector space may or may not be convergent to the identical vector space.

To examine this point, let us consider an infinite set $\{\boldsymbol{e}_i\}$ $(i = 1, 2, \cdots)$ of orthonormal vectors all belonging to a Hilbert space V. We take any vector $\boldsymbol{x} \in V$ and form the set of vectors

$$\boldsymbol{x}_n = \sum_{i=1}^n c_i \boldsymbol{e}_i, \tag{4.16}$$

where the complex number c_i is the inner product of $\boldsymbol{e}_i$ and $\boldsymbol{x}$ expressed by

$$c_i = (\boldsymbol{e}_i, \boldsymbol{x}).$$

For the pair of vectors $\boldsymbol{x}$ and $\boldsymbol{x}_n$, the Schwarz inequality (4.2) gives

$$|(\boldsymbol{x}, \boldsymbol{x}_n)|^2 \leq \|\boldsymbol{x}\|^2 \|\boldsymbol{x}_n\|^2 = \|\boldsymbol{x}\|^2 \left(\sum_{i=1}^{n} |c_i|^2 \right). \tag{4.17}$$

On the other hand, taking the inner product of (4.16) with $\boldsymbol{x}$ yields

$$(\boldsymbol{x}, \boldsymbol{x}_n) = \sum_{i=1}^{n} c_i (\boldsymbol{x}, \boldsymbol{e}_i) = \sum_{i=1}^{n} |c_i|^2 . \tag{4.18}$$

From (4.17) and (4.18), we have

$$\sum_{i=1}^{n} |c_i|^2 \leq |\boldsymbol{x}|^2 .$$

This conclusion is true for arbitrarily large n and can be stated as shown below.

♠ **Bessel inequality:**

Let $\{\boldsymbol{e}_i\}$ $(i = 1, 2, \cdots)$ be an infinite set of orthonormal vectors in a Hilbert space V. Then for any $\boldsymbol{x} \in V$ with $c_i = (\boldsymbol{e}_i, \boldsymbol{x})$, we have

$$\sum_{i=1}^{\infty} |c_i|^2 \leq \|\boldsymbol{x}\|^2,$$

which is known as the **Bessel inequality**.

The Bessel inequality shows that the limiting vector

$$\lim_{n\to\infty} \sum_{i=1}^{n} c_i \boldsymbol{e}_i \equiv \sum_{i=1}^{\infty} c_i \boldsymbol{e}_i \tag{4.19}$$

has a finite norm, which means that the vector (4.19) is convergent. However, we still do not know whether it converges to $\boldsymbol{x}$. To make such a statement, the set $\{\boldsymbol{e}_i\}$ should be equipped with the completeness property defined below.

♠ **Complete orthonormal vectors:**

An infinite set of orthonormal vectors $\{\boldsymbol{e}_i\}$ in a Hilbert space V is called **complete** if the only vector in V that is orthogonal to all the $\boldsymbol{e}_i$ is the zero vector.

The following is an immediate consequence of the above statement.

♠ **Parseval identity:**

Let $\{e_i\}$ be an infinite set of orthonormal vectors in a Hilbert space V. Then for any $x \in V$,

$$\{e_i\} \text{ is complete} \iff \|x\|^2 = \sum_{i=1}^{\infty} |c_i|^2 \text{ with } c_i = (e_i, x). \tag{4.20}$$

Proof Suppose that the set $\{e_i\}$ is complete and consider the vector defined by

$$y = x - \sum_{i=1}^{\infty} c_i e_i,$$

where $x \in V$ and $c_i = (e_i, x)$. It follows that for any e_j,

$$(e_j, y) = (e_j, x) - \sum_{i=1}^{\infty} c_i (e_j, e_i) = c_j - \sum_{i=1}^{\infty} c_i \delta_{ji} = 0. \tag{4.21}$$

In view of the definition of the completeness of $\{e_i\}$, (4.21) means that y is the zero vector. Hence, we have

$$x = \sum_{i=1}^{\infty} c_i e_i,$$

which implies

$$\|x\|^2 = \sum_{i=1}^{\infty} |c_i|^2.$$

We now consider the converse. Suppose x to be orthogonal to all the $\{e_i\}$, which means

$$(e_i, x) = c_i = 0 \text{ for all } i. \tag{4.22}$$

It follows from (4.20) to (4.22) that $\|x\|^2 = 0$, which in turn gives $x = \mathbf{0}$, because only the zero vector has a zero vector. This completes the proof. ♣

We close this section by providing precise terminology for the basis of a Hilbert space.

♠ **Basis of a Hilbert space:**

A complete orthonormal set $\{\boldsymbol{e}_i\}$ $(i = 1, 2, \cdots)$ in a Hilbert space V is called a **basis** of V.

Remark.

1. The concept of *completeness* of an orthonormal set of vectors is distinct from the concept of *completeness* of the Hilbert space, but they are mutually related.
2. In order to define **generalized Fourier coefficients** $c_i = (\boldsymbol{e}_i, \boldsymbol{x})$ for $\boldsymbol{x} \in V$ (see Sect. 4.3.4), it suffices for the set $\{\boldsymbol{e}_i\}$ to be only orthonormal, nor necessarily complete.

4.3 Hilbert Spaces of ℓ^2 and L^2

4.3.1 Completeness of the ℓ^2 Spaces

In this subsection, we examine the completeness property of the space ℓ^2 on the field $\boldsymbol{F}$ (here $\boldsymbol{F} = \boldsymbol{R}$ or $\boldsymbol{C}$). As already noted, the completeness of a given vector space V is characterized by the fact that every Cauchy sequence $(\boldsymbol{x}_n)$ involved in the space converges to an element $\boldsymbol{x} \in V$ such that $\lim_{n\to\infty} \|\boldsymbol{x} - \boldsymbol{x}_n\| = 0$. Hence, to prove the completeness of the ℓ^2 space, we show in turn that (1) every Cauchy sequence $(\boldsymbol{x}_n)$ in the ℓ^2 space converges to a limit $\boldsymbol{x}$, and (2) the limit $\boldsymbol{x}$ belongs to ℓ^2.

We consider Statement (1). Assume a set of infinite-dimensional vectors

$$\boldsymbol{x}^{(n)} = \left(x_1^{(n)}, x_2^{(n)}, \cdots\right),$$

wherein $x_i^{(n)} \in \boldsymbol{F}$, and let the sequence of vectors $\left\{\boldsymbol{x}^{(1)}, \boldsymbol{x}^{(2)}, \cdots\right\}$ be a Cauchy sequence in the sense of the norm

$$\|\boldsymbol{x}\| = \left(\sum_{i=1}^{\infty} |x_i|^2\right)^{1/2} < \infty.$$

Then, for any $\varepsilon > 0$, there exists an integer N such that

$$m, n > N \;\Rightarrow\; \left\|\boldsymbol{x}^{(m)} - \boldsymbol{x}^{(n)}\right\| = \left(\sum_{i=1}^{\infty} \left|x_i^{(m)} - x_i^{(n)}\right|^2\right)^{1/2} < \varepsilon. \tag{4.23}$$

This implies that

$$\left|x_i^{(m)} - x_i^{(n)}\right| < \varepsilon \tag{4.24}$$

for every i and every $m, n > N$. Furthermore, since (4.23) is true in the limit $m \to \infty$, we find

$$\left\|\boldsymbol{x} - \boldsymbol{x}^{(n)}\right\| < \varepsilon \tag{4.25}$$

for arbitrary $n > N$. The inequalities (4.24) and (4.25) mean that $\boldsymbol{x}^{(n)}$ converges to the limiting vector expressed by $\boldsymbol{x} \equiv (x_1, x_2, \cdots)$, in which the component $x_i \in \boldsymbol{F}$ is defined by

$$x_i = \lim_{n\to\infty} x_i^{(n)}. \tag{4.26}$$

(That the limit (4.26) belongs to $\boldsymbol{F}$ is guaranteed by the completeness of $\boldsymbol{F}$.)

The remaining task is to show that the limiting vector $\boldsymbol{x}$ belongs to the original space ℓ^2. By the triangle inequality, we have

$$\|\boldsymbol{x}\| = \left\|\boldsymbol{x} - \boldsymbol{x}^{(n)} + \boldsymbol{x}^{(n)}\right\| \le \left\|\boldsymbol{x} - \boldsymbol{x}^{(n)}\right\| + \left\|\boldsymbol{x}^{(n)}\right\|.$$

Hence, for every $n > N$ and for every $\varepsilon > 0$, we obtain

$$\|\boldsymbol{x}\| < \varepsilon + \left\|\boldsymbol{x}^{(n)}\right\|.$$

As the Cauchy sequence $(\boldsymbol{x}^{(1)}, \boldsymbol{x}^{(2)}, \cdots)$ is bounded, $\|\boldsymbol{x}\|$ cannot be greater than

$$\varepsilon + \limsup_{i\to\infty} \left\|\boldsymbol{x}^{(i)}\right\|$$

and is therefore finite. This implies that the limit vector $\boldsymbol{x}$ belongs to $\ell^2(\boldsymbol{F})$. Consequently, we have proven that the space $\ell^2(\boldsymbol{F})$ is complete.

Remark. Among the various kinds of Hilbert spaces, the space ℓ^2 has a significant importance in mathematical physics, mainly because it provides the groundwork for the theory of quantum mechanics. In fact, any element $\boldsymbol{x}$ of the space ℓ^2 satisfying the normalized conditions $\|\boldsymbol{x}\| = \sum_{i=1}^{\infty} |x_i|^2 = 1$ works as a possible state vector of quantum systems. In the Heisenberg formulation of quantum mechanics, the infinite-dimensional matrices corresponding to physical observables act on these state vectors.

4.3.2 Completeness of the L^2 Spaces

We next consider another important class of Hilbert spaces, called L^2 **spaces**, which are spanned by square-integrable functions $\{f_n(x)\}$ on a closed interval, say $[a, b]$. To prove the completeness of the L^2 space, we show that every

Cauchy sequence $\{f_n\}$ in the L^2 space converges to a limit function $f(x)$, and then verify that the f belongs to L^2.

Let $\{f_1(x), f_2(x), \cdots\}$ be a Cauchy sequence in L^2. Then for any small $\varepsilon > 0$, we can find an integer N such that

$$m, n > N \quad \Rightarrow \quad \|f_n - f_m\| = \sqrt{\int_a^b |f_n(x) - f_m(x)|^2 \, dx} < \varepsilon.$$

Then, it is always possible to find an integer n_1 such that

$$n > n_1 \quad \Rightarrow \quad \|f_{n_1}(x) - f_n(x)\| < \frac{1}{2}.$$

By mathematical induction, after finding $n_{k-1} > n_{k-2}$, we find $n_k > n_{k-1}$ such that

$$n > n_k \quad \Rightarrow \quad \|f_n(x) - f_{n_k}(x)\| < \left(\frac{1}{2}\right)^k.$$

In this way, we obtain a sequence (f_{n_k}) that is a subsequence such that

$$\|f_{n_{k+1}}(x) - f_{n_k}(x)\| < \left(\frac{1}{2}\right)^k \quad \text{for } k = 1, 2, \cdots,$$

or equivalently,

$$\|f_{n_1}\| + \sum_{k=1}^{\infty} \|f_{n_{k+1}} - f_{n_k}\| < \|f_{n_1}\| + \sum_{k=1}^{\infty} \left(\frac{1}{2}\right)^k = \|f_{n_1}\| + 1 \equiv A,$$

where A is a finite constant. Let

$$g_k = |f_{n_1}| + |f_{n_2} - f_{n_1}| + \cdots + |f_{n_{k+1}} - f_{n_k}| \ \ (k = 1, 2, \cdots).$$

Then, by the Minkowski inequality, we have

$$\begin{aligned}\int_a^b [g_k(x)]^2 \, dx &= \int_a^b \left[|f_{n_1}| + |f_{n_2} - f_{n_1}| + \cdots + |f_{n_{k+1}} - f_{n_k}|\right]^2 dx \\ &\le \left(\|f_{n_1}\| + \sum_{i=1}^{k} \|f_{n_{i+1}} - f_{n_i}\|\right)^2 \le A^2 < \infty. \qquad (4.27)\end{aligned}$$

Let $g(x) = \lim g_k(x)$. Then $[g(x)]^2 = \lim [g_k(x)]^2$, and

$$\int_a^b [g(x)]^2 dx = \int_a^b \lim_{k\to\infty} [g_k(x)]^2 \, dx = \lim_{k\to\infty} \int_a^b [g_k(x)]^2 \, dx. \qquad (4.28)$$

[See the remark below for the interchangeability of the limit and integral signs in (4.28).] It follows from (4.27) and (4.28) that

$$\int_a^b [g(x)]^2 dx < \infty,$$

or equivalently,

$$\int_a^b \left(|f_{n_1}| + \sum_{k=1}^{\infty} |f_{n_{k+1}} - f_{n_k}| \right)^2 dx < \infty.$$

This implies that the infinite sum

$$\|f_{n_1}\| + \sum_{k=1}^{\infty} \|f_{n_{k+1}} - f_{n_k}\| \tag{4.29}$$

converges to a function, denoted by $f \in L^2$, in the sense of the norm in L^2.

We next show that the limit function $f(x)$ expressed by (4.29) is an element of L^2 such as

$$\|f_n(x) - f(x)\| \to 0 \quad (n \to \infty). \tag{4.30}$$

We first note that

$$f(x) - f_{n_j}(x) = \sum_{k=j}^{\infty} \left[f_{n_{k+1}} - f_{n_k}(x) \right].$$

It follows that

$$\|f - f_{n_j}\| \leq \sum_{k=j}^{\infty} \|f_{n_{k+1}} - f_{n_k}\| < \sum_{k=j}^{\infty} \left(\frac{1}{2}\right)^k = \frac{1}{2^{j-1}},$$

so we have

$$\lim_{j\to\infty} \|f - f_{n_j}\| = 0.$$

Observe that

$$\|f_n - f\| \leq \|f_n - f_{n_k}\| + \|f_{n_k} - f\|,$$

where $\|f_n - f_{n_k}\| \to 0$ as $n \to \infty$ and $k \to \infty$; thus

$$\lim_{n\to\infty} \|f_n - f\| = 0,$$

which shows that the Cauchy sequence (f_n) converges to $f \in L^2$.

Remark. The interchangeability of limit and integral signs in (4.28) is justified by the following three facts:

(i) The sequence $([g_k(x)]^2)$ is a sequence of square-integrable functions in $[a, b]$,

(ii) $[g_k(x)]^2 \geq 0$ for all k, and

(iii) The integral $\int_a^b [g_k]^2 dx$ for each k has a common bound A^2 as shown in (4.27). The proof of this point is based on the theory of the Lebesgue integral, which we discuss in Chap. 6.

4.3.3 Mean Convergence

Before proceeding, comments on a new class of convergence that is relevant to the argument on the completeness of the L^2 space are in place. Observe that the expression (4.30) is rephrased in the following sentence: For any small $\varepsilon > 0$, it is possible to find N such that

$$n > N \Rightarrow \|f(x) - f_n(x)\| < \varepsilon. \tag{4.31}$$

Hence, we can say that the infinite sequence (f_n) converges to $f(x)$ in the norm of the L^2 space. Convergence of the type (4.31) is called **the convergence in the mean** or **the mean convergence**, which is inherently different from the uniform convergence and the pointwise convergence. The point is the fact that in the mean convergence, the quantitative deviation between $f_n(x)$ and $f(x)$ is measured not by the difference $f(x) - f_n(x)$, but by the norm in the L^2 space based on the integration procedure:

$$\|f(x) - f_n(x)\| = \left[\int_a^b |f(x) - f_n(x)dx|^2 \, dx\right]^{1/2}.$$

Hence, when $f(x)$ is convergent in the mean to $f_n(x)$ on the interval $[a, b]$, there may exist a finite number of **isolated points** such that $f(x) \neq f_n(x)$. Obviously, this situation is not allowed in cases of uniform or pointwise convergence.

4.3.4 Generalized Fourier Coefficients

Having clarified the completeness property of the two specific Hilbert spaces, ℓ^2 and L^2, we introduce two important concepts: **generalized Fourier coefficients** and **generalized Fourier series**. We shall see that they play a crucial role in revealing the close relationship between the two distinct Hilbert spaces ℓ^2 and L^2.

♠ **Generalized Fourier coefficients:**

Suppose that a set of square-integrable functions $\{\phi_i\}$ is orthonormal (not necessarily complete) in the norm of the L^2 space. Then, the numbers

$$c_k = (f, \phi_k) \tag{4.32}$$

are called the **Fourier coefficients** of the function $f \in L^2$ relative to the orthonormal set $\{\phi_i\}$, and the series

$$\sum_{k=1}^{\infty} c_k \phi_k \tag{4.33}$$

is called the **Fourier series** of f with respect to the set $\{\phi_i\}$.

Remark.

1. In general, the Fourier series shown in (4.33) may or may not be convergent; its convergence property is determined by the features of the functions f and the associated orthonormal set of functions $\{\phi_k\}$.
2. Some readers may be familiar with the Fourier series associated with trigonometric functions or imaginary exponentials. Notably, however, the concepts of Fourier series and Fourier coefficients introduced above are more general concepts than those associated with trigonometric series.

The importance of the Fourier coefficients (4.32) becomes apparent when we see that they consist of the ℓ^2 space. In fact, since c_k is the inner product of f and ϕ_k, it yields the **Bessel inequality** in terms of c_k and f:

$$\sum_{k=1}^{\infty} |c_k|^2 \leq \|f\|. \tag{4.34}$$

From the hypothesis of $f \in L^2$, the norm $\|f\|$ remains finite. Hence, the inequality (4.34) ensures the convergence of the infinite series $\sum_{k=1}^{\infty} |c_k|^2$, which consists of the Fourier coefficients defined by (4.32). This convergence means that the sequence of Fourier coefficients $\{c_k\}$ is an element of the space ℓ^2, whichever orthonormal set of functions $\phi_k(x)$ we choose. In this context, the two elements $f \in L^2$ and $\mathbf{c} = (c_1, c_2, \cdots) \in \ell^2$ are connected via the Fourier coefficient (4.32).

4.3.5 Riesz–Fisher Theorem

Recall that every Fourier coefficient satisfies the Bessel inequality (4.34). Hence, in order for a given set of complex numbers (c_i) to constitute the Fourier coefficients of a function $f \in L^2$, it is necessary that the series

$$\sum_{k=1}^{\infty} |c_k|^2$$

converge. As a matter of fact, this condition is not only necessary, but also sufficient as stated in the theorem below.

♠ **Riesz–Fisher theorem:**

Given any set of complex numbers (c_i) such that

$$\sum_{k=1}^{\infty} |c_k|^2 < \infty, \tag{4.35}$$

there exists a function $f \in L^2$ such that

$$c_k = (f, \phi_k) \quad \text{and} \quad \sum_{k=1}^{\infty} |c_k|^2 = \|f\|^2, \tag{4.36}$$

where $\{\phi_i\}$ is a complete orthonormal set.

Proof Set linear combinations of $\phi_k(x)$ as

$$f_n(x) = \sum_{k=1}^{n} c_k \phi_k(x), \tag{4.37}$$

where the c_k are arbitrary complex numbers satisfying condition (4.35). Then, for a given integer $p \geq 1$, we obtain

$$\|f_{n+p} - f_n\|^2 = \|c_{n+1}\phi_{n+1} + \cdots + c_{n+p}\phi_{n+p}\|^2 = \sum_{k=n+1}^{n+p} |c_k|^2. \tag{4.38}$$

Let $p = 1$ and $n \to \infty$. Then, from condition (4.35), we have

$$\|f_{n+1} - f_n\| = |c_{n+1}|^2 \to 0 \ \ (n \to \infty).$$

This tells us that the infinite sequence $\{f_n\}$ defined by (4.37) associated with a given set of complex numbers $\{c_i\}$ always converges in the mean to a function $f \in L^2$.

Our remaining task is to show that this limit function $f(x)$ satisfies condition (4.36), so we consider the inner product

$$(f, \phi_i) = (f_n, \phi_i) + (f - f_n, \phi_i), \tag{4.39}$$

where we assume $n \geq i$. It follows from (4.37) that the first term on the right-hand side is equal to c_i. The second term vanishes as $n \to \infty$, since

$$|(f - f_n, \phi_i)| \leq \|f - f_n\| \cdot \|\phi_i\| \to 0 \quad (n \to \infty),$$

where we used the mean convergence of $\{f_n\}$ to f. In addition, the left-hand side of (4.39) is independent of n. Hence, taking the limit $n \to \infty$ on both sides of (4.39), we obtain

$$(f, \phi_i) = c_i, \tag{4.40}$$

which means that c_i is the Fourier coefficient of f relative to ϕ_i. From our assumption, the set $\{\phi_i\}$ is complete and orthonormal. Hence, the Fourier coefficients (4.40) satisfy the **Parseval identity**:

$$\sum_{k=1}^{\infty} |c_k|^2 = \|f\|^2. \tag{4.41}$$

The results (4.40) and (4.41) are identical to condition (4.36), thus proving the theorem. ♣

4.3.6 Isomorphism between ℓ^2 and L^2

The Riesz–Fisher theorem results immediately in the isomorphism between the Hilbert spaces L^2 and ℓ^2. An isomorphism is a one-to-one correspondence that preserves the entire algebraic structure. For instance, two vector spaces U and V (over the same number field) are isomorphic if there exists a one-to-one correspondence between the vectors $\boldsymbol{x}_i$ in U and $\boldsymbol{y}_i$ in V, say $\boldsymbol{y}_i = f(\boldsymbol{x}_i)$, such that

$$f(\alpha_1 \boldsymbol{x}_1 + \alpha_2 \boldsymbol{x}_2) = \alpha_1 f(\boldsymbol{x}_1) + \alpha_2 f(\boldsymbol{x}_2).$$

The isomorphism between L^2 and ℓ^2 is closely related to the theory of quantum mechanics, which originally consisted of two distinct theories: Heisenberg's matrix mechanics, based on infinite-dimensional vectors, and Schrödinger's wave mechanics, based on square-integrable functions. From the mathematical point of view, the difference between the two theories reduces to the fact that the former uses the space ℓ^2, whereas the latter uses the space L^2. Hence, the isomorphism between the two spaces verifies the equivalence of the two theories describing the nature of quantum mechanics.

Let us prove the above point. Choose an arbitrary complete orthonormal set $\{\phi_n\}$ in L^2 and assign to each function $f \in L^2$ the sequence $(c_1, c_2, \cdots, c_n, \cdots)$ of its Fourier coefficients with respect to this set. Since

$$\sum_{k=1}^{\infty} |c_k|^2 = \|f\|^2 < \infty,$$

the sequence $(c_1, c_2, \cdots, c_n, \cdots)$ is an element of ℓ^2. Conversely, in view of the Riesz–Fisher theorem, for every element $(c_1, c_2, \cdots, c_n, \cdots)$ of ℓ^2 there is a function $f(x) \in L^2$ whose Fourier coefficients are $c_1, c_2, \cdots, c_n, \cdots$. This correspondence between the elements of L^2 and ℓ^2 is one-to-one. Furthermore, if

$$f(x) \longleftrightarrow (c_1, c_2, \cdots, c_n, \cdots)$$

and

$$g(x) \longleftrightarrow (d_1, d_2, \cdots, d_n, \cdots),$$

then

$$f(x) + g(x) \longleftrightarrow (c_1 + d_1, \cdots, c_n + d_n, \cdots)$$

and

$$kf(x) \longleftrightarrow (kc_1, kc_2, \cdots, kc_n, \cdots),$$

which readily follows from the definition of Fourier coefficients (the reader should prove it). That is, addition and multiplication by scalars are preserved by the correspondence. Furthermore, in view of Parseval's identity, it follows that

$$(f, g) = \sum_{i=1}^{\infty} c_i^* d_i. \tag{4.42}$$

All of these facts ensure the isomorphism between the spaces L^2 and ℓ^2, i.e., the one-to-one correspondence between the elements of L^2 and ℓ^2 that preserves the algebraic structures of the space. In this context, we may say that every element $\{c_i\}$ in an ℓ^2 space serves as a *coordinate system* of the L^2 space, and vice versa.

Exercises

1. Prove the inequality $\sum_{k=1}^{\infty} |c_k|^2 \le \|f\|$ given in (4.34).

Solution: Suppose a partial sum $S_n(x) = \sum_{k=1}^{n} \alpha_k \phi_k(x)$, where α_k is a certain number (real or complex). Since the set $\{\phi_i\}$ is orthonormal,

$$\|f(x) - S_n(x)\|^2 = \left(f - \sum_{j=1}^{n} \alpha_j \phi_j(x),\ f - \sum_{k=1}^{n} \alpha_k \phi_k(x) \right)$$

$$= \|f\|^2 - \sum_{k=1}^{n} |c_k|^2 + \sum_{k=1}^{n} (\alpha_k - c_k)^2 . \quad (4.43)$$

The minimum of (4.43) is assumed if $\alpha_k = c_k$. In that case, the equation (4.43) reads

$$\|f(x) - \sum_{k=1}^{n} c_k \phi_k(x)\|^2 = \|f\|^2 - \sum_{k=1}^{n} |c_k|^2,$$

which implies $\sum_{k=1}^{n} |c_k|^2 \le \|f\|^2$. Since the right-hand side is independent of n, the value of n can be taken arbitrarily large. Hence, by taking the limit $n \to \infty$, we attain the desired result: $\sum_{k=1}^{\infty} |c_k|^2 \le \|f\|^2$. ♣

2. Verify the equation $(f, g) = \sum_{i=1}^{\infty} c_i^* d_i$ given in (4.42).

Solution: This equality is verified because of the relations $(f, f) = \sum_{i=1}^{\infty} |c_i|^2$ and $(g, g) = \sum_{i=1}^{\infty} |d_i|^2$, and their consequences:

$$(f+g, f+g) = (f,f) + 2(f,g) + (g,g) = \sum_{i=1}^{\infty} |c_i + d_i|^2$$

$$= \sum_{i=1}^{\infty} |c_i|^2 + 2 \sum_{i=1}^{\infty} c_i^* d_i + \sum_{i=1}^{\infty} |d_i|^2 . \quad ♣$$

5

Orthonormal Polynomials

Abstract The theory of Hilbert spaces we dealt with in Chap. 4 can be used to construct a number of polynomial functions that are orthonormal and complete in the sense of the L^p space. In this chapter we present three important approaches for the construction of orthonormal polynomials, based, respectively, on the Weierstrass theorem (Sect. 5.1.1), the Rodrigues formula (Sect. 5.2.1), and generating functions (Sect. 5.2.7). We shall find that various orthonormal polynomials relevant to mathematical physics can be effectively classified by adopting these methods.

5.1 Polynomial Approximations

5.1.1 Weierstrass Theorem

There are a number of special polynomials that play a significant role in various aspects of mathematical physics: Legendre, Laguerre, Hermite, and Chebyshev polynomials are well known. For instance, Legendre and Laguerre polynomial expansions are often used to solve second-order differential equations having spherical symmetry. The point is that many of these special polynomials form a **complete orthonormal set of polynomials**; the origin of their orthonormality and completeness can be accounted for in terms of the theory of the Hilbert space L^2. Owing to completeness, these special polynomials enable us to produce polynomial approximations of fairly arbitrary functions with desired accuracy, which serves as a useful device in manipulating square-integrable functions.

The validity of polynomial approximations is based on the famous **Weierstrass approximation theorem**, which states that from the set of powers of a real variable x one can construct a sequence of polynomials that converges uniformly to any continuous function within a finite interval $[a, b]$. From this result, we shall see that it is possible to find various kinds of complete orthonormal sets of polynomials on any interval $[a, b]$.

In what follows, for simplicity we focus on polynomial approximations only of real-valued functions of a real variable. In the case of a complex-valued function, the separate validity of the theorem for each of its real and imaginary parts ensures the validity of the theorem.

♠ **Weierstrass approximation theorem:**
If a function $f(x)$ is continuous on the closed interval $[a,b]$, there exists a polynomial such as

$$G_n(x) = \sum_{k=0}^{n} c_k x^k \tag{5.1}$$

that converges uniformly to $f(x)$ on $[a,b]$.

The proof will is be given in Appendix C. Several remarks on this theorem are given below.

- In the polynomial approximation based on (5.1), the values of coefficients $c_m^{(n)}$ depend on n for fixed m. Thus, in order to improve the accuracy of the approximation by going to polynomials of higher degree, the earlier coefficients must change. For instance, when the approximating polynomial (5.1) is replaced by
$$G_{n+1}(x) = \sum_{k=0}^{n+1} d_k x^k,$$
we have in general
$$c_k \neq d_k \text{ for all } k(\leq n).$$
This situation is in contrast to the case of our familiar **Taylor series expansions**, in which the earlier coefficients remain unchanged.
- The Weierstrass theorem requires only that the continuity of functions be approximated. This condition is much weaker than Taylor's theorem for expansion in power series, in which the derivatives of all orders must exists (i.e., it must be **analytic**; see Sect. 7.1.2 for the definition of **analytic functions**). Furthermore, the former theorem can apply to polynomial approximations outside the radius of convergence (see Sect. 7.4.1) of a Taylor series.
- The Weierstrass theorem may be extended to functions of more than one variable. By a straightforward generalization of the proof (see Appendix C), it can be shown that if a function $f(x_1, x_2, \cdots, x_m)$ is continuous in each variable x_i located within $[a_i, b_i]$ $(i = 1, 2, \cdots, m)$, it may be approximated uniformly by the polynomials
$$G_n(x_1, x_2, \cdots, x_m) = \sum_{k_1=0}^{N_1} \sum_{k_2=0}^{N_2} \cdots \sum_{k_m=0}^{N_m} c_{k_1 k_2 \cdots k_m} x^{k_1} x^{k_2} \cdots x^{k_m}.$$
The special cases of $m = 2$ and $m = 3$ are considered in Sects. 5.1.4 and 5.1.5.

5.1.2 Existence of Complete Orthonormal sets of Polynomials

It must be emphasized that the Weierstrass theorem requires that the set of polynomials $\{G_n\}$ be neither orthogonal nor complete. Nevertheless, the theorem ensures indirectly the existence of a variety of *complete orthonormal* sets of polynomials in terms of the L^2 space. The proof of their existence is based on the **Gram-Schmidt orthogonalization method** shown below.

♠ **Gram-Schmidt orthogonalization method:**

Given any set of linearly independent functions $\{\varphi_i\}$ normalizable on a closed interval, it is possible to construct an orthonormal set of functions $\{Q_i\}$ through the recursion formula

$$Q_i(x) = \frac{u_i(x)}{\|u_i(x)\|}, (i = 1, 2, \cdots),$$

with the definitions:

$$u_1(x) = \varphi_1(x), \quad u_i(x) = \varphi_i(x) - \sum_{k=1}^{i} (u_k, \varphi_{i+1}) u_k(x).$$

Here, (u_k, φ_{i+1}) means the inner product in terms of the L^2 space. Let us apply the Gram-Schmidt orthogonalization process to a set of powers $\{x^n\}$ that is linearly independent. We then obtain an orthonormal set $\{Q_i\}$ given by

$$Q_i(x) = \sum_{m=0}^{i} {b'}_m^{(i)} x^m. \tag{5.2}$$

Owing to the orthogonality of the set $\{Q_i\}$, the original functions x^m are expressed conversely by linear combinations of $\{Q_i\}$ such as

$$x^m = \sum_{i=0}^{m} b_i^{(m)} Q_i(x). \tag{5.3}$$

Substituting (5.3) into (5.1), we obtain

$$G_n(x) = \sum_{m=0}^{n} a_m^{(n)} \sum_{i=0}^{m} b_i^{(m)} Q_i(x). \tag{5.4}$$

The superscripts (n) and (m) attached to the coefficients $a_m^{(n)}$ and $b_i^{(m)}$, respectively, remind us that the values of the terms contained in the finite sequences,

$$\left\{a_0^{(n)}, a_1^{(n)}, \cdots, a_n^{(n)}\right\} \quad \text{and} \quad \left\{b_0^{(m)}, b_1^{(m)}, \cdots, b_m^{(m)}\right\},$$

depend on n or m: as n (or m) increases, all the earlier terms in the sequence must be altered.

Now let us show the completeness of the orthonormal set $\{Q_n(x)\}$ given by (5.2), which was deduced from the orthogonalization process; this is achieved by proving that Parseval's identity,

$$\sum_{n=1}^{\infty} |(f, Q_n)|^2 = \|f\|^2,$$

holds for any $f \in L^2$, or equivalently, by proving that

$$(f, Q_n) = 0 \text{ for all } n \iff \|f\| = 0. \tag{5.5}$$

The sentence "$\|f\| = 0$ *implies* $(f, Q_n) = 0$ *for all* n" immediately follows from the **Bessel inequality**,

$$\sum_{k=0}^{\infty} |(f, Q_n)|^2 \leq \|f\|^2.$$

To prove the converse, we note that if $(f, Q_n) = 0$ for all n, we have

$$(f, G_n) = 0 \text{ for all } n, \tag{5.6}$$

since the G_n are linear combinations of the Q_n. In addition, we recall that the Weierstrass theorem guarantees the uniform convergence of the sequence (G_n) to f. Since uniform convergence implies a mean convergence, we obtain

$$\|f - G_n\| \to 0. \tag{5.7}$$

From (5.6) and (5.7), it follows that

$$\|f - G_n\|^2 \equiv (f - G_n, f - G_n) = \|f\|^2 + \|G_n\|^2 \to 0,$$

which implies that $\|f\| = 0$ as well as $\|G_n\|^2 \to 0$. (This is because $\|f\|^2$ is independent of n and $\|G_n\|^2$ is nonnegative for all n.) As summarized, we attain the desired conclusion (5.5), which indicates that the orthonormal set $\{Q_n\}$ is complete in terms of the L^2 space.

The completeness of the set $\{Q_i\}$ means that there exists a set of constants $\{c_i\}$ such that any function $g \in L^2$ can be approximated in the mean by the following sequence of partial sums:

$$g_n(x) = \sum_{i=0}^{n} c_i Q_i(x). \tag{5.8}$$

The reader should appreciate a crucial difference between (5.1) and (5.8). In the latter, the c_i are independent of n in contrast to the case of (5.1). Thus as we extend the sum to infinity, the approximation improves without changing the earlier c_i. Therefore, we may say that there exists an infinite series

$$\lim_{n\to\infty} g_n(x) = \sum_{i=0}^{\infty} c_i Q_i(x)$$

that converges to g in the mean. The expansion coefficients $c_i = (g, Q_i)$ in the infinite series are the **Fourier coefficients** we introduced in Sect. 4.3.4.

5.1.3 Legendre Polynomials

The previous discussion revealed that the orthonormal set $\{Q_i\}$ constructed from the orthogonalization process based on the set of powers $\{x^m\}$ is complete, so that the linear combination $\sum_{i=0}^{n} c_i Q_i$ converges in the mean to $f \in L^2$. Let us employ this result to find an explicit function form of a complete orthonormal set of functions $\{P_n\}$ defined on the interval $[-1, 1]$. The first member of such a complete orthogonal set is $P_0(x) = 1$ (For convenience, the normalization constant is omitted temporarily). Using the Gram-Schmidt orthogonalization process, we have

$$P_1(x) = \frac{x - (x, P_0)\, P_0}{\|x - (x, P_0)\, P_0\|} = x,$$

$$P_2(x) = \frac{x^2 - \left(x^2, P_0\right) P_0 - \left(x^2, P_1\right) P_1}{\|x^2 - (x^2, P_0)P_0 - (x^2, P_1)P_1\|} = \frac{1}{2}\left(3x^2 - 1\right),$$

where we use the notation

$$(x^m, P_n) = \int_{-1}^{1} x^m P_n(x) dx.$$

Successive procedures give

$$P_3(x) = \frac{1}{2}\left(5x^3 - 3x\right), \quad P_4(x) = \frac{1}{8}\left(35x^4 - 30x^2 + 3\right),$$
$$P_5(x) = \frac{1}{8}\left(63x^5 - 70x^3 + 15x\right), \cdots .$$

Eventually, we obtain the complete orthonormal set of polynomials $\{P_n\}$ known as the **Legendre polynomial**. The x dependence of each function is plotted in Fig. 5.1. Note that $P_n(x)$ has exactly $n - 1$ distinct zeros in the open interval $[-1, 1]$.

A general formula for $P_n(x)$ is given by

$$P_n(x) = \frac{1}{2^n} \sum_{k=0}^{[n/2]} (-1)^k \frac{(2n-2k)!}{k!\,(n-k)!\,(n-2k)!} x^{n-2k}, \tag{5.9}$$

where we used the **Gauss notation**:

$$\left[\frac{n}{2}\right] = \begin{cases} \dfrac{n}{2} & \text{if } n \text{ is even,} \\[2ex] \dfrac{n-1}{2} & \text{if } n \text{ is odd.} \end{cases}$$

Equation (5.9) is rewritten in a simpler form as

$$\begin{aligned} P_n(x) &= \frac{1}{2^n} \sum_{k=0}^{[n/2]} \frac{(-1)^k}{k!\,(n-k)!} \frac{d^n}{dx^n} x^{2n-2k} \\ &= \frac{1}{2^n n!} \frac{d^n}{dx^n} \sum_{k=0}^{n} \frac{(-1)^k n!}{k!\,(n-k)!} x^{2n-2k} \\ &= \frac{1}{2^n n!} \frac{d^n}{dx^n} (x^2-1)^n. \end{aligned} \tag{5.10}$$

The last line is known as the **Rodrigues formula** for Legendre polynomials. This is a special form of the more general Rodrigues formula that is applicable to any orthonormal polynomial function. The derivations of (5.9) and

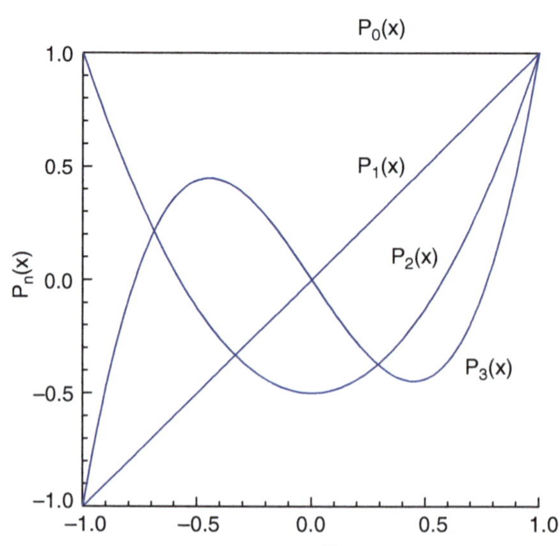

Fig. 5.1. Profiles of the first three terms of the Legendre polynomial $P_n(x)$

(5.10), as well as that of general Rodrigues formula are given in Sects. 5.2.1 and 5.2.2.

The orthogonality of the Legendre polynomials follows from the Rodrigues formula (5.10). To see this, we denote d^n/dx^n by d_n, and assume that $n \geq m$. Dropping constant factors, we have

$$\begin{aligned}\int_{-1}^{1} P_n(x)P_m(x)dx &= \int_{-1}^{1} \left[d_n(x^2-1)^n\right]\left[d_m(x^2-1)^m\right]dx \\ &= \left[d_{n-1}(x^2-1)^n\right]\left[d_m(x^2-1)^m\right]\Big|_{-1}^{1} \\ &\quad - \int_{-1}^{1} \left[d_{n-1}(x^2-1)^n\right]\left[d_{m+1}(x^2-1)^m\right]dx, \end{aligned} \tag{5.11}$$

where we employed integration by parts. Since

$$d_{n-1}(x^2-1)^n = (x^2-1) \times (\text{a polynomial}),$$

the first term in the last line of (5.11) vanishes upon putting in the limits ± 1, leaving the second term alone. Therefore, after n partial integrations, we have

$$\int_{-1}^{1} P_m(x)P_n(x)dx = (-1)^n \int_{-1}^{1} (x^2-1)^n d_{m+n}(x^2-1)^m dx.$$

Now, if $n > m$, then $n + m > 2m$ so that $d_{n+m}(x^2-1)^m = 0$. Therefore,

$$\int_{-1}^{1} P_n(x)P_m(x)dx = 0 \quad \text{for} m \neq n.$$

If $m = n$, then we have

$$\int_{-1}^{1} P_n(x)^2 dx = \frac{(-1)^n}{2^{2n}(n!)^2} \int_{-1}^{1} (x^2-1)^n d_{2n}(x^2-1)^n dx, \tag{5.12}$$

where a normalization constant is explicitly attached. Since $(x^2-1)^n$ is a polynomial of degree $2n$, its $(2n)$th derivative is just $(2n)!$. Hence, the integral (5.12) reads

$$\int_{-1}^{1} P_n(x)^2 dx = \frac{(2n)! \cdot (-1)^n}{2^{2n}(n!)^2} \int_{-1}^{1} (x^2-1)^n dx = \frac{2}{2n+1}. \tag{5.13}$$

As summarized, the orthogonal property of Legendre polynomial functions is given by

$$\int_{-1}^{1} P_m(x)P_n(x)dx = \begin{cases} 0 & (m \neq n) \\ \dfrac{2}{2n+1} & (m = n). \end{cases}$$

Remark. Equation (5.13) follows from the identity

$$\begin{aligned}\int_{-1}^{1}(1-x^2)^n dx &= 2^{2n+1}\int_0^1 t^n(1-t)^n dt &= 2^{2n+1}B(n+1,n+1)\\ &= 2^{2n+1}\frac{\Gamma(n+1)^2}{\Gamma(2n+2)} &= 2^{2n+1}\frac{(n!)^2}{(2n+1)!}.\end{aligned}$$

Here, we have changed the variable by setting $x = 2t - 1$ to obtain the **beta function** $B(x,y)$ and the **gamma function** $\Gamma(x)$ defined, respectively, by

$$B(x,y) = \int_0^1 t^{x-1}(1-t)^{y-1}dt = \frac{\Gamma(x)\Gamma(y)}{\Gamma(x+y)},$$

$$\Gamma(x) = \int_0^\infty e^{-t}t^{x-1}dt.$$

5.1.4 Fourier Series

We next consider the application of the Weierstrass theorem to functions of two variables. Through earlier discussions, we have the proof of the completeness properties of the set of trigonometric functions $\sin n\theta$ and $\cos n\theta$ $(n = 0, 1, \cdots, \infty)$.

The Weierstrass theorem tells us that any function $g(x,y)$ that is continuous in both variables on finite closed intervals may be approximated uniformly by the sequence of functions

$$g_N(x,y) = \sum_{n,m=0}^{N} a_{nm}^{(N)} x^n y^m. \tag{5.14}$$

Employ polar coordinates and restrict the domain of definition to the unit circle $x = \cos\theta$ and $y = \sin\theta$ to find

$$g_N(\cos\theta, \sin\theta) = f_N(\theta) = \sum_{n,m=0}^{N} a_{nm}^{(N)} \cos^n\theta \sin^m\theta. \tag{5.15}$$

Clearly, $f_N(\theta)$ should be periodic with periodicity 2π. Using **Euler's equation**,

$$e^{i\theta} = \cos\theta + i\sin\theta,$$

we obtain expressions for the nth powers of $\sin\theta$ and $\cos\theta$:

$$\cos^n\theta = \left[\frac{1}{2}\left(e^{i\theta} + e^{-i\theta}\right)\right]^n, \quad \sin^n\theta = \left[\frac{1}{2i}\left(e^{i\theta} - e^{-i\theta}\right)\right]^n.$$

We then rewrite (5.15) in the form

$$f_M(x) = \sum_{n=-M}^{M} \frac{c_n^{(M)}}{(2\pi)^{1/2}} e^{inx} \text{ with } M = 2N. \tag{5.16}$$

where we have inserted the factor $(2\pi)^{1/2}$ for later convenience and have replaced the variable θ by x to emphasize the generality of the result.

The superscript M attached to $c_n^{(M)}$ in (5.16) suggests the possibility that the values of $c_n^{(M)}$ are dependent of M. However, this is not the case. In fact, the values of the coefficients c_n are determined independently of M owing to the completeness of the orthonormal set of functions

$$F_n(x) = \frac{e^{inx}}{(2\pi)^{1/2}}, \quad n = 0, \pm 1, \cdots,$$

defined on the interval $[-\pi, \pi]$. The completeness of the set $\{F_n\}$ allows us to approximate an arbitrary function f in the mean by an infinite series of the F_n, and we write

$$f(x) \simeq \sum_{n=-\infty}^{\infty} c_n F_n(x) = \sum_{n=-\infty}^{\infty} \frac{c_n}{(2\pi)^{1/2}} e^{inx}, \tag{5.17}$$

where the expansion coefficients are given by

$$c_n = (F_n, f) = \frac{1}{(2\pi)^{1/2}} \int_{-\pi}^{\pi} f(x) e^{-inx} dx. \tag{5.18}$$

The series (5.17) with the coefficients (5.18) is known as the **trigonometric Fourier series**. The completeness of the set $\{F_n\}$ can be verified in a discussion similar to that in Sect. 5.1.2.

5.1.5 Spherical Harmonic Functions

We have derived the sets of Legendre polynomials and trigonometric functions from the Weierstrass approximation theorem in one and two variables, respectively. We now derive the set of spherical harmonics from a three-variable generalization. It tells us that a function g of x, y, z (i.e., $\boldsymbol{r}$) can be approximated uniformly by a sequence of partial sums given by

$$g_M(\boldsymbol{r}) = \sum_{j,k,n=0}^{M} a_{jkn}^{(M)} x^j y^k z^n. \tag{5.19}$$

We may also use an alternative coordinate system such as

$$\begin{aligned} u &\equiv x + iy = r\sin\theta e^{i\phi}, \\ v &\equiv x - iy = r\sin\theta e^{-i\phi}, \\ w &\equiv z = r\cos\theta, \end{aligned}$$

which yields

$$g_M(\boldsymbol{r}) = \sum_{\alpha,\beta,\gamma=0}^{M} b_{\alpha\beta\gamma}^{(M)} u^\alpha v^\beta w^\gamma \tag{5.20}$$

$$= \sum_{l=0}^{3M} r^l \sum_{(\alpha,\beta,\gamma)} b_{\alpha\beta\gamma}^{(M)} e^{i(\alpha-\beta)\phi} \sin^{\alpha+\beta}\theta \cos^\gamma\theta. \tag{5.21}$$

In (5.21), the symbol $\sum_{(\alpha,\beta,\gamma)}$ indicates taking the sums over combinations of α, β, γ subject to the condition $\alpha + \beta + \gamma = l$. [Note that the sum over all l in effect removes the restriction on α, β, γ and gives the same results as the original unrestricted sum in (5.20).]

We now restrict $\boldsymbol{r}$ to the unit sphere by requiring that $|\boldsymbol{r}| = 1$, and introduce an index $m = \alpha - \beta$. The expression (5.21) is then rewritten in the form

$$g_M(\theta,\phi) = \sum_{l=0}^{3M} \sum_{(\alpha,\beta,\gamma)} b_{\alpha\beta\gamma}^{(M)} e^{im\phi} \sin^{\alpha+\beta-|m|}\theta \cos^\gamma\theta \sin^{|m|}\theta.$$

A trigonometric identity gives

$$\sin^{\alpha+\beta-|m|}\theta \cos^\gamma\theta = (1-\cos^2\theta)^{(\alpha+\beta-|m|)/2} \cos^\gamma\theta,$$

which is a polynomial in $\cos\theta$ of maximum degree $\alpha + \beta + \gamma - |m| = l - |m|$, since $\alpha + \beta - |m|$ is even (see the remark below). Denoting this polynomial by $f_{lm}(\cos\theta)$, we get

$$g_M(\theta,\phi) = \sum_{l=0}^{3M} \sum_{m} b_{lm}^{(M)} e^{im\phi} \sin^{|m|}\theta \, f_{lm}(\cos\theta). \tag{5.22}$$

Remark. That $\alpha + \beta - |m|$ is even is seen by observing the identity

$$\alpha + \beta - |m| = m - |m| + 2\beta.$$

On the right-hand side, 2β is even and

$$m - |m| = \begin{cases} 0 & \text{if } m \geq 0, \\ -2m & \text{if } m < 0. \end{cases}$$

The range of the summation over m still has to be specified. Recall that all the α, β, γ are nonnegative integers subject to the condition that $\alpha+\beta+\gamma = \ell \geq 0$. This is illustrated schematically in Fig. 5.2, in which the point (α, β, γ) must lie on the oblique face of the tetrahedron depicted in the $\alpha\beta\gamma$ space. The line $\alpha - \beta = m$ on the γ plane is shown as a solid line. In order for it to intersect the oblique face, m must satisfy the condition that

$$-\ell \leq m \leq \ell.$$

Therefore, the sum over m in (5.22) is restricted to $|m| \leq l$, and the last equation becomes

$$g_M(\theta, \phi) = \sum_{l=0}^{3M} \sum_{m=-l}^{l} b_{lm}^{(M)} Y_{lm}(\theta, \phi). \tag{5.23}$$

Here the sequence of functions

$$Y_{lm}(\theta, \phi) \equiv e^{im\phi} \sin^{|m|} \theta \, f_{lm}(\cos\theta), \tag{5.24}$$

where $f_{lm}(\cos\theta)$ is a polynomial in $\cos\theta$ of degree $l - |m|$, provides a uniform approximation to any continuous function defined on the unit sphere. The functions Y_{lm} are called **spherical harmonics**. Note that for a given l, there are $2l+1$ functions Y_{lm}.

The orthonormality of the set $\{Y_{lm}\}$ is characterized by the relation

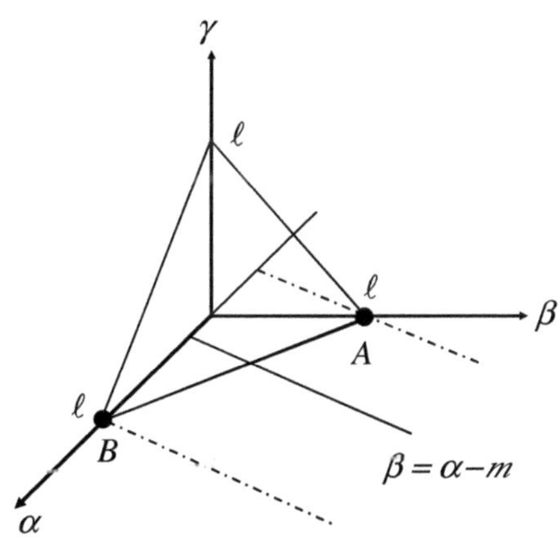

Fig. 5.2. The solid and dashed-dotted lines shown on the γ-plane indicate the relation $\alpha - \beta = m$ for $-\ell < m < \ell$ and $m = \pm\ell$, respectively. In order for the point (α, β, γ) be on the oblique face of the tetrahedron, the condition $-\ell \leq m \leq \ell$ should be satisfied so that the solid line intersects the line segment AB on the γ-plane

$$\int_0^{2\pi} d\phi \int_0^{\pi} \sin\theta d\theta \; Y_{l'm'}^*(\theta, \phi) Y_{lm}(\theta, \phi) = \delta_{ll'}\delta_{mm'},$$

which determines the functions Y_{lm} uniquely up to a phase factor.

General equations for the Y_{lm} are

$$Y_{lm}(\theta,\phi) = (-1)^m \left[\frac{2l+1}{4\pi}\frac{(l-m)!}{(l+m)!}\right]^{1/2} P_l^m(\cos\theta)e^{im\phi}, \quad m \geq 0$$
$$Y_{l,-m}(\theta,\phi) = (-1)^m Y_{lm}^*(\theta,\phi), \quad m \geq 0,$$

where

$$P_l^m(x) = (1-x^2)^{m/2}\frac{d_m}{dx^m}P_l(x)$$
$$= \frac{1}{2^\ell \ell!}(1-x^2)^{m/2}\frac{d^{\ell+m}}{dx^{\ell+m}}(x^2-1)^\ell, \quad m \geq 0, \tag{5.25}$$

are called the **associated Legendre functions**.

Remark.

1. The normalization constant of the Y_{lm} follows immediately from the orthonormality relations for the associated Legendre functions:
$$\int_{-1}^{1} P_l^m(x)P_{l'}^m(x)dx = \frac{(l+m)!}{(l-m)!}\frac{2}{2l+1}\delta_{ll'}.$$
There is, of course, a free choice of phase factor; ours is a common choice in the physics literature. However, one must be careful because different authors choose different phase factors for the spherical harmonics.
2. We should note that the associated Legendre functions $P_l^m(x)$ are not another orthonormal set of polynomials on $[-1,1]$. In fact, they are not polynomials at all as is clearly seen in equation (5.25).

Exercises

1. Find the normalized Legendre polynomials $\tilde{P}_n(x)$.

   Solution: Using equation (5.13), we write the normalized Legendre polynomials $\tilde{P}_n(x)$ as
$$\tilde{P}_n(x) = \sqrt{\frac{2n+1}{2n}}P_n(x), \quad n = 0,1,2,\cdots. \clubsuit$$

2. Derive the explicit form of each function: Y_{00}, Y_{11}, Y_{10}, and $Y_{1,-1}$.
   Solution: It follows from (5.24) that for $l = m = 0$, we obtain $Y_{00} = \sqrt{1/4\pi}$. If $l = 1$, then m can equal -1, 0, or $+1$. Recalling

that $f_{lm}(\cos\theta)$ is a polynomial in $\cos\theta$ of degree $l-|m|$, we obtain $Y_{10} = c_1\cos\theta + c_2$, $Y_{11} = c_3 e^{i\phi}\sin\theta$, $Y_{1,-1} = c_4 e^{-i\phi}\sin\theta$. The constants c_1, c_2, c_3, c_4 are determined by imposing orthonormality. For instance,

$$\begin{aligned}
\int_0^{2\pi} d\phi \int_0^{\pi} \sin\theta Y_{00}^* Y_{10} d\theta &= \sqrt{\pi}\int_0^{\pi} d\theta \sin\theta (c_1\cos\theta + c_2) \\
&= \delta_{01}\delta_{00} = 0, \\
\int_0^{2\pi} d\phi \int_0^{\pi} \sin\theta \left|Y_{10}\right|^2 d\theta &= 2\pi \int_0^{\pi} d\theta \; \left[\sin\theta(c_1\cos\theta + c_2)\right]^2 \\
&= \delta_{10}\delta_{10} = 1,
\end{aligned}$$

which result in $c_1 = \sqrt{3/(4\pi)}$ and $c_2 = 0$. Similarly, it follows that $c_3 = -c_4 = -\sqrt{3/8\pi}$. We choose the minus sign with the convention to be adopted later. Therefore, the first few members of the set $\{Y_{lm}\}$ are

$$\begin{aligned}
Y_{00} &= \sqrt{\frac{1}{4\pi}}, & Y_{11} &= -\sqrt{\frac{3}{8\pi}} e^{i\phi}\sin\theta, \\
Y_{10} &= \sqrt{\frac{3}{4\pi}}\cos\theta, & Y_{1,-1} &= \sqrt{\frac{3}{8\pi}} e^{-i\phi}\sin\theta.
\end{aligned}$$ ♣

3. From the generating function of Legendre polynomials determine that

(i) $P_n(1) = 1,\; P_n(-1) = (-1)^n$,

(ii) $P_{2n}(0) = (-1)^n \dfrac{(2n-1)!!}{(2n)!!}, \quad P_{2n+1}(0) = 0 \quad \text{with } (-1)!! = 1$,

(iii) $\displaystyle\int_0^1 x^n P_n(x)dx = \frac{2^n(n!)^2}{(2n+1)!}$.

Solution: We use the equation

$$(1-2tx+t^2)^{-1/2} = \sum_{n=0}^{\infty} P_n(x)t^n.$$

(i) For $x = 1$, we have

$$\frac{1}{1-t} = \sum_{n=0}^{\infty} t^n = \sum_{n=0}^{\infty} P_n(1)t^n,$$

which yields $P_n(1) = 1$. Similarly for $x = -1$, we obtain

$$\frac{1}{1+t} = \sum_{n=0}^{\infty} (-1)^n t^n = \sum_{n=0}^{\infty} P_n(-1) t^n,$$

which gives $P_n(-1) = (-1)^n$.

(ii) For $x = 0$, we have

$$\left(\frac{1}{1+t^2}\right)^{1/2} = \sum_{n=0}^{\infty} (-1)^n \frac{(2n-1)!!}{(2^n n!)} t^{2n} = \sum_{n=0}^{\infty} P_n(0) t^n.$$

Then, we have the desired result.

(iii) We get the relation by performing an integration in parts n times. ♣

4. Show that the Coulomb potential at $\boldsymbol{r} = \boldsymbol{r}_0$ experienced from the unit charge at $z = a$ on the z-axis is given by

$$V(\boldsymbol{r}_0) = \frac{1}{4\pi a} \sum_{n=0}^{\infty} \left(\frac{r_0}{a}\right)^n P_n(\cos\theta),$$

where θ is the angle between the z axis and the vector $\boldsymbol{r}_0$ and a satisfies the condition $r_0 < a$.

Solution: Using the generating function of Legendre polynomials, we have

$$\begin{aligned} V(\boldsymbol{r}_0) &= \frac{1}{4\pi\varepsilon_0} \frac{1}{|\boldsymbol{r}_0 - \boldsymbol{a}|} = \frac{1}{4\pi\varepsilon_0} \frac{1}{\sqrt{r_0^2 + a^2 - 2ar_0\cos\theta}} \\ &= \frac{1}{4\pi\varepsilon_0} \sum_{n=0}^{\infty} \left(\frac{r_0}{a}\right)^n P_n(\cos\theta). \end{aligned}$$

This series converges because $r_0 < a$ and $|P_n(\cos\theta) \leq 1|$. ♣

5.2 Classification of Orthonormal Functions

5.2.1 General Rodrigues Formula

In the previous section we saw that several kinds of orthonormal polynomials can be produced through the Gram-Schmidt orthogonalization process by starting with $1, x, x^2, \cdots$. However, there is a more elegant approach that applies to most polynomials of interest to physicists. This section describes this approach, which is based on the **Rodrigues formula** and classifies various orthogonal polynomials in terms of the parameters involved in the formula.

♠ **General Rodrigues formula:**

$$Q_n(x) = \frac{1}{K_n w(x)} \frac{d^n}{dx^n} [w(x)s^n(x)] \quad (n = 0, 1, 2, \cdots), \tag{5.26}$$

where it is assumed that

1. $Q_1(x)$ is a first-degree polynomial in x.
2. $s(x)$ is a polynomial in x of degree no more than 2 with real roots.
3. $w(x)$ is real, positive, and integrable in the interval $[a, b]$ and satisfies the boundary condition

$$w(a)s(a) = w(b)s(b) = 0.$$

Equation (5.26) under the three conditions noted above provides the sequence of functions $(Q_0(x), Q_1(x), Q_2(x), \cdots)$ that forms an orthogonal set of polynomials on the interval $[a, b]$ with a weight function $w(x)$, which can be normalized by a suitable choice of constants K_n. For historical reasons, different polynomial functions are normalized differently, which is why K_n is introduced here. In the meantime, we omit denoting K_n without loss of generality.

♠ **Theorem:**

The function $Q_n(x)$ defined by (5.26) is a polynomial in x of the nth degree and satisfies the orthogonality relation on the interval $[a, b]$ with weight $w(x)$:

$$\int_a^b p_m(x) Q_n(x) w(x) dx = 0 \quad (m < n), \tag{5.27}$$

where $p_m(x)$ is an arbitrary polynomial of degree $m < n$.

Proof From hypothesis, we have

$$\left. \frac{d^m}{dx^m} [w(x)s^n(x)] \right|_{x=a \text{ or } b} = 0 \quad (\text{if } m < n) \tag{5.28}$$

and

$$\frac{d^m}{dx^m} \left[w(x)s^n(x)p_{(\le k)}(x) \right] = w(x)s^{n-m}(x)p_{(\le k+m)}, \tag{5.29}$$

where the symbol $p_{(\le k)}(x)$ denotes an arbitrary polynomial in x of degree $\le k$. Then, integrating (5.27) by parts n times, we obtain for $m < n$,

$$\int_a^b p_m(x)Q_n(x)w(x)dx = \int_a^b p_m(x)\frac{d^n}{dx^n}\left[w(x)s^n(x)\right]dx$$
$$= \int_a^b w(x)s^n(x)\frac{d^n}{dx^n}p_m(x)dx \; = 0, \quad (5.30)$$

where we used (5.26) and (5.28). Next we examine whether or not Q_n is a polynomial of degree n. Set $n = m$ and $k = 0$ in (5.29) to obtain

$$\frac{1}{w(x)}\frac{d^n}{dx^n}\left[w(x)s^n(x)\right] = Q_n(x) = p_{(\leq n)}(x),$$

which indicates that $Q_n(x)$ is a polynomial of degree no more than n. We thus tentatively write

$$Q_n(x) = p_{(\leq n-1)}(x) + a_n x^n, \quad (5.31)$$

and would like to show that $a_n \neq 0$. Multiplying both parts of (5.31) by $Q_n(x)w(x)$ followed by integrating on $[a, b]$ yields

$$\int_a^b \left[Q_n(x)\right]^2 w(x)dx = \int_a^b p_{(\leq n-1)}(x)Q_n(x)w(x)dx + a_n\int_a^b x^n Q_n(x)w(x)dx$$
$$= a_n\int_a^b x^n Q_n(x)w(x)dx,$$

where we used (5.30). This clearly proves that $a_n \neq 0$, i.e., that $Q_n(x)$ is a polynomial of the nth degree. ♣

5.2.2 Classification of the Polynomials

In what follows, we classify the orthogonal polynomials that are derived from the Rodrigues formula (5.26) the three conditions according to noted earlier. By the condition **1** associated with (5.26), $Q_1(x)$ is a first-degree polynomial, and we can define it as

$$Q_1(x) = -\frac{x}{K_1}. \quad (5.32)$$

Then the Rodrigues formula (5.26) reads

$$\frac{1}{w}\frac{dw}{dx} = -\frac{x + (ds/dx)}{s}. \quad (5.33)$$

Recall that $s(x)$ can be the zeroth-, first-, or second-degree polynomial. In each case, we can find an appropriate weight function $w(x)$ that satisfies the differential equation (5.33) as well as the boundary condition **3**:

$$w(a)s(a) = w(b)s(b) = 0. \quad (5.34)$$

Such discussions determine the explicit forms of possible functions $s(x)$ and $w(x)$ under conditions **1**–**3** in Sect. 5.2.1 and then allow classification of all of the orthogonal polynomials provided by the general Rodrigues formula described below.

Hermite polynomials:

We first consider the case that $s(x)$ is a zeroth-degree polynomial, i.e., a constant given by

$$s(x) = \alpha.$$

Equation (5.33) takes the form

$$\frac{1}{w}\frac{dw}{dx} = -\frac{x}{\alpha}$$

and has the solution

$$w(x) = A \exp\left(-\frac{x^2}{2\alpha}\right) \quad \text{with a constant } A. \tag{5.35}$$

The product $w(x)s(x)$ vanishes only at $x = \pm\infty$, provided that $\alpha > 0$. To satisfy the conditions in (5.34), we have to set

$$a = -\infty, \quad b = +\infty.$$

The constants A and α affect only the multiplicative factor in front of each polynomial. Thus, without loss of generality, we can take $\alpha = 1$ and $A = 1$, which yields

$$w = e^{-x^2}.$$

The complete orthonormal polynomials corresponding to this case are known as Hermite polynomials, designated by $H_n(x)$, and satisfy the orthonormal condition

$$\int_{-\infty}^{\infty} e^{-x^2} H_m(x) H_n(x) dx = \delta_{mn}.$$

Laguerre polynomials:

Next we let $s(x)$ be a polynomial of the first degree, such as

$$s(x) = \beta(x - \alpha).$$

The Rodrigues formula (5.26) now becomes

$$\frac{1}{w}\frac{dw}{dx} = -\frac{x + \beta}{\beta(x - \alpha)},$$

which has the solution

$$w(x) = \text{const.} \times (x-\alpha)^\nu e^{-x/\beta},$$

where

$$\nu \equiv -\frac{\alpha+\beta}{\beta}.$$

If $\beta > 0$ and $\nu > -1$, then $s(x)w(x)$ vanishes at $x = \alpha$ and $x = +\infty$, and $w(x)$ is integrable in the interval $[\alpha, +\infty)$. The simplest choice is therefore to take $\alpha = 0$ and $\beta = 1$, which yields

$$w = x^\nu e^{-x}, \quad a = 0, \ \ b = +\infty.$$

These choices result in the Laguerre polynomials, commonly denoted by $L_n^\nu(x)$, whose orthonormality relation is given by

$$\int_0^\infty x^{-\nu} L_m^\nu(x) L_n^\nu(x) dx = \delta_{mn} \quad \text{with } \nu > -1.$$

Jacobi polynomials:

Finally, let us take

$$s(x) = \gamma(x-\alpha)(\beta-x), \quad \beta > \alpha.$$

Here we assume that $s(x)$ has two distinct roots. [If $s(x)$ has a double root, the boundary condition (5.34) cannot be satisfied, since in this case the function $s(x)w(x)$ cannot vanish at more than one point.] The Rodrigues formula (5.26) now reads

$$\frac{1}{w}\frac{dw}{dx} = -\frac{x+\gamma(\beta-x)-\gamma(x-\alpha)}{\gamma(x-\alpha)(\beta-x)},$$

which has the solution

$$w(x) = \text{const.} \times (x-\alpha)^\mu (\beta-x)^\nu,$$

with

$$\mu \equiv -\frac{\alpha+\beta}{\beta} \quad \text{and} \quad \nu \equiv \frac{1-\gamma}{\gamma} - \frac{\alpha}{\gamma(\beta-\alpha)}.$$

If $\mu > -1$ and $\nu > -1$, then $s(x)w(x)$ vanishes at $x = \alpha$ and $x = \beta$, and $w(x)$ is integrable on the interval $[\alpha, \beta]$. With the replacement

$$\frac{2x-\alpha-\beta}{\beta-\alpha} \to x,$$

apart from multiplicative factors, we obtain

$$w = (1-x)^\nu (1+x)^\mu \text{ with } \nu, \mu > -1, \quad a = -1, \ b = +1.$$

The corresponding complete orthonormal polynomials are called the Jacobi polynomials $G_n^{\mu,\nu}(x)$, and satisfy the relation

$$\int_{-1}^{1} (1-x)^{-\nu}(1-x)^{-\mu} G_m^{\mu,\nu}(x) G_n^{\mu,\nu}(x) dx = \delta_{mn} \text{ with } \nu, \mu > -1.$$

Remark. Jacobi polynomials can be divided into subcategories depending on the values of μ and ν. The most common and widely used in mathematical physics are collected in Table 5.1.

Table 5.1. Special cases of Jacobi polynomials

μ	ν	$w(x)$	Polynomial
$\lambda - 1/2$	$\lambda - 1/2$	$(1-x^2)^{\lambda-1/2}$	Gegenbauer, $C_n^\lambda(x)$. $\lambda > -1/2$
0	0	1	Legendre, $P_n(x)$
$-1/2$	$-1/2$	$(1-x^2)^{-1/2}$	Chebyshev of the first kind, $T_n(x)$
$1/2$	$1/2$	$(1-x^2)^{1/2}$	Chebyshev of the second kind, $U_n(x)$

5.2.3 The Recurrence Formula

We now show that all the orthogonal polynomials derived from the Rodrigues formula (5.26) satisfy the following relation:

♠ **Recurrence formula:**

$$Q_{n+1}(x) = (a_n x + b_n) Q_n(x) - c_n Q_{n-1}(x), \quad (n = 1, 2, \cdots) \tag{5.36}$$

where the constants a_n, b_n, and c_n depend on the class of polynomials considered.

Proof The only property needed for the proof of (5.36) is the orthogonality relation:

$$\int_a^b Q_n(x) p_{(<n)}(x) w(x) dx = 0, \tag{5.37}$$

where the symbol $p_{(<n)}(x)$ denotes an arbitrary polynomial in x of degree less than n. For convenience, we introduce the following notation:

$$\xi_n = \text{coefficient of } x^n \text{ in } Q_n(x),$$

$$\eta_n = \text{coefficient of } x^{n-1} \text{ in } Q_n(x), \tag{5.38}$$

$$I_n = \int_a^b Q_n^2(x) w(x) dx. \tag{5.39}$$

It then follows that

$$Q_{n+1}(x) - \frac{\xi_{n+1}}{\xi_n} x Q_n(x) = \sum_{i=0}^{n} r_i^{(n)} Q_i(x)$$

because the left-hand side is a polynomial of degree $\leq n$; $r_i^{(n)}$ are appropriate constants determined by the left-hand side. Multiplying both sides by wQ_m, taking m equal to $0, 1, 2, \cdots, n-2$ successively, and using the orthogonality relation (5.37), we obtain

$$r_m^{(n)} = 0 \quad \text{for} \ \ m = 0, 1, 2, \cdots, n-2.$$

Thus

$$Q_{n+1}(x) - \frac{\xi_{n+1}}{\xi_n} x Q_n(x) = r_n^{(n)} Q_n(x) + r_{n-1}^{(n)} Q_{n-1}(x), \tag{5.40}$$

which is the recurrence formula we are looking for. ♣

5.2.4 Coefficients of the Recurrence Formula

We now have to find the constants $r_n^{(n)}$ and $r_{n-1}^{(n)}$ in (5.40). In view of the orthogonality relation (5.37), we have

$$I_n = \int_a^b Q_n^2(x) w(x) dx = \xi_n \int_a^b Q_n(x) x^n w(x) dx. \tag{5.41}$$

Multiplying (5.40) by wQ_{n-1} and integrating, we obtain

$$\begin{aligned} I_{n-1} r_{n-1}^{(n)} &= -\frac{\xi_{n+1}}{\xi_n} \int_a^b Q_n(x) Q_{n-1}(x) x w(x) dx \\ &= -\frac{\xi_{n+1}}{\xi_n} \cdot \frac{\xi_{n-1}}{\xi_n} \int_a^b Q_n(x) \xi_n x^n w(x) dx \\ &= -\frac{\xi_{n+1}\xi_{n-1}}{\xi_n^2} I_n. \end{aligned}$$

Therefore,

$$r_{n-1}^{(n)} = -\frac{I_n}{I_{n-1}} \cdot \frac{\xi_{n+1}\xi_{n-1}}{\xi_n^2}. \tag{5.42}$$

Substituting this into (5.40) and comparing the coefficients of x^n on both sides, yields

$$r_n^{(n)} = -\frac{\eta_{n+1}}{\xi_n} \cdot \frac{\xi_{n+1}\eta_n}{\xi_n^2}. \tag{5.43}$$

Finally, it follows from (5.40)–(5.43) that the coefficients a_n, b_n, and c_n defined in (5.40) become

$$a_n = \frac{\xi_{n+1}}{\xi_n},$$

$$b_n = \frac{\xi_{n+1}}{\xi_n}\left(\frac{\eta_{n+1}}{\xi_{n+1}} - \frac{\eta_n}{\xi_n}\right),$$

$$c_n = \frac{I_n}{I_{n-1}} \cdot \frac{\xi_{n+1}\xi_{n-1}}{\xi_n^2}. \tag{5.44}$$

The constants ξ_n and η_n can, in principle, be found from the Rodrigues formula once the functions $s(x)$ and $w(x)$ as well as the constants K_n have been fixed. The constants I_n, which determine the normalization of the polynomials, are given by

$$I_n = \frac{(-1)^n \xi_n n!}{K_n} \int_a^b s(x)^n w(x) dx.$$

This follows immediately from the Rodrigues formula if we integrate n times by parts the integral

$$\begin{aligned} I_n &= \int_a^b Q_n(x)^2 w(x) dx = \xi_n \int_a^b Q_n(x) x^n w(x) dx \\ &= \frac{\xi_n}{K_n} \int_a^b x^n \frac{d^n}{dx^n} \left[s(x)^n w(x)\right] dx. \end{aligned}$$

Although the explicit form of the coefficients given in (5.44) seems rather complicated, the corresponding recurrence relation for a specific orthogonal polynomial simplifies it considerably.

5.2.5 Roots of Orthogonal Polynomials

Consider the recurrence formula (5.36) in which the polynomials $Q_n(x)$ are normalized, and from (5.39) $I_n = 1$ $(n = 0, 1, 2, \cdots)$. After some rearrangement, the equation takes the form

$$xQ_{n-1}(x) = \frac{\xi_{n-1}}{\xi_n} Q_n(x) + \frac{\xi_{n-2}}{\xi_{n-1}} Q_{n-2}(x) + \beta_{n-1} Q_{n-1}(x),$$

where

$$\beta_{n-1} = \frac{\eta_{n-1}}{\xi_{n-1}} - \frac{\eta_n}{\xi_n}.$$

The matrix form is given by

$$x\begin{pmatrix} Q_0 \\ Q_1 \\ Q_2 \\ \cdots \\ \cdots \\ Q_{N-1} \end{pmatrix} = \begin{pmatrix} \beta_0 & \xi_0/\xi_1 & 0 & \cdots\cdots & 0 \\ \xi_0/\xi_1 & \beta_1 & \xi_1/\xi_2 & \cdots\cdots & 0 \\ 0 & \xi_1/\xi_2 & \beta_2 & \cdots\cdots & 0 \\ \cdots & \cdots & \cdots & \cdots\cdots & \cdots \\ \cdots & \cdots & \cdots & \cdots\cdots & \cdots \\ 0 & 0 & 0 & \cdots\cdots & \beta_{N-1} \end{pmatrix} \begin{pmatrix} Q_0 \\ Q_1 \\ Q_2 \\ \cdots \\ \cdots \\ Q_{N-1} \end{pmatrix}$$

$$+ \begin{pmatrix} 0 \\ 0 \\ 0 \\ \cdots \\ \cdots \\ (\xi_{N-1}/\xi_N)Q_N \end{pmatrix},$$

which gives the eigenvalue equations provided that $\{x_i\}$ are the roots of the polynomial equation $Q_N(x) = 0$ such that

$$J\boldsymbol{R}(x_i) = x_i \boldsymbol{R}(x_i),$$

where the column vector $\boldsymbol{R}(x_i)$ is defined by

$$\boldsymbol{R}(x_i) = [Q_0(x_i), Q_1(x_i), Q_{N-1}(x_i)] \, .$$

Thus, the eigenvalues of the $N \times N$ matrix J are the zeros of $Q_N(x)$. The matrix is called the **Jacobi matrix** associated with the sequence $\{Q_n(x)\}$. Since J is symmetric, the eigenvalues $\{x_i\}$ are real. We thus have proved the following theorem:

♠ **Theorem:**
The eigenvalues $\{x_i\}$ $(i = 1, 2, \cdots, N)$ of the matrix J are the zeros of $Q_N(x)$. The eigenvector belonging to x_i is $\boldsymbol{R}(x_i) = [Q_0(x_i), Q_1(x_i), Q_{N-1}(x_i)]$.

5.2.6 Differential Equations Satisfied by the Polynomials

Historically, most orthogonal polynomials were discovered as solutions of differential equations. Here we give a single generic differential equation that is satisfied by all the polynomials Q_n.

♠ **Theorem:**

All of the orthogonal polynomials $Q_n(x)$ derived from the general Rodrigues formula (5.26) satisfy the differential equation

$$\frac{d}{dx}\left(sw\frac{dQ_n}{dx}\right) = -\lambda_n w Q_n,$$

with the constant

$$\lambda_n = -n\left(K_1\frac{dQ_1}{dx} + \frac{n-1}{2}\frac{d^2 s}{dx^2}\right).$$

Proof Since $dQ_n(x)/dx$ is a polynomial of degree $\leq (n-1)$, it follows from (5.29) that the function

$$\frac{1}{w}\frac{d}{dx}\left[s(x)w(x)\frac{dQ_n}{dx}\right]$$

is a polynomial of degree $\leq n$. Thus, we can write

$$\frac{1}{w}\frac{d}{dx}\left[s(x)w(x)\frac{dQ_n}{dx}\right] = -\sum_{i=1}^{n}\lambda_n^{(i)}Q_i(x), \tag{5.45}$$

where the $\lambda^{(i)}$ are undetermined constants. Multiplying both sides of (5.45) by wQ_m and integrating, we get

$$\int_a^b Q_m(x)\frac{d}{dx}\left[s(x)w(x)\frac{dQ_n}{dx}\right]dx = -\lambda_n^{(m)}I_m. \tag{5.46}$$

Here I_m is an integral given by (5.39). Integrating by parts, for $m < n$ the left-hand side of (5.46) yields

$$\begin{aligned}
&\int_a^b Q_m(x)\frac{d}{dx}\left[s(x)w(x)\frac{dQ_n}{dx}\right]dx \\
&= -\int_a^b s(x)w(x)\frac{dQ_n}{dx}\frac{dQ_m}{dx}dx \\
&= \int_a^b w(x)Q_n(x)\left[\frac{1}{w}\frac{d}{dx}\left(s(x)w(x)\frac{dQ_m}{dx}\right)\right]dx \\
&= 0.
\end{aligned}$$

We have used the condition that $s(a)w(a) = s(b)w(b) = 0$, which is assumption **3** in Sect. 5.2.1. We also used the fact that $Q_n(x)$ is orthogonal to any polynomial of degree $< n$. Consequently, we arrive at the result

$$\lambda_n^{(m)} = 0, \quad \text{for } m < n.$$

Setting

$$\lambda_n^{(n)} = \lambda_n,$$

for simplicity, we can rewrite (5.45) in the form

$$\frac{d}{dx}\left[s(x)w(x)\frac{dQ_n}{dx}\right] = -w(x)\lambda_n Q_n(x), \tag{5.47}$$

which is the differential equation satisfied by a polynomial $Q_n(x)$. The constant λ_n can be found by setting $m = n$ in (5.46) and integrating, as we demonstrate later in Exercise **4**.

5.2.7 Generating Functions (I)

As a matter of fact, all the orthogonal polynomials $Q_n(x)$ discussed thus far can be generated from a single function $g(t, x)$ of two variables by repeated differentiation with respect to t. Called a **generating function**, it plays a significant role in many areas of mathematics. Here we study the essence of generating functions together with several examples by which we can derive specific orthogonal polynomials.
A formal definition of generating functions is given below.

♠ **Generating function:**

Assume a (finite or infinite) convergent power series

$$\gamma(t) \equiv \sum_k f_k t^k.$$

The $\gamma(t)$ is called a **generating function** for the sequence of coefficients $f_1, f_2, \cdots, f_n, \cdots$.

Clearly, all the coefficients f_n are obtained from differentiating $\gamma(t)$ as given by

$$f_n = \frac{1}{n!}\frac{d^n\gamma(t)}{dt^n}.$$

For orthogonal polynomials, generating functions are assumed to take the form

$$g(t, x) = \sum_{n=0}^{\infty} A_n Q_n(x) t^n, \tag{5.48}$$

where $Q_n(x)$ is an orthogonal polynomial associated with $g(t, x)$, and the A_n are appropriate constants. The explicit form of $g(t, x)$ can often be derived using the Rodrigues formula and **Cauchy's integral formula** (see Sect. 7.3.1).

Remember that the latter formula determines an nth-order derivative of a function $f(z)$ as

$$\frac{d^n}{dz^n} f(z) = \frac{n!}{2\pi i} \int_C \frac{f(\zeta) d\zeta}{(\zeta - z)^{n+1}},$$

where $f(z)$ is **analytic** within the closed contour C. (See Sect. 7.1.2 for a definition of **analytic functions**.) Applying this to the Rodrigues formula for, say, Hermite polynomials $H_n(x)$, we obtain

$$H_n(x) = (-1)^n e^{x^2/2} \frac{d^n}{dx^n} e^{-x^2/2} = (-1)^n e^{x^2/2} \frac{n!}{2\pi i} \oint_C \frac{e^{-\zeta^2/2} d\zeta}{(\zeta - x)^{n+1}}.$$

We then try to sum the series as

$$\sum_{n=0}^{\infty} \frac{H_n(x)}{n!} t^n = \frac{e^{x^2/2}}{2\pi i} \oint_C e^{-\zeta^2/2} \left[\sum_{n=0}^{\infty} \frac{(-1)^n t^n}{(\zeta - x)^{n+1}} \right] d\zeta = \frac{e^{x^2/2}}{2\pi i} \oint_C \frac{e^{-\zeta^2/2} d\zeta}{\zeta - x + t},$$

where we require that the point $x - t$ be inside the contour. Finally we evaluate the above integral and find

$$e^{tx - (t^2/2)} = \sum_{n=0}^{\infty} \frac{H_n(x) t^n}{n!}.$$

Comparing this last equation with (5.48), we see that $e^{tx-(t^2/2)}$ is the generating function associated with Hermite polynomials $H_n(x)$. Similarly, we can derive the generating function for Laguerre polynomials as

$$\frac{e^{-tx/(1-t)}}{(1-t)^{1+\alpha}} = \sum_{n=0}^{\infty} L_n^{\alpha}(x) t^n.$$

5.2.8 Generating Functions (II)

There is an alternative way to determine a generating function, which is based on the recurrence formula for a particular polynomial. To see this, we try to find the generating function of the Legendre polynomials that satisfies the following recursion formula:

$$(n+1)P_{n+1}(x) - (2n+1)xP_n(x) + nP_{n-1}(x) = 0,$$

with $P_0(x) = 1$, $P_1(x) = x$, and for convenience we set $P_{-1}(x) = 0$. We seek an expression in closed form for

$$g(t, x) = \sum_{n=0}^{\infty} P_n(x) t^n.$$

First we note that

$$\begin{aligned}\frac{\partial g}{\partial t} &= \sum_{n=0}^{\infty} nP_n(x)t^{n-1} = \sum_{n=0}^{\infty}(n+1)P_{n+1}(x)t^n \\ &= \sum_{n=0}^{\infty}\left[(2n+1)xP_n(x) - nP_{n-1}(x)\right]t^n.\end{aligned}$$

By straightforward rearrangement we find that

$$\frac{\partial g}{\partial t} = xg(t,x) + 2tx\frac{\partial g}{\partial t} - tg(t,x) - t^2\frac{\partial g}{\partial t},$$

which leads to the partial differential equation

$$\frac{1}{g}\frac{\partial g}{\partial t} = \frac{x-t}{1-2tx+t^2}.$$

Coupled with the initial condition $g(0,x) = 1$ we finally have

$$\frac{1}{\sqrt{1-2tx+t^2}} = \sum_{n=0}^{\infty} P_n(x)t^n.$$

Generating functions for other orthogonal polynomials are given in Appendix D.

Exercises

1. Find the recurrence formula for normalized polynomials $\tilde{Q}_n(x)$.

Solution: When the polynomials are normalized, we have $I_n = 1$ $(n = 0, 1, 2, \cdots)$ from (5.39). The recurrence formula (5.36) is

$$\tilde{Q}_{n+1}(x) = (a_n x + b_n)\tilde{Q}_n(x) - \frac{a_n}{a_{n-1}}\tilde{Q}_{n-1}(x). \quad \clubsuit$$

2. Assume that a sequence of orthogonal polynomials satisfies

$$Q_{n+1}(x) = \left[(n+1)x + 1\right]Q_n(x) - 3(n+1)Q_{n-1}(x).$$

Find the normalized constants for $Q_n(x)$ defined by $\tilde{Q}_n(x) = \lambda Q_n(x)$, where $\tilde{Q}_n(x)$ are normalized polynomials.

Solution: We denote the normalized polynomials as $\tilde{Q}_n(x) = \lambda Q_n(x)$, where the constants λ_n $(n = 0, 1, 2, \cdots)$ are to be found. Substituting $\tilde{Q}_n(x)$ into the given formula, we have

$$\tilde{Q}_{n+1}(x) = \frac{\lambda_{n+1}}{\lambda_n}\left[(n+1)x + 1\right]\tilde{Q}_n(x) - 3(n+1)\frac{\lambda_{n+1}}{\lambda_n}\tilde{Q}_{n-1}(x).$$

Comparing this with the normalized recurrence formula from Exercise **1**, we have the relation $(3\lambda_n)/\lambda_{n-1} = \lambda_{n-1}/(n\lambda_n)$, which yields $\lambda_n = \lambda_{n-1}/\sqrt{3n}$. This relation gives the normalization constants of the form

$$\lambda_n = \frac{1}{3^{n/2}}(n!)^{1/2}\lambda_0. \quad ♣$$

3. Find the recurrence formula for Hermite and Legendre polynomials.
Solution: For Hermite and Legendre polynomials, (5.36) reads

$$H_{n+1}(x) = 2xH_n(x) - 2nH_{n-1}(x) \tag{5.49}$$

and

$$(n+1)P_{n+1}(x) = (2n+1)xP_n(x) - nP_{n-1}(x), \tag{5.50}$$

respectively. See Appendix D for the recurrence relations associated with the other polynomials we have discussed. ♣

4. Determine the constants λ_n given in (5.47).
Solution: Setting $m = n$ on the left-hand side of (5.46), we obtain

$$\int_a^b Q_n(x)\frac{d}{dx}\left[s(x)w(x)\frac{dQ_n}{dx}\right]dx \tag{5.51}$$

$$= \int_a^b Q_n(x)\left[\frac{d(sw)}{dx}\frac{dQ_n}{dx} + s(x)w(x)\frac{d^2Q_n}{dx^2}\right]dx$$

$$= \int_a^b w(x)Q_n(x)\left[K_1Q_1(x)\frac{dQ_n}{dx} + s(x)\frac{d^2Q_n}{dx^2}\right]dx. \tag{5.52}$$

Here we used the relation $d(sw)/dx = wK_1Q_1$ [set $n = 1$ in the general Rodrigues formula (5.26).] The orthogonality of $Q_n(x)$ means that only the nth power of x in the square brackets contributes to the integral in the last line of (5.52). [See (5.30) for details of the orthogonal property of $Q_n(x)$.] We then set up the following expressions:

$$\begin{aligned} s(x) &= ax^2 + bx + c, \\ Q_n(x) &= \xi_n x^n + \xi_{n-1} x^{n-1} + \cdots, \\ Q_1(x) &= \eta_1 + \eta_0, \end{aligned}$$

which result in

$$K_1 Q_1 \frac{dQ_n}{dx} = K_1 \eta_1 n \xi_n x^n + (\text{const.}) \times x^{n-1} + \cdots$$

and

$$s \frac{d^2 Q_n}{dx^2} = an(n-1)\xi_n x^n + (\text{const.}) \times x^{n-1} + \cdots .$$

Thus the relevant terms in the square brackets in the last line in (5.52) become

$$\left[nK_1 \frac{dQ_1}{dx} + \frac{1}{2} \frac{ds^2}{dx^2} n(n-1) \right] \xi_n x^n,$$

where we used $\eta_1 = dQ_1/dx$ and $a = (1/2)(ds^2/dx^2)$, and we get

$$\begin{aligned} &\int_a^b Q_n(x) \frac{d}{dx} \left[s(x) w(x) \frac{dQ_n}{dx} \right] dx \\ &= \left[nK_1 \frac{dQ_1}{dx} + \frac{n}{2}(n-1) \frac{d^2 s}{dx^2} \right] \int_a^b w(x) Q_m(x) \left(\xi_n x^n \right) dx \\ &= n \left(K_1 \frac{dQ_1}{dx} + \frac{n-1}{2} \frac{d^2 s}{dx^2} \right) I_n. \end{aligned}$$

Comparing this with (5.46), gives us

$$\lambda_n = -n \left(K_1 \frac{dQ_1}{dx} + \frac{n-1}{2} \frac{d^2 s}{dx^2} \right). \quad ♣$$

5.3 Chebyshev Polynomials

5.3.1 Minimax Property

Thus far we have seen that every real function $f(x)$ defined in a certain interval (finite or infinite) can be approximated in the mean by appropriate orthogonal polynomial $\{Q_n(x)\}$ as

$$f(x) \simeq \sum_{i=0}^{n} c_i Q_i(x). \tag{5.53}$$

The coefficients c_i are determined formally by using the orthogonality of the polynomials in question. The striking advantage of such polynomial approximations is that an improvement in the approximation through addition of an extra term $c_{n+1} Q_{n+1}(x)$ does not affect the previously obtained coefficients, $c_0, c_1, \cdots, c_n$.

In principle, any polynomial that we discussed in Sect. 5.2 can be approximated using (5.53). From the point of view of numerical analysis, however, the Chebyshev polynomial $\{T_n(x)\}$ is the best choice, primarily because at any point x within the domain $[-1,1]$, the function $T_n(x)$ has the smallest maximum deviation from the true function $f(x)$ to be approximated. This property, which is unique to Chebyshev polynomials, is known as the **minimax property**. In general, polynomials endowed with the minimax property are very difficult to find, but fortunately, the Chebyshev polynomials fall into this category an, moreover, are easy to compute.

To show the minimax property of Chebyshev polynomials, we have to be aware of two of their other properties. The first is a concise formula for $T_n(x)$ that is an alternative to those based on the Rodrigues formula.

♠ **Concise formula for Chebyshev polynomials:**

$$T_n(x) = \cos\left(n \cos^{-1} x\right) \quad (n = 0, 1, \cdots). \tag{5.54}$$

The derivation of (5.54) requires some lengthy calculations, so we put it in the next subsection (see Sect. 5.3.2). Equation (5.54) implies that each $T_n(x)$ has n zeros in the interval $[-1,1]$, which are located at the points

$$x = \cos\left[\frac{\pi}{n}\left(k - \frac{1}{2}\right)\right] \quad (k = 1, 2, \cdots, n). \tag{5.55}$$

In this same interval, there are $n+1$ extrema (maxima and minima), located at

$$x = \cos\left(\frac{\pi}{n} k\right) \quad (k = 0, 1, \cdots, n).$$

Note that $T_n(x) = 1$ at all of the maxima, whereas $T_n(x) = -1$ at all of the minima. This feature of T_n is exactly what makes the Chebyshev polynomials so useful in polynomial approximation of functions

Remark. Equation (5.54) combined with trigonometric identities can yield explicit expressions for $T_n(x)$:

$$T_0(x) = 1, \quad T_1(x) = x, \quad T_2(x) = 2x^2 - 1, \quad T_3(x) = 4x^3 - 3x, \cdots,$$

and more generally,

$$T_{n+1}(x) = 2x T_n(x) - T_{n-1}(x) \quad (n \geq 1).$$

The last expression is a special case of the general recurrence formula (5.40) derived in Sect. 5.2.3.

The second property of Chebyshev polynomials to be noted is the discrete orthogonality relation described below. (The proof is given in Sect. 5.3.3.)

♠ **Discrete orthogonal relation:**

If x_k $(k = 1, \cdots, n)$ are the m zeros of $T_n(x)$ given by (5.55) and $i, j < n$, then

$$\sum_{k=1}^{n} T_i(x_k) T_j(x_k) = \begin{cases} 0, & i \neq j, \\ n/2, & i = j \neq 0, \\ n, & i = j = 0. \end{cases} \tag{5.56}$$

From (5.54) and (5.56), we obtain the following theorem:

♠ **Theorem:**

Suppose $f(x)$ to be an arbitrary function in the interval $[-1, 1]$ and define c_j $(j = 1, \cdots, n)$ by

$$c_j = \frac{2}{N} \sum_{k=1}^{n} f(x_k) T_{j-1}(x_k), \tag{5.57}$$

where x_k is the kth zero of $T_n(x)$ given by (5.55). We then have

$$f(x) = \sum_{k=1}^{n} c_k T_{k-1}(x) - \frac{c_1}{2} \quad \text{for all } x = x_k. \tag{5.58}$$

What is remarkable is the fact that for $x = x_k$, the finite sum in (5.58) is equal to $f(x)$ *exactly*. For $x \neq x_k$, the sum in (5.58) just *approximates* $f(x)$; nevertheless the error can be reduced by increasing the degree n of the sum. Moreover, for practical use, we can truncate the sum in (5.58) to a much lower degree, for even if we do so, the approximation (5.58) is sufficiently accurate over the whole interval $[-1, 1]$, not only at the zeros of $T_n(x)$. This is in contrast to the case of approximations based on other polynomials, where the degree of summation n should be taken as large as possible to obtain high accuracy. In fact, this truncation capability is the reason Chebyshev polynomial expansion is far better than the other choices.

To examine the above statement, let us suppose that n is so large that (5.58) is virtually a perfect approximation of $f(x)$. We then consider the truncated approximation

$$f(x) \simeq \sum_{k=1}^{m} c_k T_{k-1}(x) - \frac{c_1}{2} \quad \text{with } m \ll n, \tag{5.59}$$

where the coefficients c_k are given in (5.57). The difference between (5.58) and (5.59) is given by

$$\sum_{k=m+1}^{n} c_k T_{k-1}(x), \tag{5.60}$$

which can be no larger than the sum of the neglected c_k's as the $T_n(x)$'s are all bounded between ± 1.

Now we consider the magnitude of the sum (5.60). We know that in general the c_k's decrease rapidly with k, which follows intuitively from the definition (5.57). Hence, the magnitude of (5.60) is dominated by the term $c_{m+1}T_m(x)$, which is much less than unity for all $x \in [-1, 1]$. In addition, $c_{m+1}T_m(x)$ is an oscillatory function with $m+1$ equal extrema distributed almost uniformly over the interval $[-1, 1]$. These two features of the dominant term $c_{m+1}T_m(x)$ result in smooth spreading out of the error of the approximation (5.59). This context implies that the Chebyshev approximation (5.59) is very nearly the same as the minimax polynomial that has the smallest maximum deviation from the true function $f(x)$.

5.3.2 A Concise Representation

The aim here is to derive the alternative representation of Chebyshev polynomials given in (5.54):

$$T_n(x) = \cos\left[n \cos^{-1}(x)\right].$$

We know that Chebyshev polynomials satisfy the relation

$$(1-x^2)\frac{d^2}{dx^2}T_n(x) - x\frac{d}{dx}T_n(x) + n^2 T_n(x) = 0,$$

which can be rewritten in the form

$$\frac{d}{dx}\left(\sqrt{1-x^2}\frac{d}{dx}T_n(x)\right) + \frac{n^2}{\sqrt{1-x^2}}T_n(x) = 0. \tag{5.61}$$

We now apply the following lemma:

♠ **Lemma:**

Let $p(x)$ and $q(x)$ be two positive, continuously differentiable functions that satisfy the differential equation

$$\frac{d}{dx}\left[p(x)\frac{d}{dx}y(x)\right] + q(x)y(x) = 0. \tag{5.62}$$

If the product $p(x)q(x)$ is nonincreasing (or nondecreasing), then the relative maxima of $[y(x)]^2$ form a nondecreasing (nonincreasing) set.

(The proof of this lemma is outlined in Exercise **1**.) We can see that if

$$p(x) = \sqrt{1-x^2} \quad \text{and} \quad q(x) = \frac{n^2}{\sqrt{1-x^2}},$$

(5.61) corresponds to (5.62), which implies that the product pq is constant. Thus, according to the lemma, all relative maxima of $T_n^2(x)$ must assume the same value.

Now we seek a polynomial $T_n(x)$ of degree n that satisfies the condition

$$T_n^2(x) = 1 \ \text{ whenever } \ T_n'(x) = 0.$$

That is, $T_n^2(x) = 1$ at all x where $T_n^2(x)$ has a relative maximum equal to 1. Clearly at these points, both $T_n^2(x) - 1$ and $\left[T_n'(x)\right]^2$ have double zeros. Then the function

$$\frac{T_n^2(x) - 1}{\left[T_n'(x)\right]^2} \tag{5.63}$$

is a rational function and all the zeros of the denominator also occur in the numerator. Is we compare the degree of the polynomials in the denominator and in the numerator, it follows that (5.63) is a quadratic, and without loss of generality we have

$$\frac{T_n^2(x) - 1}{\left[T_n'(x)\right]^2} = \alpha(x^2 - 1). \tag{5.64}$$

The constant α can be determined by dividing both sides by x^2 and letting x approach infinity. Then, inserting a polynomial of degree n for $T_n(x)$, we obtain

$$\frac{1}{n^2} = \alpha \quad \text{so that} \quad T_n(x) = \cos\left[n\cos^{-1} x + c\right],$$

which yields

$$\frac{x^2 - 1}{n^2}\left(\frac{dT_n}{dx}\right)^2 = T_n^2 - 1. \tag{5.65}$$

Equation (5.65) is a differential equation for $T_n(x)$ that determines the explicit form of our desired $T_n(x)$. To solve it, we set

$$T_n(x) = \cos\theta, \quad x = \cos\phi,$$

where θ and ϕ are functions of x. We then have

$$T_n^2(x) - 1 = -\sin^2\theta$$

and

$$\frac{d}{dx}T_n(x) = \left(\frac{d}{d\phi}\cos\theta\right)\frac{d\phi}{dx} = \frac{\sin\theta}{\sin\phi}\frac{d\theta}{d\phi}.$$

Substituting these in (5.64) yields

$$\left(\frac{d\theta}{d\phi}\right)^2 = n^2 \quad \text{so that} \quad \theta = \pm n\phi + c,$$

and we get

$$T_n(x) = \cos\left(n\cos^{-1}x + c\right).$$

To determine c, we note that

$$T_n^2(\pm 1) = 1 = \cos(c).$$

Hence, $c = 0$ and we eventually obtain

$$T_n(x) = \cos\left(n\cos^{-1}x\right). \tag{5.66}$$

5.3.3 Discrete Orthogonality Relation

We close this section by proving the discrete orthogonality relation (5.56) for Chebyshev polynomials.

Proof (of the discrete orthogonality relation): Let x_k $(k = 1, 2, \cdots, n)$ be the n zeros of $T_n(x)$, which is given by

$$x_k = \cos\left[\frac{\pi}{n}\left(k - \frac{1}{2}\right)\right] \quad (k = 1, 2, \cdots, n).$$

Then the value of $T_\ell(x)$ at $x = x_k$, in which $\ell < n$ is assumed, reads

$$T_\ell(x_k) = \cos\left[\ell\cos^{-1}(x_k)\right] = \cos\left[\frac{\pi\ell}{n}\left(k - \frac{1}{2}\right)\right].$$

Using the trigonometric identity, we have for $\ell, m < n$,

$$\begin{aligned} &T_\ell(x_k)T_m(x_k) \\ &= \frac{1}{2}\cos\left[\frac{\pi(\ell+m)}{2n}(2k-1)\right] + \frac{1}{2}\cos\left[\frac{\pi(\ell-m)}{2n}(2k-1)\right]. \end{aligned} \tag{5.67}$$

If $\ell = m = 0$, this equals 1 so that we obtain

$$\sum_{k=1}^{n} T_\ell(x_k)^2 = \sum_{k=1}^{n} 1 = n. \tag{5.68}$$

Otherwise, if $\ell = m \neq 0$, the second term in the last line of (5.67) equals 1/2 and we have

$$\begin{aligned} \sum_{k=1}^{n} T_\ell(x_k)^2 &= \frac{n}{2} + \frac{1}{2}\sum_{k=1}^{n}\cos\left[\frac{\ell\pi}{n}(2k-1)\right] \\ &= \frac{n}{2} + \frac{\sin(2\ell\pi)}{4\sin(\ell\pi/n)} = \frac{n}{2}, \end{aligned} \tag{5.69}$$

where we used the equation (see Exercise **2**)

$$\sum_{k=1}^{n} \cos(2k-1)x = \frac{\sin 2nx}{2\sin x} \quad (\text{for } \ x \neq 0).$$

In a similar manner, for the case $\ell \neq m$ we find that

$$\sum_{k=1}^{n} T_\ell(x_k) T_m(x_k) = 0. \tag{5.70}$$

Equations (5.68), (5.69), and (5.70) together are identical to the desired result given in (5.56). ♣

Exercises

1. Prove the lemma associated with the differential equation (5.62).

Solution: The proof is based on the nondecreasing property of the function defined by

$$f(x) = [y(x)]^2 + \frac{[p(x)y'(x)]^2}{p(x)q(x)},$$

in which the functions $y(x)$, $p(x)$, and $q(x)$ are assumed to satisfy the differential equation (5.62). The nondecreasing property of $f(x)$ is verified by seeing its derivative:

$$f'(x) = 2yy' + \frac{2py'}{pq}(py')' + \left(\frac{1}{pq}\right)'(py')^2 = -\frac{(pq)'}{(pq)^2}(py')^2,$$

where we used the condition (5.62). From hypothesis, pq is nonincreasing, which implies $(pq)' \leq 0$. Hence, it is readily seen that $f' \geq 0$, i.e., that f is nondecreasing.

Now we realize that, y' must vanish wherever $y(x)^2$ has a relative maximum so that $f(x) = y^2$. Suppose that x_1 and x_2 are two successive zeros of y', such that $x_1 < x_2$. Since $f(x)$ is nondecreasing, we have $f(x_2) \geq f(x_1)$, or equivalently, $y^2(x_2) \geq y^2(x_1)$, which means that the relative maxima of y^2 form a nondecreasing set. This completes the proof of the lemma. ♣

2. Prove that $\displaystyle\sum_{k=1}^{n} \cos(2k-1)x = \frac{\sin 2nx}{2\sin x}$ (for $x \neq 0$).

Solution: This equation is obtained by considering the sum

$$\sum_{k=1}^{N} e^{i(2k-1)x} = e^{-ix}\sum_{k=1}^{N} e^{2ikx} = e^{-ix}\left[\frac{1-e^{2i(N+1)x}}{1-e^{2ix}} - 1\right]$$

$$= e^{i(N-1)x} \cdot \frac{\sin(N+1)x}{\sin x} - e^{-ix}.$$

Taking the real part of both sides yields

$$\sum_{k=1}^{N} \cos(2k-1)x = \cos(N-1)x \cdot \frac{\sin(N+1)x}{\sin x} - \cos x$$
$$= \frac{\sin 2Nx + \sin 2x}{2\sin x} - \frac{2\sin x \cos x}{2\sin x} = \frac{\sin 2Nx}{2\sin x}. \quad \clubsuit$$

3. Derive the formula for Chebyshev polynomials:

$$\frac{1-t^2}{1-2tx+t^2} = T_0(x) + 2\sum_{m=1}^{\infty} T_m(x)t^m,$$

where $|t| < 1$. Then, using this equation, prove that

$$\int_0^{2\pi} \frac{\cos\theta}{1-2t\cos\theta+t^2} d\theta = \frac{2\pi t^n}{1-t^2},$$

where $n \geq 0$.

Solution: It follows that

$$1 + \sum_{m=1}^{\infty} 2t^m \cos m\theta = -1 + 2\text{Re}\sum_{m=0}^{\infty} e^{im\theta}t^m = -1 + 2\text{Re}1/(1-te^{i\theta})$$
$$= (1-t^2)/(1-2tx+t^2),$$

which the desired result. The next equation is found in the Fourier cosine series, where the coefficients can be obtained from

$$b_n = \frac{1}{\pi}\int_0^{2\pi} \frac{1-t^2}{1-2tx+t^2} \cos n\theta d\theta = 2t^n. \quad \clubsuit$$

5.4 Applications in Physics and Engineering

5.4.1 Quantum-Mechanical State in an Harmonic Potential

We now consider the application of **Hermite polynomials** $H_n(x)$ to physical systems in the theory of **quantum mechanics**. We know that $H_n(x)$ satisfies the following second-order differential equation:

$$H_n''(x) - xH_n'(x) + nH_n(x) = 0.$$

Let us introduce the related function

$$U_n(x) = e^{-x^2/4}H_n(x). \tag{5.71}$$

A simple calculation shows that

$$U_n''(x) + \left(n + \frac{1}{2} - \frac{x^2}{4}\right) U_n(x) = 0. \tag{5.72}$$

This equation is similar in form to the **Schrödinger equation** for a quantum particle whose motion is confined to an harmonic potential well. In fact, the Schrödinger equation is given by

$$\psi''(x) + \left(E - \frac{x^2}{2}\right) \psi(x) = 0, \tag{5.73}$$

where $\psi(x)$ is the quantum wave function whose squared value at the position $x = a$, namely, $|\psi(a)|^2$, represents the **probability density** of the quantum particle being observed at $x = a$. The similarity between (5.72) and (5.73) implies that the product of the function defined by (5.71), i.e., $H_n(x)$, and $e^{-x^2/4}$ behaves as a wave function that describes the quantum particle in the potential well.

However, it should be noted that solutions of (5.73) do not always satisfy the condition

$$\int_{-\infty}^{\infty} |\psi(x)|^2 dx < \infty, \tag{5.74}$$

which must be satisfied for the solutions to be physically meaningful. By comparing (5.73) with (5.72), we see that whenever

$$E = E_n = 2n + 1, \tag{5.75}$$

we have

$$\psi_n(x) = c_n e^{-x^2/2} H_n\left(\sqrt{2}x\right),$$

which clearly satisfies the condition (5.74) if the constants c_n are chosen appropriately. Furthermore, the **uniqueness theorem** for solutions of ordinary differential equations (see Sect. 15.2.4) guarantees that the values of E given in (5.75) are the only ones for which (5.73) has solutions satisfying (5.75). These specific values of E are called the **eigenenergies** of the system, and the corresponding solutions $psi_n(x)$ are called **eigenfunctions**.

5.4.2 Electrostatic potential generated by a multipole

Next, we briefly discuss the use of **Legendre polynomials** in describing the electrostatic potential field generated by a **multipole**. For simplicity, we first consider an electric dipole, i.e., a pair of positive and negative charges separated by an infinitesimal distance h. We choose our coordinate system such that both charges are located on the x-axis with the negative charge at the origin. The magnitude of the charges is taken to be $\pm(1/h)$. Then, the electrostatic potential field $\Phi_2(P)$ with respect to a point P on the sphere $x^2 + y^2 + z^2 = r^2$ is represented as

$$\Phi_2(P) = \lim_{h\to 0} \frac{1}{h} \left(\frac{1}{\sqrt{(x-h)^2+y^2+z^2}} - \frac{1}{\sqrt{x^2+y^2+z^2}} \right)$$
$$= \frac{\partial}{\partial x}\left(\frac{1}{r}\right) = -\frac{x}{r^3}.$$

Therefore, when $r = 1$, we have

$$\Phi_2(P)|_{r=1} = -x = -P_1(x) \cdot 1!,$$

where $P_n(x)$ is a Legendre polynomial.

Similar descriptions can be presented for high-degree multipoles. The potential $\Phi_4(P)$ of a quadrupole is determined as follows: Consider a double negative charge $-(2/h^2)$ located at the origin and two positive charges $1/h^2$ located at the points $(x, y, z) = (\pm h, 0, 0)$. Then, the associated potential $\Phi_4(P)$ at a point on a sphere of radius r is given by

$$\Phi_4(P) = \lim_{h\to 0} \frac{1}{h^2} \left(\frac{1}{\sqrt{(x+h)^2+y^2+z^2}} - \frac{2}{\sqrt{x^2+y^2+z^2}} + \frac{1}{\sqrt{(x-h)^2+y^2+z^2}} \right)$$
$$= \frac{\partial^2}{\partial x^2}\left(\frac{1}{r}\right) = -\frac{r^2-3x^2}{r^5},$$

so for $r = 1$,

$$\Phi_4(P)|_{r=1} = -1 + 3x^2 = P_2(x) \cdot 2!.$$

Similarly, for an octapole, we get

$$\Phi_8(P)|_{r=1} = \left.\frac{\partial^3}{\partial x^3}\left(\frac{1}{r}\right)\right|_{r=1} = -15x^3 + 9x = -P_3(x) \cdot 3!,$$

and in general

$$\Phi_{2^n}(P)|_{r=1} = \left.\frac{\partial^n}{\partial x^n}\left(\frac{1}{r}\right)\right|_{r=1} = (-1)^n P_n(x) \cdot n!.$$

The final result tells us that the potential of a 2^n-pole is described by the product of the Legendre polynomial $P_n(x)$ and the factor $(-1)^n \cdot n!$. By solving the previous equation for $P_n(x)$, we obtain the following expression for the nth Legendre polynomial:

$$P_n(x) = \frac{(-1)^n}{n!} \cdot \left.\frac{\partial^n}{\partial x^n}\left(\frac{1}{r}\right)\right|_{r=1}.$$

6

Lebesgue Integrals

Abstract The concept of "measure" (Sect. 6.1.2) is important for an understanding of the theory of the Lebesgue integral. A measure is a generalization of the concept of length that allows us to quantify the length of a set that is composed of, for instance, an infinite number of infinitesimal points with a highly discontinuous distribution. Thus, the Lebesgue integral is an effective tool for integrating highly discontinuous functions that cannot be integrated using conventional Riemann integrals.

6.1 Measure and Summability

6.1.1 Riemann Integral Revisited

It is certain that the **Riemann integral** is adequate for practical applications to most problems in physics and engineering, as the functions that we usually encounter are continuous (piecewise, at least) so that they are integrable by the Riemann procedure. In advanced subjects in mathematical physics, however, we come to a class of *highly irregular* functions where the concept of an ordinary Riemann integral is not applicable. In order to treat such functions, we have to employ another, more flexible integral than the Riemann integral. In this chapter, we present a concise description of the **Lebesgue integral**. The Lebesgue integral not only overcomes many of the difficulties inherent in the use of the Riemann integral, but its study has also generated new concepts and techniques that are extremely valuable in practical problems in modern physics and engineering.

At first, the cultivation of an intuitive feeling for the Lebesgue integral as an adjunct to formal manipulations and calculations is important, and we achieve this by comparing it with the Riemann integral. When defining the Riemann integral of a function $f(x)$ on an interval $I = [a, b]$, we divide the entire interval $[a, b]$ into small subintervals $\Delta x_k = [x_k, x_{k+1}]$ such that

$$a = x_1 < x_2 < \cdots < x_{n+1} = b.$$

The finite set $\{x_i\}$ of numbers is called a **partition** P of the interval I. Using this notation P, let us define, e.g., the sums

$$S_P(f) = \sum_{k=1}^{n} M_k(x_{k+1} - x_k), \quad s_P(f) = \sum_{k=1}^{n} m_k(x_{k+1} - x_k),$$

where M_k and m_k are the supremum and infimum of $f(x)$ on the interval $\Delta x_k = [x_k, x_{k+1}]$, respectively, given by

$$M_k = \sup_{x \in \Delta x_k} f(x), \quad m_k = \inf_{x \in \Delta x_k} f(x). \tag{6.1}$$

Evidently, the relation $S_P(f) \geq s_P(f)$ holds if the function $f(x)$ is bounded on the interval $I = [a, b]$. We take the limit inferior (or limit superior) of the sums,

$$S(f) = \liminf_{n \to \infty} S_P, \quad s(f) = \limsup_{n \to \infty} s_P, \tag{6.2}$$

where all possible choices of the partition P are taken into account. The $S(f)$ and $s(f)$ are called the **upper** and **lower Riemann–Darboux integrals** of f over I, respectively. If the relation holds, i.e., if

$$S(f) = s(f) = A,$$

the common value A is called the **Riemann integral** and the function $f(x)$ is called **Riemann integrable** such that

$$A = \int_a^b f(x)dx.$$

We note without proof that the following conditions ensure the existence of the Riemann integral of a function $f(x)$.

1. $f(x)$ is continuous in $I = [a, b]$.
2. $f(x)$ has only a finite number of discontinuities in $I = [a, b]$.

On the other hand, when the function $f(x)$ exhibits too many points of discontinuity, the above definition is of no use in forming the integral. An illustrative example is given below.

Examples Assume an enumeration $\{z_n\}$ $(n = 1, 2, \cdots)$ of the rational numbers between 0 and 1 and let

$$f(x) = \begin{cases} 1 & (x = z_1, z_2, \cdots, z_n) \\ 0 & \text{otherwise.} \end{cases}$$

That is, the function $f(x)$ has the value unity if x is rational and the value zero if x is irrational. In any subdivision of the interval $\Delta x_k \subset [0, 1]$,

$$m_k = 0, \quad M_k = 1,$$

and

$$s_P = 0, \quad S_P = 1.$$

Therefore, the upper and lower Darboux integrals are 1 and 0, respectively, whence $f(x)$ has no Riemann integral.

6.1.2 Measure

The shortcoming of the Riemann procedure demonstrated above can be successfully overcome by employing Lebesgue's procedure. The latter requires a systematic way of assigning a **measure** $\mu(X_i)$ to each subset of points X_i. In the remainder of this section, we learn about the basic properties of measure and its relevant materials, which serve as preliminaries to introduce the precise definition of Lebesgue integrals given in Sect. 6.2.

The measure for a subset of points is a generalization of the concepts of the length, area, and volume. Intuitively, it follows that the length of an interval $[a, b]$ is $b - a$. Similarly, if we have two disjoint intervals $[a_1, b_1]$ and $[a_2, b_2]$, it is natural to interpret the length of the set consisting of these two intervals as the sum $(b_1 - a_1) + (b_2 - a_2)$. However, the 'length' of a set of points of rational (or irrational) numbers on the line is not obvious. This context requires a rigorous mathematical definition of a measure of a point set, as shown below.

♠ **Measure of a set of points:**

A measure $\mu(X)$ defined on a set of points X is a function with the following two properties:

1. If the set X is empty or consists of a single point, $\mu(X) = 0$; otherwise, $\mu(X) > 0$.
2. The measure of the sum of two **nonoverlapping** sets is equal to the sum of the measures of these sets expressed by

$$\mu(X_1 + X_2) = \mu(X_1) + \mu(X_2) \quad \text{for } X_1 \cap X_2 = 0. \tag{6.3}$$

In the above statement, $X_1 + X_2$ denotes the set containing both elements of X_1 and X_2, wherein each element is counted only once. If X_1 and X_2 overlap, (6.3) is replaced by

$$\mu(X_1 + X_2) = \mu(X_1) + \mu(X_2) - \mu(X_1 \cap X_2)$$

so that the points common to X_1 and X_2 will be counted only once.

Various kinds of measures have been thus far introduced in mathematics. Among them, is the following important example of measure that plays a central role in the subsequent discussions. Consider a monotonic increasing function $\alpha(x)$ and let I be an interval (open or closed) with endpoints a and b. We define the **α-measure** of I denoted by $\mu_\alpha(I)$, which takes different values depending on the types of endpoints a and b as shown below.

♠ **α-measure of intervals:**

α-measure of intervals are defined by

- $\mu_\alpha(\,[a, b]\,) = \alpha(b^+) - \alpha(a^-)$ for the closed interval $[a, b]$,
- $\mu_\alpha(\,(a, b]\,) = \alpha(b^+) - \alpha(a^+)$ for the semiclosed interval $(a, b]$,

- $\mu_\alpha(\,[a,b)\,) = \alpha(b^-) - \alpha(a^-)$ for the semiclosed interval $[a,b)$,
- $\mu_\alpha(\,(a,b)\,) = \alpha(b^-) - \alpha(a^+)$ for the open interval (a,b),

where $\alpha(a^-) = \lim_{\varepsilon\to 0} \alpha(a-\varepsilon)$ and $\alpha(a^+) = \lim_{\varepsilon\to 0} \alpha(a+\varepsilon)$.

By definition, the open interval (a,a) is an **empty set**, so that $\mu_\alpha((a,a)) = 0$ for any $a \in \boldsymbol{R}$. The other cases of intervals $(a,a]$ and $[a,a)$ are also empty sets. Note that $\mu_\alpha(I) \geq 0$ since $\alpha(x)$ is a monotonically increasing function.

Examples Let $\alpha(x)$ be the monotonically increasing function (see Fig. 6.1)

$$\alpha(x) = \begin{cases} 0, & x < 1, \\ \frac{1}{2}, & x = 1, \\ 1, & x > 1. \end{cases} \tag{6.4}$$

We then have

$$\mu_\alpha(\,[0,1)\,) \;=\; \alpha(1^-) - \alpha(0^-) \;=\; 0 - 0 \;=\; 0$$

and

$$\mu_\alpha(\,[0,1]\,) \;=\; \alpha(1^+) - \alpha(0^-) \;=\; 1 - 0 \;=\; 1.$$

Similarly,

$$\begin{aligned} \mu_\alpha(\,[1,2]\,) &= \mu_\alpha(\,[1,2)\,) \;=\; 2 - 0 \;=\; 2, \\ \mu_\alpha(\,(1,2]\,) &= \mu_\alpha(\,(1,2)\,) \;=\; 2 - 1 \;=\; 1. \end{aligned}$$

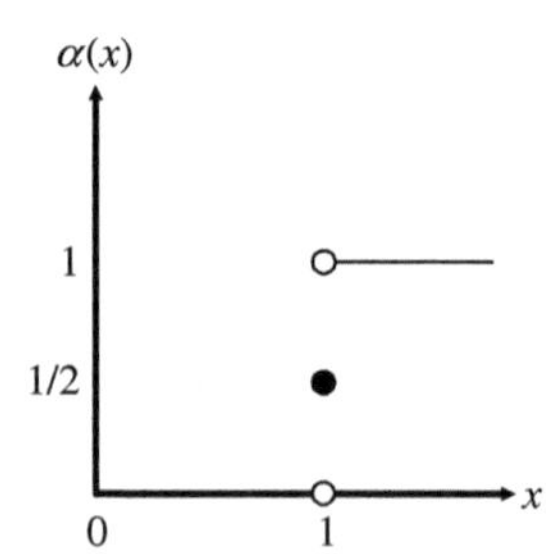

Fig. 6.1. The function $\alpha(x)$ defined in (6.4)

6.1.3 The Probability Measure

The significance of measure is understood by illustrating the probability theory as an example. Probability theory deals with statistical properties of a

random variable x associated with an event occurring sequentially or simultaneously, where it is assumed that the average of x approaches a constant value as the number of observations increases.

Given a random variable x, its **expected** (or **mean**) **value** is defined by the integral

$$E\{x\} = \int_{-\infty}^{\infty} xp(x)dx, \tag{6.5}$$

where $p(x) \geq 0$ is the **probability density function** of the random variable x defined by

$$p(x) = \frac{dP(x)}{dx},$$

with the **probability distribution function** $P(x)$. The function $P(x)$ describes the probability that the event labeled x occurs. It follows intuitively that

$$P\{x_1 < x \leq x_2\} = \int_{x_1}^{x_2} p(x)dx \tag{6.6}$$

and

$$\int_{-\infty}^{\infty} p(x)dx = 1.$$

Examples For a discrete random variable $\{x_i\}$, the integral of (6.5) can be written as a sum:

$$E\{x\} = \sum_i x_i p_i.$$

In an experiment with dice, e.g., the probability of each event is given by

$$p_1 = p_2 = \cdots = p_6 = \frac{1}{6},$$

which yields

$$E\{x_i\} = \sum_{i=1}^{6} x_i p_i = \frac{7}{3}.$$

In probability theory, the probability distribution function $P(x)$ plays the role of measure. Assume a set of continuous real numbers, $X = \{x \leq a\}$ and let the function $\alpha(a)$ be the probability that x has a value no greater than a. The function $\alpha(a)$ then reads

$$\alpha(a) = P(x \leq a), \tag{6.7}$$

where $\alpha(-\infty^+) = 0$ and $\alpha(\infty^-) = 1$. Note that $\alpha(a)$ is a monotonically increasing function. We have as well

$$P\{x_1 < x \leq x_2\} = \alpha(x_2) - \alpha(x_1),$$

since

$$P\{x \le x_2\} = P\{x \le x_1\} + P\{x_1 < x \le x_2\}.$$

Therefore, we see that the probability distribution function $P(x \in I)$ corresponds to the α-measure for any interval I, as expressed by

$$\mu_\alpha(I) = P(x \in I),$$

which behaves as $0 \le \mu_\alpha(I) \le 1$ for any I.

Remark. The mean value (6.5) of a random variable x can be interpreted as a **Riemann–Stieltjes integral**, rather than as an ordinary Riemann integral. To see this, we observe that the Riemann integral (6.5) can be expressed by the Riemann sum as

$$\int_{-\infty}^{\infty} xp(x)dx = \sum_{k=-\infty}^{\infty} \xi_k p(\xi_k)(x_{k+1} - x_k), \tag{6.8}$$

where ξ_k is any point on Δx_k. Since $p(x_k)(x_{k+1} - x_k) = \Delta P\{x_k < x \le x_{k+1}\}$ from (6.46), the mean value is written in the form

$$E\{x\} = \int_{-\infty}^{\infty} xdP = \int_{-\infty}^{\infty} xd\mu(x), \tag{6.9}$$

which is called the Riemann–Stieltjes integral of x with respect to $\mu(x)$.

6.1.4 Support and Area of a Step Function

What follows is an important concept that we use together with the concept of measure to introduce the definition of the Lebesgue integral. Let I_i be any interval, and suppose that the **step function** $\theta(x)$ given by

$$\theta(x) = \begin{cases} c_i, & x \in I_i, \ i = 1, 2, \cdots, n, \\ 0, & \text{otherwise}, \end{cases}$$

where a set $\{c_1, c_2, \cdots, c_n\}$ consists of finite and real numbers. We see that θ is constant on each interval I_i, and zero elsewhere. We now introduce the following concept:

♠ **Support of a step function:**
The disjoint set $S = I_1 \cup I_2 \cup \cdots \cup I_n \subseteq I$ on which θ is nonzero is called the **support** of $\theta(x)$.

An example of the support $\theta(x)$ is depicted in Fig. 6.2. When the support of a step function θ has a *finite* total length, we associate it with the *area* $A(\theta)$ between the graph of θ and the x-axis, with the usual rule that areas below the x-axis have a *negative* sign. We refer to $A(\theta)$ as the **area under the graph** of θ.

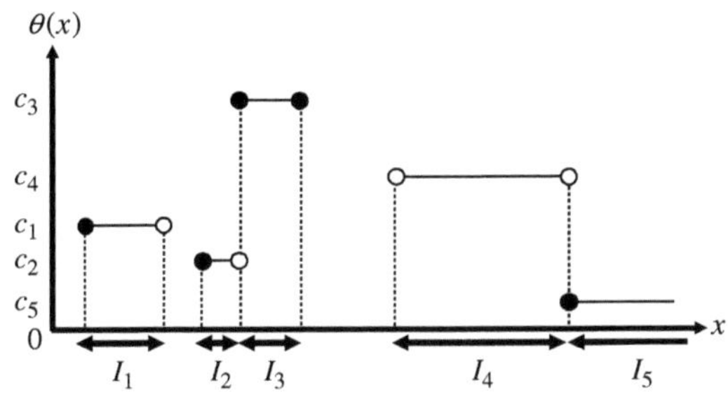

Fig. 6.2. The disjoint set $S = I_1 \cup I_2 \cup \cdots$ that serves as the support of $\theta(x)$

Concepts such as support and area can apply to a linear combination of step functions. Suppose that $\theta_1, \theta_2, \cdots, \theta_n$ are step functions on the same interval I, all with supports of finite total length, and that $a_1, a_2, \cdots, a_n$ are finite real numbers. Then, the function $\Theta(x)$ defined by

$$\Theta(x) = \sum_{j=1}^{n} a_j \theta_j(x) \quad \text{for } x \in I$$

is also a step function on I. The support of $\Theta(x)$ has a finite length and the area under the graph of $\Theta(x)$ is given by

$$A(\Theta) = \sum_{j=1}^{n} a_j A(\theta_j).$$

Examples Let $\theta_1, \theta_2 : [0,3) \to \boldsymbol{R}$ be defined by

$$\theta_1(x) = \begin{cases} 1 & \text{for } [0,2), \\ 2 & \text{for } [2,3), \end{cases} \quad \theta_2(x) = \begin{cases} -1 & \text{for } [0,1], \\ 1 & \text{for } (1,3). \end{cases} \tag{6.10}$$

Let $\Theta = 2\theta_1 - \theta_2$. Then

$$\Theta(x) = \begin{cases} 3 & \text{for } [0,1], \\ 1 & \text{for } (1,2), \\ 3 & \text{for } [2,3). \end{cases} \tag{6.11}$$

These are plotted in Fig. 6.3. Clearly Θ is a step function. Note also that the areas are

$$A(\theta_1) = 2(1) + 1(2) = 4, \quad A(\theta_2) = -1(1) + 2(1) = 1,$$

and

$$A(\Theta) = 1(3) + 1(1) + 1(3) = 7 \quad = 2A(\theta_1) - A(\theta_2).$$

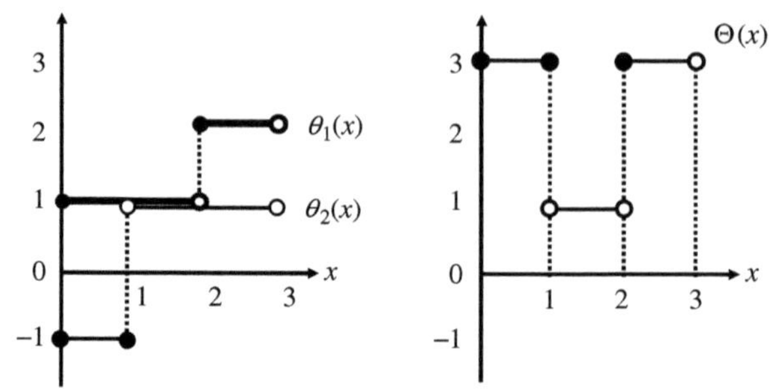

Fig. 6.3. The functions $\theta_1(x)$, $\theta_2(x)$, $\Theta(x)$ given in (6.10) and (6.11), respectively

6.1.5 α-Summability

Now, we combine the concepts of α-measure and support of a step function. Let $\alpha(x)$ be a monotonically increasing function, I be any interval, and $\theta(x)$ be a step function. We further assume that the support of θ is a **simple set**, i.e., the union of a finite collection of disjoint intervals. For example, the set $S = \bigcup_{k=1}^n I_k$ is a simple set if $I_1, I_2, \cdots, I_n$ are disjoint intervals. Then, the α-**measure** of S is given by

$$\mu_\alpha(S) = \sum_{k=1}^{n} \mu_\alpha(I_k).$$

Observe that the value of $\mu_\alpha(S)$ is independent of the way in which the set S is subdivided. Note also that

(i) $\mu_\alpha(S) \geq 0$ for any simple set S, and

(ii) if S and T are simple sets such that $S \subseteq T$, then $\mu_\alpha(S) \leq \mu_\alpha(T)$.

We are now ready to present the following statement:

♠ **α-summability:**
A step function $\theta(x)$ is α-**summable** if the support of θ has a finite α-measure with respect to a given monotonically increasing function $\alpha(x)$.

Given an α-summable step function $\theta(x)$, we associate it with a real number $A_\alpha(\theta)$ defined by

$$A_\alpha(\theta) = \sum_{k=1}^{n} c_k \mu_\alpha(I_k), \tag{6.12}$$

where c_k is the amplitude of step function $\theta(x)$ for $x \in I_k$. In general, $A_\alpha(\theta)$ can be thought of as a generalized *area*. For example, when setting $\alpha(x) = x$, the measure $\mu_\alpha(I_k)$ turns out to be just the ordinary length of the interval I_k, then $A_\alpha(\theta)$ is just the area $A(\theta)$ under the graph of θ_j as defined in Sect. 6.1.4. However, if $\alpha(x)$ has a more complicated function form, we get a different value of $A_\alpha(\theta)$ from the above since in that case a *length* along the x-axis should be measured by the α-measure rather than by ordinary length. An example of an actual calculation of $A_\alpha(\theta)$ is provided in Exercise **2**.

Remark. We shall see in Sect. 6.2.2 that the Lebesgue integral is defined by the limit $n \to \infty$ of the sum in (6.12).

6.1.6 Properties of α-summable functions

We list some basic properties of α-summable step functions without proof.

- If $\theta(x)$ is a nonnegative α-summable step function with respect to a given $\alpha(x)$, then $A_\alpha(\theta) \geq 0$ and $A_\alpha(0) = 0$.
- If θ_1 and θ_2 are α-summable step functions on the same interval I such that $\theta_1 \leq \theta_2$ on I, then $A_\alpha(\theta_1) \leq A_\alpha(\theta_2)$.
- Let a set $\{\theta_m\}$ be α-summable step functions on the same interval I, and let $\{a_m\}$ be finite real numbers. By defining $\theta : I \to \boldsymbol{R}$ as

$$\theta(x) = \sum_{j=1}^{m} a_j \theta_j(x)$$

for all $x \in I$ (θ is also an α-summable step function on I), we have

$$A_\alpha(\theta) = \sum_{j=1}^{m} a_j A_\alpha(\theta_j).$$

Exercises

1. Assume a monotonically increasing function $\alpha(x)$ defined by

$$\alpha(x) = \begin{cases} 0, & x \in (-\infty, 1), \\ x^2 - 2x + 2, & x \in [1, 2), \\ 3, & x = 2, \\ x + 2, & x \in (2, \infty). \end{cases}$$

Calculate $A_\alpha(\theta)$ for each of the two step functions:

$$\theta_1(x) = \begin{cases} -1, & x \in [0,1), \\ 2, & x \in [1,3], \end{cases}$$

and

$$\theta_2(x) = \begin{cases} -1, & x \in [0,1], \\ 2, & x \in (1,3]. \end{cases}$$

Solution: Since

$$\mu_\alpha(\,[0,1)\,) = \alpha(1^-) - \alpha(0^-) = 0 - 0 = 0,$$
$$\mu_\alpha(\,[1,3]\,) = \alpha(3^+) - \alpha(1^-) = 5 - 0 = 5,$$

we have

$$A_\alpha(\theta_1) = (-1)0 + 2(5) = 10.$$

For θ_2, on the other hand, we have a different result since

$$\mu_\alpha(\,[0,1]\,) = \alpha(1^+) - \alpha(0^-) = 1 - 0 = 1,$$
$$\mu_\alpha(\,(1,3]\,) = \alpha(3^+) - \alpha(1^+) = 5 - 1 = 4,$$

which yields

$$A_\alpha(\theta_2) = (-1)1 + 2(4) = 7.$$

It is noteworthy that the values of $A_\alpha(\theta_1)$ and $A_\alpha(\theta_2)$ are different, although the area $A(\theta)$ for them is the same. The difference comes from the fact that α has a discontinuity at the single point where θ_1 and θ_2 have different values. ♣

2. Evaluate $A_\alpha(\theta)$ of the step function:

$$\theta(x) = \begin{cases} 2, & x \in (-\infty,0], \\ 1, & x \in (0,\infty), \end{cases}$$

which is associated with the α-measure:

$$\alpha(x) = \begin{cases} 0, & x < 0, \\ \frac{1}{2}, & x = 0, \\ 1, & x > 0. \end{cases}$$

Solution: Since

$$\mu_\alpha((-\infty,0]) = \alpha(0^+) - \alpha(-\infty^+) = \frac{1}{2} - 0 = \frac{1}{2},$$

$$\mu_\alpha((0,\infty)) = \alpha(\infty^-) - \alpha(0^+) = 1 - \frac{1}{2} = \frac{1}{2},$$

we have

$$A_\alpha(\theta_1) = 1\left(\frac{1}{2}\right) + 2\left(\frac{1}{2}\right) = \frac{3}{2}. \quad ♣$$

3. Show that the function

$$f(x) = \lim_{n\to\infty} \lim_{m\to\infty} (\cos 2\pi m!x)^n ,$$

called **Dirichlet's function**, takes the form

$$f(x) = \begin{cases} 1 & \text{for all rational numbers } x, \\ 0 & \text{otherwise.} \end{cases}$$

Solution: When x is a rational number, it is expressed by a fraction p/q with relatively prime integers p and q. Hence, for sufficiently large m, the product $m!x$ becomes an integer since

$$m!x = m \cdot (m-1) \cdots (q+1) \cdot p \cdot (q-1) \cdots 2 \cdot 1.$$

Thus we have $\cos 2\pi m!x = 1$. Otherwise, if x is an irrational number, $m!x$ is also an irrational for any m, so that $|\cos 2\pi m!x| < 1$. As a result, we obtain

$$\lim_{n\to\infty} \lim_{m\to\infty} (\cos 2\pi m!x)^n = \begin{cases} 1 : x \text{ is a rational,} \\ 0 : x \text{ is an irrational.} \end{cases} \quad ♣$$

6.2 Lebesgue Integral

6.2.1 Lebesgue Measure

The Lebesgue integral procedure essentially reduces to *finding a measure for sets of arguments.* In particular if a set consists of too many points of discontinuity, we need a way to define its measure that is known as the **Lebesgue measure**. It this subsection, we explain how to construct the Lebesgue measure of a point set.

As a simple example, let us consider a finite interval $[a, b]$ of length L. This can be decomposed into two sets: a set X consisting of some of the points $x \in [a, b]$ and its **complementary set** X' consisting of all points $x \in [a, b]$ that do not belong to X. A schematic view of X and X' is shown in Fig. 6.4. Both X and X' may be sets of several continuous line segments or sets of **isolated points**.

We would like to evaluate the measure of X. To do this, we cover the set of points X by nonoverlapping intervals $\Lambda_i \subset [a, b]$ such as

$$X \subset (\Lambda_1 + \Lambda_2 + \cdots).$$

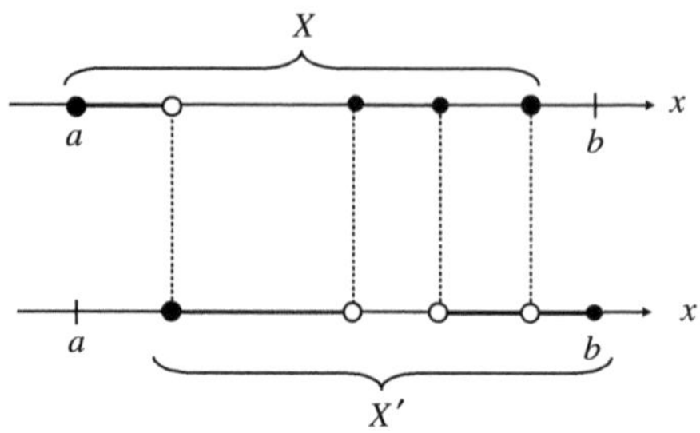

Fig. 6.4. A set X and its complementary set X'

If we denote the length of Λ_k by ℓ_k, the sum of ℓ_k must satisfy the inequality

$$0 \le \sum_k \ell_k \le L.$$

In particular, the smallest value of the sum $\sum_i \ell_i$ is referred to as the **outer measure** of X and is denoted by

$$\mu_{\text{out}}(X) = \inf\left(\sum_k \ell_k\right).$$

In the same manner, we can find intervals $\Lambda_k{}' \subset [a, b]$ of lengths $\ell_1', \ell_2', \cdots$ that cover the complementary set X' such that

$$X' \subset (\Lambda_1' + \Lambda_2' + \cdots), \quad 0 \le \sum_k \ell_k' \le L.$$

Here we define another kind of measure denoted by

$$\mu_{\text{in}}(X) \equiv L - \mu_{\text{out}}(X') \quad = L - \inf\left(\sum_k \ell_k'\right), \tag{6.13}$$

which is called the **inner measure** of X. Note that the inner measure of X is defined by the outer measure of X', not of X. It is a straightforward matter to prove the inequality

$$0 \le \mu_{\text{in}}(X) \le \mu_{\text{out}}(X). \tag{6.14}$$

Specifically, if

$$\mu_{\text{in}}(X) = \mu_{\text{out}}(X),$$

it is called the **Lebesgue measure** of the point set X, denoted by $\mu(X)$. Clearly, when X contains all the points of $[a, b]$, the smallest interval that covers $[a, b]$ is $[a, b]$ itself, and thus $\mu(X) = L$.

Our results are summarized below.

♠ **Lebesgue measure:**
A set of points X is said to be **measurable** with the **Lebesgue measure** $\mu(X)$ if and only if $\mu_{\text{in}}(X) = \mu_{\text{out}}(X) \equiv \mu(X)$.

Remark. An unbounded point set X is measurable if and only if $(-c, c) \cap X$ is measurable for all $c > 0$. In this case, we define $\mu(X) = \lim_{c\to\infty} \mu\left[(-c, c) \cap X\right]$, which may or may not be finite.

6.2.2 Definition of the Lebesgue Integral

We are now in a position to define the **Lebesgue integral**. Let the function $f(x)$ be defined on a set X that is bounded:

$$0 \le f_{\text{min}} \le f(x) \le f_{\text{max}}.$$

We partition the ordinate axis by the sequence $\{f_k\}$ $(1 \le k \le n)$ so that $f_1 = f_{\text{min}}$ and $f_n = f_{\text{max}}$. Owing to the one-to-one correspondence between x and $f(x)$, there should exist sets X_i of values x such that

$$f_k \le f(x) < f_{k+1} \quad \text{for } x \in X_k \quad (1 \le k \le n-1), \tag{6.15}$$

as well as a set X_n of values x such that $f(x) = f_n$. Each set X_k assumes a measure $\mu(X_k)$. Thus we form the sum of products $f_k \cdot \mu(X_k)$ of all possible values of f, called the **Lebesgue sum**:

$$\sum_{k=1}^{n} f_k \cdot \mu(X_k). \tag{6.16}$$

If the sum (6.16) converges to a finite value when taking the limit $n \to \infty$ such that

$$\max |f_k - f_{k+1}| \to 0,$$

then the limiting value of the sum is called the **Lebesgue integral** of $f(x)$ over the set X.

The formal definition of the Lebesgue integral is given below.

♠ **Lebesgue integral:**
Let $f(x)$ be a nonnegative function defined on a **measurable set** X and divide X into a finite number of subsets such as

$$X = X_1 + X_2 + \cdots + X_n. \tag{6.17}$$

Let $f_k = \inf_{x \in X_k} f(x)$ to form the sum

$$\sum_{k=1}^{n} f_k \mu(X_k). \tag{6.18}$$

Then the **Lebesgue integral** of $f(x)$ on X is defined by

$$\int_X f d\mu \equiv \lim_{\max|f_k - f_{k-1}| \to 0} \left[\sum_{k=1}^{n} f_k \mu(X_k) \right],$$

where all possible choices of partition (6.17) are considered.

Figure 6.5 is a schematic illustration of the Lebesgue procedure. Obviously, the value of the Lebesgue sum (6.16) depends on our choice of partition. If we take an alternative partition instead of (6.17), the value of the sum also changes. Among the infinite variety of choices, the partition that maximizes the sum (6.17) gives the Lebesgue integral of $f(x)$. That a function is **Lebesgue integrable** means that the limit superior of the sum (6.18) is determined independently of our choice of the partition of the x-axis.

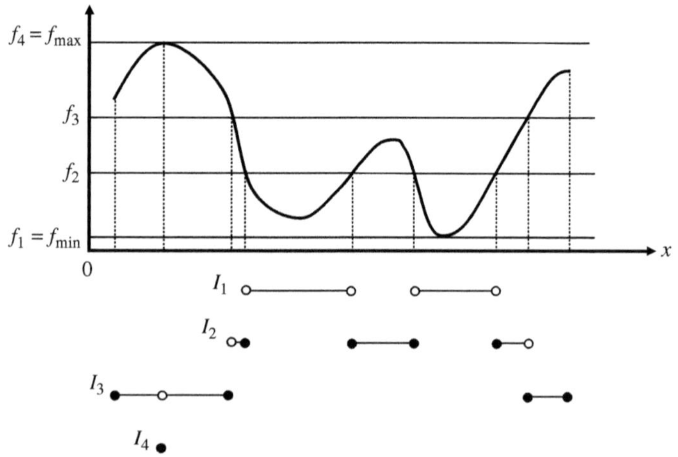

Fig. 6.5. An illustration of the Lebesgue procedure

6.2.3 Riemann Integrals vs. Lebesgue Integrals

Before proceeding further with this discussion, we compare the definitions of Riemann and Lebesgue integrals for a better understanding of the significance of the latter. In the language of measure, the **Riemann integral** of a function

$f(x)$ defined on the set X is obtained by dividing X into nonoverlapping subsets X_i as

$$X = X_1 + X_2 + \cdots + X_n, \quad X_i \cap X_j = 0, \quad \text{for any } i, j,$$

followed by setting the **Riemann sum**

$$\sum_{k=1}^{n} f(\xi_k)\mu(X_k). \tag{6.19}$$

Here, the measure $\mu(X_k)$ is identified with the length of the subset X_k, and ξ_k assumes any point that belongs to X_k. We increase the number of subsets $n \to \infty$ such that

$$\mu(X_k) \to 0 \quad \text{for any } X_k,$$

and if the limit of the sum (6.19) exists and is independent of the subdivision process, it is called the **Riemann integral** of $f(x)$ over X. Obviously, the Riemann integral can be defined under the condition that all values of $f(x)$ defined over X_k tend to a common limit as $\mu(X_k) \to 0$. Such a requirement excludes any possibility of defining the Riemann integral for functions having too many points of discontinuity.

Remark. In view of the analogy between the sum (6.12) and (6.18), we may say that, in a sense, the Lebesgue integral is the limit $n \to \infty$ of the quantity $A_\alpha(\theta)$.

Although the Lebesgue sum given in (6.16) is apparently similar to the Riemann sum given in (6.19), they are intrinsically different. In the Riemann sum (6.19), $f(\xi_i)$ is the value of $f(x)$ at an *arbitrary* point $\xi_i \in X_i$. Thus the value of ξ_i is allowed to vary within each subset, which causes an indefiniteness in the value of $f(\xi_i)$ within each subset. On the other hand, in the Lebesgue sum (6.16), the value of f_i corresponding to each subset X_i has a definite value. Therefore, for the existence of the Lebesgue integral, we no longer need local smoothness of $f(x)$. As a result, the conditions imposed on the integrated function become very *weak* compared with the case of the Riemann integral.

6.2.4 Properties of the Lebesgue Integrals

Several properties of the Lebesgue integral are given below without proof.

1. If $f(x)$ is the Lebesgue integrable on X and if $X = X_1 + X_2 + \cdots + X_n$, then

$$\int_X f d\mu = \sum_{i=1}^{n} \int_{X_i} f d\mu.$$

2. If two functions $f(x)$ and $g(x)$ are both Lebesgue integrable on X and if $f(x) \le g(x)$ for any $x \in X$, then

$$\int_X f d\mu \le \int_X g d\mu.$$

3. If $\mu(X) = 0$, then $\int_X f(x)dx = 0$.

4. If the integral $\int_X f(x)dx$ is finite, then the subset of X defined by

$$X' = \{x \mid f(x) = \pm\infty\}$$

has zero measure. This means that in order for the integral to converge, the measure of a set of points x at which $f(x)$ diverges is necessarily zero.

5. Suppose that $\int_X f(x)dx$ is finite and that $X' \subset X$. If we make $\mu(X') \to 0$, then

$$\int_{X'} f d\mu \to 0.$$

6. When $f(x)$ on X takes both positive and negative values, its Lebesgue integral is defined by

$$\int_X f d\mu = \int_X f^+ d\mu + \int_X f^- d\mu \tag{6.20}$$

and

$$\int_X |f| d\mu = \int_X f^+ d\mu - \int_X f^- d\mu, \tag{6.21}$$

where

$$f^+(x) = \begin{cases} f(x) & \text{for } \{x;\ f(x) \ge 0\}, \\ 0 & \text{for } \{x;\ f(x) < 0\}, \end{cases}$$

and

$$f^-(x) = \begin{cases} 0 & \text{for } \{x;\ f(x) \ge 0\}, \\ -f(x) & \text{for } \{x;\ f(x) < 0\}. \end{cases}$$

Definition (6.21) is justified except when both integrals on the right-hand side diverge.

6.2.5 Null-Measure Property of Countable Sets

Let us show that any countable set has a Lebesgue measure equal to zero. A rigorous definition of countable sets is given herewith.

♠ **Countable set:**

A finite or infinite set X is **countable** (or **enumerable**) if and only if it is possible to establish a reciprocal one-to-one correspondence between its elements and the elements of a set of real integers.

It follows that every finite set is countable and that every subset of a countable set is also countable. Any countable set is associated with a specific number, called the **cardinal number**, defined below.

♠ **Cardinal numbers:**
Two sets X_1 and X_2 are said to have the same cardinal number if and only if there exists a reciprocal one-to-one correspondence between their respective elements.

Remark. A set X is called an **infinite set** if it has the same cardinal number as one of its subsets; otherwise, X is called a **finite set**.

It should be stressed that an infinite set may or may not be countable. When a given infinite set is countable, then its cardinal number is denoted by $\aleph_0$, which is the same as the cardinal number of the set of the positive real integers. Furthermore, the cardinal number of every noncountable set is denoted by $\aleph$, which is identified with the cardinal number of the set of all real numbers (or the set of points on a continuous line). Cardinal numbers of infinite sets, $\aleph_0$ and $\aleph$, are called **transfinite numbers**.

The most important property of countable sets in terms of measure theory is given below.

♠ **Theorem:**
Any countable set (finite or infinite) has a Lebesgue measure of zero, namely, **null measure**.

Examples An illustrative example is the set of rational numbers that has measure zero as shown earlier. The countability of this set follows from the fact that it can be arranged in a sequence of proper fractions as

$$0,\ 1,\ \frac{1}{2},\ \frac{1}{3},\ \frac{2}{3},\ \frac{1}{4},\ \frac{3}{4},\ \frac{1}{5},\ \frac{2}{5},\ \frac{3}{5},\ \frac{4}{5},\ \cdots .$$

Accordingly, since the set of all rational numbers in the interval $[0,1]$ has zero measure, the Lebesgue integral of Dirichlet's function $\chi(x)$ over this interval is well defined and equal to zero.

Another well-known example of the set of measure zero is the **Cantor set**, which is demonstrated in Exercise **2**.

6.2.6 The Concept of Almost Everywhere

We have observed that sets of measure zero make no contribution to Lebesgue integrals. This fact provides a concept of an **equality almost everywhere** for measurable functions, which plays an important role in developing the theory of function analysis.

♠ **Equality almost everywhere:**
Two functions $f(x)$ and $g(x)$ defined on the same set X are said to be equal almost everywhere with respect to a measure $\mu(X)$ if

$$\mu\{x \in X;\ f(x) \neq g(x)\} = 0.$$

We extend this terminology to other circumstances as well. In general, a property is said to hold **almost everywhere** on X if it holds at all points of X except on a set of measure zero. Thus two functions $f(x)$ and $g(x)$ are said to be equivalent (written $f \sim g$) if they coincide almost everywhere. For example, Dirichlet's function mentioned earlier is equivalent almost everywhere to the function $g(x) \equiv 0$.

Since the behavior of functions on sets of measure zero is often unimportant, it is natural to introduce the following generalization of the ordinary notion of the convergence of a sequence of functions:

♠ **Convergence almost everywhere:**
A sequence of functions $\{f_n(x)\}$ defined on a set X is said to **converge almost everywhere** to a function $f(x)$ if

$$\lim_{n\to\infty} f_n(x) = f(x) \tag{6.22}$$

for all $x \in X$ except for points of measure zero.

Examples A typical example is the sequence

$$\{f_n(x)\} = \{(-x)^n\}$$

defined on $[0, 1]$. It converges almost everywhere to the function $f(x) \equiv 0$; in fact it converges everywhere except at the point $x = 1$.

Exercises

1. Show that the set of all rational numbers in the interval $[0, 1]$ has a Lebesgue measure equal to zero.

Solution: Denote by X' the set of irrational numbers that is complementary to X and the entire interval $[0, 1]$ by I. Since $\mu(I) = 1$, the outer measure of X' reads

$$\mu_{\text{out}}(X') = \mu_{\text{out}}(I - X) = \mu_{\text{out}}(I) - \mu_{\text{out}}(X) = 1 - \mu_{\text{out}}(X).$$

By definition, the inner measure of X is given by

$$\mu_{\rm in}(X) \;=\; \mu_{\rm in}(I) - \mu_{\rm out}(X') \;=\; 1 - [1 - \mu_{\rm out}(X)] \;=\; \mu_{\rm out}(X).$$

The last equality asserts that the set X is **Lebesgue measurable**). The remaining task is to evaluate the value of $\mu(X) = 0$.

Let x_k $(k = 1, 2, \cdots, n, \cdots)$ denote the points of rational numbers in the interval I. We cover each point $x_1, x_2, \cdots, x_n, \cdots$ by an open interval of length $\varepsilon/2$, $\varepsilon/2^2$, $\cdots$, $\varepsilon/2^n, \cdots$, respectively, where ε is an arbitrary positive number. Since these intervals may overlap, the entire set can be covered by an open set of measure not greater than

$$\sum_{n=1}^{\infty} \frac{\varepsilon}{2^n} = \frac{\varepsilon}{2\left(1 - \frac{1}{2}\right)} = \varepsilon.$$

Since ε can be made arbitrarily small, we find that $\mu_{\rm out}(X) = 0$. Hence, from (6.18) we immediately have $\mu(X) = 0$. ♣

2. Evaluate the measure of a **Cantor set**, an infinite set constructed as follows: **(i)** From the closed interval $[0, 1]$, delete the open interval $(1/3, 2/3)$ that forms its middle third; **(ii)** from each of the remaining intervals $[0, 1/3]$ and $[2/3, 1]$ delete the middle third; **(iii)** continue this process of deleting the middle thirds indefinitely to obtain the point set on the line that remains after all these open intervals.

Solution: Observe that at the kth step, we have thrown out 2^{k-1} adjacent intervals of length $1/3^k$. Thus the sum of the lengths of the intervals removed is equal to

$$\frac{1}{3} + \frac{2}{9} + \frac{4}{27} + \cdots + \frac{2^{n-1}}{3^n} + \cdots = \lim_{n\to\infty} \frac{\frac{1}{3}\left[1 - \left(\frac{2}{3}\right)^n\right]}{1 - \frac{2}{3}} = 1.$$

This is just the measure of the open set P' that is the **complement** of P. Therefore, the Cantor set P itself has null measure

$$\mu(P) = 1 - \mu(P') = 1 - 1 = 0. \quad ♣$$

3. Show that if $f(x)$ is nonnegative and integrable on X, then

$$\mu\,[\,x \in X, f(x) \geq c\,] \leq \frac{1}{c} \int_X f d\mu,$$

which is known as, **Chebyshev's inequality**.

Solution: Set $X' = \{x \in X, f(c) \geq c\}$ to observe that

$$\int_X f d\mu = \int_{X'} f d\mu + \int_{X-X'} f d\mu \geq \int_{X'} f d\mu \geq c\mu(X'). \quad ♣$$

4. Show that if $\int_X |f| d\mu = 0$, then $f(x) = 0$ **almost everywhere.**

Solution: By Chebyshev's inequality,

$$\mu \left[x \in X, |f(x)| \geq \frac{1}{n} \right] \leq n \int_X |f| d\mu = 0$$

for all $n = 1, 2, \cdots$. Therefore, we have

$$\mu \left[x \in X, f(x) \neq 0 \right] \leq \sum_{n=1}^{\infty} \mu \left[x \in X, |f(x)| \geq \frac{1}{n} \right] = 0. \quad ♣$$

6.3 Important Theorems for Lebesgue Integrals

6.3.1 Monotone Convergence Theorem

Our current task is to examine whether or not the equality

$$\lim_{n \to \infty} \int_a^b f_n(x) dx = \int_a^b f(x) dx \tag{6.23}$$

is valid under the Lebesgue procedure. This problem can be clarified by referring to two important theorems concerning the convergence property of Lebesgue integrals; the **monotone convergence theorem** and the **dominated convergence theorem**. Neither theorem is valid if we restrict our attention to Riemann integrable functions. We observe that, owing to the two convergence theorems, Lebesgue theory offers a considerable improvement over Riemann theory with regard to convergence properties.

In what follows, we assume that X is a set of real numbers, and that $\{f_n\}$ is a sequence of functions defined on X.

♠ **Monotone convergence theorem:**

If (f_n) is a sequence such that $0 \leq f_n \leq f_{n+1}$ for all $n \geq 1$ in X and $f = \lim_{n \to \infty} f_n$, then

$$\lim_{n \to \infty} \int_X f_n d\mu = \int_X \lim_{n \to \infty} f_n d\mu = \int_X f d\mu.$$

Remark. The monotone convergence theorem states that in the case of Lebesgue integrals, the conditions to reverse the order of limit and integration are much weaker than in the case of Riemann integrals; i.e., only the **pointwise convergence** of $f_n(x)$ to $f(x)$ is required in the Lebesgue case, whereas in the Riemann case we must have **uniform convergence** of $f_n(x)$ to $f(x)$.

Proof (of the monotone convergence theorem): The hypothesis $0 \le f_n \le f_{n+1}$ implies that

$$0 \le \int_X f_n d\mu \le \int_X f_{n+1} d\mu,$$

which indicates that the sequence $\{\int_X f_n d\mu\}$ increases monotonically with respect to n; thus its limit $n \to \infty$ exists as a we denote it by M (possibly equal to ∞). In addition, by hypothesis

$$\int_X f_n d\mu \le \int_X f d\mu, \quad \text{for all } n. \tag{6.24}$$

Since (6.24) is true for arbitrary n, we have

$$M = \lim_{n\to\infty} \left[\int_X f_n d\mu \right] \le \int_X f d\mu.$$

Therefore, if we can verify the opposite inequality

$$M \ge \int_X f d\mu, \tag{6.25}$$

we will get the desired result,

$$M = \lim_{n\to\infty} \int_X f_n d\mu = \int_X \lim_{n\to\infty} f_n d\mu = \int_X f d\mu.$$

To show (6.25), let c be a number such that $c \in (0,1)$ and introduce the point set

$$X_n = \{x : \; cf(x) \le f_n(x)\}.$$

Owing to the monotonically increasing property of the sequence $\{f_n(x)\}$ with regard to n, the set X_n satisfies the inclusion relation

$$X_1 \subset X_2 \subset X_3 \subset \cdots \quad \text{and} \quad \bigcup_{n=1}^{\infty} X_n = X.$$

In addition, the increasing property of the sequence $\{\int_{X_n} f_n d\mu\}$ yields

$$c \int_{X_n} f d\mu \le \int_{X_n} f_n d\mu \le \lim_{n\to\infty} \left[\int_{X_n} f_n d\mu \right] = M. \tag{6.26}$$

Since (6.26) must hold for any n, we have

$$c \int_X f d\mu \le M. \tag{6.27}$$

Furthermore, since (6.27) is true for all $c \in (0,1)$, we have

$$\int_X f d\mu \le M. \tag{6.28}$$

Note that the substitution $c = 1$ into (6.27) is allowed because the symbol $\le$, not $<$, is involved in (6.27). ♣

6.3.2 Dominated Convergence Theorem (I)

In the previous argument, we saw that the order of limit and integration can be reversed when considering monotonically increasing sequences of functions. In practice, however, the requirement in the monotone convergence theorem, i.e., that the sequence $\{f_n(x)\}$ must be monotone increasing, is sometimes very inconvenient. In this subsection, we examine the same issue for more general sequences of functions, i.e., nonmonotone sequences satisfying some looser conditions and their limit passage. Our current objective is to prove the theorem below.

♠ **Dominated convergence theorem:**
Let $\{f_n\}$ be a sequence of functions for almost everywhere on X such that **(a)** $\lim_{n\to\infty} f_n(x) = f(x)$, and **(b)** there exists a nonnegative g such that $|f_n| \le g$ for all $n \ge 1$. Then, we have

$$\lim_{n\to\infty} \int_X f_n d\mu = \int_X f d\mu.$$

Remark. Note that the condition imposed on the theorem above is that the sequences $\{f_n\}$ should be bounded almost everywhere. This condition is clearly looser than that imposed in the monotone convergence theorem. Hence, the monotone convergence theorem can be regarded as a special case of the dominated convergence theorem.

6.3.3 Fatou Lemma

The proof of the dominated convergence theorem requires the lemma given below.

♠ **Fatou lemma:**
If $f_n(x) \ge 0$ for all n and for almost everywhere in a bounded measurable set X and if $\lim_{n\to\infty} f_n(x) = f(x)$, then

$$\int_X \left[\liminf_{n\to\infty} f_n\right] d\mu = \int_X f d\mu \le \liminf_{n\to\infty} \left[\int_X f_n d\mu\right],$$

where the definition is

$$\liminf_{n\to\infty} f_n = \lim_{n\to\infty} \left[\inf_{k\ge n} f_k\right].$$

Proof Let $g_n = \inf_{k\geq n} f_k$. Since the sequence $g_n(x)$ is nonnegative and nondecreasing, we have

$$\lim_{n\to\infty} g_n = \liminf_{n\to\infty} f_n.$$

(See Sect. 2.1.4 for the precise definition of lim inf.) In addition, the monotone convergence theorem implies that

$$\lim_{n\to\infty} \int_X g_n d\mu = \int_X \lim_{n\to\infty} g_n d\mu = \int_X \liminf_{n\to\infty} f_n d\mu. \tag{6.29}$$

It also follows that

$$g_n(x) \leq f_k(x) \quad \text{for any } k \geq n.$$

Hence,

$$\int_X g_n d\mu \leq \int_X f_k d\mu \ \text{ for any } \ k \geq n,$$

that is,

$$\int_X g_n d\mu \leq \inf_{k\geq n} \int_X f_k d\mu.$$

Taking the limit $n \to \infty$ and applying the monotone convergence theorem, we get

$$\lim_{n\to\infty} \int_X g_n d\mu \leq \lim_{n\to\infty} \left[\inf_{k\geq n} \int_X f_k d\mu \right] = \liminf_{n\to\infty} \int_X f_n d\mu. \tag{6.30}$$

From (6.29) and (6.30), we conclude that

$$\int_X \liminf_{n\to\infty} f_n d\mu = \int_X f d\mu \leq \liminf_{n\to\infty} \int_X f_n d\mu. \quad \clubsuit$$

6.3.4 Dominated Convergence Theorem (II)

Our next task is to prove the dominated convergence theorem.

Proof Observe that f_n and f are Lebesgue integrable on X. From hypothesis, it follows that $f_n + g \geq 0$ and $g - f_n \geq 0$ almost everywhere. Thus by Fatou's lemma, we have

$$\int_X \liminf_{n\to\infty} (f_n + g)\, d\mu \leq \liminf_{n\to\infty} \int_X (f_n + g)\, d\mu$$

or

$$\int_X \liminf_{n\to\infty} f_n d\mu \leq \liminf_{n\to\infty} \int_X f_n d\mu \tag{6.31}$$

by the linearity of the Lebesgue integral. It is also true that $g - f_n \geq 0$ on X; thus also by Fatou's lemma we have

$$\int_X \liminf_{n\to\infty} (g - f_n)\, d\mu \leq \liminf_{n\to\infty} \int_X (g - f_n)\, d\mu,$$

or equivalently,

$$-\int_X \liminf_{n\to\infty} f_n d\mu \leq \liminf_{n\to\infty} \left[-\int_X f_n\right] d\mu.$$

The latter inequality can be rewritten as

$$\int_X \liminf_{n\to\infty} f_n d\mu \geq \limsup_{n\to\infty} \int_X f_n d\mu. \tag{6.32}$$

From (6.31) and (6.32) we set

$$\begin{aligned}\int_X \liminf_{n\to\infty} f_n d\mu &\leq \liminf_{n\to\infty} \int_X f_n d\mu \\ &\leq \limsup_{n\to\infty} \int_X f_n d\mu \ \leq \int_X \liminf_{n\to\infty} f_n d\mu,\end{aligned}$$

which clearly indicates that

$$\liminf_{n\to\infty} \int_X f_n d\mu = \limsup_{n\to\infty} \int_X f_n d\mu,$$

so that the limit $\lim_{n\to\infty} \int_X f_n d\mu$ exists and is equal to $\int_X \lim_{n\to\infty} f_n d\mu = \int_X f d\mu$. This completes the proof of the theorem. ♣

6.3.5 Fubini Theorem

For a function of several variables, we may define the Lebesgue integral by exactly the same process as for a function of one variable. In cases of two variables, for instance, a rectangle $S = [a, b] \times [c, d]$ takes on the role of intervals, and we need only to imitate the definitions and methods that we used for functions of a single variable. We can develop the theory for the entire plane $\mathbf{R}^2$ analogously to that for the real axis $\mathbf{R}$. In fact, all the consequences in Sect. 6.2 for Lebesgue integrable functions on a closed interval $[a, b]$ are easily carried over to the corresponding propositions for the double integral on the rectangle S without modifying the actual proofs in Sect. 6.2, except for replacing $f(x)$ by $f(x, y)$.

However, an important new problem arises here. If f is integrable on the rectangle $S = [a, b] \times [c, d]$, we have to determine whether the value of the integral

$$\iint_S f(x, y) dx dy \tag{6.33}$$

is equal to that of the repeated integrals

$$\int_c^d \left[\int_a^b f(x,y)dx \right] dy \quad \text{and} \quad \int_a^b \left[\int_c^d f(x,y)dy \right] dx.$$

This is true for continuous functions on S. But it is far from obvious that the existence of the double integral (6.33) guarantees the existence of either repeated integral.

The following example may lead the reader to consider the point mentioned above.

Examples Assume the function

$$f(x,y) = \begin{cases} \frac{x^2-y^2}{(x^2+y^2)^2} & \text{for } (x,y) \neq (0,0), \\ 0 & \text{for } (x,y) = (0,0), \end{cases} \tag{6.34}$$

and compute the repeated integrals

$$I_{yx} = \int_0^1 dy \left[\int_0^1 f(x,y)dx \right] \quad \text{and} \quad I_{xy} = \int_0^1 dx \left[\int_0^1 f(x,y)dy \right].$$

Straightforward calculations yield

$$I_{xy} = \int_0^1 dx \int_0^1 \frac{\partial}{\partial y}\left(\frac{y}{x^2+y^2} \right) dy = \int_0^1 \frac{dx}{x^2+1} dx = \frac{\pi}{4}$$

and

$$I_{yx} = \int_0^1 dy \int_0^1 \frac{\partial}{\partial x}\left(\frac{-x}{x^2+y^2} \right) dx = \int_0^1 \frac{-dy}{y^2+1} dy = -\frac{\pi}{4}.$$

Hence, we conclude that

$$I_{xy} \neq I_{yx},$$

which indicates that the order of integrations with respect to x and y cannot be changed.

We now present the main theorem of this subsection.

♠ **Fubini theorem:**

Let the function $f(x)$ be integrable on a rectangle $S = [a,b] \times [c,d]$. Then the following equalities hold:

$$\int\int_S f(x,y)dxdy = \int_c^d \left[\int_a^b f(x,y)dx \right] dy = \int_a^b \left[\int_c^d f(x,y)dy \right] dx.$$

According to Fubini's theorem, a double integral $\int\int_S f(x,y)dxdy$ is computed by integrating first with respect to x and then with respect to y, or vice versa. We omit an exact proof of the Fubini theorem, since it requires rather lengthy arguments regarding the existence and the convergence of the double integrals. Instead, we present some applications of the theorem.

The following is an extension of the Fubini theorem:

♠ **Fubini–Hobson–Tonelli theorem:**

Let the function $f(x)$ be defined on $S = [a,b] \times [c,d]$. Then, if either of the repeated integrals

$$\int_a^b \left[\int_c^d |f(x,y)|dy\right] dx \quad \text{or} \quad \int_c^d \left[\int_a^b |f(x,y)|dx\right] dy$$

exists, f is integrable on S and, hence,

$$\int\int_S f(x,y)dxdy = \int_a^b \left[\int_c^d f(x,y)dy\right] dx = \int_c^d \left[\int_a^b f(x,y)dx\right] dy.$$

Both the Fubini theorem and the Fubini–Hobson–Tonelli theorem for integrals on a rectangle S may be easily extended to integrals on all of $\mathbf{R}^2$ or to the integrals on any measurable subsets of $\mathbf{R}^2$.

Exercises

1. Suppose that the function

$$g_n(x) = -2k^2 x e^{-k^2x^2} + 2(k+1)^2 x e^{-(k+1)^2x^2}$$

is defined on $[0,\infty)$, and form the sum

$$f_n(x) = \sum_{k=1}^{n} g_k(x) = -2xe^{-x^2} + 2(n+1)^2 x e^{-(n+1)^2x^2}.$$

Show that $\int_0^\infty \lim_{n\to\infty} f_n(x)dx \neq \lim_{n\to\infty}\int_0^\infty f_n(x)dx$.

Solution: We have

$$\int_0^\infty \lim_{n\to\infty} f_n(x)dx = \int_0^\infty \left(-2xe^{-x^2}\right) dx = \left[e^{-x^2}\right]_0^\infty = -1,$$

whereas

$$\lim_{n\to\infty}\int_0^\infty f_n(x)dx = \left[e^{-x^2} - e^{-(n+1)^2x^2}\right]_0^\infty = 0.$$

Therefore, (6.23) is not valid. ♣

2. Given the function:

$$f_n(x) = \begin{cases} n \sin nx & \text{for } 0 \leq x \leq \pi/n, \\ 0 & \text{for } \pi/n \leq x \leq \pi, \end{cases}$$

show that

$$\lim_{n\to\infty} \int_0^1 f_n(x)dx \neq \int_0^1 \lim_{n\to\infty} f_n(x)dx.$$

Solution: We have $\lim_{n\to\infty} f_n(x) = 0$ for every x in $[0, \pi]$ and $\lim_{n\to\infty} \int_0^\pi f_n(x)dx = 2$. Hence, we obtain the desired result. ♣

3. Suppose that the nonnegative functions $\{f_n(x) : n \in \boldsymbol{N}\}$ are each summable over a measurable set X, and $f_n \leq f_{n+1}$ on X. Show that the limit function $f = \lim_{n\to\infty} f_n$ is summable over X and that

$$\lim_{n\to\infty} \int_X f_n d\mu = \int_X f d\mu.$$

Solution: Let $g_n = f_1 - f_n$, so that $0 = g_1 \leq g_2 \leq \cdots \leq f_1$. Thus, the dominated convergence theorem ensures that $\lim_{n\to\infty} g_n = f_1 - f$ is integrable, and we have $\lim_{n\to\infty} \int_X (f_1 - f_n)d\mu = \int_X (f_1 - f)d\mu$, which gives

$$\int_X f_1 d\mu - \lim_{n\to\infty} f_n d\mu = \int_X (f_1 - f)d\mu.$$

Further, as f is integrable since $0 \leq f \leq f_1$, we have

$$\int_X (f_1 - f)d\mu = \int_X f_1 d\mu - \int_X f d\mu,$$

so that

$$\int_X f_1 d\mu - \lim_{n\to\infty} \int_X f_n d\mu = \int_X f_1 d\mu - \int_X f d\mu,$$

which gives

$$\lim_{n\to\infty} \int_X f_n d\mu = \int_X f d\mu. \quad ♣$$

4. Examine the applicability to integrals $\int_0^\infty f_n(x)dx$ of dominated and monotone convergence theorems for the following: **(i)** $f_n(x) = 2n^2 e^{-n^2 x^2}$; **(ii)** $f_n(x) = nxe^{-nx^2}$

Solution:

(i) Setting $y = nx$, we have

$$\int_0^\infty f_n(x)dx = \int_0^\infty 2n^2 e^{-n^2x^2} dx = \int_0^\infty 2ne^{-y^2} dy = n\sqrt{\pi},$$

where the last term diverges as $n \to \infty$. Hence, the $\lim_{n\to\infty} \int_0^\infty f_n(x)dx$ does not exist. Next, we observe that for $x \neq 0$,

$$\lim_{n\to\infty} f_n(x) = \lim_{n\to\infty} \left(2n^2 e^{-n^2x^2}\right) = 0,$$

whereas for $x = 0$,

$$\lim_{n\to\infty} f_n(0) = \lim_{n\to\infty} \left(2n^2\right) = \infty.$$

Thus, there is no limiting function $f = \lim_{n\to\infty} f_n$ that satisfies the inequality $f(x) \geq 2n^2 e^{-n^2x^2}$ for all n in X, and we can conclude that neither the dominated nor the monotone convergence theorem is applicable.

(ii) It is found that

$$\int_0^\infty f_n(x)dx = \int_0^\infty nxe^{-nx^2} dx = \left[-\frac{1}{2}e^{-nx^2}\right]_0^\infty = \frac{1}{2},$$

and that $nxe^{-nx^2} \to 0$ pointwise as $n \to \infty$. Therefore, the limiting function $f(x)$ satisfying the inequality $f(x) \geq nxe^{-nx^2}$ does not exist. Hence, neither the dominated nor the monotone convergence theorem is applicable. ♣

5. Using Fubini's theorem, derive the formula

$$\int_0^1 \frac{x^b - x^a}{\log x} dx = \log \frac{1+b}{1+a} \quad \text{for } a, b > 0. \tag{6.35}$$

Solution: Note that the integral in the left-hand-side is beyond elementary calculus, so that it is impossible to achieve (6.35) by straightforward calculations. Instead, we observe that

$$\int_0^1 dx \int_a^b x^y dy = \int_0^1 \frac{x^b - x^a}{\log x} dx$$

and

$$\int_a^b dy \int_0^1 x^y dx = \int_a^b \frac{dy}{y+1} = \log \frac{1+b}{1+a}.$$

Thus, when we apply the Fubini theorem to the double integral

$$\int\int_{[0\le x\le 1,\ a\le y\le b]} x^y dxdy,$$

we obtain the desired result (6.35). ♣

6. Show that the function $f(x,y)$ given in (6.34) in Sect. 6.3.2 is not integrable on $[0,1]\times[0,1]$.

Solution: It follows that

$$\int\int_{0\le x,y\le 1}\left|\frac{x^2-y^2}{(x^2+y^2)^2}\right| dxdy = 2\int\int_{0\le x\le y\le 1}\frac{y^2-x^2}{(x^2+y^2)^2}dxdy$$

$$= 2\int_0^1 dy\int_0^y \frac{y^2-x^2}{(x^2+y^2)^2}dxdy = \int_0^1 \frac{dy}{y} = \infty.$$

This means that the existence and equality of two repeated integrals do *not* guarantee the existence of the double integral. ♣

6.4 The Lebesgue Spaces L^p

6.4.1 The Spaces of L^p

We close this chapter by demonstrating the relevance of the Lebesgue integral theory to the functional analysis that we discussed in Chap. 4. The Lebesgue theory on integration enables us to introduce certain spaces of functions that have properties that are of great importance in analysis as well as in mathematical physics, in particular, quantum mechanics. These are the so-called L^p spaces of complex-valued functions f such that $|f|^p$ is integrable.

We have already dealt with the concept of Hilbert space. In fact, L^2 for any measure μ satisfies the conditions for a Hilbert space. We begin with a short review of the definition of L^p spaces in terms of measure, and follow this by examining how the spaces possess vector space properties owing to the use of the Lebesgue integral.

Let p be a positive real number and let X be a measurable set in $\mathbf{R}$. The L^p space is defined as follows:

♠ **Definition of L^p space:**

The L^p space is a set of complex-valued Lebesgue measurable functions $f(x)$ on X that satisfy

$$\int_X |f|^p d\mu < \infty$$

for $p\ge 1$.

When the integral $\int_X |f(x)|^p dx$ exists, we call it the p-**norm** of f and denote it by

$$\|f\|_p = \left(\int_X |f|^p d\mu \right)^{1/p}.$$

Clearly for $p = 2$, the present definition reduces to our earlier definition of L^2.

6.4.2 Hölder Inequality

The following two inequalities are fundamentals that demonstrate the relations between the norms of functions involved in L^p.

♠ **Hölder inequality:**
For any $f, g \in L^p$ under the conditions

$$p, q > 1 \quad \text{and} \quad \frac{1}{p} + \frac{1}{q} = 1,$$

we have

$$fg \in L^1 \quad \text{and} \quad \|fg\|_1 \leq \|f\|_p \|g\|_p.$$

Proof We assume that neither f nor g is zero almost everywhere (otherwise, the result is trivial). To proceed with the proof, we first observe the inequality

$$a^{1/p} b^{1/q} \leq \frac{a}{p} + \frac{b}{q} \quad \text{for } a, b \geq 0, \tag{6.36}$$

which we is justify by rewriting it as

$$t^{1/p} \leq \frac{t}{p} + \frac{1}{q},$$

where we set $t = a/b$. Then, we note that the function given by

$$f(t) = t^{1/p} - \frac{t}{p} - \frac{1}{q} \leq 0$$

has a maximum at $t = 1$, namely,

$$\max f(t) = f(1) = 1 - \frac{1}{p} - \frac{1}{q} = 0,$$

which results in the inequality (6.36), which we use to obtain

$$\frac{|f(x)g(x)|}{AB} \leq \frac{A^{-p}|f(x)|^p}{p} + \frac{B^{-q}|g(x)|^q}{q}, \tag{6.37}$$

where

$$A = \left[\int_X |f|^p d\mu \right]^{1/p} \quad \text{and} \quad B = \left[\int_X |g|^q d\mu \right]^{1/q}.$$

The right-hand side of (6.37) is integrable from the hypothesis that $f, g \in L^p$. Therefore, using (6.37) we obtain

$$\frac{1}{AB} \int_X |fg| d\mu \leq \frac{A^{-p}}{p} \int_X |f|^p d\mu + \frac{B^{-q}}{q} \int_X |g|^q d\mu$$
$$= \frac{1}{p} + \frac{1}{q} = 1.$$

Consequently, we have

$$\int_X |fg| d\mu \leq AB,$$

which proves the inequality. ♣

6.4.3 Minkowski Inequality

The other inequality of interest is stated below.

♠ **Minkowski inequality:**
If $f, g \in L^p$ with $p \geq 1$, then

$$f + g \in L^p \quad \text{and} \quad \|f + g\|_p \leq \|f\|_p + \|g\|_p. \tag{6.38}$$

Proof For $p = 1$, the inequality is readily obtained by integrating the triangle inequality for real numbers. For $p > 1$, it follows that

$$\int_X |f + g|^p d\mu = \int_X |f + g|^{p-1} |f| d\mu$$
$$+ \int_X |f + g|^{p-1} |g| d\mu.$$

Let $q > 0$ be such that

$$\frac{1}{p} + \frac{1}{q} = 1.$$

Applying the Hölder inequality to each of these last two integrals and noting that $(p-1)q = p$, gives us

$$\int_X |f(x) + g(x)|^p dx \leq M \left[\int_X |f + g|^{(p-1)q} d\mu \right]^{1/q}$$
$$= M \left[\int_X |f + g|^p d\mu \right]^{1/q}, \tag{6.39}$$

where M denotes the right-hand side of the inequality (6.38) that we would like to prove. Now divide the extreme ends of the relation (6.39) by

$$\left[\int_X |f+g|^p d\mu\right]^{1/q}$$

to obtain the desired result. ♣

Remark. It should be noted that neither the Hölder inequality nor the Minkowski inequality holds for $0 < p < 1$ if $\mu(X) > 0$, which is why we restrict ourselves to $p \geq 1$.

6.4.4 Completeness of L^p Spaces

By virtue of the two inequalities discussed above, we can show the completeness properties of L^p spaces, which is crucially important for developing **Hilbert space theory** for Lebesgue measurable functions.

♠ **Completeness of L^p spaces:**
The space L^p is complete: i.e., for any $f_n \in L^p$ satisfying

$$\lim_{n,m\to\infty} \|f_n - f_m\|_p = 0,$$

there exists $f \in L^p$ such that

$$\lim_{n\to\infty} \|f_n - f\|_p = 0.$$

Proof Let $\{f_n\}$ be a **Cauchy sequence** in L^p. Then, there is a natural number n_1 such that for all $n > n_1$, we have

$$\|f_n - f_{n_1}\| < \frac{1}{2}.$$

By induction, after finding $n_{k-1} > n_{k-2}$, we find $n_k > n_{k-1}$ such that for all $n > n_k$ we have

$$\|f_n - f_{n_k}\| < \frac{1}{2^k}.$$

Then $\{f_{n_k}\}$ is a subsequence of $\{f_n\}$ that satisfies

$$\|f_{n_{k+1}} - f_{n_k}\| < \frac{1}{2^k}$$

or

$$\|f_{n_1}\| + \sum_{k=1}^{\infty} \|f_{n_{k+1}} - f_{n_k}\| = A < \infty.$$

Let

$$g_k = |f_{n_1}| + |f_{n_2} - f_{n_1}| + \cdots + |f_{n_{k+1}} - f_{n_k}|, \quad k = 1, 2, \cdots .$$

Then, by the Minkowski inequality,

$$\begin{aligned}\int_X g_k^p(x) d\mu &= \int_X \left(|f_{n_1}| + |f_{n_2} - f_{n_1}| + \cdots + |f_{n_{k+1}} - f_{n_k}|\right)^p d\mu \\ &\le \left(\|f_{n_1}\|_p + \sum_{k=1}^{\infty} \|f_{n_{k+1}} - f_{n_k}\|\right)^p \\ &\le A^p < \infty.\end{aligned}$$

Let $g = \lim g_k$. Then $g^p = \lim g_k^p$. By the monotone convergence theorem given in Sect. 6.2.1, we have

$$\int_X g^p d\mu = \lim_{k\to\infty} \int_X g_k^p d\mu < \infty,$$

which shows that g is in L^p, and hence

$$\int_X \left(|f_{n_1}| + \sum_{k=1}^{\infty} |f_{n_{k+1}} - f_{n_k}|\right)^p dx < \infty,$$

implying that

$$|f_{n_1}| + \sum_{k=1}^{\infty} |f_{n_{k+1}} - f_{n_k}|$$

converges almost everywhere to a function $f \in L^p$.

It remains to prove that $\|f_{n_k} - f\| \to 0$ as $k \to \infty$. We first note that

$$f(x) - f_{n_j}(x) = \sum_{k=j}^{\infty} \left[f_{n_{k+1}}(x) - f_{n_k}(x)\right].$$

It then follows that

$$\|f - f_{n_j}\| \le \sum_{k=j}^{\infty} \|f_{n_k+1} - f_{n_k}\|_p < \sum_{k=j}^{\infty} \frac{1}{2^k} = \frac{1}{2^{j-1}}.$$

Therefore, $\|f - f_{n_j}\|_p \to 0$ as $j \to \infty$. Now

$$\|f_n - f\|_p \le \|f_n - f_{n_k}\|_p + \|f_{n_k} - f\|_p,$$

where $\|f_n - f_{n_k}\|_p \to 0$ as $n \to \infty$ and $k \to \infty$ and thus $\|f_n - f\|_p = 0$ as $n \to \infty$. This shows that the Cauchy sequence $\{f_n\}$ converges to f in L^p. ♣

Before closing this chapter, we must emphasize that if we employ the Riemann integral to construct L^p spaces, the theorem mentioned above breaks down so that we can no longer expect completeness of the resulting function space. To illustrate this point, we temporarily define the 'L^1 space' by a set of Riemann integrable functions under the '1-norm':

$$\|f\|_1^{[R]} \equiv \int_0^1 |f(x)|dx < \infty.$$

We then consider a function

$$f_n(x) = \begin{cases} 1 & \text{for } x \in \{a_n\}, \\ 0 & \text{otherwise,} \end{cases}$$

where $\{a_n\}$ $(n = 1, 2, \cdots)$ is an infinite sequence of all rational numbers in $[0, 1]$. It readily follows that a function $f_n(x) - f_\mu(x) \in L^1$ is Riemann integrable and reads

$$\|f_n - f_m\|_1^{[R]} = \int_0^1 |f_n(x) - f_\mu(x)|dx = 0.$$

Nevertheless, $f_n(x)$ converges to Dirichlet's function $\chi(x)$, which is not Riemann integrable as noted earlier. As it is impossible to examine the quantity

$$\|f_n - \chi\|_1^{[R]},$$

using Reimann integrals, we cannot establish the complete function space based on that method.

6.5 Applications in Physics and Engineering

6.5.1 Practical Significance of Lebesgue Integrals

From a practical viewpoint, what makes Lebesgue integrals so important is the fact that they allow us to interchange the order of integration and other limiting procedures under very weak conditions, which is not possible in the case of Riemann integrals. In fact, in the case of Riemann integrals, the identities

$$\lim_{n\to\infty} \int_{-\infty}^{\infty} f_n(x)dx = \int_{-\infty}^{\infty} \lim_{n\to\infty} f_n(x)dx$$

and

$$\sum_{n=1}^{\infty} \int_{-\infty}^{\infty} f_n(x)dx = \int_{-\infty}^{\infty} \sum_{n=1}^{\infty} f_n(x)dx$$

are valid only if the integrands on the right-hand side, i.e., $\lim f_n$ and $\sum f_n$, are uniformly convergent. Such a restriction can be removed by using a Lebesgue integral since with the latter, only pointwise convergence of the integrand is needed. We saw in Sect. 6.3 that the **Lebesgue convergence theorem** and **Fubini's theorem** markedly weaken the conditions necessary for the validity of an interchange of the order of integration. As a result, we need not monitor the order of the limiting procedure, which is very useful in the practical calculations encountered in physics and engineering.

6.5.2 Contraction Mapping

Another important consequence of Lebesgue integral theory is the **completeness** of the **function space** L^p spanned by Lebesgue integrable functions. L^p spaces have a wide range of applications in physics, statistics, engineering, and other disciplines. For instance, they serve as a basis in the development of a rigorous theory of Fourier transformation, in which the mappings between two different L^p spaces are considered. Moreover, the theory of quantum mechanics is established on the basis of the L^2 space, a specific class of L^p spaces with $p = 2$. In both applications, the completeness property of the L^p space plays a crucial role in making the theory self-contained. In order for the reader to learn more about this issue, we present the **contraction mapping theorem** (or **Banach's fixed point theorem**) below. This theorem proves the existence of a unique solution to a certain kind of equation associated with Lebesgue integrable functions, which makes the theory based on L^p spaces self-contained.

A preliminary terminology is defined below.

♠ **Contraction mapping:**

A contraction mapping T is a mapping from L^p onto L^p that satisfies the relation

$$\|T(f) - T(g)\| \leq c\|f - g\| \quad (0 \leq c < 1) \tag{6.40}$$

for any $f, g \in L^p$ (see Fig. 6.6).

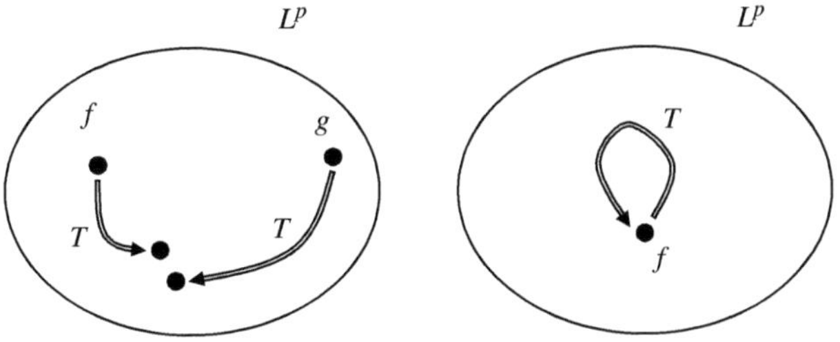

Fig. 6.6. Sketch of a contraction mapping T acting on $f, g \in L^p$

Remark. If T is regarded as a differential operator acting on a Lebesgue integrable function f, then we can say that 'a contraction mapping is a mapping that satisfies the **Lipschitz condition**' (see Sect. 15.2.3).

We should keep in mind that the norm $\|\cdots\|$ used in (6.40) is in terms of L^p spaces, so that $\|f-g\|=0$ means $f=g$ **almost everywhere**. In plain words, a contraction mapping reduces the distance between two elements in the L^p space.

We are now ready to move on to the main theorem.

♠ **Contraction mapping theorem:**

Let T be a contraction mapping and I be an identity mapping. Then the equation

$$(T-I)f=0 \tag{6.41}$$

has one and only one solution f that belongs to L^p.

Remark. The solution f of the equation (6.41) is called a **fixed point** in L^p.

The contraction mapping theorem guarantees the existence and uniqueness of fixed points of certain self-mappings and provides a constructive method for finding those fixed points. It should be emphasized that the theorem allows us to prove the existence (and uniqueness) of solutions of **ordinary differential equations** with respect to Lebesgue integrable functions, as intuitively understood if T is set to be a differential operator.

Proof (of the contraction mapping theorem): For arbitrary $f_0 \in L^p$, we introduce a sequence of functions $\{f_n\}$ defined by

$$f_1=T(f_0),\ f_2=T(f_1),\ \cdots,\ f_n=T(f_{n-1}),\ \cdots.$$

We shall see below that the sequence $\{f_n\}$ is a **Cauchy sequence** and thus has a limit $f\equiv\lim_{n\to\infty}f_n$. It follows from the definition of T that

$$\begin{aligned}\|f_n-f_{n+j}\| &= \|T(f_{n-1})-T(f_{n-1+j})\|\\ &\le c\|f_{n-1}-f_{n-1+j}\|\\ &\le\ \cdots\ \le c^n\|f_0-f_j\|\end{aligned} \tag{6.42}$$

and

$$\begin{aligned}\|f_0-f_j\| &\le \|f_0-f_1\|+\cdots+\|f_{j-1}-f_j\|\\ &\le \left(1+c+\cdots+c^{j+1}\right)\|f_0-f_1\|\\ &\le (1-c)^{-1}\|f_0-f_1\|,\end{aligned} \tag{6.43}$$

where we used the **Minkowski inequality** (6.38) with respect to the p-norm designated by $\|\cdots\|$. From (6.42) and (6.43), we set

$$\|f_n - f_{n+j}\| \le \frac{c^n}{1-c}\|f_0 - f_1\| \to 0 \quad (n \to \infty).$$

This indicates that $\{f_n\}$ is a Cauchy sequence and thus converges to a limit (denoted by f) regardless of the choice of f_0. Furthermore, the limit f always belongs to L^p since the space L^p is complete. Hence, the converging behavior of $\{f_n\}$ to f can be expressed by using the concept of the norm of L^p as

$$\lim_{n\to\infty} \|f_n - f\| = 0. \tag{6.44}$$

We then obtain

$$\begin{aligned} \|T(f) - f\| &\le \|T(f) - f_n\| + \|f_n - f\| \\ &= \|T(f) - T(f_{n-1})\| + \|f_n - f\| \\ &\le \|f - f_{n-1}\| + \|f_n - f\| \quad \to 0 \quad (n \to \infty), \end{aligned} \tag{6.45}$$

which means that $T(f) = f$ almost everywhere. Consequently, equation (6.41) has at least one solution that is a limit f of the sequence $\{f_n\}$ that we introduced.

The uniqueness of the solution f is readily understood. Suppose $g \in L^p$ such that $T(g) = g$. We then have

$$\|g - f\| = \|T(g) - T(f)\| \le c\|g - f\|.$$

This means that $\|g-f\| = 0$ since $0 \le c < 1$, so we have $g = f$ almost everywhere. ♣

Remark. Note that it is our use of the Lebesgue integral (instead of the Riemann integral) that guarantees the validity of the contraction mapping theorem. In fact, if we restrict ourselves to the Riemann integral, the limit f of the sequence $\{f_n\}$ may not belong to L^p, and we can no longer obtain the result (6.45).

6.5.3 Preliminaries for the Central Limit Theorem

The effectiveness of Lebesgue integrals is also observed in probability theory, particularly in the derivation of the **central limit theorem**, which plays a fundamental role in statistical mechanics and in the statistical analysis of experimental data. Later, we shall see that employing Lebesgue integrals is necessary for proving the central limit theorem, where the **Lebesgue convergence theorem** and **Fubini's theorem** are used time and again.

In order to prove the central limit theorem, we introduce a **random variable** x (see Sect. 6.1.3); for instance, x may be the number of spots we get when shooting a pair of dice or a real number that we randomly pick from an interval on the real axis. Suppose that x lies in a set X on the real axis. (Here, X may be a continuous interval, a set of discrete points, or a union of the two.) In modern probability theory, measures characterizing the statistical properties of the system considered are defined in terms of the Lebesgue integral. For instance, the **probability** (or **distribution**) that x is found in subset $X_0 \subset X$ is given by

$$P\left(x:\ x \in X_0\right) = \int_{X_0} p d\mu, \tag{6.46}$$

where μ is the Lebesgue measure of X_0 and p is the **probability density** associated with x. In general, p is assumed to satisfy the normalization condition $\int_X p d\mu = 1$. We can state that the random variables x and y are **independent** if

$$P(x, y) = P(x)P(y).$$

Moreover, the variables x and y are said to be **identically distributed** if

$$P(x) = P(y).$$

We also define the **expected** (or **mean**) **value** of x and the variance of x by the integrals

$$E\{x\} = \int_X x p d\mu \quad \text{and} \quad V\{x\} = \int_X \left(x - E\{x\}\right)^2 p d\mu,$$

respectively, where μ is the Lebesgue measure of X. In particular, the expected value of an imaginary exponent e^{izx}, where z is real, is known as the **characteristic function**.

♠ **Characteristic function:**
The characteristic function $\varphi_x(z)$ of a random variable x is defined by

$$\varphi_x(z) = E\{e^{izx}\}.$$

It can be shown that

$$E\{e^{iz(x+y)}\} = \varphi_x(z)\varphi_y(z)$$

if and only if the random variables x and y are independent. Furthermore, we obtain

$$\varphi_x(z) = \varphi_y(z)$$

if and only if the variables x and y are identically distributed. The latter condition is known as the **uniqueness theorem** for characteristic functions, and the proof, which involves **Fubini's theorem**, can be found in advanced texts on probability theory.

6.5.4 Central Limit Theorem

We are now ready to state the key theorem.

♠ **Central limit theorem:**
Assume a series of random variables $\{x_n\}$ in which the x_n are independent and identically distributed. For arbitrary a and b $(b > a)$, we have

$$\lim_{n\to\infty} P\left(a \le \frac{\sum_{j=1}^{n} x_j - nm}{\sigma\sqrt{n}} \le b\right) = \frac{1}{\sqrt{2\pi}} \int_a^b e^{-\xi^2/2} d\xi, \tag{6.47}$$

where $m = E\{x_n\}$ and $\sigma^2 = V\{x_n\}$.

Briefly, the theorem states that the probability that the average of n random variables equals α is proportional to $e^{-\alpha^2/2}$. (Note that α is the *average* of n variables and not a variable itself.) A random variable with the probability density $e^{-\xi^2/2}$ is said to be **normally distributed.**

Remark. The central limit theorem is very effective in describing various stochastic phenomena in nature since it can be applied regardless of the distribution of the n random variables; i.e., almost all classes of random variables obey the theorem as long as they are independent and identically distributed.

An illustrative example of the central limit theorem in physics is the **Maxwell – Boltzmann distribution** of an ideal gas. For a given temperature T, the distribution $f(v)$ of the velocity of gas molecules $v = |\boldsymbol{v}|$ is known to satisfy the equation

$$f(v) = \left(\frac{m}{2\pi k_{\rm B} T}\right)^{3/2} \exp\left(-\frac{mv^2}{2k_{\rm B}T}\right), \tag{6.48}$$

where m is the mass of a gas molecule and $k_{\rm B}$ is the Boltzmann constant. Here, the velocity $v(t_i)$ as a function of discrete time t_i $(i = 1, 2, \cdots, n)$ serves as n random variables. In general, in an equilibrium state, $v(t_i)$ for different t_i is independent and identically distributed and, thus, if n is sufficiently large, the time average of $v(t_i)$ obeys the normal distribution described by (6.48). Figure 6.7 shows the distribution of the squared velocity of gas molecules, which is determined from the formula $4\pi v^2 f(v^2)$, for various values of T; we set $k_{\rm B} = 1.38 \times 10^{-23}$ kg · m^2/s^2 · K and $m = 6.6 \times 10^{-27}$ kg by considering ^{4}He molecules. We observe that the mean value of v^2 shifts to the right with an increase in the temperature, which can be intuitively understood to be due to the acceleration of the molecules at high temperatures.

It is important to emphasize that the central limit theorem holds good for any kind of distribution of the n variables $\{x_i\}$ as long as they are independent and identically distributed. For example, let us consider n variables that obey

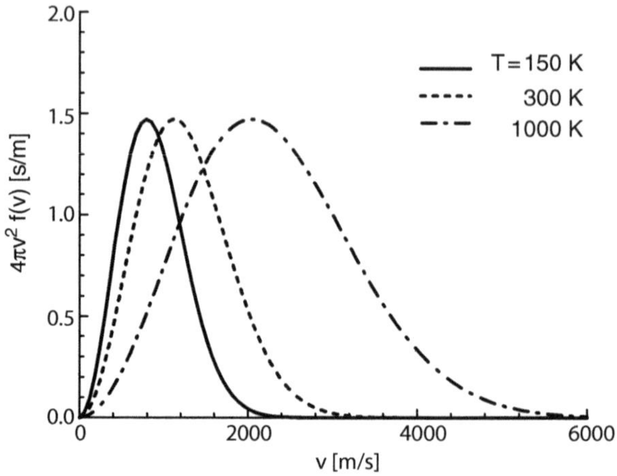

Fig. 6.7. Distribution of square velocity v^2 of ^{4}He molecules

the distribution $P(x)$ shown in Fig. 6.8. The average of these variables shows the distribution depicted in Fig. 6.8, all of which converge to the normal distribution as n increases. The fact that the distribution of $\{x_i\}$ can be disregarded is the reason the normal distribution is so universally observed in a wide variety of stochastic phenomena.

6.5.5 Proof of the Central Limit Theorem

As some further points have to be discussed in order to prove the central limit theorem, we present below only an outline and not a rigorous proof. Let us emphasize that the use of Lebesgue integrals is necessary for proving the central limit theorem, and the **Lebesgue convergence theorem** and **Fubini's theorem** are used time and again.

Proof We have only to consider the case of $m = 0$ and $\sigma = 1$; otherwise, the new variable $\tilde{x}_n \equiv (x_n - m)/\sigma$ is introduced to yield $E\{\tilde{x}_n\} = 0$ and $V\{\tilde{x}_n\} = 1$. The characteristic function $\varphi_{y_n}(z)$ for the variable

$$y_n = \frac{\sum_{j=1}^{n} x_j}{\sqrt{n}}$$

is given by

$$\varphi_{y_n}(z) = E\left\{\exp\left(\frac{iz}{\sqrt{n}}\sum_{j=1}^{n} x_j\right)\right\} = \prod_{j=1}^{n} \varphi_{x_j}\left(\frac{z}{\sqrt{n}}\right),$$

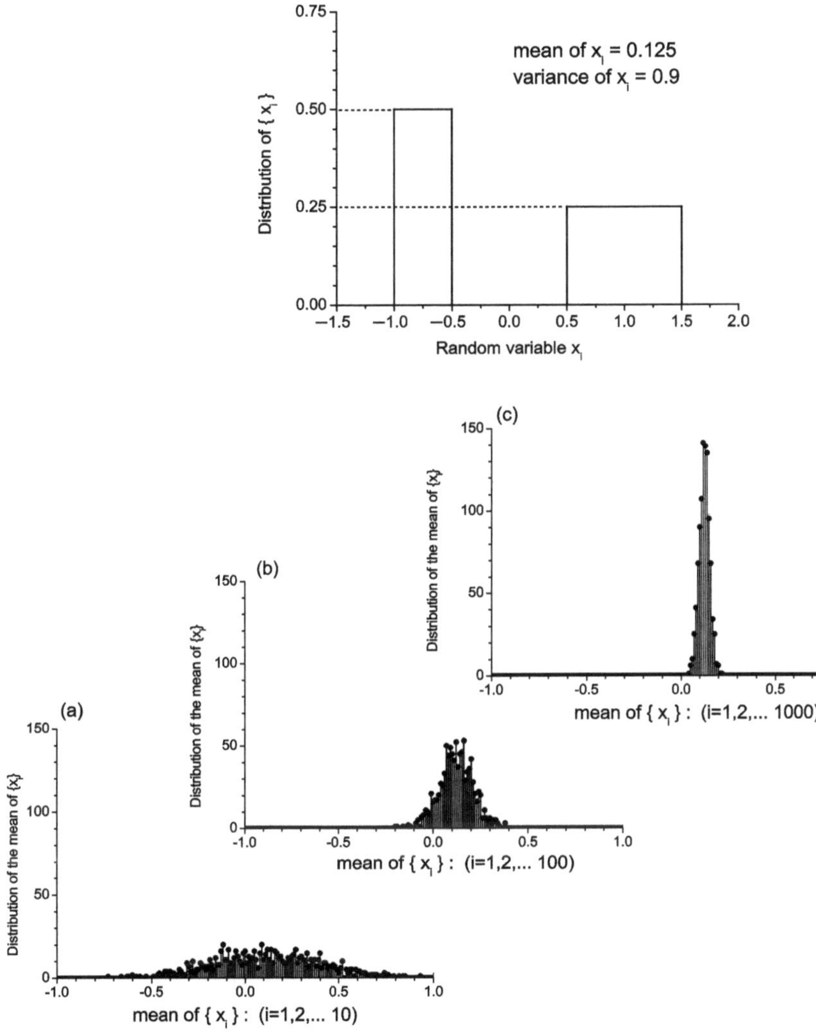

Fig. 6.8. *Top*: Distributions of a random variable x. *Bottom*: (**a**)–(**c**) Distributions of the average value α of n random variables $x_1, x_2, \cdots, x_n$ with $n = 10$ for (**a**), $n = 100$ for (**b**), and $n = 1000$ for (**c**). For each, 1000 α's are sampled to create the distribution. With increasing n, the distribution of α converges to the normal distribution around the center of 0.125 as expected

in which the condition that all x_n are independent allows us to obtain the last expression. Furthermore, since all x_n are identically distibuted, we have

$$\prod_{j=1}^{n} \varphi_{x_j}\left(\frac{z}{\sqrt{n}}\right) = \left[\varphi_{x_1}\left(\frac{z}{\sqrt{n}}\right)\right]^n,$$

which is ensured by the **uniqueness theorem** discussed in Sect. 6.5.5.

We want the limit of $\varphi_{y_n}(z)$ at $n \to \infty$, so we use the formula (see the lemma below)

$$\lim_{n\to\infty} \varphi_{y_n}(z) = e^{-z^2/2}, \tag{6.49}$$

in which the right-hand side is the characteristic function of a normal distribution. The result of (6.49) together with the **continuity theorem** (see below) states that

$$\lim_{n\to\infty} P(a \le y_n \le b) = P(a \le y \le b) = \frac{1}{\sqrt{2\pi}} \int_a^b e^{-y^2/2} dy. \quad ♣ \tag{6.50}$$

The following theorem forms the basis for the proof of the central limit theorem.

♠ **Continuity theorem:**
Let x and x_n be random variables such that

$$\lim_{n\to\infty} \varphi_{x_n}(z) = \varphi_x(z).$$

We then obtain

$$\lim_{n\to\infty} P(a \le x_n \le b) = P(a \le x \le b)$$

for arbitrary a, $b(b>a)$ satisfying $P(x = a) = P(x = b) = 0$.

This theorem states that the convergence of characteristic functions implies the convergence of the corresponding distribution functions. Since the proof requires the use of Fubini's theorem as well as the Lebesgue convergence theorem and is quite complicated, we do not present it.

♠ **Lemma:**
If $E\{x\} = 0$ and $E\{x^2\} = 1$ for a random variable x, then the characteristic function $\varphi_x(z)$ satisfies the relation

$$\lim_{n\to\infty} \left[\varphi_x\left(\frac{z}{\sqrt{n}}\right)\right]^n = \exp\left(-\frac{z^2}{2}\right). \tag{6.51}$$

Proof The assumption that $E\{x^2\} = 1 < \infty$ implies that $\varphi_x(z)$ is twice differentiable. In fact, we obtain

$$\varphi_x(z) = E\{e^{izx}\},$$
$$\varphi_x{}'(z) = E\left\{\frac{d}{dz}e^{izx}\right\} = E\left\{ixe^{izx}\right\},$$
$$\varphi_x{}''(z) = E\left\{\frac{d^2}{dz^2}e^{izx}\right\} = E\left\{-x^2e^{izx}\right\},$$

where the **Lebesgue convergence theorem** was used to interchange the order of differentiation d/dz and integration $\int dx$ associated with calculation of $E\{\cdots\}$. The twice differentiability of $\varphi(z)$ allows us to expand it around $z = 0$ as

$$\varphi_x\left(\frac{z}{\sqrt{n}}\right) = \varphi_x(0) + \frac{z}{\sqrt{n}}\varphi_x{}'(0) + \frac{z^2}{2n}\varphi_x{}''(\eta),$$

where η is small enough to be $|\eta| \leq |z|/\sqrt{n}$. Since $\varphi_x(0) = 1$ and $\varphi_x{}'(0) = E\{ix\} = 0$, we have

$$\begin{aligned}\log\left[\varphi\left(\frac{z}{\sqrt{n}}\right)\right]^n &= n\log\varphi\left(\frac{z}{\sqrt{n}}\right)\\ &= n\log\left(1 + \frac{z^2}{2n}\varphi''(\eta)\right)\\ &= \frac{z^2}{2}\varphi''(\eta) - \frac{z^4}{8n}\varphi''(\eta)^2 + \cdots \quad (n \gg 1),\end{aligned}$$

where we used the inequality $|\varphi''(\eta)| \leq 1$ to expand the logarithmic term for $n \gg 1$. As a result, we set

$$\lim_{n\to\infty}\log\left[\varphi\left(\frac{z}{\sqrt{n}}\right)\right]^n = \frac{z^2}{2}\varphi''(0) = -\frac{z^2}{2},$$

which is equivalent to (6.51). ♣.

Part III

Complex Analysis

7

Complex Functions

Abstract Differentiation and integration of complex functions are significantly different from those of real functions. In this chapter, we show that two very important theorems—the Cauchy theorem (Sect. 7.2.2) and the Taylor series expansion (Sect. 7.4.3)—result in a broad range of mathematical consequences that are highly relevant and useful in mathematical physics. However, before moving on to the principal discussion, we deal with the underlying concepts of analytic functions (Sect. 7.1.2) and the geometric meaning of analyticity (Sect. 7.1.5).

7.1 Analytic Functions

7.1.1 Continuity and Differentiability

This chapter describes the theory of functions of a complex variable. Let $\boldsymbol{C}$ denote the set of all elements z of the form

$$z = x + iy,$$

where $x, y \in \boldsymbol{R}$ and i is a familiar symbol defined by $i^2 = -1$. Let D be a domain in $\boldsymbol{C}$. Then, a **complex function** defined by

$$f : D \to \boldsymbol{C}$$

is a rule that assigns a complex-valued function $f(z)$ to each $z \in D$. This $f(z)$ is equivalent to an ordered pair of real-valued functions $u(z)$ and $v(z)$. Thus, $f(z)$ can be written in the form

$$w = f(z) = u(z) + iv(z).$$

The real-valued functions $u(z)$ and $v(z)$ are called the **real** and **imaginary parts** (or **components**) of $f(z)$ (see Fig. 7.1). We may write $u = \mathrm{Re}f$ and $v = \mathrm{Im}f$.

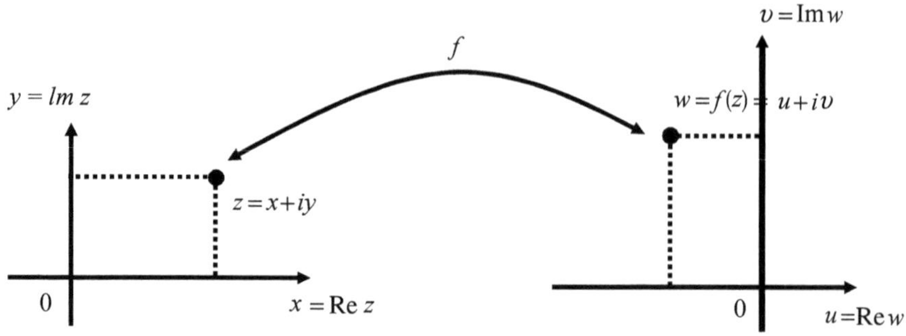

Fig. 7.1. A complex function $w = f(z)$ that assigns a point on the w-plane to each point on the z-plane

Once we introduce complex functions, the concepts of differentiation and integration encountered in ordinary real calculus acquire new depth and significance. When $f(z)$ has its derivative in D, it is referred to as an **analytic function** in D. (More precise definitions of analytic functions are given in Sect. 7.1.3.) We shall see that the conditions for a complex-valued function $f(z)$ to be differentiable with respect to a complex variable z is much stronger than that for a real-valued function $f(x)$ with respect to a real variable x. This restriction forces a great deal of the structure of $f(z)$.

An exact definition of an analytic function is obtained by considering its derivative with respect to a complex variable z. Therefore, our first task is to determine the necessary and sufficient conditions for a complex function $f(z)$ to have a derivative with respect to z. Before stating what is meant by the derivative $f'(z)$, we begin with the definition of continuity for $f(z)$.

♠ **Continuity of complex functions:**

Let $f : D \to \boldsymbol{C}$ be a complex function and z_0 a point in D. Then, a function $w = f(z) \in \boldsymbol{C}$ is continuous at the point z_0 if

$$\lim_{z \to z_0} f(z) = f(z_0). \tag{7.1}$$

In the limit of (7.1), the complex variable z may approach z_0 from any direction in D (see Fig. 7.2). Hence, if we say the limit (7.1) exists, it means that a unique quantity $f(z_0)$ must result from the limiting process regardless of how the limit $z \to z_0$ is taken.

A similar feature is found in the definition of the derivative of $f(z)$.

♠ **Derivatives of complex functions:**

A complex function $f(z)$ is said to be **differentiable** at the point z_0 if and only if the limit

$$\lim_{z \to z_0} \frac{f(z) - f(z_0)}{z - z_0}, \quad z \in D \tag{7.2}$$

exists and is uniquely determined regardless of the manner in which z approaches z_0. When the limit exists, we denote it by $f'(z_0)$, the **derivative** of $f(z)$ at z_0.

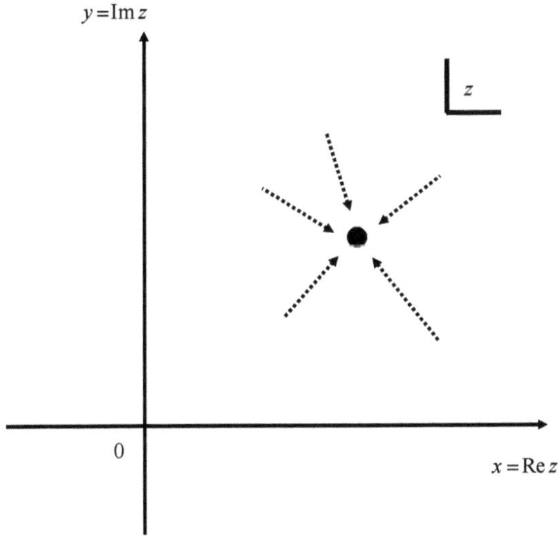

Fig. 7.2. Approaching direction of z to z_0

The definition(7.2) requires that the ratio $[f(z_0 + \Delta z) - f(z_0)]/\Delta z$ always tend to a unique limiting value, no matter the path along which z approaches z_0. This is an extremely strict condition; in fact, a number of theorems in the theory of analytic functions are derived from this requirement.

Keep in mind that a function $f(z)$ may be differentiable only at a point, or on a curve, or through a region. An example for a differentiable function at single point is presented in Example 3 in Sect. 7.1.2.

7.1.2 Definition of an Analytic Function

Among many differentiable functions, some specific kinds of functions form the class of analytic functions as stated below.

♠ **Analytic functions:**

A function $f(z)$ is said to be **analytic** at the point $z = z_0$ if and only if it is differentiable throughout a neighborhood of $z = z_0$.

Remark. There are some synonyms for the term analytic: **holomorphic**, **regular**, and **regular analytic**.

We offer some comments on the distinction between **differentiability** and **analyticity**. As noted above, the conditions for $f(x)$ to be analytic are more stringent than those for it to be differentiable; in fact, a function $f(z)$ is said to be analytic at a point z_0 if it has a derivative at z_0 and *at all points in some neighborhood* z_0. In this context, if we say that a function is analytic on a curve, we mean that it has a derivative at all points on a two-dimensional narrow strip containing the curve. If a function is differentiable only at a point or only along a curved line, then it is not analytic so that we say it is **singular** there. A typical example of $f(x)$ that is differentiable only at a point is demonstrated in Example **3** below.

Examples **1.** The function $f(z) = z^n$ is differentiable and analytic everywhere. In fact, the limit

$$\lim_{\Delta z\to 0}\frac{(z_0+\Delta z)^n - z_0^n}{\Delta z}$$
$$= \lim_{\Delta z\to 0}\left[nz_0^{n-1} + \frac{n(n-1)}{2}z_0^{n-2}\Delta z + \cdots + (\Delta z)^{n-1}\right] = nz_0^{n-1}$$

exists for arbitrary z_0, and is clearly independent of the path along which $\Delta z \to 0$. This means that any polynomial in z is differentiable and analytic everywhere.

2. The function $f(z) = z^*$ is neither differentiable nor analytic anywhere, since the limit yields

$$\lim_{\Delta z\to 0}\frac{(z_0+\Delta z)^* - z_0^*}{\Delta z} = \lim_{\Delta z\to 0}\frac{\Delta z^*}{\Delta z}. \tag{7.3}$$

If $\Delta z \to 0$ parallel to the real axis, then $\Delta z = \Delta z^* = \Delta x$ so that the limit equals 1. However, if $\Delta z \to 0$ parallel to the imaginary axis, then $\Delta z = i\Delta y = -i\Delta z^*$ so that the limit equals -1. Therefore, the quantity (7.3) depends on the path $\Delta z \to 0$, which means that it is neither differentiable nor analytic anywhere.

3. The function $f(z) = |z|$ is differentiable only at the origin. In fact,

$$\begin{aligned} f(z_0+\Delta z) - f(z_0) &= (z_0+\Delta z)(z_0^*+\Delta z_0^*) - z_0 z_0^* \\ &= z_0\Delta z^* + z_0^*\Delta z - \Delta z\Delta z^*, \end{aligned}$$

which yields

$$\begin{aligned} \lim_{\Delta z\to 0}\frac{f(z_0+\Delta z) - f(z_0)}{\Delta z} &= \lim_{\Delta z\to 0}\frac{z_0\Delta z^* + z_0^*\Delta z - \Delta z\Delta z^*}{\Delta z} \\ &= cz_0 + z_0^*, \end{aligned}$$

where $c = \lim_{\Delta z \to 0}(\Delta z^* / \Delta z)$ is a complex-valued constant that depends on the path of $\Delta z \to 0$. Hence, the limit noted above is uniquely determined only when $z_0 = 0$, which means that the function $f(z) = |z|$ is differentiable only at a point $z = 0$.

7.1.3 Cauchy–Riemann Equations

Let $f : D \to \boldsymbol{C}$ with $f(z) = u(z) + iv(z)$ as usual. We give the necessary and sufficient conditions for a function $f(z) = u(x,y) + iv(x,y)$ to be differentiable at a point $z_0 \in D$. Let us assume that $f(z)$ is differentiable at $z_0 \in D$. Then we have

$$f'(z_0) = \lim_{\Delta z \to 0} \frac{\Delta f}{\Delta z} = \lim_{\Delta z \to 0} \left(\frac{\Delta u}{\Delta z} + i \frac{\Delta v}{\Delta z} \right).$$

Since $f'(z_0)$ exists, it is independent of the path $\Delta z \to 0$; i.e., it is independent of the ratio $\Delta y / \Delta x$. If the limit is taken parallel to the real axis, $\Delta y = 0$ and $\Delta z = \Delta x$, we have

$$f'(z_0) = \lim_{\Delta x \to 0} \left(\frac{\Delta u}{\Delta x} + i \frac{\Delta v}{\Delta x} \right) = \frac{\partial u}{\partial x} + i \frac{\partial v}{\partial x}.$$

On the other hand, if the limit approaches the point z_0 along the line parallel to the imaginary axis, $\Delta x = 0$ and $\Delta z = i\Delta y$, then

$$f'(z_0) = \lim_{\Delta y \to 0} \left(\frac{\Delta v}{\Delta y} - i \frac{\Delta u}{\Delta y} \right) = \frac{\partial v}{\partial y} - i \frac{\partial u}{\partial y}.$$

From the initial assumption, these two limits must be equal, so equating real and imaginary parts gives us

$$\frac{\partial u}{\partial x} = \frac{\partial v}{\partial y} \quad \text{and} \quad \frac{\partial u}{\partial y} = -\frac{\partial v}{\partial x}. \tag{7.4}$$

Equations (7.4) are known as the **Cauchy–Riemann relations** (abbreviated by CR relations), and they are a necessary condition for differentiability.

However, alone they are not sufficient, as they provide only necessary condition. This is because they were determined from special cases of the requirement of differentiability as demonstrated above. In fact, the sufficient conditions for the differentiability of $f(z)$ at z_0 consist of the following two statements:

♠ **Theorem:**
A function $f(z)$ is differentiable at $z_0 \in D$ if and only if

(i) the first-order partial derivatives of $u(x,y)$ and $v(x,y)$ exist and are continuous at z_0, and
(ii) those derivatives at z_0 satisfy the CR equations.

Proof We prove that conditions **(i)** and **(ii)** imply the differentiability of $f(z)$ at $z_0 \in D$. (The converse was proven implicitly in the beginning of this subsection.) From hypothesis **(i)**, the functions u, $\partial u/\partial x$, and $\partial u/\partial y$ are all continuous at the point $z_0 = x_0 + iy_0$, so we have

$$\begin{aligned}\Delta u &= u(x_0 + \Delta x, y_0 + \Delta y) - u(x_0, y_0)\\ &= \frac{\partial u}{\partial x}\Delta x + \frac{\partial u}{\partial y}\Delta y + \varepsilon_1 \Delta x + \varepsilon_2 \Delta y,\end{aligned} \tag{7.5}$$

in approximation of the order Δx and Δy. In (7.5), the partial derivatives are equated at the point (x_0, y_0), and the real numbers ε_1 and ε_2 vanish as $\Delta x, \Delta y \to 0$. Using a similar formula for $v(x, y)$, we have

$$\begin{aligned}\Delta f = f(z_0 + \Delta z) - f(z_0) &= \Delta u + i\Delta v\\ = \frac{\partial u}{\partial x}\Delta x + \frac{\partial u}{\partial y}\Delta y + \varepsilon_1 \Delta x + \varepsilon_2 \Delta y\\ +i\left(\frac{\partial v}{\partial x}\Delta x + \frac{\partial v}{\partial y}\Delta y + \varepsilon_3 \Delta x + \varepsilon_4 \Delta y\right).\end{aligned}$$

Using the CR equations that are supposed to hold at the point (x_0, y_0) from assumption **(ii)** above gives us

$$\Delta f = \left(\frac{\partial u}{\partial x} + i\frac{\partial v}{\partial x}\right)(\Delta x + i\Delta y) + \Delta x(\varepsilon_1 + i\varepsilon_3) + \Delta y(\varepsilon_2 + i\varepsilon_4).$$

Dividing the both sides by $\Delta z = \Delta x + i\Delta y$ yields

$$\frac{\Delta f}{\Delta z} = \frac{\partial u}{\partial x} + i\frac{\partial v}{\partial x} + (\varepsilon_1 + i\varepsilon_3)\frac{\Delta x}{\Delta z} + (\varepsilon_2 + i\varepsilon_4)\frac{\Delta y}{\Delta z}. \tag{7.6}$$

Since $|\Delta z| = \sqrt{(\Delta x)^2 + (\Delta y)^2}$, we have

$$|\Delta x| \le |\Delta z| \quad \text{and} \quad |\Delta y| \le |\Delta z|,$$

so that

$$\left|\frac{\Delta x}{\Delta z}\right| \le 1 \quad \text{and} \quad \left|\frac{\Delta y}{\Delta z}\right| \le 1. \tag{7.7}$$

Hence, it follows from (7.7) that the last two terms in (7.6) tend to zero with $\Delta z \to 0$ because $\lim_{\Delta z \to 0} \varepsilon_n = 0$ $(1 \le n \le 4)$. As a result, the limit

$$\lim_{\Delta z \to 0} \frac{\Delta f}{\Delta z} = \frac{\partial u}{\partial x} + i\frac{\partial v}{\partial x} \tag{7.8}$$

is independent of the path of $\Delta z \to 0$, so the derivative $f'(z_0)$ exists. We thus have verified that $f(z)$ is differentiable at z_0 if conditions **(i)** and **(ii)** are satisfied. This completes the proof of the analyticity of $f(z)$. ♣

Examples **1.** Regarding the function

$$f(z) = z^2 = (x^2 - y^2) + i(2xy) \equiv u + iv, \tag{7.9}$$

we have

$$\frac{\partial u}{\partial x} = 2x = \frac{\partial v}{\partial y}, \text{ and } \frac{\partial v}{\partial x} = 2y = -\frac{\partial u}{\partial y}. \tag{7.10}$$

These equations mean that everywhere in the complex plane the CR relations hold and the partial derivatives are continuous. Hence, the function (7.9) is analytic in the entire complex plane. Such analytic functions are called **entire functions**.

2. We saw in Sect. 7.1.1 that the function $f(z) = |z|^2 = x^2 + y^2$ is not analytic anywhere since it is differentiable only at the origin. In fact, it yields

$$\frac{\partial u}{\partial x} = 2x, \quad \frac{\partial u}{\partial y} = 2y, \quad \frac{\partial v}{\partial x} = \frac{\partial u}{\partial y} = 0,$$

which satisfy the CR relations only at the origin.

7.1.4 Harmonic Functions

The CR relations immediately provide one remarkable result that points to connections with physics. Provided that the CR relations hold in a region, we set

$$\frac{\partial}{\partial x}\frac{\partial u}{\partial x} = \frac{\partial}{\partial x}\frac{\partial v}{\partial y} = \frac{\partial}{\partial y}\frac{\partial v}{\partial x} = -\frac{\partial}{\partial y}\frac{\partial u}{\partial y}. \tag{7.11}$$

Here we assume the continuity of the second-order partial derivatives of $u(x, y)$ and $v(x, y)$, which allows us to interchange the orders of differentiation in the mixed partial derivatives in (7.11). (This qualification, however, can be dropped since the second-order partial derivatives of an analytic function are necessarily continuous as we prove later.) Equation (7.11) yields the **Laplace equation**:

$$\frac{\partial^2 u}{\partial x^2} + \frac{\partial^2 u}{\partial y^2} = \nabla^2 u = 0.$$

In the same way, it follows that

$$\nabla^2 v = 0.$$

Thus we set the following theorem:

♠ **Theorem:**
Each of the real and imaginary parts of analytic functions satisfies the two-dimensional Laplace equation.

Any function ϕ satisfying $\nabla^2\phi = 0$ is called an **harmonic function**. Accordingly, if $f = u + iv$ is an analytic function, then u and v are called **conjugate**

harmonic functions since $\nabla^2 u = \nabla^2 v = 0$ holds. The fact that real and imaginary components of analytic functions satisfy the Laplace equation plays a crucial role in solving applied second-order partial differential equations. Detail discussions on this point are presented in Sect. 9.4.3.

7.1.5 Geometric Interpretation of Analyticity

To gain in-depth insight into the nature of analytic functions, we reveal the geometric meaning of "**analyticity**." We know that the analyticity of $f(z)$ within a domain D ensures the existence of the derivative $f'(z) = df/dz$ defined by

$$f'(z) = \lim_{h\to 0} \frac{f(z+h) - f(z)}{h}.$$

This suggests that at a point z_0 within D,

$$f(z_0 + h) - f(z_0) \simeq f'(z_0)h \tag{7.12}$$

for an arbitrary complex number h the magnitude $|h|$ is sufficiently small.

Let us consider the geometrical meaning of (7.12). For the discussion to be concrete, we assume, for the moment, that the derivative $f'(z)$ takes the values

$$f'(z_0) = 1 + i \quad \text{and} \quad f'(z_1) = \frac{-1+\sqrt{3}i}{2}$$

at the points z_0 and z_1 in D. It then follows that

$$f'(z_0)h = (1+i)h = \sqrt{2}\left(\frac{1}{\sqrt{2}} + \frac{i}{\sqrt{2}}\right)h = \sqrt{2}e^{i\pi/4}h, \tag{7.13}$$

where $h = |h|(\cos\theta + i\sin\theta)$ is a complex number having a certain argument θ. Equation (7.13) means that $f'(z_0)h$ is obtained through the rotation of the vector h by $\pi/4$ followed by multiplication by $\sqrt{2}$. (Note that any complex number can be regarded as a vector on the two-dimensional complex plane.) Similarly, we have

$$f'(z_1)h = e^{2\pi i/3}h, \tag{7.14}$$

which states that $f'(z_1)h$ is obtained through the rotation of h by $2\pi/3$. The processes are schematically illustrated in Fig. 7.3. The vector h is depicted by thin arrows and the corresponding vectors $f'(z)h$ by thick arrows. Noteworthy is that the magnitude $|f'(z)h|$ at both z_0 and z_1, is invariant no matter what direction the vector h takes; indeed it follows from (7.13) and (7.14) that

$$|f'(z_0)h| = \sqrt{2}|h| \quad \text{and} \quad |f'(z_1)h| = |h|.$$

Hence, when the direction of h is shifted by increasing θ, $|f'(z)h|$ remains unchanged so that the front edge of the vector $f'(z)h$ moves along a circle centered at the origin.

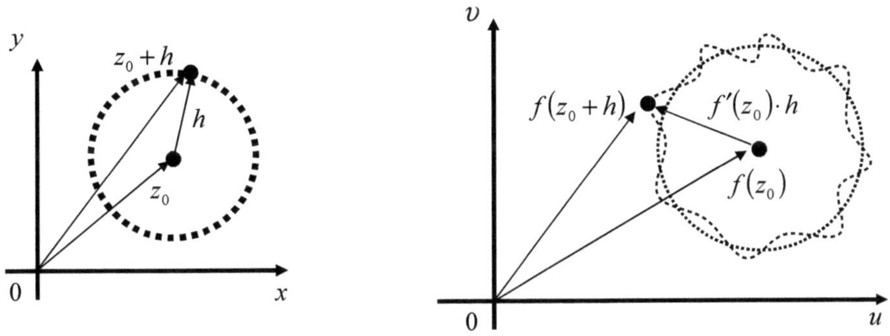

Fig. 7.3. Illustration of analyticity of $f(z)$ at z_0. An infinitesimal circle on the z-plane centered at an analytic point is mapped to a circle on the w-plane with slight modulation

Now we go back to (7.12), which says that if $f(z)$ is analytic at z_0, the acquired vectors $f'(z_0)h$ given above are almost equal to the vectors $f(z_0+h)-f(z_0)$. This implies that the magnitude $|f(z_0+h)-f(z_0)|$ is almost invariant to the change in the direction of h characterized by θ. Thus as θ increases, the front edge of $f(z_0+h)-f(z_0)$ should trace a circle centered at the origin. (To be precise, the radius may be subjected to a slight fluctuation, as shown in Fig. 7.3, owing to contributions from higher-order terms than h^2.) In other words, since $f(z_0)$ is fixed, an increase in θ from 0 to 2π results in movement of $f(z_0+h)$ along the circle centered at $f(z_0)$. This means that for analytic functions $f(z)$, the change in the magnitude of f for an infinitesimal change in z is isotropic. This isotropy is the geometric interpretation of the analyticity of $f(z)$.

Better understanding can be attained by considering the case of nonanalytic functions. Let us use the same argument for the function

$$f(z) = x^2 + iy, \tag{7.15}$$

where $u = x^2$ and $v = y$. This function is not analytic, since it does not satisfy the CR relations. Indeed,

$$\frac{\partial u}{\partial x} = 2x \neq 1 = \frac{\partial v}{\partial y}$$

except at $x = 1/2$. For such a nonanalytic function, the isotropy regarding the magnitude of the difference $|f(z+h)-f(z)|$ for infinitesimal h breaks down, as is shown below. Once we set $h = |h|(\cos + i\sin\theta)$ with $|h| = \varepsilon = \text{const}$, we have

$$\begin{aligned} f(z_0+h) &= (x_0+\varepsilon\cos\theta)^2 + i(y_0+\varepsilon\sin\theta) \\ &\simeq x_0^2 + 2\varepsilon\cos\theta\cdot x_0 + iy_0 + i\varepsilon\sin\theta \\ &= f(z_0) + 2\varepsilon\cos\theta\cdot x_0 + i\varepsilon\sin\theta, \end{aligned} \tag{7.16}$$

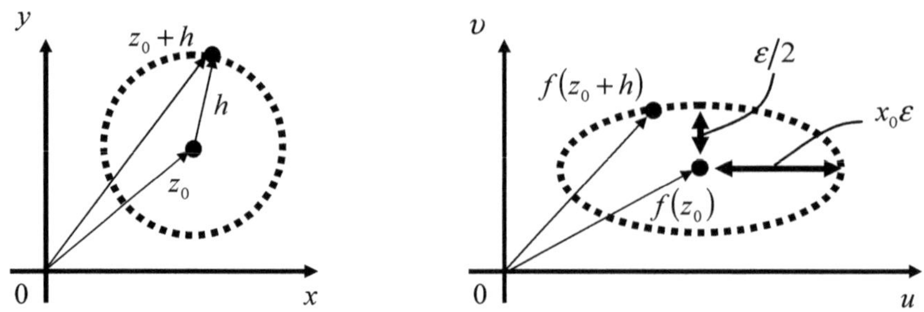

Fig. 7.4. Schematic illustration of nonanalyticity. When $f(z)$ is not analytic at $z = z_0$, then an infinitesimal circle centered at z_0 is mapped to an ellipse so the isotropy breaks out

up to the order of ε. Equation (7.16) indicates that when θ increases, the front edge of the vector $f(z_0+h)$ moves along an ellipse that has a major axis of $2x_0\varepsilon$ and a minor axis ε (see Fig. 7.4). That is, the magnitude $|f(z_0+h)-f(z_0)|$ is no longer isotropic, but depends on the direction of h (except for the particular case of $x_0 = 1/2$).

Exercises

1. Show that $f(z)$ is continuous at z_0 if it is analytic there.

   Solution: From the identity, we have

$$f(z) - f(z_0) = f(z_0 + \Delta z) - f(z_0) = \Delta z \cdot \frac{f(z_0 + \Delta z) - f(z_0)}{\Delta z}$$

   and with the definition $\Delta z = z - z_0$, we set

$$\lim_{\Delta z \to 0} [f(z_0 + \Delta z) - f(z_0)] = \left(\lim_{\Delta z \to 0} \Delta z\right) f'(z_0) = 0.$$

   Moreover, if we write $f(z) = u(z) + iv(z)$, it follows that $u(z)$ and $v(z)$ are both continuous. ♣

2. Express the Cauchy–Riemann relations in polar coordinates (r, θ).

   Solution: By imposing $z = x + iy = re^{i\theta}$, we transform the partial derivatives in terms of x into $\partial/\partial x = (\partial r/\partial x)(\partial/\partial r) + (\partial\theta/\partial x)(\partial/\partial\theta)$. After some algebra, we obtain $\partial/\partial x = \cos\theta(\partial/\partial r) - (\sin\theta/r)(\partial/\partial\theta)$, which, together with the same procedure with respect to $\partial/\partial y$, yields the polar form of the CR relations as

$$\frac{\partial u}{\partial r} = \frac{1}{r}\frac{\partial v}{\partial \theta}, \quad \frac{\partial u}{\partial \theta} = -r\frac{\partial v}{\partial r}.$$

   Their abbreviated forms read $u_r = v_\theta/r$ and $u_\theta = -rv_r$. ♣

3. If $f(z)$ is analytic in a region D and if $|f(z)|$ is constant there, then $f(z)$ is constant. Prove it.

Solution: If $|f| = 0$, the proof is immediate. Otherwise we have

$$u^2 + v^2 \equiv c \neq 0. \tag{7.17}$$

Taking the partial derivatives with respect to x and y, we have $uu_x + vv_x \equiv 0$ and $uu_y + vv_y \equiv 0$. Using the CR relations, we obtain $uu_x - vu_y \equiv 0$ and $vu_x + uu_y \equiv 0$, so that

$$(u^2 + v^2)u_x \equiv 0. \tag{7.18}$$

From (7.17) and (7.18), and from the CR relations, we conclude that $u_x = v_y \equiv 0$. We can obtain $u_y = v_x \equiv 0$ in a similar manner. Therefore, f is constant. ♣

4. Let $\phi(x,y)$ and $\psi(x,y)$ be harmonic functions in a domain D. Show that if we set $u = \phi_y - \psi_x$ and $v = \phi_x - \psi_y$, the function $f(z) = u + iv$ with the variable $z = x + iy$ becomes analytic in D.

Solution: It follows that $u_x - v_y = (\phi_{yx} - \psi_{xx}) - (\phi_{xy} - \psi_{yy}) = -\nabla^2\psi$, where $\psi_{yx} = \psi_{xy}$ was used. Since $\nabla^2\psi = 0$, we have $u_x = v_y$. Similarly, we obtain $u_y = -v_x$. Hence, u and v satisfy the CR relations in D, which indicates the analyticity of f on D. ♣

7.2 Complex Integrations

7.2.1 Integration of Complex Functions

We now turn to the integration of functions $f(z)$ with respect to a complex variable z. The theory of integration in the complex plane is just the theory of the line integral as defined by

$$\int_{\alpha_1}^{\alpha_2} f(z)dz = \lim_{N\to\infty, \Delta z_i \to 0} \sum_{i=1}^{N} f(z_i)\Delta z_i.$$

Here (Δz_i) is a sequence of small segments situated at z_i of the curve that connects the complex number α_1 to the other number α_2 in the z-plane. Since there are infinitely many choices for connecting α_1 to α_2, it is possible to obtain different values for the integral for different paths.

Examples Assume the contour integral

$$I = \oint_{C_i} z^* dz$$

from $z = 1$ to $z = -1$ along the three paths (see Fig. 7.5): **(i)** the unit circle centered at the origin in the counterclockwise direction, designated by C_1; **(ii)** that in the clockwise direction, denoted by C_2; and **(iii)** the real axis, C_3.

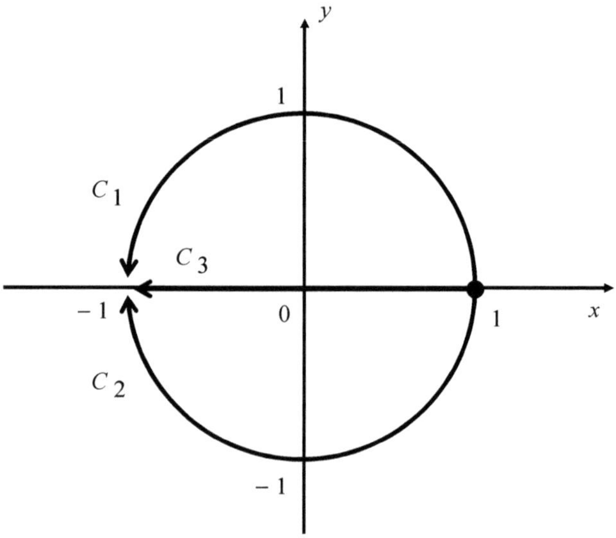

Fig. 7.5. Three paths. C_1, C_2, and C_3

(i) The values of z on the circle are given by $z = e^{i\theta}$, so $dz = ie^{i\theta}d\theta$. Thus,

$$I(C_1) = \oint_{C_1} z^* dz = \int_0^{\pi} e^{-i\theta} i e^{i\theta} d\theta = \pi i. \tag{7.19}$$

(ii) In a similar manner as in **(i)**, we have

$$I(C_2) = \oint_{C_2} z^* dz = \int_0^{-\pi} e^{-i\theta} i e^{i\theta} d\theta = -\pi i. \tag{7.20}$$

(iii) On the real axis, $z = x$ and $dz = dx$ so that

$$I(C_3) = \oint_{C_3} z^* dz = \int_1^{-1} x dx = -2. \tag{7.21}$$

In general, complex integrals on the path C possess the following property:

♠ **Darboux's inequality:**
Contour integrals on a path C satisfy the relation

$$\left| \int_C f(z) dz \right| \le ML, \tag{7.22}$$

where $M = \max |f(z)|$ on C and L is the length of C.

This property is very useful because in working with complex line integrals it is often necessary to establish upper bounds on their absolute values.

Proof Recall the original definition of complex integrals:

$$\int_C f(z)dz = \lim_{n\to\infty} \sum_{k=1}^{n} f(z_k)\Delta z_k.$$

It follows that

$$\left|\sum_{k=1}^{n} f(z_k)\Delta z_k\right| \le \sum_{k=1}^{n} |f(z_k)|\ |\Delta z_k| \le M \sum_{k=1}^{n} |\Delta z_k| \le ML,$$

where we have used the facts that $|f(z)| \le M$ for all points z on C, that the $\sum |\Delta z_k|$ represents the sum of all the chord lengths joining z_{k-1} and z_k, and that this sum is not greater than the length of C. Taking the limit of both sides, we obtain the desired inequality (7.22). ♣

7.2.2 Cauchy Theorem

We are now in a position to proceed with the key theorem in the theory of functions of a complex variable. Consider the complex integral

$$I(C_i) = \oint_{C_i} \sin z dz$$

along the closed paths C_i $(i = 1, 2, 3)$ shown in Fig. 7.6: **(a)** $C_1 = OP$, **(b)** $C_2 = OQ + QP$, **(c)** $C_3 = OR + RP$. After some algebra, we obtain

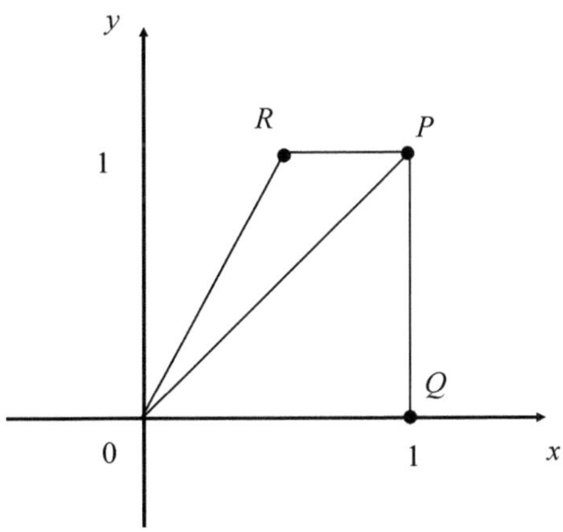

Fig. 7.6. Three paths: OQP, OP, and ORP

$$I(C_1) = I(C_2) = I(C_3) = ...,$$

which gives us the possibility that the integral from O to B remains invariant in quantity for our choices of integration paths. Actually, this is entirely true; it depends only on the two endpoints O and B. This peculiarity of integration comes from the fact that the integrand $\sin z$ is analytic on the integration paths in question. (In fact, it is analytic everywhere on the complex plane.) This result can be generalized to the following statement, called **Cauchy's theorem**, which is pivotal in the theory of complex function analysis.

♠ **Cauchy's theorem:**
If $f(z)$ is analytic within and on a closed contour C, then

$$\oint_C f(z)dz = 0. \tag{7.23}$$

The somewhat lengthy discussions that are needed for a proof of Cauchy's theorem, are beyond the scope of this textbook, but two immediate corollaries of the theorem are listed below.

♠ **Path independence:**
If $f(z)$ is analytic in the region R and if contours C_1 and C_2 lie in R and have the same endpoints, then

$$\int_{C_1} f dz = \int_{C_2} f dz.$$

The proof readily follows by applying Cauchy's theorem to the closed contour consisting of C_2 and $-C_1$ as shown in Fig. 7.7:

$$\int_{C_2} + \int_{-C_1} = 0 \Rightarrow \int_{C_2} = -\int_{-C_1} = \int_{C_1}.$$

Intuitively, the symbol $-C$ denotes the contour C traced in the opposite direction. A discussion on the path independence follows the theorem below.

♠ **Uniqueness of the integral:**
If $f(z)$ is analytic within a region bounded by a closed contour C, then the integration $\int_{z_1}^{z_2} f(z)dz$ along any contour within C depends only on z_1 and z_2.

This theorem states that an analytic function $f(z)$ has a unique integral not only a unique derivative. From a practical viewpoint, this theorem is frequently

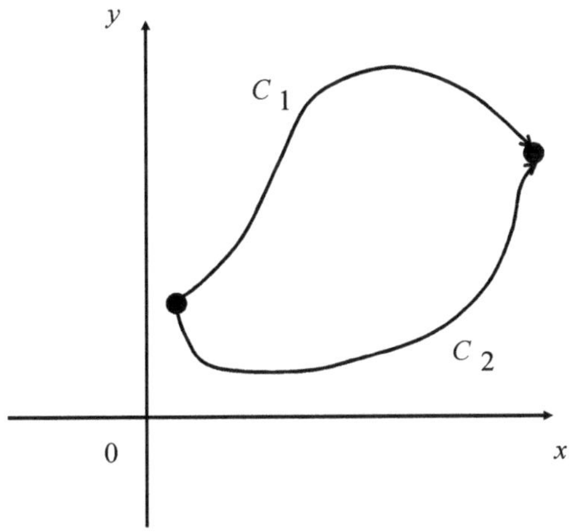

Fig. 7.7. Two integration paths: C_1 and $C_2 = -C_1$

used in the evaluation of contour integrals, since it allows us to choose an appropriate contour.

Remark. When integrating along a closed contour, we agree to move along the contour in such a way that the enclosed region lies to our left. An integration that follows this convention is called integration in the *positive* sense. Integration performed in the opposite direction acquires a minus sign.

7.2.3 Integrations on a Multiply Connected Region

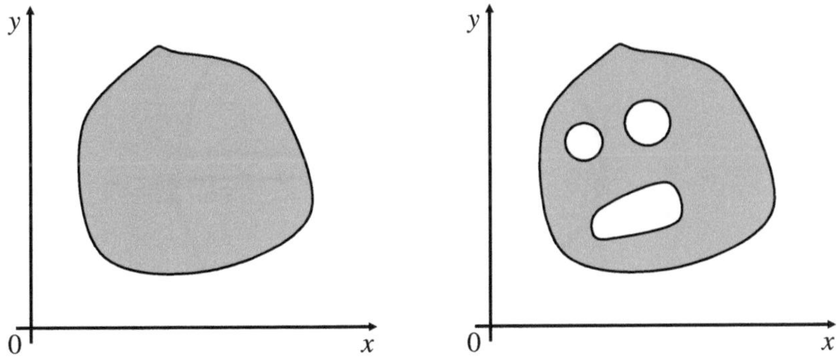

Fig. 7.8. *Left*: A simply connected region. *Right*: A multiply connected region

We should note that Cauchy's theorem applies in a direct way only to simply connected regions. A region R is said to be **simply connected** if every closed curve in R can be continuously contracted into a point without leaving R. Otherwise, it is said to be **multiply connected**; (see Fig. 7.8). The physical reason for this restriction is easy to find. The important fact is that Cauchy's theorem is a restatement that no singular point is included within the region bounded by the contour C. If the region R bounded by C is multiply connected, it becomes possible to put on singular points *within the closed contour* C but surely *outside the region* R in question. In this case, Cauchy's theorem no longer holds even though the integrand $f(z)$ is analytic everywhere in the region.

Nevertheless, there is still a way to apply Cauchy's theorem to multiply connected regions, which is based on allowing the deformation of contours as described below.

Suppose that $f(z)$ is analytic in the region that lies between two closed contours C and C', where C encloses C'. Draw two lines AB and EF close together, so as to connect the two contours. Then $ABDEFGA$ described as shown in Fig. 7.9 is a closed contour, which we denote by S and $f(z)$ is analytic within it. Then, we have

$$\oint_S f(z)dz = 0.$$

Now let the lines AB and FE approach infinitely close to one another. The contribution from the part BDE tends toward the integral around C in the positive (i.e., counterclockwise) direction. Similarly, the contribution from FGA tends toward that around C' in the negative (clockwise) direction, thus minus that around C' in the positive direction. The contributions from AB

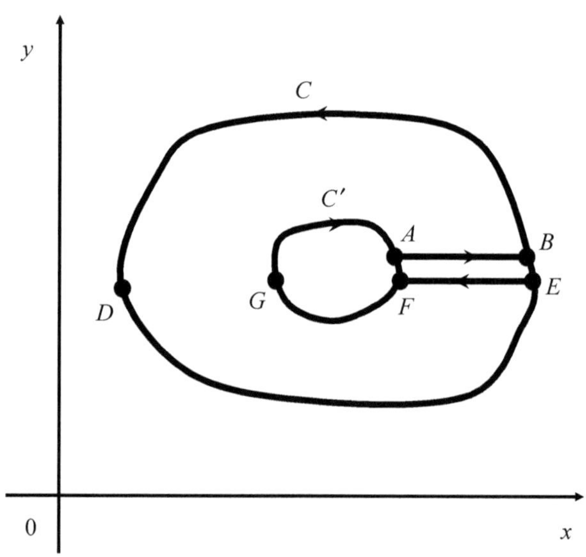

Fig. 7.9. Closed contour of ABDEFGA that consists of C and C′

and EF approach equal and opposite values since they ultimately become the same path described in opposite directions. We thus come to the conclusion that

$$\oint_C f(z)dz = \oint_{C'} f(z)dz.$$

This means that, *if a function is analytic between two contours, its integrals around both contours have the same value.*

Remark. There is an immediate extension to the case where C encloses several closed paths $C_1, C_2, \cdots$, all external to one another. Because of Cauchy's theorem, an integration contour can be moved across any region of the complex plane over which the integrand is analytic without changing the value of the integral. It cannot be moved across a hole (the shaded area) or a singularity (the dot), but it can be made to collapse around either, as shown in Fig. 7.10. As a result, an integration contour C enclosing n holes or singularities can be replaced by n separated closed contours C_i, each enclosing a hole or a singularity as given by

$$\oint_C f(z)dz = \sum_{i=1}^{n} \oint_{C_i} f(z)dz.$$

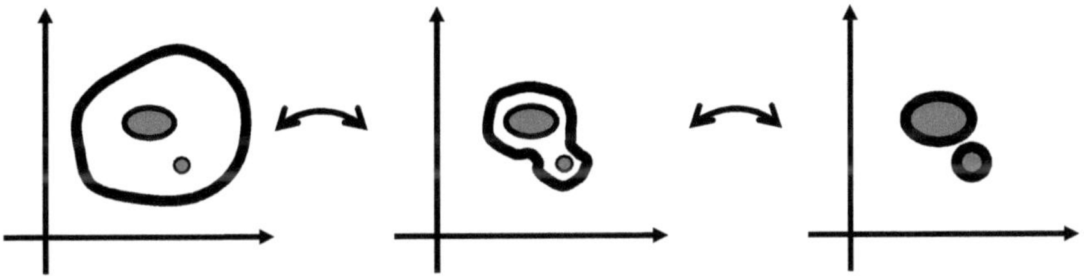

Fig. 7.10. Collapse of an integration path onto the boundaries of a hole (*a large shaded region*) and singularity (*a small shaded dot*)

7.2.4 Primitive Functions

Here is a definition of the primitive function of a complex function.

♠ **Primitive function:**

Let $f(z)$ be a function that is continuous in a domain D and has the property $\oint_C f(z)dz = 0$ for every closed path C in D. Then, the primitive function $F(z)$ of $f(z)$ is defined by

$$F(z) = \int_{z_0}^{z} f(z')dz' \quad (z_0, z \in D),$$

which is analytic in D with the derivative

$$\frac{dF(z)}{dz} = f(z).$$

Proof Consider the differential

$$F(z+\Delta z)-F(z) = \int_{z_0}^{z+\Delta z} f(z')dz' - \int_{z_0}^{z} f(z')dz' = \int_{z}^{z+\Delta z} f(z')dz', \quad (7.24)$$

where we make use of the path-independence property. If we write

$$\begin{aligned}\int_{z}^{z+\Delta z} f(z')dz' &= f(z)\int_{z}^{z+\Delta z} f(z')dz', + \int_{z}^{z+\Delta z} [f(z') - f(z)]dz' \\ &= f(z)\Delta z + \int_{z}^{z+\Delta z} [f(z') - f(z)]dz',\end{aligned}$$

then (7.24) becomes

$$F(z+\Delta z) - F(z) - f(z)\Delta z = \int_{z}^{z+\Delta z} [f(z') - f(z)]dz'. \quad (7.25)$$

Since $f(z)$ is continuous, corresponding to an arbitrary small positive number ε, there is a number δ such that

$$|z - z'| < \delta \ \Rightarrow \ |f(z) - f(z')| < \varepsilon.$$

Now choose $|\Delta z| < \delta$, which ensures $|z - z'| < \delta$ for z' on the path C in question. Therefore, we have

$$\left| \int_{z}^{z+\Delta z} [f(z') - f(z)]dz', \right| \le \int_{z}^{z+\Delta z} |f(z') - f(z)|\,|dz'| < \varepsilon|\Delta z|$$

and (7.25) can be written as

$$\left| \frac{F(z+\Delta z) - F(z)}{\Delta z} - f(z) \right| < \varepsilon \ \text{ for } |\Delta z| < \delta.$$

Since ε can be arbitrarily small, we conclude that

$$\lim_{\Delta z \to 0} \frac{F(z+\Delta z) - F(z)}{\Delta z} = f(z),$$

or equivalently,

$$\frac{dF(z)}{dz} = f(z).$$

This result is obtained for any point in D, so $F(z)$ is analytic in D. ♣

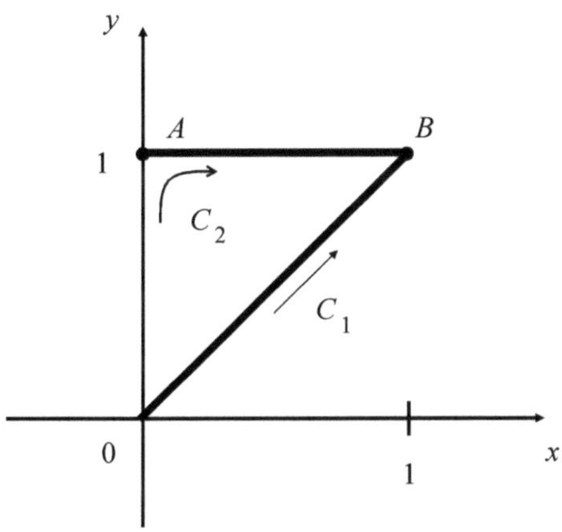

Fig. 7.11. Integration paths used in Exercise **1**

Exercises

1. Evaluate the integral

$$I(C) = \int_C \sin z dz \tag{7.26}$$

along the two paths shown in Fig. 7.11: **(a)** $C_1 = OB$, **(b)** $C_2 = OA+AB$.

Solution: Since $\sin z = \sin(x+iy) = \cosh y \sin x + i \sinh y \cos x$ and $dz = dx + idy$, we can divide (7.26) into real and imaginary parts as

$$\begin{aligned} I(C) &= \int_C (\cosh y \sin x dx - \sinh y \cos x dy) \\ &\quad + \int_C (\cosh y \sin x dy + \sinh y \cos x dx). \end{aligned}$$

Noting $x = y$ along the curve C_1, we have

$$\begin{aligned} I(C_1) &= (1+i)\int_0^1 \cosh x \sin x dx - (1-i)\int_0^1 \sinh x \cos x dx \\ &= [\cosh x \cos x]_0^1 + [\sinh x \sin x]_0^1 \\ &= (1 - \cosh 1 \cos 1) + i(\sinh 1 \sin 1), \end{aligned} \tag{7.27}$$

where we employ partial integrations. Next we evaluate I along C_2. Along the path from O to A, $x = 0$ and $dx = 0$, and along the path from A to B, $y = 1$ and $dy = 0$. Therefore,

$$\begin{aligned} I(C_2) &= \int_{C_2} \sin z dz \\ &= -\int_0^1 \sinh y dy + \int_0^1 \cosh x \sin x dx + i\int_0^1 \sinh x \cos x dx \\ &= (1 - \cosh 1 \cos 1) + i\,(\sinh 1 \sin 1)\,. \end{aligned}$$

Observe that $I(C_1) = I(C_2)$. ♣

2. Set $C : |z| = r$, and calculate the following integrals:

$$\text{(i)} \ \left| \oint_C \frac{dz}{z} \right|, \quad \text{(ii)} \ \oint_C \frac{dz}{|z|}, \quad \text{(iii)} \ \oint_C \frac{|dz|}{z}.$$

Solution: Let $z = re^{i\theta}$, which yields $dz = ire^{i\theta}d\theta$ and $|dz| = rd\theta$. Hence, we have the results: (i) $|\oint_C dz/z| = |\int_0^{2\pi}(ire^{i\theta})/(re^{i\theta})d\theta| = 2\pi$, (ii) $\oint_C dz/|z| = \int_0^{2\pi}(ire^{i\theta})/rd\theta = 0$, (iii) $\oint_C |dz|/z = \int_0^{2\pi} r/(re^{i\theta})d\theta = 0$. ♣

3. Let $f(z)$ be analytic on a unit circle D about the origin. For any two points z_1 and z_2 on D, there exists two points ξ_1 and ξ_2 on the line segment $[z_1, z_2]$ that satisfy the relation

$$f(z_2) - f(z_1) = \{\text{Re}\,[f'(\xi_1)] + i\text{Im}\,[f'(\xi_2)]\}\,(z_2 - z_1). \tag{7.28}$$

Prove it. (This is a generalization of the **mean value theorem** that is valid for real functions.)

Solution: From assumptions, we have

$$\begin{aligned} f(z_2) - f(z_1) &= \int_{z_1}^{z_2} f'(z)dz \quad = (z_2 - z_1)\int_0^1 f'[z_1 + t(z_2 - z_1)]dt \\ &= (z_2 - z_1)\left\{\int_0^1 \text{Re}\,[f'(z_1 + t(z_2 - z_1))]\,dt \right. \\ &\quad \left. + i\int_0^1 \text{Im}\,[f'(z_1 + t(z_2 - z_1))]\,dt\right\}. \end{aligned} \tag{7.29}$$

Note that the integrals in the last line are both real. Hence, they satisfy the mean value theorem for integrals of real-valued functions $g(t)$ that are expressed by

$$\int_0^1 g[z_1 + t(z_2 - z_1)]dt = g[z_1 + c(z_2 - z_1)] \ \text{ when } \ 0 < c < 1.$$

Setting $\xi_k = z_1 + c_k(z_2 - z_1)$ with $0 < c_k < 1$ $(k = 1, 2)$, we get the desired equation (7.28). ♣

7.3 Cauchy Integral Formula and Related Theorem

7.3.1 Cauchy Integral Formula

We now turn to the famous integral formula that is the chief tool in the application of the theory of analytic functions in physics.

♠ **Cauchy integral formula:**
If $f(z)$ is analytic within and on a closed contour C, we have

$$\oint_C \frac{f(z)}{z-a}dz = \begin{cases} 2\pi i f(a) & \text{if } a \text{ is interior to } C, \\ 0 & \text{if } a \text{ is exterior to } C. \end{cases} \tag{7.30}$$

Proof The latter case is trivial; when $z = a$ is exterior to C, the integrand in (7.30) becomes analytic within C so that we have at once $\oint [f(z)/(z-a)]dz = 0$ by virtue of the Cauchy theorem. Hence, we consider below only the case where $z = a$ is within C.

Suppose that the integral

$$J(a) = \oint_C \frac{f(z)}{z-a}dz \tag{7.31}$$

around a closed contour C within and on which $f(z)$ is analytic. In view of the discussion in Sect. 7.2.3, the contour C may be deformed into a small circle of radius r about the point a. Accordingly, the variable z is expressed by $z = a + re^{i\theta}$.

Now, we rewrite (7.31) as

$$J(a) = f(a)\oint_C \frac{dz}{z-a}dz + \oint_C \frac{f(z)-f(a)}{z-a}dz. \tag{7.32}$$

The first integral on the right-hand side becomes

$$\oint_C \frac{dz}{z-a} = \int_0^{2\pi} \frac{ire^{i\theta}}{re^{i\theta}}d\theta = 2\pi i. \tag{7.33}$$

Hence, (7.30) is confirmed if the second integral of (7.32) vanishes for some choice of the radius r of the circle C. To show this, we note the continuity of $f(z)$ at a, which tells us that for all $\varepsilon > 0$ there exists an appropriate quantity δ such that

$$|z-a| < \delta \ \Rightarrow \ |f(z)-f(a)| < \varepsilon.$$

This implies that for any arbitrarily small ε, we can find $r = |z-a|$ that satisfies the relation

$$\left|\oint_C \frac{f(z)-f(a)}{z-a}dz\right| \le \oint_C \frac{|f(z)-f(a)|}{|z-a|}|dz| < \frac{\varepsilon}{\delta}2\pi\delta = 2\pi\varepsilon. \tag{7.34}$$

Thus by taking r small enough, but still greater than zero, the absolute value of the integral can be made smaller than any preassigned number. From (7.32 to 7.34), we obtain the desired equation:

$$\oint \frac{f(z)}{z-a} dz = 2\pi i f(a) \text{ if } a \text{ is within } C. \quad \clubsuit \tag{7.35}$$

Remark. If a is a point located just on the contour C, the integral (7.30) will have the **principal value integral** (see Sect. 9.4.1).

The Cauchy integral formula gives us another hint by which to comprehend the rigid structure of analytic functions: If a function is analytic within and on a closed contour C, its value at every point inside C is determined by its values on the bounding curve C.

7.3.2 Goursat Formula

A remarkable consequence of the Cauchy's integral formula is the fact that, when $f(z)$ is analytic at $z = a$, all of its derivatives are also analytic. Furthermore, the region of analyticity for those derivatives is identical with that of $f(z)$. To prove the theorem, we use the integral representation (7.35) to evaluate the derivative,

$$\begin{aligned} &2\pi i f'(a) \\ &= 2\pi i \lim_{h\to 0} \frac{f(a+h)-f(a)}{h} = \lim_{h\to 0}\left[\frac{1}{h}\oint f(z)\left(\frac{1}{z-a-h}-\frac{1}{z-a}\right)dz\right] \\ &= \lim_{h\to 0}\oint \frac{f(z)}{(z-a)(z-a-h)}dz = \oint \frac{f(z)}{(z-a)^2}dz. \end{aligned} \tag{7.36}$$

The last equality in (7.36) is verified from

$$\begin{aligned} &\oint \left[\frac{f(z)}{(z-a-h)(z-a)} - \frac{f(z)}{(z-a)^2}\right] dz \\ &= h\oint \frac{f(z)}{(z-a)^2(z-a-h)}dz \le \frac{hML}{b^2(b-|h|)}, \end{aligned} \tag{7.37}$$

where M is the maximum value of $|f(z)|$ on the contour, L is the length of the contour, and b is the minimum value of $|z-a|$ on the contour. The right-hand side of the inequality in (7.37) approaches zero as $h \to 0$, so we have

$$\lim_{h\to 0}\oint \left[\frac{f(z)}{(z-a)(z-a-h)} - \frac{f(z)}{(z-a)^2}\right] dz = 0,$$

which ensures the equality of the last part of (7.36).

We can continue with the same process to obtain higher derivatives, arriving at the general formula for the nth derivative of f at $z = a$:

♠ **Goursat formula:**
If $f(z)$ is analytic within and on a closed contour C, we have

$$f^{(n)}(a) = \frac{n!}{2\pi i} \oint \frac{f(z)}{(z-a)^{n+1}} dz \quad (n = 0, 1, 2, \cdots). \tag{7.38}$$

Note that equation (7.38) guarantees the existence of all the derivatives $f'(a), f''(a), \cdots$ and the analyticity at all a's within C.

Remark. The Goursat formula (7.38) is valid only within the contour, and thus gives no information as to the existence of the derivatives *just on* the contour.

7.3.3 Absence of Extrema in Analytic Regions

An additional noteworthy fact associated with **Cauchy's integral formula** (7.30) is that it points up the absence of either a maximum or a minimum of an analytic function within a **region of analyticity**.

For example, if $z = a$ is a point within C, from (7.30) we see that

$$f(a) = \frac{1}{2\pi} \int_0^{2\pi} f(a + \rho e^{i\phi}) d\phi, \tag{7.39}$$

which means that $f(a)$ is the arithmetic average of the values of $f(z)$ on any circle centered at a. We thus have $|f(a)| \leq M$, where M is the maximum value of $|f|$ *just on* the circle. (Equality can occur only if f is constant on the contour.)

The above argument applies to arbitrary points *within* the circle and, further, to a region bounded by any contour C (not necessary a circle). We thus conclude that the inequality $|f(z)| \leq M$ holds for all z within C, which means that $|f(z)|$ has no maximum within the region of analyticity.

Similarly, if $f(z)$ has no zero within C, then $1/f(z)$ is an analytic function inside C and $|1/f(z)|$ has no maximum within C, taking its maximum value on C. Therefore $|f(z)|$ does not have a minimum within C but does have one on the contour C. We thus arrive at the following important theorem.

♠ **Absolute maximum theorem:** If a nonconstant function $f(z)$ is analytic within and on a closed contour C, then $|f(z)|$ can have no maximum within C.

♠ **Absolute minimum theorem:**

If a nonconstant function $f(z)$ is analytic within and on a closed contour C, and if $f(z) \neq 0$ there, then $|f(z)|$ can have no minimum within C.

Accordingly, points at which $df/dz = 0$ are saddle points, rather than true maxima or minima.

We further observe that the theorems apply not only to $|f(z)|$ but also to the real and imaginary parts of an analytic function. To see this, we rewrite (7.39) as

$$2\pi f(a) = 2\pi(u_a + iv_a) \quad \text{and} \quad 2\pi f(a) = \int_0^{2\pi} f(x+iy)d\phi = \int_0^{2\pi} (u+iv)d\phi, \tag{7.40}$$

where u_a and v_a are the values of $u(x,y)$ and $v(x,y)$ at $z = x + iy = a$. Equating the last terms of the two equations in (7.40), we obtain

$$u_a = \frac{1}{2\pi}\int_0^{2\pi} u d\phi \text{ and } v_a = \frac{1}{2\pi}\int_0^{2\pi} v d\phi,$$

so that u_a and v_a are the arithmetic averages of the values of $u(x,y)$ and $v(x,y)$, respectively, on the boundary of the circle. Hence, based on the same reasoning as above, we see that both of u and v take on their minimum and maximum values on the boundary curve of a region within which f is analytic.

7.3.4 Liouville Theorem

We saw in the previous discussion that $|f(z)|$ has its maximum M on the boundary of the region of analyticity of $f(z)$. In certain cases, the maximum of $|f(z)|$ bounds the absolute value of derivatives $|f^{(n)}(z)|$, as stated in the theorem below.

♠ **Cauchy inequality:**

If $f(z)$ is analytic within and on a circle C with a radius r, and $M(r)$ is the maximum of $|f(z)|$ on C, then we have

$$\left|f^{(n)}(z)\right| \leq \frac{n!}{r^n} M(r) \quad \text{within and on } C.$$

This is called the **Cauchy inequality**.

Proof Goursat's formula reads

$$f^{(n)}(z_0) = \frac{n!}{2\pi i}\oint_C \frac{f(z)}{(z-z_0)^{n+1}}dz.$$

Take $|z - z_0| = r$ and use the Darboux inequality to get the desired result:

$$\left|f^{(n)}(z_0)\right| \leq \left|\frac{n!}{2\pi i}\right| \oint_C \frac{|f(z)|}{|(z-z_0)^{n+1}|}|dz| \leq \frac{n!}{2\pi r^{n+1}} M(r) \cdot 2\pi r$$
$$= \frac{n! M(r)}{r^n}. \quad \clubsuit \tag{7.41}$$

If the $f(z)$ we have considered is analytic at all points on the complex plane, i.e., if it is an **entire function**, the above result reduces to the following theorem:

♠ **Liouville theorem:**

If $f(z)$ is an **entire function** and $|f(z)|$ is bounded for all values of z, then $f(z)$ is a constant.

Proof Let $n = 1$ and $M(r) = M$ in (7.41) to obtain

$$|f'(z_0)| < \frac{M}{r}.$$

Since $f(z)$ is an entire functions we may take r as large as we like. Thus we can make $|f'(z_0)| < \varepsilon$ for any preassigned ε. That is, $|f'(z_0)| = 0$, which implies that $f'(z_0) = 0$ for all z_0, so $f(z_0) = \text{const.}$ ♣

Liouville's theorem is a very powerful statement about analytic functions over the complex plane. In fact, if we restrict our attention to the real axis, then it becomes possible to find many *real* functions that are entire and bounded but are *not* constant; $\cos x$ and e^{-x^2} are cases in point. In contrast, there is no such freedom for complex analytic functions; any analytic function is either not bounded (goes to infinity somewhere on the complex plane) or not entire (is not analytic at some points of the complex plane).

7.3.5 Fundamental Theorem of Algebra

The next theorem follows easily from Liouville's theorem and provides a remarkable tie-up between analysis and algebra. In what follows, the points z at which $f(z) = 0$ are called the **zeros** of $f(z)$ or roots of $f(z)$.

♠ **Fundamental theorem of algebra:**

Every nonconstant polynomial of degree n with complex coefficients has n zeros in the complex plane.

Proof Let $P(z)$ be any polynomial. If $P(z) \neq 0$ for all z, then the function $f(z) = 1/P(z)$ is entire. Moreover, if P is nonconstant, then $P \to \infty$ as $z \to \infty$ so that f is bounded. Hence, in view of Liouville's theorem, f must be a constant. This result means that P is also a constant, which is contrary to our assumption that P is a nonconstant polynomial. We thus conclude that $P(z)$ has at least one zero in the complex plane.

Furthermore, an induction argument shows that an nth-degree polynomial has n zeros (counting multiplicity; see Remark **1** below). If we assume that every kth-degree polynomial can be written

$$P_k(z) = A(z - \alpha_1) \cdots (z - \alpha_k),$$

it follows that

$$P_{k+1}(z) = A(z - \alpha_0)(z - \alpha_1) \cdots (z - \alpha_k). \quad ♣$$

Remark.

1. The point α is called a **zero of order** k (or **zero of multiplicity** k) of the function $P(z)$ if it reads

$$P(z) = (z - \alpha)^k Q(z),$$

where $Q(z)$ is a polynomial with $Q(\alpha) \neq 0$. Equivalently, α is a zero of order k if

$$P(\alpha) = P'(\alpha) = \cdots = P^{(k-1)}(\alpha) = 0 \quad \text{and} \quad P^{(k)}(\alpha) \neq 0.$$

2. It can be shown that if $f_1(z)$ and $f_2(z)$ are analytic within and on C and if $|f_2(z)| < |f_1(z)| \neq 0$ on C, then $f_1(z)$ and $f_1(z) + f_2(z)$ have the same number of zeros within C. This is called **Rouché's theorem**, which is verified in Sect. 9.3.4.

7.3.6 Morera Theorem

The final important theorem is called Morera's theorem and, is in a sense the converse of Cauchy's theorem.

♠ **Morera theorem:**
Let $f(z)$ be a continuous function on some domain D and suppose that

$$\oint_C f(z)dz = 0$$

for every simple closed curve C in D whose interior also lies in D. Then f is analytic in D.

Proof For some fixed point z_0 in D, define the function

$$F(z) = \int_{z_0}^{z} f(z')dz', \quad z \in D,$$

where the path is along the line segment in D from z_0 to z. From this, we have

$$\frac{F(z+\Delta z) - F(z)}{\Delta z} = \frac{f(z)}{\Delta z}\int_{z}^{z+\Delta z} dz' + \frac{1}{\Delta z}\int_{z}^{z+\Delta z} \{f(z') - f(z)\}\, dz' = f(z),$$

where Darboux's inequality is used in the second term in the limit $\Delta z \to 0$. As a result, we get

$$F'(z) = f(z),$$

which indicates the existence of the first derivative of $F(z)$, so $F(z)$ is analytic in D and $f(z)$ is also analytic. ♣

Exercises

1. Let $f(z)$ be analytic within a circle $D : z = |R|$, and let it satisfy the relations $|f(z)| \le M$ and $f(0) = 0$.

   (i) Prove that

   $$|f(z)| \le \frac{M}{R}|z| \text{ for } z \in D. \tag{7.42}$$

   (ii) Prove that the equality in (7.42) holds at $z = z_0$ if and only if there exists a complex number c that yields $|c| = 1$ and

   $$f(z_0) = c\frac{M}{R}z. \tag{7.43}$$

   Statements **(i)** and **(ii)** constitute the **Schwarz lemma**.

   Solution: **(i)** Equation in (7.42) holds trivially for $z = 0$. For considering the case of $z \neq 0$, we specify the circle $D' : |z| = \rho < R$ and set the function $g(z) = f(z)/z$. Since g is analytic within and on D', it follows from the theorem in Sect. 7.3.3 that

   $$|g(z)| \le \max_{z \in D'} |g(z)| \le \frac{M}{\rho},$$

   which means that

   $$|f(z)| \le \frac{M}{\rho}|z| \text{ for } z \in D'.$$

   By fixing z within D' and taking the limit of ρ to R, we get to (7.42).

   (ii) If the equality in (7.42) holds at some $z_0 \in D$ except at the origin, we have

$$|g(z_0)| = \frac{M}{R} \geq |g(z)| \text{ for } z \in D.$$

It follows again from the theorem in Sect. 7.3.3 that $g(z)$ must be constant within D. Hence, we have

$$g(z) = c\frac{M}{R} \text{ with } |c| = 1.$$

This reduces to the desired result (7.43). ♣

2. Let $f(z)$ be analytic on a domain D and $f(z) \not\equiv 0$. Show that if $f(a) = 0$ with $a \in D$, then it is always possible to find small $\rho > 0$ such that

$$0 < |z - a| < \rho \Rightarrow f(z) \neq 0.$$

This means that zeros of $f(z)$ are necessarily isolated from each other.

Solution: Suppose that $z = a$ is an nth zero of $f(z)$. From the definition of zero of a complex function, there exists an $n \in \boldsymbol{N}$ such that

$$p < n \Rightarrow f^{(n)}(a) \neq 0 \text{ and } f^{(p)}(a) = 0.$$

Hence, the Taylor series of $f(z)$ around $z = a$ reads

$$f(z) = \sum_{p=0}^{\infty} \frac{f^{(n+p)}(a)}{(n+p)!}(z-a)^{n+p} = (z-a)^n g_n(z),$$

where

$$g_n(z) = \sum_{p=0}^{\infty} \frac{f^{(n+p)}(a)}{(n+p)!}(z-a)^p, \text{ so } g_n(a) = \frac{f^{(n)}(a)}{n!} \neq 0.$$

Since $g_n(z)$ is analytic at a, it is continuous there. Thus we can find $\rho > 0$ such that

$$|z - a| < \rho \Rightarrow |g_n(z) - g_n(a)| < \frac{1}{2}\frac{|f^{(n)}(a)|}{n!}.$$

It follows from the triangular inequality that

$$|g_n(z)| > |g_n(a)| - \frac{1}{2}\frac{|f^{(n)}(a)|}{n!} = \frac{1}{2}\frac{|f^{(n)}(a)|}{n!} > 0.$$

This implies that for our choice of ρ,

$$0 < |z - a| < \rho \Rightarrow f(z) = (z-a)^n g_n(z) \neq 0. \clubsuit$$

3. Obtain an alternative form of Cauchy's integral formula expressed by

$$f(z) = f(re^{i\phi}) = \frac{R^2 - r^2}{2\pi} \int_0^{2\pi} \frac{f(re^{i\theta})}{R^2 - 2rR\cos(\theta - \phi) + r^2} d\theta$$

that is valid for $|z| < R$ if $f(z)$ is analytic for $|z| \le R$. This is called **Poisson's integral formula.**

Solution: Consider the function

$$g(\zeta) = \frac{z^*}{R^2 - z^*\zeta} f(\zeta),$$

which is analytic for $|\zeta| \le R$. Hence, for the contour $C : |\zeta| = R$, we have $\oint_C g(\zeta)d\zeta = 0$. Furthermore, Cauchy's integral formula tells us that $\oint_C f(\zeta)/(\zeta - z)d\zeta = 0$. From these two results, we obtain

$$\frac{1}{2\pi i} \oint_C \left(\frac{1}{\zeta - z} + \frac{z^*}{R^2 - z^*\zeta} \right) f(\zeta)d\zeta = \frac{R^2 - |z|^2}{2\pi i} \oint_C \frac{f(\zeta)}{(\zeta - z)(R^2 - z^*\zeta)} d\zeta = 0. \tag{7.44}$$

Setting $z = re^{i\phi}$ and $\zeta = Re^{i\theta}$, we have

$$\begin{aligned}(\zeta - z)(R^2 - z^*\zeta) &= \left(Re^{i\theta} - re^{i\phi}\right)\left(R^2 - re^{-i\phi}Re^{i\theta}\right) \\ &= R^2 e^{i\theta}\left[R^2 - 2rR\cos(\theta - \phi) + r^2\right].\end{aligned}$$

Substituting in (7.44), we arrive at the desired formula. ♣

7.4 Series Representations

7.4.1 Circle of Convergence

We now turn to a very important notion: series representations of complex analytic functions. To begin with, we note (without proof) that most of the definitions and theorems in connection with the convergence of series of *real* numbers and real functions presented in Chap. 2 and 3 can be applied to complex counterparts with little or no change. Here we give a basic theorem regarding the convergence property of infinite power series consisting of complex numbers.

♠ **Theorem:**
If the power series

$$\sum_{n=0}^{\infty} a_n z^n \tag{7.45}$$

converges at $z = z_0 \neq 0$, then it converges absolutely at every point of $|z| < |z_0|$ and, furthermore, it converges uniformly for $|z| \le \rho$ where $0 < \rho < |z_0|$.

Proof We first prove the statement regarding absolute convergence. From hypothesis, we see that the series $\sum_{n=0}^{\infty} a_n z_0^n$ converges. We set

$$s_n = \sum_{k=0}^{n} a_k z_0^k,$$

to obtain

$$|s_n - s_{n-1}| = |a_n z_0^n| \to 0 \quad (n \to \infty).$$

Hence, there exists an integer $M > 0$ that satisfies

$$|a_n z_0^n| \leq M \quad \text{for all } n,$$

which implies

$$\sum_{n=0}^{\infty} |a_n z^n| = \sum_{n=0}^{\infty} |a_n z_0^n| \left|\frac{z}{z_0}\right| \leq M \sum_{n=0}^{\infty} \left|\frac{z}{z_0}\right|.$$

Therefore, if $|z| < |z_0|$, the right-hand side converges so that the series (7.45) converges absolutely.

Next we consider uniform convergence. For every z satisfying the relation $|z| \leq \rho < |z_0|$, we have

$$\sum_{n=0}^{\infty} |a_n z^n| \leq M \sum_{n=0}^{\infty} \frac{\rho^n}{|z_0|^n},$$

since $0 < \rho/|z_0| < 1$. In view of the Weierstrass M-test, we conclude that the series (7.45) converges uniformly on the region of $|z| \leq \rho$. ♣

This theorem states that converging behavior of power series

$$\sum_{n=0}^{\infty} a_n z^n \tag{7.46}$$

can be classified into the following three types:

1. It converges at all z.
2. It converges (ordinary and thus absolutely) at $|z| < R$, but diverges at $|z| > R$, in which the real constant R depends on the feature of the series.
3. It diverges at all z except the origin.

This classification leads us to introduce the concept of **radius of convergence** R of the power series (7.46). For the above three cases, it becomes

1. $R = 0$, **2.** R itself, **3.** $R = \infty$,

respectively. The circle C with the radius R about the origin is called the **circle of convergence** associated with the series. Note that just on C, converging behavior of the corresponding series is inconclusive—it may or may not converge.

The following theorems provide us with a clue for finding the radius of convergence of a given power series.

♠ **Theorems:**
Given a power series $\sum_{n=0}^{\infty} a_n z^n$, its radius of convergence R equals

(i) $R = \lim_{n\to\infty} \left| \frac{a_n}{a_{n+1}} \right|$, if the limit exists;

(ii) $R = \frac{1}{\limsup_{n\to\infty} \sqrt[n]{|a_n|}}$.

7.4.2 Singularity on the Radius of Convergence

Given a complex-valued power series, the convergence criterion based on the radius of convergence discussed in the previous subsection does not provide us with any information about the convergence property of the series just on the circle of convergence. We present below two important theorems regarding the latter point.

♠ **Theorem:**
If the power series $\sum_{n=0}^{\infty} a_n z^n$ has a radius of convergence R, then it has at least one singularity on the circle $|z| = R$.

Proof Set

$$f(z) = \sum_{n=0}^{\infty} a_n z^n.$$

If $f(z)$ were analytic at every point on the circle of convergence, then for each z with $|z| = R$, there would exist some maximal ε_{z_0} such that $f(z)$ could be **continued analytically** to a circular region $|z - z_0| < \varepsilon_{z_0}$ where z_0 is located on the circle $|z| = R$. (See Sect. 8.3 for details of **analytic continuation**.) Here ε_{z_0} would depend on z_0 and we define

$$\varepsilon \equiv \min_{|z_0|=R} \varepsilon_{z_0} > 0.$$

By performing continuations successfully for all possible z_0, we obtain a function $g(z)$ that is analytic for $|z| < R + \varepsilon$. Clearly for $|z| < R$, g must be identical to f. In addition, g must have a power series representation,

$$g(z) = \sum_{n=0}^{\infty} b_n z^n, \tag{7.47}$$

that is convergent for $|z| < R + \varepsilon$. Yet since for $|z| < R$

$$g(z) = f(z) = \sum_{n=0}^{\infty} a_n z^n,$$

we conclude that

$$a_n \equiv b_n.$$

This implies that the radius of convergence of (7.47) would be R, which clearly gives us a contradiction. We thus conclude that $f(z)$ has at least one singularity on the circle $|z| = R$. ♣

In general, it is difficult to determine when a function has a singularity at a particular point on the circle of convergence of its power series. The following theorem is one of the few results we have in this direction.

♠ **Theorem:**
Suppose that a power series $\sum_{n=0}^{\infty} a_n z^n$ has a radius of convergence $R < \infty$ and that $a_n \geq 0$ for all n. Then the series has a singularity at $z = R$ on the real axis.

Proof By the previous theorem, the function

$$f(z) = \sum_{n=0}^{\infty} a_n z^n$$

has a singularity at some point $Re^{i\alpha}$. If we consider the power series for f about a point $\rho e^{i\alpha}$ with $0 < \rho < R$, we have

$$f(z) = \sum_{n=0}^{\infty} b_n \left(z - \rho e^{i\alpha}\right) = \sum_{n=0}^{\infty} \frac{f^{(n)}(\rho e^{i\alpha})}{n!} \left(z - \rho e^{i\alpha}\right)^n,$$

where the radius of convergence is $R - \rho$. (If it were larger, the power series would define an **analytic continuation** of f beyond $Re^{i\alpha}$.) Note, however, that for any nonnegative integer j, the derivative $f^{(j)}$ reads

$$f^{(j)}(\rho e^{i\alpha}) = \sum_{n=j}^{\infty} n(n-1)\cdots(n-j+1) a_n (\rho e^{i\alpha})^{n-j}.$$

Since $a_n \geq 0$, we have

$$\left| f^{(j)}(\rho e^{i\alpha}) \right| \leq f^{(j)}(\rho).$$

This implies that the power series representation of f around, $z = \rho$, expressed by

$$f(z) = \sum_{n=0}^{\infty} \frac{f^{(n)}(\rho)}{n!} (z - \rho)^n$$

must have a radius of convergence $R-\rho$. On the other hand, if f were analytic at $z=R$, the above power series would converge on a disc of radius greater than $R-\rho$. Therefore, f is singular at $z=R$. ♣

7.4.3 Taylor Series

Below is the one of the main theorems of this section, which states that any analytic function can be expanded into a power series around its analytic point.

♠ **Taylor series expansion:**

If $f(z)$ is analytic within and on the circle C of radius r around $z=a$, then there exists a unique and uniformly convergent series in powers of $(z-a)$,

$$f(z)=\sum_{k=0}^{\infty} c_k (z-a)^k \quad (|z-a| \le r), \tag{7.48}$$

with

$$c_k = \frac{f^{(k)}(a)}{k!} = \frac{1}{2\pi i}\oint_C \frac{f(\zeta)}{(\zeta-a)^{k+1}} d\zeta.$$

The largest circle C for which the power series (7.48) converges is called the **circle of convergence** of the power series and its radius is called the **radius of convergence**.

Proof Let $f(z)$ be analytic within and on a closed contour C. From Cauchy's integral formula, we have

$$f(a+h) = \frac{1}{2\pi i}\oint_C \frac{f(z)}{z-a-h} dz, \tag{7.49}$$

where a is inside a contour C. The contour is taken to be a circle about a, inasmuch as the region of convergence of the resulting series is circular. We employ the identity

$$\left[1+\frac{h}{z-a}+\frac{h^2}{(z-a)^2}+\cdots+\frac{h^{N-1}}{(z-a)^{N-1}}\right]\left(\frac{z-a-h}{z-a}\right) = 1-\frac{h^N}{(z-a)^N}$$

to obtain the exact expression

$$\frac{1}{z-a-h} = \sum_{n=0}^{N-1}\left[\frac{h^n}{(z-a)^{n+1}}\right] + \frac{h^N}{(z-a-h)(z-a)^N}.$$

Substituting this into (7.49), we have

$$f(a+h) = \sum_{n=0}^{N-1} \frac{h^n}{2\pi i}\oint_C \frac{f(z)}{(z-a)^{n+1}} dz + \frac{h^N}{2\pi i}\oint_C \frac{f(z)}{(z-a)^N (z-a-h)} dz. \tag{7.50}$$

Since the first integral can be replaced by the nth derivative of f at $z = a$, we have

$$f(a+h) = \sum_{n=0}^{N-1} \frac{h^n}{n!} f^{(n)}(a) + R_N, \tag{7.51}$$

where, R_N is the second term on the right-hand side of (7.50). It follows from (7.51) that if $\lim_{N\to\infty} R_N = 0$, the Taylor series expansion of $f(z)$ around $z = a$ is obtained successfully. This is indeed the case. As $f(z)$ is analytic within and on the contour C, the absolute value of R_N is bounded as

$$|R_N| = \left| \frac{h^N}{2\pi i} \oint_C \frac{f(z)}{(z-a)^N (z-a-h)} dz \right| \le \frac{|h|^N M r}{r^N (r - |h|)}, \tag{7.52}$$

where r is the radius of the circle and M is the maximum value of $|f|$ on the contour. Within the radius r, $|h| < r$ so that

$$\lim_{N\to\infty} R_N = 0.$$

Hence, we have

$$f(a+h) = \sum_{n=0}^{\infty} \frac{h^n}{n!} f^{(n)}(a), \tag{7.53}$$

which holds at any point $z = a + h$ within the circle of radius r. ♣

We note that the series (7.53) converges for large h as far as $|h| < r_c$, since R_N vanishes as $N \to \infty$ for any value of $|h|$ smaller than r_c. Furthermore, as the inequality (7.52) holds whenever $f(z)$ is analytic within and on the circle of radius r_c, the radius of convergence, r, can extend up to the singularity is nearest neighbor to $z = a$. When the extending circle goes beyond the nearest singular point, the inequality becomes invalid so that the Taylor series expansion fails.

7.4.4 Apparent Paradoxes

We have seen that the radius of convergence is determined by the distance to the nearest singularity. Interestingly, this explains some apparent paradoxes that which occur if we restrict our attention only to values of the series along the real axis of z.

A familiar example is the Taylor expansion of $f(z) = 1/(1-z)$ around the origin:

$$\frac{1}{1-z} = 1 + z + z^2 + \cdots . \tag{7.54}$$

Obviously, both sides of (7.54) "blow up" at $z = 1$. At $z = -1$, on the other hand, the right-hand side diverges, whereas the left-hand side has a finite value of $1/2$. Notably, this apparent paradox occurs at all points represented by $z = e^{i\phi}$, i.e., at any point on a unit circle surrounding the origin. The reason for this is clear from the point of view of the radius of convergence. (We leave it to the reader.)

Another example is

$$f(z) = e^{-1/z^2}.$$

Observe that $f^{(n)}(0) = 0$ for any $n = 0, 1, \cdots$, so if one puts this result blindly into the Taylor formula around $z = 0$, one obtains apparent nonsense as $e^{-1/z^2} = 0$. The point here is that $z = 0$ is a singularity, where the Taylor series expansion is prohibited.

These two examples suggest the importance of realizing the difference between the series representing a function and "the function itself." A power series, such as a Taylor series, has only a limited range of representation characterized by the radius of convergence. Beyond this range, the power series is unable to represent the function. For example, the function considered in (7.54),

$$f(z) = \frac{1}{1-z}, \tag{7.55}$$

exists and is analytic everywhere except at $z = 1$, but its power series around $z = 0$, given by

$$1 + z + z^2 + \cdots$$

exists and represents f *only* within the unit circle centered at the origin (i.e., $|z| < 1$). The region in which a power series reproduces its original function is dependent on the explicit form of the series expansion. In fact, an alternative series expansion of (7.55) around $z = 3$ is given by

$$-\frac{1}{2} + \frac{1}{4}(z-3) - \frac{1}{8}(z-3)^2 + \cdots,$$

which exists and represents (7.55) only within the circle of radius 2 centered at $z = 3$. We thus conclude that power series (including Taylor's, Laurent's, and others) are not regarded as pieces of a versatile mold by means of which one can cast a copy of the function. Each piece of the mold can reproduce the behavior of f only within the region where the series converges, but gives no indication of the shape of f beyond its range.

7.4.5 Laurent Series

When expanding a function $f(z)$ around its singular point $z = a$, Taylor's expansion is obviously not suitable but we can obtain an alternative expansion that is valid for a singular point. The latter kind of expansion is called a **Laurent series expansion**. Laurent series enter quite often in mathematical analyses of physical problems, where functions to be considered have a finite number of singularities.

♠ **Laurent series expansions:**

Let $f(z)$ be analytic within and on a closed contour C except at a point $z = a$ enclosed by C. Then, $f(z)$ can be expanded around $z = a$ as

$$f(z) = \sum_{n=-\infty}^{\infty} c_n (z-a)^n, \tag{7.56}$$

with the definition

$$c_n = \frac{1}{2\pi i} \oint_C \frac{f(\zeta)}{(\zeta - a)^{n+1}} d\zeta. \tag{7.57}$$

The series (7.56) with the constants (7.57) is called the Laurent series expansion of $f(z)$.

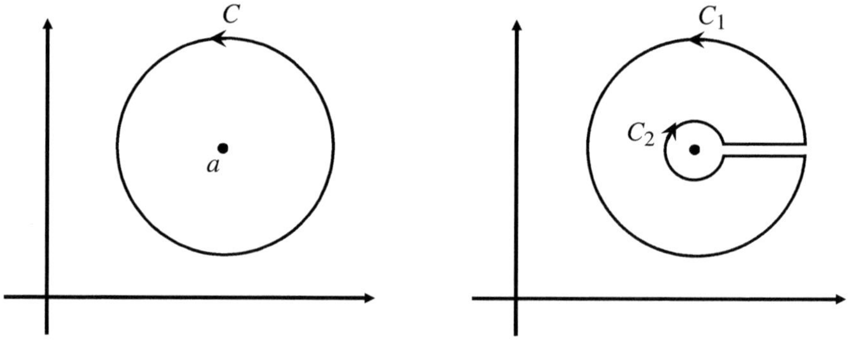

Fig. 7.12. Conversion of a closed contour C into $C_1 + C_2$ so as not to involve the singularity of $f(z)$ at $z = a$ in it

Proof The trick to deriving a, Laurent series expansion is to use the contour $C_1 + C_2$ illustrated in Fig. 7.12 such that its interior does not contain the singular point of $f(z)$ at $z = a$ (i.e., f is analytic within and on the contour). As is indicated, the original contour C can be reduced to two circular contours C_1 and C_2 encircling $z = a$ counterclockwise and clockwise, respectively. Applying Cauchy's theorem, we have

$$\begin{aligned} f(a+h) &= \frac{1}{2\pi i} \oint_C \frac{f(z)}{z-a-h} dz \\ &= \frac{1}{2\pi i} \oint_{C_1} \frac{f(z)}{z-a-h} dz - \frac{1}{2\pi i} \oint_{C_2} \frac{f(z)}{z-a-h} dz. \end{aligned} \tag{7.58}$$

Note that $|z-a| > |h|$ on the contour C_1 and $|z-a| < |h|$ on C_2. We thus have

$$\frac{1}{z-a-h} = \frac{1}{z-a} \cdot \frac{1}{1-\frac{h}{z-a}} = \frac{1}{z-a} \sum_{n=0}^{\infty} \left(\frac{h}{z-a} \right)^n \quad \text{on } C_1 \tag{7.59}$$

and

$$\frac{1}{z-a-h} = \frac{1}{h} \cdot \frac{1}{\frac{z-a}{h}-1} = \frac{-1}{h} \sum_{n=0}^{\infty} \left(\frac{z-a}{h} \right)^n \quad \text{on } C_2. \tag{7.60}$$

The substitution of these two expressions into (7.58) yields

$$f(a+h) = \frac{1}{2\pi i} \left[\oint_{C_1} \sum_{n=0}^{\infty} \frac{h^n}{(z-a)^{n+1}} f(z)dz + \oint_{C_2} \sum_{n=1}^{\infty} \frac{(z-a)^{n-1}}{h^n} f(z)dz \right]. \tag{7.61}$$

The order of integration and summation within the square brackets can be reversed since the infinite series involved in the integrals converge. Eventually, we obtain

$$f(a+h) = \sum_{n=-\infty}^{\infty} c_n h^n; \quad c_n = \frac{1}{2\pi i} \oint_C \frac{f(z)}{(z-a)^{n+1}} dz. \tag{7.62}$$

Here, the contour for the coefficients c_n should be C_1 in the positive direction for $n \geq 0$ and C_2 in the negative direction for $n < 0$. The series (7.62) is what we call the Laurent series expansion of $f(z)$ around the singular point $z = a$. Note that C_1 can be taken as the contour for all values of n with the reverse direction for negative n's. This is because the integrand is analytic in the region between C_1 and C_2, which allows us to expand the size of the contour C_2 until it coincides with the larger contour C_1. ♣

7.4.6 Regular and Principal Parts

An important property of Laurent series is the series resolution. To see this, we rewrite (7.62) as follows:

$$f(a+h) = \sum_{n=0}^{\infty} c_n h^n + \sum_{n=1}^{\infty} c_{-n} h^{-n}. \tag{7.63}$$

The first term in (7.63) converges everywhere within the outer circle of convergences, whereas the second term converges anywhere outside the inner circle. This means that the Laurent series expansion resolves the original function $f(z)$ into two parts: one that is analytic within the outer circle of convergence, and the other that is analytic outside the inner circle of convergence. Obviously, each part is analytic over different portions of the complex plane.

The part of the Laurent series consisting of positive powers of h is called the **regular part**. The other part, consisting of negative powers, is called the **principal part**. Either part (or both) may terminate at a finite degree of the sum or be identically zero. Particularly when the principal part is identically zero, then $f(z)$ is analytic at $z = a$, and the Laurent series is identical with the Taylor series.

Remark. At first glance, the regular part exhibited in (7.63) resembles the Taylor series. However, this is not the case; the nth coefficient cannot generally be associated with $f^{(n)}(a)$ because the latter may not exist. In most applications, $f(z)$ is not analytic at $z = a$.

7.4.7 Uniqueness of Laurent Series

Taylor and Laurent series allow us to express an analytic function as a power series. For a Taylor series of $f(z)$, the expansion is routine because the coefficient of its n term is simply $f^{(n)}(z_0)/n!$, where z_0 is the center of the circle of convergence. In contrast, for the case of a Laurent series expansion, the nth coefficient is not (in general) easy to evaluate. It can usually be found by inspection and certain manipulations of other known series, but if we use such an intuitive approach to determine the coefficients, we cannot be sure that the result we obtain is correct. The following theorem addresses this issue.

♠ **Theorem:**

If the series

$$\sum_{n=-\infty}^{\infty} a_n(z-z_0)^n \tag{7.64}$$

converges to $f(z)$ at all points in some annular region around z_0, then it is the unique Laurent series expansion of $f(z)$ in that region.

Proof Multiply both sides of (7.64) by

$$\frac{1}{2\pi i(z-z_0)^{k+1}},$$

integrate the result along a contour C in the annular region, and use the easily verifiable fact that

$$\frac{1}{2\pi i}\oint_C \frac{dz}{(z-z_0)^{k-n+1}} = \delta_{kn}$$

to obtain

$$\frac{1}{2\pi i}\oint_C \frac{f(z)}{(z-z_0)^{k+1}} = a_k.$$

Thus, the coefficient a_k in the power series (7.64) is precisely the coefficient in the Laurent series of $f(z)$ given in (7.57), and the two must be identical. ♣

Remark. A Laurent series is unique only for a specified annulus. In general, a function $f(z)$ can possess two or more entirely different Laurent series about a given point, valid for different (nonoverlapping) regions; For instance,

$$f(z) = \frac{1}{z(1-z)} = \begin{cases} \dfrac{1}{z} + 1 + z + z^2 + \cdots, & 0 < |z| < 1, \\ -\dfrac{1}{z^2} - \dfrac{1}{z^3} - \dfrac{1}{z^4} - \cdots, & 1 < |z| < \infty. \end{cases}$$

7.4.8 Techniques for Laurent Expansion

The following examples illustrate several useful techniques for the construction of Taylor and Laurent series.

(a) Use of geometric series

Suppose that a function

$$f(z) = \frac{1}{z-a} \tag{7.65}$$

fails to be analytic at $z = a$. We would like to obtain the Laurent series of $f(z)$ around $z = a$. First we note that for $|z| < |a|$, $f(z)$ reads

$$\frac{1}{z-a} = -\frac{1}{a}\frac{1}{1-(z/a)} = -\frac{1}{a}\sum_{n=0}^{\infty}\left(\frac{z}{a}\right)^n. \tag{7.66}$$

This is obviously the Taylor series expansion of $f(z)$ around the point $z = 0$. That is, for $|z| < |a|$, the Laurent series of $f(z)$ given in (7.65) becomes identical to its Taylor series. Nevertheless this is not the case for $|z| > |a|$, since its radius of convergence is $R = |a|$. Hence, we should also evaluate the Laurent series around $z = a$ that is valid for $|z| > |a|$. In a similar manner as above, we obtain

$$\frac{1}{z-a} = \frac{1}{z}\sum_{n=0}^{\infty}\left(\frac{a}{z}\right)^n = \sum_{n=0}^{\infty}\frac{a^n}{z^{n+1}} \quad \text{for } |z| > |a|. \tag{7.67}$$

Expansions (7.66) and (7.67) both serve as the Laurent series expansions of $f(z)$, although the regions of convergence are different from one another.

Remark. The function $f(z)$ given in (7.65) can be expanded by this method about any point $z = b$; Indeed, write

$$f(z) = \frac{1}{z-a} = \frac{1}{(z-b)-(a-b)} \quad (b \neq a).$$

Then, either

$$f(z) = -\frac{1}{a-b}\sum_{n=0}^{\infty}\frac{(z-b)^n}{(a-b)^n} \quad (|z-b| < |a-b|)$$

or

$$f(z) = \sum_{n=0}^{\infty}\frac{(a-b)^n}{(z-b)^{n+1}} \quad (|z-b| > |a-b|)\,.$$

(b) Rational fraction decomposition

Next we assume a function

$$f(z) = \frac{1}{z^2-(2+i)z+2i}.$$

The roots of the denominator are $z = i$ and $z = 2$, which are the only points at which $f(z)$ fails to be analytic. Hence, $f(z)$ has a Taylor series about $z = 0$ that is valid for $|z| < 1$ and two Laurent series about $z = 0$ that are valid for $1 < |z| < 2$ and $|z| > 2$. To obtain them, we use the identities

$$z^2-(2+i)z+2i = (z-i)(z-2)$$

and

$$f(z) = \frac{1}{(z-i)(z-2)} = \frac{1}{2-i}\left(\frac{1}{z-2}-\frac{1}{z-i}\right).$$

When we want the Laurent series of $f(z)$ around $z = 0$ that is valid for $1 < |z| < 2$, it suffices to expand the function $1/(z-2)$ in the Taylor series about $z = 0$ [see **(a)** above] and then expand $1/(z-i)$ in the Laurent series about $z = 0$ that is valid for $|z| > 1$. (The latter series is also valid for $1 < |z| < 2$.) If these two series are subtracted, we obtain a series for $f(z)$ that is valid for $1 < |z| < 2$, which is the desired Laurent series.

(c) Differentiation

The method used in **(b)** fails for functions with a double root in the denominator such that

$$f(z) = \frac{1}{(z-1)^2}.$$

Among alternative methods, the simplest one is the differentiation

$$\frac{1}{(z-1)^2} = \frac{d}{dz}\left(\frac{1}{1-z}\right).$$

From the discussions regarding the earlier case **(a)**, the function $1/(1-z)$ is seen to be represented by

$$\frac{1}{1-z} = \begin{cases} \displaystyle\sum_{n=0}^{\infty} z^n, & |z| < 1, \\ \displaystyle -\sum_{n=0}^{\infty} \frac{1}{z^{n+1}}, & |z| > 1. \end{cases}$$

Hence, term-by-term differentiations yield

$$\frac{1}{(z-1)^2} = \begin{cases} \displaystyle\sum_{n=0}^{\infty} (n+1)z^n, & |z| < 1, \\ \displaystyle -\sum_{n=0}^{\infty} (n+1)z^{-(n+2)}, & |z| > 1. \end{cases}$$

Exercises

1. Let $f(z)$ be an entire function. Employ the Taylor series expansion to show that the function defined by

$$g(z) = \begin{cases} \dfrac{f(z)-f(a)}{z-a}, & z \neq a, \\ f'(a), & z = a. \end{cases}$$

is also entire.

Solution: For $z \neq a$, we employ the Taylor series expansion of $f(z)$ to obtain

$$g(z) = f'(a) + \frac{f''(a)}{2!}(z-a) + \frac{f^{(3)}(a)}{3!}(z-a)^2 + \cdots. \tag{7.68}$$

By the definition of g, the representation (7.68) is valid at $z = a$. Hence, g is equal to an everywhere-convergent power series and is thus an entire function. ♣

2. If f is entire and if for some integer $k \geq 0$ there exist positive constants A and B such that

$$|f(z)| \leq A + B|z|^k,$$

then f is a polynomial of degree k at most. Prove it

Solution: Note that the case $k = 0$ is the original Liouville theorem. To prove the case of $k > 0$, we employ mathematical induction, and consider

$$g(z) = \begin{cases} \dfrac{f(z) - f(0)}{z}, & z \neq 0, \\ f'(0), & z = 0, \end{cases} \tag{7.69}$$

where $f(z)$ is assumed to obey the conditions noted above. By Exercise **1**, g is entire. In addition, by hypothesis on f we have

$$|g(z)| \leq C + D|z|^{k-1}.$$

Hence, by induction, g is a polynomial of degree $k - 1$ at most, then f is polynomial of degree k at most owing to the definition (7.69). This completes the proof. ♣

3. Find the Laurent series of the multivalued logarithmic function given by

$$f(z) = \log(1+z) = \log|1+z| + i\arg(1+z).$$

Solution: The **branch cut** (see Sect. 8.2.3) is set so as to extend from $-\infty$ to -1 along the real axis. Hence, $\log(1+z)$ is analytic within the circle $|z| = 1$. Since

$$\frac{d}{dz}\log(1+z) = \frac{1}{1+z},$$

we may expand

$$\frac{1}{1+z} = 1 - z + z^2 - z^3 + \cdots = \sum_{n=0}^{\infty}(-1)^n z^n \quad (|z| < 1).$$

Then, term-by-term integration yields

$$\int^z \frac{d\xi}{1+\xi} = z - \frac{z^2}{2} + \frac{z^3}{3} - \cdots + C \quad (|z| < 1),$$

where C is the constant of integration. Since $\log 1 = 0$, it follows that $C = 0$ and

$$\log(1+z) = z - \frac{z^2}{2} + \frac{z^3}{3} - \cdots = \sum_{n=1}^{\infty}(-1)^{n+1}\frac{z^n}{n} \quad (|z| < 1).$$

Other branches of $\log(1+z)$ have the same series except for different values of the constant C. ♣

4. Find the power series representation of $f(z)$ about $z = 0$ that satisfies the differential equation

$$f'(z) + f(z) = 0 \quad \text{with} \quad f(0) = 1. \tag{7.70}$$

Solution: Let $f(z) = 1 + \sum_{n=1}^{\infty} a_n z^n$. Then we have $f'(z) = \sum_{n=1}^{\infty} n a_n z^{n-1}$ $= a_1 + \sum_{n=1}^{\infty} (n+1) a_{n+1} z^n$. Substitute this into (7.70) to obtain

$$1 + a_1 = 0 \quad \text{and} \quad a_n + (n+1)a_{n+1} = 0 \ \text{ for } \ n \geq 1.$$

The latter result yields

$$a_n = (-1)\frac{1}{n} a_{n-1} = (-1)^2 \frac{1}{n(n-1)} a_{n-2} = \cdots = (-1)^n \frac{1}{n!} a_1.$$

Hence, we have $a_n = (-1)^n / n!$, so that

$$f(z) = 1 + \sum_{n=1}^{\infty} \frac{(-1)^n}{n!} z^n = e^{-z}. \quad \clubsuit$$

5. Let $f(z) = \sum_{n=0}^{\infty} c_n (z-a)^n$ be analytic for $|z - a| < R$. Prove that

$$\frac{1}{2\pi} \int_0^{2\pi} \left| f(a + re^{i\theta}) \right|^2 d\theta = \sum_{n=0}^{\infty} |c_n|^2 r^{2n} \quad \text{for any} \ \ r < R.$$

Then show that

$$\sum_{n=0}^{\infty} |c_n|^2 r^{2n} \leq M(r)^2, \tag{7.71}$$

in which $M(r) = \max_{|z-a|=r} |f(z)|$. The result (7.71) is called **Gutzmer's theorem**.

Solution: From assumption, it follows that

$$|f(z)|^2 = \left[\sum_{n=0}^{\infty} c_n (re^{i\theta})^n \right] \left[\sum_{m=0}^{\infty} c_m^* (re^{-i\theta})^m \right] = \sum_{n,m=0}^{\infty} c_n c_m^* r^{m+n} e^{i(n-m)\theta}.$$

This infinite series converges uniformly on the circle $|z - a| = r < R$, which allows us to interchange the order of integration and summation as expressed by

$$\int_0^{2\pi} |f(a + re^{i\theta})|^2 d\theta = \sum_{n,m=0}^{\infty} c_n c_m^* r^{m+n} \int_0^{2\pi} e^{i(n-m)\theta} d\theta.$$

The right-hand side vanishes when $n \neq m$ since the integral equals zero. Hence, we have

$$\int_0^{2\pi} |f(a + re^{i\theta})|^2 d\theta = \sum_{n=0}^{\infty} |c_n|^2 r^{2n} \times 2\pi,$$

which is equivalent to the desired equation. Furthermore, since $|f(a + re^{i\theta})| \le M(r)$, we have

$$\sum_{n=0}^{\infty} |c_n|^2 r^{2n} = \frac{1}{2\pi} \int_0^{2\pi} |f(a+re^{i\theta})|^2 d\theta \le \frac{1}{2\pi} \int_0^{2\pi} M(r)^2 d\theta = M(r)^2. \quad ♣$$

7.5 Applications in Physics and Engineering

7.5.1 Fluid Dynamics

This section demonstrates the effectiveness of using complex function theory for analyzing fluid dynamics in a two-dimensional plane. The primary aim is to derive the **Kutta–Joukowski theorem** (see Sect. 7.5.2), which describes the **lift force** exerted on a solid material placed in a uniform flow. Before proceeding, we introduce terminologies and several basic concepts that pertain to fluid dynamics.

The fundamental quantities that characterize a two-dimensional fluid flow are velocity $\boldsymbol{v} = u\boldsymbol{e}_x + v\boldsymbol{e}_y$ and vorticity $\boldsymbol{\omega} = \nabla \times \boldsymbol{v}$, both of which are vector-valued functions of the position $\boldsymbol{r}$. Here, we restrict our attention to the case of an **irrotational** ($\boldsymbol{\omega} = \boldsymbol{0}$) and **incompressible** ($\nabla \cdot \boldsymbol{v} = 0$) fluid. The assumption $\boldsymbol{\omega} = \nabla \times \boldsymbol{v}$ allows us to define an appropriate function $\Phi(x, y)$ such that

$$\boldsymbol{v} = \nabla \Phi, \tag{7.72}$$

since $\nabla \times (\nabla f) = \boldsymbol{0}$ for any analytic function $f(x, y)$ in the x-y plane. The function $\Phi(x, y)$ defined by (7.72) is called the **velocity potential**. Further, our assumption of $\nabla \cdot \boldsymbol{v} = 0$ implies that

$$\frac{\partial u}{\partial x} + \frac{\partial v}{\partial y} = 0,$$

which in turn suggests the presence of an analytic function $\Psi(x, y)$ defined by

$$u \equiv \frac{\partial \Psi}{\partial y}, \quad v \equiv -\frac{\partial \Psi}{\partial x} \tag{7.73}$$

that satisfies the two-dimensional Laplace equation $\nabla^2 \Psi = 0$. Such a function $\Psi(x, y)$ is called a **stream function**.

Remark. The name *stream function* originates from the fact that the curves of $\Psi(x, y) = \text{const.}$ in the x-y plane represent streamline flow. This is shown by noting that if $d\Psi = 0$, we have

$$d\Psi = \frac{\partial \Psi}{\partial x} dx + \frac{\partial \Psi}{\partial y} dy = -v dx + u dy = 0,$$

so that $dx/u = dy/v$, which implies that $d\boldsymbol{r}$ is parallel to $\boldsymbol{v}$.

From (7.72) to (7.73), it follows that the components of the velocity $\boldsymbol{v}$ are expressed as

$$u = \frac{\partial \Phi}{\partial x} = \frac{\partial \Psi}{\partial y}, \quad v = \frac{\partial \Phi}{\partial y} = -\frac{\partial \Psi}{\partial x}.$$

This allows us to introduce the concept of a **complex velocity potential** $f(z)$ in the complex plane:

$$f(z) = \Phi(z) + i\Psi(z) \quad \text{with} \quad z = x + iy. \tag{7.74}$$

Note that since $f(z)$ is analytic,

$$\frac{\partial f}{\partial x} = \frac{df}{dz} = u - iv = |\boldsymbol{v}| e^{-i\theta},$$

i.e., the absolute value of the derivative $|df/dz|$ gives the magnitude of the velocity $|\boldsymbol{v}|$. Furthermore, the contour integral of $f(z)$ has important physical implications. Given a closed contour C placed on a two-dimensional flow, we have

$$\oint_C df = \Gamma(C) + iQ(C),$$

where

$$\Gamma(C) = \oint_C d\Phi = \oint_C (u dx + v dy) = \oint_C \boldsymbol{v} \cdot d\boldsymbol{r},$$

$$Q(C) = \oint_C d\Psi = \oint_C (u dy - v dx) = \oint_C |\boldsymbol{v} \times d\boldsymbol{r}|.$$

Hence, the integrals $\Gamma(C)$ and $Q(C)$ represent the **circulation** (or **rotation**) and the **fluid flow**, respectively.

7.5.2 Kutta–Joukowski Theorem

We are now ready to study the Kutta–Joukowski theorem, which describes the lift force in a two-dimensional flow. The lift force is a component of the fluid dynamic force that is perpendicular to the flow direction. It is the lift force that makes it possible for airplanes, helicopters, sail boats, etc. to move against the gravitational force or water currents.

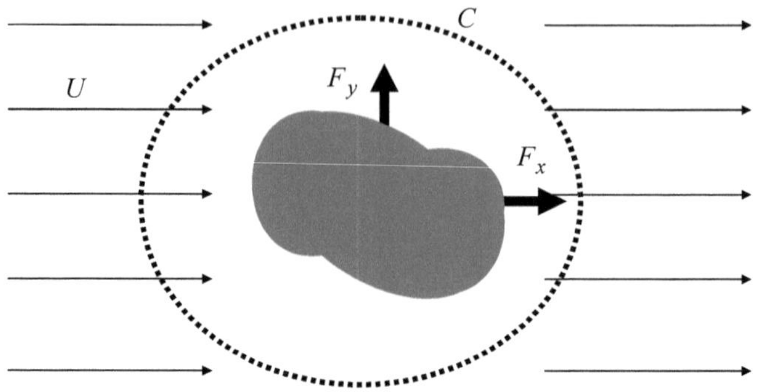

Fig. 7.13. Spatial configuration of material placed into a two-dimensional uniform flow with speed U showing the components F_x, F_y of the flow-induced force $\mathbf{F}$ acting on the material

♠ **Kutta–Joukowski theorem:**

The lift force F_y that acts on a material placed in a uniform flow U in the x-direction is given by

$$F_y = -\rho U \Gamma(C), \tag{7.75}$$

where ρ and $\Gamma(C)$ are the mass density and the circulation of the fluid, respectively, within a closed contour C surrounding the material (see Fig. 7.13).

The lift force is generated in accordance with **Bernoulli's theorem** and the **law of conservation of momentum**. Both of these principles are used to explain the mechanism responsible for the occurrence of the lift force in a uniform flow, which is given by the **Blasius formula** (see 7.5.3):

$$\boldsymbol{F} = \frac{i\rho}{2} \oint_C w^2 dz,$$

which plays a key role in the proof of the Kutta–Joukowski theorem, as shown below.

Proof (of the Kutta–Joukowski theorem). Assume a uniform flow oriented to the x-axis. Then the function $w = df/dz$ is analytic and satisfies the relation

$$\lim_{z\to\infty} w = U = \text{const.}$$

Hence, w can be expanded at points sufficiently far from the origin:

$$w = \frac{df}{dz} = U + \frac{k_0}{z} - \frac{c_1}{z^2} - \frac{2c_2}{z^3} - \cdots \quad (z \to \infty), \tag{7.76}$$

which implies that

$$f = Uz + k_0 \log z + c_0 + \frac{c_1}{z} + \frac{c_2}{z^2} + \cdots \quad (z \to \infty) \tag{7.77}$$

and

$$w^2 = U^2 + \frac{2Uk_0}{z} + \left(k_0^2 - 2Uc_1\right)\frac{1}{z^2} + \cdots \quad (z \to \infty). \tag{7.78}$$

From (7.77) we have

$$\oint_C f dz = 2\pi i k_0 = \Gamma(C) + iQ(C), \tag{7.79}$$

and substituting (7.78) into the **Blasius formula** expressed by $\boldsymbol{F} = (i\rho/2) \oint_C w^2 dz$, we obtain

$$\boldsymbol{F} = F_x + iF_y = \frac{i\rho}{2} \cdot 2\pi i \cdot 2Uk_0 = -2\pi\rho U k_0. \tag{7.80}$$

Combining (7.79) and (7.80) yields

$$F_x + iF_y = \rho U(-Q + i\Gamma),$$

i.e.,

$$F_x = -\rho UQ, \quad F_y = -\rho U\Gamma. \tag{7.81}$$

of the two results above, it is the second one regarding F_y that states the theorem. ♣

Remark. The first equation in (7.81) indicates that $F_x = 0$ if $Q = 0$; i.e. no force in the direction of the stream is relevant to a material inside the closed contour C if no source is located interior to C. This is precisely the case for an ideal flow without any viscosity.

7.5.3 Blasius Formula

We conclude this section by explaining the **Blasius formula**, which is important for the proof of the Kutta–Joukowski theorem discussed above. Consider a two-dimensional flow of irrotational and incompressible fluid and assume that a solid material is placed inside a closed contour C encircling a portion of the fluid. Apparently, a force $\boldsymbol{F}$ from the flow is exerted on the material. Hence, the **law of the conservation of momentum** within the contour C is written as

$$\boldsymbol{F} + \oint_C d\boldsymbol{G} = 0,$$

where $d\boldsymbol{G}$ represents the sum of momentums that pass through a line element ds of the closed contour C per unit time. It is given by

$$d\boldsymbol{G} = p\boldsymbol{n}ds + \rho\boldsymbol{v}v_n ds, \tag{7.82}$$

where p is the fluid pressure, $\boldsymbol{n}$ is a basis vector normal to the contour C, ρ is the density of the fluid, and $v_n = \boldsymbol{v} \cdot \boldsymbol{n}$. The first and second terms on the right-hand side of (7.82) represent the impulse transmitted to the interior of C through ds and the volume of fluid passing through ds, respectively. Using the stream potential Ψ, we rewrite as (7.82)

$$d\boldsymbol{G} = p\boldsymbol{n}ds + \rho\boldsymbol{v}d\Psi, \tag{7.83}$$

since $d\Psi = v_n ds$.

In order to obtain the complex-number representation of (7.83), we denote by dz an infinitesimal vector having length ds and a direction normal to $\boldsymbol{n}$. We then have

$$dz = i(n_x + in_y)ds.$$

when we apply this relation to (7.82), the quantity $d\boldsymbol{G}$ is expressed as

$$dG_x + idG_y = -ipdz + \rho\frac{df^*}{dz^*} \cdot \frac{df - df^*}{2i}, \tag{7.84}$$

where we consider $d\Psi$ to be the imaginary part of df. The pressure p is known to correlate with f via **Bernoulli's theorem**, which is expressed by

$$p = p_0 - \frac{\rho}{2}\left|\frac{df}{dz}\right|^2 = p_0 - \frac{\rho}{2}\frac{df}{dz}\frac{df^*}{dz^*}, \tag{7.85}$$

where p_0 is the pressure at a position far from the material (i.e., $z \to \infty$). It then follows from (7.84) to (7.85) that

$$dG_x + idG_y = -ip_0dz + \frac{i\rho}{2}\frac{df}{dz}\frac{df^*}{dz^*}dz - \frac{i\rho}{2}\frac{df^*}{dz^*}\left(\frac{df}{dz}dz - \frac{df^*}{dz^*}dz^*\right) \tag{7.86}$$

$$= -ip_0dz - \frac{i\rho}{2}\left(\frac{df}{dz}\right)^2 dz. \tag{7.87}$$

Since $\oint_C dz = 0$, we finally obtain

$$\boldsymbol{F} = F_x + iF_y = \frac{i\rho}{2}\oint_C \left(\frac{df}{dz}\right)^2 dz, \tag{7.88}$$

which is known as the **Blasius formula**.

8

Singularity and Continuation

Abstract We devote the first half of this chapter to the essential properties and classification of singularities, which are nonanalytic points in a complex plane. We then describe analytic continuation, which is a most important concept from a theoretical as well as an applied point of view. Through analytic continuations, we observe the interesting fact that the functional form of a complex function may undergo various changes depending on the defining region in the complex plane.

8.1 Singularity

8.1.1 Isolated Singularities

A **singularity** of a complex function $f(z)$ is any point where it is not analytic. In particular, the point $z = a$ is called an **isolated singularity** if and only if $f(z)$ is analytic in some neighborhood but not at $z = a$. Most singularities we have encountered so far in this text were isolated singularities. However, we will see later that there are singularities that are not isolated.

When $z = a$ is an isolated singularity of $f(z)$, it is classified as follows:

1. A **removable singularity** if and only if $f(z)$ is finite throughout a neighborhood of $z = a$, except possibly at $z = a$ itself.
2. A **pole of order** m $(m = 1, 2, \cdots)$ if and only if $(z-a)^m f(z)$ but not $(z-a)^{m-1} f(z)$ is analytic at $z = a$. In this case, $\lim_{z \to a} |f(z)| = \infty$ no matter how z approaches $z = a$.
3. An **essential singularity** if and only if the Laurent series of $f(z)$ around $z = a$ has an infinite number of terms involving negative powers of $(z - a)$.

Remark. There is an alternative definition of a pole: the point $z = a$ is a pole of mth order of $f(z)$ if and only if $1/f(z)$ is analytic and has a **zero of order** m at $z = a$.

The three types of isolated singularities described above can be distinguished by the degree of expansion of the Laurent series of $f(z)$ being considered. Let $f(z)$ have an isolated singularity at $z = a$. Then there is a real number $\delta > 0$ such that $f(z)$ is analytic for $0 < |z - a| < \delta$ but not for $z = a$, which means that $f(z)$ can be represented by the Laurent series

$$f(z) = \sum_{n=0}^{\infty} c_n (z-a)^n + \sum_{n=0}^{M} c_{-n} \frac{1}{(z-a)^n}. \tag{8.1}$$

Thus, it suffices to examine the expansion degree M of the principal part, the second sum in (8.1), in order to determine the type of the isolated singularity $z = a$.

Case 1. Removable singularities ($M = 0$)

In this case, the principal part is absent so that the Laurent series around $z = a$ reads

$$f(z) = c_0 + c_1(z-a) + c_2(z-a)^2 + \cdots \quad (z \neq a).$$

Observe that $\lim_{z\to a} f(z) = c_0$ as is consistent with statement **1** above, which says that $f(z)$ is finite in a neighborhood of $z = a$. This kind of singularity can be eliminated by redefining $f(a)$ as c_0, which is why we call it *removable*.

Examples Consider the function

$$f(z) = \frac{\sin z}{z}. \tag{8.2}$$

This yields $\lim_{z\to 0} f(z) = 1$, but the value of $f(0)$ is not defined. Hence, $z = 0$ is a removable singularity of (8.2). In a similar sense, the functions

$$e^{\sin z/z} \quad \text{and} \quad \frac{1}{z} - \frac{1}{\tan z}$$

are regarded as analytic at $z = 0$, since this point is the removable singularity for each.

Case 2. Isolated poles (M is finite)

The second type of isolated singularity, for which the principal part reads

$$\sum_{n=1}^{M} c_{-n} h^{-n} \quad (c_{-M} \neq 0,\ M \geq 1),$$

is called a **pole** of order M. Order M is the minimum of the integer that makes the quantity

$$\lim_{z\to z_0} (z - z_0)^M f(z)$$

a finite, nonzero complex number.

Examples **1.** The function $f(z) = 1/\sin z$ has Laurent series valid for $0 < |z| < \pi$;

$$\frac{1}{\sin z} = \frac{1}{z} + \frac{z}{6} + \frac{7}{360}z^3 + \frac{31}{15120}z^5 + \cdots,$$

from which it follows that it has a simple pole at the origin.

2. The function $f(z) = 1/z$ has a simple pole at $z = 0$, which is easily seen by noting that $\lim_{z\to 0} zf(z) = 1$.

Case 3: Essential singularities ($M = \infty$)

The third type of isolated singularity, **essential singularity**, gives rise to an infinite principal part.

Examples The function $f(z) = e^{1/z}$ has the Laurent series

$$e^{1/z} = 1 + \frac{1}{z} + \frac{1}{2!\, z^2} + \frac{1}{3!\, z^3} + \cdots,$$

which is valid for $|z| > 0$. Since the principal part is infinite, the function has essential singularity at $z = 0$.

Remark. An infinite principal part in the Laurent series implies essential singularity *only when* the series is valid for all points in a neighborhood $|z - a| < \varepsilon$ except $z = a$. For example, the series

$$f(z) = \frac{1}{(z-1)^2} + \frac{1}{(z-1)^3} + \frac{1}{(z-1)^4} + \cdots$$

does not mean that $z = 1$ is an essential singularity of $f(z)$, since the series converges only if $|z-1| > 1$. It actually represents the function $f(z) = 1/(z^2 - 3z + 2)$ in the annulus $1 < |z-1| < R$, which evidently has a simple pole at $z = 1$.

8.1.2 Nonisolated Singularities

As noted earlier, there are other kinds of singular points that are neither poles nor essential singularities. For example, neither $\sqrt{z}$ nor $\log z$ can be expanded near $z = 0$ in Laurent series; both of them are discontinuous along an entire line (say, the negative real axis) so that the singular point $z = 0$ is not isolated. Singularities of this kind, called **branch points**, are discussed in the next subsection.

Another type of singular behavior of an analytic function occurs when it possesses an infinite number of isolated singularities converging to some limit point. Consider, for instance,

$$f(z) = \frac{1}{\sin(1/z)}.$$

The denominator has simple zeros whenever

$$z = \frac{1}{n\pi} \qquad (n = \pm 1, \pm 2, \cdots).$$

The function $f(z)$ has simple poles at these points and the sequence of these poles converges toward the origin. The origin cannot be regarded as an isolated singularity because every one of its neighborhoods contains at least one pole (actually an infinite number of poles).

8.1.3 Weierstrass Theorem for Essential Singularities

The behavior of a function in the neighborhood of an isolated essential singularity is different from the cases of other isolated singularities such as poles and removable singularities. Most remarkable is the fact that $f(z)$ can be made to take any arbitrary complex value by choosing an appropriate path of $z \to a$. For instance, if z approaches zero along the negative real semiaxis, then the function $f(z) = e^{1/z}$ yields $|f(z)| \to 0$. However, if z approaches zero along the positive real semiaxis, then $|f(z)| \to \infty$. Finally, if z approaches zero along the imaginary axis, then $|f(z)|$ remains constant but $\arg f(z)$ oscillates, and so on. The character of a function near an essential singularity is described by the following theorem:

♠ **Weierstrass theorem:**
In any neighborhood of an isolated essential singularity, an analytic function approaches any given value arbitrarily closely.

Proof We use the contraposition method to prove our theorem. Let $z = a$ be an isolated essential singularity of $f(z)$. We assume for the moment that for $|z-a| < \varepsilon$, $|f(z) - \gamma|$ with a given complex number γ does not become arbitrarily small. Then, the function $[f(z) - \gamma]^{-1}$ is bounded in the region of $|z - a| < \varepsilon$ so that it is possible to find a constant M such that

$$\left| \frac{1}{f(z) - \gamma} \right| < M \text{ for } |z - a| < \varepsilon.$$

Hence, $[f(z) - \gamma]^{-1}$ is analytic for $|z - a| < \varepsilon$ (or at worst has a removable singularity) and can be expanded by

$$\frac{1}{f(z) - \gamma} = b_0 + b_1(z-a) + b_2(z-a)^2 + \cdots. \tag{8.3}$$

If $b_0 \neq 0$, then

$$\lim_{z \to a} \frac{1}{f(z) - \gamma} = b_0 \quad \text{so that} \quad \lim_{z \to a} f(z) = \gamma + \frac{1}{b_0}.$$

This means that $z = a$ is not a singularity of $f(z)$, which contradicts our assumption. Otherwise, if $b_0 = 0$, we have

$$f(z) = \gamma + \frac{1}{(z-a)^k \left[b_k + b_{k+1}(z-a) + \cdots\right]},$$

where b_k is the first nonzero coefficient in the series (8.3). This clearly shows that $z = a$ is a pole of $f(z)$ of kth degree, which again is inconsistent with our assumption. Therefore, we conclude that $|f(z) - \gamma|$ with a given γ can be arbitrarily small in the vicinity of an essential singularity $z = a$. Furthermore, since γ is arbitrary, the function $f(z)$ approaches any given complex value arbitrarily closely. ♣

Remark. The above theorem becomes invalid if the **point at infinity** is taken into account; the point at infinity $z = \infty$ is defined as the point $\overline{z}$ that is mapped onto the origin $z = 0$ by the transformation $\overline{z} = 1/z$. For instance, the function $f(z) = e^z$ has an essential singularity at $z = \infty$ but never approaches zero there.

8.1.4 Rational Functions

In comparisons with the previous case, the behavior of an analytic function near a pole is easy to describe. We now derive the following result:

♠ **Theorem:**
A **rational function** has no singularities other than poles. Conversely, an **analytic function** that has no singularities other than poles is necessarily a rational function.

A **rational function** $f(z)$ is of the form

$$f(z) = \frac{p(z)}{q(z)}, \tag{8.4}$$

where

$$p(z) = \alpha_0 + \alpha_1 z + \alpha_2 z^2 + \cdots + \alpha_n z^n$$

and

$$q(z) = \beta_0 + \beta_1 z + \beta_2 z^2 + \cdots + \beta_m z^m.$$

Observe that the polynomials $p(z)$ and $q(z)$ are analytic at all finite points on the complex plane.

Proof In what follows, we assume that $p(z)$ and $q(z)$ have no common zeros; if they do have a common zero at $z = z_0$, it is always possible to write $f(z)$ in (8.4) as the quotient of two polynomials with no common zeros by canceling a suitable number of the $(z - z_0)$-factors.

Obviously, the only possible singularities of $f(z)$ are situated at the zeros of $q(z)$. Since the zeros of $p(z)$ do not coincide with those of $q(z)$, $f(z)$ necessarily diverges at the zeros of $q(z)$. Such points can be poles but not essential singularities in view of the Weierstrass theorem given in Sect. 8.1.3. We have thus proved that all singularities of rational functions $f(z)$ are necessarily poles.

To prove the converse, suppose that all the singularities of an analytic function $f(z)$ are poles at the points $a_1, a_2, \cdots, a_n$. The orders of these poles are denoted by $m_1, m_2, \cdots, m_n$, respectively. In the vicinity of the point a_ν, the function $f(z)$ has a Laurent series expansion of the form

$$f(z) = \frac{c_{-m_\nu}^{(\nu)}}{(z - a_\nu)^{m_\nu}} + \cdots + \frac{c_{-1}^{(\nu)}}{(z - a_\nu)} + \sum_{\mu=0}^{\infty} c_\mu^{(\nu)} (z - a_\nu)^\mu,$$

where the superscripts (ν) on $c^{(\nu)}$ indicate that they are the coefficients that belong to the νth poles, $z = a_\nu$. Denote the principal part by

$$g_\nu(z) = \frac{c_{-m_\nu}^{(\nu)}}{(z - a_\nu)^{m_\nu}} + \cdots + \frac{c_{-1}^{(\nu)}}{(z - a_\nu)} \tag{8.5}$$

and consider the expression

$$h(z) = f(z) - g_1(z) - g_2(z) - \cdots - g_n(z).$$

Since $f(z) - g_\nu(z)$ is analytic at $z = a_\nu$, and $g_\nu(z)$ is analytic everywhere except at $z = a_\nu$, it follows that $h(z)$ is analytic at all points of the complex plane, including the **point at infinity**. In view of Liouville's theorem such a function is necessarily a constant. Thus we have identically $h(z) \equiv \gamma_0$, whence

$$f(z) = \gamma_0 + \sum_{\nu=1}^{n} g_\nu(z), \tag{8.6}$$

which implies that $f(z)$ can be brought into the form (8.4). This completes the proof of our theorem. ♣

Exercises

1. Find the poles and their order of the following functions:

$$\text{(a) } f(z) = \frac{\sin(z+1)}{z^3}, \quad \text{(b) } f(z) = \frac{\sin z}{z^3}.$$

Solution: **(a)** Clearly, $\lim_{z\to 0} z^2 f(z) = \infty$ and $\lim_{z\to 0} z^3 f(z) = \sin(1) \neq 0$. Hence, f has a third-order pole at $z = 0$ arising from the factor $1/z^3$. **(b)** Since $\lim_{z\to 0} z^3 f(z) = 0$, the pole of $f(z)$ is not a third-order pole. Instead, noting the asymptotic behavior of $\sin z$ near $z = 0$, we obtain

$$\lim_{z\to 0} z^2 f(z) = \lim_{z\to 0} z^2 \frac{z - (z^3/3!) + \cdots}{z^3} = 1.$$

Hence, $f(z)$ has a second-order pole at $z = 0$. ♣

2. Show that a function $f(z)$ cannot be bounded in the neighborhood of its isolated singular point $z = a$.

Solution: Use the contraposition method; if $|f(z)| < M$ for $|z - a| \leq r$, then the expansion coefficients read

$$|c_{-n}| = \left| \frac{1}{2\pi i} \oint_C (\zeta - a)^{n-1} f(\zeta) d\zeta \right| \leq M r^n \text{ for any } n,$$

where C is the circle given by $|z - a| = r$. Since r may be taken as small as desired, we have

$$c_{-1} = c_{-2} = \cdots = 0,$$

which means that the Laurent series reduces to a Taylor series. Hence, $f(z)$ should be analytic at $z = a$, which contradicts the assumption that $z = a$ is a singular point. ♣

3. Let both $f(z)$ and $g(z)$ be analytic in the vicinity of $z = a$ and have a zero of mth order at $z = a$. Prove that

$$\lim_{z\to a} \frac{f(z)}{g(z)} = \frac{f^{(m)}(a)}{g^{(m)}(a)}. \tag{8.7}$$

This result is called **l'Hôpital's rule.**

Solution: In the vicinity of $z = a$, we have

$$f(z) = (z-a)^m \left[\frac{f^{(m)}(a)}{m!} + (z-a) \frac{f^{(m+1)}(a)}{(m+1)!} + (z-a)^2 \frac{f^{(m+2)}(a)}{(m+2)!} + \cdots \right],$$

and we also have a form similar to $g(z)$. These expressions immediately yield the desired equation (8.7). ♣

4. Prove that if $f(z)$ has an essential singularity at $z = a$, $1/f(z)$ also has an essential singularity.

Solution: Suppose that f has an essential singularity at $z = a$ but that $1/f$ does not. If this is true, $1/f$ will at most have a pole there (of order N, for instance) and is expressed in terms of the series as

$$\frac{1}{f} = \sum_{n=-N}^{\infty} b_n h^n.$$

Rewrite this to obtain

$$f = \frac{h^N}{\sum_{m=0}^{\infty} b_{m-N} h^m}.$$

Note that the denominator $\sum b_{m-N} h^m$ is analytic within C_1, and thus the fraction $1/\sum b_{m-n} h^m$ is as well. As a result, the function f would be expanded into a power series in h starting with h^N; this result contradicts our assumption that $f(z)$ has an essential singularity at $z = a$. Therefore, wherever $f(z)$ has an essential singularity, $1/f$ also necessarily has one. ♣

Remark. The above result sounds intriguing when compared with the behavior of an $f(z)$ that has a pole. If $f(z)$ has a pole of order N at $z = a$, $1/f$ obviously has no pole but does have a zero of order N; i.e., $1/f \propto (z-a)^N$.

8.2 Multivaluedness

8.2.1 Multivalued Functions

Up to this point, our concern has been limited to single-valued functions, i.e., functions whose values are uniquely specified once z is given. When we consider multivalued functions, many important theorems must be reformulated.

The necessary concepts are best illustrated by considering the behavior of the function $f(z) = z^{1/2}$ in a graphical manner. Figure 8.1 gives a contour of a

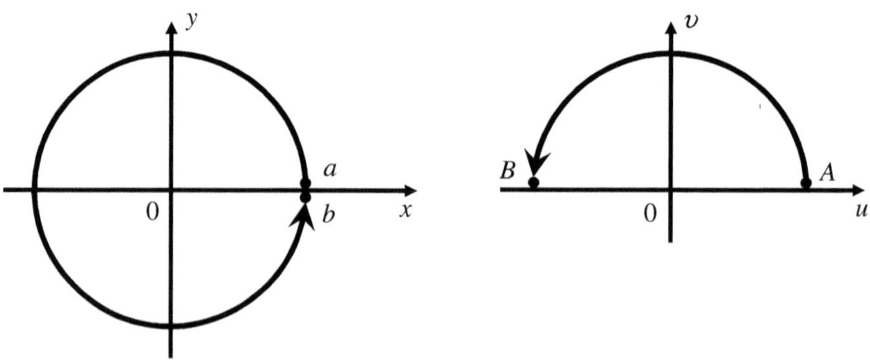

Fig. 8.1. Mapping of a circle on the z-plane onto an upper-half circle on the w-plane through $f(z) = z^{1/2}$

unit circle $a \to b$ on the z-plane. Through the transformation $w = f(z) = z^{1/2}$, the circle is mapped onto a semicircle $A \to B$ on the w-plane such that

$$\begin{aligned} z = 1 \quad &\to \quad w = 1, \\ z = i = e^{\pi i/2} \quad &\to \quad w = \left(e^{\pi i/2}\right)^{1/2} = e^{\pi i/4}, \\ z = -1 = e^{\pi i} \quad &\to \quad w = \left(e^{\pi i}\right)^{1/2} = e^{\pi i/2}, \\ z = -i = e^{3\pi i/2} \quad &\to \quad w = \left(e^{3\pi i/2}\right)^{1/2} = e^{3\pi i/4}. \end{aligned}$$

Of importance is the fact that the images of the points a and b, i.e., A and B, respectively, are not equal but are distinct on the w-plane. This suggests that the value of $z^{1/2}$ for $z = 1$ is not uniquely determined. Furthermore, a similar phenomenon occurs for any circular contour $a \to b$ with an arbitrarily large (or small) radius. We thus see that the function $f(z) = z^{1/2}$ is multivalued, at least along the positive real axis; one point on the positive real axis of the z-plane is associated with two distinct points on the w-plane.

As a matter of fact, the multivaluedness of the function $f(z) = z^{1/2}$ noted above occurs at all points on the whole z-plane (except at the origin). To see this, we observe again that the circular contour $a \to b$ may have any radius. As a result, all the points on the z-plane are correlated with only half of the points on the w-plane, those for which Im $[w] = v > 0$. The remaining values of w are generated if a second circuit $a \to b$ is made. Namely, the values of w with $v < 0$ will be correlated with those values of z whose arguments lie between 2π and 4π. As a consequence, all values for $z^{1/2}$ represented by on the w-plane may be divided into two independent sets: the set of values of w generated on the first circuit of the z-plane $0 < \phi < 2\pi$ and those generated on the second circuit $2\pi < \phi < 4\pi$. These two independent sets of values for $z^{1/2}$ are called the **branches** of $z^{1/2}$.

The concept of branch allows us to apply the theory of analytic functions to many-valued functions, where each branch is defined as a single-valued continuous function throughout its region of definition.

8.2.2 Riemann Surfaces

For the case $z^{1/2}$, the notion that the regions $0 < \phi < 2\pi$ and $2\pi < \phi < 4\pi$ correspond to two different regions of the w-plane is awkward geometrically, since each of these two regions covers the z-plane completely. To re-establish the single-valuedness and continuity of $f(z)$, it is desirable to give separate geometric meanings to two z-plane regions. This is achieved through the use of the notion of **Riemann surfaces**.

A Riemann surface is an ingenious device for representing both branches by means of a single continuous mapping. Suppose that two separate z-planes are cut along the positive real semiaxis from $+\infty$ to 0 (see Fig. 8.2), and that the planes are superimposed on each other but retain their separate identities.

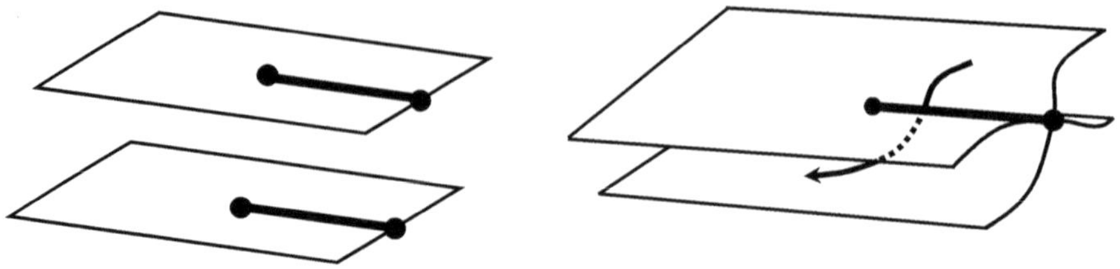

Fig. 8.2. A Riemann surface composed of two separated z-planes

Now suppose that the first quadrant of the upper sheet is joined along the cut to the fourth quadrant of the lower sheet to form a continuous surface. It is now possible to start a curve C in the first quadrant of the upper sheet, go around the origin, and cross the positive real semiaxis into the first quadrant of the lower sheet in a continuous motion. The curve can be continued on the lower sheet around the origin into the first quadrant of the lower sheet. This process of cutting and cross-joining two planes leads to the formation of a Riemann surface, which is thought of as a single continuous surface formed of two Riemann sheets.

Several important remarks are in order.

1. According to this model, the positive real semiaxis appears as a line where all four edges of our cuts meet. However, the Riemann surface has no such property. This results in the line between the first quadrant of the upper sheet and the fourth quadrant of the lower sheet being considered distinct from the line between the first quadrant of the lower sheet and the fourth quadrant of the upper one. There are two real positive semiaxes on the Riemann surface just as there are two real negative semiaxes. Hence, the entire Riemann surface is mapped one-to-one onto the w-plane. (The origin $z = 0$ belongs to neither branch since the polar angle θ is not defined for $z = 0$.)

2. The splitting of a multivalued function into branches is arbitrary to a great extent. For instance, we can define the following two functions, both of which may be treated as branches of $f(z) = \sqrt{z}$:

$$\text{BranchA}: \quad f_A(z) = \begin{cases} \sqrt{r}e^{i\theta/2} & \text{for} \quad 0 < \theta \le \pi, \\ \sqrt{r}e^{i(\theta+2\pi)/2} & \text{for} \quad -\pi < \theta \le 0. \end{cases}$$

$$\text{BranchB}: \quad f_B(z) = \begin{cases} \sqrt{r}e^{i(\theta+2\pi)/2} & \text{for} \quad 0 < \theta \le \pi, \\ \sqrt{r}e^{i\theta/2} & \text{for} \quad -\pi < \theta \le 0. \end{cases}$$

Note that branch A is continuous on the negative real semiaxis but is discontinuous on the positive real semiaxis (so is branch B). These two branches together, constitute, the double-valued function $f(z) = \sqrt{z}$, and this representation is no better and no worse than the previous one.

3. The above-mentioned technique can be extended to other multivalued functions that require more than two Riemann sheets (for instance, $f(z) = \sqrt[3]{z}$ requires three). There are functions requiring an infinite number of Riemann sheets, such as $f(z) = z^{\alpha}$ with an irrational α.

8.2.3 Branch Point and Branch Cut

We so back to the behavior of the multivalued function $w = f(z) = z^{1/2}$ to introduce other important concepts referred to as **branch point** and **branch cut**. Let us consider a certain closed curve C without self-intersections in the z-plane. Specify a point z_0 to which we assign a definite value of the argument θ_0. Through the mapping $w = z^{1/2}$, we will find two distinct points: $w_0(z_0)$ and $w_1(z_0)$.

In what follows, we examine the variation of the functions $w_0(z)$ and $w_1(z)$ as the point z moves continuously along the curve C. Since the argument of the point z on the curve C varies continuously, the functions $w_0(z)$ and $w_1(z)$ are continuous functions of z on the curve C.

Here, two different cases are possible. In the first case, the curve C does not contain the point $z = 0$ within it. Then, after traveling the curve C, the argument of the point z_0 returns to the original value $\arg z_0 = \theta_0$. Hence, the values of the functions $w_0(z)$ and $w_1(z)$ are also equal to their original values at the point $z = z_0$ after traveling the curve C. Thus, in this case, two distinct single-valued functions of the complex variable z are defined on C:

$$w_0 = r^{1/2} e^{i\theta/2} \quad \text{and} \quad w_1 = r^{1/2} e^{i/2(\theta+2\pi)}.$$

Obviously, if the domain D of the z-plane has the property that any closed curve in the domain does not contain the point $z = 0$, then two distinct single-valued continuous functions, $w_0(z)$ and $w_1(z)$, are defined in D. We call the functions $w_0(z)$ and $w_1(z)$ **branches** of the multivalued function $w(z) = z^{1/2}$.

In the second case, the curve C contains the point $z = 0$ within it. Then, after traversing C in the positive direction, the value of the argument of the point z_0 does not return to the original value θ_0 but changes by 2π as expressed by

$$\arg z_0 = \theta_0 + 2\pi.$$

Therefore, as a result of their continuous variation after traversing the curve C, the values of the functions $w_0(z)$ and $w_1(z)$ at the point z_0 are no longer be equal to the original values. More precisely, we obtain

$$\tilde{w}_0(z_0) = w_0(z_0) e^{i\pi} \quad \text{and} \quad \tilde{w}_1(z_0) = w_1(z_0) e^{i\pi},$$

which indicate that the function $w_0(z)$ goes into the function $w_1(z)$ and vice versa. This recurrence phenomenon stems from the fact that $z = 0$ is the branch point of the multivalued function $f(z) = z^{1/2}$. A formal definition of branch point is given below.

♠ **Branch point:**
Suppose that several of branches of $f(z)$ are analytic in the neighborhood of $z = a$ but not at $z = a$. Then, the point $z = a$ is a **branch point** if and only if $f(z)$ passes from one of these branches to another when z moves along a closed circuit around $z = a$.

Remark. The point at infinity, $z = \infty$, is a branch of $f(z)$ if and only if the origin is a branch point of $f(1/z)$.

It is important to note that the branch points for a given multivalued function, always occur pairwise so that they are connected by a simple curve called the **branch cut** (**cut** or **branch line**). Branch cuts bound the regions within which the individual single-valued branches are defined. For instance, in the case of $f(z) = z^{1/2}$, the branch cut ran from the branch point at $z = 0$ to another branch point at $z = \infty$ along the positive real axis. It should be emphasized here that any curve joining the origin ($z = 0$) and the point of infinity ($z = \infty$) would have done just as well. For example, we could have used the negative real axis as the branch cut, for which the regions

$$-\pi < \phi < \pi \quad \text{and} \quad \pi < \phi < 3\pi$$

(instead of $0 < \phi < 2\pi$ and $2\pi < \phi < 4\pi$) serve as the defining regions for the first and second branch. On the w-plane, these two would correspond to Re $v > 0$ and $v < 0$, respectively. We therefore may choose the branch cut that is most convenient for the problem at hand.

Remark. The choice of branches and branch cuts for a given multivalued function is not unique; however, the branch points and the number of branches are uniquely determined once a function is given.

Exercises

1. Examine the multivaluedness of a logarithm function $\ln z$.

   Solution: Expressing z in polar form, $\ln z = \ln\left(re^{i\phi}\right) = \ln r + i\phi$, and changing ϕ by $2\pi k$ results in

$$\ln z(r, \phi + 2\pi k) = \ln r + i(\phi + 2\pi k) = \ln z(r, \phi) + 2\pi ik. \quad (8.8)$$

It follows from (8.8) that there is no nonzero value of k for which $\ln z(r,\phi+2\pi k)$ and $\ln z(r,\phi)$ are equal. Therefore, the logarithm function is an infinite-valued function. ♣

2. Evaluate $\log e$, $\log(-1)$, $\log(1+i)$ according to the expression (8.8).

Solution: $\log e = \log|e| + i\arg e = 1 + 2n\pi i$,
$\log(-1) = \log|-1| + i\arg(-1) = (2n+1)\pi i$,
$\log(1+i) = \log|1+i| + i\arg(1+i) = \frac{\log 2}{2} + (2n+\frac{1}{4})\pi i$. ♣

3. Evaluate 1^i and i^i according to the definition of power functions: $z^a = e^{a\log z}$, where $z(\neq 0)$ and a are complex numbers.

Solution:
$1^i = e^{i\log 1} = e^{i\cdot 2n\pi i} = e^{2n\pi}$,
$i^i = e^{i\log i} = e^{i(2n+\frac{1}{2})\pi i} = e^{(2n-\frac{1}{2})\pi}$. ♣

4. Show that a power function $z^{m/n}$ with an irreducible rational number m/n $(n \geq 2)$ is an n-valued function.

Solution: The multiple values of $z(r,\phi)^{m/n} = r^{m/n}e^{im\phi/n}$ are found by varying the integer k in the expression:

$$z(r,\phi+2\pi k)^{m/n} = r^{m/n}e^{im\phi/n}e^{i2\pi km/n} = e^{i2\pi km/n}z(r,\phi)^{m/n}.$$

Substituting $k = n$ yields

$$z(r,\phi+2\pi n)^{m/n} = e^{i2\pi m}z(r,\phi)^{m/n} = z(r,\phi)^{m/n},$$

wherein $e^{i2\pi m} = 1$ for arbitrary $m \in \boldsymbol{N}$. Hence, all multiple values of $z^{m/n}$ at a given z are found with a value of k in the range $0 \leq k \leq n-1$. Since there are n different values of k in this range, $z^{m/n}$ is an n-valued function. ♣

8.3 Analytic Continuation

8.3.1 Continuation by Taylor Series

It is often the case that a complex function is defined only in a limited region in the complex plane. For instance, a series representation of a function is of use only within its radius of convergence, but provides no direct information about the function outside this radius of convergence. An illustrative example is a function $f(z)$ defined by

$$f(z) = 1 + z + z^2 + \cdots . \tag{8.9}$$

Obviously, this function is identified with $1/(1-z)$ for $|z| < 1$, whereas it diverges for $|z| > 1$ and thus is no longer equivalent to $1/(1-z)$. Nevertheless, a sophisticated technique makes it possible to identify the function $f(z)$ given in (8.9) with $1/(1-z)$ even for the region $|z| > 1$. This technique, by which the defined region of a function is extended to an 'uncultivated' region, is called **analytic continuation**. The resultant function may often be defined by sequential continuation over the entire complex plane without reference to the original region of definition.

To see an actual process of analytic continuation, we suppose that a function f is given as a power series around $z = 0$, with a radius of convergence R and a singular point of f being on the **circle of convergence**. We show that it is possible to extend the function outside R. We first note that at any point $z = a$ within the circle ($|z| < R$), we can evaluate not only the value of the series but all its derivatives at that point as well because the function f is analytic and the series representation has the same radius of convergence. Therefore, we can obtain a Taylor series of $f(z)$ around $z = a$ as

$$f(z) = \sum_{n=0}^{\infty} \frac{f^{(n)}(a)}{n!}(z-a)^n. \tag{8.10}$$

The radius of convergence of this series is the distance to the nearest singular point, say $z = z_s$ (see Fig. 8.3a). The resultant circle of convergence with radius $R_0 = |z_s - z_0|$ is indicated by the solid circle in the figure. One may setup this process using a new point, e.g., $z = b$, not necessarily within the original circle of convergence (see Fig. 8.3b), about which a new series such as (8.10) can be set up (see Fig. 8.3c). Continuing on in this way, it is apparently possible by means of such a series of overlapping circles to obtain values for f for every point in the complex plane excluding the singular points.

Our current discussion can be summarized as follows:

1. Let $f(z)$ be defined by its Taylor series expansion around $z = a$ within some circle $|z-a| = r$.
2. Specify a certain point $z = b$ within the circle and evaluate $f(b), f'(b), \cdots$ to obtain a Taylor series of $f(z)$ around $z = b$.
3. Observe that the latter series converges within a circle $|z-b| = r'$ that intersects the first circle but may contain a region that is not within the first circle.
4. Specify again another point $z = c$ within the circle $|z-b| = r'$ and repeat the process described above.

8.3.2 Function Elements

We know that the term 'analytic continuation' refers to a method that allows us to extend the defining region of a complex function. Alternatively, this term can refer to the function that is newly found through analytic continuation of some other function. The formal definition is given below.

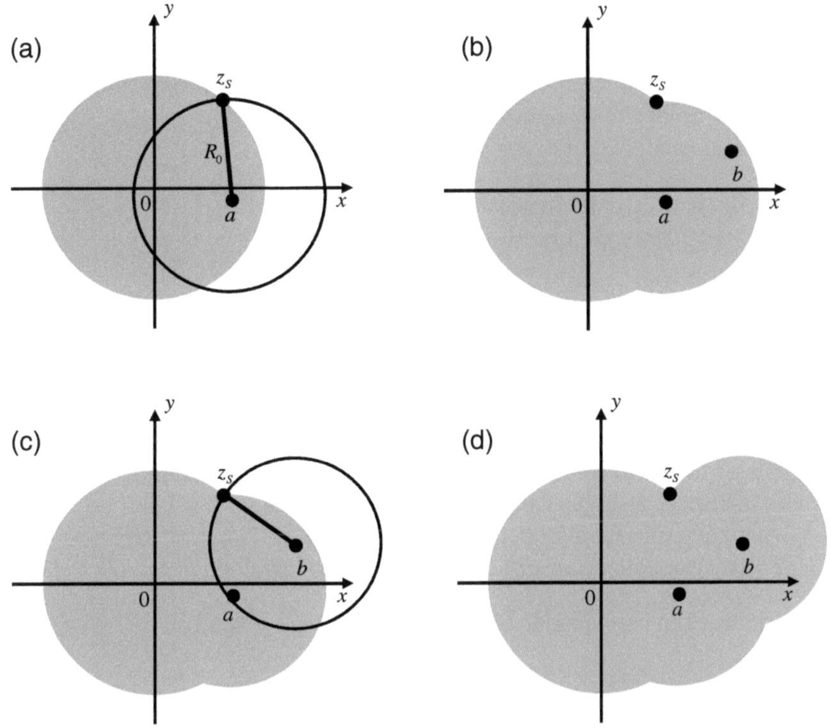

Fig. 8.3. Illustration of an analytic continuation procedure

♠ **Analytic continuation:**

Given a single-valued analytic function $f_1(z)$ defined on a region D_1, the analytic function $f_2(z)$ defined on D_2 is called an **analytic continuation** of $f_1(z)$ to D_2 if and only if the intersection $D_1 \cap D_2$ contains a simply connected open region where $f_1(z) \equiv f_2(z)$.

If the two analytic functions $f_1(z)$ and $f_2(z)$ defined on D_1 and D_2, respectively, are analytic continuations of one another, then it is evident that an analytic function $f(z)$ can be defined on $D_1 \cup D_2$ by setting

$$f(z) = \begin{cases} f_1(z) & \text{in } D_1, \\ f_2(z) & \text{in } D_2. \end{cases}$$

Here, f_1 and f_2 are called **function elements** of f. More generally, we can consider a sequence of function elements $(f_1, f_2, \cdots, f_n)$ such that f_k is an analytic continuation of f_{k-1}. The elements of such a sequence are called **analytic continuations of each other**. Relevant terminology for this point is given below.

♠ **General analytic function:**

A general analytic function f is a nonvoid collection of function elements f_k in which any two elements are analytic continuations of each other by way of a chain whose links are members of f.

♠ **Complete analytic function:**

A complete analytic function f is a general analytic function that contains all the analytic continuations of any one of its elements.

A **complete analytic function** is evidently *maximal* in the sense that it cannot be further extended. Moreover, it is clear that every function element belongs to a unique complete analytic function. **Incomplete general analytic functions** are more arbitrary, and there are many cases in which two different collections of function elements should be regarded as defining the same function. For instance, a single-valued function $f(z)$ defined in D can be identified either with the collection that consists of the single function element defined on D or with the collection of all function elements defined on $D' \subset D$.

Examples **1.** Let us consider the functions

$$f_1(z) = \sum_{n=0}^{\infty} z^n \text{ defined on } |z| < 1 \tag{8.11}$$

and

$$f_2(z) = \sum_{n=0}^{\infty} \left(\frac{3}{5}\right)^{n+1} \left(z + \frac{2}{3}\right)^n \text{ defined on } \left|z + \frac{2}{3}\right| < \frac{5}{3}. \tag{8.12}$$

Both series converge to $1/(1-z)$; Particularly the latter converges since

$$f_2(z) = \frac{3}{5} \sum_{n=0}^{\infty} \left[\frac{3}{5}\left(z + \frac{2}{3}\right)\right]^n = \frac{\frac{3}{5}}{1 - \frac{3}{5}(z + \frac{2}{3})} = \frac{1}{1-z}.$$

Therefore, the two functions represent the same function $f(z) = 1/(1-z)$ in the two overlapping regions (see Fig. 8.4), although they have different series representations. In this context, we can write

$$f(z) = \begin{cases} f_1(z) & \text{when for } z \in D_1,\ D_1 = \{z : |z| < 1\}, \\ f_2(z) & \text{when for } z \in D_2,\ D_2 = \{z : \left|z + \frac{2}{3}\right| < \frac{5}{3}\}. \end{cases}$$

2. Another illustrative example is given by

$$f_1(z) = \int_0^{\infty} e^{-zt} dt \text{ defined on } \mathrm{Re} z > 0 \tag{8.13}$$

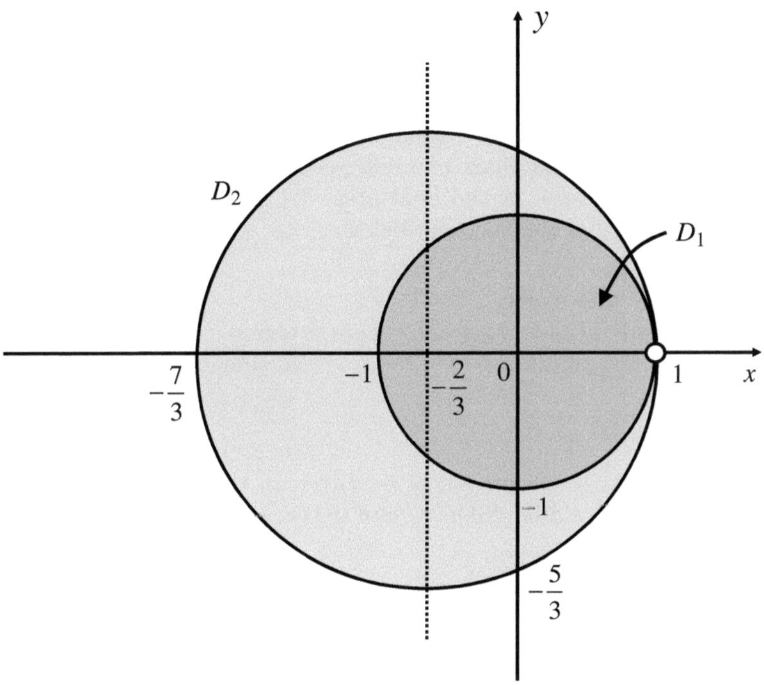

Fig. 8.4. Both functions $f_1(z)$ in (8.11) and $f_2(a)$ in (8.12) represent the same function $f(z) = 1/(1-z)$ in the overlapping region $D_1 \cap D_2$

and

$$f_2(z) = i \sum_{n=0}^{\infty} \left(\frac{z+i}{i} \right)^n \quad \text{defined on } |z+i| < 1.$$

Observe that each f_1 and f_2 reads $1/z$ for the respective defining region. Thus, we have

$$\frac{1}{z} = \begin{cases} f_1(z) & \text{for } z \in D_1, \ \ D_1 = \{z : \mathrm{Re} z > 0\}, \\ f_2(z) & \text{for } z \in D_2, \ \ D_2 = \{z : |z+i| < 1\}. \end{cases}$$

The two functions are analytic continuations of one another, and $f(z) = 1/z$ is the analytic continuation of both f_1 and f_2 for all z except $z = 0$.

Remark. In some cases, it is impossible to extend the function outside of a finite region because an infinite number of singularities are located densely on the boundary of the region. In that event, the boundary of this region is called the **natural boundary** of the function and the region within this boundary is called the **region of the existence** of the function.

8.3.3 Uniqueness Theorem

Having introduced the concept of analytic continuation, we may ask a question as to whether the function resulting from an analytic continuation process is uniquely determined, independent of the continuing path; i.e., whether a function that is continued along two different routes from one area to another will have the same value in the final area. We now attempt to answer this question by examining the theorem below.

♠ **Uniqueness theorem:**

Let $f_1(z)$ and $f_2(z)$ be analytic within a region D. If the two functions coincide in the neighborhood of a point $z \in D$, then they coincide throughout D.

Proof The theorem to be proven is rewritten in the following statement: *If both $f(z)$ and $g(z)$ are analytic at z_0 and if $f(z_n) = g(z_n)$ with $n = 1, 2, \cdots$ at points z_n that satisfy $\lim_{n\to\infty} z_n = z_0$ but $z_n \neq z_0$ for all n, then $f(z) \equiv g(z)$ throughout D.* We now prove it.

Let $h(z) = f(z) - g(z)$. Here, f and g are assumed to satisfy the conditions given in the statement above, so that $h(z_n) = 0$ for all n and $h(z)$ is analytic at z_0. Owing to the analyticity of $h(z)$ at z_0, we have the expansion

$$h(z) = a_0 + a_1(z - z_0) + a_2(z - z_0)^2 + \cdots ,$$

which converges in a certain circle around z_0. Since $h(z)$ is continuous at z_0, we have

$$h(z_0) = \lim_{n\to\infty} h(z_n) = 0,$$

which means that the coefficient a_0 is zero. Then, since $h'(z)$ is also continuous at z_0, we set

$$h'(z_0) = \lim_{n\to\infty} h'(z_n) = 0,$$

which means that $a_1 = 0$. Continuing in this fashion, we find successively that all the coefficients vanish. In its circle of convergence, the function $h(z)$ is therefore identically zero. This completes the proof. ♣

This remarkable theorem demonstrates the strong correlation between the behaviors of analytic functions on different parts of the complex plane. For example, if two functions agree in value over a small arc (arbitrarily small as long as it is not a point), then they are identical in their common region of analyticity.

8.3.4 Conservation of Functional Equations

An important consequence of the uniqueness theorem is the so-called principle of the conservation of a functional equation.

♠ **Conservation of functional equations:**

Let $F(p,q,r)$ be an analytic function for all values of the three variables p,q,r, and let $f(z)$ and $g(z)$ be analytic functions of z. If a relation $F[f(z),g(z),z]=0$ between function elements $f(z)$ and $g(z)$ holds on a domain, then this relation is also true for all analytic continuations of these function elements.

Remark. In plain words, this theorem states that analytic continuations of $f(z)$ satisfy every functional (and differential) equation satisfied by the original $f(z)$.

This theorem can easily be generalized to cases of functional equations involving more than two functions. We illustrate this by two examples.

Examples **1.** From elementary trigonometry, we know that the real function $\sin x$ has the additional theorem

$$\sin(x+u) = \sin x \cos u + \cos x \sin u,$$

where u is an arbitrary real value. Since $\sin z$, $\cos z$, and $\sin(z+u)$ are analytic for all finite values of z, and since the relation

$$\sin(z+u) = \sin z \cos u + \cos z \sin u$$

is satisfied if z is any point on the real axis, it follows by analytic continuation that the same relation must hold for all values of z. If we report the same argument with respect to the real variable u, we find that u may be replaced by a complex variable w without invalidating the relation in question. Hence, the **addition theorem** of the function $\sin z$ is true for arbitrary complex values of z and w.

2. Another important example is afforded by functions satisfying differential equations. To take a simple case, we consider the function

$$f(z) = \log(1+z).$$

This is represented for $|z|<1$ by the power series

$$f(z) = z - \frac{z^2}{2} + \frac{z^3}{3} - \cdots. \tag{8.14}$$

which yields

$$f'(z) = 1 - z + z^2 - z^3 + \cdots = \frac{1}{1+z}.$$

In this context, the identity

$$f'(z) = \frac{1}{1+z} \tag{8.15}$$

appears to be valid for $|z| < 1$. However, it follows that the identity (8.15) must hold for all analytic continuations of the power series (8.14).

8.3.5 Continuation Around a Branch Point

The uniqueness theorem given in Sect. 8.3.3 also gives us the following corollary:

♠ **Theorem:**
If D_1 and D_2 are regions into which $f(z)$ has been continued from D, yielding the corresponding functions f_1 and f_2, and if $D_3 = D_1 \cap D_2$ also overlaps D, then $f_1 = f_2$ throughout D_3.

It is important to note that the validity of this theorem is due to the condition that D_3 and D have a common region. If this condition is not satisfied, the uniqueness of analytic continuation may break down. Instead, one can say: If analytic continuation of a function f along two different routes from z_0 to z_1 yields two different values at z_1, then $f(z)$ must have a certain kind of singularity between the two routes. This seems obvious by recalling the fact that the radius of convergence of a power series extends up to the next singularity of the function; if there were no singularities between the two routes, then it would be possible to fill in the region between the two routes by means of analytic continuation based on the power series. Then we would obtain sufficient overlapping so that the uniqueness theorem would be satisfied. In that event $f(z_1)$ for the two different routes would be identical, in contradiction to our hypothesis. There must therefore be a singularity between the two routes.

Note that the last discussion does not state that different values must be obtained if there is any kind of singularity between the two routes. It must be a particular type of singularity to cause a discrepancy, and we call it a **branch point**, as we introduced earlier. An analytic function involving branch points is said to be multivalued and the various possible sets of values generated by the process of analytic continuation are known as **branches**. Intuitively, all the possible values of a function at a given point may be obtained by the process of analytic continuation if one winds about the branch point as many times as necessary.

8.3.6 Natural Boundaries

In all the examples considered so far, the singularities were isolated points. It is, however, easy to construct functions for which this is not the case. Consider, say, the function

$$f(z) = \frac{1}{\sin(1/z)}.$$

The denominator vanishes for $1/z = n\pi$ with an integer n. Hence, the points $z = (1/n\pi)$ are singular points of $f(z)$, but are clearly isolated in the vicinity of the origin. It is further possible for the singular points of a function to fill a whole arc of a continuous curve; in this case, we speak of a **singular line** of the function.

Particularly interesting is a situation in which a function $f(z)$ has a closed singular line C. In this case, it is obviously impossible to continue $f(z)$ analytically across C. The entire domain of definition of $f(z)$ is therefore the interior of C, and we say that C is a **natural boundary** of $f(z)$.

Such an occurrence is not as unusual as it may seem. Consider, for instance, the analytic function $f(z)$ defined by the power series

$$f(z) = z + z^2 + z^4 + z^8 + \cdots = \sum_{n=0}^{\infty} z^{2^n}. \tag{8.16}$$

By the root test given in Sect. 2.4.3, the circle of convergence of this series turns out to be $|z| < 1$. Thus $f(z)$ must have at least one singularity on $|z| = 1$. For the sake of simplicity, we assume that this singularity is situated at the point $z = 1$; a different location will cause a minor change in the argument. From the definition of $f(z)$, it follows that

$$f(z^2) = z^2 + z^4 + z^8 + \cdots = \sum_{n=1}^{\infty} z^{2^n} = f(z) - z.$$

By the principle of conservation (see Sect. 8.3.4), the functional equation

$$f(z) = z + f(z^2) \tag{8.17}$$

is true for all analytic continuations of $f(z)$. Observe that (8.17) gives

$$f'(z) = 1 + 2zf'(z^2),$$

which means that $f(z)$ cannot have a derivative at $z = -1$ since from hypothesis $f(1)$ does not exist. Thus, $z = -1$ is also a singular point of $f(z)$. In the same way, from the relation

$$f(z) = z + f(z^2) = z + z^2 + f(z^4)$$

it follows that the points z for which $z^4 = 1$ are singularities of $f(z)$. Continuing in this fashion, we conclude that all points z for which $z^{2^n} = 1$ are singularities of $f(z)$. But these are the points $e^{2\pi i/(2^n)}$ that divide the circumference $|z| = 1$ into 2^n equal parts. Since, for $n \to \infty$, all points on $|z| = 1$ are limits of these points and since the limit point of singular points is also a singularity, it follows that all points on $|z| = 1$ are singular points of $f(z)$. We have thus proven that the unit circle is the natural boundary of the analytic function (8.16).

8.3.7 Technique of Analytic Continuations

The uniqueness theorem is the fundamental theorem in the theory of analytic continuation. However, in practice, the most relevant method would be one that tells us whether a function f_2 is the analytic continuation of a function f_1.

Let us describe two possible methods of analytic continuation: The first is based on the **Schwarz principle of reflection**, which essentially makes use of the functional relation $f(z^*) = f(z)^*$.

♠ **Schwarz principle of reflection:**
If $f(z)$ is analytic within a region D intersected by the real axis and is real on the real axis, then we have $f(z^*) = f(z)^*$.

Proof Expand $f(z)$ in a Taylor series about a point a on the real axis. The coefficients of the Taylor series are real by virtue of the hypothesis that $f(z)$ is real on the real axis. Hence, we have

$$f(z) = \sum_n c_n (z-a)^n, \tag{8.18}$$

where c_n is real. Then

$$f(z)^* = \sum_n c_n \left(z^* - a\right)^n = f(z^*), \tag{8.19}$$

proving the theorem. ♣

The above theorem holds for any point within the circle of convergence of the power series. By the methods of analytic continuation, therefore, it may be extended to include any nonsingular point conjugate to a point in D. As a result, the function in question can be continued from a region above the real axis to a region below.

A second method employs explicit functional relations such as **addition formulas** or **recurrence relations**. A simple example is provided by the addition formula

$$f(z + z_1) = f(z) f(z_1).$$

If f were known only in a given region, it would be continued outside that region to any point given by the addition of the coordinates of any two points within the region. A less trivial example occurs in the theory of gamma functions. The **gamma function** is defined by the integral

$$\Gamma(z) = \int_0^\infty e^{-t} t^{z-1} dt. \tag{8.20}$$

This integral converges only for Re $z > 0$, so that it defines $\Gamma(z)$ for only the right half of the complex plane. From (8.20), one may readily derive (by integrating by parts) a functional relationship between $\Gamma(z)$ and $\Gamma(z+1)$:

$$z\Gamma(z) = \Gamma(z+1). \tag{8.21}$$

We may now use (8.21) to continue $\Gamma(z)$ into the Re $z < 0$ part of the complex plane. As first, we assume that $\Gamma(z)$ is known for $x > 0$. Then using recurrence relation (8.21), the points in the strip $-1/2 < x < 1/2$ can be computed in terms of the values of $\Gamma(z)$ for $x > 0$. The function so defined and the original function have an overlapping region of convergence so that it is the analytic continuation into the negative x-region.

8.3.8 The Method of Moment

Suppose that we are given a power series $f(z) = \sum_{n=0}^{\infty} a_n z^n$ where the coefficients a_n are the *moments* of a given continuous function. For example, suppose that there exists a continuous function g on $[0,1]$ such that

$$a_n = \int_0^1 g(t) t^n dt.$$

Then

$$f(z) = \sum_{n=0}^{\infty} \left[\int_0^1 g(t) t^n dt \right] z^n = \sum_{n=0}^{\infty} \left[\int_0^1 g(t)(zt)^n dt \right],$$

and interchanging the order of summation and integration, we find that

$$f(z) = \int_0^1 \left[\sum_{n=0}^{\infty} g(t)(zt)^n \right] dt = \int_0^1 \frac{g(t)}{1-zt} dt.$$

(The interchange of summation and integration is easy to justify if $|z| < 1$.) Moreover, this integral form serves to define an analytic extension of the original power series.

Examples Consider

$$f(z) = \sum_{n=0}^{\infty} \frac{z^n}{n+1} \quad (|z| < 1). \tag{8.22}$$

Since

$$\frac{1}{n+1} = \int_0^1 t^n dt,$$

we set $g(t) = 1$ to obtain

$$f(z) = \int_0^1 \frac{dt}{1 - zt} \quad \text{for } |z| < 1.$$

The integral above is the analytic continuation of the original representation (8.22), so that the latter is analytic throughout the complex plane except for the semi-infinite line $[1, \infty)$. [In fact, the analytic continuation has a discontinuity at every point of the interval $[1, \infty)$.]

Exercises

1. Suppose $f(z) = \sum_{k=0}^{\infty} c_k z^{n_k}$ with $\liminf_{k \to \infty} \frac{n_{k+1}}{n_k} > 1$. Prove that the circle of convergence of $f(z)$ above is a natural boundary for f.

Solution: Since the result is independent of c_k, we may assume without loss of generality that the radius of convergence is 1. In addition, neglecting finitely many terms if necessary, we assume that for some $\delta > 0$ and for all k, $n_{k+1}/n_k = 1 + \delta$. Finally, it suffices to show that f is singular at the point $z = 1$. The same result applied to the series $\sum_{k=0}^{\infty} c_k (ze^{-i\theta})^{n_k}$ shows that f is singular at any point $z = e^{i\theta}$.

Choose an integer $m > 0$ such that $(m+1)/m < 1 + \delta$ and consider the power series $g(w)$ obtained by setting $z = (w^m + w^{m+1})/2$. We then find that

$$\begin{aligned} g(w) &= f\left(\frac{w^m + w^{m+1}}{2}\right) \\ &\frac{c_0}{2^{n_0}} w^{mn_0} + \frac{c_0 n_0}{2^{n_0}} w^{mn_0+1} + \cdots + \frac{c_0}{2^{n_0}} w^{mn_0+n_0} \\ &+ \frac{c_1}{2^{n_1}} w^{mn_1} + \frac{c_1 n_1}{2^{n_1}} w^{mn_1+1} + \cdots + \frac{c_1}{2^{n_1}} w^{mn_1+n_1} + \cdots . \end{aligned}$$

Note that in this expression no two terms involve the same power of w, since the inequality $mn_{k+1} > mn_k + n_k$ holds whenever $n_{k_1}/n_k > (m+1)/m$. If $|w| < 1$, then $(|w|^m + |w|^{m+1})/2 < 1$, and since $f(z)$ is absolutely convergent for $|z| < 1$, the series $\sum_{k=0}^{\infty} |c_k| [(|w|^m + |w|^{m+1})/2]^{n_k}$ converges. Hence, for $|w| < 1$, $g(w)$ is absolutely convergent. On the other hand, if we take w real and greater than 1, then $(w^m + w^{m+1})/2 > 1$, so the series $\sum_{k=0}^{\infty} c_k [(w^m + w^{m+1})/2]^{n_k}$ diverges. Note, though, that the jth partial sums s_j of the above series are exactly the $n_j(m+1)$th partial sums of the power series of g. Hence, the series for $g(w)$

diverges and g, too, has a radius of convergence of 1. This means that $g(w)$ must have a singularity at some point w_0 with $|w_0| = 1$. If $w_0 \neq 1$, then $|(w^m + w^{m+1})/2| < 1$ and since f is analytic in $|z| < 1$, g is analytic at w_0. Thus g must have a singularity at $w_0 = 1$ and since $g(w) = f[(w^m + w^{m+1})/2]$, $f(z)$ must have a singularity at $z = 1$. ♣

2. Define an analytic continuation of: **(i)** $\sum_{n=1}^{\infty} \frac{z^n}{\sqrt[3]{n}}$, **(ii)** $\sum_{n=0}^{\infty} \frac{z^n}{n^2+1}$.

Solution:
(i) Since $\frac{1}{n^{1/3}} = \Gamma(1/3) \int_0^\infty e^{-nt} t^{-2/3} dt$, we have

$$\sum_{n=1}^{\infty} \frac{z^n}{n^{1/2}} = \Gamma\left(\frac{1}{3}\right) \int_0^\infty \sum_{n=1}^{\infty} \left(ze^{-t}\right)^n t^{-2/3} dt$$
$$= \Gamma\left(\frac{1}{3}\right) \int_0^\infty \frac{z}{t^{2/3}(e^t - z)} dt,$$

which is analytic outside of the interval $[1, \infty)$.

(ii) Since $\frac{1}{n^2+1} = \int_0^\infty e^{-nt} \sin t dt$,

$$\sum_{n=0}^{\infty} \frac{z^n}{n^2+1} = \int_0^\infty \sum_{n=0}^{\infty} \left(ze^{-t}\right)^n \sin t dt = \int_0^\infty \frac{e^t \sin t}{e^t - z} dt,$$

which is analytic outside of the interval $[1, \infty)$. ♣

3. Suppose that f is bounded and analytic in $\text{Im}z \geq 0$ and real on the real axis. Prove that f is constant.

Solution: By the Schwarz reflection principle, f can be extended to the entire plane and would then be a bounded entire function. Hence, f is constant. ♣

4. Given an entire function that is real on the real axis and imaginary on the imaginary axis, prove that it is an odd function; i.e., $f(z) = -f(-z)$.

Solution: Set $f(z) = f(x, y) = u(x + iy) + iv(x + iy)$. The Schwarz reflection principle implies that $f(z^*) = f(x - iy) = u(x - iy) + iv(x - iy) = u(x + iy) - iv(x + iy) = -f(z)$. In a similar way, we have $f(-z) = f(-x - iy) = u(-x - iy) + iv(-x - iy) = -u(x + iy) - iv(x + iy) = -f(z)$. ♣

9

Contour Integrals

Abstract In this chapter, we show that singularities do not interfere with the analysis of complex functions but are useful in extracting complex integrals along closed contours. This utility of singularities is based on the residue theorem (Sect. 9.1.1), argument principle (Sect. 9.4), and principal value integrals (Sect. 9.5.1), all of which correlate the nature of singularities within and/or on the contour with the relevant complex integrals.

9.1 Calculus of Residues

9.1.1 Residue Theorem

In the preceding two chapters, we provided the theoretical bases of complex functions. This chapter deals with more practical matters that are relevant to computations of contour integrations on a complex plane. The theorem below is central to the development of this topic.

♠ **Residue theorem:**

If a function $f(z)$ is analytic everywhere within a closed contour C except at a finite number of poles, its contour integral along C yields

$$\oint_C f(z)dz = 2\pi i \sum_j \operatorname{Res}(f, a_j). \tag{9.1}$$

Here, $\operatorname{Res}(f, a_j)$ is called the **residue** of $f(z)$ at the pole $z = a_j$. When the pole is mth order, it reads

$$\operatorname{Res}(f, a_j) = \frac{1}{(m-1)!} \lim_{z \to a_j} \frac{d^{(m-1)}}{dz^{(m-1)}} \left[(z - a_j)^m f(z) \right]. \tag{9.2}$$

Once the residue is evaluated, the integral $\oint_C f(z)dz$ around the contour C surrounding the pole $z = a$ can be determined by the above theorem. Notably, this theorem enables us to evaluate various kinds of integrals of real functions that are unfeasible by means of elementary calculus.

Before demonstrating the utility of the residue theorem, we present a short review of the nature of residues. Originally, the residue of $f(z)$ is defined in association with a particular coefficient of the Laurent series expansion. We know that $f(z)$ around its pole at $z = a$ may be expressed by a Laurent series expansion such as

$$f(a+h) = \sum_{n=-\infty}^{\infty} c_n h^n, \quad c_n = \frac{1}{2\pi i}\oint_C \frac{f(z)}{(z-a)^{n+1}}dz.$$

Then, the specific coefficient

$$c_{-1} = \frac{1}{2\pi i}\oint_C f(z)dz \tag{9.3}$$

is called the residue of $f(z)$ at $z = a$. In fact, the result (9.3) immediately reduces to the form of (9.1) as

$$\frac{1}{2\pi i}\oint_C f(z)dz = 2\pi i c_{-1}.$$

The equivalence of the two quantities, $\text{Res}(f, a)$ in (9.2) and c_{-1} in (9.3), is verified as follows.

Proof (of the residue theorem). Suppose that $f(z)$ has a pole of order m at a. Then $f(z)$ can be written as

$$f(z) = \frac{c_{-m}}{(z-a)^m} + \frac{c_{-m+1}}{(z-a)^{m-1}} + \cdots + \frac{c_{-1}}{(z-a)} + \sum_{n=0}^{\infty} c_n (z-a)^n. \tag{9.4}$$

Now we introduce the quantity

$$g(z) \equiv (z-a)^m f(z) \quad = c_{-m} + c_{-m+1}(z-a) + \cdots$$

$$= \sum_{n=0}^{\infty} c_{n-m}(z-a)^n. \tag{9.5}$$

Since $g(z)$ is analytic everywhere in a neighborhood around a, it can be expanded in terms of a Taylor series as

$$g(z) = \sum_{n=0}^{\infty} \frac{g^{(n)}(a)}{n!}(z-a)^n. \tag{9.6}$$

The residue c_{-1} is the coefficient of the $n = m-1$ term in (9.5). Hence, comparing (9.5) with (9.6), we have

$$c_{-1} = \frac{1}{(m-1)!}g^{(m-1)}(a) = \frac{1}{(m-1)!}\lim_{z\to a}\frac{d^{(m-1)}}{dz^{(m-1)}}\left[(z-a)^m f(z)\right], \quad (9.7)$$

which is simply equation (9.2). ♣

9.1.2 Remarks on Residues

The reason that only the particular coefficient c_{-1} plays a role in evaluating the contour integral is clarified by integraing both sides of (9.4) along the contour containing the mth-order pole a. For convenience, we rewrite (9.4) as

$$f(z) = \sum_{n=1}^{m} \frac{c_{-n}(a)}{(z-a)^n} + \Psi_A(z), \quad (9.8)$$

where

$$\Psi_A(z) = \sum_{n=0}^{\infty} c_n(a)(z-a)^n$$

is the regular part of the series (9.8), thus being analytic everywhere in a region within a closed contour C containing a. By integrating $f(z)$ along the contour C, we set

$$\oint_C f(z)dz = \sum_{n=1}^{m} c_{-n} \oint_C \frac{1}{(z-a)^n}dz \quad (9.9)$$

because of the analyticity of $\Psi_A(z)$. The integral of (9.9) can be easily evaluated by letting the contour be a circle of radius ρ centered at a. Since any point on the contour can be expressed as $z = a + \rho e^{i\phi}$, we have

$$\oint_C \frac{1}{(z-a)^n}dz = \int_0^{2\pi} \frac{i\rho e^{i\phi}}{\rho^n e^{in\phi}}d\phi = i\rho^{-(n-1)} \int_0^{2\pi} e^{-i(n-1)\phi}d\phi. \quad (9.10)$$

Note that the integral (9.10) vanishes for all $n \neq 1$, and it is only when $n = 1$ that it has a nonzero value:

$$\oint_C \frac{1}{z-a}dz = i\int_{\phi_0}^{\phi_0+2\pi} d\phi = 2\pi i.$$

Therefore, all the terms in the sum of (9.9) are zero except the $n = 1$ term, and Goursat's formula takes the form

$$\oint_C f(z)dz = 2\pi i\, c_{-1}. \quad (9.11)$$

In short, once we integrate the function $f(z)$ in (9.8), only the term involving c_{-1} survives, whereas the other terms vanish. This results in the fact that the contour integral $\oint_C f(z)dz$ around a pole is determined by the value of the specific coefficient c_{-1}.

9.1.3 Winding Number

To evaluate $\oint_C f(z)dz$ when C is a general closed curve (and when f may have isolated singlarities), we introduce the following concept:

♠ **Winding number:**
Suppose that C is a closed curve and that the point $z = a$ is not located on C. Then the number

$$n(C,a) = \frac{1}{2\pi i}\oint_C \frac{dz}{z-a}$$

is called the **winding number** of C around a.

Note that if C represents the boundary of a circle (traversed counterclockwise), then the winding number reads

$$n(C,a) = \begin{cases} 0 \text{ if } a \text{ is inside the circle,} \\ 1 \text{ if } a \text{ is outside the circle.} \end{cases}$$

Both identities have already been proven in the context of Cauchy's theorem. In addition, if the curve C encloses k times the point a, then we have

$$n(C,a) = \frac{1}{2\pi i}\int_0^{2k\pi} i d\theta = k,$$

which explains the terminology "winding number."

♠ **Theorem:**
For any closed curve C and point $a \notin C$, the winding number $n(C,a)$ is an integer.

Proof Suppose that C is parametrized by $z(t)$, $0 \le t \le 1$, and set

$$f(s) = \int_0^s \frac{z'(t)}{z(t)-a}dt \quad (0 \le s \le 1).$$

Then, it follows from

$$f'(s) = \frac{z'(s)}{z(s)-a}$$

that the quantity

$$[z(s)-a]e^{-f(s)}$$

is a constant, and setting $s = 0$, we have

$$[z(s)-a]e^{-f(s)} = z(0)-a.$$

Hence,

$$e^{f(s)} = \frac{z(s)-a}{z(0)-a} \text{ and } e^{f(1)} = \frac{z(1)-a}{z(0)-a} = 1,$$

since C is closed, i.e., $z(1) = z(0)$. Thus

$$f(1) = 2\pi k i \text{ for some integer } k$$

and

$$n(C,a) = \frac{1}{2\pi i} f(1) = k. \clubsuit$$

In terms of the winding number, the residue theorem given in Sect. 9.1.1 can be restated as follows:

♠ **Residue theorem (restated):**

Suppose $f(z)$ is analytic in a simply connected domain D except for isolated singularities at $z_1, z_2, \cdots, z_m$. Let C be a closed curve that does not intersect any of the singularities. Then

$$\oint_C f(z)dz = 2\pi i \sum_{k=1}^{m} n(C, z_k)\text{Res}(f, z_k). \tag{9.12}$$

The proof is left to the reader.

9.1.4 Ratio Method

We saw in Sect. 7.4.5 that a function having a pole of order m can be expressed by the ratio of two polynomials such as

$$f(z) = \frac{p(z)}{q(z)}. \tag{9.13}$$

In this case, it is possible to formulate an alternative equation that determines the residue of $f(z)$. Employing such an equation to evaluate the residue is referred to as a **ratio method**.

To derive these equations, we first recall the fact that if a function $R(z)$ satisfies

$$p(a) = p'(a) = \cdots = p^{(m-1)}(a) = 0 \quad \text{and} \quad p^{(m)}(a) \neq 0,$$

the Taylor series for $R(z)$ is given by

$$p(z) = \frac{p^{(m)}(a)}{m!}(z-a)^m + h.o.,$$

where $h.o.$ means the terms of higher order. Such a function, for which the lowest power of $(z-a)$ is m, is said to have an m**th-order zero** at a.

Now we present the equation for the residue of $f(z)$ at a simple pole a. As seen from (9.13), a simple pole of $f(z)$ arises from the fact that $p(z)$ has a zero of $(m-1)$th order and $q(z)$ has a zero of order m. Then,

$$f(z) = \frac{\dfrac{p^{(m-1)}(a)}{(m-1)!}(z-a)^{m-1} + h.o.}{\dfrac{q^{(m)}(a)}{m!}(z-a)^m + h.o.}.$$

For such a function, we obtain the residue of f at the simple pole a as

$$c_{-1} = \lim_{z\to a}(z-a)f(z) = m\frac{p^{(m-1)}(a)}{q^{(m)}(a)}. \tag{9.14}$$

By means of 9.14, we can compute the residue of $f(z)$ at a simple pole a quite easily.

Next we consider the equation for a second-order pole of $f(z)$ at a. Such a pole arises when $p(z)$ has a zero of order m and $q(z)$ has a zero of order $(m+2)$ at a. Then,

$$f(z) = \frac{\dfrac{p^{(m)}(a)}{m!}(z-a)^m + \dfrac{p^{(m+1)}(a)}{(m+1)!}(z-a)^{m+1} + h.o.}{\dfrac{q^{(m+2)}(a)}{(m+2)!}(z-a)^{m+2} + \dfrac{q^{(m+3)}(a)}{(m+3)!}(z-a)^{m+3} + h.o.},$$

from which we set

$$\begin{aligned} c_{-1} &= \lim_{z\to a}\frac{d}{dz}\left[(z-a)^2 f(z)\right] \\ &= \frac{m+2}{m+3}\cdot\frac{(m+3)p^{(m+1)}(a)q^{(m+2)}(a) - (m+1)p^{(m)}(a)q^{(m+3)}(a)}{\left[q^{(m+2)}(a)\right]^2}. \end{aligned} \tag{9.15}$$

For example, if the second-order pole of a function arises from a second-order zero of $q(z)$, then $m=0$. The residue of such a pole is given by (9.15) as

$$c_{-1} = \frac{2}{3}\frac{3p'(a)q''(a) - p(a)q^{(3)}(a)}{[q''(a)]^2}. \tag{9.16}$$

9.1.5 Evaluating the Residues

In what follows, we demonstrate actual procedures to evaluate the residue by means of the three methods discussed in the previous subsections. As an instructive example, we consider the function

$$f(z) = \frac{e^z}{z(z+2)^2},$$

which has a simple pole at $z = 0$ and a second-order pole at $z = -2$.

Using a Laurent expansion:

The present purpose is to evaluate the coefficient c_{-1} of the Laurent series expansion of $f(z)$ around the poles at $z = 0$ and $z = -2$. In order to do this we first determine the Taylor series for the factor $e^z/(z+2)^2$ around $z = 0$. Since the expressions

$$e^z = 1 + z + \frac{z^2}{2!} + \cdots$$

and

$$\frac{1}{(z+2)^2} = \frac{1}{4}\left[\frac{1}{1+(z/2)}\right]^2 = \frac{1}{4}\left(1 - z + \frac{3}{4}z^2 - \cdots\right)$$

hold around $z = 0$, we have

$$\frac{e^z}{z(z+2)^2} = \frac{1}{4z}\left(1 + z + \frac{z^2}{2!} + \cdots\right)\left(1 - z + \frac{3}{4}z^2 - \cdots\right) = \frac{1}{4z} + \frac{z}{16} + \cdots.$$

Thus, we immediately obtain

$$c_{-1}(0) = \frac{1}{4}.$$

Similarly we have

$$c_{-1}(-2) = -\frac{3}{4}e^{-2} \quad \text{(see Exercise 1)}.$$

Using Goursat's formula:

The residue of the simple pole at $z = 0$ is given by

$$c_{-1}(0) = \lim_{z\to 0}\left[z\,\frac{e^z}{z(z+2)^2}\right] = \frac{1}{4}$$

and that of the second-order pole at $z = -2$ is given by

$$c_{-1}(-2) = \frac{1}{1!}\lim_{z\to -2}\frac{d}{dz}\left[(z+2)^2\,\frac{e^z}{z(z+2)^2}\right] = -\frac{3}{4}e^{-2}.$$

Using the ratio method:

For this example, the numerator and denominator functions can be chosen in different ways. For the residue at $z = 0$, we could take

$$p(z) = e^z, \quad q(z) = z(z+2)^2$$

or, alternatively,

$$p(z) = \frac{e^z}{(z+2)^2}, \quad q(z) = z.$$

For either choice, the residue for the simple pole is given by

$$c_{-1}(0) = \frac{p(0)}{q'(0)} = \frac{1}{4}.$$

The residue $c_{-1}(-2)$ can be obtained in a similar manner as above (see Exercise **2**).

Exercises

1. Evaluate the residue of

$$f(z) = \frac{e^z}{z(z+2)^2}$$

at $z = -2$ by using a Laurent expansion.

Solution: The residue of $f(z)$ at $z = -2$ is found by using the expression

$$e^z = e^{-2}e^{z+2} = e^{-2}\sum_{n=0}^{\infty}\frac{(z+2)^n}{n!} = e^{-2}\left[1 + (z+2) + \frac{(z+2)^2}{2!} + \cdots\right],$$

and the Taylor series expansion for $1/z$ around $z = -2$ as

$$\frac{1}{z} = -\frac{1}{2}\left[\frac{1}{1-(z+2)/2}\right] = -\sum_{m=0}^{\infty}\frac{(z+2)^m}{2^{m+1}} = -\frac{1}{2} - \frac{z+2}{4} - \frac{(z+2)^2}{8} - \cdots.$$

Thus, the Laurent series for $f(z)$ around $z = -2$ is

$$\frac{e^z}{z(z+2)^2} = -\frac{1}{2}e^{-2}\left[\frac{1}{(z+2)^2} + \frac{3}{2}\frac{1}{z+2} + \frac{5}{4} + \cdots\right],$$

from which we have

$$c_{-1}(-2) = -\frac{3}{4}e^{-2}. \clubsuit$$

2. Evaluate the residue of

$$f(z) = \frac{e^z}{z(z+2)^2}$$

at $z = -2$ using the ratio method.

Solution: For the pole at $z = -2$, we can choose either

$$p(z) = e^z, \quad q(z) = z(z+2)^2$$

as before or

$$p(z) = \frac{e^z}{z}, \quad q(z) = (z+2)^2.$$

Then, regardless of how the numerator and denominator are chosen, we refer to (9.16) to obtain

$$c_{-1}(-2) = \frac{2}{3}\frac{3p'(-2)q''(-2) - p(-2)q^{(3)}(-2)}{[q''(-2)]^2} = -\frac{3}{4}e^{-2}. \ ♣$$

9.2 Applications to Real Integrals

9.2.1 Classification of Evaluable Real Integrals

Using the residue theorem, we can evaluate the five types of real integrals listed below.

1. $\displaystyle\int_0^{2\pi} f(\cos\theta, \sin\theta)d\theta$, where $f(x,y)$ is a rational function without a pole on the circle $x^2 + y^2 = 1$.

2. $\displaystyle\int_{-\infty}^{\infty} f(x)dx$, where $f(z)$ is a rational function without a real pole and is subject to the condition that $\displaystyle\lim_{|x|\to\infty} xf(x) = 0$.

3. $\displaystyle\int_{-\infty}^{\infty} f(x)e^{ix}dx$, where $f(z)$ is an analytic function in the upper-half plane $\text{Im}z \geq 0$ except at a finite number of points.

4. $\displaystyle\int_0^{\infty} f(x)/x^{\alpha}dx$, where α denotes a real number such that $0 < \alpha < 1$ and $f(z)$ is a rational function with no pole on the positive real axis $x \geq 0$, which satisfies the condition $f(z)/z^{\alpha-1} \to 0$ as $z \to 0$ and $z \to \infty$.

5. $\displaystyle\int_0^{\infty} f(x)\log x dx$, where $f(z)$ is a rational function with no pole on the positive real axis $x \geq 0$ and satisfies the condition $\displaystyle\lim_{x\to+\infty} xf(x) = 0$.

In Sect. 9.2.2–9.2.6 we demonstrate actual processes for evaluating the above integrals.

9.2.2 Type 1: Integrals of $f(\cos\theta, \sin\theta)$

Consider an integral of the form

$$\int_0^{2\pi} f(\cos\theta,\ \sin\theta)d\theta.$$

Setting $z = e^{i\theta}$ makes it a contour integral around the unit circle, and thus the evaluation of the residues within the circle completes the integration.

Example We evaluate the integral

$$I = \int_0^{2\pi} \frac{d\theta}{1 - 2p\cos\theta + p^2} \quad (p < 1). \tag{9.17}$$

If we express $\cos\theta$ in terms of $z = e^{i\theta}$, 9.17 becomes a contour integral,

$$I = \oint_C \frac{1}{1 - p(z + z^{-1}) + p^2} \frac{dz}{iz} = \frac{1}{i} \oint_C \frac{dz}{(1 - pz)(z - p)}, \tag{9.18}$$

where C is a unit circle centered at the origin. The integrand in (9.18) has a simple (first-order) pole at $z = p$ within C. Hence, we obtain

$$I = \frac{1}{i} \times 2\pi i \lim_{z \to p} \left(\frac{1}{1 - pz} \right) = \frac{2\pi}{1 - p^2}. \quad ♣$$

9.2.3 Type 2: Integrals of Rational Function

Next consider the integral

$$I = \int_{-\infty}^{\infty} f(x)dx, \tag{9.19}$$

where $f(x)$ is a rational function subject to the condition

$$\lim_{|x| \to \infty} xf(x) = 0,$$

which is a necessary and sufficient condition for the integral to be convergent. To evaluate (9.19), we consider the integral of $f(z)$ along a closed contour consisting of the real axis from $-R$ to $+R$ and a semicircle $\Gamma(R)$ in the upper half-plane. The contour integral is expressed as

$$\oint_C f(z)dz = \int_{-R}^{R} f(x)dx + \int_{\Gamma(R)} f(z)dz. \tag{9.20}$$

From the Lemma below, it follows that the second term in (9.20) vanishes in the limit $R \to \infty$. Hence, we obtain

$$\lim_{R\to\infty} \oint_C f(z)dz = \int_{-\infty}^{\infty} f(x)dx, \tag{9.21}$$

and applying the residue theorem yields

$$\int_{-\infty}^{\infty} f(x)dx = 2\pi i \sum_j \mathrm{Res}(f, a_j),$$

where a_j is the jth pole of $f(z)$ in the upper half-plane. Therefore, the evaluation of the residues located within the upper half-plane completes the integration.

Example We prove the equation

$$I = \int_{-\infty}^{\infty} \frac{dx}{1+x^2} = \pi.$$

Since $x/(1+x^2)$ vanishes as $|x| \to \infty$, we may follow a process similar to the one discussed above. Since $z = i$ is the only pole of $1/(1+z^2) = 1/(z+i)(z-i)$ involved in the upper half-plane, we have

$$I = (2\pi i) \cdot \mathrm{Res}(i) = 2\pi i \frac{1}{2i} = \pi.$$

Less simple examples will be found in Exercises Sect. 9.2. ♣

As was noted earlier, our result (9.21) is based on the following lemma:

♠ **Lemma:**
Let $f(z)$ be continuous in the sector $\theta_1 < \arg z < \theta_2$. If

$$\lim_{|z|\to\infty} z f(z) = 0 \ \text{ for } \ \theta_1 < \arg z < \theta_2, \tag{9.22}$$

then the integral $\int f(z)dz$ extended over the arc of the circle $|z| = r$ contained in the sector tends to 0 as $r \to \infty$.

Proof Let $M(r)$ be the upper bound of $|f(z)|$ on the arc of the circle $|z| = r$. Then we have

$$\left| \int f(z)dz \right| \le M(r) \int_{\theta_1}^{\theta_2} r d\theta = M(r) \cdot r(\theta_2 - \theta_1). \tag{9.23}$$

In view of the condition (9.22), the right-hand side of (9.23) vanishes as $r \to \infty$. This completes the proof. ♣

9.2.4 Type 3: Integrals of $f(x)e^{ix}$

We now study integrals of the form

$$\int_{-\infty}^{\infty} f(x)e^{ix}dx,$$

where f is analytic on the upper half-plane $\mathrm{Im}z \geq 0$ except at a finite number of singularities (if they exist). We first consider the case when the singularities are not on the real axis. Then, the integral

$$\int_{-R}^{R} f(x)e^{ix}dx$$

has a meaning, which can be seen from the following theorem:

♠ **Theorem:**
If $\lim_{|z|\to\infty} f(z) = 0$ for $\mathrm{Im}z \geq 0$, then

$$\lim_{R\to+\infty} \int_{-R}^{R} f(x)e^{ix}dx = 2\pi i \sum \mathrm{Res}\left[f(z)e^{ix}\right],$$

the summation extending over the singularities of $f(z)$ contained in the upper half-plane $y > 0$.

Before starting the proof, we note that $|e^{iz}| \leq 1$ in the half-plane $y \geq 0$. This leads us to integrate on the half-plane $y \geq 0$ along the contour used above for an integral of type **2**. To prove the theorem, thus it suffices to show that the integral $\int_{\Gamma(R)} f(z)e^{iz}dz$ tends to 0 as r tends to ∞.

If we know in advance that $\lim_{|z|\to\infty} zf(z) = 0$, then it would be sufficient to apply the lemma in Sect. 9.2.3. To prove that $\int_{\Gamma(R)} f(z)e^{iz}dz$ tends to 0 with only the hypothesis of the theorem above, we use the following lemma:

♠ **Jordan Lemma:**
Let $f(z)$ be a function defined in a sector of the half-plane $y \geq 0$. If $\lim_{|z|\to\infty} f(z) = 0$, the integral $\int f(z)e^{iz}dz$ extended over the arc of the circle $|z| = r$ contained in the sector tends to 0 as r tends to ∞.

Proof Let us put $z = re^{i\theta}$ and let $M(r)$ be the upper bound of $|f(re^{i\theta})|$ as θ varies, the point $e^{i\theta}$ remaining in the sector. Then,

$$\left|\int f(z)e^{iz}dz\right| \leq M(r) \int_0^{\pi} e^{-r\sin\theta} r\theta = 2M(r) \int_0^{\pi/2} e^{-r\sin\theta} r\theta. \qquad (9.24)$$

Since

$$\frac{\pi}{2} \le \frac{\sin\theta}{\theta} \le 1 \text{ for } 0 \le \theta \le \frac{\pi}{2},$$

we have

$$\int_0^{\pi/2} e^{-r\sin\theta} r\theta \le \int_0^{\pi/2} e^{-\frac{2}{\pi}r\theta} r\theta \le \int_0^{\infty} e^{-\frac{2}{\pi}r\theta} r\theta = \frac{\pi}{2}. \tag{9.25}$$

From (9.24) and (9.25), it follows that

$$\left| \int f(z) e^{iz} dz \right| \le \pi M(r). \tag{9.26}$$

In view of our assumption $\lim_{|z|\to\infty} f(z) = 0$, the right-hand side of (9.26) vanishes as $r \to \infty$, which completes the proof. ♣

Remark.

1. If we have to calculate an integral

$$\int_{-\infty}^{\infty} f(x) e^{-ix} dx$$

that involves a *negative* imaginary exponential e^{-ix}, it would be necessary to integrate in the *lower* half-plane instead of the upper one because the function $|e^{-iz}|$ is bounded in the lower half-plane $y \le 0$. More generally, an integral of the form $\int_{-\infty}^{\infty} f(x) e^{ax} dx$ (where a is complex constant) can be evaluated by integrating in the half-plane where $|e^{az}| \le 1$.
2. Remember that $\sin z$ and $\cos z$ are not bounded in any half-plane. To evaluate integrals of the form

$$\int_{-\infty}^{\infty} f(x) \sin^n x dx \quad \text{and} \quad \int_{-\infty}^{\infty} f(x) \cos^n x dx,$$

we always express the trigonometric functions in terms of complex exponentials so that the preceding methods can be applied.

9.2.5 Type 4: Integrals of $f(x)/x^\alpha$

Consider integrals of the form

$$\int_0^{\infty} \frac{f(x)}{x^\alpha} dx,$$

where α denotes a real number such that $0 < \alpha < 1$, and $f(x)$ is a rational function with no pole on the positive real axis $x \ge 0$. In addition, we assume $f(z)$ such that $f(z)/z^{\alpha-1} \to 0$ in the limits $z \to 0$ and $z \to \infty$.

To calculate such an integral, we consider the function

$$g(z) = \frac{f(z)}{z^\alpha}$$

of the complex variable z, defined in the plane with the positive real axis $x \geq 0$ excluded. Let D be the open set thus defined. It is necessary to specify the branch of z^α chosen in D, so we take the branch of the argument of z between 0 and 2π. With this convention, we integrate $g(z)$ along the closed path $C(r,\varepsilon)$ as follows: we first trace the real axis from $\varepsilon > 0$ to $r > 0$, then the circle $\Gamma(r)$ of centered at the origin and radius r in the positive sense, then the real axis from r to ε, and finally, the circle $\gamma(\varepsilon)$ of center 0 and radius ε in the negative sense. The integral

$$\int_{C(r,\varepsilon)} \frac{f(z)}{z^\alpha} dz$$

is equal to the sum of the residues of the poles of $f(z)/z^\alpha$ contained in D if r has been chosen sufficiently large and ε sufficiently small. We have

$$\int_{C(r,\varepsilon)} \frac{f(z)}{z^\alpha} dz = \int_{\Gamma(r)} \frac{f(z)}{z^\alpha} dz + \int_{\gamma(\varepsilon)} \frac{f(z)}{z^\alpha} dz + \left(1 - e^{-2\pi i\alpha}\right) \int_\varepsilon^r \frac{f(x)}{x^\alpha} dx$$

because when the argument of z is equal to 2π,

$$z^\alpha = e^{2\pi i\alpha} |z|^\alpha.$$

From assumption, $f(z)/z^{\alpha-1}$ tends to 0 when z tends to 0 or when $|z|$ tends to infinity. Thus the integrals along $\Gamma(r)$ and $\gamma(\varepsilon)$ tend to 0 as $r \to \infty$ and $\varepsilon \to 0$. On the limit, we have

$$\left(1 - e^{-2\pi i\alpha}\right) \int_0^\infty \frac{f(x)}{x^\alpha} dx = 2\pi i \sum \mathrm{Res}\left[\frac{f(z)}{z^\alpha}\right]. \tag{9.27}$$

This relation allows us to calculate the original integral.

Example Try to evaluate the integral

$$I = \int_0^\infty \frac{dx}{x^\alpha(1+x)} \quad (0 < \alpha < 1).$$

Here we have

$$f(z) = \frac{1}{1+z},$$

where there is only one pole at $z = -1$. As the branch of the argument of z is equal to π at this point, the residue of $f(z)/z^\alpha$ at this pole is equal to $1/e^{\pi i\alpha}$. Relation (9.27) then gives

$$I = \frac{\pi}{\sin \pi\alpha}.$$

9.2.6 Type 5: Integrals of $f(x) \log x$

The final type of integral to be noted is a class of the form

$$\int_0^\infty f(x) \log x dx,$$

where f is a rational function with no pole on the positive real axis $x \geq 0$ and

$$\lim_{x \to \infty} x f(x) = 0.$$

This last condition ensures that the integral is convergent.

We consider the same open set D as for integrals of **Type 4** and the same path of integration. Here again, we must specify the branch chosen for $\log z$, and we choose the argument of z between 0 and 2π. For a reason that will soon be apparent, we integrate the function $f(z)(\log z)^2$ instead of $f(z) \log z$. Here again the integrals along the circles $\Gamma(r)$ and $\gamma(\varepsilon)$ tend to 0 as $r \to \infty$ and $\varepsilon \to 0$, respectively.

When the argument z is equal to 2π, we have

$$\log z = \log x + 2\pi i.$$

Thus we have the relation

$$\int_0^\infty f(x)(\log x)^2 dx - \int_0^\infty f(x)(\log x + 2\pi i)^2 dx = 2\pi i \sum \text{Res}\left[f(z)(\log z)^2\right].$$

and, hence,

$$-2\int_0^\infty f(x) \log x dx - 2\pi i \int_0^\infty f(x) dx = \sum \text{Res}\left[f(z)(\log z)^2\right]. \quad (9.28)$$

By taking the imaginary part of the relation (9.28), we obtain the desired result:

$$\int_0^\infty f(x) \log x dx = -\frac{1}{2}\text{Im}\left\{\sum \text{Res}\left[f(z)(\log z)^2\right]\right\}.$$

Example Consider the integral

$$I = \int_0^\infty \frac{\log x}{(1+x)^3} dx.$$

As the residue of $(\log z)^2/(1+z)^3$ at the pole $z = -1$ is equal to $1 - i\pi$, we find

$$I = -\frac{1}{2}.$$

Exercises

1. Evaluate the integral defined by $I = \int_0^{2\pi} \frac{d\theta}{(1 - a\cos\theta)^2}$ $(0 < a < 1)$.

Solution: Let $z = e^{i\theta}$ and set $C: |z| = 1$. Then

$$I = \frac{4}{ia^2} \oint_C \frac{zdz}{[z^2 + (2z/a) + 1]^2}.$$

The integrand has two poles of second order at $z = z_1, z_2$ ($|z_1| < |z_2|$), which are the solutions of the equation $g(z) = z^2 + (2z/a) + 1 = 0$. Since $0 < a < 1$, only the pole $z_1 = (-1 + \sqrt{1 - a^2})/a$ is found within C. The residue at z_1 is given by

$$\begin{aligned}
\mathrm{Res}(z_1) &= \lim_{z \to z_1} \frac{d}{dz}\left[(z - z_1)^2 \frac{z}{g(z)^2}\right] = \lim_{z \to z_1} \frac{d}{dz} \frac{z}{(z - z_2)^2} \\
&= -\frac{z_1 + z_2}{(z_1 - z_2)^3} = \frac{2/a}{\left(2\sqrt{1 - a^2}/a\right)^3},
\end{aligned}$$

and thus we obtain

$$I = \frac{4}{ia^2} \times 2\pi i \mathrm{Res}(z_1) = \frac{2\pi}{\left(1 - a^2\right)^{3/2}}. \quad ♣$$

2. Evaluate the integral $I = \frac{1}{2\pi i} \oint_C \frac{e^z}{z^n} dz$ $(C: |z| = 1)$ for integer n.

Solution: For integers $n \leq 0$, it is apparent that $I = 0$ since the integrand is analytic within and on C. For integers $n > 0$, $f(z) = e^z z^{-n}$ has a pole of order n at $z = 0$. Using the residue theorem, we have $I = 1/(n-1)!$. ♣

3. Calculate the integral $I = \int_{-\infty}^{\infty} \frac{dx}{(1 + x^2)^{n+1}}$.

Solution: Define the function $f(z) = 1/(1 + z^2)^{n+1}$, and set the semicircle C as shown in Fig. 9.1. Within C, $f(z)$ has the pole of $(n+1)$th order at $z = i$, and its residue reads

$$\begin{aligned}
\frac{1}{n!}\left[\frac{d^n}{dz^n} \frac{(z - i)^{n+1}}{(1 + z^2)^{n+1}}\right]_{z=i} &= \frac{1}{n!}\left[\frac{d^n}{dz^n}(z + i)^{-(n+1)}\right]_{z=i} \\
&= \frac{(-1)^n (n+1)(n+2)\cdots 2n}{n!}(2i)^{-(2n+1)} \\
&= \frac{(2n)!}{2^{2n}(n!)^2} \frac{1}{2i}.
\end{aligned}$$

Hence, in view of Cauchy's theorem, we have

$$\oint_C f(z)dz = \frac{\pi(2n)!}{2^{2n}(n!)^2}. \tag{9.29}$$

We now observe that

$$\oint_C f(z)dz = \int_{-R}^{R} \frac{dx}{(1+x^2)^{n+1}} + \int_\Gamma \frac{dz}{(1+z^2)^{n+1}}, \tag{9.30}$$

where Γ denotes the upper half-circle. Since $|1+z^2| \geq R^2 - 1$ on C, the second integral in the limit $R \to \infty$ yields

$$\left| \int_\Gamma \frac{dz}{(1+z^2)^{n+1}} \right| \leq \frac{\pi R}{(R^2-1)^{n+1}} \to 0. \tag{9.31}$$

From (9.29)–(9.31), we conclude that

$$I = \frac{\pi(2n)!}{2^{2n}(n!)^2}. \quad ♣$$

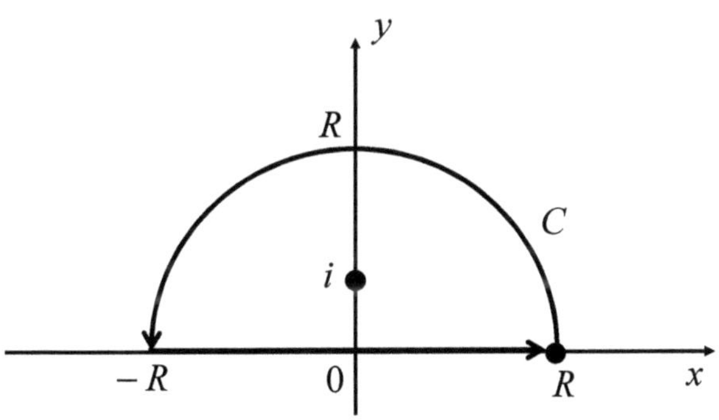

Fig. 9.1. The integration path used in Exercise **3**

4. Calculate the integral $I = \displaystyle\int_0^\pi \log(1 - 2r\cos\theta + r^2)d\theta$, where $r \neq 1$.

Solution: First we assume that $r < 1$. Observe that the function $\log(1-z)/z = -1-(z/2)-(z^2/3)-\cdots$ is analytic for $|z| \leq r < 1$. Hence, if we set the circle $C : |z| = r$, we have

$$\oint_C \frac{\log(1-z)}{z}dz = i\int_0^{2\pi} \log(1-z)d\theta = 0. \tag{9.32}$$

Since $|1-z|^2 = 1-2r\cos\theta+r^2$ on C, the real component of the second integral in (9.32) reads $(i/2)\int_0^{2\pi} \log\left(1 - 2r\cos\theta + r^2\right) d\theta = 0$, so we get

$$I = 0 \quad \text{for } r < 1.$$

Next we consider the case of $r > 1$. Set $s = 1/r < 1$ to obtain

$$0 = \int_0^\pi \log\left(1 - 2s\cos\theta + s^2\right) d\theta = \int_0^\pi \log\left(1 - \frac{2}{r}\cos\theta + \frac{1}{r^2}\right) d\theta$$
$$= \int_0^\pi \left[\log\left(1 - 2r\cos\theta + r^2\right) - \log r^2\right] d\theta.$$

Hence, we conclude that

$$I = 2\pi \log r \quad \text{for } r > 1. \quad \clubsuit$$

5. Calculate the integral $I = \displaystyle\int_0^\infty \frac{x^{\alpha-1}}{1+x} dx$, where $0 < \alpha < 1$.

Solution: Consider the power function

$$z^\beta = e^{\beta \log z} = e^{\beta(\log|z| + i\arg z)}$$

with $-1 < \beta < 0$. Its branch for $0 < \arg z < 2\pi$ is single-valued on the domain D enclosed by the contour $C = AB + \Gamma + B'A' + \gamma$ depicted in Fig. 9.2. Let the radius r of the circle γ be sufficiently small and that R of Γ be sufficiently large. Then, the pole $z = -1$ of the function $f(z) = z^\beta/(1+z)$ is located within C so that we have

$$\oint_C f(z)dz = \int_r^R \frac{x^\beta}{1+x} + \int_\Gamma f(z)dz - \int_r^R \frac{x^\beta e^{\beta \cdot 2\pi i}}{1+x} + \int_\gamma f(z)dz. \tag{9.33}$$

Observe that

$$\left|\int_\Gamma f(z)dz\right| \le \int_\Gamma \frac{|z^\beta|}{|1+z|}|dz| \le \frac{2\pi R^{\beta+1}}{R-1} \to 0 \quad (R \to \infty)$$

and

$$\left|\int_\gamma f(z)dz\right| \le \frac{2\pi r^{\beta+1}}{1-r} \to 0 \quad (r \to 0).$$

Take the limits $R \to \infty$ and $r \to 0$ on both sides of (9.33) to yield

$$\left(1 - e^{\beta \cdot 2\pi i}\right)\int_0^\infty \frac{x^\beta}{1+x} dx = \text{Res}\left[\frac{z^\beta}{1+z}, -1\right] = 2\pi i \lim_{z \to -1} z^\beta = 2\pi i e^{\beta \cdot \pi i},$$

which then gives us

$$\int_0^\infty \frac{x^\beta}{1+x} dx = 2\pi i \frac{e^{\beta \cdot \pi i}}{1 - e^{\beta \cdot 2\pi i}}.$$

Since $\beta = \alpha - 1$, the above result is equivalent to

$$I = \frac{\pi}{\sin \alpha\pi}. \quad \clubsuit$$

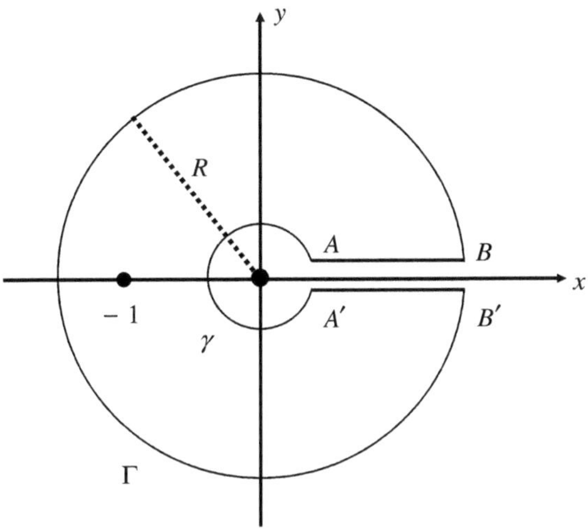

Fig. 9.2. Integration path $C = AB + \Gamma + B'A' + \gamma$ used in Exercise **5**

9.3 More Applications of Residue Calculus

9.3.1 Integrals on Rectangular Contours

The integrals discussed so far are evaluated using the residue theorem based on a circular (or semicircular) contour whose radius is eventually made to be infinitely large or infinitely small. However, there are other integrals that can be evaluated by the residue theorem that do not have to be closed with a circle. Several examples are given below.

Let us consider the integral

$$I = \int_{-\infty}^{\infty} \frac{xe^x}{(1+e^{2x})^2} dx.$$

To evaluate it, we examine the contour integral

$$J = \oint_C \frac{ze^z}{(1+e^{2z})^2} dz \tag{9.34}$$

around the rectangular contour shown in Fig. 9.3. Beginning at the lower left hand corner of the rectangle,

$$\begin{aligned} J = &\int_{-L}^{L} \frac{xe^x}{(1+e^{2x})^2} dx + \int_0^{\pi} \frac{(L+iy)e^{L+iy}}{\left(1+e^{2(L+iy)}\right)^2} idy + \int_L^{-L} \frac{(x+i\pi)e^{x+i\pi}}{\left(1+e^{2(x+i\pi)}\right)^2} dx \\ &+ \int_{\pi}^{0} \frac{(-L+iy)e^{-L+iy}}{\left(1+e^{2(-L+iy)}\right)^2} idy. \end{aligned} \tag{9.35}$$

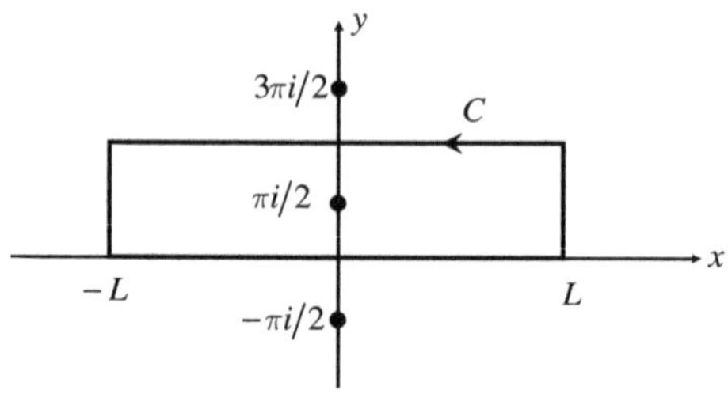

Fig. 9.3. Rectangular contour surrounding the path $z = \pi i/2$

In the limit $L \to \infty$, the second and fourth integral of (9.35) go to zero, since in this limit the magnitude of $e^{2(L+iy)}$ and $e^{2(-L+iy)}$ become very large or very small, respectively, compared to unity. Hence, we have

$$\begin{aligned}\lim_{L\to\infty} J &= \int_{-\infty}^{\infty} \frac{xe^x}{(1+e^{2x})^2}dx + \int_{\infty}^{-\infty} \frac{(x+i\pi)e^{x+i\pi}}{\left(1+e^{2(x+i\pi)}\right)^2}dx, \\ &= I + \int_{-\infty}^{\infty} \frac{(x+i\pi)e^{x+i\pi}}{\left(1+e^{2(x+i\pi)}\right)^2}dx \;=\; 2I - i\pi \int_{-\infty}^{\infty} \frac{e^x}{(1+e^{2x})^2}dx, \quad (9.36)\end{aligned}$$

where we have used the expressions $e^{x+i\pi} = -e^x$ and $e^{2(x+i\pi)} = e^{2x}$. As a result, the integral I to be evaluated is expressed in terms of J as

$$I = \frac{1}{2}\lim_{L\to\infty} J + \frac{i\pi}{2}\int_{-\infty}^{\infty} \frac{e^x}{(1+e^{2x})^2}dx. \quad (9.37)$$

The contour integral J is readily evaluated by employing the residue theorem. Looking back to the definition (9.34), we see that J has second-order poles at the values of z for which $e^{2z} = -1$. These values are

$$z = \pm\frac{i\pi}{2}, \pm\frac{3i\pi}{2}, \cdots, \pm i\left(N+\frac{1}{2}\right)\pi,$$

where N is a nonnegative integer. Note that only the pole at $z = i\pi/2$ is enclosed in the rectangle (see Fig. 9.3). Hence, using the ratio method (see Sect. 9.1.4) we have

$$J \;= 2\pi i \mathrm{Res}\left(\frac{i\pi}{2}\right) = 2\pi i \cdot 2\frac{p'(i\pi/2)}{q''(i\pi/2)} = \frac{-\pi(2+i\pi)}{4}, \quad (9.38)$$

where $p(z) = ze^z$ and $q(z) = (1+e^{2z})^2$ are constituents of the integrand in (9.34).

The latter integral of (9.37) is evaluated by substituting $w = e^x$, and it follows that

$$\int_{-\infty}^{\infty} \frac{e^x}{(1+e^{2x})^2}dx = \int_0^{\infty} \frac{dw}{(1+w^2)^2} = \frac{1}{2}\int_{-\infty}^{\infty} \frac{dw}{(1+w^2)^2}, \quad (9.39)$$

Thus, applying the residue theorem yields

$$\frac{1}{2}\int_{-\infty}^{\infty}\frac{dw}{(1+w^2)^2} = \frac{1}{2}\cdot 2\pi i\cdot \mathrm{Res}(i) = \pi i \lim_{w\to i}\frac{d^2}{dw^2}\frac{1}{(w+i)^2} = \frac{\pi}{4}. \quad (9.40)$$

From (9.38) and (9.40), we finally obtain

$$I = -\frac{\pi}{4}(2+i\pi) + \frac{\pi}{4} = -\frac{\pi}{4}(1+i\pi).$$

9.3.2 Fresnel Integrals

We would like to derive the equations

$$\int_0^{\infty}\cos(kx^2)dx = \int_0^{\infty}\sin(kx^2)dx = \frac{1}{2}\sqrt{\frac{\pi}{2k}}$$

with a real positive constant k. These are known as the **Fresnel cosine integral** and **Fresnel sine integral**. Integrals of this type are encountered in the study of a phenomenon called diffraction, which is exhibited by all types of waves such as light and sound.

In this connection we consider the integral

$$I = \oint_C e^{ikz^2}dz \quad (k>0) \quad (9.41)$$

around the contour shown in Fig. 9.4. The integral variable z becomes $z=x$ on the segment along the real axis, $z = Re^{i\phi}$ $(0\le\phi\le\pi/4)$ along the large (ultimately infinite) arc, and $z = x(1+i)$ along the slanted segment defined by $y=x$. Therefore, with $(1+i)^2 = 2i$, we have

$$\lim_{R\to\infty}\oint_C e^{ikz^2}dz = \int_0^{\infty}e^{ikx^2}dx + \lim_{R\to\infty}\left[\int_0^{\pi/4}e^{ikR^2e^{2i\phi}}iRd\phi\right] + (1+i)\int_{\infty}^{0}e^{-2kx^2}dx. \quad (9.42)$$

Our objective is to evaluate the real and imaginary parts of the first integral on the right-hand side of (9.42). Then, evaluations of the other integrals shown in (9.42) complete the computation.

First, we readily obtain

$$\oint_C e^{ikz^2}dz = 0, \quad (9.43)$$

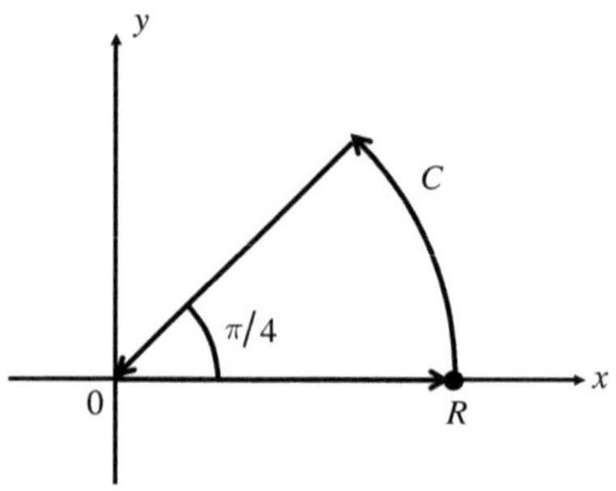

Fig. 9.4. Contour for evaluating the integral (9.41)

since there are no poles within the contour of Fig. 9.4.

Second, we consider the integral along the arc, which is given in the second term on the right-hand side of (9.42). On the large arc, we have

$$\left| Re^{ikR^2 e^{2i\phi}} \right| = \left| Re^{ikR^2 \cos(2\phi)} e^{-kR^2 \sin(2\phi)} \right| \le Re^{-kR^2 \sin(2\phi)},$$

where the sign of $\sin(2\phi)$ is always nonnegative in the range $0 \le \phi \le \pi/4$. Hence,

$$\lim_{R\to\infty} Re^{-kR^2 \sin(2\phi)} = 0, \tag{9.44}$$

so that the integral along the arc vanishes in the limit $R \to \infty$. In fact, l'Hôpital's rule states that for $a \ge 0$,

$$\lim_{R\to\infty} \frac{R}{e^{aR^2}} = \lim_{R\to\infty} \frac{1}{2aRe^{aR^2}} = 0.$$

Finally we examine the integral along the slanted segment, i.e., the third term on the right-hand side of (9.42). To evaluate it, we consider the quantity

$$J \equiv \left(\int_{-\infty}^{\infty} e^{-2kx^2} dx \right)^2 = \int_{-\infty}^{\infty} e^{-2kx^2} dx \cdot \int_{-\infty}^{\infty} e^{-2ky^2} dy$$
$$= \int_{-\infty}^{\infty} dx \int_{-\infty}^{\infty} dy\, e^{-2k(x^2+y^2)}.$$

In terms of the polar coordinates, it yields

$$J = \int_0^{\infty} dr \int_0^{2\pi} d\theta\, re^{-2kr^2} = \int_0^{\infty} \frac{d(r^2)}{2} \int_0^{2\pi} d\theta\, e^{-2kr^2} = \frac{\pi}{2k}.$$

and we have the Gaussian integral given by

$$\int_{-\infty}^{\infty} e^{-2kx^2} dx = \sqrt{\frac{\pi}{2k}},$$

so that

$$(1+i)\int_{\infty}^{0} e^{-2kx^2}dx = \frac{1+i}{2}\sqrt{\frac{\pi}{8k}}. \tag{9.45}$$

Substituting the results of (9.43), (9.44), and (9.45) into (9.42), we find that

$$\int_0^{\infty} e^{ikx^2}dx = \frac{1+i}{2}\sqrt{\frac{\pi}{2k}}. \tag{9.46}$$

Writing the exponential in trigonometric form and equating the real and imaginary parts of both sides of (9.42), we obtain the Fresnel integral:

$$\int_0^{\infty}\cos(kx^2)dx = \int_0^{\infty}\sin(kx^2)dx = \frac{1}{2}\sqrt{\frac{\pi}{2k}}. \quad ♣$$

9.3.3 Summation of Series

Our final application of the residue theorem is the summation of a series $\sum_{n=-\infty}^{\infty} f(n)$. Using this method, we can convert a certain type of series to simple forms such as

$$\sum_{n=-\infty}^{\infty}\frac{1}{(a+n)^2} = \frac{\pi^2}{\sin^2(\pi a)} \tag{9.47}$$

and

$$\sum_{n=1}^{\infty}\frac{2x}{x^2+n^2\pi^2} = \coth x - \frac{1}{x}.$$

This technique is particularly useful, for instance, to express a power series solution of a differential equation in a simple closed form. In fact, this device is generalized for various series summations as shown below.

♠ **Theorem:**

An infinite series of functions $f(n)$ with respect to an integer n is given by

$$\sum_{n=-\infty}^{\infty} f(n) = -\sum_{n=-\infty}^{\infty}\text{Res}\,(g, a_n), \tag{9.48}$$

where Res (g, a_n) is the residue of the specific function

$$g(z) = \frac{\pi f(z)}{\tan(\pi z)}$$

at the nth pole of $f(z)$ located at $z = a_n$.

According to this theorem, we see that if the number of poles of $f(z)$ is finite and the values of Res (g, a_n) are readily obtained, the series on the left-hand side of (9.48) is written in a simple form.

Proof The key point is to use a function given by $\pi/\tan(\pi z)$. This function has simple poles at $z = 0, \pm 1, \pm 2, \cdots$, each with residue 1 evaluated as

$$\lim_{z\to n} \frac{\pi}{\tan(\pi z)} \cdot (z-n) = \lim_{z\to n} \frac{\pi}{\pi/\cos^2(\pi z)} = 1,$$

where we used **l'Hôpital's rule** (see Exercise **3** in Sect. 8.1). In addition, the function $\pi/\tan(\pi z)$ is bounded at infinity except on the real axis. To derive (9.48), let us consider the contour integral

$$\oint_{C_1} \frac{\pi f(z)}{\tan(\pi z)} dz \tag{9.49}$$

around the contour C_1 shown in Fig. 9.5. Here $f(z)$ is assumed to have no branch points or essential singularities anywhere. Since only the pole at $z = 0$ is found within C_1, the contour integral equals $2\pi i$ times the residue of the integrand at $z = 0$, which is $f(0)$, i.e.,

$$\oint_{C_1} \frac{\pi f(z)}{\tan(\pi z)} dz = 2\pi i f(0).$$

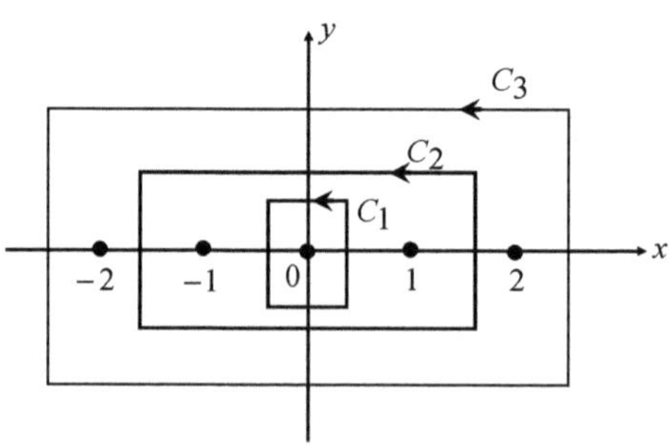

Fig. 9.5. A sequence of rectangular contours to derive equation (9.48)

Next, the integral around contour C_2 is

$$\oint_{C_2} \frac{\pi f(z)}{\tan(\pi z)} dz = 2\pi i \left[f(0) + f(1) + f(-1) + \text{Res}(g, a_1) \right],$$

where $\text{Res}(g, a_1)$ stems from the contribution of the pole of $f(z)$ located at $z = a_1$. Finally, for a contour at infinity, the integral must be

$$\oint_{C_\infty} \frac{\pi f(z)}{\tan(\pi z)} dz = 2\pi i \left\{ \sum_{n=-\infty}^{\infty} \left[f(n) + \text{Res}(g, a_n) \right] \right\}. \tag{9.50}$$

If $|zf(z)| \to 0$ as $|z| \to \infty$, the infinite contour integral is zero so that we successfully obtain the equation:

$$\sum_{n=-\infty}^{\infty} f(n) = -\sum_{n=-\infty}^{\infty} \text{Res}(g, a_n). \quad \clubsuit \tag{9.51}$$

9.3.4 Langevin and Riemann zeta Functions

Our present aim is to establish the equivalence between **Langevin's function**,

$$\coth x - (1/x),$$

and the sum

$$\sum_{n=1}^{\infty} \frac{2x}{x^2 + n^2\pi^2}.$$

Letting $f(z) = 2x/(x^2 + z^2\pi^2)$, and using the above equation, we obtain

$$\sum_{m=-N}^{N} \frac{2x}{x^2 + n^2\pi^2} = \frac{1}{2\pi i} \oint_C \pi \cot \pi z f(z) dz - \sum_{\text{poles}} \text{Res}\left[\pi \cot(\pi z) f(z)\right],$$

where C is a closed contour, say, a rectangle, enclosing the points $z = 0, \pm 1, \cdots$. Now let the length and width of the rectangle C approach ∞. As this happens,

$$\left| \frac{1}{2\pi i} \oint_C \pi \cot \pi z f(z) dz \right| \le \frac{1}{2} \oint_C \pi |\cot \pi z| \left| \frac{2x}{x^2 + n^2\pi^2} \right| |dz| \to 0. \tag{9.52}$$

Hence, we have

$$\begin{aligned} \sum_{m=-\infty}^{\infty} \frac{2x}{x^2 + n^2\pi^2} &= -\text{Res}\left[\frac{(\pi \cot \pi z) 2x}{x^2 + z^2\pi^2}\right]_{z=\pm ix/\pi} \\ &= -\frac{2x}{\pi}\left[\frac{\cot(ix)}{2ix/\pi} + \frac{\cot(-ix)}{-2ix/\pi}\right] \\ &= 2i \cot(ix) = 2 \coth x. \end{aligned}$$

This result can be rewritten as

$$2 \sum_{m=1}^{\infty} \frac{2x}{x^2 + n^2\pi^2} + \frac{2}{x} = 2 \coth x$$

or

$$\coth x - \frac{1}{x} = \sum_{m=1}^{\infty} \frac{2x}{x^2 + n^2\pi^2}, \tag{9.53}$$

which establishes the result we stated at the outset.

Remark. To see that the integral in (9.52) vanishes as $z \to \infty$, we observe that

$$|\cot \pi z| = \frac{|\cos \pi z|}{|\sin \pi z|} = \sqrt{\frac{\cos^2 \pi x + \sinh^2 \pi y}{\sin^2 \pi x + \sinh^2 \pi y}}.$$

If we choose the rectangle whose vertical sides cross the x-axis at a large enough half-integer, say, $x = 10^5 + \frac{1}{2}$ so that $\cos \pi x = 0$ and $\sin \pi x = 1$, then over these sides of the rectangle

$$|\cot \pi z| = \sqrt{\frac{\sinh^2 \pi y}{1 + \sinh^2 \pi y}} = |\tanh \pi y| \leq 1.$$

Over the horizontal sides of the rectangle, $\lim_{z\to\infty} |\cot \pi z| = 1$. Thus the integrand goes as $|1/z^2|$ as $|z| \to \infty$, and the integral vanishes.

If we integrate both sides of (9.53) from 0 to x, we get

$$\sum_{m=1}^{\infty} \ln\left(1 + \frac{x^2}{m^2\pi^2}\right) = \ln\left[\prod_{m=1}^{\infty}\left(1 + \frac{x^2}{m^2\pi^2}\right)\right] = \ln\left(\frac{\sinh x}{x}\right).$$

Hence,

$$\frac{\sinh x}{x} = \prod_{m=1}^{\infty}\left(1 + \frac{x^2}{m^2\pi^2}\right).$$

We may extend this result to all z in the complex plane by **analytic continuation**. Then setting $x = i\theta$ with θ real, we obtain

$$\sin \theta = \theta \prod_{n=1}^{\infty}\left(1 - \frac{\theta^2}{n^2\pi^2}\right).$$

This infinite product formula displays all the zeros of $\sin \theta$ explicitly. It represents the complete factorization of the Taylor series and can, in fact, be taken as the definition of the sine function.

By equating coefficients of the θ^3 term of both sides of the above equation, we obtain a useful sum:

$$\sum_{n=1}^{\infty} \frac{1}{n^2} = \frac{\pi^2}{6},$$

which is a special value of the **Riemann zeta function**,

$$\zeta(z) = \sum_{n=1}^{\infty} \frac{1}{n^z}.$$

Exercises

1. Evaluate $\sum_{n=-\infty}^{\infty} \frac{1}{(a+n)^2}$ by considering the contour integral:

$$I = \oint_C \frac{\pi}{\tan(\pi z)} \frac{1}{(a+z)^2} dz,$$

where a is not an integer and C is a circle of large radius,

Solution: In order to use equation (9.48), we define

$$f(z) = \frac{1}{(a+z)^2} \quad \text{and} \quad g(z) = \frac{\pi}{\tan(\pi z)} \cdot \frac{1}{(a+z)^2}.$$

Since the integrand $g(z)$ has simple poles at $z = 0, \pm 1, \pm 2 \cdots$ and a double pole at $z = -a$, evaluation of $\mathrm{Res}(g, -a)$ completes the problem [see (9.48)]. To find the residue at $z = -a$, set $z = -a + \xi$ for small ξ and determine the coefficient of ξ^{-1}:

$$\begin{aligned} \frac{\pi}{\tan(\pi z)} \frac{1}{(a+z)^2} &= \frac{\pi}{\xi^2} \frac{1}{\tan(-a\pi + \xi\pi)} \\ &= \frac{\pi}{\xi^2} \left\{ \frac{1}{\tan(-a\pi)} + \xi \left[\frac{d}{dz} \frac{1}{\tan(\pi z)} \right]_{z=-a} + \cdots \right\}. \end{aligned}$$

It follows from (9.54) that the residue at the double pole $z = -a$ is

$$\pi \left[\frac{d}{dz} \frac{1}{\tan(\pi z)} \right]_{z=-a} = \pi \left[\frac{-\pi}{\sin^2(\pi z)} \right]_{z=-a} = -\frac{\pi^2}{\sin^2(\pi a)}.$$

Therefore, it is readily seen from (9.48) that

$$\sum_{n=-\infty}^{\infty} \frac{1}{(a+n)^2} = \frac{\pi^2}{\sin^2 \pi a}. \quad ♣$$

9.4 Argument Principle

9.4.1 The Principle

It may occur that a function $f(z)$ has several zeros and poles simulateously in a domain D. If we denote the number of such zeros and poles by N_0 and N_∞, respectively, these numbers are related to one another as stated below.

♠ **Argument principle:**

Let $f(z)$ be an analytic function within a closed contour C except at a finite number of poles. If $f(z) \neq 0$ on C, then

$$\frac{1}{2\pi i} \oint_C \frac{f'(z)}{f(z)} dz = N_0 - N_\infty, \tag{9.54}$$

where N_0 and N_∞ are the numbers of zeros and poles of $f(z)$ in C, respectively. Both zeros and poles are to be counted with their multiplicities.

Proof By the residue theorem, the integral

$$\frac{1}{2\pi i} \oint_C \frac{f'(z)}{f(z)} dz$$

is equal to the sum of the residues of the logarithmic derivative of $f(z)$ in D, i.e.,

$$g(z) = \frac{f'(z)}{f(z)} = \frac{d[\log f(z)]}{dz}.$$

The only possible singularities of $g(z)$ in D coincide with the zeros and poles of $f(z)$. In order to determine the residue of $g(z)$ at a zero of $f(z)$, we observe that in the neighborhood of a zero a of the nth order, $f(z)$ has an expansion

$$f(z) = (z-a)^n \left[c_1 + c_2(z-a) + \cdots\right], \quad c_1 \neq 0.$$

We therefore have

$$f(z) = (z-a)^n f_1(z),$$

where $f_1(z) \not\equiv 0$ in a certain neighborhood of $z = a$. Hence,

$$\log f(z) = n \log(z-a) + \log f_1(z),$$

and

$$\frac{f'(z)}{f(z)} = \frac{n}{z-a} + \frac{f_1'(z)}{f_1(z)},$$

where the last term is analytic at $z = a$. It follows that the residue of $g(z)$, which is called the **logarithmic residue** of $f(z)$ at $z = a$ is n, i.e., it is equal to the order of the zero of $f(z)$ at $z = a$. If the zeros of $f(z)$ in D are counted with their multiplicities, the sum of the logarithmic residues of $f(z)$ at the zeros of $f(z)$ in D will be equal to the number of zeros.

We now turn to the poles of $f(z)$ in D. If $z = b$ is a pole of order m, we have near it an expansion

$$\begin{aligned} f'(z) &= \frac{c_1}{(z-b)^m} + \cdots + \frac{c_m}{z-b} + c_{m+1} + \cdots \\ &= \frac{1}{(z-b)^m}\left[c_1 + c_2(z-b) + \cdots\right] \\ &= \frac{f_2(z)}{(z-b)^m}, \end{aligned}$$

where $f_2(z)$ is analytic at $z = b$ and $f_2(z) \neq 0$. Hence,

$$\frac{f'(z)}{f(z)} = -\frac{m}{z-b} + \frac{f_2'(z)}{f_2(z)},$$

which shows that the logarithmic residue of $f(z)$ at a pole of $f(z)$ of order m is $-m$. If the poles of $f(z)$ in D are counted with their multiplicities, the sum of the logarithmic residues of $f(z)$ at the points of $f(z)$ in D will be equal to minus the number of these poles. Since $g(z)$ has no singularities in D except at the zeros and poles of $f(z)$, we have proven our theorem. ♣

Remark. If we replace $f(z)$ in (9.54) by $f(z) - a$, this formula will yield the difference between the number of zeros and the poles of $f(z) - a$. Since the latter are identical with the poles of $f(z)$, we find that

$$\frac{1}{2\pi i}\oint_C \frac{f'(z)}{f(z)-a}dz = N_a - N_\infty,$$

where N_a indicates how often the value of a is taken by $f(z)$ in D.

Examples **1.** For $f(z) = z^2$ and $C : |z| = 1$, $N_0 = 2$ and $N_\infty = 0$ so that we have

$$\frac{1}{2\pi i}\oint_C \frac{f'(z)}{f(z)}dz = 2.$$

In fact, the integral reads

$$\frac{1}{2\pi i}\oint_C \frac{f'(z)}{f(z)}dz = \frac{1}{2\pi i}\oint_C \frac{2z}{z^2}dz = \frac{1}{2\pi i} \times 2 \times 2\pi i = 2.$$

2. For $f(z) = z/(z-a)$ and $C : |z| = R$, $N_0 = 1$ and

$$N_\infty = \begin{cases} 1 & \text{if } R > a, \\ 0 & \text{if } R < a. \end{cases}$$

Hence, we have

$$\frac{1}{2\pi i}\oint_C \frac{f'(z)}{f(z)}dz = \begin{cases} 0 & \text{if } R > a, \\ 1 & \text{if } R < a. \end{cases} \tag{9.55}$$

Indeed, $f(z) = 1 + [a/(z-a)]$, $f'(z) = -1/(z-a)^2$, $f'/f = (1/z) - [1/(z-a)]$, which yields (9.55).

9.4.2 Variation of the Argument

Equation (9.54) can be brought into a different form in which its geometric character becomes more apparent. If we write

$$\varphi = \arg f(z), \quad f(z) = |f(z)|e^{i\varphi},$$

we obtain

$$\begin{aligned}\frac{1}{2\pi i}\oint_C \frac{f'(z)}{f(z)}dz &= \frac{1}{2\pi i}\oint_C d\log f(z)\\ &= \frac{1}{2\pi i}\oint_C [d\log|f(z)| + id\varphi]\\ &= \frac{1}{2\pi i}\oint_C d\log|f(z)| + \frac{1}{2\pi}\oint_C d\varphi.\end{aligned}$$

Recall that $\log w(z)$ is a many-valued function of w. If $\log w$ is continued along a closed curve that surrounds to origin, we shall not return to the value of $\log w$ with which we started. However, this many-valuedness is confined to $\mathrm{Im}(\log w) = \arg w$, i.e., $\mathrm{Re}(\log w) = \log|w|$ is single-valued. If we write $w = f(z)$, it follows that

$$\oint_C d\log|f(z)| = 0.$$

In fact,

$$\int_{z_1}^{z_2} d\log|f(z)| = \log|f(z_2)| - \log|f(z_1)|,$$

and if the integration is performed over a closed contour, the terminals z_1 and z_2 of the integration coincide; moreover, owing the single-valuedness of $\log|f(z)|$, the value of the integral is zero. Hence, we have

$$\frac{1}{2\pi i}\oint_C \frac{f'(z)}{f(z)}dz = \frac{1}{2\pi}\oint_C d\varphi, \tag{9.56}$$

where $\varphi = \arg f(z)$.

To interpret (9.56), we observe that

$$\int_{z_1}^{z_2} d\varphi = \varphi(z_2) - \varphi(z_1) = \arg f(z_2) - \arg f(z_1)$$

is the quantitiative change in the argument of $f(z)$, which is called the **variation of the argument** of $f(z)$. The integral $\oint_C d\varphi$ is therefore the total variation of $\arg f(z)$ if z describes the entire boundary C of the domain

D. It is clear that the value of this integral must be an integral multiple of 2π. If z describes C, the point $f(z)$ describes a closed curve C', and if C' surrounds the origin m times in the positive (counterclockwise) direction, the increase in $\arg f(z)$ along C' will be $2m\pi$. In view of (9.54) and (9.56), we obtain the theorem below.

♠ **Theorem:**
Let the domain D be bounded by one or more closed contours C and let a function $f(z)$ be single-valued and analytic apart from a finite number of poles. If N_0 and N_∞ denote the number of zeros and poles of $f(z)$ in D, respectively, and $f(z) \neq 0$ on C, then

$$\frac{1}{2\pi}\Delta_c = N_0 - N_\infty,$$

where Δ_c denotes the **total variation** of $\arg f(z)$.

9.4.3 Extentson of the Argument Principle

The argument principle can be extended to the case in which $f(z)$ has zeros or poles on the boundary C of the domain D. Suppose that $f(z_0) = 0$, where z_0 is situated on C. Let $f(z)$ be analytic at z_0; then we have

$$f(z) = (z - z_0)^m f_1(z), \quad f_1(z_0) \neq 0,$$

if m is the multiplicity of the zero. In view of the relation

$$\log f(z) = m \log(z - z_0) + \log f_1(z),$$

it follows that

$$\arg f(z) = m\arg(z - z_0) + \arg f_1(z).$$

At $z = z_0$, $f_1(z) \neq 0$ and $\log f(z)$ is analytic. Hence, $\arg f_1(z)$ will vary continuously if z varies along C and passes through $z = z_0$, but the expression $\arg(z - z_0)$ shows a different behavior. Since this is the angle between the parallel to the positive axis through z_0 and the linear segment drawn from z_0 to z, it is clear that if z_0 is passed $\arg(z - z_0)$ jumps by the amount π. The contribution of this zero to $\arg f(z)$ will be $m\pi$, i.e., one-half of what it would have been if the zero were situated in the interior of D. If $z = z_0$ is a pole of order m, its contribution to $\arg f(z)$ will be $-m\pi$. This follows immediately from the fact that $f(z)^{-1}$ has a zero of order m at z_0 and that

$$\log[f(z)^{-1}] = -\log f(z).$$

We therefore have the following extension of the argument principle.

♠ **Extended argument principle:**
The argument principle remains valid if $f(z)$ has poles and zeros on the boundary, provided that these poles and zeros are counted with half their multiplicities.

9.4.4 Rouché Theorem

As an application of the argument principle, we prove the following result, known as the **Rouché theorem.**

♠ **Rouché theorem:**
If the function $f(z)$ and $g(z)$ are analytic and single-valued in a domain D and on its boundary C and if $|g(z)| < |f(z)|$ on C, then the number of zeros of the function $f(z)+g(z)$ within D is equal to that of zeros of $f(z)$.

Proof We have

$$\log\left[f(z)+g(z)\right] = \log f(z) + \log\left[1+\frac{g(z)}{f(z)}\right],$$

whence

$$\arg\left[f(z)+g(z)\right] = \arg f(z) + \arg\left[1+\frac{g(z)}{f(z)}\right]. \tag{9.57}$$

On the contour C, we have

$$\left|\frac{g(z)}{f(z)}\right| < 1.$$

It thus follows that the points

$$w = 1+\frac{g(z)}{f(z)}, \quad z \in C \tag{9.58}$$

are all situated in the interior of the circle $|1-w| < 1$. Since this circle does not contain the origin, the curve (9.58) cannot surround that point. As a result, the total variation of the argument of (9.58) along C is zero. Hence, by (9.57), we have

$$\Delta_c\left[f(z)+g(z)\right] = \Delta_c\left[f(z)\right].$$

Since neither $f(z)$ nor $f(z)+g(z)$ has poles in D, it follows from (9.54) that these two functions have the same number of zeros in D. ♣

The application of Rouché's theorem is illustrated by the following short proof of the maximum principle. If $f(z)$ is analytic in $D + C$ and there is a point z_0 in D such that

$$|f(z)| < |f(z_0)| \text{ for } z \in C,$$

then it follows from Rouché's theorem that the function $f(z_0) - f(z)$ and $f(z_0)$ have the same number of zeros in D and the function $f(z) - f(z_0)$ has at least one zero there, namely, at $z = z_0$. The assumption that $|f(z)| < |f(z_0)|$ for $z \in C$ thus leads to a contradiction.

Exercises

1. Let z_j be the zeros of a function $f(z)$ that is analytic in a circular domain D and let $f(z) \not\equiv 0$. Each zero is counted as many times as its multiplicity. Prove that for every closed curve C in D that does not pass through a zero, the sum of winding numbers yields

$$\sum_j n(C, z_j) = \frac{1}{2\pi i} \oint_C \frac{f'(z)}{f(z)} dz. \tag{9.59}$$

Solution: From hypothesis, we can write $f(z) = (z - z_1)(z - z_2) \cdots (z - z_n) g(z)$, where $g(z)$ is analytic and $g(z) \neq 0$ in D. Forming the logarithmic derivative, we obtain

$$\frac{f'(z)}{f(z)} = \frac{1}{z - z_1} + \frac{1}{z - z_2} + \cdots + \frac{1}{z - z_n} + \frac{g'(z)}{g(z)}$$

for $z \neq z_j$, and particularly on C. Since $g(z) \neq 0$ in D, Cauchy's theorem yields $\oint_C g'(z)/g(z) dz = 0$. Recalling the definition of $n(C, z_j)$, we set the desired result (9.59). ♣

2. Show that an analytic function in a domain D that takes only real values on the boundary C of D reduces to a constant.

Solution: Let $\xi = a + ib$, $b \neq 0$, be a nonreal complex number and consider the values of $f(z) - \xi$ for $z \in C$. If $b > 0$, say, we have $\text{Im}[f(z) - \xi] = b > 0$ since $f(z)$ is real on C. The vales of $f(z) - \xi$ are thus confined to the upper half-plane so that the curve described by $f(z) - \xi$ cannot surround the origin. Hence, we have $\Delta_C[f(z) - \xi] = 0$. Furthermore, since $f(z) - \xi$ is analytic in $D + C$, it follows from the argument principle that $f(z) - \xi \neq 0$, i.e., $f(z) \neq \xi$ in D. The same reasoning also applies to values ξ for which $b < 0$. We thus conclude that $f(z)$ does not take nonreal values in D.

Next we show that the above result means that $f(z)$ reduces to a constant. Since $f(z)$ is analytic in D, we have

$$f'(z) = \lim_{h\to 0} \frac{f(z+h)-f(z)}{h} = \lim_{h\to 0} \frac{f(z+ih)-f(z)}{ih},$$

where $h \to 0$ through positive values. Since $f(z)$ is real throughout D, the first limit is real and the second limit is imaginary. They can therefore be equal only if they are both zero. Since z is arbitrary, it follows that $f'(z) = 0$ throughout D; hence, $f(z) = \text{const.}$ ♣

3. Show that all zeros of polynomials

$$p(z) = z^n + a_{n-1}z^{n-1} + \cdots + a_1 z + a_0$$

are located within the region $|z| \le R_0$, where

$$R_0 = \max\{1+|a_{n-1}|, 1+|a_{n-2}|, \cdots, 1+|a_1|, |a_0|\}.$$

Solution: Let $f(z) = z^n$, $g(z) = a_{n-1}z^{n-1} + \cdots + a_1 z + a_0$, and let $R_k = R_0 + (1/k)$ for an arbitrary fixed $k \in \boldsymbol{N}$. Observe that

$$|a_j| \le R_0 - 1 < R_k - 1 \;\text{ for } j = 1, 2, \cdots, n-1$$

and $|a_0| \le R_0 < R_k$. Then, if $|z| = R_k$, we have

$$\begin{aligned} |g(z)| &\le |a_{n-1}||z|^{n-1} + \cdots + |a_1||z| + |a_0| \\ &\le (R_k - 1)R_k^{n-1} + \cdots + (R_k - 1)R_k + R_k = R_k^n = |f(z)|. \end{aligned}$$

In view of Rouche's theorem, $f(z)$ and $f(z) + g(z) = p(z)$ have the same number of zeros within the region $|z| < R_k$. Since $f(z)$ has n zeros and $p(z)$ is an nth-order polynomial, we conclude that all the zeros of $p(z)$ have to be located within the region $|z| < R_k$. Finally, we take the limit $k \to \infty$ (since k is arbitrary) to find that all the zeros of $p(z)$ have to be located within $|z| < R_0$. ♣

4. Show that the equation $z^3 + 3z + 1 = 0$ has solutions whose absolute values are less than 2.

Solution: Let z be on the circle $|z| = 2$. Then we have

$$|z^3| = 8 > 3 \cdot 2 + 1 > 3|z| + 1 \ge |3z + 1|.$$

This means that there are three solutions to the equation $z^3 + 3z + 1 = 0$ and that all of them satisfy $|z| < 2$. ♣

9.5 Dispersion Relations

9.5.1 Principal Value Integrals

The previous sections treated contour integrals whose integrand has no pole on the contour C. If a pole is located on C, the integrand diverges at the pole so that we cannot use ordinary integration methods. This difficulty is overcome by introducing a new concept called the **principal value integral**. To derive it, we consider an integral

$$I = \oint_C \frac{f(z)}{z-\alpha} dz \tag{9.60}$$

with the integration contour depicted in Fig. 9.6. In (9.60), α is assumed to be real without loss of generality. In addition, we assume that $f(z)$ is analytic at $\text{Im} z > 0$, and behaves as $z^\beta |f(z)| \to A$ ($\beta > 0$) as $|z| \to \infty$ there. In order for the integral (9.60) to be defined, the contour C must be traversed in such a way as to avoid the pole at $z = \alpha$. Then, since both $f(z)$ and $1/(z-\alpha)$ are analytic within and on C, (9.60) equals zero. Therefore, by breaking it up, we obtain the following expression:

$$\begin{aligned} &\oint_C \frac{f(z)}{z-\alpha} \\ &= \int_{-R}^{\alpha - r} \frac{f(x)}{x-\alpha} dx + \int_\gamma \frac{f(z)}{z-\alpha} dz + \int_{\alpha+r}^{R} \frac{f(x)}{x-\alpha} dx + \int_\Gamma \frac{f(z)}{z-\alpha} dz \\ &= 0. \end{aligned} \tag{9.61}$$

Here r is the radius of the small semicircle γ centered at $x = \alpha$ and R is the radius of the large semicircle Γ centered at the origin. The radius r can be chosen as small as we please and R can be chosen as large as we please.

Our current interest is to determine where the sum of the four integrals appearing in the second line of (9.61) converges in the limits of $r \to 0$ and $R \to \infty$. This is seen by evaluating the integrals along γ and Γ given in (9.61). First, once we set $z = Re^{i\theta}$, the integral along the large semicircle Γ yields

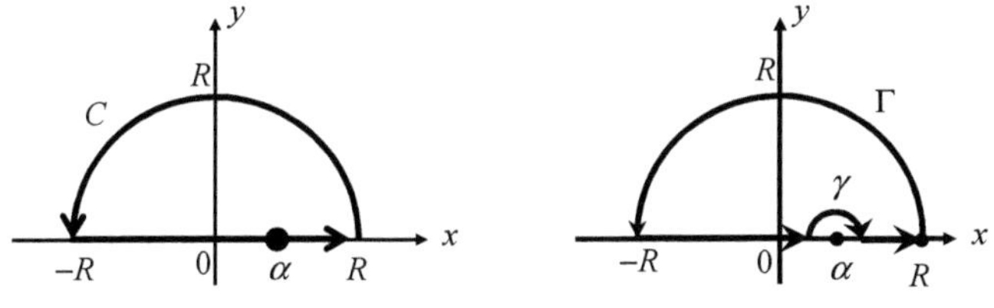

Fig. 9.6. Integration contour on which the pole of the integrand is located

$$\int_{\Gamma} \frac{f(z)}{z-\alpha} dz = i \int_0^{\pi} \frac{f(Re^{i\theta})}{Re^{i\theta}-\alpha} Re^{i\theta} d\theta;$$

hence,

$$\left| \int_{\Gamma} \frac{f(z)}{z-\alpha} dz \right| \leq \frac{R}{|R-\alpha| R^{\beta}} \int_0^{\pi} \left| f(Re^{i\theta}) \right| d\theta, \tag{9.62}$$

where we have used the inequality

$$|Re^{i\theta} - \alpha| = \sqrt{R^2 + \alpha^2 - 2R\alpha\cos\theta} \geq \sqrt{R^2 + \alpha^2 - 2R\alpha} = |R - \alpha|.$$

In the limit $R \to \infty$, the right-hand side of (9.62) vanishes since $\beta > 0$. Therefore, the integral over the semicircle Γ can be made arbitrarily small by choosing R sufficiently large.

Next, we write the integral along γ as

$$\int_{\gamma} \frac{f(z)}{z-\alpha} dz = f(\alpha) \int_{\gamma} \frac{1}{z-\alpha} dz + \int_{\gamma} \frac{f(z)-f(\alpha)}{z-\alpha} dz. \tag{9.63}$$

By setting $z - \alpha = re^{i\theta}$, the first integral on the right-hand side is evaluated as

$$f(\alpha) \int_{\gamma} \frac{1}{z-\alpha} dz = i f(\alpha) \int_{\pi}^{0} d\theta = -i\pi f(\alpha).$$

In addition, the Taylor series expansion of $f(z)$ around $z = \alpha$ yields

$$\frac{f(z)-f(\alpha)}{z-\alpha} dz = f'(\alpha) \cdot i\varepsilon e^{i\theta} d\theta + \frac{f''(\alpha)}{2} \cdot \varepsilon e^{i\theta} \cdot i\varepsilon e^{i\theta} d\theta + \cdots = O(\varepsilon),$$

which means that the second integral in (9.63) vanishes in the limit $r \to 0$. Equation (9.61) thus yields

$$\lim_{R\to\infty} \lim_{r\to 0} \left[\int_{-R}^{\alpha-r} \frac{f(x)}{x-\alpha} dx + \int_{\alpha+r}^{R} \frac{f(x)}{x-\alpha} dx \right] - i\pi f(\alpha) = 0. \tag{9.64}$$

Now we introduce a new notation as shown below.

♠ **Principal value integral:**
The notation

$$\mathcal{P} \int_{-R}^{R} \frac{f(x)}{x-\alpha} dx \equiv \lim_{r\to 0} \left[\int_{-R}^{\alpha-r} \frac{f(x)}{x-\alpha} dx + \int_{\alpha+r}^{R} \frac{f(x)}{x-\alpha} dx \right]$$

provides the **principal value integral** (or the **Cauchy principal value**) of $f(z)/(z-\alpha)$ for real α.

with this notation, (9.64) reads

$$\lim_{R\to\infty} \mathcal{P}\int_{-R}^{R} \frac{f(x)}{x-\alpha}dx = i\pi f(\alpha),$$

where $f(x)$ is a complex-valued function of a real variable x. For the sake of brevity, we write this simply as

$$\mathcal{P}\int_{-\infty}^{\infty} \frac{f(x)}{x-\alpha}dx = i\pi f(\alpha). \tag{9.65}$$

This result provides a way to evaluate the contour integrals involving singularities on the integration path. When we decompose $f(x)$ in (9.65) as $f(x) = f_R(x) + if_I(x)$ and equate the real and imaginary parts, we obtain an important relation between f_R and f_I:

♠ **Hilbert transform pair:**
A pair of functions f_R and f_I that satisfies the relations

$$\begin{aligned} f_R(\alpha) &= \frac{1}{\pi}\mathcal{P}\int_{-\infty}^{\infty} \frac{f_I(x)}{x-\alpha}dx, \\ f_I(\alpha) &= -\frac{1}{\pi}\mathcal{P}\int_{-\infty}^{\infty} \frac{f_R(x)}{x-\alpha}dx. \end{aligned} \tag{9.66}$$

is called a **Hilbert transform pair.**

It readily follows from (9.66) that if $f_I(x) \equiv 0$, then $f_R(x) \equiv 0$.

9.5.2 Several Remarks

The principal value integral is seen as a way of avoiding singularities on a path of integration; we integration to the point just before the singularity in question, skip over the singularity, and begin integrating again immediately beyond the singularity. This prescription enables us to make sense out of integrals such as

$$\int_{-R}^{R} \frac{dx}{x}. \tag{9.67}$$

Apparently, this integral seems to be zero, since an odd function is integrated over a symmetric domain. However, the singularity at the origin makes the integral meaningless unless we insert a symbol $\mathcal{P}$ in front of it. Following the prescription for principal value integrals, we can easily evaluate the principal value of (9.67):

$$\mathcal{P}\int_{-R}^{R} \frac{dx}{x} = \lim_{r\to 0}\left[\int_{-R}^{-r} \frac{dx}{x} + \int_{r}^{R} \frac{dx}{x}\right].$$

In the first integral on the right-hand side, we set $x = -y$. Then

$$\mathcal{P}\int_{-R}^{R}\frac{dx}{x} = \lim_{r\to 0}\left[\int_{R}^{r}\frac{dy}{y} + \int_{r}^{R}\frac{dx}{x}\right],$$

where the two integrals within the brackets obviously cancel out. Consequently, we have

$$\mathcal{P}\int_{-R}^{R}\frac{dx}{x} = 0. \tag{9.68}$$

We emphasize again that the integral (9.68) is completely different from the meaningless quantity in (9.67).

As a further step, we evaluate the principal value integral defined by

$$\mathcal{P}\int_{-R}^{R}\frac{f(x)}{x-\alpha}dx.$$

It follows from the result of (9.83) that

$$\mathcal{P}\int_{-R}^{R}\frac{f(x)}{x-\alpha}dx = \mathcal{P}\int_{-R}^{R}\left[\frac{f(\alpha)}{x-\alpha} + \frac{f(x)-f(\alpha)}{x-\alpha}\right]dx$$

$$= f(\alpha)\ln\left(\frac{R-\alpha}{R+\alpha}\right) + \mathcal{P}\int_{-R}^{R}\frac{f(x)-f(\alpha)}{x-\alpha}dx. \tag{9.69}$$

It often happens that the second integral in the second equation in (9.69) is not be singular at $x = \alpha$; for instance, as in the case where $f(x)$ is differentiable at $x = \alpha$. In this case, the symbol P there can be dropped

Particularly interesting is the behavior of (9.69) in the limit $R \to \infty$, which yields

$$\mathcal{P}\int_{-\infty}^{\infty}\frac{f(x)}{x-\alpha}dx = \mathcal{P}\int_{-\infty}^{\infty}\frac{f(x)-f(\alpha)}{x-\alpha}dx. \tag{9.70}$$

Hence, substituting (9.70) into (9.65), we obtain

$$\begin{aligned} f_R(\alpha) &= \frac{1}{\pi}\mathcal{P}\int_{-\infty}^{\infty}\frac{f_I(x)-f_I(\alpha)}{x-\alpha}dx, \\ f_I(\alpha) &= -\frac{1}{\pi}\mathcal{P}\int_{-\infty}^{\infty}\frac{f_R(x)-f_R(\alpha)}{x-\alpha}dx, \end{aligned} \tag{9.71}$$

which are complementary expressions of a Hilbert transform pair (9.66).

Remark. Equation (9.70) is equivalent to

$$\mathcal{P}\int_{-\infty}^{\infty}\frac{f(a)}{x-a}dx = 0, \quad \text{and thus} \quad \mathcal{P}\int_{-\infty}^{\infty}\frac{dx}{x-a} = 0,$$

which readily follows from the result (9.68).

9.5.3 Dispersion relations

Mathematical arguments given so far are interesting in their own right, but their applications to physical sciences are also significant. In the following discussions, we show that general physical quantities associated with response phenomena satisfy the Hilbert transform relations given in (9.66) and (9.71). In the language of physics, the relation between corresponding parts of Hilbert transform pairs referred to as a **dispersion relation**, plays an important role in describing the properties of response functions.

We begin by considering a physical system for which an input $I(t)$ is related to a response $R(t)$ in the following linear manner:

$$R(t) = \frac{1}{\sqrt{2\pi}} \int_{-\infty}^{\infty} G(t-t')I(t')dt'. \tag{9.72}$$

For example, $I(t')$ might be the electric field acting on a physical object at a time t' and $R(t)$ is the resulting polarization field at time t. We have assumed that G depends only on the difference $t-t'$ because we want the system to respond to a sharp input at t_0 as expressed by $I(t') = I_0\delta(t'-t_0)$. In the same way, it would respond to a sharp input at $t_0+\tau$, i.e., at a time τ later. For the first case, we have

$$R_1(t) = \frac{1}{\sqrt{2\pi}} \int_{-\infty}^{\infty} G(t-t')I_0\delta(t'-t_0)dt' = \frac{I_0}{\sqrt{2\pi}}G(t-t_0). \tag{9.73}$$

and for the second,

$$R_2(t) = \frac{1}{\sqrt{2\pi}} \int_{-\infty}^{\infty} G(t-t')I_0\delta(t'-t_0-\tau)dt' = \frac{I_0}{\sqrt{2\pi}}G(t-t_0-\tau),$$

or, in other words,

$$R_2(t+\tau) = \frac{I_0}{\sqrt{2\pi}}G(t-t_0) = R_1(t).$$

Thus if we shift the input by τ, the response is also shifted by τ.

Now, in order to derive the dispersion relation for the physical systems of interest, we consider the Fourier transform of (9.72). Using the convolution theorem, we find that

$$r(\omega) = g(\omega)j(\omega),$$

where

$$r(\omega) = \frac{1}{\sqrt{2\pi}} \int_{-\infty}^{\infty} R(t)e^{i\omega t}dt, \qquad g(\omega) = \frac{1}{\sqrt{2\pi}} \int_{-\infty}^{\infty} G(t)e^{i\omega t}dt,$$

$$\text{and} \qquad j(\omega) = \frac{1}{\sqrt{2\pi}} \int_{-\infty}^{\infty} I(t)e^{i\omega t}dt.$$

Notably, it is possible to extend $g(\omega)$ into the complex z-plane, based on the assumptions that

(i) $g(z)$ is analytic for Im$z > 0$, and
(ii) $g(z) \to 0$ as $z \to \infty$.

Observe that **(i)** and **(ii)**, are the conditions under which we derived the Hilbert transform pair (see Sect. 9.4.1). After some discussion, we see that $g(z)$ arising from a $G(t)$ that satisfies the necessary assumptions yields

$$g_R(\omega) = \frac{1}{\pi}\mathcal{P}\int_{-\infty}^{\infty}\frac{g_I(\omega')}{\omega'-\omega}d\omega',$$

$$g_I(\omega) = -\frac{1}{\pi}\mathcal{P}\int_{-\infty}^{\infty}\frac{g_R(\omega')}{\omega'-\omega}d\omega'. \tag{9.74}$$

These relations between g_R and g_I are called the **dispersion relations** for g. The validity of assumptions **(i)** and **(ii)** that the function $g(z)$ must satisfy is demonstrated in Sect. 9.5.6.

9.5.4 Kramers–Kronig Relations

The term "dispersion relation" is often restricted to mean a relation between two functions whose arguments are quantitatively treatable experimentally. For instance, in (9.74) only a positive frequency ($\omega \geq 0$) should actually be accessible, so they are not directly practical as they stand. In the following, we derive an alternative expression of the dispersion relations that involve only positive, experimentally meaningful frequencies.

We first assume that $G(t)$ is real, which is obvious from (9.73), where R_1 and I_0 are real. Hence, we may proceed as follows:

$$\begin{aligned} g(z) &= \frac{1}{\sqrt{2\pi}}\int_0^{\infty} G(t)e^{izt}dt, \\ g^*(z) &= \frac{1}{\sqrt{2\pi}}\int_0^{\infty} G^*(t)e^{-iz^*t}dt = \frac{1}{\sqrt{2\pi}}\int_0^{\infty} G(t)e^{-iz^*t}dt \\ &= g(-z^*). \end{aligned} \tag{9.75}$$

As a consequence, we have

$$g^*(z) = g(-z^*),$$

which is referred to as the **reality condition**.

Next let us assume z to be real ($z = \omega$) in order to discuss the behavior of $g(z)$ on the real axis. It follows from the reality condition (9.75) that

$$g_R(\omega) - ig_I(\omega) = g_R(-\omega) + ig_I(-\omega)$$

or

$$g_R(\omega) = g_R(-\omega) \quad \text{and} \quad g_I(\omega) = -g_I(-\omega). \tag{9.76}$$

That is, g_R and g_I are even and odd functions of ω, respectively. Note that if the conditions in (9.76) are satisfied, the function

$$G(t) = \frac{1}{\sqrt{2\pi}} \int_{-\infty}^{\infty} g(\omega) e^{-i\omega t} d\omega$$

becomes a real function. (The proof is left to the reader).

Now we rewrite the first part of (9.74) as

$$g_R(\omega) = \frac{1}{\pi} \mathcal{P} \int_{-\infty}^{0} \frac{g_I(\omega')}{\omega' - \omega} d\omega' + \frac{1}{\pi} \mathcal{P} \int_{0}^{\infty} \frac{g_I(\omega')}{\omega' - \omega} d\omega'.$$

we rewrite $\omega' \to -\omega'$ in the first integral and use (9.76) to obtain

$$g_R(\omega) = \frac{2}{\pi} \mathcal{P} \int_{0}^{\infty} \frac{\omega' g_I(\omega')}{\omega'^2 - \omega^2} d\omega' \tag{9.77}$$

and an identical procedure yields

$$g_I(\omega) = -\frac{2\omega}{\pi} \mathcal{P} \int_{0}^{\infty} \frac{g_R(\omega')}{\omega'^2 - \omega^2} d\omega'. \tag{9.78}$$

Eventually, the expressions (9.77) and (9.78) involve only positive, experimentally accessible frequencies. These equations are referred to as the **Kramers–Kronig relations**.

9.5.5 Subtracted Dispersion Relation

In deriving dispersion relations, it often happens that the quantity of interest, say $g(z)$, does not tend toward zero as $|z| \to \infty$. Furthermore, we are not usually fortunate enough to know the precise behavior of the quantity as $|z|$ tends to infinity. Nevertheless, if we at least know that the quantity is bounded for large values of $|z|$, the dispersion relation can be reformulated in the following way:

Suppose that $f(z)$ is analytic in the upper half-plane, and let α_0 be some point on the real axis at which $f(z)$ is analytic. Our aim is to derive the dispersion relation for $f(x)$ under the condition that the asymptotic behavior of $f(z)$ for $z \to \infty$ is unknown. Then, instead of $f(z)$, we consider the function

$$\frac{f(z) - f(\alpha_0)}{z - \alpha_0} \equiv \phi(z),$$

which is also analytic in the upper half-plane and not singular at $z = \alpha_0$, and $|\phi(z)| \to 0$ as $|z| \to \infty$ owing to the boundedness of $|f(z)|$ for $z \to \infty$. Thus in a manner, similar to the case in (9.65), we can write

$$i\pi\phi(x) = \mathcal{P} \int_{-\infty}^{\infty} \frac{\phi(x')}{x' - x} dx'.$$

In actuality, we have

$$\begin{aligned} & i\pi\left[\frac{f(x)-f(\alpha_0)}{x-\alpha_0}\right] \\ &= \mathcal{P}\int_{-\infty}^{\infty}\frac{f(x')-f(\alpha_0)}{(x'-x)(x'-\alpha_0)}dx' \\ &= \mathcal{P}\int_{-\infty}^{\infty}\frac{f(x')}{(x'-x)(x'-\alpha_0)}dx' - \frac{f(\alpha_0)}{x-\alpha_0}\mathcal{P}\int_{-\infty}^{\infty}\left(\frac{1}{x'-x}-\frac{1}{x'-\alpha_0}\right)dx', \end{aligned}$$

so

$$\begin{aligned} i\pi f(x) = \quad & i\pi f(\alpha_0) + (x-\alpha_0)\mathcal{P}\int_{-\infty}^{\infty}\frac{f(x')}{(x'-x)(x'-\alpha_0)}dx' \\ & - f(\alpha_0)\mathcal{P}\int_{-\infty}^{\infty}\frac{dx'}{x'-x} + f(\alpha_0)\mathcal{P}\int_{-\infty}^{\infty}\frac{dx'}{x'-\alpha_0}. \end{aligned}$$

The last two principal value integrals are equal to zero as we demonstrate later in (9.83). Hence, separating the real and imaginary parts, we finally obtain

$$\begin{aligned} f_R(x) &= f_R(\alpha_0) + \frac{x-\alpha_0}{\pi}\mathcal{P}\int_{-\infty}^{\infty}\frac{f_I(x')}{(x'-x)(x'-\alpha_0)}dx', \\ f_I(x) &= f_I(\alpha_0) - \frac{x-\alpha_0}{\pi}\mathcal{P}\int_{-\infty}^{\infty}\frac{f_R(x')}{(x'-x)(x'-\alpha_0)}dx'. \end{aligned} \tag{9.79}$$

Relations of the type of (9.79) are referred to as **once-subtracted dispersion relations**. Emphasis is placed on the fact that the relations (9.79) are free from the assumption that $|f(z)|$ should vanish in the limit $z \to \infty$. For them to be of use in a particular physical problem, we must have a means of determining, say, $f_R(\alpha_0)$ for some α_0.

9.5.6 Derivation of Dispersion Relations

This subsection provides a proof of the dispersion relation (9.74). We shall see that by making a few very reasonable assumptions about the system in question, we can show that the real and imaginary parts of the physical quantity $g(\omega)$ are intimately related to one another for real values of ω (i.e., a dispersion relation). The key assumption is the **causality requirement**: we may say that causality of the function $G(t)$ implies the analytic properties of $g(z)$ in the upper half-plane and thus verifies the dispersion relations with respect to $g(\omega)$ on the real axis.

Toward this end, let us consider what can be said about $G(\tau)$ on general physical grounds. First to be noted is that an input at t should not give rise to a response at times prior to t, i.e., $G(\tau)=0$ for $\tau<0$. Thus we have

$$R(t) = \int_{-\infty}^{t} G(t-t')I(t')dt', \tag{9.80}$$

which shows that the response at t is the weighted linear superposition of all inputs prior to t, which is the causality requirement.

Secondly, the possibility that $G(\tau)$ is singular for any finite τ is excluded because, on physical grounds, the response from a sharp input given by

$$R(t) = \frac{I_0}{\sqrt{2\pi}} G(t - t_0), \quad t > t_0$$

must always be finite.

Finally, it is assumed that the effect of an input in the remote past does not appreciably influence the present. This may be stated as the requirement that $G(\tau) \to 0$ as $\tau \to \infty$, since the response to any impulse dies down after a sufficiently long time (i.e., any system has some dissipative mechanism). Furthermore, $G(\tau)$ should vanish faster than τ^{-1} so that it becomes integrable. Recall that $g(z)$ is defined through an integration of $G(t)$ with respect to t.

The following three points summarise our physically motivated assumptions on $G(\tau)$:

(i) $G(\tau) = 0$ for $\tau < 0$,
(ii) $G(\tau)$ is bounded for all τ, and
(iii) $|G(\tau)|$ is integrable, so $G(\tau) \to 0$ faster than $1/\tau$ as $\tau \to \infty$.

We demonstrate below that these three assumptions for $G(t)$ lead naturally to the two conditions for $g(z)$ under which we have derived the dispersion relation of $g(\omega)$.

First, we show that these three conditions require that $|g(z)| \to 0$ at $z \to \infty$ on the upper half-plane. It is possible to write

$$g(\omega) = \frac{1}{\sqrt{2\pi}} \int_0^\infty G(t) e^{i\omega t} dt.$$

We extend this relation into the complex plane by using the definition

$$g(z) = \frac{1}{\sqrt{2\pi}} \int_0^\infty G(t) e^{izt} dt = \frac{1}{\sqrt{2\pi}} \int_0^\infty G(t) e^{i\omega t} e^{-\eta t} dt,$$

where we have written $z = \omega + i\eta$. We now restrict our attention to the upper half-plane ($\eta > 0$), where the term $e^{-\eta t}$ is a decaying exponential. For $0 < \theta < \pi$, it reads

$$|g(z)| \le \frac{1}{\sqrt{2\pi}} M \int_0^\infty e^{-(|z| \sin \theta) t} dt,$$

where we have replaced $G(t)$ by its maximum value M in view of assumption **(ii)** above. Hence, we have

$$|g(z)| \le \frac{M_G}{\sqrt{2\pi} |z| \sin \theta}.$$

This means that for $0 < \theta < \pi$, $|g(z)| \to 0$ as $|z| \to \infty$. On the other hand, when $\theta = 0$ or π, we have

$$g(\omega, \eta = 0) = \frac{1}{\sqrt{2\pi}} \int_0^\infty G(t) e^{i\omega t} dt.$$

This results in Parseval's identity:

$$\int_{-\infty}^\infty |g(\omega, \eta = 0)|^2 d\omega = \frac{1}{2\pi} \int_{-\infty}^\infty |G(t)|^2 dt,$$

where both sides of improper integrals converge. (See Sect. 3.4.2 for the convergence conditions of an improper integral.) Thus $|g(\omega, \eta = 0)|$ vanishes as $\omega \to \infty$. As a result, $|g(z)| \to 0$ as $|z| \to \infty$ in the whole region of $0 \le \theta \le \pi$, i.e., in any direction in the upper half-plane.

Now we want to show that $g(z)$ is analytic in the upper half-plane. Using

$$g(z) = \frac{1}{\sqrt{2\pi}} \int_0^\infty G(t) e^{izt} dt = \frac{1}{\sqrt{2\pi}} \int_0^\infty G(t) e^{i\omega t} e^{-\eta t} dt, \tag{9.81}$$

we see that for $\eta > 0$,

$$\frac{d^n g}{dz^n} = \frac{1}{\sqrt{2\pi}} \int_0^\infty G(t) \frac{d^n}{dz^n} e^{izt} dt = \frac{i^n}{\sqrt{2\pi}} \int_0^\infty t^n G(t) e^{i\omega t} e^{-\eta t} dt. \tag{9.82}$$

The integrals in (9.82) are uniformly convergent owing to the term $e^{-\eta t}$ ($\eta > 0$, $t > 0$). Thus $g(z)$ is analytic in the upper half-plane ($\eta > 0$). Hence, for any $g(z)$ arising from a $G(t)$ that satisfies assumptions **(i)**, **(ii)**, and **(iii)**, we can proceed according to the argument in Sect. 9.4.1, and we finally obtain the dispersion relation (9.74).

Exercises

1. Prove that

$$\mathcal{P} \int_{-R}^R \frac{dx}{x-a} = \ln\left(\frac{R-a}{R+a}\right)$$

when $-R < a < R$.

Solution: We write

$$\mathcal{P} \int_{-R}^R \frac{dx}{x-a} = \lim_{\varepsilon \to 0} \left[\int_{-R}^{a-\varepsilon} \frac{dx}{x-\varepsilon} + \int_{\varepsilon+a}^R \frac{dx}{x-a} \right].$$

Setting $x = -y$ in the first integral on the right-hand side, we find that

$$\begin{aligned} \mathcal{P} \int_{-R}^R \frac{dx}{x-a} &= \lim_{\varepsilon \to 0} \left[\int_R^{\varepsilon - a} \frac{dy}{y+a} + \ln(R-a) - \ln\varepsilon \right] \\ &= \lim_{\epsilon \to 0} \left[\ln\epsilon - \ln(R+a) + \ln(R-a) - \ln\epsilon \right] \\ &= \ln\left(\frac{R-a}{R+a}\right) \quad (-R < a < R). \quad ♣ \end{aligned} \tag{9.83}$$

2. By using the formula (9.71), prove that

$$\int_{-\infty}^{\infty} \frac{\sin x}{x} dx = \pi.$$

Solution: Consider the function $f(z) = e^{iz}$. This function is analytic everywhere, and if we write $z = Re^{i\theta}$, then $|f(z)| \to 0$ as $R \to \infty$ for all θ such that $0 < \theta < \pi$. In this case, $f_R(x) = \cos x$ and $f_I(x) = \sin x$, so using (9.71), we obtain

$$\cos \alpha = (1/\pi) \int_{-\infty}^{\infty} (\sin x - \sin \alpha)/(x - \alpha) dx.$$

Since $\sin x - \sin \alpha = 2 \sin[(x-\alpha)/2] \cos[(x+\alpha)/2]$, there is no singularity of the integrand at $x = \alpha$. For the special case $\alpha = 0$, we find that $1 = (1/\pi) \int_{-\infty}^{\infty} (\sin x/x) dx$, i.e., $\int_{-\infty}^{\infty} (\sin x/x) dx = \pi$. From this result, we also obtain $\int_0^{\infty} (\sin x/x) dx = \pi/2$ by symmetry. ♣

3. Show that the integral $S(t) = \dfrac{1}{2\pi i} \lim_{\varepsilon \to 0} \displaystyle\int_{-\infty}^{\infty} \frac{e^{ixt}}{x - i\varepsilon} dx$ reads

$$S(t) = \begin{cases} 1, & t > 0, \\ 0. & t < 0. \end{cases}$$

Solution: Taking the contours $\mathrm{Im}(z) > 0$ for $t > 0$, and $\mathrm{Im}(z) < 0$ for $t < 0$, we have the desired result, which is the integral representation of Heaviside's step function. ♣

10

Conformal Mapping

Abstract Conformal mapping refers to transformation from one complex plane to another such that the local angles and shapes of infinitesimally small figures are preserved. This special class of mapping is indispensable for solving physics and engineering problems that are expressed in terms of complex functions with inconvenient geometries. In this chapter we show that a problem can be drastically simplified by choosing an appropriate mapping, which allows us to evaluate the solution using elementary calculus.

10.1 Fundamentals

10.1.1 Conformal Property of Analytic Functions

We are concerned here the mapping properties of an **analytic function** $w = f(z)$ in a domain D on the z-plane into the w-plane. Through the mapping, any line drawn on the z-plane results in a line on the w-plane. Particularly when $f = u + iv$ is analytic, the transformation is **angle-preserving** or **conformal**. This means that through the transformation from (x, y) to (u, v), the angle between the crossing lines on the w-plane is equal to the angle between the crossing lines on the z-plane (see Fig. 10.1). In physics and engineering, the subject derives its usefulness from the possibility of transforming a problem that occurs naturally in a rather difficult setting into another simpler one.

Let D be a domain on the z-plane, and let Γ_1 and Γ_2 be two differentiable arcs lying in D and intersecting at a point $z = a$ in D. If $f(z)$ is an analytic function in D, the images $f(\Gamma_1)$ and $f(\Gamma_2)$ are differentiable arcs lying in a domain $D' = f(D)$ and intersecting at a point $a' = f(a)$. Then we say the following:

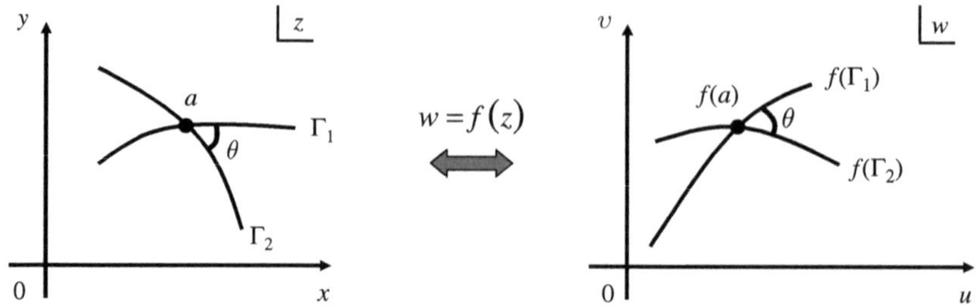

Fig. 10.1. Angle-preserving property of a conformal mapping $w = f(z)$

♠ **Conformal mapping:**

The mapping $w = f(z)$ is **conformal** at $z = a$ if for every such pair of arcs, the angle between the arcs Γ_1 and Γ_2 intersecting at $z = a$ on the z-plane is equal to the angle between the arcs $f(\Gamma_1)$ and $f(\Gamma_2)$ at their intersecting point $f(a)$ on the w-plane.

The mapping is said to be conformal in D if it is conformal at each point in D. We shall see that if a function $w = f(z)$ is analytic, it is necessarily conformal except at a finite number of specific points; this fact is formally stated below.

♠ **Theorem:**

Given an analytic function $f(z)$, the mapping $w = f(z)$ is conformal at $z = a$ if and only if $f'(a) \neq 0$.

Proof For proving sufficiency, we consider the arcs Γ_1 and Γ_2 given parametrically by

$$z_1 = \Psi_1(t) \text{ and } z_2 = \Psi_2(t) \quad (0 \leq t \leq 1)$$

and assume that z_1, z_2 are points on Γ_1, Γ_2 at a short distance ℓ from $z = a$. Then, from the relation

$$z_1 - a = \ell e^{i\alpha}, \quad z_2 - a = \ell e^{i\beta},$$

we have the ratio

$$\frac{z_2 - a}{z_1 - a} = e^{i(\beta - \alpha)}.$$

As $\ell \to 0$, $\beta - \alpha$ must approach the angle θ between the curves on the z-plane. That is,

$$\theta = \lim_{\ell \to 0} \text{ang}\left(\frac{z_2 - a}{z_1 - a}\right).$$

For the angle $\tilde{\theta}$ between the arcs of $f(\Gamma_1)$ and $f(\Gamma_2)$ at $f(a)$, we have

$$\tilde{\theta} = \lim_{\ell \to 0} \text{ang} \left[\frac{f(z_2) - f(a)}{f(z_1) - f(a)} \right]$$

$$= \lim_{\ell \to 0} \text{ang} \left[\frac{\dfrac{f(z_2) - f(a)}{z_2 - a} \cdot (z_2 - a)}{\dfrac{f(z_1) - f(a)}{z_1 - a} \cdot (z_1 - a)} \right]$$

$$= \lim_{\ell \to 0} \text{ang} \left[\frac{f'(a) \cdot (z_2 - a)}{f'(a) \cdot (z_1 - a)} \right] = \theta, \quad \text{if } f'(a) \neq 0. \tag{10.1}$$

Thus, the condition $f'(a) \neq 0$ is necessary. Conversely if $f^{(n)}(a) = 0$ with $n = 1, 2, \cdots$ and $f^{(p)}(a) \neq 0$, near $z = a$ we have

$$f(z) = f(a) + O\left[(z - a)^p\right].$$

Thus, we get

$$\tilde{\theta} = \lim_{\ell \to 0} \arg \left[\frac{f(z_2) - f(a)}{f(z_1) - f(a)} \right]$$

$$= \lim_{\ell \to 0} \arg \left[\frac{(z_2 - a)^p}{(z_1 - a)^p} \right]$$

$$= p \lim_{\ell \to 0} \arg \left(\frac{z_2 - a}{z_1 - a} \right) = p\theta,$$

which shows that the angle is magnified by p. Therefore, if the mapping $w = f(z)$ is conformal, we necessarily have $p = 1$, which completes the proof of the sufficiency of the condition. ♣

10.1.2 Scale Factor

There is another important geometric property that analytic functions possess: whenever $f(z)$ is analytic, any infinitesimal figure plotted on the z-plane is transformed into a similar figure on the w-plane with a change in size but with the proportions (and angles) preserved. We prove this by considering the length of an infinitesimally small quantity df given by

$$df = du + idv = \left(\frac{\partial u}{\partial x} dx + \frac{\partial u}{\partial y} dy \right) + i \left(\frac{\partial v}{\partial x} dx + \frac{\partial v}{\partial y} dy \right). \tag{10.2}$$

Its square length reads

$$
\begin{aligned}
|df|^2 &= \left(\frac{\partial u}{\partial x}dx + \frac{\partial u}{\partial y}dy\right)^2 + \left(\frac{\partial v}{\partial x}dx + \frac{\partial v}{\partial y}dy\right)^2 \\
&= \left[\left(\frac{\partial u}{\partial x}\right)^2 + \left(\frac{\partial v}{\partial x}\right)^2\right](dx)^2 + \left[\left(\frac{\partial u}{\partial y}\right)^2 + \left(\frac{\partial v}{\partial y}\right)^2\right](dy)^2 \\
&\quad +2\left(\frac{\partial u}{\partial x}\frac{\partial u}{\partial y} + \frac{\partial v}{\partial x}\frac{\partial v}{\partial y}\right)dxdy. \qquad (10.3)
\end{aligned}
$$

Substituting the Cauchy–Riemann relations into (10.3), we obtain

$$
|df|^2 = h^2|dz|^2, \quad \text{where } h = \sqrt{\left(\frac{\partial u}{\partial x}\right)^2 + \left(\frac{\partial u}{\partial y}\right)^2} = \sqrt{\left(\frac{\partial v}{\partial x}\right)^2 + \left(\frac{\partial v}{\partial y}\right)^2}. \qquad (10.4)
$$

The quantity h is known as a **scale factor** and measures a magnification ratio of the elementary lines through the transformation $w = f(z)$. From (10.4), it readily follows that

$$
h = \left|\frac{df}{dz}\right|. \qquad (10.5)
$$

We see from (10.5) that since df/dz is isotropic, the scale factor h is also isotropic (i.e., independent of the direction of dz) for any analytic function f. This means that any infinitesimal figures on the z-plane are transformed into similar figures on the w-plane with a change in their size by $h = |df/dz|$.

Note that the magnitude of h depends on points z and may vanish at points where $f'(z) = 0$. Points where $f'(z) = 0$ are called **critical points** of the transformation $w = f(z)$, and at these points, the transform becomes non conformal. The simplest example is

$$
f(z) = z^2
$$

for which we have

$$
h = |f'(0)| = 0.
$$

In fact, when two line elements passing through $z = 0$ make an angle $\beta - \alpha$ with respect to one another, the corresponding lines on the w-plane make an angle of $2(\beta - \alpha)$. Thus mapping is not conformal at $z = 0$. In general, the region in the neighborhood of the point at which $h = 0$ on the w-plane becomes greatly compressed. In contrast, the corresponding region on the z-plane is tremendously expanded.

10.1.3 Mapping of a Differential Area

The scale factor h given in (10.4) can be derived in a different way by considering the conformal mapping of a differential area. Let $f(z)$ be a conformal mapping that transforms any points in D of the z-plane onto a region S of the

w-plane. In the domain D, we define a rectangular differential area element with sides of the rectangle parallel to the x and y-axes. These sides are given by

$$dz_1 = dx \text{ and } dz_2 = idy,$$

The images of dz_1 and dz_2 are differential curves in the w-plane given by

$$dw_1 = du_1 + idv_1 \text{ and } dw_2 = du_2 + idv_2.$$

Note that the differential area element of the rectangle in the z-plane reads $dA_z = dxdy$ and that of the parallelogram in the w-plane is

$$dA_w = |\text{Im}(dw_1^* dw_2)|.$$

Since $dz_1 = dx$ and $dz_2 = idy$, the images of these line elements can be written as

$$dw_1 = \frac{\partial f}{\partial x}dx = \left(\frac{\partial u}{\partial x} + i\frac{\partial v}{\partial x}\right) dx$$

and

$$dw_2 = \frac{1}{i}\frac{\partial f}{\partial y} idy = \left(\frac{\partial u}{\partial y} + i\frac{\partial v}{\partial y}\right) dy.$$

Therefore, dA_w is given by

$$dA_w = |\text{Im}(dw_1^* dw_2)| = \left|\frac{\partial u}{\partial x}\frac{\partial v}{\partial y} - \frac{\partial u}{\partial y}\frac{\partial v}{\partial x}\right| dxdy = \frac{\partial(u,v)}{\partial(x,y)} dA_z, \tag{10.6}$$

where

$$\frac{\partial(u,v)}{\partial(x,y)} \equiv \left|\frac{\partial u}{\partial x}\frac{\partial v}{\partial y} - \frac{\partial u}{\partial y}\frac{\partial v}{\partial x}\right| = \begin{vmatrix} \dfrac{\partial u}{\partial x} & \dfrac{\partial u}{\partial y} \\ \dfrac{\partial v}{\partial x} & \dfrac{\partial v}{\partial y} \end{vmatrix}$$

is called the **Jacobian determinant** of the transformation. Since $f(z)$ is analytic, u and v satisfy the Cauchy–Riemann relations over the region R, so the Jacobian determinant can be written as

$$\frac{\partial(u,v)}{\partial(x,y)} = \left(\frac{\partial u}{\partial x}\right)^2 + \left(\frac{\partial u}{\partial y}\right)^2 = \left(\frac{\partial v}{\partial x}\right)^2 + \left(\frac{\partial v}{\partial y}\right)^2. \tag{10.7}$$

This provides a physical interpretation of the Jacobian determinant $\partial(u,v)/\partial(x,y)$; namely, it is identical to the square of the same factor h introduced in (10.4).

10.1.4 Mapping of a Tangent Line

We consider the mapping of a tangent line. Let C be a curve in the z-plane and Γ be the image of C in the w-plane (see Fig. 10.2). A differential segment dw along Γ is related to the differential segment dz along C by

$$dw = \frac{df}{dz} dz = f'(z) dz. \tag{10.8}$$

We suppose w_0 to be a point on Γ that is the image of z_0 on C. Then from (10.8), the tangent to Γ at w_0, denoted by $\tau(w_0)$, is related to the tangent to C at z_0, denoted by $t(z_0)$:

$$\tau(w_0) \equiv \left. \frac{dw}{d\lambda} \right|_{w=w_0} = f'(z_0) \left. \frac{dz}{d\lambda} \right|_{z=z_0} \equiv f'(z_0) t(z_0), \tag{10.9}$$

where λ parametrizes the curve of Γ on the w-plane.

An immediate consequence of equation (10.9) is that if $f'(z_0) = 0$, the tangent $t(z_0)$ on the z-plane cannot be related to the tangent $\tau(w_0)$ on the w-plane. The point z_0 that satisfies $f'(z) = 0$ is called a **critical point** on the curve. For simplicity in the following discussion, we assume that the curve C does not contain any critical points.

The characteristics of the mapping (10.9) become clear by employing the polar form.

$$\tau(w_0) = |\tau(w_0)| e^{i\psi(w_0)}, \quad f'(z_0) = |f'(z_0)| e^{i\phi(z_0)}, \quad \text{and} \quad t(z_0) = |t(z_0)| e^{i\theta(z_0)}. \tag{10.10}$$

The first equation shows that $\tau(w_0)$ is oriented at an angle $\psi(w_0)$ to the u-axis; similarly, the third one shows that $t(z_0)$ makes an angle $\theta(z_0)$ with the x-axis. It follows from (10.9) that

$$|\tau(w_0)| e^{i\psi(w_0)} = |f'(z_0)||t(z_0)| e^{i[\phi(z_0)+\theta(z_0)]}.$$

Thus the magnitude of $\tau(w_0)$ and its argument read

$$|\tau(w_0)| = |f'(z_0)||t(z_0)| \quad \text{and} \quad \psi(w_0) = \phi(z_0) + \theta(z_0).$$

Each equation gives us the properties of the conformal mapping of a tangent line as follows:

(i) The magnitude of the tangent $|t(z_0)|$ is modified by the scale factor $|f'(z_0)|$, thus being enlarged or shrunk by the mapping. Since $|f'(z_0)|$ depends on z_0, the magnification varies from point to point on C.

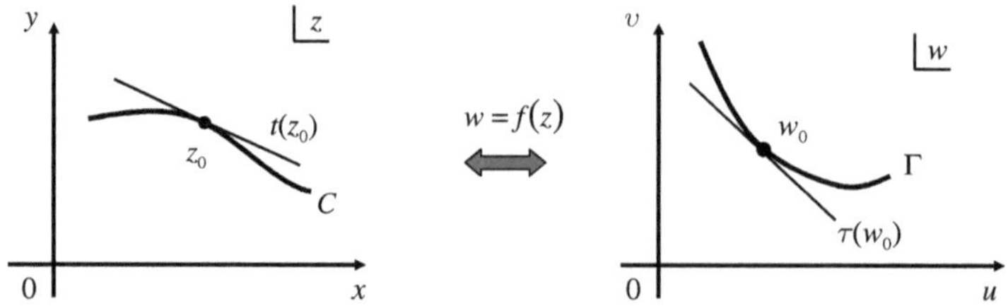

Fig. 10.2. Conformal mapping of a tangential line

(ii) The angle between the tangent $t(z)$ and the x-axis at z_0 differs from the angle between the tangent $\tau(w)$ and the u-axis at w_0. The difference is determined by the argument of $f'(z_0)$, denoted by ϕ, called the argument of the mapping; ϕ also depends on z_0 and thus varies from point to point on C.

10.1.5 The Point at Infinity

For later use, we introduce a few concepts that are at the basis of further investigations on conformal mapping. Our aim is to understand the way in which the entire spherical curved surface is mapped conformally onto the entire flat plane with a one-on-one correspondence. This is achieved with the help of a stereographic projection between the complex plane and an artificial sphere as described below.

Let us consider a sphere of radius R (for convenience, R is taken as $1/2$) such that the complex plane is tangential to it at the origin, as shown in Fig. 10.3. The point P on the sphere opposite the origin (called the north pole, for convenience) is used as the "eye" of the stereographic projection. We draw straight lines through P that intersect both the sphere and the plane. These lines permit a mapping of point z on the plane onto the point ζ on the sphere (see Fig. 10.3). In this fashion the entire complex plane is mapped onto the sphere (called a **Riemann** or a **complex sphere**).

As to the properties of the Riemann sphere, the following statements can be verified without much difficulty.

1. Straight lines in the z-plane are mapped onto circles on the sphere that pass through P.

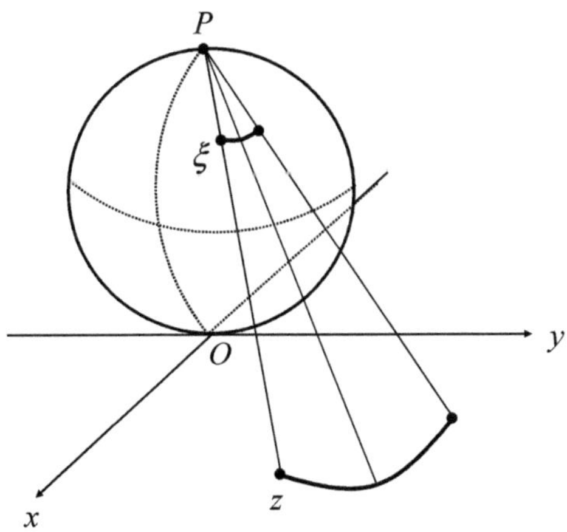

Fig. 10.3. Riemann sphere

2. The images of intersecting straight lines on the plane have two common points on the Riemann sphere, one of which is P.
3. The images of parallel straight lines on the z-plane have only the point P in common, and they have a common tangent at P.
4. The exterior of a circle $|z| = R$ with $R \gg 1$ is mapped onto the interior of a small spherical cap around point P. As $R \to \infty$ the cap shrinks to P.

Note that the point P itself has no counterpart on the z-plane. Nevertheless, it has been found convenient to adjoin an extra point to the z-plane, known as the **point at infinity**, in such a way that a curve passing through P on the Riemann sphere is the image of a curve on the z-plane that approaches the point at infinity.

♠ **Point at infinity:**
The point at infinity $z = \infty$ is defined as the point $\tilde{z}$ that is mapped onto the origin $z = 0$ by the transformation $\tilde{z} = 1/z$.

The importance of the point at infinity is greatly enhanced once we appreciate the conformal property of the stereographic projection: i.e., if two curves intersect on the z-plane at an angle γ, then their images on the sphere intersect at the same angle. This conformal property permits the definition of the angle between two *parallel* straight lines on the z-plane, i.e., the angle that their images make on the sphere at point P. (Indeed this angle is equal to zero as noted in **3** above.)

10.1.6 Singular Point at Infinity

The concept of a point at infinity is closely interwoven with the study of singularities of analytic functions. The notion of analyticity can be extended to a point at infinity by the following device: A function $f(z)$ is considered to be analytic at infinity if the function

$$g(z) = f\left(\frac{1}{z}\right)$$

is analytic at $z = 0$. A more precise statement on this mater is given below.

♠ **Extended definition of conformal mappings:**
A function $w = f(z)$ is said to transform the neighborhood of a point z_0 conformally into a neighborhood of $w = \infty$ if the function $\eta = 1/f(z)$ transforms the neighborhood of z_0 conformally into a neighborhood of $\eta = 0$.

Example The mapping $w = 1/z$ is conformal at the origin $z = 0$. Initially, the function $f(z) = 1/z$ is not defined at $z = 0$; however, the subterfuge based on the Riemann sphere makes the mapping $w = 1/z$ meaningful (and, furthermore, conformal) at $z = 0$. Note that it is also conformal at $z = \infty$ even though the derivative $f'(z)$ approaches zero as $z \to \infty$.

Owing to the above convention, it becomes possible to introduce the concept of a **pole at infinity**, a **branch at infinity**, and so on, through the corresponding behavior of $g(z)$ at the origin. In fact, owing to our convention, a function $f(z) = e^z$ that has no singularities in the original z-plane comes to possess an **essential singularity at infinity**. Other functions that have no singularities (e.g., all the polynomials in z) are also found to have a breakdown of analyticity at infinity. In contrast, functions that are **analytic at infinity** possess at least one singularity for some finite value of z. The natural conjecture is that there may not be a perfectly analytic function. This problem has actually been resolved and is embodied in the theorem below.

♠ **Entire function:**

A function $f(z)$ whose only singularity is an isolated singularity at the point at infinity $z = \infty$ is called an **entire function** (or **integral function**). If this singularity is a pole of mth order, then $f(z)$ must be a polynomial of degree m.

♠ **Liouville theorem:**

The only function $f(z)$ that is analytic in the entire complex plane as well as at the point at infinity is the constant function $f(z) = \text{const}$.

Remark. In some texts the term "complex plane" is tacitly assumed to mean the extended complex plane with the **point at infinity** included. Certain theorems may then be stated more conveniently. However, one should never forget that while there is a point at infinity, there is still no such thing as a complex number "infinity" in the sense that it possesses the algebraic properties shared by other complex numbers.

Exercises

1. Suppose that two differential curves on the z-plane, meet at a point z_0 at which $f'(z_0) = f''(z_0) = \cdots = f^{(m-1)}(z_0) = 0$ and $f^{(m)}(z_0) \neq 0$. Show that the angle θ between the two curves is magnified by m times through the conformal mapping $w = f(z)$.

   Solution: From hypothesis, $f(z)$ can be expanded in the neighborhood of the point z_0 as

$$f(z) = f(z_0) + c_m(z - z_0)^m + c_{m+1}(z - z_0)^{m+1} + \cdots,$$

where $c_m \neq 0$. Then, by the same scenario as we used in deriving (10.1), the angle $\tilde{\theta}$ between the mapped arcs at $f(z_0)$ reads

$$\tilde{\theta} = \lim_{\ell \to 0} \arg \frac{f(z_2) - f(z_0)}{f(z_1) - f(z_0)} = \lim_{\ell \to 0} \arg \left(\frac{z_2 - z_0}{z_1 - z_0} \right)^m$$
$$= m \lim_{\ell \to 0} \arg \frac{z_2 - z_0}{z_1 - z_0} = m\theta. \quad ♣$$

2. We say that the mapping $w = f(z)$ is **locally one-to-one** at z_0 if $f(z_1) \neq f(z_2)$ for any two distinct points z_1 and z_2 within the circle $|z - z_0| < \delta$ with some $\delta > 0$. Show that $w = f(z)$ is locally one-to-one at z_0 if $f(z)$ is analytic at z_0 and $f'(z_0) \neq 0$.

Solution: Let $f(z_0) = \alpha$ and take $\delta > 0$ small enough so that $f(z) - \alpha$ has no other zero in $|z - z_0| < \delta$. In view of the theorem regarding the isolated property of zeros, such a δ can always be found. The argument principle says that

$$1 = \frac{1}{2\pi i} \oint_C \frac{f'(z)}{f(z) - \alpha} dz,$$

where C is a circle $|z - z_0| = \delta$. Denoting $\Gamma = f(C)$, we have

$$1 = \frac{1}{2\pi i} \oint_\Gamma \frac{dw}{w - \alpha} = \frac{1}{2\pi i} \oint_\Gamma \frac{dw}{w - \beta}$$

for any β satisfying $|\beta - \alpha| < \varepsilon$ with sufficiently small ε. If we take $\delta' \leq \delta$ so that

$$D = \{z; |z - z_0| < \delta'\} \subset f^{-1} \left[D^* = \{w; |w - \alpha| < \varepsilon\} \right],$$

it follows that for any $z_1, z_2 \in D$,

$$1 = \frac{1}{2\pi i} \oint_\Gamma \frac{dw}{w - f(z_1)} = \frac{1}{2\pi i} \oint_\Gamma \frac{dw}{w - f(z_2)},$$

or equivalently,

$$1 = \frac{1}{2\pi i} \oint_\Gamma \frac{f'(z)}{f(z) - f(z_1)} dz = \frac{1}{2\pi i} \oint_\Gamma \frac{f'(z)}{f(z) - f(z_2)} dz.$$

This means that each function $f(z) - f(z_1)$ and $f(z) - f(z_2)$ has only one zero inside the circle $|z - z_0| = \delta$. Therefore, we conclude that $f(z_1) \neq f(z_2)$ if $z_1 \neq z_2$. ♣

10.2 Elementary Transformations

10.2.1 Linear Transformations

The most simple conformal mapping $w = f(z)$ would be the following:

♠ **Linear transformation:**

$$w = \alpha z + \beta, \tag{10.11}$$

where α and β are complex numbers.

A linear transformation generates a translation plus a magnification and a rotation of a polygon, but does not affect its shape. Thus, for example, a line maps to a line, a rectangle maps to a rectangle, a circle maps to a circle, etc.

To appreciate the above statement, we first consider the particular case of $\alpha = 1$. From (10.11), we have

$$w = z + \beta, \tag{10.12}$$

which describes a translation by the constant β of the points being mapped. Obviously, a translation does not modify the length of a line or its orientation, only changes its position with respect to the coordinate axes. Since a polygon is constructed from three or more lines, the size and orientation of a polygon are not affected by a translation; only the position of the polygon is changed.

Next we consider the case of $\beta = 0$. When we express α in polar form, the linear transformation becomes

$$w = |\alpha| e^{i\gamma} z$$

with a constant argument γ. Then, the line between two points transforms as

$$w_1 - w_2 = |\alpha| e^{i\gamma}(z_1 - z_2) = |\alpha| \cdot |z_1 - z_2| e^{i(\gamma+\theta)}.$$

Therefore, the length of a line in the z-plane, $|z_1 - z_2|$, becomes magnified by a constant factor $|\alpha|$ and the line is rotated through an angle γ. Thus, the lengths of the sides of a polygon and the orientation of the polygon with respect to the axes is modified. Nevertheless, its shape remains unchanged by the linear transformation with $\beta = 0$.

We have seen that the values of α and β straightforwardly determine the image of a polygon in the z-plane under a particular linear transformation. Conversely, if one knows the coordinates of two points on the original polygon in the z-plane and the images of those two points in the w-plane, one can determine α and β and thus the linear transformation.

10.2.2 Bilinear Transformations

There is another important conformal mapping referred to as the **bilinear transformation** (or the **fractional** or **Möbius transformation**):

♠ **Bilinear transformation:**

$$w = \frac{\alpha z + \beta}{\gamma z + \delta}, \tag{10.13}$$

where α, β, γ and δ are complex numbers satisfying the relation $\alpha\delta - \beta\gamma \neq 0$.

The condition $\alpha\delta - \beta\gamma \neq 0$ ensures that

$$\frac{df}{dz} = \frac{\alpha\delta - \beta\gamma}{(\gamma z + \delta)^2}$$

is nonzero at any finite point of the plane. Accordingly, the bilinear transformation (10.13) possesses the one-to-one property because if $f(z_1) = f(z_2)$, then

$$\frac{\alpha z_1 + \beta}{\gamma z_1 + \delta} = \frac{\alpha z_2 + \beta}{\gamma z_2 + \delta},$$

which implies $(\alpha\delta - \beta\gamma)(z_1 - z_2) = 0$, and thus $z_1 = z_2$.

Remark.

1. If $\gamma = 0$, the bilinear transformation (10.13) reduces to a linear transformation, which has already been discussed. Thus, we require that $\gamma \neq 0$ in what follows.
2. The function $f(z) = (\alpha z + \beta)/(\gamma z + \delta)$ serves as a **general solution** (see Sect. 15.1.4) of the differential equation:

$$\left(\frac{f''}{f'}\right)' - \frac{1}{2}\left(\frac{f''}{f'}\right)^2 = 0,$$

   which is called the **Schwarz differential equation**.

Observe that the mapping (10.13) has two apparent exceptional points: $z = \infty$ and $z = -\delta/\gamma$ at which w diverges. It is possible to weed out these exceptions by extending the definition of conformal representation such that the point at infinity is included. With such an extension, the conformal property of the transformation (10.13) at the two points is recovered, even though the function $f(z)$ itself diverges. Similarly, we cay say that $w = f(z)$ transforms the neighborhood of $z = \infty$ conformally into that of a point w_0 if $w = \phi(\xi) = f(1/\xi)$ transforms the neighborhood of $\xi = 0$ conformally into that of the point w_0.

A particularly interesting example of the bilinear transformation is

$$w = f(z) = \frac{z - z_0}{z - z_0^*}, \tag{10.14}$$

where $\mathrm{Im}(z_0) \neq 0$. This transformation maps the upper half-plane of the z-plane including the x-axis, onto the unit circle centered at the origin of the w-plane. This is demonstrated in Exercise **1**.

10.2.3 Miscellaneous Transformations

In what follows, we note several elementary transformations that facilitate a better understanding of the conformal nature of analytic functions. We shall see that any conformal transformation may be regarded as a transformation from Cartesian to **orthogonal curvilinear coordinates**.

Example 1. $w = z^2$, $w = \sqrt{z}$
Assume a conformal mapping defined by

$$w = z^2. \tag{10.15}$$

Setting $z = x + iy$ and separating the real and imaginary parts, we have

$$x^2 - y^2 = u, \quad 2xy = v. \tag{10.16}$$

Thus, the straight lines parallel to the x- and y-axes in the z-plane denoted by

$$x = a \quad \text{and} \quad y = b$$

are mapped onto rectangular hyperbolas in the w-plane given by

$$u = a^2 - \frac{v^2}{4a^2} \quad \text{and} \quad u = \frac{v^2}{4b^2} - b^2,$$

respectively. This is shown schematically in Fig. 10.4.

Another important feature of the mapping (10.15) is found by expressing z and w in polar coordinates:

$$z = \rho e^{i\phi}, \quad w = re^{i\theta}.$$

On substitution in (10.15), we obtain

$$r = \rho^2, \quad \theta = 2\phi. \tag{10.17}$$

Hence, the upper half of the z-plane, $0 \leq \phi \leq \pi$, goes into the entire w-plane, $0 \leq \theta \leq 2\pi$; the lower half also goes into the entire w-plane. In other words, points z and $-z$ in the z-plane obviously go into the same point in the

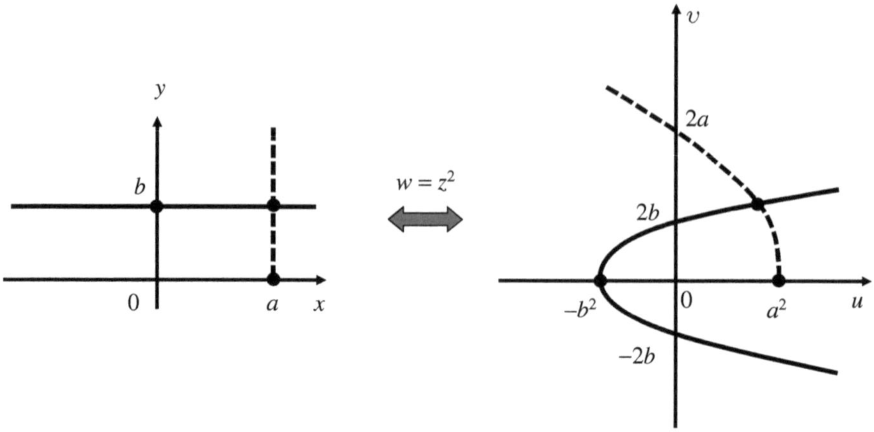

Fig. 10.4. Mapping $w = z^2$

w-plane. This suggests the possibility that some distinct geometric figures in the z-plane may go into coincident figures in the w-plane.

Next we consider the transformation: $w = \sqrt{z}$. In terms of polar forms, it reads

$$\sqrt{z} = \rho^{1/2} e^{i\phi/2} e^{in\pi},$$

so that we have

$$r = \sqrt{\rho}, \quad \theta = \frac{\phi}{2} + n\pi. \tag{10.18}$$

Owing to the additional term $n\pi$ in the latter equation in (10.18), a half revolution in the z-plane corresponds to one complete revolution in the w-plane. This is obviously a manifestation of the multivaluedness of the root function. The mapping of the upper half of the z-plane onto the w-plane is illustrated schematically in Fig. 10.5.

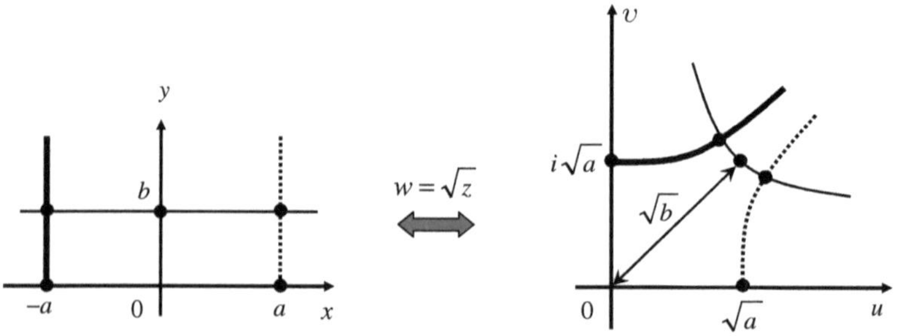

Fig. 10.5. Mapping $w = \sqrt{z}$

Example 2. $w = e^z$, $w = \log z$
In the case of

$$w = e^z, \tag{10.19}$$

there are simple relationships between the Cartesian coordinates in the z-plane and the polar coordinates in the w-plane

$$re^{i\theta} = e^{x+iy} = e^x(\cos y + i \sin y); \text{ i.e.,} \quad r = e^x, \ \ \theta = y.$$

The lines $x =$ const., parallel to the y-axis, become concentric circles in the w-plane; the lines $y =$ const., parallel to the x-axis, become rays emerging from the origin. Accordingly, a strip of the z-plane bounded by $y = y_0$ and $y = y_0 + 2\pi$ goes into the entire w-plane.

In the inverse of (10.19)

$$z = \log w, \ \ x = \log r, \ \ y = \theta + 2n\pi,$$

which is an infinitely many-valued function since all points for different values of n correspond to the same point in the w-plane.

Example 3. $w = \cosh z$
Next let us consider the following functions:

$$w = \cosh z.$$

The Cartesian coordinates in the two planes are related as follows:

$$\begin{aligned} u + iv = \cosh(x + iy) \ &= \cosh x \cos y + i \sinh x \sin y, \\ u = \cosh x \cos y, \ \ &v = \sinh x \sin y. \end{aligned} \tag{10.20}$$

Dividing the first equation by $\cosh x$, the second by $\sinh x$, squaring and adding, we have an ellipse in the w-plane that corresponds to the straight line $x =$ const. in the z-plane. Similarly, $y =$ const. goes into a hyperbola in the w-plane. The equations of the ellipses and hyperbolas are

$$\frac{u^2}{\cosh^2 x} + \frac{v^2}{\sinh^2 x} = 1, \quad \frac{u^2}{\cos^2 y} - \frac{v^2}{\sin^2 y} = 1. \tag{10.21}$$

The semimajor and semiminor axes of the ellipses are $\cosh x$ and $\sinh x$; the semifocal distance is unity. The semiaxes of the hyperbolas are $\cos y$ and $\sin y$; the semifocal distance is unity. Hence, equations (10.21) represent families of confocal ellipses and hyperbolas. This transformation may be regarded as a transformation from Cartesian to elliptic coordinates.

Example 4. $w = 1/z$
Consider the function

$$w = \frac{1}{z} \tag{10.22}$$

and use rectangular coordinates to obtain

$$(u + iv)(x + iy) = 1.$$

By equating real and imaginary parts, we set

$$ux - vy = 1, \quad vx + uy = 0.$$

By an algebraic elimination first of x and then of y, we arrive at the two families of circles:

$$u^2 + \left(v + \frac{1}{2y}\right)^2 = \frac{1}{4y^2}, \quad \left(u - \frac{1}{2x}\right)^2 + v^2 = \frac{1}{4x^2}. \tag{10.23}$$

The degenerate cases $x = 0$ and $y = 0$ cannot be handled by (10.23), but from (10.22) we find that respectively, they give the two axes $u = 0$ and $v = 0$.

The transformation is shown in Fig. 10.6. Note that through the transformation, the edge of the z-plane at infinity ($z = \infty$) is pulled into the origin of the w-plane ($w = 0$), whereas the center of the z-plane is stretched out in all directions to infinity in the w-plane. It is possible to visualize this process by introducing an artificial concept, called "the **point at infinity**"; see Sect. 10.1.5 for details.

Remark. The mapping $w = 1/z$ reverses the orientation of the circumference of the circle to be mapped: $\arg(w) = -\arg(z)$. For example, the circumference of $|w| = 1$ is described in the negative since if $|z| = 1$ is described in the positive sense.

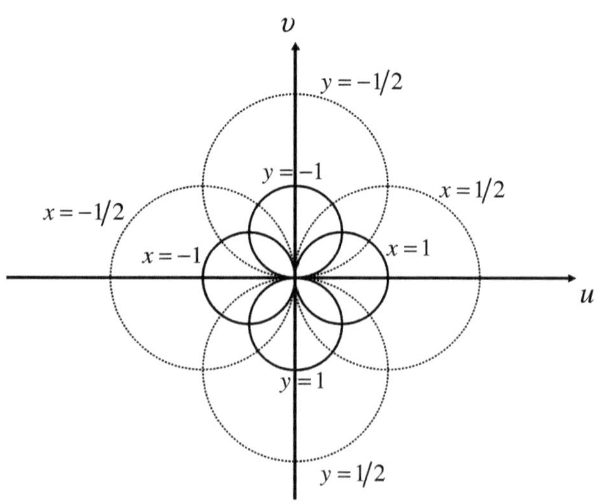

Fig. 10.6. Mapping $w = 1/z$

10.2.4 Mapping of Finite-Radius Circle

Remember that the **analyticity** of functions is characterized by the isotropy of their derivatives. Owing to the isotropy, infinitely small circles on the z-plane are transformed into infinitely small circles an the w-plane through any analytic function $w = f(z)$. Of course, this shape-preserving behavior disappears when the circle has a finite radius; because the scale factor h generally depends on z. Nevertheless, there exist a class of nontrivial analytic functions that transform a finite circle on the z-plane onto the w-plane, which is simply a bilinear transformation.

♠ **Theorem:**
Bilinear transformations $w = f(z)$ map circles (or straight lines) on the z-plane onto circles (or straight lines) on the w-plane.

Proof Our proof is based on the fact that the bilinear transformation formula (10.11) can be rewritten as

$$w = f(z) = \frac{\alpha}{\gamma} + \frac{\beta\gamma - \alpha\delta}{\gamma}\frac{1}{\gamma z + \delta}.$$

This is composed of a sequential transformation of the following:

1. $w = z + b$, a simple translation of the plane by the complex vector b.
2. $w = az$, a rotation of the plane through the angle $\arg a$, followed by an expansion (or contraction) by $|a|$.
3. $w = 1/z$, an **inversion** that takes the interior of the unit circle to the exterior and vice versa.

Since these transformations are all conformal, their composition surely maps circles (or straight lines) onto circles (or straight lines). ♣

Remark. Statement **3** above regarding the inversion $w = 1/z$ is followed by considering the equation

$$\alpha(x^2 + y^2) + \beta x + \gamma y + \delta = 0,$$

which represents a circle ($\alpha \neq 0$) or straight line ($\alpha = 0$) in the z-plane. This can be written as

$$\alpha|z|^2 + \frac{\beta}{2}(z + z^*) + \frac{\gamma}{2i}(z - z^*) + \delta = 0. \qquad (10.24)$$

Then, the transformation $w = 1/z$ maps it onto

$$\delta|w|^2 + \frac{\beta}{2}(w + w^*) - \frac{\gamma}{2i}(w - w^*) + \alpha = 0,$$

which is a circle ($\delta \neq 0$) or a straight line ($\delta = 0$).

10.2.5 Invariance of the Cross ratio

The following peculiarity of a Möbius transformation serves as a useful device in applications of conformal mapping.

♠ **Invariance of the cross ratio:**
Any Möbius transformation $w = f(z)$ that maps the four points z_i $(i = 1, 2, 3, 4)$ into w_i $(i = 1, 2, 3, 4)$, respectively, satisfies

$$\frac{(w_1 - w_4)(w_3 - w_2)}{(w_1 - w_2)(w_3 - w_4)} = \frac{(z_1 - z_4)(z_3 - z_2)}{(z_1 - z_2)(z_3 - z_4)} \equiv \lambda.$$

The constant λ is called the **cross ratio** (or **anharmonic ratio**).

Proof Let z_i $(i = 1, 2, 3, 4)$ be four distinct finite points on the z-plane and let w_i $(i = 1, 2, 3, 4)$ be their corresponding images through a Möbius transformation. Then, for any two of the points, we have

$$w_k - w_i = \frac{\alpha z_k + \beta}{\gamma z_k + \delta} - \frac{\alpha z_i + \beta}{\gamma z_i + \delta} = \frac{\alpha\delta - \beta\gamma}{(\gamma z_k + \delta)(\gamma z_i + \delta)}(z_k - z_i),$$

and, consequently, for all four,

$$\frac{(w_1 - w_4)(w_3 - w_2)}{(w_1 - w_2)(w_3 - w_4)} = \frac{(z_1 - z_4)(z_3 - z_2)}{(z_1 - z_2)(z_3 - z_4)}. \tag{10.25}$$

This clearly ensures the invariance of the cross ratio λ under the Möbius transformation. ♣

Remark. If one of the points of w_i, say w_1, is the **point at infinity**, the corresponding result is obtained by letting $w_1 \to \infty$ in (10.25). The left-hand side then takes the form

$$\frac{w_3 - w_2}{w_3 - w_4}.$$

This expression is to be regarded as the cross ratio of the points ∞, w_2, w_3, w_4. A similar remark applies if one of the points z_i is the point at infinity.

If z_4 is taken to be a variable z, then the corresponding image w_4 on the w-plane becomes a function of z that obeys the relation

$$\frac{(w_1 - w)(w_3 - w_2)}{(w_1 - w_2)(w_3 - w)} = \frac{(z_1 - z)(z_3 - z_2)}{(z_1 - z_2)(z_3 - z)}. \tag{10.26}$$

By solving (10.26) for w, we can verify that it transforms the three points z_1, z_2, z_3 into the corresponding points w_1, w_2, w_3. In this context, the expression (10.26) turns out to show that a Möbius transformation is uniquely determined by three correspondences. Since a circle is uniquely determined by three points on its circumference, (10.26) can be used to find Möbius transformations that map a given circle determined by $z_i (i = 1, 2, 3)$ onto a second given circle (or straight line) determined by $w_i (i = 1, 2, 3)$.

Example If we take $z_1 = 1, z_2 = i, z_3 = -1$ and $w_1 = 0, w_2 = 1, w_3 = \infty$, we obtain the transformation

$$w = i\,\frac{1-z}{1+z}.$$

This maps the circle $|z| = 1$ on the real axis and the interior $|z| < 1$ of the unit circle on the upper half of the w-plane.

Exercises

1. Consider the function $w = f(z) = (z - z_0)/(z - z_0^*)$ with $\text{Im}(z_0) \neq 0$. Show that it maps the region $\text{Im} z > 0$ onto $|w| < 1$.

Solution: Set $z = x$ to obtain

$$|w|^2 = \left(\frac{x-z_0}{x-z_0^*}\right)\left(\frac{x-z_0}{x-z_0^*}\right)^* = \left(\frac{x-z_0}{x-z_0^*}\right)\left(\frac{x-z_0^*}{x-z_0}\right) = 1.$$

That is, the image on the x-axis is the circumference of the unit circle centered at the origin of the w-plane.

Next we evaluate the image of a point off the x-axis in the upper half of the z-plane. Expressing z and z_0 in polar form, we have

$$|w|^2 = \frac{\left(re^{i\theta} - r_0e^{i\theta_0}\right)\left(re^{-i\theta} - r_0e^{-i\theta_0}\right)}{\left(re^{i\theta} - r_0e^{-i\theta_0}\right)\left(re^{-i\theta} - r_0e^{i\theta_0}\right)} = \frac{\xi_1 - \xi_2}{\xi_1 + \xi_2}, \qquad (10.27)$$

where

$$\xi_1 = r^2 + r_0^2 - 2rr_0 \cos\theta \cos\theta_0 \quad \text{and} \quad \xi_2 = 2rr_0 \sin\theta \sin\theta_0.$$

Since $-1 \leq \cos\theta \cos\theta_0 < 1$, we have

$$(r - r_0)^2 \leq r^2 + r_0^2 - 2rr_0 \cos\theta \cos\theta_0 = \xi_1, \quad \text{i.e.,} \;\; \xi_1 \geq 0.$$

In addition, since z and z_0 are in the upper half-plane, both $\sin\theta$ and $\sin\theta_0$ are positive, so $\xi_2 > 0$. Consequently, we have

$$|w|^2 < 1,$$

which means that the images of points in the upper half of the z-plane are located in the interior of the unit origin-centered circle. ♣

Remark. If z_0 were real, all points z would be mapped onto the single point $w = 1$, which is the reason we assumed $\mathrm{Im}(z_0) \neq 0$ in the first place.

2. Show that $w = (z - z_0)/(z_0^* z - 1)$ in which $|z_0| < 1$ maps $|z < 1|$ onto $|w| < 1$ and $z = z_0$ onto $w = 0$.

   Solution: Observe that

$$1 - |w|^2 = 1 - \frac{|z - z_0|^2}{|z_0^* z - 1|^2} = \frac{|z_0|^2|z|^2 - z^2 - (z_0^*)^2 + 1}{|z_0^* z - 1|^2}$$
$$= \frac{(1 - |z|^2)(1 - |z_0|^2)}{|z_0^* z - 1|^2}.$$

   Hence, $|z| = 1$ corresponds to $|w| = 1$. In addition, $z = z_0$ corresponds to $w = 0$. These mean that $|z| < 1$ is transformed onto $|w| < 1$. ♣

3. Let C and C^* be two simple closed contours in the z- and the w-plane, respectively, and let $w = f(z)$ be analytic within and on C. If $w = f(z)$ maps C onto C^* in such a way that C^* is traversed by w exactly once in the positive sense under the condition that z describes C in the positive sense, then $w = f(z)$ maps the domain bounded by C onto the domain bounded by C^*.

   Solution: We denote the domains bounded by C and C^* by D and D^*, respectively. Then it suffices to prove that every point of D^* is taken exactly once if z is in D. Recall that the number n of zeros of the function $w_0 - f(z)$ in D is given by

$$n = \frac{1}{2\pi i} \oint_C \frac{f'(z)}{f(z) - w_0} dz.$$

   With the substitution $w = f(z)$, $f'(z)dz = dw$, this is rewritten as

$$n = \frac{1}{2\pi i} \oint_{C^*} \frac{dw}{w - w_0},$$

   where the integration has to be extended over the contour C^* into which C is transformed by $w = f(z)$. By the residue theorem, the value of this expression is 1 if w_0 is within C^* and 0 if w_0 is outside C^*. This shows that every point in D^* is taken exactly once and that a value outside D^* is not taken at all. This completes the proof. ♣

4. Find a conformal mapping $w = f(z)$ of the region between the two circles $|z| = 1$ and $|z - (1/4)| = 1/4$ onto an annulus $a < |z| < 1$.

   Solution: To solve this, we have to find a bilinear transformation that simultaneously maps $|z| < 1$ onto $|z| < 1$ and $|z-(1/4)| < 1/4$ onto a disc of the form $|z| < a$. Note that

$$w = \frac{z-\alpha}{1-\alpha^* z}$$

maps $|z|<1$ onto $|z|<1$, and that

$$g(z) = a\frac{4z-1-\beta}{1-\beta^*(4z-1)}$$

maps $|z|<1$ and $|z-(1/4)|<1/4$ onto a disc of the form $|z|<a$. Equating coefficients leads us to $\alpha = 2-\sqrt{3}$. ♣

5. Find the bilinear transformation that maps $z = 0, i, -1$ onto $w = 1, -1, 0$, respectively.

   Solution: Set $[z, 0, i, -1] = [w, 1, -1, 0]$ to obtain $w = -(z+i)/(3z-i)$. ♣

6. Show that four distinct arbitrary points on the z-plane can be mapped through the bilinear transformation onto $w = 1, -1, c, -c$ on the w-plane, where c is a complex number depending on the cross ratio λ of the mapping. Determine an explicit form of c as a function of λ.

   Solution: Let $[z_1, z_2, z_3, z_4] = [1, -1, c, -c]$ to obtain $c = (1+\lambda \pm 2\sqrt{2})/(1-\lambda)$ and $c_1 c_2 = 1$. ♣

10.3 Applications to Boundary-Value Problems

10.3.1 Schwarz–Christoffel Transformation

In the preceding section, we discussed rich properties of the bilinear transformation that can transform the upper half of the z-plane onto the unit circle of the w-plane. Now we turn to a similar kind of important mappings called the **Schwarz–Christoffel transformation** (abbreviated SC transformation), which transforms the upper (or lower) half of the z- plane onto the inside of a n-sided polygon drawn on the w-plane. This transformation is defined by the following integral:

$$w(z) = \beta + \alpha \sum_{i=1}^{n} \int^{z} (z' - x_i)^{-\theta_i/\pi} dz'. \qquad (10.28)$$

Here x_i $(1 \le i \le n)$ are n distinct fixed points along the x-axis, and the angle θ_j is defined as shown in Fig. 10.7, being either positive or negative according to whether we follow the boundary of the polygon counterclockwise or clockwise. (For example, θ_1 and θ_2 are positive, but θ_3 is negative in Fig. 10.7.) The constant α gives rise to a magnification of that image by a factor $|\alpha|$ and a rotation of that image by an angle $\arg(\alpha)$. The constant β generates a translation of the magnified and rotated image.

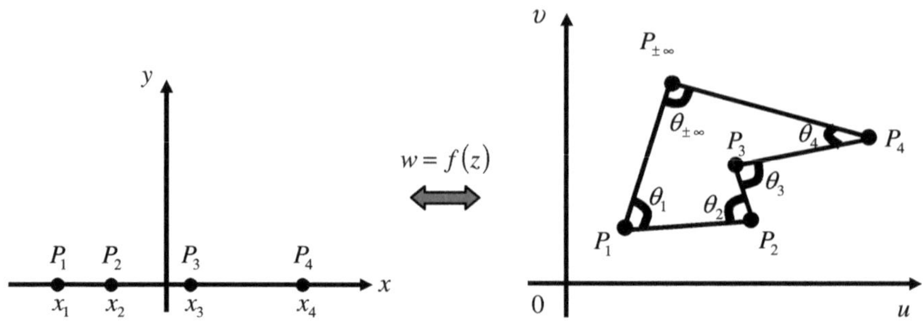

Fig. 10.7. Schwarz–Christoffel transformation of the real axis of the z-plane to a polygon on the w-plane

Remark. If we wish to transform the upper half of the z-plane into the exterior of the polygon in the w-plane, it suffices to define

$$w(z) = \beta + \alpha \int^{z} (z' - x_1)^{\theta_1/\pi}(z' - x_2)^{\theta_2/\pi} \cdots (z' - x_N)^{\theta_N/\pi} dz',$$

where the θ's are assigned the same values as in the preceding case.

Example The function

$$w = f(z) = \int^{z} \frac{d\xi}{\sqrt{(1-\xi^2)(1-k^2\xi^2)}} \quad (0 < k < 1) \tag{10.29}$$

maps the upper half of the z-plane ($\text{Im}z > 0$) into the interior of a rectangle on the w-plane. In fact, (10.29) is obtained by putting $n = 4$, $\theta_i = \pi/2$ for all $i = 1, 2, 3, 4$ in the definition (10.28), followed by setting $x_1 = 1$, $x_2 = -1$, $x_3 = 1/k$ and $x_4 = -1/k$, all of which are located on the real axis. The integral in (10.29) is called an **elliptic integral of the first kind**.

10.3.2 Derivation of the Schwartz–Christoffel Transformation

In order to derive equation (10.28) for the Schwarz–Christoffel transformation, we let

$$x_1 < x_2 < \cdots < x_n$$

be points on the real axis and consider the function $f(z)$ whose derivative is

$$f'(z) = \alpha(z - x_1)^{-k_1}(z - x_2)^{-k_2} \cdots (z - x_n)^{-k_n}. \tag{10.30}$$

For this function we have

$$\arg f'(z) = \arg \alpha - k_1 \arg(z - x_1) - k_2 \arg(z - x_2) - \cdots - k_n \arg(z - x_n).$$

Now, visualize the point z as moving from left to right along the real axis, starting to the left of the point x_1. When $z < x_1$, we have

$$\arg(z - x_1) = \arg(z - x_2) = \cdots = \arg(z - x_n) = \pi,$$

whereas for $x_1 < z < x_2$, $\arg(z - x_1) = 0$, the others remaining at π. Hence, as z crosses a_1 from left to right, $\arg f'(z)$ increases by $k_1\pi$. It remains constant for $x_1 < z < x_2$ and increases by $k_2\pi$ as z crosses x_2, etc. As a result, the image of the segment $-\infty < z < a_1$ becomes a straight line, the image of $x_1 < z < x_2$ being another whose argument exceeds that of the first by $k_1\pi$, and so on.

If we constrain the numbers $k_1, \cdots, k_n$ to lie between -1 and 1, then the increments in the argument of $f'(z)$ will lie between $-\pi$ and π. Further, for $k_1 < 1, k_2 < 1, \cdots, k_n < 1$, it is obvious that the function $f(z)$ whose derivative is (10.30) is actually continuous at each of the points $x_1, x_2, \cdots, x_n$. Therefore, the image of the moving point z will be a polygonal line. Finally, integrate (10.30) to set the equation

$$f(z) = \beta + \alpha \int^z (z' - a_1)^{-k_1} (z' - a_2)^{-k_2} \cdots (z' - a_n)^{-k_n} dz', \qquad (10.31)$$

which maps the x-axis onto a polygonal line.

Remark.

1. The sum of the exterior angles of this polygonal line is
$$k_1\pi + k_2\pi + \cdots + k_n\pi = \pi \sum_{i=1}^{n} k_i.$$
Hence, in order for the polygon to be closed, it is necessary that $\sum_{i=1}^{n} k_i = 2$. Particularly when $k_i > 0$ for all i, then the polygon becomes convex.
2. The complex constants, α and β, control the position, size, and orientation of the polygon. Thus β may be so chosen that one of the vertices of the polygon will coincide with some specified point e.g., the origin. Then α may be chosen so that one side of the polygon will be of given size and parallel to a given direction.

10.3.3 The Method of Inversion

The Schwarz–Christoffel transformation itself is applicable to polygons composed of straight lines, but not to those of circular ones. Nevertheless, combining the **method of inversion**, the former transformation can be extended to regions bounded by circular arcs.

♠ **Inversion with respect to a circle:**

An inversion transformation $w = f(z)$ with respect to a circle $|z| = a$ is defined by

$$w = \frac{a^2}{z^*}. \tag{10.32}$$

through which the interior points of the circle are mapped onto exterior points, and vice versa.

The inversion preserves the magnitude of the angle between two intersecting curves, but it reverses the sign of the angle. This is attributed to the fact that (10.32) consists of two successive transformations: the first a^2/z, and the second a reflection with respect to the real axis. The first of these is conformal, whereas the second maintains the angle but reverses its sign.

For the purpose of this section, we investigate the inversion of a circle of radius $|z_0|$ centered at $z = z_0 \neq 0$. This circle is expressed by

$$|z - z_0| = |z_0| \tag{10.33}$$

or

$$z^* z - z_0(z + z^*) = 0. \tag{10.34}$$

Note that this circle passes through the origin, i.e., the center of an inversion circle. Through the inversion (10.32), the circle (10.34) is mapped onto

$$\frac{a^4}{ww^*} - z_0\left(\frac{a^2}{w} + \frac{a^2}{w^*}\right) = 0.$$

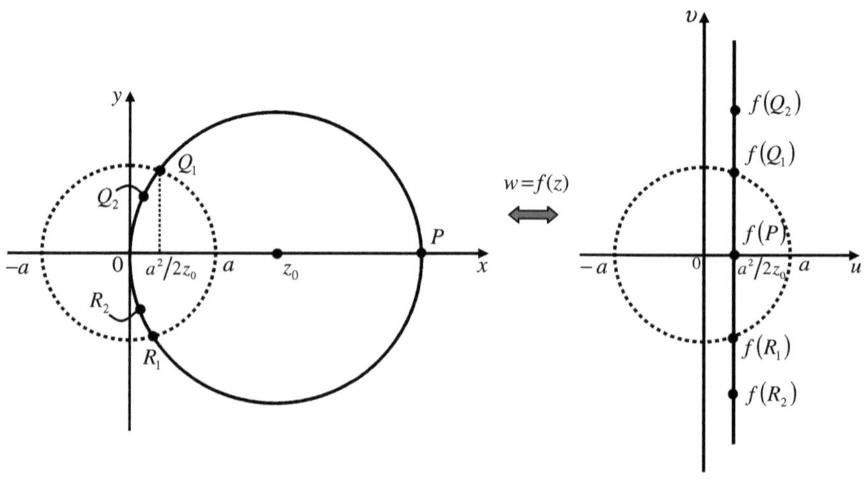

Fig. 10.8. Inversion of the circle $|z - z_0| = |z_0|$ in (10.33) with respect to a circle $|z| = a$ through the mapping $w = a^2/z^*$ given in (10.32)

By multiplying ww^* on both sides and putting $w = u + iv$, we have

$$a^4 - 2a^2 z_0 u = 0,$$

or equivalently,

$$u = \frac{a^2}{2z_0}.$$

This means that by the inversion, the circle (10.33) is mapped onto a straight line parallel to the imaginary axis of the w-plane (see Fig. 10.8).

The role that inversion plays in extending the Schwarz–Christoffel transformation should now be clear. Assume two interesting circular arcs such as P and Q in Fig. 10.9 and a circle R of radius a whose center is the intersection of the two circular arcs. Then, by an inversion with respect to R, the point at the intersection is transformed into the **point at infinity**, the arcs themselves being transformed into the solid portions of the lines P' and Q'. As a result, the Schwarz–Christoffel transformation may now be applied to these two straight lines, whereas it may not be applied to the original circular arcs.

Exercises

1. Find a transformation that maps the upper half of the z-plane onto the triangular region shown in Fig. 10.10 in such a way that the points $x_1 = -1$ and $x_2 = 1$ are mapped onto the points $w = -a$ and $w = a$, respectively, and the point $x_3 = \pm\infty$ is mapped onto $w = ib$.

   Solution: Let us denote the angles at w_1 and w_2 in the w-plane by $\phi_1 = \phi_2 = \phi$, where $\phi = \tan^{-1}(b/a)$. Since x_3 is taken at infinity we may omit the corresponding factor in (10.28) to obtain

$$w = \beta + \alpha \int_0^z (\xi + 1)^{-\phi/\pi} (\xi - 1)^{-\phi/\pi} d\xi = \beta + \alpha \int_0^z (\xi^2 - 1)^{-\phi/\pi} d\xi. \tag{10.35}$$

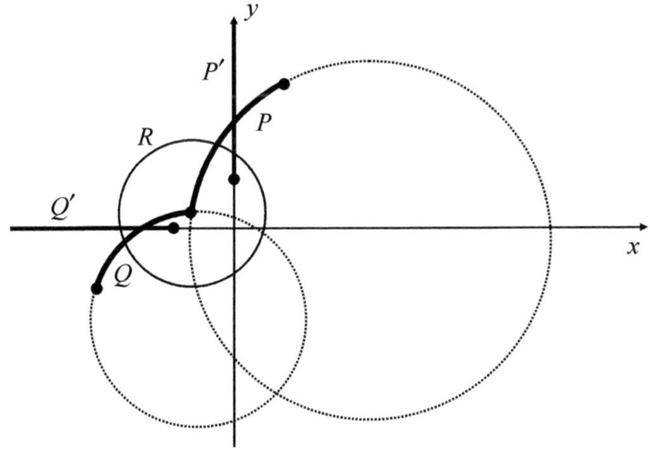

Fig. 10.9. Inversion of circular arcs P and Q with respect to the circle R

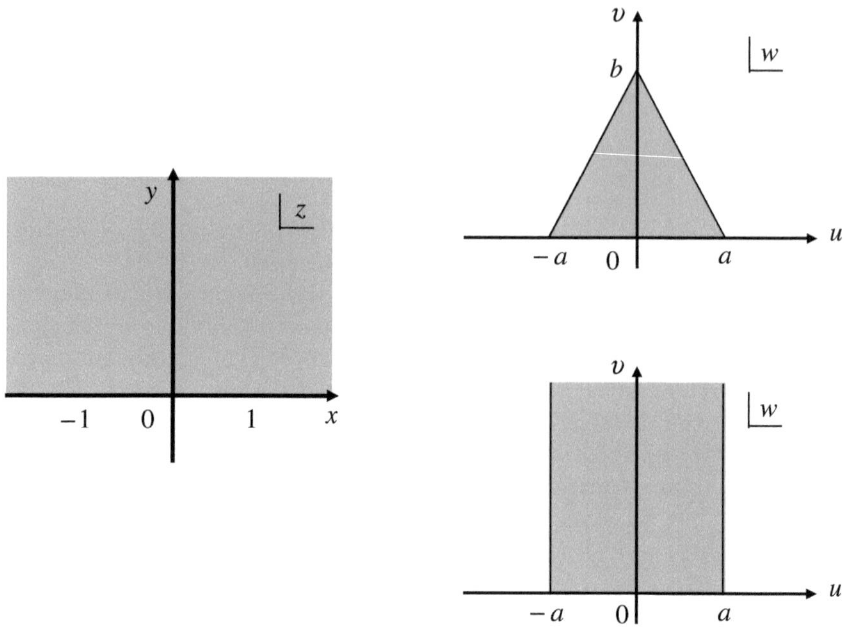

Fig. 10.10. Mapping of the upper half of the z-plane onto a certain limited region of the w-plane

The required transformation may then be found by fixing the constants α and β as follows. Since the point $z = 0$ lies on the line segment x_1x_2 it will be mapped onto the line segment w_1w_2 in the w-plane, and by symmetry must be mapped onto the point $w = 0$. Thus setting $z = 0$ and $w = 0$ in (10.35), we obtain $\beta = 0$. An expression for α can be found by considering the region in the w-plane in Fig. 10.10 to be the limiting case of the triangular region with the vertex w_3 at infinity. Thus we may use the above, but with the angles at w_1 and w_2 set to $\phi = \pi/2$. From (10.35), we obtain $w = \alpha \int_0^z (1/\sqrt{\xi^2 - 1})d\xi = i\alpha \sin^{-1} z$. By setting $z = 1$ and $w = a$, we find $i\alpha = 2a/\pi$, so the required transformation is $w = (2a/\pi)\sin^{-1} z$. ♣

2. Find the conformal mapping that transforms the interior of the circle $|z| < 1$ to the interior of a polygon on the w-plane, subject to the condition that the points $z_1, z_2, \cdots, z_n$ lying on the circle $|z| = 1$ are mapped, respectively, onto the vertex $w_1, w_2, \cdots, w_n$ of the polygon.

Solution: Consider first the transformation

$$\tau(z) = \frac{-iz + 1}{z - i}, \qquad (10.36)$$

which maps $|z| < 1$ onto $\text{Im}\tau > 0$. It yields

$$\frac{d\tau}{dz} = -\frac{2}{(z-i)^2} \text{ and } \tau - \tau_j = \frac{2(z_j - z)}{(z-i)(z_j - i)} \quad (j = 1, 2, \cdots, n). \tag{10.37}$$

Next, we assume that through (10.36), the points $z_1, z_2, \cdots z_n$ are mapped, respectively, onto the points $\tau_1, \tau_2, \cdots, \tau_n$ that are located on the line $\text{Im}\tau = 0$. Then, the transformation that maps $\text{Im}\tau > 0$ onto the interior of a polygon on the w-plane is given by $w = w(\tau)$, whose derivative reads

$$\frac{dw}{d\tau} = \alpha(\tau - \tau_1)^{(k_1/\pi)-1}(\tau - \tau_2)^{(k_2/\pi)-1} \cdots (\tau - \tau_n)^{(k_n/\pi)-1}. \tag{10.38}$$

Here k_i is the internal angle of the polygon at the ith vertex, which satisfies $\sum_{i=1}^{n} k_i = (n = 2)\pi$. From (10.37) and (10.38), we have

$$\frac{dw}{dz} = -\frac{2\alpha}{(z-i)^2} \cdot \frac{1}{2^2} \cdot \frac{(z_1 - z)^{(k_1/\pi)-1} \cdots (z_n - z)^{(k_n/\pi)-1}}{(z-i)^{-2}(z_1 - i)^{(k_1/\pi)-1} \cdots (z_n - i)^{(k_n/\pi)-1}}.$$

Replace $(\alpha/2)(z_1 - i)^{1-(k_1/\pi)} \cdots (z_n - i)^{1-(k_n/\pi)}$ by α to obtain the final result:

$$w = f(z) = \alpha \int_{z_0}^{z} (z_1 - \zeta)^{(k_1/\pi)-1} \cdots (z_n - \zeta)^{(k_n/\pi)-1} d\zeta + \beta,$$

where $\alpha(\neq 0), \beta$ are complex constants and $z_0 \neq z_1, \cdots, z_n$. ♣

3. Prove that the function

$$w = f(z) = \int_0^z \frac{1}{\sqrt[3]{1 - \xi^6}} d\xi \tag{10.39}$$

maps the unit circle on the z-plane onto a regular hexagon on the w-plane.

Solution: Observe that $\xi^6 - 1 = (\xi - \xi_1) \cdots (\xi - \xi_6)$ with $|\xi_j| = 1$ $(j = 1, \cdots, 6)$. Similarly to Exercise **2** above, we map $|z| < 1$ onto $\text{Im}\tau > 0$, and then let the points $\tau_1, \cdots \tau_6$ located on the line $\text{Im}\tau = 0$ correspond to $\xi_1, \cdots \xi_6$. Then, by setting $n = 6$ and $k_j = (2/3)\pi$ for all j, we see that the transformation (10.39) maps $|z| < 1$ onto a regular hexagon on the w-plane, $(\sqrt[3]{2}/6)\Gamma(1/3)$ on a side.

4. Suppose that $\phi(z)$ satisfies the Laplace equation and let $w = f(z)$ be a conformal mapping. Then, show that the function

$$\phi(w) = \phi(u, v)$$

also satisfies Laplace's equation in the w-plane; i.e.,

$$\frac{\partial^2 \phi}{\partial u^2} + \frac{\partial^2 \phi}{\partial v^2} = 0. \tag{10.40}$$

Solution: Since $x = x(u, v)$, the partial derivative $\partial/\partial x$ can be rewritten as $\partial/\partial x = u_x(\partial/\partial u) + v_x(\partial/\partial v)$, where $u_x = \partial u/\partial x$ and $v_x = \partial v/\partial x$. It yields

$$\begin{aligned} \frac{\partial^2 \phi}{\partial x^2} &= \left(u_x \frac{\partial}{\partial u} + v_x \frac{\partial}{\partial v} \right) \left(u_x \frac{\partial}{\partial u} + v_x \frac{\partial}{\partial v} \right) \phi \\ &= (u_x)^2 \frac{\partial^2 \phi}{\partial u^2} + (v_x)^2 \frac{\partial^2 \phi}{\partial v^2} + 2u_x v_x \frac{\partial^2 \phi}{\partial u \partial v}. \end{aligned} \tag{10.41}$$

Similarly, we have

$$\begin{aligned} \frac{\partial^2 \phi}{\partial y^2} &= (u_y)^2 \frac{\partial^2 \phi}{\partial u^2} + (v_y)^2 \frac{\partial^2 \phi}{\partial v^2} + 2u_y v_y \frac{\partial^2 \phi}{\partial u \partial v} \\ &= (v_x)^2 \frac{\partial^2 \phi}{\partial u^2} + (u_x)^2 \frac{\partial^2 \phi}{\partial v^2} - 2v_x u_x \frac{\partial^2 \phi}{\partial u \partial v}, \end{aligned} \tag{10.42}$$

where we have used the Cauchy–Riemann relations: $u_x = v_y$, $u_y = -v_x$. Adding up the sides of the second lines of (10.41) and (10.42), we obtain

$$\frac{\partial^2 \phi}{\partial x^2} + \frac{\partial^2 \phi}{\partial y^2} = \left[(u_x)^2 + (u_y)^2 \right] \left(\frac{\partial^2 \phi}{\partial u^2} + \frac{\partial^2 \phi}{\partial v^2} \right).$$

The quantity inside the square brackets is equal to $|f'(z)|^2$, which is nonzero for analytic functions $f(z)$. As a consequence, we conclude that

$$\frac{\partial^2 \phi}{\partial x^2} + \frac{\partial^2 \phi}{\partial y^2} = 0 \quad \Longleftrightarrow \quad \frac{\partial^2 \phi}{\partial u^2} + \frac{\partial^2 \phi}{\partial v^2} = 0. \quad ♣$$

10.4 Applications in Physics and Engineering

10.4.1 Electric Potential Field in a Complicated Geometry

The Schwarz–Christoffel transformation is useful in mathematical physics, since it can be used to solve two-dimensional **Laplace equations** under certain boundary conditions. In fact, there are many physical systems that are described by Laplace's equation subject to **Dirichlet** or **Neumann boundary conditions**. For example, Laplace's equation can be used to describe heat conduction in a uniform medium, nonturbulent fluid flow, and an electrostatic field in a uniform system. In this subsection, we demonstrate how the Schwarz–Christoffel transformation works efficiently to solve such two-dimensional Laplace equations. It should be emphasized that our method is independent of the physical system being described. In the meantime, we apply the transformation to problems in electrostatics in order to illustrate the method of solution, bearing in mind that that these techniques are also applicable to problems involving other physical systems.

The general procedure for determining the electrostatic potential by using conformal mapping methods involves transforming a complicated charge-distribution geometry in the z-plane into a simple geometry in the w-plane. After solving the problem for the simpler geometry, the inverse transformation to the z-plane is applied to obtain the potential for the original geometry.

As a concrete example, we consider a metal block with a cut out wedge of angle γ as shown in Fig. 10.11. There is a vacuum inside the wedge. The block extends to $\pm\infty$ in the direction perpendicular to the plane of the page. Since charge moves freely inside a metal, all of the charge placed in the conductor is distributed in such a way that the potential at all points along these edges is the same. We denote this potential by ϕ_0, i.e., the system under consideration is subject to the Dirichlet boundary conditions given by

$$\phi(r, \theta = 0) \; = \; \phi(r, \theta = \gamma) \; = \; \phi_0. \tag{10.43}$$

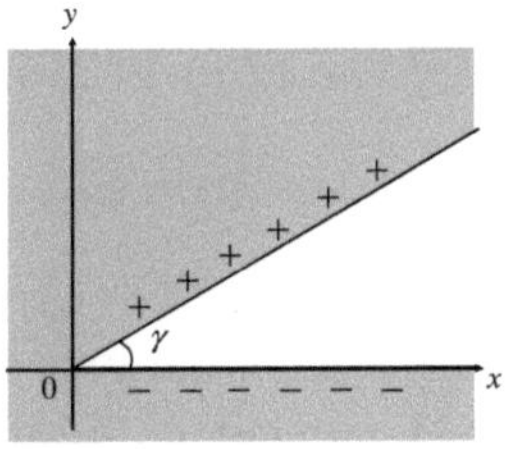

Fig. 10.11. Wedge cut of a metal

Our objective is to evaluate the potential $\phi(z)$ at points in the vacuum region inside the wedge (defined by $0 \le \arg(z) \le \gamma$). This potential satisfies the Laplace equation, and thus, it can be determined by conformal mapping methods. For this purpose, we attempt to find the mapping that transforms the wedge in the z-plane onto the real axis of the w-plane. We know that the transformation of the real axis in the z-plane onto the wedge shown in Fig. 10.11 is given by the Schwarz–Christoffel transformation:

$$w = \beta + \alpha(z - x_1)^{-\theta_1/\pi}.$$

Therefore, the inverse mapping

$$z = x_1 + \left[\frac{1}{\alpha}(w - \beta)\right]^{-\pi/\theta_1} \tag{10.44}$$

transforms the wedge in the w-plane with an internal angle $-\theta_1$ onto the real axis of the z-plane. By interchanging z and w in (10.44), we obtain the mapping

$$w = u_1 + \left[\frac{1}{\alpha}(z - \beta)\right]^{-\pi/\theta_1}, \tag{10.45}$$

which transforms the wedge in the z-plane onto the real axis of the w-plane. In order to apply this mapping to the configuration shown in Fig. 10.11, we set

$$\gamma = -\theta_1, \quad u_1 = \beta = 0, \quad \text{and} \quad \frac{1}{\alpha} = 1,$$

where α is real. Then, the mapping in (10.45) becomes

$$w = z^{\pi/\gamma}. \tag{10.46}$$

Remark. It immediately follows that the mapping in (10.46) transforms the space within the wedge onto the upper half of the w plane. This is because points within the wedge that satisfy the condition $0 < \arg(z) = \theta < \gamma$ are mapped onto $w = r^{\pi/\gamma} e^{i\pi\theta/\gamma}$, whose argument $\pi\theta/\gamma$ takes values in the interval $(0, \pi)$.

Remember that the Dirichlet boundary condition is invariant under conformal mappings. Hence, the boundary condition of (10.43) is mapped to

$$\phi(u, v = 0) = \phi_0, \tag{10.47}$$

where $v = 0$ is the image of the wedge. As noted earlier, the mapping in (10.46) transforms the problem of finding the potential in the region within the wedge in Fig. 10.11 to that of finding the potential in the upper half of the w plane due to a flat metal surface that extends along the entire u-axis and is maintained at a potential ϕ_0 by a uniformly distributed charge.

We now consider the "mapped" Laplace equation for the w-plane. Since all points on the surface of the flat plane are at the same potential, the potential all points (u, v) located at the same distance v above the plate is the same. Thus, the potential at any point must be independent of the value of u and the Laplace equation in the w-plane becomes

$$\frac{d^2\phi}{dv^2} = 0.$$

Integration of this differential equation followed by application of the boundary condition (10.47) yields

$$\phi(v) = \phi_0 + cv. \tag{10.48}$$

The constant c is obtained by using the property that the derivative of the potential (i.e., the electrostatic field) is a constant for a charged flat plate. Similar to ϕ_0, the value of this constant field E_0 depends on how much charge is distributed over a given area on the plate. With reference to (10.48),

$$\frac{\partial \phi}{\partial v} = c = -E_0, \quad \text{so that} \quad \phi(v) = \phi_0 - E_0 v.$$

In order to complete the analysis, the potential must be expressed in terms of the coordinates in the z-plane. From this expression

$$v = \operatorname{Im}(w) = \operatorname{Im}\left(r^{\pi/\gamma} e^{i\pi\theta/\gamma}\right) = r^{\pi/\gamma} \sin(\pi\theta/\gamma),$$

the potential is given by

$$\begin{aligned}\phi &= \phi_0 - E_0 r^{\pi/\gamma} \sin(\pi\theta/\gamma) \\ &= \phi_0 - E_0 \left(x^2 + y^2\right)^{\pi/(2\gamma)} \sin\left[\frac{\pi}{\gamma} \tan^{-1}\left(\frac{y}{x}\right)\right]. \end{aligned} \tag{10.49}$$

This is the final solution to the problem in question. We see from (10.49) that $\phi = \phi(r, \theta)$ is constant when

$$r^{\pi/\gamma} \sin(\pi\theta/\gamma) = \text{const.}$$

This is the equation for a family of equipotential curves.

10.4.2 Joukowsky Airfoil

Our final discussion related to the applications of conformal mappings concerns the **Joukowsky transformation**, which is an important conformal

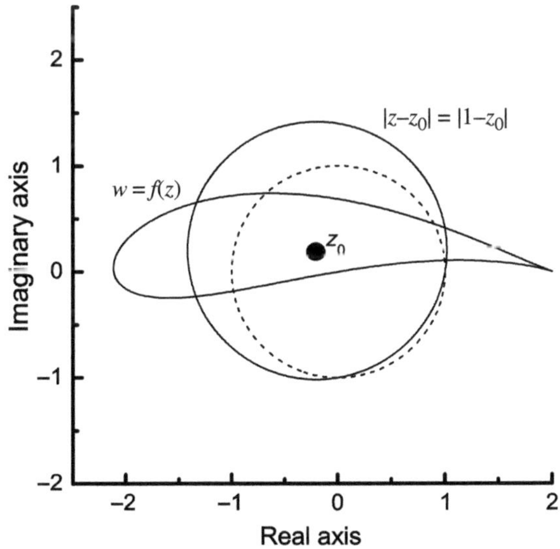

Fig. 10.12. The Joukowsky transformation (10.50) of the *circle* $|z - z_0| = |1 - z_0|$ with $z_0 = (-0.2, 0.2)$ to the airfoil indicated by the *thick curve*

mapping that has been historically employed in the theory of **airfoil** design. Here, the term "airfoil" refers to the cross-sectional shape of a wing (or a propeller or a turbine). According to the literature on airfoil theory, any object with an angle of attack in a moving fluid generates a **lift**, a force perpendicular to the flow. Airfoils are designed as efficient shapes that increase the lift that the object generates. The Joukowsky transformation maps a circle on the complex plane into a family of airfoil shapes called **Joukowsky airfoils**, which simplify the analysis of two-dimensional fluid flows around an airfoil with a complicated geometry.

The Joukowsky transformation $w = f(z)$ is defined by

$$w = f(z) = z + \frac{1}{z}, \tag{10.50}$$

where z is located on a circle C that passes through the point $z = 1$ and encloses the point $z = -1$ as well as the origin $z = 0$. Note that the center of the circle, denoted by z_0, does not coincide with the origin, but is located close to the origin. In fact, the coordinates of z_0 are variables, and changes in these variables alter the geometry of the resulting airfoil. An example of an airfoil generated by the transformation (10.50) is shown in Fig. 10.12, where $z_0 = (-0.2, 0.2)$. We see that the circle $C: \ |z - z_0| = |1 - z_0|$ is mapped onto an airfoil indicated by a thick curve. The stream lines for a flow around the airfoil can be obtained by applying an inverse transformation to the streamlines for a flow around the circle and the latter can be easily evaluated.

Part IV

Fourier Analysis

11

Fourier Series

Abstract A Fourier series is an expansion of a periodic function in terms of an infinite sum of sines and cosines. The use of a Fourier series allows us to break up an arbitrary periodic function into a set of simple terms that can be solved individually and then recombined in order to obtain the solution to the original problem with the desired level of accuracy. In this chapter, we place particular emphasis on the mean convergence property of a Fourier series (Sect. 11.2.1) and the conditions that are necessary for the series to be uniformly convergent (Sect. 11.3.1). Better understanding of convergence properties clarifies the reasons for the utility and the limit of validity of Fourier series expansion in mathematical physics.

11.1 Basic Properties

11.1.1 Definition

Fourier series are infinite series consisting of trigonometric functions with a particular definition of expansion coefficients. They can be applied to almost all periodic functions whether the functions are continuous or not. With these expansion, physical phenomena involving some periodicity are reduced to a superposition of simple trigonometric functions, which helps us a great deal in arithmetic and practical aspects. I section We begin this with a description of the basic properties of Fourier series. We follow this by considering the convergence theory of Fourier series, which is the issue in the next section.

First of all, it is important to clarify the distinction between the following two concepts: trigonometric series and Fourier series.

♠ **Trigonometric series:**
The series

$$\frac{A_0}{2} + \sum_{n=1}^{\infty} (A_n \cos nx + B_n \sin nx)$$

is called a **trigonometric series.**

Here the set of coefficients $\{A_n\}$ and $\{B_n\}$ can be taken arbitrarily. (The expression $A_0/2$ instead of A_0 is just due to our convention.) Among the infinite choices of $\{A_n\}$ and $\{B_n\}$, a specific definition of the coefficients noted below provides the Fourier series of a given function $f(x)$.

♠ **Fourier series:**
The series

$$\frac{a_0}{2} + \sum_{n=1}^{\infty} (a_n \cos nx + b_n \sin nx) \tag{11.1}$$

is called a **Fourier series** of a function $f(x)$ if and only if the coefficients are given by the **Euler–Fourier formula** expressed by

$$a_n = \frac{1}{\pi}\int_{-\pi}^{\pi} f(x) \cos nx dx,$$
$$b_n = \frac{1}{\pi}\int_{-\pi}^{\pi} f(x) \sin nx dx. \tag{11.2}$$

Accordingly, a Fourier series is a specific kind of trigonometric series whose coefficients bear a definite relation (11.2) to some function $f(x)$. In (11.1) we have written the constant term as $a_0/2$ rather than a_0, so that the expression for a_0 is given by taking $n = 0$ in (11.2). There is no b_0 for $\sin(0 \cdot x) = 0$.

By definition, every Fourier series is a trigonometric series. However, the converse is not true, as demonstrated below.

Example It is known that the trigonometric series given by

$$\sum_{n=2}^{\infty} \frac{\sin nx}{\log n}$$

is not a Fourier series. Indeed, no function can be related to the coefficient $1/\log n$ via (11.2).

11.1.2 Dirichlet Theorem

Emphasis should be placed on the fact that the definition of Fourier series provides no information as to its convergence; thus the infinite series (11.1) may converge or diverge depending on the behavior of the function $f(x)$. This leads us to discuss which functions $f(x)$ make the series (11.1) convergent. This issue is clarified in part by the following theorem (and by referring Fig. 11.1):

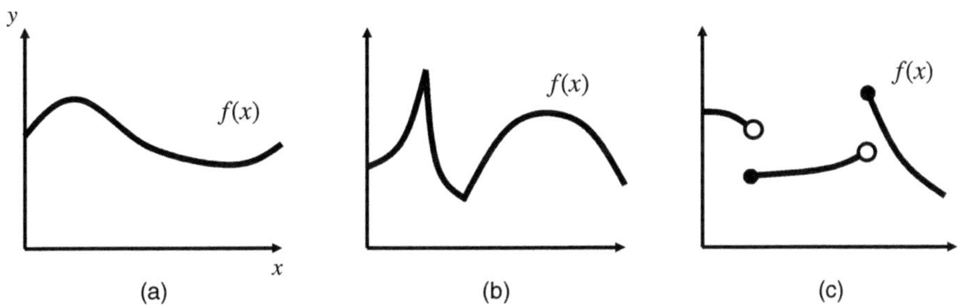

Fig. 11.1. **(a)** Continuous and smooth function. **(b)** Continuous but nonsmooth function. **(c)** Function with a finite number of discontinuities

♠ **Dirichlet theorem:**

If $f(x)$ is periodic with the period 2π and if $f'(x)$ is continuous or at most have a finite number of discontinuity in $[0, 2\pi]$, then its Fourier series converges to

1. $f(x)$, if x is a point of continuity, or
2. $\dfrac{f(x+0)+f(x-0)}{2}$, if x is a point of discontinuity.

The set of conditions noted above is called **Dirichlet's conditions**. It is worthy to note that the Dirichlet conditions are *sufficient* but not *necessary*. That is, if the conditions are satisfied, the convergence of the series is guaranteed; but if they are not satisfied, the series may or may not converge. An exact proof of Dirichlet's theorem requires rather complicated calculations, which will be demonstrated in the next section.

Remarks.

1. The Dirichlet conditions do not require the continuity of $f(x)$ within $[0, 2\pi]$.
2. Almost all periodic functions that we encounter in physical problems satisfy the Dirichlet conditions; therefore, the Fourier series expansion can be used almost regardless of its convergence.

It follows that if $f(x)$ is continuous within $[0, 2\pi]$ and satifies Dirichlet's conditions, then the Fourier series of $f(x)$ converges to $f(x)$ at all the points within $[0, 2\pi]$. This means that the Fourier series of $f(x)$ converges *uniformly* to $f(x)$. Once uniform convergence is ensured, we generally write

$$f(x) = \frac{a_0}{2} + \sum_{n=1}^{\infty} (a_n \cos nx + b_n \sin nx) \tag{11.3}$$

with the definition (11.2) for the coefficients. Consequently, if we form the Fourier series of $f(x)$ without first examining its convergence to $f(x)$, we should write

$$f(x) \sim \frac{a_0}{2} + \sum_{n=1}^{\infty} (a_n \cos nx + b_n \sin nx) \tag{11.4}$$

instead of (11.3). The symbol "$\sim$" in (11.4) means that the series on the right-hand side only *corresponds* to the function $f(x)$ and can be replaced by the equality "=" only if we succeed in proving that the infinite series converges uniformly to $f(x)$.

11.1.3 Fourier Series of Periodic Functions

Preceding arguments were limited to the case of periodic functions with period 2π. But Fourier series expansions can apply to periodic functions whose periods differ from 2π. This is seen by replacing x in (11.3) by $(2\pi/\lambda)x$, which transforms a series convergent in the interval $[0, 2\pi]$ to another series convergent to $[0, \lambda]$. The resulting Fourier series is

$$f(x) = \frac{a_0}{2} + \sum_{n=1}^{\infty} (a_n \cos nkx + b_n \sin nkx), \tag{11.5}$$

where $k = 2\pi/\lambda$ and

$$a_n = \frac{2}{\lambda}\int_0^{\lambda} f(x)\cos nkx dx \text{ and } b_n = \frac{2}{\lambda}\int_0^{\lambda} f(x)\sin nkx dx. \tag{11.6}$$

Obviously, these latter expressions can be reduced to the original definitions (11.1) and (11.2) by setting $\lambda = 2\pi$.

The expressions (11.5) and (11.6) become more concise by imposing the relations

$$\cos(nkx) = \frac{e^{inkx} + e^{-inkx}}{2}, \quad \sin(nkx) = \frac{e^{inkx} - e^{-inkx}}{2i}.$$

Then the Fourier series reads

$$f(x) = \frac{a_0}{2} + \sum_{n=1}^{\infty} \left(\frac{a_n - ib_n}{2}\right) e^{inkx} + \sum_{n=1}^{\infty} \left(\frac{a_n + ib_n}{2}\right) e^{-inkx}. \tag{11.7}$$

We rewrite the index n in the second sum by $-n'$ to find

$$\begin{aligned}
\sum_{n=1}^{\infty} \left(\frac{a_n + ib_n}{2}\right) e^{-inkx} &= \sum_{-n'=1}^{\infty} \left(\frac{a_{-n'} + ib_{-n'}}{2}\right) e^{in'kx} \\
&= \sum_{n'=-1}^{-\infty} \left(\frac{a_{n'} - ib_{n'}}{2}\right) e^{in'kx},
\end{aligned}$$

where the identities $a_{-n} = a_n$ and $b_{-n} = -b_n$ were used. As a result, we obtain a complex form of the Fourier series as

$$f(x) \sim \frac{a_0}{2} + \sum_{n=1}^{\infty} \left(\frac{a_n - ib_n}{2} \right) e^{inkx} + \sum_{n=-1}^{-\infty} \left(\frac{a_n - ib_n}{2} \right) e^{inkx}$$

$$= \sum_{n=-\infty}^{\infty} c_n e^{inkx}, \tag{11.8}$$

with the definition

$$c_n = \frac{a_n - ib_n}{2}. \tag{11.9}$$

An explicit form of c_n is given by substituting the definition of a_n and b_n, given by (11.6), into (11.9) as

$$c_n = \frac{1}{2} \left\{ \frac{2}{\lambda} \int_0^{\lambda} f(x) \cos(nkx) dx - \frac{2i}{\lambda} \int_0^{\lambda} f(x) \sin(nkx) dx \right\}$$

$$= \frac{1}{\lambda} \int_0^{\lambda} f(x) e^{-inkx} dx. \tag{11.10}$$

11.1.4 Half-range Fourier Series

Fourier series expansions sometimes involve only sine or cosine terms. This actually occurs when the function being expanded is either even $[f(-x) = f(x)]$ or odd $[f(-x) = -f(x)]$ over the interval $[-\lambda/2, \lambda/2]$. When a given function is even or odd, unnecessary work in determining Fourier coefficients can be avoided. For instance, for an odd function $f_o(x)$, we have

$$a_n = \frac{2}{\lambda} \int_{-\lambda/2}^{\lambda/2} f_o(x) \cos(nkx) dx$$

$$= \frac{2}{\lambda} \left\{ \int_{-\lambda/2}^{0} f_o(x) \cos(nkx) dx + \int_0^{\lambda/2} f_o(x) \cos(nkx) dx \right\}$$

$$= \frac{2}{\lambda} \left\{ - \int_0^{\lambda/2} f_o(x) \cos(nkx) dx + \int_0^{\lambda/2} f_o(x) \cos(nkx) dx \right\}$$

$$= 0 \quad (n = 0, 1, 2, \cdots,) \tag{11.11}$$

and

$$b_n = \frac{2}{\lambda} \left\{ \int_{-\lambda/2}^{0} f_o(x) \sin(nkx) dx + \int_0^{\lambda/2} f_o(x) \sin(nkx) dx \right\}$$

$$= \frac{4}{\lambda} \int_0^{\lambda/2} f_o(x) \sin(nkx) dx \quad (n = 0, 1, 2, \cdots,). \tag{11.12}$$

Here we used the identities $\cos(-nkx) = \cos(nkx)$ and $\sin(-nkx) = -\sin(nkx)$. Accordingly, we have

$$f_o(x) \sim \sum_{n=1}^{\infty} b_n \sin(nkx),$$

which is called the **Fourier sine series.**

Similarly, in the Fourier series corresponding to an even function $f_e(x)$, the same process yields

$$a_n = \frac{4}{\lambda} \int_0^{\lambda/2} f_e(x) \cos(nkx) dx \quad (n = 0, 1, 2, \cdot) \tag{11.13}$$

and $b_n = 0$ for all n. Accordingly, the Fourier series becomes

$$f_e(x) \sim \frac{a_0}{2} + \sum_{n=1}^{\infty} a_n \cos(nkx),$$

which is called the **Fourier cosine series.**

Note that a_n and b_n given in (11.12) and (11.13) are computed in the interval $[0, \lambda/2]$, whose width is half of the period λ. Thus, the Fourier sine or cosine series of an odd or even function, respectively, is often called a **half-range Fourier series.** As discussed later, half-range Fourier series expansion is important from a practical viewpoint because it enables us to expand a nonperiodic function within its domain.

♠ **Theorem:**

If $f(x)$ is an even or odd function and it is periodic with period λ, then the Fourier coefficients a_n and b_n become

$$a_n = \frac{4}{\lambda} \int_0^{\lambda/2} f(x) \cos(nkx) dx, \quad b_n = 0 \quad \text{if } f(x) \text{ is even}$$

and

$$a_n = 0, \quad b_n = \frac{4}{\lambda} \int_0^{\lambda/2} f(x) \sin(nkx) dx \quad \text{if } f(x) \text{ is odd.}$$

11.1.5 Fourier Series of Nonperiodic Functions

A problem that arises quite often in applications is how to apply a Fourier series expansion to a function $f(x)$ that is defined only on the interval $[0, L]$. In this case, nothing is said about the periodicity of $f(x)$. However, this does not prevent us from writing the Fourier series of $f(x)$, since the Euler–Fourier formulas (11.2) involve only the finite interval.

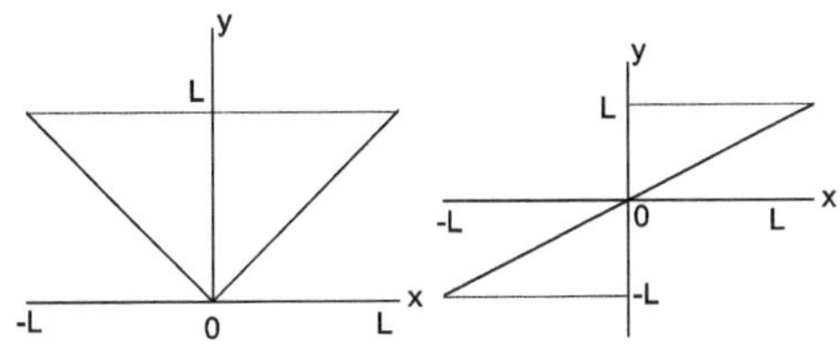

Fig. 11.2. Functions $f_e(x)$ and $f_o(x)$ defined in (11.14) and (11.15), respectively

As an example, we try to expand the function

$$f(x) = x \text{ for } [0, L]$$

as a Fourier series. In this case, $f(x)$ is not periodic, but we can make it a periodic function by extending it as an even or odd function over $[-L, L]$ and periodic with period $2L$. The respective definitions of $f_e(x)$ and $f_o(x)$ in $[-L, L]$ are

$$f_e(x) = \begin{cases} -x & \text{for } -L \leq x < 0, \\ x & \text{for } \ 0 \leq x \leq L \end{cases} \tag{11.14}$$

and

$$f_o(x) = x \quad \text{for } \ -L \leq x \leq L, \tag{11.15}$$

whose profiles are shown in Fig. 11.2.

First, we consider the case of the even function $f_e(x)$. In terms of the Fourier cosine expansion, the coefficients a_0 and a_n are given by

$$a_n = \frac{2}{L}\int_0^L f_e(x)\cos(nkx)dx = \frac{2L}{\pi^2}\frac{[(-1)^n - 1]}{n^2} = \begin{cases} \dfrac{-4L}{n^2\pi^2}, \ n = 1, 3, \cdots, \\ 0, \quad n = 2, 4, \cdots \end{cases},$$

$$a_0 = \frac{2}{L}\int_0^L x dx = L.$$

Here we have used $kL = \pi$. Hence, the cosine series becomes

$$f(x) = \frac{L}{2} - \frac{4L}{\pi^2}\sum_{n=1}^{\infty}\frac{1}{(2n-1)^2}\cos\frac{(2n-1)\pi x}{L}. \tag{11.16}$$

The partial sums of the series given in (11.16) are illustrated in Fig. 11.3. Although the original function $f(x)$ is defined only within the interval $[0, L]$, the resulting Fourier series produces not only $f(x)$ in $[0, L]$, but also the even extension $f_e(x)$ with $f_e(x) = f_e(x + 2L)$.

Second, we look at the sine series of f_o given by (11.15). In this case,

$$b_n = \frac{2}{L}\int_0^L f_o(x)\sin(nkx)dx = \frac{2L}{\pi}\frac{(-1)^{n+1}}{n}$$

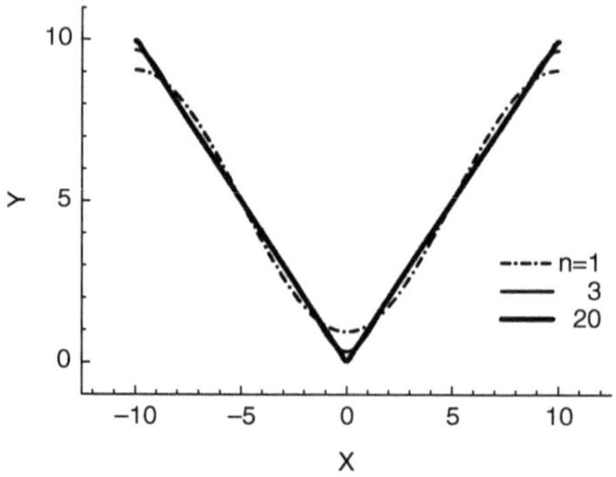

Fig. 11.3. A partial sum on the right-hand side of (11.16)

and the sine series is

$$f(x) = \frac{2L}{\pi} \sum_{n=1}^{\infty} \frac{(-1)^{n+1}}{n} \sin(nkx). \tag{11.17}$$

Figure 11.4 shows some partial sums of (11.17). As in the case of even extension, the Fourier series produces the odd extension $f_o(x)$ with $f_o(x) = f_e(x + 2L)$.

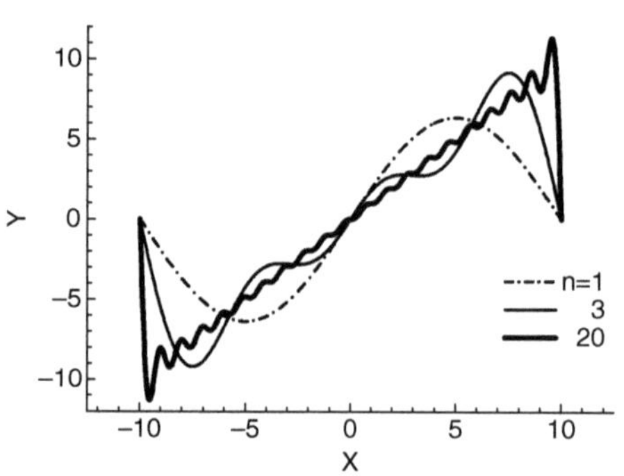

Fig. 11.4. A partial sum on the right-hand side of (11.17)

11.1.6 The Rate of Convergence

We have had two kinds of Fourier series representations for $f(x) = x$ in the interval $[0, L]$. This poses the following question: Does it make any difference which kind of Fourier series, (11.16) or (11.17), we use to represent $f(x) = x$ in the interval $[0, \lambda/2)$? Yes, it does. In the above-mentioned case, the even extension $f_e(x)$ is more suitable than the odd extension $f_o(x)$ for following two reasons.

The first reason concerns the rate of convergence of the resulting Fourier series. The coefficients given in (11.16) go as $1/(2n-1)^2$, whereas those in (11.17) go as $1/n$. Thus, the former series converges more quickly than the latter. The difference in the rate of convergence is due to the fact that the periodic extension of $f_e(x)$ is continuous, but that of $f_o(x)$ has discontinuities at odd multiples of L. In general, the Fourier coefficients of discontinuous functions decay as $1/n$, whereas those of continuous functions decay at least as rapidly as $1/n^2$. These observations as to the rate of convergence of the coefficient with respect to n can be formulated as follows:

♠ **Theorem:**
If $f(x)$ and its first k derivatives satisfy the Dirichlet conditions on the interval $[0, \lambda]$ and if the periodic extensions of $f(x), f'(x), \cdots, f^{(k-1)}(x)$ are all continuous, then the Fourier coefficients of $f(x)$ decay at least as rapidly as $1/n^{k+1}$.

The second reason is that the Fourier series representation corresponding to the odd extension $f_o(x)$ exhibits a small discrepancy from the original function $f(x)$ around points of discontinuity of $f_o(x)$. This discrepancy is a **Gibbs phenomenon**, illustrated in Sect. 11.3.5. When an extension generates points of discontinuities, a **Gibbs phenomenon** will inevitably occur, which makes the resulting Fourier series representation highly unreliable in the vicinity of the discontinuity. Consequently, when performing half-range expansions of nonperiodic functions, the way of extension that renders the resulting function continuous (and smooth) over its domain is preferred.

11.1.7 Fourier Series in Higher Dimensions

It is important to generalize the Fourier series to more than one dimension. This generalization is especially useful in crystallography and solid-state physics, which deal with the three-dimensional periodic structures of atoms and molecules. To generalize to N dimensions, we first consider a special case in which an N-dimensional periodic function is a product of N one-dimensional periodic functions. That is, we take the N functions $f^{(j)}(x)$ $[j = 1, 2, \cdots, N]$ with period L_j:

$$f^{(j)}(x) = \sum_{n=-\infty}^{\infty} c_n^{(j)} e^{2\pi i n x / L_j}, \quad j = 1, 2, \cdots, N, .$$

Let us define $F(\boldsymbol{r})$ by the product of all the N functions $f^{(j)}(x_j)$

$$\begin{aligned}
F(\boldsymbol{r}) &= f^{(1)}(x_1)f^{(2)}(x_2)\cdots f^{(N)}(x_N) \\
&= \sum_{n_1}\sum_{n_2}\cdots\sum_{n_N} c_{n_1}^{(1)}c_{n_2}^{(2)}\cdots c_{n_N}^{(N)}\cdot e^{2\pi i(n_1x_1/L_1+\cdots+n_Nx_N/L_N)} \\
&= \sum_{\boldsymbol{k}} C_{\boldsymbol{k}} e^{i\boldsymbol{k}\cdot\boldsymbol{r}},
\end{aligned} \tag{11.18}$$

where we have used the following new notation:

$$\begin{aligned}
C_{\boldsymbol{k}} &= c_{n_1}^{(1)}c_{n_2}^{(2)}\cdots c_{n_N}^{(N)}, \\
\boldsymbol{k} &= 2\pi(n_1/L_1, n_2/L_2, \cdots n_N/L_N), \\
\boldsymbol{r} &= (x_1, x_2, \cdots, x_N).
\end{aligned}$$

We take (11.18) as the definition of the Fourier series for any periodic function of N variables. The definition of the coefficient $C_{\boldsymbol{k}}$ can be developed for a general periodic function $F(\boldsymbol{r})$ of N variables:

$$F(\boldsymbol{r}) = \sum_{\boldsymbol{k}} C_{\boldsymbol{k}} e^{i\boldsymbol{k}\cdot\boldsymbol{r}} \quad \Longleftrightarrow \quad C_{\boldsymbol{k}} = \frac{1}{V}\int_V F(\boldsymbol{r})e^{-i\boldsymbol{k}\cdot\boldsymbol{r}}d^Nx, \tag{11.19}$$

where $V \equiv L_1L_2\cdots L_N$ determines the smallest region of periodicity in N dimensions. When $N = 1$ (11.19) obviously reduces to the Fourier series in one dimension.

Remark. The application of (11.19) requires some clarification regarding the region V of the integral. In one dimension, the shape of the smallest region of periodicity is unique, being simply a line segment of length L. In two or more dimensions, however, such regions can have a variety of shapes. For instance, in two dimensions, they can be rectangles, pentagons, hexagons, and so forth. Thus, we let V in (11.18) stand for a primitive cell of the N-dimensional lattice. This cell in three dimensions, which is important in solid-state physics, is called the **Wigner–Seitz cell.**

Recall that $F(\boldsymbol{r})$ is a periodic function of $\boldsymbol{r}$. This means that when $\boldsymbol{r}$ is changed by $\boldsymbol{R}$, where $\boldsymbol{R}$ is a vector describing the boundaries of a cell, then we should get the same function: $F(\boldsymbol{r}+\boldsymbol{R}) = F(\boldsymbol{r})$. This implies that the periodicity of

$F(\boldsymbol{r})$ requires the vector $\boldsymbol{k}$ to take only restricted directions and magnitudes. In fact, when replacing $\boldsymbol{r}$ in (11.19) by $\boldsymbol{r}+\boldsymbol{R}$, we have

$$F(\boldsymbol{r}+\boldsymbol{R}) = \sum_{\boldsymbol{k}} C_{\boldsymbol{k}} e^{i\boldsymbol{k}\cdot(\boldsymbol{r}+\boldsymbol{R})} = \sum_{\boldsymbol{k}} \left(e^{i\boldsymbol{k}\cdot\boldsymbol{R}} \cdot C_{\boldsymbol{k}} e^{i\boldsymbol{k}\cdot\boldsymbol{r}} \right),$$

which is equal to $F(\boldsymbol{r})$ if

$$e^{i\boldsymbol{k}\cdot\boldsymbol{R}} = 1, \quad \text{i.e.,} \quad \boldsymbol{k}\cdot\boldsymbol{R} = 2\pi \times (\text{integer}). \tag{11.20}$$

Equation (11.20) is a key relation in determining the allowed directions and magnitudes of the vector $\boldsymbol{k}$. In one-dimensional cases, the inner product reduces to $\boldsymbol{k}\cdot\boldsymbol{R} = (2\pi n/L)\cdot L = 2\pi n$; thus (11.20) obviously holds true. In three dimensions, the vector $\boldsymbol{R}$ is represented as $\boldsymbol{R} = m_1\boldsymbol{a}_1 + m_2\boldsymbol{a}_2 + m_3\boldsymbol{a}_3$, where m_1, m_2, and m_3 are integers and $\boldsymbol{a}_1$, $\boldsymbol{a}_2$, and $\boldsymbol{a}_3$ are crystal axes, which are not generally orthogonal. Hence, condition (11.20) is satisfied when $\boldsymbol{k} = n_1\boldsymbol{b}_1 + n_2\boldsymbol{b}_2 + n_3\boldsymbol{b}_3$, where n_1, n_2, and n_3 are integers and $\boldsymbol{b}_1$, $\boldsymbol{b}_2$, and $\boldsymbol{b}_3$ are the reciprocal lattice vectors defined by

$$\boldsymbol{b}_1 = \frac{2\pi(\boldsymbol{a}_2\times\boldsymbol{a}_3)}{\boldsymbol{a}_1\cdot(\boldsymbol{a}_2\times\boldsymbol{a}_3)}, \quad \boldsymbol{b}_2 = \frac{2\pi(\boldsymbol{a}_3\times\boldsymbol{a}_1)}{\boldsymbol{a}_1\cdot(\boldsymbol{a}_2\times\boldsymbol{a}_3)}, \quad \boldsymbol{b}_3 = \frac{2\pi(\boldsymbol{a}_1\times\boldsymbol{a}_2)}{\boldsymbol{a}_1\cdot(\boldsymbol{a}_2\times\boldsymbol{a}_3)}.$$

In fact,

$$\boldsymbol{k}\cdot\boldsymbol{R} = \left(\sum_{i=1}^{3} n_i\boldsymbol{b}_i\right)\cdot\left(\sum_{j=1}^{3} m_j\boldsymbol{a}_j\right) = \sum_{i,j}^{3} n_i m_j \boldsymbol{b}_i\cdot\boldsymbol{a}_j,$$

and the reader may verify that $\boldsymbol{b}_i\cdot\boldsymbol{a}_j = 2\pi\delta_{ij}$. Thus we obtain

$$\boldsymbol{k}\cdot\boldsymbol{R} = 2\pi\sum_{j=1}^{3} m_j n_j = 2\pi\times(\text{integer}).$$

Exercises

1. Expand the following functions in Fourier series:

(i) $f(x) = \sin ax$ on $[-\pi,\pi]$, where a is *not* an integer.

(ii) $f(x) = \sin ax$ on $[0,\pi]$, where a is *not* an integer.

Solution: It is straightforward to obtain the results:

$$\text{(i)} \quad \sin ax = \frac{2}{\pi}\sin a\pi \sum_{n=1}^{\infty}(-1)^n \frac{n\sin nx}{a^2-n^2}.$$

(ii) $\sin ax = \dfrac{1-\cos a\pi}{\pi}\left[\dfrac{1}{a} + 2a\sum_{n=1}^{\infty}\dfrac{\cos 2nx}{a^2-4n^2}\right]$

$$+\ 2a\frac{1+\cos a\pi}{\pi}\sum_{n=0}^{\infty}\frac{\cos(2n+1)x}{a^2-(2n+1)^2}. \clubsuit$$

2. Expand the functions $f(x) = \cos x$ on $[0, \pi]$ in a Fourier sine series.

Solution: $\cos x = \dfrac{8}{\pi}\displaystyle\sum_{n=1}^{\infty}\dfrac{n\sin 2nx}{4n^2-1}.$ ♣

3. **(i)** Find the Fourier series of $f(x) = x$ on the interval $[-\pi, \pi]$.

(ii) Prove that the identity $\dfrac{\pi}{4} = 1 - \dfrac{1}{3} + \dfrac{1}{5} - \dfrac{1}{7} + \cdots$.

Solution:

(i) $f(x) = \displaystyle\sum_{n=1}^{\infty}\dfrac{2(-1)^{n+1}}{n}\sin nx.$

(ii) If we substitute $x = \pi/2$ in the series, we obtain

$$\frac{\pi}{2} = \sum_{n=1}^{\infty}\frac{2(-1)^{n+1}}{n}\sin\frac{n\pi}{2} = 2\left(1 - \frac{1}{3} + \frac{1}{5} - \frac{1}{7} + \cdots\right),$$

which obviously gives the desired result. ♣

4. Expand the function $f(x) = x^2$ into the Fourier cosine series on the domain $[-\pi, \pi]$ and then prove that $\displaystyle\sum_{n=1}^{\infty}\frac{1}{n^2} = \frac{\pi^2}{6}$ and $\displaystyle\sum_{n=1}^{\infty}\frac{4(-1)^n}{n^2} = -\frac{\pi^2}{3}$.

Solution: Straightforward calculations yield $x^2 = \displaystyle\sum_{n=1}^{\infty}\frac{4(-1)^n}{n^2}\cos nx$.
By substituting $x = \pi$ and $x = 0$, we obtain the desired equations. ♣

5. Determine both the cosine and sine series of $f(x) = x^3 - x$ defined on the interval $[0, 1]$. Which series do you suppose converges more quickly?

Solution: We may set the even and odd extensions of $f(x)$ over $[-1, 1]$, respectively, as

$$f_o(x) = x^3 - x \quad \text{for } -1 \le x \le 1,$$

and

$$f_e(x) = \begin{cases} -x^3 + x & \text{for } -1 \le x < 0, \\ x^3 - x & \text{for } 0 \le x \le 1. \end{cases}$$

It follows that $f_o(x)$ is smoother than $f_e(x)$; namely, $f_e(x)$ has a discontinuity in its derivative at $\pm n$. This implies that the sine series converges more rapidly than the cosine series. In fact, straightforward calculations yield the sine series

$$f_o(x) = \frac{12}{\pi^3} \sum_{n=1}^{\infty} \frac{(-1)^n}{n^3} \sin(n\pi x)$$

and the cosine series

$$f_e(x) = -\frac{1}{4} + \frac{2}{\pi^2} \sum_{n=1}^{\infty} \left\{ \frac{1+(-1)^n 2}{n^2} + \frac{6}{n^4\pi^2}[1-(-1)^n] \right\} \cos(n\pi x).$$

The continuity of $f_o(x)$ and its first derivatives leads to Fourier coefficients that decay as $1/n^3$, whereas the continuity of $f_e(x)$ coupled with the discontinuity in $f'_e(x)$ leads to Fourier coefficients that decay as $1/n^2$. ♣

11.2 Mean Convergence of Fourier Series

11.2.1 Mean Convergence Property

We know that Fourier series are endowed with a specific class of convergence called **mean convergence** (or **convergence in the mean**). This converging behavior is expressed by an integral:

$$\lim_{N\to\infty} \int_0^{\lambda} \left| f(x) - \sum_{n=-N}^{N} c_n e^{inkx} \right|^2 dx = 0. \tag{11.21}$$

Equation (11.21) applies regardless of the continuity and smoothness of the function $f(x)$, as far as $f(x)$ is square-integrable.

Remark. From the viewpoint of **Hilbert space theory**, the relation (11.21) comes from the completeness property of the set of functions $\{e^{inkx}\}$ in the sense of the **norm in the L^2 space**. The L^2 space is a specific kind of Hilbert space that is composed of a set of square-integrable functions $f(x)$ expressed by

$$\int_a^b |f(x)|^2 dx < \infty.$$

The **inner product** (f, g) and the norm $\|f\|$ of elements $f, g \in L^2$, respectively, are given by

$$(f,g) = \int_a^b f(x)^* g(x) dx \quad \text{and} \quad \|f\| = (f,f) = \int_a^b |f(x)|^2 dx.$$

The mean convergence of the Fourier series [i.e., the equality in (11.21)] holds even when the integrand in (11.21) has a nonzero value at discrete points of x. This comes from the fact that the definition of the mean convergence is determined through integration, and that a finite number of discontinuities of the integrand do not contribute to the result of its integration. This is explained schematically in Fig. 11.5, in which we find

$f(x)$: a continuous function,
$g_n^{(1)}(x)$: a series that converges uniformly to $f(x)$ except at a point of discontinuity $x = a$.
$g_n^{(2)}(x)$: a series that converges uniformly to $f(x)$ except at points of discontinuity $x = a_1, a_2, a_3, \cdots$.

As shown in Fig. 11.5, these three functions are distinct from one another. However, if we integrate the squared deviation between two of them followed by taking the limit $n \to \infty$, we have

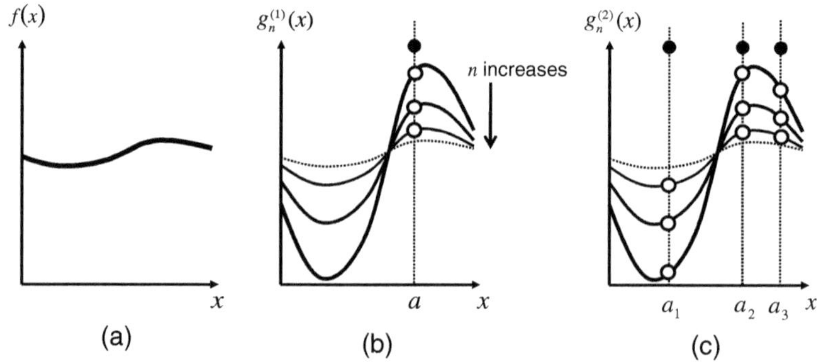

Fig. 11.5. Sketches of a continuous function $f(x)$, a series of functions $g_n^{(1)}(x)$ converging to $f(x)$ except at a discontiuity, and a similar series of functions $g_n^{(2)}(x)$ having several discontinuities. Series $\{g_n^{(1)}(x)\}$ and $\{g_n^{(1)}(x)\}$ both converge *in the mean* to $f(x)$

$$\lim_{n\to\infty}\int_0^\lambda |f(x)-g_n^{(1)}(x)|^2dx = \lim_{n\to\infty}\int_0^\lambda |f(x)-g_n^{(2)}(x)|^2dx = 0. \quad (11.22)$$

This is because the area surrounded by two of them vanishes with $n \to \infty$, i.e., the area right below (or above) points of discontinuity are zero owing to their discreteness. Thus, the series $g_n^{(1)}(x)$ and $g_n^{(2)}(x)$ both converge to $f(x)$ in the mean regardless of their discrepancy from $f(x)$ at points of discontinuity.

11.2.2 Dirichlet and Fejér Integrals

It is pedagogical to give an alternative exposition of mean convergence of Fourier series, which is based on the two important concepts: **Dirichlet's integral** and **Fejér's integral**.

Consider the partial sum $S_N(x)$ of the Fourier series of $f(x)$ expressed by

$$S_N(x) = \sum_{n=-N}^{N} c_n e^{inkx}$$

and its arithmetic mean

$$\sigma_N(x) = \frac{1}{N+1}(S_0 + S_1 + \cdots + S_N). \quad (11.23)$$

After some algebra, we obtain their integral representations as given below (see Exercises **1** and **2** in 11.2.2 for references).

♠ **Dirichlet integral:**

$$S_N(x) = \frac{1}{\lambda}\int_{-x}^{\lambda-x} f(t+x)\frac{\cos(Nkt)-\cos\{(N+1)kt\}}{1-\cos(kt)}dt. \quad (11.24)$$

♠ **Fejér integral:**

$$\sigma_N(x) = \frac{1}{(N+1)\lambda}\int_{-\lambda/2}^{\lambda/2} f(t+x)\frac{\sin^2\frac{N+1}{2}kt}{\sin^2\frac{1}{2}kt}dt. \quad (11.25)$$

Remark. Note the distinctive difference between the convergence of S_N and that of σ_N. Whereas $\lim_{N\to\infty} S_N = S$ implies $\sigma_N \to S$, the converse does not generally hold true; in fact, σ_N may converge even when S_N diverges. A typical example is the case of the numerical sequence $u_n = (-1)^n$, where, $S_N = \sum u_n$ does not converge because $S_{2N} = 0$ and $S_{2N+1} = 1$, whereas $\sigma_N = \sum S_n/(N+1)$ converges to $1/2$.

By putting $f(x) \equiv 1$ in (11.25), we have the following notation:

♠ **Dirichlet kernel:** The function

$$D_N(t) \equiv \frac{1}{N+1} \cdot \frac{\sin^2 \frac{N+1}{2}kt}{\sin^2 \frac{1}{2}kt} \tag{11.26}$$

is called the **Dirichlet kernel**, which satisfies the identity:

$$1 = \frac{1}{\lambda} \int_{-\lambda/2}^{\lambda/2} D_N(t)dt. \tag{11.27}$$

The derivation of the identity (11.27) is straightforward. When $f(x) \equiv 1$, we have $f(t+x) = 1$, $c_0 = 1$, and $c_n = 0$ ($|n| \geq 1$), which obviously yield $S_N = 1$ for arbitrary N and thus $\sigma_N = 1$. Substitute this into (11.25) to obtain the identity (11.27). Figure 11.6 plots the behavior of $D_N(t)$ with increasing N; it shows maxima at $t = 0, \pm\lambda, \pm 2\lambda, \cdots$, and the magnitude of the maxima become singular when $N \to \infty$.

From (11.25) and(11.27), we arrive at the key relation

$$\sigma_N(x) - f(x) = \frac{1}{\lambda} \int_{-\lambda/2}^{\lambda/2} \{f(t+x) - f(x)\} D_N(t)dt. \tag{11.28}$$

If $f(x)$ is continuous (piecewise, at least), the integral in (11.28) can be made arbitrarily small by taking a sufficiently large N (see Exercise **3** below). To be precise, there exists an m for each $\varepsilon > 0$ such that

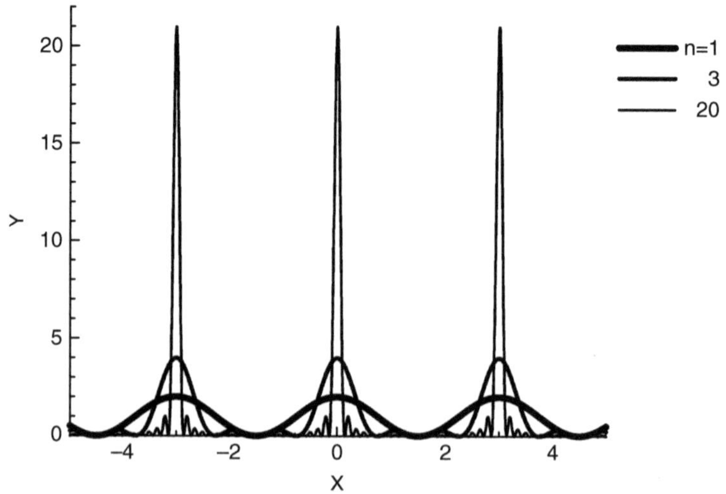

Fig. 11.6. Dirichlet's kernel $D_N(t)$ defined in (11.26)

$$N > m \Rightarrow |\sigma_N(x) - f(x)| < \varepsilon. \tag{11.29}$$

This clearly means that $\sigma_N(x)$ converges uniformly to $f(x)$ if $f(x)$ is continuous.

As is shown later, the result (11.29) immediately yields the mean convergence of the Fourier series to $f(x)$.

11.2.3 Proof of the Mean Convergence of Fourier Series

We are now in a position to prove the mean convergence property of Fourier series.

The function $\sigma_N(x)$ can be expressed as a trigonometric polynomial, since it consists of N's trigonometric polynomials $S_0, S_1, \cdots, S_N$ as given by (11.23). Hence, (11.29) implies the existence of a trigonometric series that converges uniformly to $f(x)$. [This is simply **Fejér's theorem** (see Sect. 11.3.2).] Thus we have

$$\sigma_N(x) = \sum_{n=-N}^{N} \gamma_n e^{inkx},$$

where all the coefficients $\{\gamma_n\}$ have to be determined.

We now make use of the fact that for any choice of $\{\gamma_n\}$, the inequality

$$\int_0^\lambda \left| f(x) - \sum_{n=-N}^{N} \gamma_n e^{inkx} \right|^2 dx \geq \int_0^\lambda \left| f(x) - \sum_{n=-N}^{N} c_n e^{inkx} \right|^2 dx$$

holds true with the Fourier coefficients $\{c_n\}$ of $f(x)$. (See the discussion in Sect. 11.2.4 for the proof.) Taking the limit $N \to \infty$ yields

$$\lim_{N\to\infty} \int_0^\lambda |f(x) - \sigma_N(x)|^2 dx \geq \lim_{N\to\infty} \int_0^\lambda \left| f(x) - \sum_{n=-N}^{N} c_n e^{inkx} \right|^2 dx. \tag{11.30}$$

Let $f(x)$ be continuous (piecewise, at least). Then the left-hand side vanishes owing to the uniform convergence of $\sigma_N(x)$ to $f(x)$ at continuous points x of $f(x)$ (A finite number of discontinuous points of $f(x)$ makes no contribution to the integral.) Eventually, we come to the desired conclusion:

$$\lim_{N\to\infty} \int_0^\lambda \left| f(x) - \sum_{n=-N}^{N} c_n e^{inkx} \right|^2 dx = 0,$$

which is a restatement of (11.21).

11.2.4 Parseval Identity

The mean convergence property of Fourier series can be represented by a more concise expression, called the **Parseval identity**. We first note the main conclusion of this subsection and then go on to its proof. For simplicity of notation, we use the following short form:

$$\frac{1}{\lambda}\int_0^\lambda f(x)g^*(x)dx \equiv (f,g)\,.$$

♠ **Parseval identity:**

A necessary and sufficient condition for the mean convergence of the Fourier series of $f(x)$ is given by

$$(f,f) = \sum_{n=-\infty}^{\infty} |c_n|^2\,,$$

which is called the **Parseval identity**.

To prove the above statement, we assume $f(x)$ to be square-integrable and consider the total squared error of $f(x)$ relative to the series of exponential functions:

$$\mathcal{E}_N = \frac{1}{\lambda}\int_0^\lambda \left| f(x) - \sum_{n=-N}^{N} \gamma_n e^{inkx} \right|^2 dx, \tag{11.31}$$

whose variables are N and the sequence $\{\gamma_n\}$ consisting of complex numbers. Term-by-term integration of (11.31) yields

$$\begin{aligned}
\mathcal{E}_N &= (f,f) - \sum_{n=-N}^{N} \gamma_n^* \left(f, e^{inkx}\right) - \sum_{n=-N}^{N} \gamma_n \left(f, e^{inkx}\right)^* \\
&\quad + \sum_{m,n=-N}^{N} \gamma_m^* \gamma_n \left(e^{inkx}, e^{imkx}\right) \\
&= (f,f) - \sum_{n=-N}^{N} (\gamma_n^* c_n + \gamma_n c_n^*) + \sum_{n=-N}^{N} \gamma_n^* \gamma_n \\
&= (f,f) + \sum_{n=-N}^{N} |\gamma_n - c_n|^2 - \sum_{n=-N}^{N} |c_n|^2.
\end{aligned} \tag{11.32}$$

Here we have used the orthonormality of imaginary exponentials, $\left(e^{inkx}, e^{imkx}\right) = \delta_{m,n}$, as well as the definitions of the Fourier coefficient $c_n = \left(f, e^{inkx}\right)$. Note that (f,f) appearing in (11.32) is nonnegative because

$$(f, f) = \frac{1}{\lambda} \int_0^{\lambda} |f(x)|^2 dx \geq 0.$$

Hence, $\mathcal{E}_N$ becomes minimal when $\gamma_n = c_n$ and its minimum value reads

$$\min\{\mathcal{E}_N\} = (f, f) - \sum_{n=-N}^{N} |c_n|^2. \tag{11.33}$$

We are now ready to complete our proof. Recall that the mean convergence of the Fourier series for $f(x)$ is defined by

$$\lim_{N\to\infty} \int_0^{\lambda} \left| f(x) - \sum_{n=-N}^{N} c_n e^{inkx} \right|^2 dx = 0. \tag{11.34}$$

From (11.31) and (11.33), we see that the definition of the mean convergence (11.34) is rewritten as

$$\lim_{N\to\infty} \min\{\mathcal{E}_N\} = 0, \tag{11.35}$$

or equivalently,

$$(f, f) = \sum_{n=-\infty}^{\infty} |c_n|^2. \tag{11.36}$$

Relation (11.36) is thus a necessary and sufficient condition for satisfying the mean convergence of the Fourier series to $f(x)$. Since Parseval's identity applies to any square-integrable function f, Fourier series for the functions f surely converge in the mean to $f(x)$.

11.2.5 Riemann–Lebesgue Theorem

As by-products of the argument in 11.2.4, we obtain the following two important properties regarding the Fourier series expansion. The first is the **Bessel inequality**

$$(f, f) \geq \sum_{n=-N}^{N} |c_n|^2. \tag{11.37}$$

This is obtained from the fact that $\min\{\mathcal{E}_N\}$ given in (11.33) is nonnegative. Here we can let $N \to \infty$ in (11.37), because the right-hand side of (11.37) forms a monotonically increasing sequence that is bounded by its left-hand side. Then we obtain

$$(f, f) \geq \sum_{n=-\infty}^{\infty} |c_n|^2. \tag{11.38}$$

We further note that the series on the right-hand side of (11.38) necessarily converges, since it is nondecreasing and bounded from above. Consequently, we arrive at the second important property to be noted:

$$\lim_{n\to\infty} c_n = 0. \tag{11.39}$$

Separating the real and imaginary parts in (11.39), we eventually find the second point to be noted:

♠ **Riemann–Lebesgue theorem:**
If $f(x)$ is square-integrable on the interval $[0, \lambda]$, then

$$\lim_{n\to\infty} \int_0^\lambda f(x)\cos(nkx)dx = 0, \quad \lim_{n\to\infty} \int_0^\lambda f(x)\sin(nkx)dx = 0.$$

Exercises

1. Derive the expressions (11.24) and (11.25) regarding the Dirichlet and Fejér integrals, respectively.

   Solution: From the definition of c_n, the partial sum $S_N(x)$ yields its integral form:

$$\begin{aligned} S_N(x) &= \sum_{n=-N}^{N} \left\{ \frac{1}{\lambda} \int_0^\lambda f(t) e^{-inkt} dt \right\} \cdot e^{inkx} \\ &= \frac{1}{\lambda} \int_0^\lambda f(t) \left\{ \sum_{n=-N}^{N} e^{-ink(t-x)} \right\} dt \\ &= \frac{1}{\lambda} \int_{-x}^{\lambda-x} f(t+x) \left(\sum_{n=-N}^{N} e^{-inkt} \right) dt. \qquad (11.40) \end{aligned}$$

   The finite series of exponential terms reads

$$\begin{aligned} \sum_{n=-N}^{N} e^{-inkt} &= e^{-iNkt} \sum_{n=0}^{2N} e^{inkt} = e^{-iNkt} \cdot \frac{1 - e^{i(2N+1)kt}}{1 - e^{ikt}} \\ &= \frac{\cos(Nkt) - \cos\{(N+1)kt\}}{1 - \cos(kt)}. \end{aligned}$$

   Substituting this in (11.40) yields Dirichlet's integral (11.24). Moreover, its arithmetic mean reduces to Fejér's integral (11.25) as demonstrated by

$$\begin{aligned} \sigma_N(x) &= \frac{1}{N+1} \{S_0 + S_1 + \cdots + S_N\} \\ &= \frac{1}{(N+1)\lambda} \int_{-x}^{\lambda-x} f(t+x) \frac{1 - \cos\{(N+1)kt\}}{1 - \cos(kt)} dt \\ &= \frac{1}{(N+1)\lambda} \int_{-\lambda/2}^{\lambda/2} f(t+x) \frac{\sin^2 \frac{N+1}{2} kt}{\sin^2 \frac{1}{2} kt} dt. \qquad (11.41) \end{aligned}$$

In the last line of (11.41), the interval of the integration from $[-x, \lambda - x]$ to $[-\lambda/2, \lambda/2]$ is replaced by taking account of the periodicity of the integrand. ♣

2. Prove that $\sigma_N(x)$ uniformly converges to $f(x)$ by postulating the continuity of $f(x)$.

Solution: Recall that the continuity of $f(x)$ allows us to determine a δ that satisfies

$$|x - x'| < \delta \ \Rightarrow \ |f(x) - f(x')| < \varepsilon \tag{11.42}$$

for an arbitrary small ε to be positive. Further, owing to its continuity, the function $f(x)$ is bounded as $|f(x)| < M$ with an appropriate constant M. We divide the range of integration given in (11.28) as $\int_{-\lambda/2}^{\lambda/2} = \int_{-\lambda/2}^{-\delta} + \int_{-\delta}^{\delta} + \int_{\delta}^{\lambda/2}$ and use the inequality (11.42) to obtain the middle term:

$$\left| \int_{-\delta}^{\delta} \{f(t+x) - f(x)\} D_N(t) dt \right| \le \int_{-\delta}^{\delta} |f(t+x) - f(x)| \, D_N(t) dt$$

$$\le \varepsilon \int_{-\delta}^{\delta} D_N(t) dt \ \le \ \varepsilon\lambda \tag{11.43}$$

and

$$\left| \left(\int_{-\lambda/2}^{-\delta} + \int_{\delta}^{\lambda/2} \right) \{f(t+x) - f(x)\} D_N(t) dt \right|$$

$$\le \left(\int_{-\lambda/2}^{-\delta} + \int_{\delta}^{\lambda/2} \right) \{|f(t+x)| + |f(x)|\} \frac{\sin^2 \frac{N+1}{2} kt}{(N+1)\sin^2 \frac{1}{2} kt} dt$$

$$\le \left(\int_{-\lambda/2}^{-\delta} + \int_{\delta}^{\lambda/2} \right) 2M \cdot \frac{dt}{(N+1)\sin^2(k\delta/2)}$$

$$\le \frac{2M\lambda}{(N+1)\sin^2(k\delta/2)}. \tag{11.44}$$

From (11.28), (11.43), and (11.44), we obtain

$$|\sigma_N(x) - f(x)| \ < \ \varepsilon + \frac{2M}{(N+1)\sin^2(k\delta/2)}.$$

Taking the limit $N \to \infty$ and fixing the small quantity δ, the second term vanishes. We thus conclude that

$$\lim_{N\to\infty} |\sigma_N(x) - f(x)| = 0. \ ♣ \tag{11.45}$$

11.3 Uniform Convergence of Fourier series

11.3.1 Criterion for Uniform and Pointwise Convergence

We know that the Fourier series of $f(x)$ converges in the mean to $f(x)$ as far as $f(x)$ is square-integrable. However, the mean convergence of the Fourier series provides no information as to its **uniform convergence**. In order for the Fourier series to converge (uniformly or pointwise) to the original function $f(x)$, several conditions regarding continuity and periodicity of $f(x)$ have to be satisfied. These are formally stated in the following two theorems:

♠ **Uniform convergence of Fourier series:**
The Fourier series of a continuous, piecewise smooth, and periodic function $f(x)$ converges to $f(x)$ absolutely and uniformly.

♠ **Pointwise convergence of Fourier series:**
The Fourier series of a piecewise smooth and periodic function $f(x)$ (continuous or discontinuous) converges to:

(i) $f(x)$ at any point of continuity, and

(ii) $\dfrac{f(x+0)+f(x-0)}{2}$ at any point of discontinuity.

Our main concern in this section is to prove these two theorems, and we follow this by demonstrating several important features of Fourier series that occur at discontinuous points of $f(x)$.

Remark. Observe that the above theorems are consistent with the conclusion of the **Dirichlet theorem** given in Sect. 11.1.2; the latter says that a Fourier series representation becomes identical to $f(x)$ provided that $f(x)$ is periodic, continuous, and further smooth (piecewise, at least).

11.3.2 Fejér theorem

The theorems given in the previous subsection clearly exhibit *sufficient* conditions for the Fourier series to converge. It is pedagogical to compare them with the **Fejér theorem**:

♠ **Fejér theorem:**
Any continuous and periodic function $f(x)$ with a period λ can be reproduced by an infinite trigonometric series

$$\lim_{N\to\infty}\left|f(x)-\sum_{n=-N}^{N}\gamma_n e^{inkx}\right|=0 \quad \text{for all } x, \tag{11.46}$$

with an appropriate choice of the set of expansion coefficients $\{\gamma_n\}$.

At first glance, Fejér's theorem appears to ensure the uniform convergence of the Fourier series. However, this is not the case at all; the sequence of the optimal coefficients $\{\gamma_n\}$ satisfying (11.46) cannot in general be replaced by the Fourier coefficients $\{c_n\}$ defined by

$$c_n=\frac{1}{\lambda}\int_0^{\lambda} f(x)e^{-inkx}dx.$$

In fact, even when $f(x)$ is continuous and periodic, its Fourier series may diverge at discrete points, as is expressed by

$$\lim_{N\to\infty}\left|f(x)-\sum_{n=-N}^{N}c_n e^{inkx}\right|=\infty \quad \text{at some points } x. \tag{11.47}$$

Hence, Fejér's theorem does not guarantee the uniform convergence of the Fourier series representation. Equation (11.47) also suggests that the continuity and periodicity of $f(x)$ are only necessary but not sufficient conditions, for the uniform convergence of its Fourier series to the original function $f(x)$.

11.3.3 Proof of Uniform Convergence

We are now in a position to prove the criterion for uniform convergence of the Fourier series $\sum_{n=-\infty}^{\infty} c_n e^{inx}$ to $f(x)$. The proof that is presented below is based on the mean convergence property of Fourier series. Recall that the mean convergence of Fourier series is expressed by

$$\int_{-\infty}^{\infty}\left|f(x)-\sum_{n=-\infty}^{\infty}c_n e^{inx}\right|^2 dx=0. \tag{11.48}$$

In general, the relation

$$\lim_{N\to\infty}\int_a^b\left|\sum_{n=0}^{N}u_n(x)\right|^2 dx=0$$

means

$$\lim_{N\to\infty}\left\{\sum_{n=0}^{N}u_n(x)\right\}=0$$

if and only if the infinite series $\sum_{n=0}^{\infty} u_n(x)$ converges uniformly to a certain function of x within the range of integration $[a, b]$. Therefore, in order to obtain the desired equality

$$f(x) = \sum_{n=-\infty}^{\infty} c_n e^{inkx}$$

for any $x \in [0, \lambda]$, we must seek the condition that the infinite series $\sum_{n=-\infty}^{\infty} c_n e^{inkx}$ converges uniformly to some function of x [not necessarily to $f(x)$]. Thus, we rewrite the Fourier coefficient c_n as

$$\begin{aligned} c_n &= \frac{1}{\lambda} \int_0^{\lambda} f(x) e^{-inkx} dx \\ &= \frac{1}{-ink\lambda} \left[f(x) e^{-inkx} \right]_0^{\lambda} + \frac{1}{ink\lambda} \int_0^{\lambda} f'(x) e^{-inkx} dx \\ &= \frac{1}{ink} \cdot \frac{1}{\lambda} \int_0^{\lambda} f'(x) e^{-inkx} dx \quad = \frac{c_n'}{ink}, \end{aligned} \tag{11.49}$$

where c_n' is the Fourier coefficient of the derivative $f'(x)$. Here $f(x)$ is assumed to be periodic, e.g., $f(0) = f(\lambda)$ and $k\lambda = 2\pi$. We further assume that $f(x)$ is continuous and smooth (piecewise, at least) on the interval $[0, \lambda]$. Then, $f'(x)$ is continuous (or piecewise continuous) to yield Parseval's identity:

$$\int_0^{\lambda} |f'(x)|^2 dx = \sum_{n=-\infty}^{\infty} |c_n'|^2 \equiv A,$$

where A is a constant. Observe that

$$\sum_{n=-\infty}^{\infty} |c_n| = \sum_{n=-\infty}^{\infty} \left| \frac{c_n'}{ink} \right| = \sum_{n=-\infty}^{\infty} \frac{|c_n'|}{nk}. \tag{11.50}$$

From the Schwartz inequality, it follows that

$$\begin{aligned} \sum_{n=-\infty}^{\infty} \frac{|c_n'|}{nk} &= \sum_{n=-\infty}^{\infty} \sqrt{\frac{1}{n^2 k^2}} \sqrt{|c_n'|^2} \le \sqrt{\sum_{n=-\infty}^{\infty} \frac{1}{n^2 k^2}} \sqrt{\sum_{n=-\infty}^{\infty} |c_n'|^2} \\ &= \frac{A}{k} \sqrt{\sum_{n=-\infty}^{\infty} \frac{1}{n^2}}. \end{aligned} \tag{11.51}$$

It follows that $\sum_{n=1}^{\infty}(1/n^2)$ is convergent (See the remark below). Hence, from (11.50) and (11.51), we see that $\sum_{n=-\infty}^{\infty} |c_n|$ converges. This implies that $\sum_{n=-\infty}^{\infty} c_n e^{inkx}$ converges uniformly to a certain function on $[0, \lambda]$ since $|c_n e^{inkx}| \le |c_n|$ for all n on $[0, \lambda]$. (See Sect. 3.3.1 for the criteria for uniform convergence.) This completes our proof.

Remark. That the series $\sum_{n=1}^{\infty}(1/n^2)$ is convergent is verified as follows: set $A_{2^{k+1}-1}$ to be a partial sum consisting of the first $2^{k+1}-1$ terms. Then we have

$$\begin{aligned} A_{2^{k+1}-1} &= 1 + \left(\frac{1}{2^2}+\frac{1}{3^2}\right) + \left(\frac{1}{4^2}+\cdots+\frac{1}{7^2}\right) + \cdots \\ &\quad + \left[\frac{1}{(2^k)^2}+\cdots+\frac{1}{(2^{k+1}-1)^2}\right] \\ &< 1 + \frac{1}{2^2}\times 2 + \frac{1}{4^2}\times 4 + \cdots + \frac{1}{(2^k)^2}\times 2^k \\ &< \sum_{j=0}^{k}\left(\frac{1}{2}\right)^k = \frac{1-(1/2)^{k+1}}{1-(1/2)} < 2. \end{aligned}$$

This means that $A_{2^{k+1}-1}$ for any k is bounded above. Furthermore, the sequence (A_m) is monotonically increasing. Hence, (A_m) converges in the limit of $m \to \infty$, which completes the proof.

11.3.4 Pointwise Convergence at Discontinuous Points

This subsection gives an account of the second criterion in Sect. 11.3.1, which is restated below.

♠ **Pointwise convergence at discontinuities:**
When a function $f(x)$ is piecewise continuous and piecewise smooth, its Fourier series converges pointwise to $\{f(x+0)-f(x-0)\}/2$ at a point of discontinuity.

This theorem can be proven in the following manner. It readily follows from (11.24) that the partial sum of the Fourier series $S_N(x)$ is expressed by

$$S_N(x) = \frac{1}{2i\lambda}\int_{-x}^{\lambda-x} f(x+t)\frac{e^{i\left(N+\frac{1}{2}\right)kt}-e^{-i\left(N+\frac{1}{2}\right)kt}}{\sin\left(\frac{1}{2}kt\right)}dt.$$

We rewrite this as

$$\begin{aligned} S_N(x) = &\ \frac{1}{2i\lambda}\int_{-x}^{\lambda-x} f(x+t)\frac{e^{i\frac{1}{2}kt}}{\sin\left(\frac{1}{2}kt\right)}\cdot e^{iNkt}dt \\ &-\frac{1}{2i\lambda}\int_{-x}^{\lambda-x} f(x+t)\frac{e^{-i\frac{1}{2}kt}}{\sin\left(\frac{1}{2}kt\right)}\cdot e^{-iNkt}dt \\ = &\ \frac{1}{2i\lambda}\int_{-x}^{\lambda-x}\{f(x+t)+f(x-t)\}\frac{e^{i\frac{1}{2}kt}}{\sin\left(\frac{1}{2}kt\right)}\cdot e^{iNkt}dt. \end{aligned}$$

Here we have set $t \to -t$ in the second integral in the first line. Further,

$$S_N(x) = \frac{1}{2i\lambda}\int_{-x}^{\lambda-x} g(t)e^{iNkt}dt \; + \; \frac{f(x+0)+f(x-0)}{2i\lambda}\int_{-x}^{\lambda-x} \frac{e^{i\frac{1}{2}kt}e^{iNkt}}{\sin\left(\frac{1}{2}kt\right)}dt, \tag{11.52}$$

where we have introduced the notation

$$g(t) = \{f(x+t) - f(x+0) + f(x-t) - f(x-0)\}\,\frac{e^{i\frac{1}{2}kt}}{\sin\left(\frac{1}{2}kt\right)}.$$

The second term in (11.52) can be simplified via the relation

$$\frac{1}{i\lambda}\int_{-x}^{\lambda-x} \frac{e^{i\frac{1}{2}kt}e^{iNkt}}{\sin\left(\frac{1}{2}kt\right)}dt = 1.$$

(See Exercise **3** in Sect. 11.3 for its derivation.) Substituting this into (11.52), we get

$$S_N(x) = \frac{1}{2i\lambda}\int_{-x}^{\lambda-x} g(t)e^{iNkt}dt + \frac{f(x+0)+f(x-0)}{2}. \tag{11.53}$$

If the integration term in (11.53) vanishes with $N \to \infty$, we will successfully obtain the desired result. In fact, when $g(t)$ is piecewise continuous in the interval $[-x, \lambda - x]$, the integral in (11.53) vanishes owing to the Riemann–Lebesgue theorem (see Sect. 11.2.5). The remaining task is, therefore, to prove the piecewise continuity of $g(t)$ on $[-x, \lambda - x]$, which is actually verified through the following discussion.

When $t \neq 0$, $f(t)$ is piecewise continuous and $\sin(t/2)$ and $e^{it/2}$ are bounded; thus $g(t)$ is surely piecewise continuous. When $t = 0$, we have

$$g(t) = \left\{\frac{f(x+t)-f(x+0)}{t} + \frac{f(x-t)-f(x-0)}{t}\right\}\cdot\frac{t}{2\sin\left(\frac{1}{2}kt\right)}\cdot 2e^{i\frac{1}{2}kt},$$

so

$$\lim_{t\to 0} g(t) = \left\{\lim_{t\to 0}\frac{f(t+x)-f(x+0)}{t} + \lim_{t\to 0}\frac{f(x-t)-f(x-0)}{t}\right\}\cdot 2. \tag{11.54}$$

The first and second terms in (11.54) are the derivatives of $f(x)$ on the right and left, respectively. Since $f(t)$ is assumed to be piecewise smooth, $f'(t)$ is piecewise continuous; thus both terms in (11.54) exist. This indicates that the limit $\lim_{t\to 0} g(t)$ exists, so $g(t)$ is piecewise continuous within the interval $[-x, \lambda - x]$.

Consequently, we can conclude from (11.53) that

$$\lim_{N\to\infty} S_N(x) = \frac{f(x+0)+f(x-0)}{2},$$

which implies the pointwise convergence of the Fourier series to $[f(x+0) + f(x-0)]/2$.

11.3.5 Gibbs Phenomenon

If a function $f(x)$ has discontinuities in the defining region, its Fourier series does not reproduce the behavior of $f(x)$ at points of discontinuity. In other words, the partial sums of a Fourier series cannot approach $f(x)$ uniformly in the vicinity of a point of discontinuity. Furthermore, close to discontinuous points, the Fourier series inevitably overshoots the value of the original function to be expanded. The size of the overshoot is proportional to the magnitude of the discontinuity. This overshoot is known, which as the **Gibbs phenomenon** is nicely illustrated with the Fourier series for the step function

$$f(x) = \begin{cases} +1 & \text{for } 0 \le x \le \dfrac{\lambda}{2}, \\ -1 & \text{for } \dfrac{\lambda}{2} < x < \lambda, \end{cases}$$

which is a periodic square wave with period λ. The complex Fourier coefficient c_n reads

$$\begin{aligned} c_n &= \frac{1}{\lambda}\left(\int_0^{\lambda/2} e^{-inkx}dx - \int_{\lambda/2}^{\lambda} e^{-inkx}dx\right) \\ &= \frac{1}{2in\pi}\left(1-e^{-in\pi}\right)^2 = \begin{cases} 0 & (n = \text{even}) \\ \dfrac{2}{in\pi} & (n = \text{odd}). \end{cases} \end{aligned}$$

Then we have

$$\begin{aligned} f(x) &\sim \sum_{n=\cdots,-3,-1,1,3,\cdots} \frac{2}{in\pi}e^{inkx} = \sum_{n=1,3,\cdots}\left(\frac{2}{-in\pi}e^{-inkx} + \frac{2}{in\pi}e^{inkx}\right) \\ &= \frac{4}{\pi}\sum_{n=1}^{\infty}\frac{\sin(2n-1)kx}{2n-1}. \end{aligned} \tag{11.55}$$

Figure 11.7 shows $f(x)$ for $0 \le x \le \lambda$ for the sum of four, six, and ten terms of the series. Three features deserve attention.

(i) There is a steady increase in the accuracy of the representation as the number of included terms is increased.

(ii) All the curves pass through the midpoint of $f(x) = 0$ at the points of discontinuity $x = n\lambda/2$ $(n = 0, \pm 1, \pm 2, \cdots)$.

(iii) In the vicinity of $x = n\lambda/2$, there is an overshoot that persists and shows no sign of diminishing.

As more and more terms are taken, the small oscillations along each horizontal portion get smaller and smaller and, except for the two outer terms of

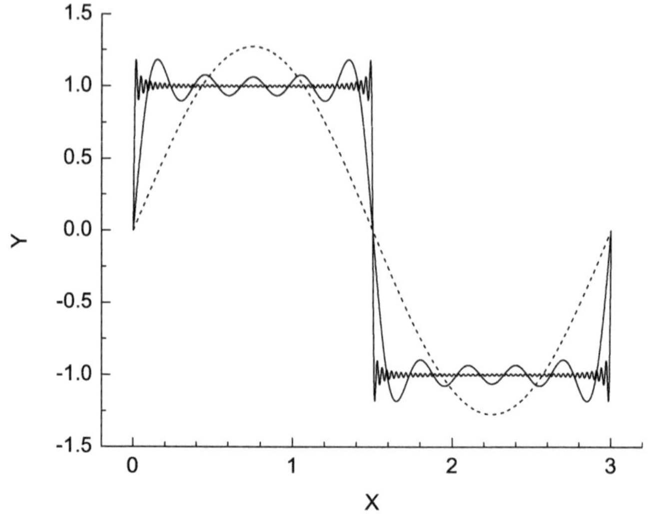

Fig. 11.7. Gibbs phenomena for the Fourier series of a step function. The partial sums of one, five, and fifty terms the right-hand side of (11.55) are given

each portion closes to the discontinuities, eventually disappear. Even in the limit of an infinite number of terms, there is still a small overshoot. This overshoot is nothing but what we call the **Gibbs phenomenon**, which results in the fact that the Fourier series cannot have uniform convergence at a point of discontinuity.

11.3.6 Overshoot at a Discontinuous Point

Owing to Gibbs phenomena, a Fourier series representation is highly unreliable in the vicinity of a discontinuity. We now consider the resulting degree of error when we represent a function $f(x)$ by a Fourier series having a discontinuity.

The maximum overshoot can be evaluated analytically through the following procedure. Let us consider a finite sum of the Fourier series in the complex form

$$S_N(x) = \sum_{n=-N}^{N} c_n e^{inkx},$$

which yields

$$S_N(x) = \frac{1}{\lambda}\int_{-x}^{\lambda-x} f(t+x)K_N(t)dt, \quad K_N(t) \equiv \frac{\sin\left[(N+\frac{1}{2})kt\right]}{\sin\left(\frac{1}{2}kt\right)}. \tag{11.56}$$

We consider the behavior of $S_N(x)$ in the vicinity of a discontinuity at $x = x_0$. We denote the jump of $f(x)$ at this discontinuity by Δf and the jump of its finite Fourier sum by ΔS_N:

$$\Delta f \equiv f(x_0+\varepsilon) - f(x_0-\varepsilon), \quad \Delta S_N \equiv S_N(x_0+\varepsilon) - S_N(x_0-\varepsilon),$$

where ε is infinitesimal. We then have

$$\Delta S_N = \frac{1}{\lambda}\int_{-x_0-\varepsilon}^{\lambda-x_0-\varepsilon} f(t+x_0+\varepsilon)K_N(t)dt - \frac{1}{\lambda}\int_{-x_0+\varepsilon}^{\lambda-x_0+\varepsilon} f(t+x_0-\varepsilon)K_N(t)dt.$$

Owing to the periodicity of the integrand $f(t+x)K_N(t)$, we replace the range of integration as follows:

$$\Delta S_N = \frac{1}{\lambda}\int_{-\varepsilon}^{\lambda-\varepsilon} f(t+x_0+\varepsilon)K_N(t)dt - \frac{1}{\lambda}\int_{\varepsilon}^{\lambda+\varepsilon} f(t+x_0-\varepsilon)K_N(t)dt.$$

Hence, we have

$$\begin{aligned}\Delta S_N &= \frac{1}{\lambda}\left(\int_{-\varepsilon}^{\varepsilon} + \int_{\varepsilon}^{\lambda-\varepsilon}\right) f(t+x_0+\varepsilon)K_N(t)dt \\ &\quad -\frac{1}{\lambda}\left(\int_{\varepsilon}^{\lambda-\varepsilon} + \int_{\lambda-\varepsilon}^{\lambda+\varepsilon}\right) f(t+x_0-\varepsilon)K_N(t)dt \\ &= \frac{1}{\lambda}\int_{-\varepsilon}^{\varepsilon} \left[f(t+x_0+\varepsilon) - f(t+x_0-\varepsilon)\right] K_N(t)dt \\ &\quad +\frac{1}{\lambda}\int_{\varepsilon}^{\lambda-\varepsilon} \left[f(t+x_0+\varepsilon) - f(t+x_0-\varepsilon)\right] K_N(t)dt. \end{aligned} \tag{11.57}$$

The integrand of (11.57) gives zero for all values of t except near $t=0$. Close to $t=0$, the integrand has a somewhat large value because of (i) the jump of $f(t+x_0)$ at $t=0$ and (ii) the significant contribution of $K_N(t)$ in the vicinity of $t=0$. Hence, we can confine the integration to the small interval $(-\delta,+\delta)$ for which the difference in the square brackets in (11.57) is simply Δf. It now follows that

$$\Delta S_N \simeq \frac{\Delta f}{\lambda}\int_{-\delta}^{\delta} \frac{\sin\left\{\left(N+\frac{1}{2}\right)kt\right\}}{\sin\frac{1}{2}kt}dt \simeq \frac{4\Delta f}{\lambda}\int_{0}^{\delta} \frac{\sin\left\{\left(N+\frac{1}{2}\right)kt\right\}}{kt}dt, \tag{11.58}$$

where the sine in the dominator was approximated by its argument because of the smallness of t.

The value of ΔS_N depends crucially on the interval δ, since the integrand in (11.58) rapidly alternates its sign as t increases. The reader may find the plot of the integrand in Fig. 11.8, where it is shown clearly that the major contribution to the integral comes from the interval $[0, \lambda/(2N+1)]$, where $\lambda/(2N+1)$ is the first zero of the integrand. Hence, if the upper limit is larger than $\lambda/(2N+1)$, the result of the integral will clearly decrease, because in each interval of length λ, the area below the horizontal axis is larger than that above. Therefore, if we are interested in the maximum overshoot of the finite

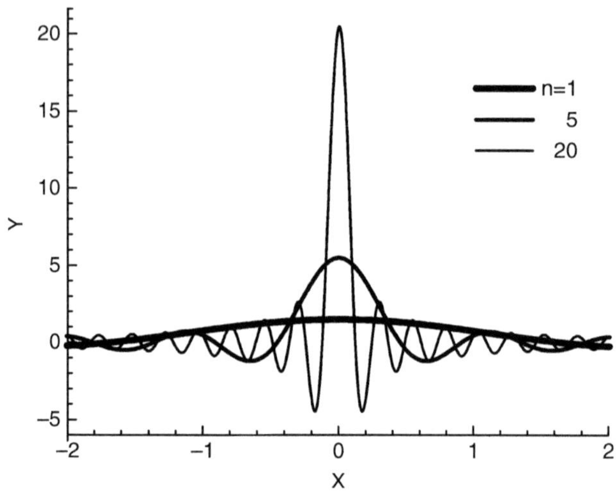

Fig. 11.8. The integrand of (11.58)

sum ΔS_N, we must set the upper limit equal to $\lambda/(2N+1)$. It follows that the maximum overshoot is

$$\begin{aligned}(\Delta S_N)_{\max} &\simeq \frac{4\Delta f}{\lambda}\int_0^{\lambda/(2N+1)} \frac{\sin(N+\frac{1}{2})kt}{kt}dt \\ &= \frac{4\Delta f}{\lambda}(N+\frac{1}{2})\int_0^{\pi} \frac{\sin x}{x}\frac{dx}{(N+\frac{1}{2})k} \quad = \frac{2\Delta f}{\pi}\int_0^{\pi}\frac{\sin x}{x}dx \\ &\simeq 1.179\Delta f.\end{aligned}$$

We thus conclude that the finite (large-N) sum approximation of the discontinuous function overshoots the function itself at a discontinuity by about 18% in this case. This means that the Fourier series tends to overshoot the positive corner by some 18% and to undershoot the negative corner by the same amount. The inclusion of more terms (increasing r) does nothing to remove this overshoot but merely moves it closer to the point of discontinuity.

Exercises

1. Let $f(x)$ be absolutely integrable and form the Fourier series of $f(x)$ in the interval $(-\pi, \pi)$. Show that the convergence of its Fourier series at a specified point x within the interval depends only on the behavior of f in the immediate vicinity of this point. (This result is referred to as the **localization theorem**.)

   Solution: We use the integral formula for the partial sums

$$S_n(x) = \frac{1}{\pi}\int_{-\pi}^{\pi} f(x+u)\frac{\sin mu}{2\sin(u/2)}du$$
$$= \frac{1}{\pi}\int_{-\delta}^{\delta} f(x+u)\frac{\sin mu}{2\sin(u/2)}du + I_1 + I_2,$$

where we have set $m = n + (1/2)$. Here δ is an arbitrarily small positive number, and I_1, I_2 are the integrals over the intervals $[\delta, \pi]$ and $[-\pi, -\delta]$, respectively. On these intervals, the function $1/[2\sin(u/2)]$ is continuous (since $|u| \geq \delta$) and, therefore, the function

$$\phi(u) = \frac{f(x+u)}{2\sin(u/2)}$$

is absolutely integrable. It then follows from the Riemann–Lebesgue theorem that the integral

$$I_1 = \frac{1}{\pi}\int_{\delta}^{\pi} \phi(u)\sin mudu$$

approaches zero as $m \to \infty$. The same is true of I_2. Thus, whether or not the partial sums of the Fourier series have a limit at the point x depends on the behavior of the integral

$$\frac{1}{\pi}\int_{-\delta}^{\delta} f(x+u)\frac{\sin mu}{2\sin(u/2)}du$$

as $m \to \infty$, which involves only the values of the function $f(x)$ in the neighborhood $[x-\delta, x+\delta]$ of the point x. This completes the proof. ♣

2. Let $f(x) = -\log|2\sin(x/2)|$, which is even and becomes infinite at $x = 2k\pi$ $(k = 0, \pm 1, \pm 2, \cdots)$.

(i) Show that $f(x)$ is integrable.

(ii) Calculate the Fourier series of $f(x)$.

(iii) Derive the identity: $\log 2 = 1 - (1/2) + (1/3) - (1/4) + \cdots$.

Solution:

(i) The given $f(x)$ equals zero at $x = \pi/3$ and is 2π-periodic. Hence, to prove the integrability of $f(x)$, it suffices to show that it is integrable on the interval $[0, \pi/3]$. Clearly we have

$$-\int_{\varepsilon}^{\pi/3} \log\left|2\sin\frac{x}{2}\right| dx = \varepsilon\log\left(2\sin\frac{\varepsilon}{2}\right) + \int_{\varepsilon}^{\pi/3}\frac{x\cos(x/2)}{2\sin(x/2)},$$

where we have dropped the absolute value sign, since $2\sin(x/2) > 1$ for $0 < x < \pi/3$. As $\varepsilon \to 0$, the quantity $\varepsilon \log[2\sin(\varepsilon/2)]$ approaches zero, which is verified by using **l'Hôpital's rule** (see Sect. 1.4.1), whereas the last integral converges since the integrand is bounded. (Recall that $\lim_{x\to 0} x/[2\sin(x/2)] = 1$.) Thus, $-\int_0^{\pi/3} \log\left|2\sin\frac{x}{2}\right| dx$ exists, i.e., $f(x)$ is integrable on the interval $[0, \pi/3]$.

(ii) Since $f(x)$ is even, we have $b_n = 0$ $(n = 1, 2, \cdots)$ and

$$a_n = -\frac{2}{\pi}\int_0^{\pi} \log\left(2\sin\frac{x}{2}\right)\cos nx dx \quad (n = 0, 1, 2, \cdots).$$

For $n \neq 0$, integrating by parts and then applying l'Hôpital's rule, we get

$$a_n = \frac{1}{n\pi}\int_0^{\pi} \frac{\sin nx\cos(x/2)}{\sin(x/2)} dx \quad (n = 1, 2, \cdots),$$

and then use the identity $2\sin nx\cos(x/2) = \sin[n+(1/2)]x + \sin[n-(1/2)]x$ to obtain

$$\begin{aligned} a_n &= \frac{1}{n\pi}\int_0^{\pi} \frac{\sin[n+(1/2)]x}{2\sin(x/2)} dx + \frac{1}{n\pi}\int_0^{\pi} \frac{\sin[n-(1/2)]x}{2\sin(x/2)} dx \\ &= \frac{1}{n}. \quad (n = 0, 1, 2, \cdots). \end{aligned}$$

For $n = 0$, we have

$$\begin{aligned} a_0 &= -\frac{2}{\pi}\int_0^{\pi} \log\left(2\sin\frac{x}{2}\right) dx = -\frac{2}{\pi}\int_0^{\pi} \left(\log 2 + \log\sin\frac{x}{2}\right) dx \\ &= \pi\log 2 + \int_0^{\pi} \log\left(\sin\frac{x}{2}\right) dx. \end{aligned}$$

The last integral, denoted by I, reads

$$\begin{aligned} I &= 2\int_0^{\pi/2} \log(\sin t)dt = 2\int_0^{\pi/2} \log\left(2\sin\frac{t}{2}\cos\frac{t}{2}\right) dt \\ &= \pi\log 2 + 2\int_0^{\pi/2} \log\left(\sin\frac{t}{2}\right) dt + 2\int_0^{\pi/2} \log\left(\cos\frac{t}{2}\right) dt. \end{aligned}$$

The substitution $t = \pi - u$ gives $\int_0^{\pi/2} \log[\cos(t/2)]dt = \int_{\pi/2}^{\pi} \log[\sin(u/2)]du$, which implies that $I = \pi\log 2 + 2I$, *i.e.*, $I = -\pi\log 2$. Consequently, $a_0 = 0$.

(iii) Since the function $f(x)$ is obviously differentiable for $x \neq 2k\pi$ $(k = 0, \pm 1, \pm 2, \cdots)$, it follows that

$$-\log \left| 2 \sin \frac{x}{2} \right| = \cos x + \frac{\cos 2x}{2} + \frac{\cos 3x}{3} + \cdots \qquad (11.59)$$

for $x \neq 2k\pi$ $(k = 0, \pm 1, \pm 2, \cdots)$. Setting $x = \pi$ in (11.59), we obtain the desired result. ♣

3. Show that

$$\int_{-x}^{\lambda - x} \frac{e^{i(N+\frac{1}{2})kt}}{\sin\left(\frac{1}{2}kt\right)} dt = i\lambda. \qquad (11.60)$$

Solution:

Recall an alternative form of $S_N(x)$ given in (11.40):

$$S_N(x) = \frac{1}{\lambda} \int_{-x}^{\lambda - x} f(t+x) \left(\sum_{n=-N}^{N} e^{-inkt} \right) dt. \qquad (11.61)$$

Setting $f(t) \equiv 1$ into (11.61) and (11.52) and comparing them, we have

$$0 + \frac{1}{i\lambda} \int_{-x}^{\lambda - x} \frac{e^{i\frac{1}{2}kt} e^{iNkt}}{\sin\left(\frac{1}{2}kt\right)} dt = \frac{1}{\lambda} \int_{-x}^{\lambda - x} \left(\sum_{n=-N}^{N} e^{-inkt} \right) dt$$

$$= \frac{1}{\lambda} \sum_{n=-N}^{N} \left(\int_{-x}^{\lambda - x} e^{-inkt} dt \right) = \frac{1}{\lambda} \sum_{n=-N}^{N} \lambda \delta_{n,0} = 1. \quad ♣$$

11.4 Applications in Physics and Engineering

11.4.1 Temperature Variation of the Ground

The most important applications of Fourier series expansions in the physical sciences are in solving **partial differential equations** that describe a wide variety of physical phenomena. In this section, two typical examples of such applications are presented, while more rigorous discussions on partial differential equations are given in Chap. 17.

First, we consider the temperature variation of the ground exposed to sunlight. The temperature at a depth of x meters at time t, denoted by $u(x,t)$, is known to be determined by the **diffusion equation**

$$\frac{\partial u}{\partial t} = \kappa \frac{\partial^2 u}{\partial x^2}. \qquad (11.62)$$

Here, the proportionality constant κ is called the **thermal conductivity** and its magnitude on the ground is roughly estimated at $\kappa = 3.0 \times 10^{-6}$ m^2/s. We will see below that the Fourier series expansion provides a means of solving equation (11.62) and clarifying the physical interpretation of its solution.

Suppose that the temperature of the land surface, $u(x = 0, t)$, changes periodically with a period T; the period T may range from a day to a year. It is then reasonable to express $u(x, t)$ by the Fourier series

$$u(x,t) = \sum_{n=-\infty}^{\infty} c_n(x) e^{in\omega t} \quad \left(\omega = \frac{2\pi}{T}\right).$$

Substitute this into (11.62) to obtain

$$in\omega c_n(x) = \kappa \frac{d^2 c_n}{dx^2},$$

which implies

$$c_n(x) \propto \begin{cases} \exp\left(-\frac{1+i}{\sqrt{2}}\sqrt{\frac{n\omega}{\kappa}}x\right) & n > 0 \\ \exp\left(-\frac{1-i}{\sqrt{2}}\sqrt{\frac{|n|\omega}{\kappa}}x\right) & n < 0. \end{cases}$$

Here we have chosen the solutions that behave as $|c_n(x)| \to 0$ in the limit of $x \to \infty$. In order to obtain the zeroth term $c_0(x)$, we note that

$$\frac{dc_0(x)}{dx^2} = 0,$$

and thus

$$c_0(x) = A_0 + B_0 x.$$

Owing to the condition that $\lim_{x\to\infty} |c_0(x)| = 0$, we see that $B_0 = 0$ and $A_0 = \text{const}$. As a result, we obtain

$$u(x,t) = A_0 + 2\sum_{n=1}^{\infty} A_n e^{-\alpha_n x} \cos\left(n\omega t - \alpha_n x + \phi_n\right), \tag{11.63}$$

where

$$\alpha_n = \sqrt{\frac{n\omega}{2\kappa}}$$

and the constants A_n and ϕ_n are determined by the t-dependence of the surface temperature $u(x = 0, t)$.

Note the presence of the parameter α_n in the general solution (11.63). It indicates that a wave component with the period T/n has the following features: (i) decay of the wave amplitude by $e^{-\alpha_n x}$ with an increase in x, and (ii) a phase shift by $\alpha_n x$ relative to the surface temperature $u(x=0,t)$.

Let us quantify the actual value of α_n. For this, we consider the case of $T = 1$ day (i.e., $60 \times 60 \times 24\,\mathrm{s}$) and assume monochromatic variation of the surface temperature given by

$$u(0,t) = 15 + 5\cos\left(\frac{2\pi}{T}t\right) \quad {}^\circ\mathrm{C}.$$

Comparing this with (11.63) with $x = 0$, we get $A_0 = 15$, $A_1 = 5/2$, and $A_n = 0$ for $n \geq 2$. Then, since

$$\alpha_1 = \sqrt{\frac{2 \times 3.14}{2 \times (3.0 \times 10^{-6}) \times (60 \times 60 \times 24)}} \simeq 3.5,$$

we have

$$u(x,t) = 15 + 5e^{-3.5x}\cos\left(\frac{2\pi}{T}t - 3.5x\right).$$

A three-dimensional plot of $u(x,t)$ in the x-t plane is shown in Fig. 11.9. We observe that at depths greater than 1 m, the temperature variation is almost in antiphase to that at the surface ($x = 0$) and the amplitude decreases considerably.

11.4.2 String Vibration Under Impact

The second example is the vibration of an elastic string subject to an impact force in a local region. Consider the case of a piano wire under an impact force applied by a hammer. Suppose that an impulse I is applied at the position $x = a$ of a suspended string with length ℓ and mass density ρ. The vibrational amplitude of the string, denoted by $u(x,t)$, is governed by the **wave equation**

$$\frac{\partial^2 u}{\partial t^2} = c^2 \frac{\partial^2 u}{\partial x^2}. \tag{11.64}$$

The string is initially assumed to be stationary, i.e.,

$$u(x, t=0) = 0. \tag{11.65}$$

The initial velocity of the line element at x is denoted by $v(x)$. Then, the **law of the conservation of momentum** states that

$$\int_0^\ell \rho v(x) dx = I, \tag{11.66}$$

where

$$v(x) = V \cdot \delta(x-a), \tag{11.67}$$

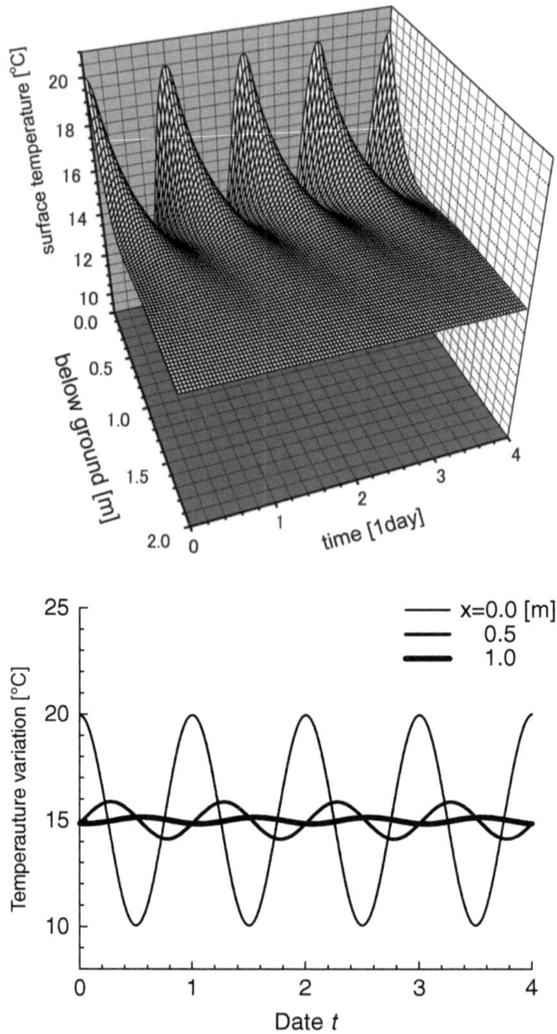

Fig. 11.9. Temperature variation $u(x, t)$ of the underground below x meters on t days

with an appropriate constant V. From (11.66) and (11.67), we have $V = I/\rho$. Furthermore, since

$$v(x) = \left.\frac{\partial u}{\partial t}\right|_{t=0},$$

we have

$$\left.\frac{\partial u}{\partial t}\right|_{t=0} = \frac{I}{\rho}\delta(x - a). \tag{11.68}$$

Under the two initial conditions (11.65) and (11.68), the general solution of (11.64) is given by

$$u(x,t) = \sum_{n=1}^{\infty} A_n \sin k_n x \sin(\omega_n t + \phi_n), \tag{11.69}$$

where

$$k_n = \frac{n\pi}{\ell}, \quad \omega_n = ck_n.$$

The constants A_n and ϕ_n in (11.69) are again determined by the initial conditions. First, imposing the condition $u(x, t = 0) = 0$ into (11.69) implies

$$A_n \sin \phi_n = 0 \quad \text{for all } n, \tag{11.70}$$

owing to the linear independence of $\{\sin k_n x\}$. Next, it follows from (11.69) that

$$\left.\frac{\partial u}{\partial t}\right|_{t=0} = \sum_{n=1}^{\infty} A_n \omega_n \cos \phi_n \sin k_n x = \frac{I}{\rho}\delta(x-a).$$

Mutiplying both sides by $\sin k_m x$ and then integrating yields

$$A_m \omega_m \cos \phi_n \int_0^{\ell} \sin^2 k_m x dx = \frac{I}{\rho} \sin k_m a \quad \text{for all } m. \tag{11.71}$$

From (11.70) and (11.71), we finally obtain

$$\phi_n = 0 \quad \text{and} \quad A_n = \frac{2I}{\rho \ell \omega_n} \sin k_n a \quad \text{for all } n. \tag{11.72}$$

The second expression in (11.72) implies that the position $x = a$ that satisfies $\sin k_n a = 0$ yields $A_n = 0$; i.e., the nth vibration mode is not excited by the impulsive force applied at $x = a$ that satisfies $\sin k_n a = 0$. In contrast, if we apply an impulsive force at $x = a$ satisfying $|\sin k_n a| = 1$, the corresponding nth mode will have a large vibrational amplitude, as is actually the case inside a piano.

12

Fourier Transformation

Abstract Fourier transformation is an effective tool for confirming the dual nature of a complex-valued function (as well as a real-valued one). Furthermore, the transformation enables us to measure certain correlations of a function with itself or with other functions; thus a Fourier transform can be applied to probability theory, signal analysis, etc. In this chapter we also provide the essence of a discrete Fourier transform (Sect. 12.3), which refers to a Fourier transform applied to a discrete complex-valued series. A discrete Fourier transform is commonly used in the numerical computation of Fourier transforms because of its computational efficiency.

12.1 Fourier Transform

12.1.1 Derivation of Fourier Transform

The properties of Fourier series that we have already developed are adequate for handling the expansion of any periodic function. Nevertheless, there are many problems in physics and engineering that do not involve periodic functions, so it is important to generalize Fourier series to include nonperiodic functions. A nonperiodic function can be considered as a limit of a given periodic function whose period becomes infinite.

Let us write Fourier series representing a periodic function $f(x)$ in complex form:

$$f(x) = \sum_{n=-\infty}^{\infty} c_n e^{ikx}, \tag{12.1}$$

with the definition $k = 2n\pi/\lambda$, in which

$$c_n = \frac{1}{\lambda} \int_{-\frac{\lambda}{2}}^{\frac{\lambda}{2}} f(x) e^{-ikx} dx.$$

We then introduce the quantity

$$\Delta k = \frac{2\pi}{\lambda} \Delta n. \tag{12.2}$$

From the definition (12.2), the adjacent values of k are obtained by setting $\Delta n = 1$, which corresponds to $(\lambda/2\pi)\Delta k = 1$. Therefore, multiplying each side of (12.1) by $(\lambda/2\pi)\Delta k$ yields

$$f(x) = \sum_{n=-\infty}^{\infty} c_\lambda(k) e^{ikx} \Delta k, \tag{12.3}$$

where

$$c_\lambda(k) \equiv \frac{\lambda}{2\pi} c_n = \frac{1}{2\pi} \int_{-\frac{\lambda}{2}}^{\frac{\lambda}{2}} f(x) e^{-ikx} dx.$$

In the limit as $\lambda \to \infty$, the ks are distributed continuously instead of discretely, i.e., $\Delta k \to dk$. Thus, the sum in (12.3) becomes exactly the definition of an integral. As a result, we arrive at the conclusion

$$c(k) = \lim_{\lambda \to \infty} c_\lambda(k) = \frac{1}{2\pi} \int_{-\infty}^{\infty} f(x) e^{-ikx} dx \tag{12.4}$$

and

$$f(x) = \int_{-\infty}^{\infty} c(k) e^{ikx} dk. \tag{12.5}$$

Further, by defining $F(k) = \sqrt{2\pi} c(k)$, equations (12.4) and (12.5) take the symmetrical form given below, known as the **Fourier transform** or **Fourier integral representation** of $f(x)$.

♠ **Fourier transform:**
The Fourier transform of $f(x)$ is defined by

$$F(k) = \frac{1}{\sqrt{2\pi}} \int_{-\infty}^{\infty} f(x) e^{-ikx} dx. \tag{12.6}$$

♠ **Inverse Fourier transform:**
The inverse Fourier transform of $F(k)$ given above is defined by

$$f(x) = \frac{1}{\sqrt{2\pi}} \int_{-\infty}^{\infty} F(k) e^{ikx} dk, \tag{12.7}$$

We often write the expressions (12.6) and (12.7) in simpler form:

$$F(k) = \mathcal{F}[f(x)] \quad \text{and} \quad f(x) = \mathcal{F}^{-1}[F(k)].$$

Observe that $F(k)$ as well as $f(x)$ are, in general, complex-valued functions of the real variables k and x, respectively. Yet, if $f(x)$ is real, then

$$F(-k) = F^*(k),$$

which gives two immediate corollaries (proofs are left to the reader):

♠ **Fourier integral theorem:**

1. If $f(x)$ is real and even, $F(k)$ is real.
2. If $f(x)$ is real and odd, $F(k)$ is purely imaginary.

12.1.2 Fourier Integral Theorem

Our derivations of the Fourier transform and its inverse transform, (12.7) and (12.6), have been ambiguous from a mathematical viewpoint. For developing exact derivations and clarifying the conditions for the infinite integrals in (12.7) and (12.6) to converge, the following theorem is of crucial importance:

♠ **Fourier integral theorem:**
If $f(x)$ is piecewise smooth and absolutely integrable, then

$$\frac{1}{\pi}\int_0^\infty \left[\int_{-\infty}^\infty f(t)\cos u(x-t)dt\right] du = \frac{f(x+0)+f(x-0)}{2}. \tag{12.8}$$

Remark. The theorem is valid for each fixed x, so x can be considered a constant insofar as the integrations are concerned.

Before starting the proof of the theorem, we note that (12.8) reduces to the form of (12.7) and (12.6) when x is a continuous point of $f(x)$. To see this, we make use of the identity

$$\int_0^\xi \cos u(x-t)du = \frac{1}{2}\int_{-\xi}^\xi e^{iu(x-t)}du. \tag{12.9}$$

Since (12.8) reads

$$f(x) = \lim_{\xi\to\infty}\frac{1}{\pi}\int_{-\infty}^\infty f(t)dt\int_0^\xi \cos u(x-u)du, \tag{12.10}$$

we substitute (12.9) and (12.10) to obtain

$$f(x) = \frac{1}{2\pi}\int_{-\infty}^\infty e^{iux}du\int_{-\infty}^\infty f(t)e^{-itu}dt = \frac{1}{2\pi}\int_{-\infty}^\infty F(u)e^{iux}du,$$

where

$$F(u) = \frac{1}{2\pi}\int_{-\infty}^\infty f(t)e^{-itu}dt.$$

These results are clearly equivalent to the forms of (12.7) and (12.6).

12.1.3 Proof of the Fourier Integral Theorem

The proof of the Fourier integral theorem is based on the following two lemmas:

♠ **Lemma 1:** If $f(x)$ is piecewise smooth for all $x \in \boldsymbol{R}$, then

$$\lim_{\xi\to\infty} \int_0^b f(x) \frac{\sin \xi x}{x} dx = \frac{\pi}{2} f(0+) \quad \text{for b} > 0.$$

♠ **Lemma 2:** If $f(x,t)$ is a continuous function of t for $a \le t \le b$ and if $\lim_{c\to\infty} \int_0^c f(x,t)dx$ exists and converges uniformly to a certain function $g(t)$ in the interval, then $g(t)$ is continuous in the interval and

$$\int_a^b g(t)dt = \int_a^b \left[\int_0^\infty f(x,t)dx \right] dt = \int_0^\infty \left[\int_a^b f(x,t)dt \right] dx.$$

Note that $f(0+)$ in Lemma **1** denotes the limiting value of $f(x)$ as x tends to zero through positive values. The proof of Lemma **1** is left to Exercise **2**. Lemma **2** follows from the fact that uniform convergence allows us to interchange the order of limiting and integration procedures (see Chapter 3 for details).

We are now ready to prove the Fourier integral theorem expressed by (12.8).

Proof (of the Fourier integral theorem): Let $f(x)$ be piecewise smooth and absolutely integrable. Consider the integral

$$\int_{-\infty}^{\infty} f(t) \cos u(x-t)dt.$$

Since $|\cos u(x-t)| \le 1$, the convergence of this integral is ensured by our hypothesis that $\int_{-\infty}^{\infty} |f(t)|dt$ converges, and since this conclusion is independent of u and x, the convergence is uniform for all u. Therefore, in view of Lemma **2**, we can interchange the order of integration in

$$I \equiv \int_0^b \left[\int_{-\infty}^{\infty} f(t) \cos u(x-t)dt \right] du$$

to obtain

$$I \equiv \int_{-\infty}^{\infty} \left[\int_0^b f(t) \cos u(x-t)du \right] dt = \int_{-\infty}^{\infty} \frac{\sin b(x-t)}{x-t} f(t)dt.$$

We now decompose this into four integrals:

$$I = \left[\int_{-\infty}^{-M} + \int_{-M}^{x} + \int_x^M + \int_M^\infty \right] \frac{\sin b(x-t)}{x-t} f(t)dt, \qquad (12.11)$$

where M is taken to be so large that the first and the last integrals in (12.11) are less in absolute value than some prescribed $\varepsilon > 0$. By changing variables, taking $u = t - x$, we can write the third integral in (12.11) as

$$\int_0^{M-x} \frac{\sin bu}{u} f(x+u)du.$$

In view of Lemma **1**, this tends to $\pi f(x+0)/2$ as $b \to \infty$. Similarly, the second integral tends to $\pi f(x-0)/2$. Therefore, by taking M sufficiently large, we obtain

$$\lim_{b\to\infty} I < \frac{\pi \left[f(x+0)+f(x-0)\right]}{2} + 2\varepsilon,$$

or equivalently,

$$\int_0^\infty \left[\int_{-\infty}^\infty f(t)\cos u(x-t)dt\right] du - \frac{\pi \left[f(x+0)+f(x-0)\right]}{2} < 2\varepsilon.$$

This completes the proof of the theorem. ♣

12.1.4 Inverse Relations of the Half-width

In practice, we often encounter functions $f(x)$ having a sharp peak at a specific point, say $x = 0$. The width of the peak of such a function is possibly correlated with the width of the peak that is exhibited by the resulting Fourier transform $F(k) = \mathcal{F}[f(x)]$. A typical example of this phenomenon is seen by considering the Fourier transform of a Gaussian function $f(x) = ae^{-bx^2}$ with $a, b > 0$, i.e.,

$$F(k) = \frac{a}{\sqrt{2\pi}} \int_{-\infty}^\infty e^{-bx^2} e^{-ikx} dx = \frac{ae^{-k^2/(4b)}}{\sqrt{2\pi}} \int_{-\infty}^\infty e^{-b[x+ik/(2b)]^2} dx.$$

We substitute $y = x + ik/(2b)$ to evaluate the integral as

$$\int_{-\infty}^\infty e^{-b[x+ik/(2b)]^2} dx = \int_{-\infty}^\infty e^{-by^2} dy = \sqrt{\frac{\pi}{b}}.$$

and we get

$$F(k) = \frac{a}{\sqrt{2b}} e^{-k^2/(4b)},$$

which is also Gaussian. It is noteworthy that the width of $f(x)$, which is proportional to $1/\sqrt{b}$, is in inverse relation to the width of $F(k)$, which is proportional to $\sqrt{b}$. Therefore, increasing the width of $f(x)$ results in a decrease in the width of $F(k)$. In the limit of infinite width (a constant function), we get infinite sharpness (the delta function). In fact, denoting the widths as Δx and Δk, we have $\Delta x \Delta k \sim 1$.

♠ **Inverse relation of the half-width:**
When $f(x)$ consists of a single peak whose width is characterized by Δx, its Fourier transform $F(k)$ is also a single-peak function with a width Δk, which yields $\Delta x \Delta k \sim 1$.

For the second example, we evaluate the Fourier transform of a box function defined by

$$f(x) = \begin{cases} b, \text{ if } |x| < a, \\ 0, \text{ if } |x| > a. \end{cases}$$

From the definition, we have

$$F(k) = \frac{1}{\sqrt{2\pi}} \int_{-\infty}^{\infty} f(x) e^{-ikx} dx = \frac{b}{\sqrt{2\pi}} \int_{-a}^{a} e^{-ikx} dx = \frac{2ab}{\sqrt{2\pi}} \left(\frac{\sin ka}{ka} \right).$$

Observe again that the width of $f(x)$, $\Delta x = 2a$, is in inverse relation to the width of $F(k)$, which is roughly the distance between its first two roots, k_+ and k_-, on either side of $k = 0$: $\Delta k = k_+ - k_- = 2\pi/a$. In addition, if $a \to \infty$, the function $f(x)$ becomes a constant function over the entire real line, and we get

$$F(k) = \frac{2b}{\sqrt{2\pi}} \lim_{a\to\infty} \frac{\sin ka}{k} = \frac{2b}{\sqrt{2\pi}} \pi \delta(k).$$

Otherwise, if $b \to \infty$ and $a \to 0$ in such a way that $2ab$ [the area under the graph of $f(x)$] remains fixed at unity, then $f(x)$ approaches the delta function and $F(k)$ becomes

$$F(k) = \lim_{a\to 0} \lim_{b\to\infty} \frac{2ab}{\sqrt{2\pi}} \frac{\sin ka}{ka} = \frac{1}{\sqrt{2\pi}}.$$

12.1.5 Parseval Identity for Fourier Transforms

If $F(k)$ and $G(k)$ are Fourier transforms of $f(x)$ and $g(x)$, respectively, we have

$$\begin{aligned}
&\int_{-\infty}^{\infty} f(x) g^*(x)\, dx \\
&= \int_{-\infty}^{\infty} \left\{ \frac{1}{\sqrt{2\pi}} \int_{-\infty}^{\infty} F(k) e^{-ikx} dk \right\} \times \left\{ \frac{1}{\sqrt{2\pi}} \int_{-\infty}^{\infty} G^*(k') e^{ik'x} dk' \right\} dx \\
&= \int_{-\infty}^{\infty} dk \int_{-\infty}^{\infty} dk' F(k) G^*(k) \left\{ \frac{1}{2\pi} \int_{-\infty}^{\infty} e^{-i(k-k')x} dx \right\} \\
&= \int_{-\infty}^{\infty} dk F(k) \int_{-\infty}^{\infty} dk' G^*(k') \delta(k'-k) \\
&= \frac{1}{2\pi} \int_{-\infty}^{\infty} F(k) G^*(k)\, dk,
\end{aligned} \tag{12.12}$$

or similarly,

$$\int_{-\infty}^{\infty} f(x)g(x)\,dx = \frac{1}{2\pi}\int_{-\infty}^{\infty} F(k)G(-k)\,dk. \tag{12.13}$$

In particular, if we set $g(x) = f(x)$ in (12.12), we have

$$\int_{-\infty}^{\infty} |f(x)|^2 dx = \frac{1}{2\pi}\int_{-\infty}^{\infty} |F(k)|^2 dk. \tag{12.14}$$

Here $|F(k)|^2$ is referred to as the **power spectrum** of the function $f(x)$. Equation (12.14), or the more general (12.13), is known as the **Parseval identity** for Fourier integrals.

Remark. A sufficient condition for interchanging the order of integration in (12.12) is the absolute convergence of the integrals: $\int_{-\infty}^{\infty} F(k)e^{-ikx}dk$ and $\int_{-\infty}^{\infty} G(k')e^{-ik'x}dk'$.

Parseval's identity is very useful for understanding the physical interpretation of the transform function $F(k)$ when the physical significance of $f(x)$ is known, as illustrated in the following example:

Examples The displacement of a **damped harmonic oscillator** as a function of time is given by

$$f(t) = \begin{cases} 0 & \text{for } t < 0, \\ e^{-t/\tau}\sin\omega_0 t & \text{for } t \geq 0. \end{cases}$$

The Fourier transform of this function is given by

$$\begin{aligned} F(\omega) &= \int_{-\infty}^{0} 0 \times e^{-i\omega t}dt + \int_{0}^{\infty} e^{-t/\tau}\sin\omega_0 t e^{-i\omega t}dt \\ &= 0 + \frac{1}{2i}\int_0^{\infty}\left[e^{-i(\omega-\omega_0)t-t/\tau} - e^{-i(\omega+\omega_0)t-t/\tau}\right]dt \\ &= \frac{1}{2}\left(\frac{1}{\omega+\omega_0-i/\tau} - \frac{1}{\omega-\omega_0-i/\tau}\right). \end{aligned}$$

The physical interpretation of $|F(\omega)|^2$ is the energy content per unit frequency interval (i.e., the energy spectrum) while $|f(t)|^2$ is proportional to the sum of the kinetic and potential energies of the oscillator. Hence, Parseval's identity, expressed by

$$\int_{-\infty}^{\infty} |f(t)|^2 dt = \frac{1}{2\pi}\int_{-\infty}^{\infty} |F(w)|^2 d\omega,$$

shows the equivalence of these two alternative specifications for the total energy to within a constant.

12.1.6 Fourier Transforms in Higher Dimensions

The concept of the Fourier transform can be extended naturally to more than one dimension. For example, in three dimensions we can define the Fourier transform of $f(x, y, z)$ as

$$F(k_x, k_y, k_z) = \frac{1}{(2\pi)^{3/2}} \iiint f(x, y, z) e^{-ik_x x} e^{-ik_y y} e^{-ik_z z} dx dy dz \quad (12.15)$$

and its inverse as

$$f(x, y, z) = \frac{1}{(2\pi)^{3/2}} \iiint F(k_x, k_y k_z) e^{ik_x x} e^{ik_y y} e^{ik_z z} dk_x dk_y dk_z. \quad (12.16)$$

Denoting the vector with components k_x, k_y, k_z by $\boldsymbol{k}$ and that with components x, y, z by $\boldsymbol{r}$, we can write the Fourier transform pair (12.15), (12.16) as follows:

♠ **Fourier transforms in three dimensions:**

$$F(\boldsymbol{k}) = \frac{1}{(2\pi)^{3/2}} \int f(\boldsymbol{r}) e^{-i\boldsymbol{k}\cdot\boldsymbol{r}} d\boldsymbol{r},$$

$$f(\boldsymbol{r}) = \frac{1}{(2\pi)^{3/2}} \int F(\boldsymbol{k}) e^{i\boldsymbol{k}\cdot\boldsymbol{r}} d\boldsymbol{k}.$$

It is pedagogical to evaluate the Fourier transform of a function $f(\boldsymbol{r})$ under the condition that the system possesses spherical symmetry, i.e., $f(\boldsymbol{r}) = f(r)$. We employ **spherical coordinates** in which the vector $\boldsymbol{k}$ of the Fourier transform lies along the polar axis ($\theta = 0$). We then have

$$d\boldsymbol{r} = r^2 \sin\theta dr d\theta d\phi \quad \text{and} \quad \boldsymbol{k}\cdot\boldsymbol{r} = kr\cos\theta,$$

where $k = |\boldsymbol{k}|$. The Fourier transform is then given by

$$\begin{aligned} F(\boldsymbol{k}) &= \frac{1}{(2\pi)^{3/2}} \int f(\boldsymbol{r}) e^{-i\boldsymbol{k}\cdot\boldsymbol{r}} d\boldsymbol{r} \\ &= \frac{1}{(2\pi)^{1/2}} \int_0^\infty dr r^t f(r) \int_0^\pi d\theta \sin\theta e^{-ikr\cos\theta}. \end{aligned}$$

The integral over θ may be straightforwardly evaluated by noting that

$$\frac{d}{d\theta} e^{-ikr\cos\theta} = ikr\sin\theta\, e^{-ikr\cos\theta}.$$

Therefore,

$$F(\boldsymbol{k}) = \frac{1}{(2\pi)^{1/2}} \int_0^\infty r^2 f(r) dr \left[\frac{e^{-ikr\cos\theta}}{ikr} \right]_{\theta=0}^{\theta=\pi}$$
$$= \frac{1}{(2\pi)^{1/2}} \int_0^\infty 2r^2 f(r) \left(\frac{\sin kr}{kr} \right) dr.$$

Remark. A similar result may be obtained for **two-dimensional Fourier transforms** in which $f(\boldsymbol{r}) = f(\rho)$, i.e., $f(\boldsymbol{r})$ is independent of the azimuthal angle ϕ. In this case, we find

$$F(\boldsymbol{k}) = \int_0^\infty \rho f(\rho) J_0(k\rho) d\rho,$$

where $J_0(x)$ is the zeroth order **Bessel function.**

Exercises

1. Show that if $f(x)$ is piecewise continuous over (a, b), then

$$\lim_{\xi \to \infty} \int_a^b f(x) \sin \xi x dx = 0.$$

Solution: If f has a continuous derivative, this is easily proved; we integrate by parts to obtain

$$\int_a^b f(x) \cos \xi x dx = \left[f(x) \frac{\sin \xi x}{\xi} \right]_a^b - \frac{1}{\xi} \int_a^b f'(x) \sin \xi x dx,$$

which tends to zero as $\xi \to \infty$ since the integral on the right-hand side is bounded. If f is not integrable, let p be a continuously differentiable function such that $\int_a^b |f(x) - p(x)| dx < \varepsilon$. Then

$$\left| \int_a^b [f(x) - p(x)] \cos \xi x dx \right| \le \int_a^b |f(x) - p(x)| \, |\cos \xi x| dx$$
$$\le \int_a^b |f(x) - p(x)| dx < \varepsilon$$

independently of ξ, and as the preceding discussion gave us $\int_a^b p(x) \cos \xi x dx \to 0$, it follows that $\int_a^b f(x) \cos \xi x dx \to 0$ as well. The proof that $\int_a^b p(x) \sin \xi x dx \to 0$ is similar. ♣

2. Show that $\displaystyle\int_0^\infty \frac{\sin x}{x}dx = \frac{\pi}{2}$.

Solution: If we substitute $\lambda = 2\pi$, $x = \pi$ into (11.60) and note that the integrand is an odd function, it follows that

$$\int_0^\pi \frac{\sin\left(\frac{2n+1}{2}u\right)}{\sin(u/2)}du = \pi. \qquad (12.17)$$

Applying the result of Exercise **1** noted above to the function $[(2/u) - 1/\sin(u/2)]$ (which is bounded in $0 < u < \pi$), we have

$$\lim_{n\to\infty}\int_0^\pi \sin\left(\frac{2n+1}{2}u\right)\left\{\frac{2}{u} - \frac{1}{\sin(u/2)}\right\}du = 0. \qquad (12.18)$$

Summing (12.17) and (12.18), we obtain

$$\lim_{n\to\infty}\int_0^\pi \frac{2\sin\frac{2n+1}{2}u}{u}du = \pi.$$

Changing variables and letting $t = (2n+1)u/2$, we set

$$\lim_{n\to\infty}\int_0^{(2n+1)\pi/2} \frac{\sin t}{t}dt = \frac{\pi}{2}.$$

We already know that $\int_0^M (\sin t)/t\, dt$ tends to a limit as $M \to \infty$ which completes our proof. ♣

3. Show that

$$\lim_{A\to\infty}\int_0^b f(x)\frac{\sin Ax}{x}dx = \frac{\pi}{2}f(0+)\ forb > 0$$

whenever f is piecewise smooth.

Solution: Observe that

$$\begin{aligned}\int_0^b f(x)\frac{\sin Ax}{x}dx &= \int_0^b f(0+)\frac{\sin Ax}{x}dx + \int_0^b \frac{f(x) - f(0+)}{x}\sin Ax dx\\ &= f(0+)\int_0^{Ab}\frac{\sin u}{u}du + \int_0^b \frac{f(x) - f(0+)}{x}\sin Ax dx.\end{aligned}$$

From the result of Exercise **1**, the last integral tends to zero as $A \to \infty$, since the integrand is piecewise smooth in the interval $0 < x < b$. It also remains bounded in this interval since, as x tends to zero, $[f(x) - f(0+)]/x$ tends to $f'(0+)$. From Exercise **2**, the other integral tends to the desired value. ♣

12.2 Convolution and Correlations

12.2.1 Convolution Theorem

In the application of the Fourier transform, we often encounter a product such as $F(k)G(k)$, where each of two functions is the Fourier transform of a function $f(x)$ and $g(x)$, respectively. Here, we are interested in finding out how the inverse Fourier transform of the product denoted by

$$\mathcal{F}^{-1}\left[F(k)G(k)\right],$$

is related to the individual inverse function

$$\mathcal{F}^{-1}[F(k)] = f(x) \quad \text{and} \quad \mathcal{F}^{-1}[G(k)] = g(x).$$

To begin with, we introduce a key concept called **convolution** and then state an important theorem that plays a central role in the discussion of the matter.

♠ **Convolution:**
The convolution of the function $f(x)$ and $g(x)$, denoted by $f * g$, is defined by

$$f * g = \frac{1}{\sqrt{2\pi}} \int_{-\infty}^{\infty} f(u)g(x-u)du. \tag{12.19}$$

The convolution obeys the **commutative**, **associative**, and **distributive** laws of algebra, i.e., if we have function f_1, f_2, f_3, then

$$\begin{aligned} &f_1 * f_2 = f_2 * f_1 && \text{(Commutative).} \\ &f_1 * (f_2 * f_3) = (f_1 * f_2) * f_3 && \text{(Associative).} \\ &f_1 * (f_2 + f_3) = (f_1 * f_2) + (f_1 * f_3) && \text{(Distributive).} \end{aligned} \tag{12.20}$$

We are now ready to prove the following important theorem regarding the product $F(k)G(k)$ of two Fourier transforms.

♠ **Convolution theorem:**
If $F(k)$ and $G(k)$ are Fourier transforms of $f(x)$ and $g(x)$, respectively, then

$$F(k)G(k) = \mathcal{F}[f * g]. \tag{12.21}$$

Proof It follows from the definition of the Fourier transform that

$$F(k) = \frac{1}{\sqrt{2\pi}} \int_{-\infty}^{\infty} f(x) e^{-ikx} dx,$$

$$G(k) = \frac{1}{\sqrt{2\pi}} \int_{-\infty}^{\infty} g(x) e^{-ikx} dx,$$

which yields

$$F(k)G(k) = \frac{1}{2\pi} \int_{-\infty}^{\infty} \int_{-\infty}^{\infty} f(x) g(x') e^{-ik(x+x')} dx dx'. \tag{12.22}$$

Let $x+x' = u$ in the double integral of (12.22) transform independent variables from (x, x') to (x, u). We thus have

$$dxdx' = \frac{\partial(x, x')}{\partial(x, u)} dudx,$$

where the Jacobian of the transformation is

$$\frac{\partial(x, x')}{\partial(x, u)} = \begin{vmatrix} \frac{\partial x}{\partial x} & \frac{\partial x}{\partial u} \\ \frac{\partial x'}{\partial x} & \frac{\partial x'}{\partial u} \end{vmatrix} = \begin{vmatrix} 1 & 0 \\ 0 & 1 \end{vmatrix} = 1.$$

Then (12.22) becomes

$$\begin{aligned} F(k)G(k) &= \frac{1}{2\pi} \int_{-\infty}^{\infty} \int_{-\infty}^{\infty} f(x) g(u-x) e^{-iku} dx du \\ &= \frac{1}{\sqrt{2\pi}} \int_{-\infty}^{\infty} e^{-iku} \left\{ \frac{1}{\sqrt{2\pi}} \int_{-\infty}^{\infty} f(x) g(u-x) dx \right\} du \\ &= \mathcal{F}[f * g]. \ \clubsuit \end{aligned} \tag{12.23}$$

12.2.2 Cross-Correlation Functions

There are several important functions related to the convolution, which are called **correlation functions** (see below) and **auto-correlation functions** (see Sect. 12.2.3).

♠ **Cross-correlation function:**
The cross-correlation of two functions f and g is defined by

$$c(z) = \frac{1}{\sqrt{2\pi}} \int_{-\infty}^{\infty} f^*(x) g(x+z) dx. \tag{12.24}$$

Despite the apparent similarity between the cross-correlation function (12.24) and the definition of convolution (12.19), their uses and interpretations are very different: the cross-correlation provides a quantitative measure of the similarity of two functions f and g since one is displaced through a distance z relative to the other.

Remark. Similar to the convolution, the cross-correlation is both **associative** and **distributive**. Unlike the convolution, however, it is not **commutative**.

We arrive at an important theorem by considering the Fourier transform of (12.24):

♠ **Wiener–Kinchin theorem:**

The Fourier transform of the cross-correlation of f and g is equal to the product of $F^*(k)$ and $G(k)$ multiplied by $\sqrt{2\pi}$, i.e.,

$$\mathcal{F}[c(z)] \equiv C(k) = F^*(k)G(k). \tag{12.25}$$

Proof

$$\begin{aligned}\mathcal{F}[c(x)] \equiv C(k) &= \frac{1}{\sqrt{2\pi}}\int_{-\infty}^{\infty} dz e^{-ikz}\left\{\frac{1}{\sqrt{2\pi}}\int_{-\infty}^{\infty} f^*(x)g(z+x)dx\right\} \\ &= \frac{1}{\sqrt{2\pi}}\int_{-\infty}^{\infty} dx f^*(x)\left\{\frac{1}{\sqrt{2\pi}}\int_{-\infty}^{\infty} g(z+x)e^{-ikz}dz\right\}.\end{aligned}$$

Making the substitution $u = z + x$ in the second integral, we obtain

$$\begin{aligned}C(k) &= \frac{1}{\sqrt{2\pi}}\int_{-\infty}^{\infty} dx f^*(x)\left\{\frac{1}{\sqrt{2\pi}}\int_{-\infty}^{\infty} g(u)e^{-ik(u-x)}du\right\} \\ &= \left\{\frac{1}{\sqrt{2\pi}}\int_{-\infty}^{\infty} f^*(x)e^{ikx}dx\right\}\left\{\frac{1}{\sqrt{2\pi}}\int_{-\infty}^{\infty} g(u)e^{-iku}du\right\} \\ &= F^*(k)G(k). \quad \clubsuit\end{aligned} \tag{12.26}$$

It readily follows from the definition (12.24) and the theorem (12.25) that

$$c(z) = \frac{1}{\sqrt{2\pi}}\int_{-\infty}^{\infty} C(k)e^{ikz}dx = \int_{-\infty}^{\infty} F^*(k)G(k)e^{ikz}dk. \tag{12.27}$$

Then, setting $z = 0$ gives us the multiplication theorem

$$\int_{-\infty}^{\infty} f^*(x)g(x)dx = \int_{-\infty}^{\infty} F^*(k)G(k)dk. \tag{12.28}$$

Further, by letting $g = f$, we arrive at the following identity:

♠ **Plancherel identity:**
A function $f(x)$ and its Fourier transform $F(k)$ are related to one another by the identity

$$\int_{-\infty}^{\infty} |f(x)|^2 dx = \int_{-\infty}^{\infty} |F(k)|^2 dk, \tag{12.29}$$

which is called the **Plancherel identity.**

Plancherel's identity is sometimes called Parseval's identity, aims to the analogy with Fourier series.

12.2.3 Autocorrelation Functions

Particularly when $g(x) = f(x)$, the cross-correlation function $c(z)$ is referred to specifically as follows:

♠ **Autocorrelation function:**
The autocorrelation function of $f(x)$ is defined by

$$a(z) = \frac{1}{\sqrt{2\pi}} \int_{-\infty}^{\infty} f^*(x) f(x+z) dx.$$

Using the Wiener–Kinchin theorem (12.26), we see that

$$\begin{aligned} a(z) &= \frac{1}{\sqrt{2\pi}} \int_{-\infty}^{\infty} A(k) e^{ikx} dk = \frac{1}{\sqrt{2\pi}} \int_{-\infty}^{\infty} \sqrt{2\pi} F^*(k) F(k) e^{ikx} dk \\ &= \frac{1}{\sqrt{2\pi}} \int_{-\infty}^{\infty} |F(k)|^2 e^{ikx} dk. \end{aligned}$$

This implies that the quantity $|F(k)|^2$, called the **power spectrum** of $f(x)$, is the Fourier transform of the autocorrelation function as formally stated below.

♠ **Power spectrum:**
Given $f(x)$, we have

$$|F(k)|^2 = \frac{1}{\sqrt{2\pi}} \int_{-\infty}^{\infty} a(z) e^{-ikx} dx,$$

where $F(k)$ and $a(z)$ are, respectively, the Fourier transform and the autocorrelation function of $f(x)$.

This result is frequently made use of in practical applications of Fourier transforms.

12.3 Discrete Fourier Transform

12.3.1 Definitions

The present section includes several topics associated with numerical computation of Fourier transforms. Generally, in computational work, we do not treat a continuous function $f(t)$, but rather $f(t_n)$ given by a discrete set of t_n's. (For now, we assume that a physical process of interest is described in the time domain.) In most common situations, the value of $f(t)$ is recorded at evenly spaced intervals. In this context, we have to estimate the Fourier transform of a function from a finite number of its sampled points.

Suppose that we have a set of measurements performed at equal time intervals of Δ. Then the sequence of sampled values is given by

$$f_k = f(t_k), \quad t_k = k\Delta \qquad (k = 0, 1, 2, \cdots, N-1). \tag{12.30}$$

For simplicity, we assume that N is even. With N numbers of input, we can produce at most N independent numbers of output. So, instead of trying to estimate the Fourier transform $F(\omega)$ in the whole range of frequency ω, we seek estimates only at the discrete values $\omega = \omega_n$ with $n = 0, 1, \cdots, N-1$. By analogy with the Fourier transform for a continuous function $f(t)$, we may define the Fourier transform for a discrete set of $f_k = f(t_k)$ $(k = 0, 1, \cdots N-1)$ as below.

♠ **Discrete Fourier transform:**

The discrete Fourier transform for a discrete set of f_k given by (12.30) is defined by

$$F_n = F(\omega_n) = \frac{1}{N}\sum_{k=0}^{N-1} f(t_k)e^{-i\omega_n t_k} = \frac{1}{N}\sum_{k=0}^{N-1} f_k e^{-2\pi i k n/N}, \tag{12.31}$$

with the definition

$$\omega_n \equiv \frac{2\pi n}{N\Delta} \qquad (n = 0, 1, \cdots, N-1). \tag{12.32}$$

Note that F_n is associated with frequency ω_n. Of importance is the fact that in (12.31), n can be any integer from $-\infty$ to ∞, whereas k in (12.31) runs from 0 to $N-1$. The latter restriction is due to the fact that F_n is periodic with a period of N terms. In fact, for any integer n such that $0 \le n \le N-1$, we have

$$F_n = F_{n\pm N} = F_{n\pm 2N} = \cdots$$

as readily follows from (12.31).

12.3.2 Inverse Transform

Given the discrete transform F_n, we can reproduce the time series f_k with the aid of the inverse relationship:

♠ **Inverse of discrete Fourier transform:**
The discrete Fourier transform of a set $\{f_k\}$ satisfies the relation

$$f_k = \sum_{n=0}^{N-1} F_n e^{2\pi ikn/N}. \tag{12.33}$$

Proof For the proof, it suffices to observe that

$$\sum_{n=0}^{N-1} e^{-2\pi in(k-k')/N} = \begin{cases} N & (k=k'), \\ 0 & \text{(otherwise)}. \end{cases} \tag{12.34}$$

(see Exercise **1** in Sect. 12.3). Then, from (12.31) and (12.34), we have

$$\begin{aligned} \sum_{n=0}^{N-1} F_n e^{2\pi ink'/N} &= \frac{1}{N} \sum_{n=0}^{N-1} \sum_{k=0}^{N-1} f_k e^{-2\pi in(k-k')/N} \\ &= \frac{1}{N} \sum_{k=0}^{N-1} f_k \cdot N\delta_{kk'} \quad = f_{k'}. \quad ♣ \end{aligned}$$

Note that the only differences between expressions (12.31) and (12.33) for F_n and f_k, respectively, are (i) changing the sign in the exponential, and (ii) dividing the answer by N. This means that a computational procedure for calculating discrete Fourier transforms can, with slight modifications, also be used to calculate the inverse transform. In addition, we see from the inverse transform that only N values of the frequency ω_n are needed and that they range from 0 to $N-1$, just as with the discrete time t_k.

12.3.3 Nyquest Frequency and Aliasing

In the above discussion, we have taken the view that the index n in (12.31) varies from 0 to N. In this convention, n in F_n and k in f_k vary over exactly the same range, so the mapping of N numbers into N numbers is manifest. Alternatively, since the quantity F_n given in (12.31) is periodic in n with period N (i.e., $F_n = F_{N+n}$), n in F_n is allowed to vary from $-N/2$ to $(N/2)-1$. In the latter convention, the discrete Fourier transform and its inverse transform read, respectively,

$$F_n = \sum_{k=-N/2}^{N/2-1} f_k e^{-2\pi ikn/N} \quad \text{and} \quad f_k = \frac{1}{N}\sum_{n=-N/2}^{N/2-1} F_n e^{2\pi ikn/N}. \tag{12.35}$$

Emphasis is placed on the fact that in (12.35), the upper bound of the summation is not $N/2$ but $(N/2)-1$. This ensures the count of ω_n to N. Indeed, the periodicity of F_n in n with the period N implies that the descretized frequency $\omega_n = 2\pi n/(N\Delta)$ is also periodic in n with N. Hence, the two extreme values of ω_n, i.e.,

$$\omega_{-N/2} = -\frac{\pi}{\Delta} \text{ and } \omega_{N/2} = \frac{\pi}{\Delta},$$

contribute to F_n as given in (12.31) in the same way. These two indistinguishable frequencies are known as the **Nyquist critical frequencies**.

♠ **Nyquist critical frequency:**
A Nyquist critical frequency is defined by

$$\omega_c \equiv \frac{\pi}{\Delta},$$

where Δ is the sampling interval: $t_k = k\Delta$ $(k = 0, 1, \cdots, N-1)$.

The Nyquist critical frequency has the following peculiarity. Suppose that we sample a sine wave of the Nyquist critical frequency, expressed by

$$f(t) = \sin(\omega_c t),$$

at the sampling interval Δ. Then we have

$$f_k = f(t_k) = \sin(\omega_c t_k + \theta) = \sin\left[\frac{\pi}{\Delta}(k\Delta + \theta)\right] = \sin(k\pi + \theta)$$
$$(k = 0, 1, \cdots, N-1),$$

where θ is determined by the initial condition: $f(0) = \sin\theta$. Then, the sampling becomes two sample points per cycle: $\sin\theta$ and $-\sin\theta$.

The above arguments further suggest that descretized frequencies ω_n above (and below) ω_c are identified with ω_{n-N} (and ω_{n+N}). This phenomenon, peculiar to discrete sampling, leads to the following important consequence:

♠ **Aliasing:**
When a continuous function $f(t)$ is sampled with an interval Δ, all of the power spectral density lying outside of the range $[-\omega_c, \omega_c)$ with $\omega_c = \pi/\Delta$ is moved into that range. Owing to a phenomenon called **aliasing.**

Through discrete sampling, therefore, any frequency component outside of the range $[-\omega_c, \omega_c)$ is falsely translated into that range.

Example Suppose that two continuous waves $\exp(i\omega_1 t)$ and $\exp(i\omega_2 t)$ are sampled with the same interval Δ. Then, if $\omega_2 = \omega_1 \pm 2\omega_c$, we obtain the same samples, since

$$\begin{aligned}\exp(i\omega_2 t_k) &= \exp(i\omega_1 t_k) \times \exp(\pm 2i\omega_c t_k) \\ &= \exp(i\omega_1 t_k) \times \exp(\pm 2k\pi i) \quad = \exp(i\omega_1 t_k),\end{aligned}$$

where $t_k = k\Delta \quad (k = 0, 1, \cdots, N-1)$. Hence, a sinusoidal wave having a frequency lying outside the range $[-\omega_c, \omega_c)$ appears the same as the sinusoidal wave whose frequency is within the range.

Remark. The way to overcome aliasing is to (i) know the natural bandwidth limit of the signal – or else enforce a known limit by analog filtering of the continuous signal, and then (ii) sample at a rate sufficiently rapid to give at least two points per cycle of the highest frequency present.

12.3.4 Sampling Theorem

We present below a famous theorem that is useful in certain applications of the discrete Fourier transform.

♠ **Sampling theorem:**
Suppose that a continuous function $f(t)$ is sampled at an interval Δ as $f_k = f(k\Delta)$. If its Fourier transform satisfies the condition that $F(\omega) = 0$ for all $|\omega| \geq \omega_c = \pi/\Delta$, then we have

$$f(t) = \sum_{k=-\infty}^{\infty} f_k \frac{\sin\left[\omega_c(t - k\Delta)\right]}{\pi(t - k\Delta)}.$$

This theorem states that if a signal $f(t)$ that is in question is bandwidth-limited (i.e., $F(\omega) = 0$ for $|\omega| \geq |\omega_0|$) with a certain preassigned frequency ω_0, then the entire information content of the signal can be recorded by sampling it at the interval $\Delta = \pi/\omega_0$.

Proof Given a continuous function $f(t)$, we express it by the inverse Fourier transform as

$$f(t) = \frac{1}{\sqrt{2\pi}} \int_{-\infty}^{\infty} F(\omega) e^{i\omega t} d\omega.$$

From hypothesis, $F(\omega)$ vanishes at $\omega \geq |\omega_c|$ so that

$$f(t) = \frac{1}{\sqrt{2\pi}} \int_{-\omega_c}^{\omega_c} F(\omega) e^{i\omega t} d\omega,$$

which yields

$$f(t_k) = \frac{1}{\sqrt{2\pi}} \int_{-\omega_c}^{\omega_c} F(\omega) e^{i\omega t_k} d\omega \quad \text{for } t_k = k\Delta \ (k \in \boldsymbol{Z}).$$

Consider the Fourier series expansion of $F(\omega)$ as

$$F(\omega) = \sum_{k=-\infty}^{\infty} c_k e^{-i\omega t_k} \quad \text{for } |\omega| \leq \omega_c, \tag{12.36}$$

where the coefficients c_k read

$$c_k = \frac{1}{\sqrt{2\pi}} \int_{-\omega_c}^{\omega_c} F(\omega) e^{i\omega t_k} d\omega = f(t_k). \tag{12.37}$$

From (12.36) and (12.37), we obtain

$$F(\omega) = \sum_{k=-\infty}^{\infty} f(t_k) e^{-i\omega t_k} \quad \text{for } |\omega| \leq \omega_c.$$

Now we define

$$H(\omega) = \sum_{k=-\infty}^{\infty} f(t_k) e^{-i\omega t_k} \quad \text{for all } \omega.$$

While the function $H(\omega)$ is a periodic function with period $2\omega_c$, the $F(\omega)$ is identically zero outside the interval $[-\omega_c, \omega_c]$. This being so, we can write

$$F(\omega) = H(\omega) S(\omega) \quad \text{with } S(\omega) = \begin{cases} 1 & |\omega| \leq \omega_c, \\ 0 & |\omega| > \omega_c. \end{cases}$$

Thus we have

$$F(\omega) = \sum_{k=-\infty}^{\infty} f(t_k) e^{-i\omega t_k} S(\omega),$$

and its inverse transform reads

$$
\begin{aligned}
f(t) &= \frac{1}{\sqrt{2\pi}} \int_{-\infty}^{\infty} \left[\sum_{k=-\infty}^{\infty} f(t_k) e^{-i\omega t_k} S(\omega) \right] e^{i\omega t} d\omega \\
&= \sum_{k=-\infty}^{\infty} f(t_k) \frac{1}{\sqrt{2\pi}} \int_{-\omega_c}^{\omega_c} e^{i\omega(t-t_k)} d\omega \\
&= \sum_{k=-\infty}^{\infty} f(t_k) \frac{1}{\sqrt{2\pi}} \int_{-\infty}^{\infty} S(\omega) e^{i\omega(t-t_k)} d\omega \\
&= \sum_{k=-\infty}^{\infty} f(t_k) \frac{\sin[\omega_c(t-t_k)]}{\omega_c(t-t_k)}. \quad \clubsuit
\end{aligned}
$$

12.3.5 Fast Fourier Transform

The **fast Fourier transform** (often abbreviated by **FFT**) is an algorithm for calculating discrete Fourier transforms and is widely known as a useful tool in computational physics. In this subsection, we demonstrate the efficiency of this computational method.

In a typical discrete Fourier transform, one has a sum of N terms expressed by

$$
F_n = \sum_{k=0}^{N-1} W^{nk} f_k, \tag{12.38}
$$

where W is a complex number defined by

$$
W \equiv e^{2\pi i/N}.
$$

Notably, the left-hand side of (12.38) can be regarded as a product of the vector consisting of the elements $\{f_k\}$ with a matrix whose (n,k)th element is the constant W to the power $n \times k$. The matrix multiplication produces a vector whose components are the F_n's. This operation evidently requires N^2 complex-number multiplications plus a smaller number of operations to generate the required powers of W. Thus, the discrete Fourier transform appears to be an $O(N^2)$ process.

The efficiency of the fast Fourier transform manifests in the fact that it enables us to compare the discrete Fourier transform in $O(N \log_2 N)$ operations. The difference between $O(N^2)$ and $O(N \log_2 N)$ is immense. With $N = 10^8$, e.g., it is the difference between, roughly, 2 s and 3 months of CPU time on a gigahertz cycle computer.

The fast Fourier transform is based on the fact that a discrete Fourier transform of length N can be rewritten as the sum of two discrete Fourier transforms, each of length $N/2$. This is easily seen from (12.38) as follows:

$$
\begin{aligned}
F_n &= \sum_{k=0}^{N-1} e^{2\pi ikn/N} f_k \\
&= \sum_{k=0}^{N/2-1} e^{2\pi in(2k)/N} f_{2k} + \sum_{k=0}^{N/2-1} e^{2\pi in(2k+1)/N} f_{2k+1} \\
&= \sum_{k=0}^{N/2-1} e^{2\pi ink/(N/2)} f_{2k} + W^n \sum_{k=0}^{N/2-1} e^{2\pi ink/(N/2)} f_{2k+1} \\
&= F_n^e + W^n F_n^o. \qquad (12.39)
\end{aligned}
$$

Here W is the same complex constant we defined in (12.38). The F_n^e denotes the nth component of the Fourier transform of the sequence (f_{2k}) with length $N/2$ expressed by

$$(f_{2k}) = (f_0, f_2, f_4, \cdots, f_{N-2}),$$

which consists of even components of the original f_k's. Similarly, the F_n^o is the corresponding transform of length $N/2$ formed from odd components. Recall that F_n is periodic in n with the period N. On the other hand, the transforms F_n^e and F_n^o are periodic in k with length $N/2$. This period-reduction property is the origin of the efficiency of the fast Fourier transform as demonstrated below.

Having decomposed F_n into F_n^e and F_n^o, we can apply the same procedure to F_n^e and F_n^o to produce $N/4$ even-numbered and odd-numbered data:

$$
\begin{aligned}
F_n^e &= \sum_{k=0}^{N/4-1} e^{2\pi ink/(N/4)} f_{4k} + W^k \sum_{k=0}^{N/4-1} e^{2\pi ink/(N/4)} f_{4k+2} \\
&= F_n^{ee} + W^n F_n^{eo}, \qquad (12.40) \\
F_n^o &= \sum_{k=0}^{N/4-1} e^{2\pi ink/(N/4)} f_{4k+1} + W^n \sum_{k=0}^{N/4-1} e^{2\pi ink/(N/4)} f_{4k+3} \\
&= F_n^{oe} + W^n F_n^{oo}. \qquad (12.41)
\end{aligned}
$$

Here, the F_k^{eo}, e.g., is the transform of the sequence (f_{4k+2}) given by

$$(f_{4k+2}) = (f_2, f_6, \cdots, f_{N-2}),$$

whose length is $N/4$. We can continue the above procedure until we obtain the transform of a single-point sequence, say,

$$F_n^{eoeeoeo\cdots oee} = f_k \quad \text{for some } k. \qquad (12.42)$$

This implies that for every pattern of $\log_2 N$ e's and o's, there is a one-point transform that is just one of the input numbers f_k. Therefore, by relating all the terms f_k $(0 \le k \le N-1)$ to $\log_2 N$ patterns of e's and o's and then tracking back to the procedures (12.39), (12.40), (12.41), and (12.42) to reproduce F_n, we will successfully obtain the discrete Fourier transform F_n $(0 \le n \le N-1)$ of the original data f_k $(0 \le k \le N-1)$.

One may ask a question as to the way we can figure out which value of k corresponds to which pattern of e's and o's in (12.42). As we demonstrate later, this can be achieved by reversing the pattern of e's and o's and setting $e = 0$ and $o = 1$. Then, we have the corresponding value of k in a binary expression. This idea of **bit reversal** can be exploited in a very clever way that makes FFTs practical.

12.3.6 Matrix Representation of FFT Algorithm

To make our discussion more concrete, we now present an actual FFT procedure to obtain the discrete Fourier transform $F(n)$ $(n = 0, 1, 2, 3)$ of the original vector data $f(k)$ $(k = 0, 1, 2, 3)$. By definition, $F(n)$ is given in the matrix representation as

$$\begin{bmatrix} F(0) \\ F(1) \\ F(2) \\ F(3) \end{bmatrix} = \begin{bmatrix} W^0 & W^0 & W^0 & W^0 \\ W^0 & W^1 & W^2 & W^3 \\ W^0 & W^2 & W^4 & W^6 \\ W^0 & W^3 & W^6 & W^9 \end{bmatrix} \begin{bmatrix} f(0) \\ f(1) \\ f(2) \\ f(3) \end{bmatrix} = \begin{bmatrix} 1 & 1 & 1 & 1 \\ 1 & W^1 & W^2 & W^3 \\ 1 & W^2 & W^0 & W^2 \\ 1 & W^3 & W^2 & W^1 \end{bmatrix} \begin{bmatrix} f(0) \\ f(1) \\ f(2) \\ f(3) \end{bmatrix}, \tag{12.43}$$

where we used the fact that

$$W^4 = \left(e^{2\pi i/4}\right)^4 = e^{2\pi i} = 1;$$

More generally, we have

$$W^{nk} = W^{nk \bmod (N)},$$

where the number

$$nk \bmod (N)$$

is the remainder when the integer nk is divided by N. The trick involved in the FFT algorithm is to decompose the product of the vector and the matrix appearing in (12.43) into that of a vector and two matrices:

$$\begin{bmatrix} F(0) \\ F(2) \\ F(1) \\ F(3) \end{bmatrix} = \begin{bmatrix} 1 & W^0 & 0 & 0 \\ 1 & W^2 & 0 & 0 \\ 0 & 0 & 1 & W^1 \\ 0 & 0 & 1 & W^3 \end{bmatrix} \begin{bmatrix} 1 & 0 & W^0 & 0 \\ 0 & 1 & 0 & W^0 \\ 1 & 0 & W^2 & 0 \\ 0 & 1 & 0 & W^1 \end{bmatrix} \begin{bmatrix} f(0) \\ f(1) \\ f(2) \\ f(3) \end{bmatrix}. \tag{12.44}$$

The equivalence between (12.43) and (12.44) is verified in a straightforward manner. Nevertheless, the reader should pay attention to the fact that in (12.44), the order of elements in the vector $F(n)$ is altered from that in the original form (12.43). As we demonstrate later, this altering property of the order of $F(n)$ enables us to compute efficiently the $F(n)$ from $f(k)$ with the help of the **bit-reversing process**.

The efficiency of FFT can be observed by counting up the number of multiplication (and additions) between matrix elements in order to complete the matrix operation given in (12.44). First we set

$$\begin{bmatrix} f_1(0) \\ f_1(1) \\ f_1(2) \\ f_1(3) \end{bmatrix} = \begin{bmatrix} 1 & 0 & W^0 & 0 \\ 0 & 1 & 0 & W^0 \\ 1 & 0 & W^2 & 0 \\ 0 & 1 & 0 & W^1 \end{bmatrix} \begin{bmatrix} f_0(0) \\ f_0(1) \\ f_0(2) \\ f_0(3) \end{bmatrix},$$

in which $f_0(k) = f(k)$ $(k = 0, 1, 2, 3)$. Then $f_1(0)$ is obtained through one complex-number multiplication and one complex-number addition, i.e.,

$$f_1(0) = f_0(0) + W^0 f_0(2). \tag{12.45}$$

We can obtain $f_1(1)$ in the same manner as above. On the contrary, to obtain $f_1(2)$, only one complex-number addition is needed due to the relation $W^2 = -W^0$. In fact,

$$f_1(2) = f_0(0) + W^2 f_0(2) \quad = f_0(0) - W^0 f_0(2),$$

in which the product $W^0 f_0(2)$ was evaluated earlier in the calculation of (12.45). Likewise, $f_1(3)$ is also computed by only one addition owing to the relation $W^3 = -W^1$. As a consequence, the vector $f_1(k)$ $(k = 0, 1, 2, 3)$ is calculated through four-times additions and two-times multiplications.

A similar scenario can apply to the remaining computation:

$$\begin{bmatrix} F(0) \\ F(2) \\ F(1) \\ F(3) \end{bmatrix} = \begin{bmatrix} f_2(0) \\ f_2(1) \\ f_2(2) \\ f_2(3) \end{bmatrix} = \begin{bmatrix} 1 & W^0 & 0 & 0 \\ 1 & W^2 & 0 & 0 \\ 0 & 0 & 1 & W^1 \\ 0 & 0 & 1 & W^3 \end{bmatrix} \begin{bmatrix} f_1(0) \\ f_1(1) \\ f_1(2) \\ f_1(3) \end{bmatrix}.$$

Calculation of each number $f_2(0)$ and $f_2(2)$ requires both one addition and one multiplication, whereas for $f_2(1)$ and $f_2(3)$ only one addition is required because of the relations $W^2 = -W^0$ and $W^3 = -W^1$. Therefore, the entire computation to yield $F(n)$ in the above context requires *four-time multiplications* and *eight-time additions*. This computational cost is significantly small compared with the direct matrix calculation given in (12.43), where *16-times multiplications* and *12-times additions* are needed. More generally, when considering the transform $F(k)$ of the length $N = 2^\gamma$, the FFT procedure requires the multiplications of $N\gamma/2$ times and the additions of $N\gamma$ times, whereas the direct matrix calculation procedure demands N^2-times multiplications and $N(N-1)$-times additions. Thus the superiority of FFT method is considerably enhanced when $N >> 1$.

12.3.7 Decomposition Method for FFT

It is still unclear as to how we can find an appropriate decomposition of general $N \times N$ matrices as performed in (12.44). To see this, we express the indices n and k in terms of two-digit expressions:

$$n = 2n_1 + n_0, \quad k = 2k_1 + k_0,$$

where each n_1, n_0, k_1, k_0 takes the value 0 or 1 [e.g., $n = 3$ corresponds to $(n_0, n_1) = (1, 0)$]. Then, the discrete Fourier transform reads

$$F(n_1, n_0) = \sum_{k_0=0}^{1} \sum_{k_1=0}^{1} f_0(k_1, k_0) W^{(2n_1+n_0)(2k_1+k_0)}.$$

Now we apply the identity

$$W^{(2n_1+n_0)(2k_1+k_0)} = W^{4n_1k_1} W^{2n_0k_1} W^{(2n_1+n_0)k_0} = W^{2n_0k_1} W^{(2n_1+n_0)k_0}$$

to obtain

$$F(n_1, n_0) = \sum_{k_0=0}^{1} \left[\sum_{k_1=0}^{1} f_0(k_1, k_0) W^{2n_0k_1} \right] W^{(2n_1+n_0)k_0}. \tag{12.46}$$

Denoting the sum in the square bracket by $f_1(n_0, k_0)$, we have

$$f_1(n_0, k_0) = \sum_{k_1=0}^{1} f_0(k_1, k_0) W^{2n_0k_1}, \tag{12.47}$$

or equivalently,

$$\begin{aligned} f_1(0,0) &= f_0(0,0) + f_0(1,0)W^0, \\ f_1(0,1) &= f_0(0,1) + f_0(1,1)W^0, \\ f_1(1,0) &= f_0(0,0) + f_0(1,0)W^2, \\ f_1(1,1) &= f_0(0,1) + f_0(1,1)W^2. \end{aligned}$$

This system of equations is expressed in matrix form as

$$\begin{bmatrix} f_1(0,0) \\ f_1(0,1) \\ f_1(1,0) \\ f_1(1,1) \end{bmatrix} = \begin{bmatrix} 1 & 0 & W^0 & 0 \\ 0 & 1 & 0 & W^0 \\ 1 & 0 & W^2 & 0 \\ 0 & 1 & 0 & W^1 \end{bmatrix} \begin{bmatrix} f_0(0,0) \\ f_0(0,1) \\ f_0(1,0) \\ f_0(1,1) \end{bmatrix}.$$

Similarly, from (12.46) and (12.47), it follows that

$$\begin{bmatrix} f_2(0,0) \\ f_2(0,1) \\ f_2(1,0) \\ f_2(1,1) \end{bmatrix} = \begin{bmatrix} 1 & W^0 & 0 & 0 \\ 1 & W^2 & 0 & 0 \\ 0 & 0 & 1 & W^1 \\ 0 & 0 & 1 & W^3 \end{bmatrix} \begin{bmatrix} f_1(0,0) \\ f_1(0,1) \\ f_1(1,0) \\ f_1(1,1) \end{bmatrix}.$$

Hence, we have

$$F(n_1, n_0) = f_2(n_0, n_1),$$

in which the order of n_0 and n_1 in the parentheses differs on the two sides. This indicates that the individual numbers $f_2(n_0, n_1)$ are in order not of $n = 2n_1 + n_0$, but of the numbers obtained by bit-reversing n, which is why the **bit-reversing process** is required to obtain the discrete Fourier transform $F(n)$ using FFT, The above discussion also clearly demonstrates the way to construct the decomposed product of matrices that makes the entire computations a fast.

Exercises

1. Show that

$$\sum_{n=0}^{N-1} e^{-2\pi i n(k-k')/N} = \begin{cases} N & \text{if k} = \text{k}', \\ 0 & \text{otherwise}, \end{cases}$$

where k and k' are integers ranging from 0 to $N - 1$.

Solution: The proof for the case of $k = k'$ is trivial. When $k = k'$, then

$$e^{-2\pi i n(k-k')/N} \neq 1 \quad \text{and} \quad e^{-2\pi i n(k-k')} = 1$$

for any choice of k and k', so that we have

$$\sum_{n=0}^{N-1} e^{-2\pi i n(k-k')/N} = \frac{1 - e^{-2\pi i n(k-k')}}{1 - e^{-2\pi i n(k-k')/N}} = 0. \quad ♣$$

12.4 Applications in Physics and Engineering

12.4.1 Fraunhofer Diffraction I

In optics, Fourier transformation is a powerful tool to describe an important class of wave diffractions, called **Fraunhofer diffraction**; this refers to the diffraction of electromagnetic radiation observed at a point far from a slit or an aperture. A Fraunhofer diffraction pattern can be described by using the wave theory of light, which predicts the areas of constructive and destructive interference.

Let us derive the diffraction pattern produced by a rectangular aperture with width a and height b. We assume that both incident and diffracted waves can be approximated as being plain waves with wavelength λ. In order to make this assumption, the diffracting obstacle and the observation point must be sufficiently far from the light source so that the curvature of the incident and diffracted light can be neglected (see Fig. 12.1). According to elementary wave optics, the amplitude of light at $\boldsymbol{R}$ on the screen is given by

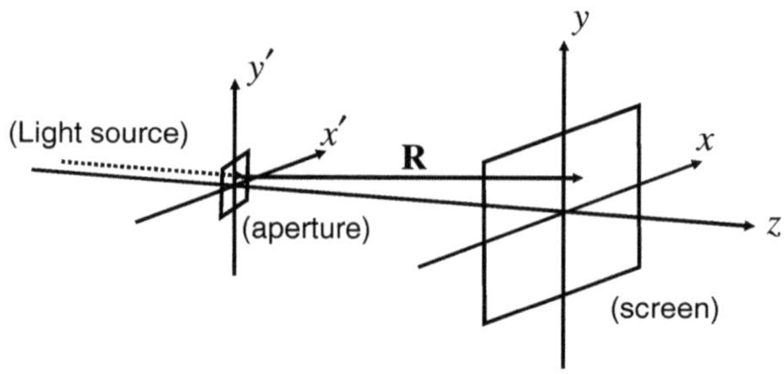

Fig. 12.1. Configurations of the light source, a recrangular aperture, and the screen

$$u(\boldsymbol{R}) = -\frac{ik}{2\pi R} \int_{\Delta S'} u(\boldsymbol{r}') e^{ik|\boldsymbol{R}-\boldsymbol{r}'|} d\boldsymbol{r}'.$$

Here, $k = 2\pi/\lambda$, $\Delta S'$ represents the area of the rectangular aperture through which light passes and $u(\boldsymbol{r}')$ is the amplitude of the incident wave at $\boldsymbol{r}'$ within the aperture:

$$u(\boldsymbol{r}') = Ae^{i\boldsymbol{k}\cdot\boldsymbol{r}'}.$$

We assume that this incident wave is oriented in the direction of the z-axis. Then, the wave vector $\boldsymbol{k}$ is perpendicular to the position vector $\boldsymbol{r}'$ so that

$$u(\boldsymbol{r}') = A = \text{const.}$$

Hence, we have

$$u(\boldsymbol{R}) = -\frac{ikA}{2\pi R} \int_{-a}^{a} dx' \int_{-b}^{b} dy' e^{ik|\boldsymbol{R}-\boldsymbol{r}'|}. \qquad (12.48)$$

Set $\boldsymbol{R} = (x, y, z)$ and $\boldsymbol{r}' = (x', y', 0)$, where the origin is located at the center of the aperture. Under the assumption that $z \gg |x|, |y|$ and $|x|, |y| \gg |x'|, |y'|$, we have

$$
\begin{aligned}
|\boldsymbol{R}-\boldsymbol{r}'| &= \sqrt{z^2+(x-x')^2+(y-y')^2} \\
&= \sqrt{R^2-2(xx'+yy')+x'^2+y'^2} \\
&\simeq R\left(1-\frac{xx'+yy'}{R^2}+\frac{x'^2+y'^2}{2R^2}\right) \simeq R\left(1-\frac{xx'+yy'}{R^2}\right).
\end{aligned}
$$

Substituting this into (12.48) yields

$$
\begin{aligned}
u(\boldsymbol{R}) &= -\frac{ikA}{2\pi R}e^{ikR}\int_{-a}^{a} e^{-\frac{ikx}{R}x'}dx' \int_{-b}^{b} e^{-\frac{iky}{R}y'}dy' \\
&= -\frac{2ikA}{\pi R}e^{ikR}\,\frac{\sin\dfrac{kax}{R}}{\dfrac{kx}{R}}\,\frac{\sin\dfrac{kby}{R}}{\dfrac{ky}{R}}.
\end{aligned}
$$

The light intensity distribution $I(\boldsymbol{R})$ on the screen is thus given by

$$
I(\boldsymbol{R}) \equiv |u(\boldsymbol{R})|^2 \propto k^2a^2b^2\left[\frac{\sin(ka\tilde{x})}{ka\tilde{x}}\frac{\sin(kb\tilde{y})}{kb\tilde{y}}\right]^2,
$$

where

$$
\tilde{x}=\frac{x}{R},\quad \tilde{y}=\frac{y}{R}.
$$

Remember that $(\sin\xi)/\xi = 0$ at $\xi = \pm n\pi$ with integers $n = 1, 2, \cdots$. In addition, since $k = 2\pi/\lambda$, we conclude that

$$
I(\boldsymbol{R}) = 0 \quad \text{at } \tilde{x} = \pm\frac{m\lambda}{2a} \quad \text{or} \quad \tilde{y} = \pm\frac{n\lambda}{2b}, \quad (m, n = 1, 2, \cdots),
$$

which describes the diffraction pattern generated on the screen.

12.4.2 Fraunhofer Diffraction II

We next consider the case of a circular aperture with radius a. For convenience, we use the polar coordinates defined by $x = r\cos\theta$, $y = r\sin\theta$. Then (12.49) reads

$$
\begin{aligned}
u(\boldsymbol{R}) &\propto e^{ikR}\int_{\Delta S'} \exp\left[\frac{-ik(xx'+yy')}{R}\right]d\boldsymbol{r}' \\
&= \int_0^a dr' \int_0^{2\pi} d\theta' r' \exp\left[\frac{-ikrr'(\cos\theta\cos\theta'+\sin\theta\sin\theta')}{R}\right] \\
&= \int_0^a dr' \int_0^{2\pi} d\theta' r' \exp\left[\frac{-ikrr'\cos(\theta'-\theta)}{R}\right].
\end{aligned}
$$

To make it consise, we use the following formulae based on the **Bessel function** $J_n(x)$:

$$\int_0^{2\pi} e^{i\zeta\cos\phi} d\phi = 2\pi J_0(\zeta), \quad \int_0^{\eta} \zeta J_0(\zeta) = \eta J_1(\eta).$$

These give us

$$u(\boldsymbol{R}) \propto 2\pi a^2 \frac{J_1\left(\frac{kar}{R}\right)}{\frac{kar}{R}},$$

where the explicit form of $J_1(x)$ is obtained from the definition of $J_\nu(x)$,

$$J_\nu(x) = \left(\frac{x}{2}\right)^\nu \sum_{\ell=0}^{\infty} \frac{(-1)^\ell (x/2)^{2\ell}}{\ell! \Gamma(\nu+\ell+1)},$$

and thus $\lim_{x\to 0} J_1(x)/x = 1/2$. The first zero of $J_1(x)$ is located at $x \simeq 1.22\pi$. Therefore, the radius r_0 of the innermost dark ring on the screen is given by

$$\frac{kar_0}{R} \simeq 1.22\pi, \quad \text{i.e.,} \quad r_0 \simeq \frac{0.61\lambda}{a} R.$$

12.4.3 Amplitude Modulation Technique

We conclude this chapter with a discussion regarding the use of Fourier transformations in an **amplitude modulation** (AM) technique. This technique is used in electronic communication, most commonly for transmitting information via a radio carrier wave. As the name indicates, AM works by modulating the vibrational amplitude of the transmitted signal according to the information being sent. This is in contrast to the **frequency modulation** (FM) technique that is also commonly used for transmitting sound, but by modulating its frequency.

For AM, we use two kinds of waves: a **carrier wave** $c(t)$ and a **message wave** $m(t)$ that contains information on the message to be transmitted. For simplicity, the carrier wave is modeled here as a simple sine wave written as

$$c(t) = C \cdot \cos(\omega_c t + \phi_c),$$

where the radio frequency (in Hertz) is given by $\omega_c/(2\pi)$. C and ϕ_c are constants representing the carrier amplitude and the initial phase, respectively, and their values are set to 1 and 0. AM is then realized by determining the product:

$$y(t) = m(t) \cdot c(t),$$

whose Fourier transform $Y(\omega)$ is expressed as

$$\begin{aligned} Y(\omega) &= \mathcal{F}[m(t)c(t)] \\ &= \frac{1}{\sqrt{2\pi}} \int_{-\infty}^{\infty} m(t) e^{-i\omega t} \frac{e^{i\omega_c t} + e^{-i\omega_c t}}{2} dt \\ &= \frac{1}{2} \left[M(\omega+\omega_c) + M(\omega-\omega_c) \right]. \end{aligned} \tag{12.49}$$

Here $M(\omega)$ is the Fourier transform of $m(t)$.

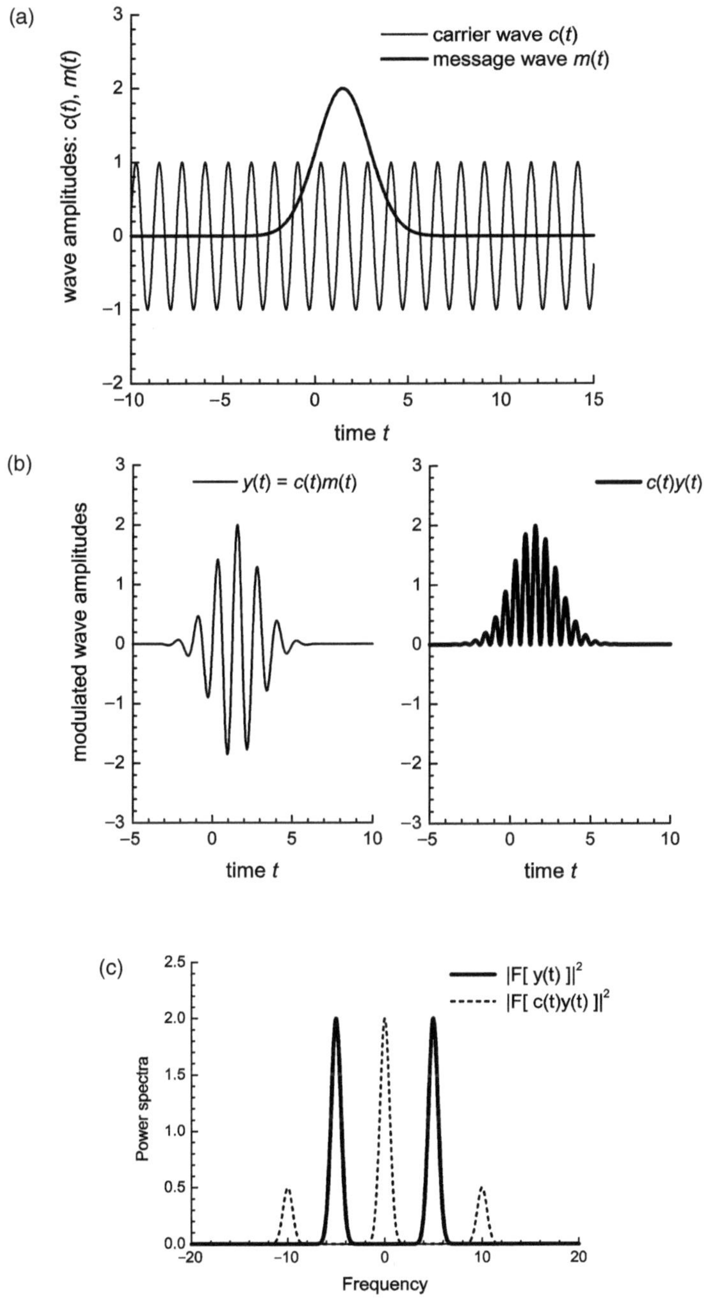

Fig. 12.2. *Top*: A carrier wave $c(t) = \sin(\omega_c t)$ with $\omega_c = 5.0$ and a message wave $m(t) = 2\exp[-(t - t_0)^2/4]$ with $t_0 = 1.5$. *Middle*: The products $c(t)m(t) \equiv y(t)$ and $c^2(t)m(t)$. *Bottom*: The power spectra $|\mathcal{F}[y(t)]|^2$ and $|\mathcal{F}[c(t)y(t)]|^2$

The result in (12.49) implies that the modulated signal $y(t)$ has two groups of components: one at positive frequencies (centered at $+\omega_c$) and one at negative frequencies (centered at $-\omega_c$). Figure 12.2 illustrates a carrier wave $c(t) = \sin(\omega_c t)$ with $\omega_c = 5.0$, a message wave $m(t) = 2\exp[-(t-t_0)^2/4]$ with $t_0 = 1.5$, and the power spectrum of $y(t) = c(t)m(t)$ [i.e., ω-dependence of $Y(\omega)$] described by (12.49), together with the associated message wave $m(t)$. The frequency shift from ω to $\omega \pm \omega_c$, which is clearly evident, facilitates the tuning of the frequency of the transmitted signal to the desired value. We are concerned only with positive frequencies. The negative ones are mathematical artifacts that carry no additional information.

In order to reproduce the original signal $m(t)$ from the modulated one $y(t)$, it is sufficient to multiply $c(t)$ by $y(t)$ and follow that with a filtering process. The Fourier transform of the product $c(t)y(t)$ is given as

$$\begin{aligned}\mathcal{F}[c(t)y(t)] &= \mathcal{F}[m(t)\cos^2(\omega_c t)] \\ &= \frac{M(\omega)}{2} + \frac{1}{4}\left[M(\omega + 2\omega_c) + M(\omega - 2\omega_c)\right].\end{aligned}$$

We pick up the first term in the last expression and take its inverse transform, thus obtaining $\mathcal{F}^{-1}[M(\omega)] = m(t)$.

13

Laplace Transformation

Abstract Using the Laplace transform for the mathematical description of a physical system considerably simplifies the analysis of its behavior Many useful applications and formulas related to Laplace transforms can be found in other textbooks, but here we focus on the theoretical background, particularly, on the convergence properties of the various forms of Laplace transforms. It is important to note that a Laplace transform exists only if the corresponding improper integral, known as the Laplace integral, converges. Hence, the convergence of the improper integral must be confirmed prior to discussing the Laplace transform of a given function. Thus we devote a portion of this chapter to an analysis of the conditions necessary for the convergence of Laplace integrals, in contrast to the standard literature that deals primarily with the practical applications of Laplace transforms.

13.1 Basic Operations

13.1.1 Definitions

The **Laplace transformation** associates a function $f(x)$ of a real variable x with a suitable function $F(s)$ of a complex variable s. This correspondence is essentially a reciprocal one-to-one and often allows us to replace a given complicated function by a simpler one. The advantage of this operation manifests particularly in applications to problems of **linear differential equations** (see Chap. 15). We shall see that the Laplace transformation allows us to reduce a linear differential equation of $f(x)$ to a certain simple algebraic equation of $F(s)$, which yields solutions of the original differential equations more readily than other techniques. Furthermore, it turns out that this reduction method can be extended to systems of differential equations (ordinary and partial) as well as to integral equations, which enhances the importance of studying and understanding the Laplace transform.

To begin with, we define the Laplace transformation operator L that maps a function $f(x)$ to a corresponding function $F(s)$:

♠ **Laplace transformation:**

The (one-sided) **Laplace transformation**, denoted by the operator L, is defined by

$$L[f(x)] = \int_0^\infty e^{-sx} f(x)dx = F(s), \tag{13.1}$$

which associates an image function $F(s)$ of the complex variable $s = \sigma + i\omega$ with a single-valued function $f(x)$ (x real) such that the integral (13.1) exists.

♠ **Laplace integral:**

The integral given in (13.1) is called the **Laplace integral.** If the Laplace integral exists for a given $f(x)$, the image function $F(s)$ is called the (one-sided) **Laplace transform** of $f(x)$.

It is important to keep in mind the difference between the Laplace *integral* and the Laplace *transform.* Namely, the Laplace transform exists only when the Laplace integral exists (i.e., converges). Convergence properties of Laplace integrals are determined by the value of s and the feature of the function $f(x)$, which is discussed fully in Sect. 13.3. In the meantime, we assume that $f(x)$ is a function that allows the Laplace integral to converge for certain s.

13.1.2 Several Remarks

Below are several important remarks regarding the properties of the Laplace transform (13.1).

1. The definition (13.1) states that for a given $F(s)$, there is at most one *continuous* function $f(x)$. Nevertheless, it does not determine a unique $f(x)$ because if $f(x)$ in (13.1) were altered at a finite number of isolated points, $F(s)$ would remain unchanged, as such discontinuous points make no contribution to the integral. For this reason, we assume in the remainder of this chapter that $f(x)$ is continuous except at isloated points.

2. In order for the integral (13.1) to exist, any discontinuity of the integrand inside the interval $(0, \infty)$ must be a finite jump so that there are right-hand and left-hand limits at those discontinuous points. An exception is a discontinuity at $x = 0$ (if it exists); for instance, the function $f(x) = 1/\sqrt{x}$ diverges at $t = 0$ but the integral (13.1) exists.

3. The **inverse Laplace transform** of $F(s)$ is a function $f(x)$ such that $L[f(x)] = F(s)$. Hence, the operation of taking an inverse Laplace transform is denoted by L^{-1} and we have

$$L^{-1}[F(s)] = f(x).$$

This expression implies the possibility of dealing with the operators L and L^{-1} algebraically, just as the equation $ax = y$ can be rewritten as $x = a^{-1}y$. At thus point, it is not clear as to how the inverse operation L^{-1} is to be performed, but actual manipulations are discussed in detail in Sect. 13.4.2.

4. Not every function $F(s)$ has an inverse Laplace transform. A sufficient condition for $F(s)$ to have its inverse transform is presented in Sect. 13.4.2.

13.1.3 Significance of Analytic Continuation

Observe that the Laplace integral (13.1) involves a complex-valued term e^{-sx} in its integrand, which makes it difficult to employ the standard methods of integration that are applicable to real integrands. One way to proceed would be to use the equation $e^{-sx} = e^{-\sigma x}\cos\omega x - ie^{-\sigma x}\sin\omega x$, which yields two real integrands. This is, however, more complicated than necessary. An easier method is to make use of the following theorem, which is verified in Sect. 13.3.7:

♠ **Analytic property of Laplace transform:**
The Laplace transform $F(s)$, which is a complex-valued function of a complex variable s, is an **analytic function** in a region of Re $(s) > \sigma_c$ with a specific real number σ_c.

Remark. Just at $\text{Re}(s) = \sigma_c$, however, no general conclusion can be drawn.

This theorem states that once the value of $F(\sigma)$ on the real axis is known, $F(s)$ on an arbitrary point of the complex plane can be obtained by simply replacing σ by s. This replacement is based on an **analytic continuation** from the semi-infinite line of the real axis, $\sigma > \sigma_c$, to the right half of the s-plane, Re $(s) > \sigma_c$, which is why we can perform the integration (13.1) as if s were a real variable. Several examples given later clearly show the efficacy of identifying s as a real parameter.

At first glance, the formality of replacing σ by s amounts simply to a change in symbol. But, without analytic continuation, we could no longer regard our replacement from σ to s as a mere formality; i.e., the concept of analytic continuation lurks in the background.

Remark. In particular, those cases in which $F(s)$ becomes multivalued cannot be treated without paying heed in detail to the difference between σ and s. The latter issue regarding multivalued $F(s)$ is discussed in Sect. 13.2.5.

13.1.4 Convergence of Laplace Integrals

Emphasis is placed on the fact that the Laplace integral (13.1) may or may not exist depending on the value of s as well as the nature of $f(x)$. A sufficient condition for the Laplace integral to converge is that the real component of s, $\mathrm{Re}(s)$, is greater than a specific value. This intuitively follows from the definition (13.1) that says if the integral (13.1) exists for

$$s_0 = \sigma_0 + i\omega_0,$$

then the integral also exists for every s such that $\mathrm{Re}\,(s) > \sigma_0$, since in the latter case

$$\left|e^{-sx}\right| < \left|e^{-s_0 x}\right| = e^{-\sigma_0 x}.$$

This is stated rigorously in the theorem below.

♠ **Convergence of Laplace integrals:**
If the Laplace integral

$$\int_0^\infty f(x)e^{-sx}dx \tag{13.2}$$

converges for $\mathrm{Re}\,(s) = \sigma_0$, then it also converges for $\mathrm{Re}\,(s) > \sigma_0$.

The proof is given in Sect. 13.3.4. This theorem implies the existence of a specific real number σ_c such that the integral (13.2) converges for $\mathrm{Re}\,(s) > \sigma_c$ and diverges for $\mathrm{Re}\,(s) < \sigma_c$ (see Fig. 13.1). The number σ_c is called the **abscissa of convergence** of the Laplace integral, whose value depends on the nature of the function $f(x)$. With this notation, we say that **the region of convergence of the Laplace integral** is a half-plane to the right of $\mathrm{Re}\,(s) = \sigma_c$. This region of convergence is of course identified with the defining region of the Laplace transform $F(s)$.

Remark. By definition, σ_c may take $-\infty$ (or ∞), which means that the integral (13.2) converges (or diverges) for all σ.

Examples Set $f(x) = 1$ for every $x \geq 0$. Then

$$\begin{aligned} L[f(x)] &= \int_0^\infty e^{-sx}dx = \lim_{X\to\infty}\int_0^X s^{-sx}dx \\ &= \lim_{X\to\infty}\left[\frac{s^{-sx}}{-s}\right]_0^X = \frac{1}{s} - \frac{1}{s}\lim_{X\to\infty} e^{-sX}. \end{aligned}$$

Hence, we have

$$L[f(x)] = \frac{1}{s} \quad \text{for} \quad s > 0.$$

For $s \leq 0$, the integral does not converge. This indicates that in this case $\sigma_c = 0$.

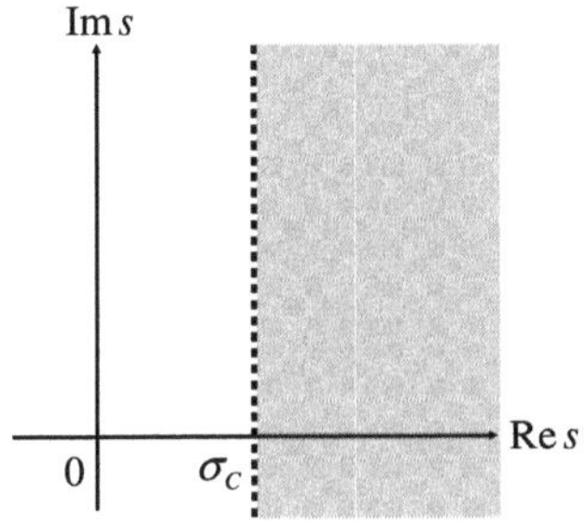

Fig. 13.1. The abscissa of convergence σ_c to the right of which the Laplace integral converges

13.1.5 Abscissa of Absolute Convergence

When the Laplace integral converges in the ordinary sense, it might converge *absolutely* in part or in all of its converging region. (Remember that the conditions for absolute convergence are more stringent than those for ordinary convergence). This leads us to define an abscissa of *absolute* convergence as follows:

♠ **Abscissa of absolute convergence:**

Suppose that the Laplace integral (13.2) converges absolutely for $\mathrm{Re}(s) = \sigma_0$ as

$$\int_0^\infty \left| f(x)e^{-sx} \right| dx = \int_0^\infty |f(x)|e^{-\sigma_0 x} dx < \infty. \tag{13.3}$$

The greatest lower bound σ_a of such a σ_0 that satisfies (13.3) is called the **abscissa of absolute convergence** of the Laplace integral (13.2).

Thus once σ_a is determined, we say that the integral (13.2) converges absolutely for $\sigma > \sigma_a$, does not converge absolutely for $\sigma < \sigma_a$, and may or may not converge absolutely at $\sigma = \sigma_a$. Since absolute convergence implies ordinary convergence, it is clear that

$$\sigma_c \leq \sigma_a.$$

The following example shows that σ_a does not generally coincide with σ_c (see Fig. 13.2).

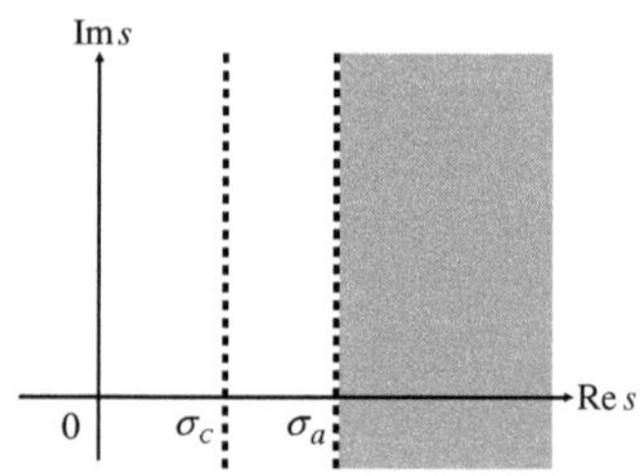

Fig. 13.2. The abscissa of convergence σ_c and the abscissa of *absolute* convergence σ_a

Example $f(x) = e^x \sin e^x$

Set $u = e^x$; then we have

$$F(s) = \int_0^\infty e^{-sx} e^x \sin e^x dx = \int_1^\infty \frac{\sin u}{u^s} du.$$

The integral converges absolutely for Re $(s) = \sigma > 1$, converges conditionally for $0 < \sigma \leq 1$, and diverges for $\sigma = 0$. Hence, we have

$$\sigma_c = 0 \text{ and } \sigma_a = 1,$$

which clearly indicates that in this case $\sigma_c \neq \sigma_a$.

13.1.6 Laplace Transforms of Elementary Functions

Let us evaluate the Laplace transforms $F(s)$ of several classes of elementary functions. We treat the complex variable s as if it were real, bearing in mind that this formalism is based on the **analyticity** of $F(s)$, as discussed in Sect. 13.1.3. The defining region of each $F(s)$ is found on the right-hand side of the equation in question.

1. $f(x) = x^n$, where n is a positive integer.

Integrating by parts, we have

$$\begin{aligned} F(s) = L[x^n] &= \int_0^\infty x^n e^{-sx} dx \\ &= \left[\frac{-x^n e^{-sx}}{s} \right]_0^\infty + \frac{n}{s} \int_0^\infty t^{n-1} e^{-sx} dx. \end{aligned} \tag{13.4}$$

Since $s > 0$ and $n > 0$, the first term in the last expression of (13.4) vanishes. Iteration of this process yields

$$L[x^n] = \frac{n(n-1)(n-2)\cdots 2 \cdot 1}{s^n} L[t^0] = \frac{n!}{s^{n+1}}$$

since $L[t^0] = L[1] = 1/s$. As a result, we have

$$F(s) = L[x^n] = \frac{n!}{s^{n+1}} \quad (\sigma > 0).$$

2. $f(x) = e^{ax}$, where a is a real constant.

$$F(s) = L[e^{ax}] = \int_0^\infty e^{-sx} e^{ax} dx = \frac{1}{s-a} \quad (\sigma > a).$$

3. $f(x) = \sin ax$, where a is a real constant.

Integrating by parts twice, we obtain

$$\begin{aligned} F(s) = L[\sin ax] &= \int_0^\infty e^{-sx} \sin ax dx \\ &= \left[-\frac{e^{-sx}}{a} \cos ax \right]_0^\infty + \frac{1}{a} \int_0^\infty (-s) e^{-sx} \cos ax dx \\ &= \frac{1}{a} - \frac{s}{a} \left\{ \left[\frac{e^{-sx}}{a} \sin ax \right]_0^\infty + \frac{s}{a} \int_0^\infty e^{-sx} \sin ax dx \right\} \\ &= \frac{1}{a} - \frac{s^2}{a^2} F(s), \end{aligned}$$

where we have used the fact that as s is positive, $e^{-sx} \to 0$ as $x \to \infty$, whereas $\sin ax$ and $\cos ax$ are bounded as $x \to \infty$. Eventually, we set to

$$F(s) = L[\sin ax] = \frac{a}{s^2 + a^2}. \quad (\sigma > 0).$$

In a simiar manner, we obtain

$$L[\cos ax] = \int_0^\infty e^{-sx} \cos ax dx = \frac{s}{s^2 + a^2} \quad (\sigma > 0).$$

4. $f(x) = \cosh ax$, where a is a real constant.

Using the linearity property of the Laplace transform operator L, we obtain

$$L[\cosh ax] = L\left[\frac{e^{ax} + e^{-ax}}{2}\right] = \frac{1}{2}L[e^{ax}] + \frac{1}{2}L[e^{-ax}]$$
$$= \frac{1}{2}\left(\frac{1}{s-a} + \frac{1}{s+a}\right) = \frac{s}{s^2 - a^2} \quad (\sigma > |a|).$$

Exercises

1. Show the linearity of the Laplace transformation operator L.

Solution: It follows from the definition of the operator L that

$$L[c_1 f(x) + c_2 g(x)] = \int_0^\infty e^{-sx}\{c_1 f(x) + c_2 g(x)\}dx$$
$$= c_1 \int_0^\infty e^{-sx} f(x)dx + c_2 \int_0^\infty e^{-sx} g(x)dx$$
$$= c_1 L[f(x)] + c_2 L[g(x)],$$

where c_1 and c_2 are arbitrary constants. This clearly shows the linearity of the operator L. ♣

2. Find the Laplace transform of the function,

$$f(x) = \begin{cases} 0, & 0 \le x < c, \\ 1, & x \ge c. \end{cases}$$

Solution: $L[f(x)] = \int_0^\infty e^{-sx} f(x)dx = \int_c^\infty e^{-sx}dx = e^{-cs}/s \ (\sigma > 0)$.
♣

3. Show that if $f(x)$ is real and $F(x) = L[f(x)]$ is single-valued, then $F(s)$ is real.

Solution: Set $s = \sigma > \sigma_c$ in the equation $F(s) = \int_0^\infty f(x)e^{-sx}dx$. Then the integrand $f(x)e^{-\sigma x}$ is real, so $F(\sigma)$ is real. This establishes that $F(s)$ is real on the real axis to the right of the point $s = \sigma_c$. In view of analytic continuation, therefore, $F(s)$ is a real-valued analytic function. ♣

13.2 Properties of Laplace Transforms

13.2.1 First Shifting Theorem

In physical applications, we are sometimes required to calculate the Laplace transform of functions multiplied by exponential factors such as

$$e^{-ax}f(x),$$

where a is real or complex. This kind of problem can be simplified by applying the theorem below.

♠ **The first shifting theorem:**

If $F(s) = L[f(x)]$ for $\sigma > \sigma_c$, then

$$F(s+a) = L[e^{-ax}f(x)]$$

for $\sigma > \sigma_c - \text{Re}\,(a)$, where a is real or complex.

Proof Suppose σ_c to be the abscissa of convergence for $F(s)$. Then the integral

$$\int_0^\infty e^{-ax}f(x)e^{-sx}dx = \int_0^\infty f(x)e^{-(s+a)t}dx \tag{13.5}$$

clearly converges for $\text{Re}\,(s+a) > \sigma_c$. Observe that the integral on the right-hand side of (13.5) is an expression for $F(s+a)$. Thus we have the general result:

$$L\left[e^{-ax}f(x)\right] = F(s+a),$$

where $F(s) = L[f(x)]$. ♣

The above theorem states that if we know the Laplace transform of any function, the transform of that function multiplied by an exponential can immediately be obtained by a simple *shift* (or translation) in the s variable.

Examples **1.** The first shifting theorem tells that

$$L[e^{-ax}x^n] = \frac{n}{(s+a)^{n+1}}, \quad \sigma > -a,$$

since

$$L[x^n] = \frac{n}{s^{n+1}}, \quad \sigma > 0.$$

2. Similarly, it follows from the first shifting theorem that

$$L[e^{-ax}\sin bt] = \frac{b}{(s+a)^2+b^2}, \qquad \sigma > -a,$$

where we use the fact that

$$L[\sin at] = \frac{a}{\sigma^2+a^2}.$$

13.2.2 Second Shifting Theorem

For the next case, assume again that a function $f(x)$ has a transform $F(s)$ and consider a shift in the x variable from x to $x - x_0$, where x_0 is a positive constant. Stipulating that the new function be zero for $x < x_0$, it can be written

$$f(x-x_0)\theta(x-x_0), \tag{13.6}$$

where

$$\theta(x) = \begin{cases} 0, \; x < 0, \\ 1, \; x > 0. \end{cases}$$

The Laplace transform of the shifted function (13.6) is thus represented by the integral

$$\int_0^\infty f(x-x_0)\theta(x-x_0)e^{-sx}dx = \int_{x_0}^\infty f(x-x_0)e^{-sx}dx.$$

Now we change the variable of integration to $t' = x - x_0$, which gives us

$$L\left[f(x-x_0)\theta(x-x_0)\right] = e^{-sx_0}\int_0^\infty f(t')e^{-st'}dx' = e^{-sx_0}F(s).$$

The result is stated formally below.

♠ **The second shifting theorem:**

If $F(s) = L[f(x)]$ for $\sigma > \sigma_c$, then

$$e^{-sx_0}F(s) = L[f(x-x_0)\theta(x-x_0)]$$

for $\sigma > \sigma_c$, where $\theta(x)$ is a unit step function and T is a real and positive constant.

Examples Consider the Laplace transform of the function

$$f(x) = \begin{cases} 0 & (x < 0), \\ 1/a & (0 \le x < a), \\ 0 & (x \ge a). \end{cases}$$

Using the step function, we express it as

$$f(x) = \frac{\theta(x) - \theta(x-a)}{a}.$$

Hence, it follows from the second shifting theorem that

$$L[f(x)] = \frac{L[1] - e^{-as}L[1]}{a} = \frac{1 - e^{-as}}{as}.$$

Note that in view of **l'Hôpital's rule**, when $a \to 0$, $L[f(x)] = 1$. The latter result means that the Laplace transform of $f(x)$ equals 1.

13.2.3 Laplace Transform of Periodic Functions

We now consider the Laplace transform of a periodic function $f(x)$ of period λ, i.e., $f(x+\lambda) = f(x)$. Assuming that the $f(x)$ is piecewise continuous, we have by definition

$$\begin{aligned} L[f(x)] &= \int_0^\infty e^{-sx} f(x)dx \\ &= \int_0^\lambda e^{-sx} f(x)dx + \int_\lambda^{2\lambda} e^{-sx} f(x)dx + \int_{2\lambda}^{3\lambda} e^{-sx} f(x)dx + \cdots . \end{aligned}$$

On the right-hand side, let $x = u + \lambda$ in the second integral, $x = u + 2\lambda$ in the third integral, and so on. We then set

$$\begin{aligned} L[f(x)] = \int_0^\lambda e^{-sx} f(x)dx &+ \int_0^\lambda e^{-s(u+\lambda)} f(u+\lambda)du \\ &+ \int_0^\lambda e^{-s(u+2\lambda)} f(u+2\lambda)du + \cdots . \end{aligned}$$

From hypothesis, $f(u+\lambda) = f(u)$, $f(u+2\lambda) = f(u)$, etc. Replacing the dummy variable u by x yields

$$\begin{aligned} L[f(x)] &= (1 + e^{-s\lambda} + e^{-2s\lambda} + \cdots) \int_0^\lambda e^{-sx} f(x)dx \\ &= \frac{1}{1 - e^{-s\lambda}} \int_0^\lambda e^{-sx} f(x)dx. \end{aligned} \tag{13.7}$$

Once we introduce the function

$$f_0(x) = \begin{cases} f(x), \ 0 \le t \le \lambda, \\ 0, \quad \text{otherwise,} \end{cases}$$

equation (13.7) becomes

$$L[f(x)] = \frac{F_0(s)}{1 - e^{-s\lambda}},$$

where

$$F_0(s) = \int_0^\infty e^{-sx} f_0(x) dx = \int_0^\lambda e^{-sx} f(x) dx,$$

So we have proven the following result:

♠ **Laplace transform of a periodic function:**

If $f(x)$ is a periodic function of period λ, its Laplace transform is given by

$$F(s) = \frac{F_0(s)}{1 - e^{-s\lambda}}, \tag{13.8}$$

where

$$F_0(s) = \int_0^\lambda e^{-sx} f(x) dx.$$

Examples Consider the Laplace transform of the periodic square wave described by $f(x + 2\lambda) = f(x)$ with

$$f(x) = \begin{cases} 1 & (0 < x < \lambda), \\ -1 & (\lambda < x < 2\lambda). \end{cases}$$

From (13.8), we obtain

$$\begin{aligned} F(s) &= \frac{1}{1 - e^{-2s\lambda}} \int_0^{2\lambda} e^{-sx} f(x) dx \\ &= \frac{1}{1 - e^{-2s\lambda}} \left(\int_0^\lambda - \int_\lambda^{2\lambda} \right) e^{-sx} f(x) dx \\ &= \frac{(1 - e^{-s\lambda})^2}{s(1 - e^{-2s\lambda})} = \frac{1}{s} \tanh\left(\frac{s\lambda}{2} \right). \end{aligned}$$

13.2.4 Laplace Transform of Derivatives and Integrals

The Laplace transform of derivatives is a most important issue in terms of applications for solving differential equations. We shall see below that through the transform, certain kinds of differential equations are reduced to algebraic equations that are easy to manipulate.

♠ **Laplace transform of derivatives:**

If $F(s) = L[f(x)]$ for $\sigma > \sigma_c$ and if

$$\lim_{t\to\infty} e^{-sx} f(x) = 0 \quad \text{for } \sigma > \sigma_c, \tag{13.9}$$

then we have

$$L[f'(x)] = sF(s) - f(0). \tag{13.10}$$

Proof Integration by parts yields

$$L[f'(x)] = \int_0^\infty e^{-sx} f'(x) dx$$

$$= \left[e^{-sx} f(x)\right]_0^\infty - \int_0^\infty (-s)e^{-sx} f(x) dx. \tag{13.11}$$

The second term on the right-hand side of (13.11) converges to $sF(s)$ for $\sigma > \sigma_c$. In addition, the first term reads $f(0)$ from the hypothesis of (13.9). Thus for $\sigma > \sigma_c$, we obtain (13.10). ♣

This result can be extended to cases of higher derivatives.

♠ **Laplace transform of higher derivatives:**

Suppose $f(x)$ to be such that $f^{(n-1)}(x)$ is continuous. If $F(s) = L[f(x)]$ for $\sigma > \sigma_c$ and if

$$\lim_{t\to\infty} e^{-sx} f^{(k)}(x) = 0$$

for $k = 0, 1, \cdots, n-1$ and $\sigma > \sigma_c$, then

$$L[f^{(n)}(x)] = s^n F(s) - \sum_{k=1}^{n} s^{n-k} f^{(k-1)}(0).$$

The above theorem is central to the use of the Laplace transform for solving differential equations with specified initial conditions (i.e., **initial value problems**).

♠ **Laplace transform of integrals:**
If $g(x) = \int_0^x f(u)du$, $L[f(x)] = F(s)$ and if

$$\lim_{t\to\infty} e^{-sx}g(x) = 0,$$

then

$$L[g(x)] = \frac{F(s)}{s}.$$

Proof From hypothesis, we have $g(0) = 0$ and $g'(s) = f(x)$, and thus

$$L[g'(x)] = L[f(x)].$$

The left-hand side becomes

$$L[g'(x)] = sL[g(x)] - g(0) = sL[g(x)].$$

As a result, we obtain

$$L[g(x)] = \frac{1}{s}L[f(x)] = \frac{F(s)}{s}. \quad ♣$$

13.2.5 Laplace Transforms Leading to Multivalued Functions

Some care should be taken when the Laplace transform results in a **multi-valued function**. A typical example is the transform of the function

$$f(x) = \frac{1}{\sqrt{x}} \quad (x \geq 0). \tag{13.12}$$

Although this function has a singularity at $x = 0$, the **improper integral** having a real integrand,

$$\int_0^\infty \frac{e^{-\sigma x}}{\sqrt{x}} dx, \tag{13.13}$$

converges for $\sigma > 0$. In what follows, we first evaluate the integral (13.13) and then **continue analytically** with the result to arrive at a suitable region of the complex s-plane where we can get a precise form of $F(s)$.

The integral (13.13) can be readily evaluated by setting $\sigma x = u^2$; then it reads

$$\frac{2}{\sqrt{\sigma}} \int_0^\infty e^{-u^2} du = \frac{\sqrt{\pi}}{\sqrt{\sigma}}. \tag{13.14}$$

Now we would like to continue analytically to take the result (13.14) to the complex s-plane. At first glance, it suffices to replace $\sqrt{\sigma}$ by $\sqrt{s}$ symbolically. However, this is not sufficient because the function $\sqrt{s}$ is double-valued (e.g., when $s = i = e^{\pi i/2}$, $\sqrt{s}$ may take the two distinct values: $e^{\pi i/4}$ and $e^{-3\pi i/4}$;

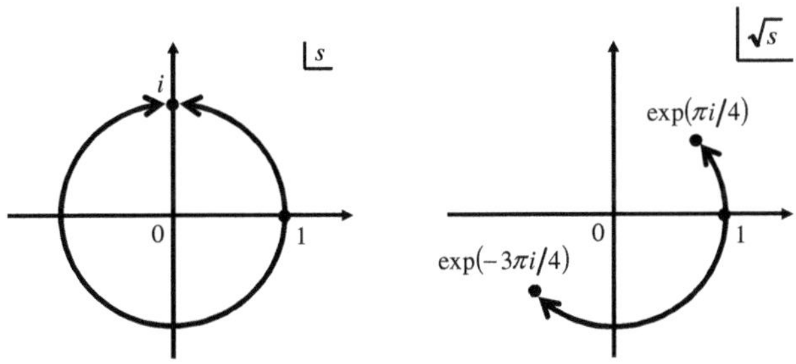

Fig. 13.3. The double-valuedness of the function $\sqrt{s}$

(see Fig. 13.3). Thus we have two possible choices (i.e., two sheets of **Riemann surfaces**) when performing analytic continuation from the real single-valued function $\sqrt{\sigma}$ to the complex double-valued function $\sqrt{s}$. We go into only one sheet of Riemann surface, the choice being the one on which the points of $\sqrt{\sigma}$ are situated [i.e., the right half of the whole s-plane, expressed by Re $(s) > 0$]. With this convention, we arrive at the result

$$F(s) = L\left[\frac{1}{\sqrt{x}}\right] = \frac{\sqrt{\pi}}{\sqrt{s}}, \tag{13.15}$$

where the symbol $\sqrt{s}$ implies the single-valued branch mentioned above.

Remark. If the above case had been treated throughout with the variable s retained, the formal variable change would have led to the factor $1/\sqrt{s}$ as in (13.15). However, we would not then have a clear meaning for $\sqrt{s}$; i.e., there would be no way to determine which branch is to be taken.

Exercises

1. Show that

$$\lim_{x\to+0} f(x) = \lim_{s\to\infty} sF(s)$$

and

$$\lim_{x\to\infty} f(x) = \lim_{s\to 0} sF(s),$$

where the Laplace integral $L[f(x)] = F(s)$ converges for $\sigma \geq 0$.

Solution: Take the limits $s \to \infty$ on both sides of equation

$$\int_0^\infty f'(x)e^{-sx}dx = sF(s) - f(0). \tag{13.16}$$

Then we have $0 = \lim_{s\to\infty} sF(s) - f(0+)$, which gives us our first result. Moreover, in the limit $s \to 0$, the left-hand side of (13.16) reads $\int_0^\infty f'(x)dx = \lim_{x\to\infty} f(x) - f(0+)$, so that we set our second result. ♣

2. Find the transform of the function

$$f(x) = \sqrt{t^k},$$

where $k \geq 1$ and is an odd integer.

Solution: This function gives convergence for $\sigma > 0$. Integration by parts yields a general recurrence equation:

$$\int_0^\infty \sqrt{t^k} e^{-\sigma x} dx = -\left[\frac{\sqrt{t^k} e^{-\sigma x}}{\sigma}\right]_0^\infty + \frac{k}{2\sigma}\int_0^\infty \sqrt{t^{k-2}} e^{-\sigma x} dx.$$

Since $k \geq 1$, the lower limit can be used in the first term on the right-hand side (and thus the integral exists). The result can be stated as

$$L\left[\sqrt{t^k}\right] = \frac{k}{2s} L\left[\sqrt{t^{k-2}}\right] \quad \text{where } k \geq 1 \text{ and odd.}$$

This yields a sequence of equations, starting with $\sqrt{t^{-1}}$, that is obtained from (13.15). Consequently, we have

$$L\left[\frac{1}{\sqrt{x}}\right] = \frac{\sqrt{\pi}}{\sqrt{s}}, \quad L\left[\sqrt{x}\right] = \frac{\sqrt{\pi}}{2\sqrt{s^3}}, \quad L\left[\sqrt{x^3}\right] = \frac{3\sqrt{\pi}}{4\sqrt{s^5}}, \quad \cdots$$

$$L\left[\sqrt{x^k}\right] = \frac{(k+1)!\,\sqrt{\pi}}{2^{k+1}[(k+1)/2]!\,\sqrt{s^{k+2}}}.$$

In these general equations, the root of a power of s is always interpreted as being on the sheet of the Riemann surface on which the values of $\sqrt{\sigma^{k+2}}$ are found. ♣

13.3 Convergence Theorems for Laplace Integrals

13.3.1 Functions of Exponential Order

The Laplace integral is **improper** by virtue of an infinite limit of integration, as shown clearly by

$$\int_0^\infty f(x) e^{-sx} dx = \lim_{R\to\infty} \int_0^R f(x) e^{-sx} dx. \tag{13.17}$$

This improper integral can be identified with the Laplace transform $F(s)$ *only* when it converges for the values of s in question. Therefore, it is important to clarify the conditions under which the Laplace integral converges. As a first step in addressing this issue, we introduce a new class of functions:

♠ **Functions of exponential order:**

A function $f(x)$ is said to be of **exponential order** α_0 if there is a real number α_0 such that

$$\lim_{x\to\infty} f(x)e^{-\alpha x} = 0 \qquad \text{for any } \alpha > \alpha_0, \tag{13.18}$$

and with the limit not existing when $\alpha < \alpha_0$.

See Fig. 13.4 for the decaying behavior of a function $f(x)$ of exponential order α. Note that condition (13.18) is not necessarily satisfied at $\alpha = \alpha_0$. The order number α_0 may take $-\infty$ if $f(x)$ is identically zero beyond some finite value of x.

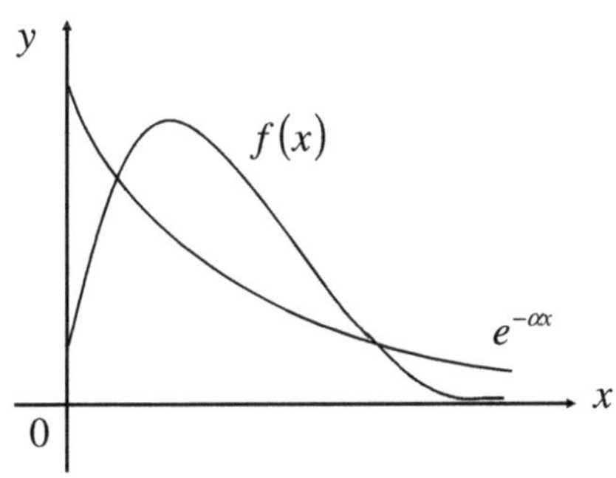

Fig. 13.4. Decaying behavior of a function $f(x)$ of exponential order α

Examples **1.** The function $f(x) = x^3$ is of exponential order zero. To see this, it suffices to check whether or not

$$\lim_{x\to\infty} \left(e^{-\alpha x} x^3\right) \tag{13.19}$$

exists. If $\alpha > 0$, then **l'Hôpital's rule** gives

$$\lim_{x\to\infty} \frac{x^3}{e^{\alpha x}} = \lim_{x\to\infty} \frac{6}{\alpha^3 e^{\alpha x}} = 0.$$

In contrast, when $\alpha < 0$, (13.19) obviously diverges. Therefore x^3 is of exponential order zero. In a similar manner, it can be shown that x^n for any integer $n \geq 0$ is of exponential order zero.

2. The function $f(x) = e^{cx}$ with any real constant c is of exponential order c, owing to the fact that

$$\lim_{x \to \infty} e^{cx} e^{-\alpha x} = 0$$

if and only if $\alpha > c$.

13.3.2 Convergence for Exponential-Order Cases

Suppose $f(x)$ to be of exponential order α_0. Then, we can show that the Laplace integral

$$\int_0^\infty f(x) e^{-sx} dx \tag{13.20}$$

converges absolutely whenever the real component of s is located within the range

$$\text{Re}\,(s) = \sigma > \alpha_0. \tag{13.21}$$

Since absolute convergence implies ordinary convergence, the inequality (13.21) serves as a sufficient condition for the Laplace integral (13.20) to converge. This result is formally stated by the theorem below.

♠ **Theorem:** (= A sufficient condition for convergence for exponential-order cases)

If $f(x)$ is of exponential order α_0, then the Laplace integral $\int_0^\infty f(x) e^{-sx} dx$ converges for

$$\text{Re}\,(s) > \alpha_0.$$

(See also Fig. 13.5.)

Proof For any σ in the range of (13.21), we can pick a number α_1 such that

$$\alpha_0 < \alpha_1 < \sigma.$$

Since $f(x)$ is of exponential order α_0, we have

$$\lim_{x \to \infty} f(x) e^{-\alpha_1 x} = 0.$$

This implies that for any given small $\varepsilon > 0$, we can find an appropriate X such that

$$|f(x)| e^{-\alpha_1 x} < \varepsilon \qquad \text{for any } x > X.$$

Hence, given any small $\varepsilon > 0$, there exists an X such that for $A, A' > X$,

$$\int_A^{A'} |f(x)|\, e^{-\sigma x} dx = \int_A^{A'} |f(x)| e^{-\alpha_1 x} e^{-(\sigma - \alpha_1) x} dx$$

$$< \varepsilon \int_A^{A'} e^{-(\sigma - \alpha_1) x} dx, \tag{13.22}$$

where the last integral in (13.22) converges to a finite value because $\sigma > \alpha_1$. This means that the leftmost integral in (13.22) can be made to approach zero by taking X sufficiently large. Thus in view of the **Cauchy's test for improper integrals** given in Sect. 3.4.4, the inequality (13.22) establishes the absolute (and thus ordinary) convergence of the integral (13.20) in the region Re $(s) > \alpha_0$. ♣

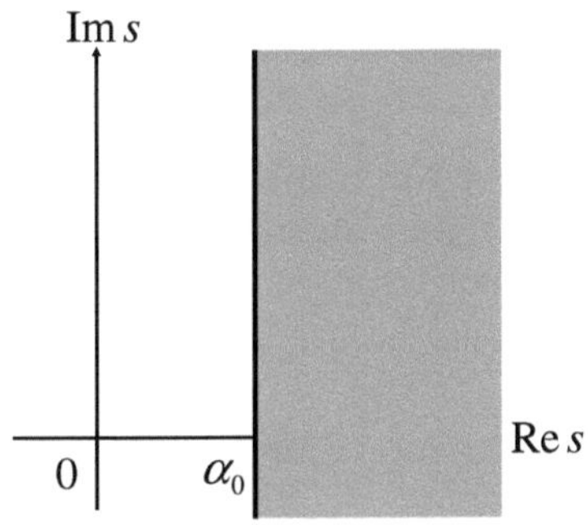

Fig. 13.5. Converging region of the Laplace integral of a function of exponential order α_0

Remark. The above theorem provides a sufficient condition for the ordinary convergence of the Laplace integral. Hence, a given Laplace integral of the function of exponential order α_0 *must* converge for Re $(s) > \alpha_0$, whereas it *may or may not* converge at Re $(s) \leq \alpha_0$. For example, $f(x) = \cos e^x$ gives $\alpha_0 = 0$, but the corresponding Laplace integral converges for Re $(s) > -1$.

13.3.3 Uniform Convergence for Exponential-Order Cases

Next we examine the condition for *uniform* convergence. Here, uniform convergence means that the improper integral (13.20) as a function of s converges uniformly to $F(s)$ over the whole defining region of the s-plane. To proceed, let α_2 be a number greater than α_0 and let σ be in the range

$$\alpha_0 < \alpha_2 \leq \sigma. \tag{13.23}$$

For any choice of α_2, we can find a number α_1 such that

$$\alpha_0 < \alpha_1 < \alpha_2.$$

The relation (13.22) is again valid by use of α_2 instead of α_1, as expressed by

$$\int_A^{A'} |f(x)|e^{-\sigma x}dx < \varepsilon \int_A^{A'} e^{-(\sigma-\alpha_2)x}dx.$$

Furthermore, by introducing α_1, we can extend this inequality to

$$\int_A^{A'} |f(x)|e^{-\sigma x}dx < \varepsilon \int_A^{A'} e^{-(\alpha_1-\alpha_2)x}dx.$$

Note that the last integral converges and is independent of σ. Therefore, in view of the **Weierstrass test for improper integrals** (see Sect. 3.4.4), the Laplace integral $\int_0^\infty f(x)e^{-sx}dx$ converges uniformly for Re $(s) \geq \alpha_2 > \alpha_0$. We have thus proved the following theorem:

♠ **Theorem:** (= A sufficient condition for uniform convergence for exponential-order cases)

If $f(x)$ is of exponential order α_0, then the Laplace integral $\int_0^\infty f(x)e^{-sx}dx$ converges uniformly to $F(s) = L[f(x)]$ for

$$\text{Re } (s) \geq \alpha_2 > \alpha_0.$$

(See also Fig. 13.6.)

Here, the constant α_2 emphasizes that the converging region guaranteed by this theorem is closed at the lower end.

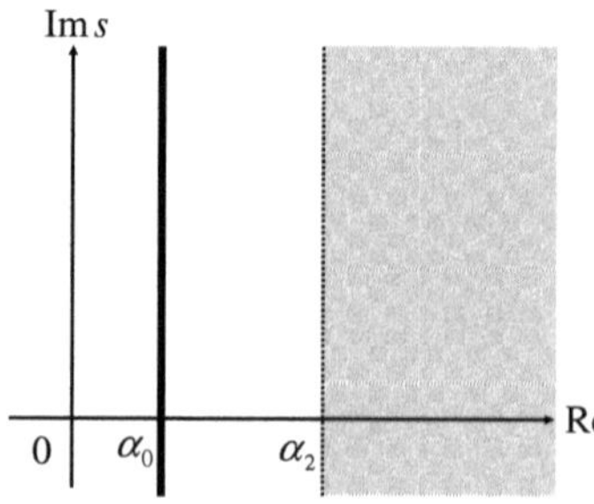

Fig. 13.6. The region of uniform convergence associated with a function of exponential order α_0

Remark. It is important to remember that the above theorem gives only a *sufficient* condition for convergence of the Laplace integral. In fact, it is possible that some functions of exponential order allow their Laplace integrals to converge uniformly to the *left* of α_0.

13.3.4 Convergence for General Cases

The previous two theorems tell a great deal about convergence of the Laplace integral for practical functions. On the other hand, for functions that are not of exponential order (but continuous within the integration interval), the following slightly different theorem applies.

♠ **Theorem:** (= A sufficient condition for convergence for general cases)

If the improper integral

$$\int_0^\infty f(x)e^{-sx}dx$$

converges for $s = s_0$, then it converges for Re $(s) >$ Re (s_0) (see also Fig. 13.7).

Proof The proof requires an auxiliary function

$$g(x) = \int_x^\infty f(\tau)e^{-s_0\tau}d\tau, \tag{13.24}$$

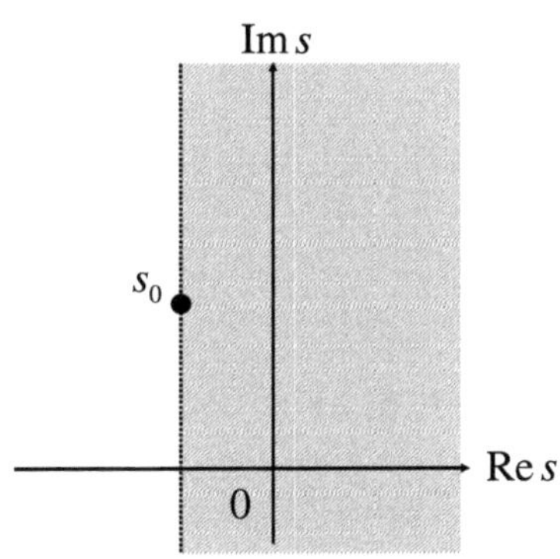

Fig. 13.7. Converging region of the Laplace integral that converges for $s = s_0$

where $f(x)$ is assumed to satisfy the conditions given above. Since $f(x)$ is continuous, $g(x)$ is also continuous and thus its derivative is given by

$$g'(x) = -f(x)e^{-s_0 x}.$$

In terms of $g(x)$, the Laplace integral can be written as

$$\int_0^\infty f(x)e^{-sx}dx = \int_0^\infty f(x)e^{-s_0 x}e^{-wx}dx = -\int_0^\infty g'(x)e^{-wx}dx, \quad (13.25)$$

where we have set $w = s - s_0$.

We now examine sufficient conditions for the rightmost integral in (13.25) to converge. **Cauchy's test for improper integrals** given in Sect. 3.4.4 says that it converges if and only if for an arbitrary small $\varepsilon > 0$, we can find an X that yields

$$\left| \int_{A'}^{A} g'(x)e^{-wx}dx \right| < \varepsilon \quad (13.26)$$

with $A',\ A > X$. Therefore, our task is to show that the relation (13.26) holds for Re $(s) >$ Re (s_0).

Integration by parts gives us

$$\int_{A'}^{A} g'(x)e^{-wx}dx = -g(A')e^{-wA'} + g(A)e^{-wA} + w\int_{A'}^{A} g(x)e^{-wx}dx, \quad (13.27)$$

which results in

$$\left| \int_{A'}^{A} g'(x)e^{-wx}dx \right| < |g(A')|e^{-xA'} + |g(A)|e^{-xA} + |z|\int_{A'}^{A} \left|g(x)e^{-wx}\right| dx. \quad (13.28)$$

From (13.24) and from the hypothesis given in the theorem, it follows that for an arbitrary small $\varepsilon' > 0$, there exists a number X such that

$$|g(x)| < \varepsilon' \text{ when } x > X.$$

Thus, if $A',\ A > X$ we have

$$|g(A')|,\ |g(A)| < \varepsilon'.$$

In addition, if

$$u \equiv \text{Re } (w) > 0,$$

then the relation (13.28) becomes

$$\left| \int_{A'}^{A} g'(x)e^{-wx}dx \right| < \varepsilon' \left[e^{-uA'} + e^{-uA} + \frac{|w|}{u}\left(e^{-uA'} - e^{-uA}\right) \right]$$
$$< \varepsilon' \left(2 + \frac{|w|}{u}\right). \quad (13.29)$$

Observe that the quantity in parentheses in (13.29) is finite for any fixed value of w with $u > 0$. Therefore, by making ε' small enough, the quantity

$$\varepsilon \equiv \varepsilon' \left(2 + \frac{|w|}{u} \right) \tag{13.30}$$

becomes arbitrarily small; this can be the ε in the relation (13.26). Consequently, the relation (13.26) holds for any $u > 0$, or equivalently, for any

$$u = \text{Re}\,(w) = \text{Re}\,(s) - \text{Re}\,(s_0) > 0.$$

This completes the proof of the theorem. (Note that if $u = 0$, the quantity in parenthesies in (13.29) diverges, and if $u < 0$, the inequality (13.29) itself does not hold.) ♣

Remark. The theorem is inconclusive for the convergence property on the line $\text{Re}(s) = \text{Re}(s_0)$ depicted on the complex s-plane. Note that we do not get convergence when $\text{Re}\,(s) = \sigma_0$. This means that even though the integral converges at a point on the line of $\text{Re}\,(s) = \sigma_0$, it does not necessarily converge all along the same line. A simple example is given by

$$f(x) = \begin{cases} 0, & 0 \le x < 1, \\ \dfrac{1}{x}, & x \ge 1. \end{cases}$$

The Laplace integral

$$\int_0^\infty f(x) e^{-s_0 x} dx = \int_1^\infty \frac{e^{-i\omega_0 x}}{x} dx = \int_1^\infty \frac{\cos \omega_0 x}{x} dx - i \int_1^\infty \frac{\sin \omega_0 x}{x} dx$$

converges for $s_0 = 0 + i\omega_0$ with $\omega_0 \neq 0$, but diverges at $s_0 = 0$.

13.3.5 Uniform Convergence for General Cases

A sufficient condition for uniform convergence is obtained in a similar way as in Sect. 13.3.4, although it is not the same as that for ordinary convergence. The difference is due to the fact that in the proof above, ε defined by (13.30) is dependent on s through $|w| = |s - s_0|$. In order to get the range of uniform convergence, we need a certain infinitesimal factor that can be taken independently of s.

To derive such a factor, let θ be the angle of $w = s - s_0$, and observe that $u = \text{Re}(w)$ satisfies the relation

$$\frac{|w|}{u} = \left| \frac{1}{\cos \theta} \right|,$$

when $u > 0$. If θ is restricted to the range

$$|\theta| < \frac{\pi}{2}, \tag{13.31}$$

we can find an angle θ' that satisfies

$$|\theta| \le \theta' < \frac{\pi}{2},$$

or equivalently,

$$\frac{|w|}{u} = \frac{1}{\cos\theta} \le \frac{1}{\cos\theta'}.$$

Inserting this into (13.30), we have

$$\varepsilon = \varepsilon' \left(2 + \frac{|w|}{u}\right) \le \varepsilon' \left(2 + \frac{1}{\cos\theta'}\right) \equiv \varepsilon'',$$

where the quantity ε'' is independent of s and becomes arbitrarily small by making ε' small enough. This is true as far as condition (13.31) is satisfied; in this context, (13.31) represents the region of uniform convergence of the Laplace integral. Rewriting θ by $\arg(s - s_0)$, we arrive at the following theorem:

♠ **Theorem** (= A sufficient condition for uniform convergence of the Laplace integrals for general cases):

If the improper integral

$$\int_0^\infty f(x)e^{-sx}dx$$

converges for $s = s_0$, then it converges uniformly to $F(s) = L[f(x)]$ for

$$|\arg(s - s_0)| \le \theta' < \frac{\pi}{2}.$$

(See also Fig. 13.8.)

Here, the θ' shows the **closedness** of the converging region. The θ' can be arbitrarily close to but not equal $\pi/2$.

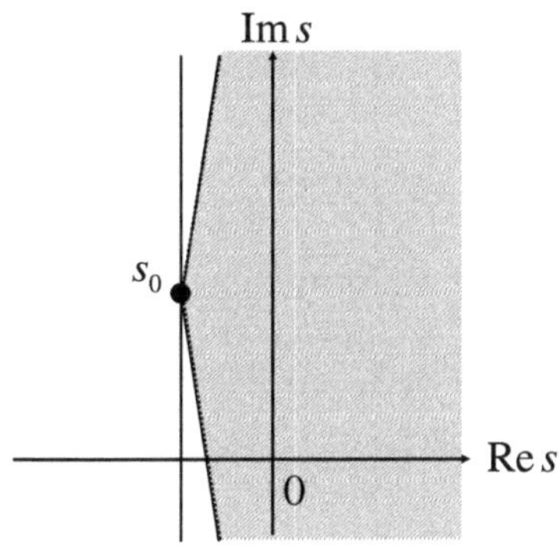

Fig. 13.8. The region of uniform convergence for the Laplace integral that converges for $s = s_0$

13.3.6 Distinction Between Exponential-Order Cases and General Cases

We have thus far presented four convergence theorems in connection with Laplace integrals, where the former two are associated with functions of exponential order and the latter two are relevant to are more general functions. The theorems for the two cases are similar to the extent that they all identify a half-plane of convergence for the Laplace integral. Moreover, the general cases that we have considered cover a wide class of functions that includes exponential-order functions as a special case. At first glance, these remarks appear to imply that each of the former two theorems for exponential-order cases is a special case of each of the latter for general cases, but, *this is not true at all.* Below we give the reasons for this not being so.

First, the theorem for ordinary convergence in the exponential-order case is intrinsically different from that in the general case. Observe that the former theorem not only tells us that the Laplace integral converges in a half-plane; it also gives a specific number (i.e., α_0) for the abscissa of a left-hand boundary of such a half-plane. (Of course $\sigma_c \leq \alpha_0$ since it gives a *sufficient* condition for convergence.) In contrast, the latter theorem merely states convergence to the right of any point at which we already know that the integral converges; it gives no information about a boundary of the region of convergence.

Second, the regions of uniform convergence are specified in a different manner for the two cases. Whereas the theorem for general cases tells us only that the Laplace integral converges uniformly in an angular sector of the right half-plane, the theorem for exponential-order cases indicates uniform convergence in a less restricted region, namely, a half-plane.

In short, the theorems for the two cases are essentially different. As well, it should be emphasized again that all the four theorems provide *sufficient* conditions for convergence of the Laplace integrals—not *necessary* or *necessary and sufficient* conditions.

13.3.7 Analytic Property of Laplace Transforms

An important consequence of uniform convergence of the Laplace integral is the fact that the corresponding Laplace transform,

$$F(s) = \int_0^\infty f(x)e^{-sx}dx, \tag{13.32}$$

is an **analytic function** on the complex s-plane. We know that if $F(s)$ is analytic, it will exist outside the range of convergence of its integral representation, which can be uniquely determined by **analytic continuation**. From a practical viewpoint, the **analyticity** of $F(s)$ plays a crucial role in evaluating the Laplace transform of a given function, since we can use it to treat the complex variable s as if it were real (see Sect. 13.1.3). We close this section by proving the analyticity of $F(s)$.

♠ **Theorem:**
The Laplace transform $F(s)$ is analytic in the region of uniform convergence of the corresponding Laplace integral (13.32).

Proof We first recall that for $F(s)$, there is a region of uniform convergence in the s-plane and then we perform a contour integration with respect to s over an arbitrary simple closed path C in this region. Owing to the uniform convergence property, the order of integration may be inverted so that we have

$$\oint_C F(s)ds = \int_0^\infty f(x)\left(\oint_C e^{-sx}ds\right)dx = 0.$$

which gives us zero because **Cauchy's integral formula** means that

$$\oint_C e^{-sx}ds = 0.$$

Since the path of C is arbitrary in the region of uniform convergence, Morera's theorem establishes that $F(s)$ is analytic inside the region of uniform convergence of its corresponding Laplace integral. ♣

13.4 Inverse Laplace Transform

13.4.1 The Two-Sided Laplace Transform

This section describes the **inverse Laplace transformation**. Intuitively understood, the inverse Laplace transform $L^{-1}[F(s)]$ of a function $F(s)$ is a function $f(x)$ whose Laplace transform is $F(s)$. Nevertheless, actual operations represented by the operator L^{-1} take some time to develop. To set to the explicit formula for manipulating the inverse transformation, we first introduce another kind of Laplace transform:

♠ **Two-sided Laplace transform:**
If the improper integral

$$\int_{-\infty}^{\infty} f(x)e^{-sx}dx \tag{13.33}$$

exists, it is called the **two-sided Laplace transform** (or **bilateral Laplace transform**), designated by $\mathcal{F}(s) = \mathcal{L}[f(x)]$.

It is easy to determine the region of convergence of such an integral. Observe that

$$\mathcal{L}[f(s)] = \int_{-\infty}^{0} f(x)e^{-sx}dx + \int_{0}^{\infty} f(x)e^{-sx}dx. \tag{13.34}$$

The second integral is an ordinary Laplace integral so that it converges on a half-plane right to a fixed point denoted by $x = \sigma_{c1}$. By the change of variable $x = -u$ the first integral becomes

$$\int_{-\infty}^{0} f(x)e^{-sx}dx = \int_{0}^{\infty} f(-u)e^{su}du.$$

Here, the latter integral is also an ordinary Laplace integral, although s has been replaced by $-s$. Hence, its region of convergence is a half-plane *left* to a point, say $x = \sigma_{c2}$. As a result, the common part of the two half-planes, $\sigma_{c1} < \mathrm{Re}(s) < \sigma_{c2}$, forms the region of convergence of the integral (13.34) as depicted in Fig. 13.9.

Remark. Typically, the range of convergence of (13.34) forms a vertical strip with a finite interval, but may be a right half plane, a left half-plane, the whole s-plane, a single point, or fail to exist.

Example We show that the function $1/(s^2+s)$ can be expressed as a two-sided Laplace integral. We readily see that

$$\frac{1}{s(s+1)} = \frac{1}{s} - \frac{1}{s+1}.$$

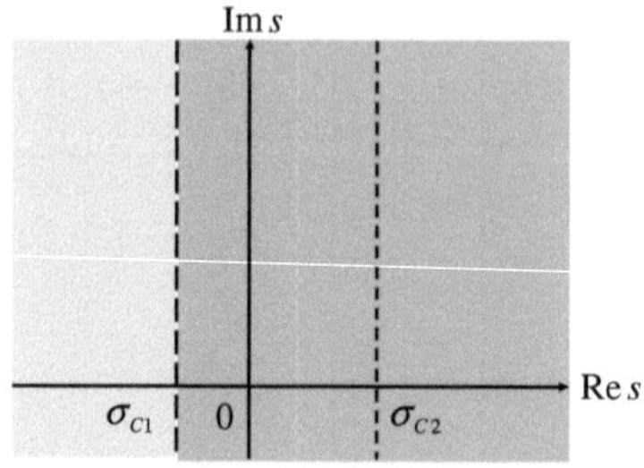

Fig. 13.9. Overlapping region: the region of convergence of the two-sided Laplace integral

We know that

$$\frac{1}{s+1} = \int_0^\infty e^{-x}e^{-sx}dx \quad \text{for } \sigma > -1 \tag{13.35}$$

and

$$\frac{1}{s} = \int_0^\infty e^{-sx}dx = \int_{-\infty}^0 e^{sx}dx \quad \text{for } \sigma > 0$$

the latter and that can be rewritten as

$$\frac{1}{s} = \int_{-\infty}^0 e^{-sx}(-1)dx \quad \text{for } \sigma < 0. \tag{13.36}$$

From (13.35) and (13.36), we obtain

$$\frac{1}{s(s+1)} = \frac{1}{s} - \frac{1}{s+1} = \int_{-\infty}^\infty f(x)e^{-sx}dx,$$

where

$$f(x) = \begin{cases} -e^{-x}, & 0 < x < \infty, \\ -1, & -\infty < x < 0, \end{cases}$$

which means that $1/(s^2+s) = \mathcal{L}[f(x)]$. The interval of convergence is seen to be $-1 < \sigma < 0$. example

13.4.2 Inverse of the Two-Sided Laplace Transform

Having introduced the two-sided Laplace transform, we are ready to undertake the inverse Laplce transformation. We first observe that the two-sided Laplace transform

$$\mathcal{F}(s) = \int_{-\infty}^\infty f(x)e^{-sx}dx \tag{13.37}$$

is identical with the **Fourier transform**

$$\mathcal{F}(\sigma + i\omega) = \int_{-\infty}^{\infty} f(x) e^{-\sigma x} e^{-i\omega x} dx$$

if we regard the real number σ as fixed. We use the **inverse Fourier transformation** to yield

$$f(x)e^{-\sigma x} = \frac{1}{2\pi} \int_{-\infty}^{\infty} \mathcal{F}(\sigma + i\omega) e^{i\omega x} d\omega,$$

or equivalently,

$$f(x) = \frac{1}{2\pi} \int_{-\infty}^{\infty} \mathcal{F}(\sigma + i\omega) e^{\sigma x} e^{i\omega x} d\omega. \tag{13.38}$$

We then replace $\sigma + i\omega$ by s, keeping in mind that s should lie on the vertical line with the abscissa $\text{Re}(s) = \sigma$. Then the integral (13.38) can be regarded as a contour integral along the vertical line $\text{Re}(s) = \sigma$. On this contour,

$$ds = i d\omega,$$

so the integral (13.38) becomes

$$f(x) = \frac{1}{2\pi i} \int_{\sigma - i\infty}^{\sigma + i\infty} \mathcal{F}(s) e^{sx} ds \quad (\text{Re}(s) = \sigma \text{ is fixed}). \tag{13.39}$$

This result provides a clue for evaluating the explicit form of $f(x)$ from its two-sided Laplace transform $\mathcal{F}(s)$.

The result (13.39) is not yet satisfying. We should recall that $f(x)$ is not determined uniquely by $\mathcal{F}(s)$ through (13.39) unless the location of the point $x = \sigma$ is specified (see Exercise **3** in Sect. 13.4). If we know in advance that σ lies in the region of convergence of the two-sided integral given by (13.37), i.e., the strip of convergence, $f(x)$ is uniquely determined by (13.39). However, if σ used in (13.39) is set outside this strip, the integral of (13.39) is altered quantitatively because the integration contour passes over one or more singular points of $\mathcal{F}(s)$. Thus for us to be able to use equation (13.39), we must know the region of convergence of the Laplace integral of $f(x)$ *before* we can fix the real number σ. If only $\mathcal{F}(s)$ is given, we will not be able to locate this region, and not be able to obtain $f(x)$ because we will not know where to put σ. These caveats lead to the following theorem:

♠ **Theorem:**
The inverse of the two-sided Laplace transform

$$f(x) = \frac{1}{2\pi i}\int_{\sigma-i\infty}^{\sigma+i\infty} \mathcal{F}(s)e^{-sx}ds \quad (\mathrm{Re}(s) = \sigma \text{ is fixed})$$

determines $f(x)$ uniquely only if we know where σ should be located.

13.4.3 Inverse of the One-Sided Laplace Transform

Let us develop the theory that correspond to the above for the one-sided transform. We compare the two-sided transform $\mathcal{L}[f(x)]$ and its one-sided counterpart $L[f(x)]$, where $f(x)$ is the same function in both cases and is defined for all x. From the definitions of the one- and two-sided transforms, it is evident that

$$F(s) = L[f(x)] = \mathcal{L}[f(x)\theta(x)],$$

where $\theta(x)$ is the step function. This implies that

$$f(x)\theta(x) = \frac{1}{2\pi i}\int_{\sigma-i\infty}^{\sigma+i\infty} F(s)e^{sx}ds \quad (\sigma \text{ is fixed}). \tag{13.40}$$

Here, σ must be to the right of all the singularities of $F(s)$ in order for the integral in (13.40) to converge. As a consequence, we have arrived at the following theorem:

♠ **The inverse Laplace transformation:**
If the function $F(s)$ defined by

$$F(s) = \int_0^\infty e^{-sx}f(x)dx$$

is analytic for $\mathrm{Re}(s) > \sigma_c$, then $f(x)$ for $x > 0$ is uniquely determined by

$$f(x) = \lim_{\omega\to\infty}\frac{1}{2\pi i}\int_{\sigma-i\omega}^{\sigma+i\omega} e^{sx}F(s)ds,$$

where σ is arbitrary for all $\sigma > \sigma_c$.

13.4.4 Useful Formula for Inverse Laplace Transformation

In contrast to the situation with the inverse Fourier transformation, the use of the inverse Laplace transformation formula is less convenient. This is primarily because the calculation of the complex integral

$$\int_{\sigma-i\infty}^{\sigma+i\infty} e^{sx}F(s)ds$$

can be rather complicated. In this subsection, we present a simple and natural method of computing integrals of this form that is based on the **residue theorem**.

Suppose that $F(s)$ is analytic on the domain $\mathrm{Re}(s) > \sigma_c$. We wish to compute

$$f(x) = \lim_{\omega\to\infty} \frac{1}{2\pi i} \int_{\sigma-i\omega}^{\sigma+i\omega} e^{sx}F(s)ds, \quad x > 0.$$

No general method for doing this exists, but it is possible to evaluate this integral under certain conditions on $F(s)$. Suppose that $F(s)$ is analytic on the entire complex plane, except at a finite number of singularities $s_1, s_2, \cdots, s_n$ satisfying

$$\mathrm{Re}(s_j) < \sigma_c, \quad j = 1, 2, \cdots, n.$$

Figure 13.10 is a sketch for this situation. Let $\sigma > \sigma_c$ and let $R > 0$ be a real number sufficiently large that the left half-circle C_R with center $s = \sigma$ and radius R encloses all the points $s_1, s_2, \cdots, s_n$. Devide C_R into the two segments:

$$I_R = \{s \in \boldsymbol{C} : \ s = \sigma + i\omega, -R < \omega < R\},$$

$$\Gamma_R = \{s \in \boldsymbol{C} : \ |s - \sigma| = R, \mathrm{Re}(s) \le \sigma\}.$$

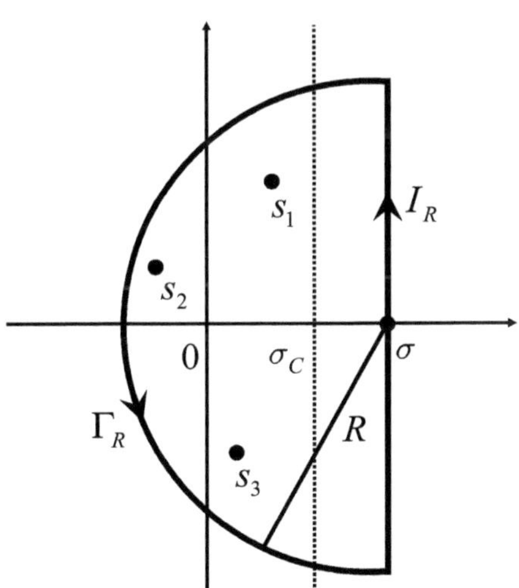

Fig. 13.10. A finite number of singularities s_j of $F(s)$ enclosed by the *left half-circle* C_R composed of Γ_R and I_R

By the residue theorem,

$$\oint_{C_R} e^{sx} F(s) ds = \sum_{j=1}^{n} \text{Res}\left[e^{sx} F(s);\ s_j\right].$$

The right-hand side is independent of R, if R is sufficiently large. From $C_R = \Gamma_R \cup I_R$, it follows that

$$\oint_{C_R} e^{sx} F(s) ds = \int_{\Gamma_R} e^{sx} F(s) ds + \int_{I_R} e^{sx} F(s) ds.$$

Clearly,

$$\lim_{M\to\infty} \int_{\sigma-iM}^{\sigma+iM} e^{sx} F(s) ds = \lim_{R\to\infty} \int_{I_R} e^{sx} F(s) ds.$$

Therefore if, by chance, we have

$$\lim_{R\to\infty} \int_{\Gamma_R} e^{sx} F(s) ds = 0, \tag{13.41}$$

then we obtain

$$f(x) = \lim_{M\to\infty} \frac{1}{2\pi i} \int_{\sigma-iM}^{\sigma+iM} e^{sx} F(s) ds = \sum_{j=1}^{n} \text{Res}\left[e^{sx} F(s);\ s_j\right].$$

Unfortunately, condition (13.41) does not hold for every F. The next theorem presents a sufficient condition on F under which (13.41) holds true.

♠ **Theorem:**

Let F be an analytic function on the complex plane except at a finite number of points (if they exist) and let Γ_R be as above. If

$$\lim_{R\to\infty} \max_{s\in\Gamma_R} |F(s)| = 0,$$

then

$$\lim_{R\to\infty} \int_{\Gamma_R} e^{sx} F(s) ds = 0$$

holds for every $x > 0$.

Proof This theorem is a reinterpretation of **Jordan's lemma** given in Sect. 9.2.4. ♣

An immediate consequence of this theorem is the following:

♠ **Theorem:**
Let F be an analytic function on the complex plane except at a finite number of points $s_1, s_2, \cdots, s_n$, satisfying $\text{Re}(s_j) < \sigma$ for all j. If

$$\lim_{R\to\infty} \max_{s\in C_R} |F(s)| = 0, \tag{13.42}$$

then the inverse Laplace transform of $F(s)$ is given by

$$f(x) = \sum_{j=1}^{n} \text{Res}\left[e^{sx}F(s);\ s_j\right]. \tag{13.43}$$

13.4.5 Evaluating Inverse Transformations

Below are several examples of actual evaluations of inverse Laplace transforms via the residue formula (13.43).

Example 1. Assume a complex-valued function

$$F(s) = \frac{1}{s^2 - 3s + 2}$$

that has two simple poles $s_1 = 1$ and $s_2 = 2$ We thus choose $\sigma = 3$ and set

$$C_R = \{s : \ |s-3| = R, \text{Re}(s) \le 3\}$$

in order to make use of equation (13.43). Before doing so, we must check that condition (13.42) is satisfied. Observe that

$$\max_{s\in C_R} |F(s)| = \max_{s\in C_R} \left| \frac{1}{(s-1)(s-2)} \right|.$$

If we let $R = |s-3|$ go to infinity, then $|s-1|$ and $|s-2|$ will also converge to infinity, so that

$$\lim_{R\to\infty} \max_{s\in C_R} \frac{1}{|(s-1)(s-2)|} = 0.$$

Thus (13.43) provides the desired result:

$$\begin{aligned} f(x) &= \text{Res}\left[\frac{e^{sx}}{(s-1)(s-2)};\ s=1\right] + \text{Res}\left[\frac{e^{sx}}{(s-1)(s-2)};\ s=2\right] \\ &= \left.\frac{e^{sx}}{s-2}\right|_{s=1} + \left.\frac{e^{sx}}{s-1}\right|_{s=2} \\ &= \frac{e^x}{1-2} + \frac{e^{2x}}{2-1} = -e^x + e^{2x}. \quad ♣ \end{aligned}$$

Remark. The above example can be solved more easily by rewriting F using partial fractions $F(s) = 1/(s-2) - 1/(s-1)$, followed by applying known equations to get

$$f(x) = L^{-1}[F(s)] = L^{-1}\left[\frac{1}{s-2}\right] - L^{-1}\left[\frac{1}{s-2}\right] = e^{2x} - e^{x}.$$

Example 2. It should be cautioned that equation (13.43) is valid only when the condition (13.42) is satisfied. As a negative example, let us consider the step function

$$\theta(x) = \begin{cases} 1, & x > c, \\ 0, & x < c, \end{cases}$$

with $c > 0$, whose Laplace transform reads

$$F(s) = L[\theta(x)] = \frac{e^{-cs}}{s} \quad (s > 0),$$

We would like to derive $\theta(x)$ from $F(s)$ through the inverse transformation given by

$$f(x) = L^{-1}[F(s)] = \lim_{M\to\infty} \frac{1}{2\pi i} \int_{\sigma - iM}^{\sigma + iM} \frac{e^{sx} e^{-cs}}{s} ds \quad (0 < x \neq c).$$

However, we cannot use (13.43) to obtain $f(x)$, since the function e^{-cs}/s does not satisfy condition (13.42). In fact, if we set $s = \sigma - R$, then

$$\max_{s \in C_R} \left| \frac{e^{-cs}}{s} \right| \geq \frac{e^{cR} e^{-c\sigma}}{|\sigma - R|} \to \infty \quad (R \to \infty)$$

since $c > 0$.

Remark. If we were to use (13.43) in Example **2**, we would obtain a wrong result. The function e^{-cs}/s has a single simple pole at $s = 0$, so

$$\text{Res}\left[\frac{e^{sx} e^{-cs}}{s};\ s = 0\right] = 1$$

for each value of x. This is, of course, not the step function $\theta(x)$.

Example 3. Next we consider the inverse Laplace transformation $L^{-1}[F(s)]$ of the function

$$F(s) = \frac{1}{s + a} \quad (a > 0).$$

The $F(s)$ has a first-order pole at $s = -a$. The residue of $F(s)e^{sx}$ at $s = -a$ reads

$$\text{Res}\,[F(s)e^{sx},\ -a] = \lim_{s\to -a}(s+a)F(s)e^{sx} = \lim_{s\to -a} e^{sx} = e^{-ax}.$$

Hence, we have

$$f(x) = L^{-1}[F(s)] = e^{-ax} \quad (x > 0).$$

13.4.6 Inverse Transform of Multivalued Functions

Some caution must be taken when considering the inverse Laplace transform of **multivalued functions**. As an example, we consider a multivalued function $F(s) = 1/\sqrt{s}$, and examine its inverse transform given by

$$f(x) = \frac{1}{2\pi i}\int_C \frac{e^{sx}}{\sqrt{s}}ds, \tag{13.44}$$

where the symbol $\sqrt{s}$ represents values of the original double-valued function $s^{1/2}$ in the same sheet of the **Riemann surface**. The function $1/\sqrt{s}$ has a **branch point** at $s = 0$, so among many choices we set its **branch cut** at $(-\infty, 0]$.

Since the function $1/\sqrt{s}$ approaches zero as $|s| \to \infty$, **Jordan's lemma** is applicable. Nevertheless, the problem becomes rather complicated owing to the presence of the branch cut. To perform the integration of (13.44), we close the path Γ by a circle to the left, bypassing the branch cut in the manner shown in Fig. 13.11. No singularities are enclosed by the closed curve consising of $C' + \Gamma + \gamma + C''$, in which C' is the vertical line, C'' is the pair of parallel

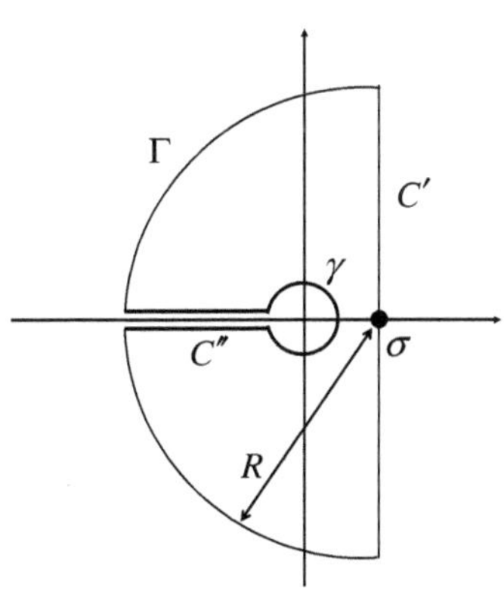

Fig. 13.11. Closed loop employed in evaluating the integral (13.44)

horizontal segments, γ is the small circle of radius δ, and Γ is a semicircle from which the infinitesimal gap at the branch cut has been omitted. Hence, we have

$$\int_{C'} \frac{e^{sx}}{\sqrt{s}} ds + \int_{\Gamma} \frac{e^{sx}}{\sqrt{s}} ds + \int_{\gamma} \frac{e^{sx}}{\sqrt{s}} ds + \int_{C''} \frac{e^{sx}}{\sqrt{s}} ds = 0.$$

In the limit $R \to \infty$, the integral over Γ vanishes and the path C' reduces to C as given in (13.44), which implies that

$$f(x) = \lim_{R\to\infty} \frac{1}{2\pi i} \int_C \frac{e^{sx}}{\sqrt{s}} ds = \lim_{R\to\infty} \frac{-1}{2\pi i} \int_{C''+\gamma} \frac{e^{sx}}{\sqrt{s}} ds. \tag{13.45}$$

Thus our remaineing task is to evaluate the last term in (13.45).

Recall that $\sqrt{s}$ is double-valued so that it is discontinuous across the branch cut. Consequently, on the parallel segments C'',

$$\sqrt{s} = i\sqrt{\rho} \quad \text{and} \quad -i\sqrt{\rho} \quad \text{with } \rho = |s|,$$

above and below the branch cut, respectively. On the small circle γ,

$$\sqrt{s} = \sqrt{\delta} e^{i\phi/2},$$

where $-\pi < \phi \le \pi$, and $ds = -d\rho$ on each of the straight lines. Thus, we have

$$\int_{C''+\gamma} \frac{e^{sx}}{\sqrt{s}} ds = -i \int_{R-\sigma_1}^{\delta} \frac{e^{-\rho x}}{\sqrt{\rho}} (-d\rho) + i \int_{\delta}^{R-\sigma_1} \frac{e^{-\rho x}}{\sqrt{\rho}} (-d\rho)$$
$$+ i \frac{\delta}{\sqrt{\delta}} \int_{\pi}^{-\pi} \frac{e^{(\delta \cos\phi)x} e^{i(\delta \sin\phi)x} e^{i\phi}}{e^{i\phi/2}} d\phi.$$

Let δ go to zero and R approach infinity; then, the first two integrals on the right-hand side combine into a single integral. The last integral on the right-hand side approaches zero. As a result, we have

$$\lim_{\delta\to 0} \lim_{R\to\infty} \int_{C''+\gamma} \frac{e^{sx}}{\sqrt{s}} ds = -2i \int_0^{\infty} \frac{e^{-\rho x}}{\sqrt{\rho}} d\rho,$$

which implies that

$$f(x) = \frac{1}{\pi} \int_0^{\infty} \frac{e^{-\rho x}}{\sqrt{\rho}} d\rho.$$

By substituting $\rho x = u^2$, the right-hand side becomes

$$\frac{1}{\pi} \int_0^{\infty} \frac{e^{-\rho x}}{\sqrt{\rho}} d\rho = \frac{2}{\pi\sqrt{x}} \int_0^{\infty} e^{-u^2} du = \frac{1}{\sqrt{\pi x}}.$$

Eventually, we obtain

$$f(x) = L^{-1}\left(\frac{1}{\sqrt{s}}\right) = \frac{1}{\sqrt{\pi x}},$$

which is consistent with the earlier result presented in (13.15).

Exercises

1. Find **(a)** $L^{-1}[5/(p+2)]$ and **(b)** $L^{-1}[1/p^s]$ where $s > 0$.

Solution: **(a)** Recall that $L[e^{ax}] = 1/(p-a)$; hence $L^{-1}[1/(p-a)] = e^{ax}$. It follows that

$$L^{-1}\left[\frac{5}{p+2}\right] = 5L^{-1}\left[\frac{1}{p+2}\right] = 5e^{-2x}.$$

(b) Recall that

$$L[x^k] = \int_0^\infty e^{-sx}x^k dx = \frac{\Gamma(k+1)}{p^{k+1}}.$$

From this we have

$$L\left[\frac{x^k}{\Gamma(k+1)}\right] = \frac{1}{p^{k+1}},$$

so

$$L^{-1}\left[\frac{1}{p^{k+1}}\right] = \frac{x^k}{\Gamma(k+1)}.$$

If we now let $k+1 = s$, then

$$L^{-1}\left[\frac{1}{p^s}\right] = \frac{x^{s-1}}{\Gamma(s)}. \clubsuit$$

2. Solve the differential equation

$$f''(x) + f(x) = 1 \tag{13.46}$$

with the initial conditions

$$f(0) = f'(0) = 0$$

using the Laplace transformation.

Solution: Taking the Laplace transform of both sides of (13.46), we obtain

$$L[f''(x)] + L[f(x)] = L[1]. \tag{13.47}$$

Substituting the result

$$L[f''(x)] = s^2 L[f(x)] - s \cdot f(0) - f'(0) \quad = s^2 L[f(x)]$$

and $L[1] = 1/s$ into (13.47) yields

$$s^2 L[f(x)] + L[f(x)] = \frac{1}{s},$$

i.e.,

$$L[f(x)] = F(s) = \frac{1}{s(s^2+1)} = \frac{1}{s} - \frac{s}{s^2+1}.$$

Thus we see that

$$\begin{aligned} f(x) = L^{-1}[F(s)] &= L^{-1}\left[\frac{1}{s}\right] - L^{-1}\left[\frac{s}{s^2+1}\right] \\ &= \begin{cases} 1 - \cos x & \text{for } x \geq 0, \\ 0 & \text{for } x < 0, \end{cases} \end{aligned}$$

which is the solution of the initial value problem originally given by (13.47). ♣

3. Derive the two-sided Laplace transform of the following three functions:

$$f_a(x) = \begin{cases} e^{-2x} - e^{-x}, & x > 0, \\ 0, & x < 0, \end{cases} \qquad f_b(x) = \begin{cases} e^{-2x}, & x > 0, \\ e^{-x}, & x < 0, \end{cases}$$

and

$$f_c(x) = \begin{cases} 0, & x > 0, \\ e^{-2x} - e^{-x}, & x < 0. \end{cases}$$

Solution: The two-sided Laplace transform read, respectively,

$$\begin{aligned} \mathcal{L}[f_a(x)] &= \frac{1}{s+2} - \frac{1}{s+1} \quad \text{for } \sigma > -1, \\ \mathcal{L}[f_b(x)] &= \frac{1}{s+2} - \frac{1}{s+1} \quad \text{for } -2 < \sigma < -1, \\ \mathcal{L}[f_c(x)] &= \frac{1}{s+2} - \frac{1}{s+1} \quad \text{for } \sigma < -2. \end{aligned}$$

Clearly, all the s functions are the same and may be labeled $\mathcal{F}(s)$ (although the region of convergence is different). This implies that the inverse of a two-sided transform is uniquely determined only after the location of σ is fixed. ♣

13.5 Applications in Physics and Engineering

13.5.1 Electric Circuits I

The most familiar applications of Laplace transformations in the physical sciences are encountered in analyses of electric circuits. Consider the RC circuit depicted in Fig. 13.12. The electric charge $q(t)$ deposited in the condenser with capacitance C is governed by the equation

$$R\frac{dq(t)}{dt} + \frac{q(t)}{C} = v(t), \quad q(t=0) = 0, \tag{13.48}$$

where R is a resistance and $v(t)$ is the external voltage. We set a rectangular voltage defined by (see Fig. 13.13)

$$v(t) = v_0 \times [\theta(t-a) - \theta(t-b)] \quad (a < b) \tag{13.49}$$

with the step function

$$\theta(t-a) = \begin{cases} 0, & t < a, \\ 1, & t \geq a. \end{cases}$$

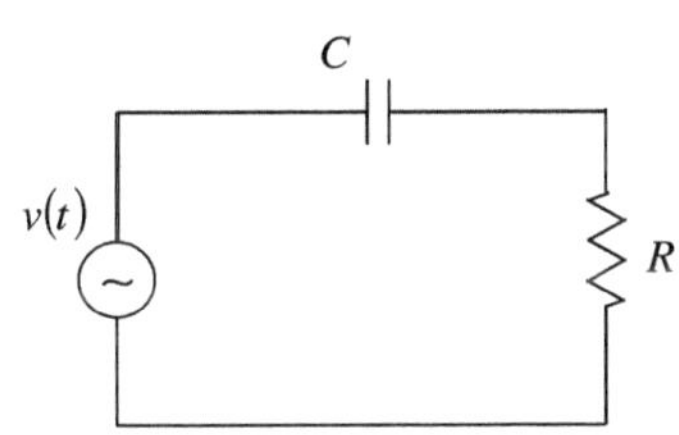

Fig. 13.12. Diagram of an RC circuit

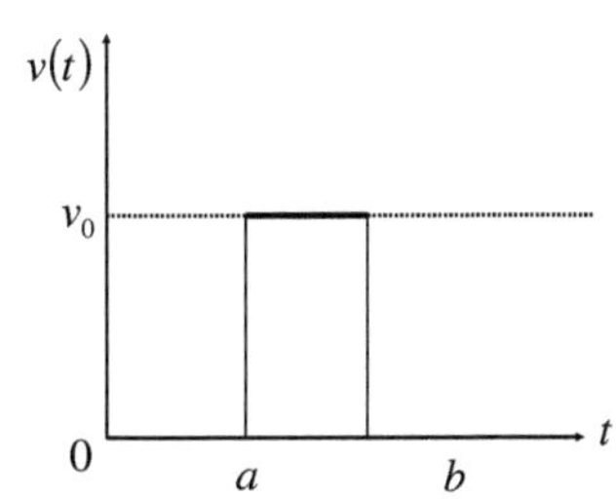

Fig. 13.13. The time dependence of a rectangular voltage applied to the RC circuit

We now want to solve the differential equation (13.48) with respect to $q(t)$. To do this, we apply the Laplace transform to both sides of (13.48) and make use of the symbol $Q(s) \equiv L[q(t)]$. Straightforward calculation yields

$$sQ(s) + \frac{Q(s)}{\tau} = \frac{v_0}{R}\left(\frac{e^{-as}}{s} - \frac{e^{-bs}}{s}\right),$$

where $\tau \equiv RC$ is called a **damping time constant**. Hence, we have

$$\begin{aligned} Q(s) &= \frac{v_0}{R}\frac{1}{s+\tau^{-1}}\left(\frac{e^{-as}}{s} - \frac{e^{-bs}}{s}\right) \\ &= Cv_0\left(\frac{1}{s} - \frac{1}{s+\tau^{-1}}\right)\left(e^{-as} - e^{-bs}\right) \\ &= Cv_0\left(\frac{e^{-as}}{s} - \frac{e^{-bs}}{s} - \frac{e^{-as}}{s+\tau^{-1}} + \frac{e^{-bs}}{s+\tau^{-1}}\right). \end{aligned}$$

Then we use the inverse transform to obtain

$$\begin{aligned} q(t) &= L^{-1}[Q(s)] \\ &= Cv_0\left[\theta(t-a) - \theta(t-b) - e^{-(t-a)/\tau}\theta(t-a) + e^{-(t-b)/\tau}\theta(t-b)\right], \\ &= \begin{cases} 0 & t < a, \\ cv_0\left[1 - e^{-(t-a)/\tau}\right] & a < t < b, \\ cv_0\left[e^{b/\tau} - e^{a/\tau}\right]e^{-t/\tau} & t > b. \end{cases} \end{aligned}$$

The explicit time-dependence of the charge $q(t)$ given by (13.50) is illustrated in Fig. 13.14, in which various separations $b - a$ are taken.

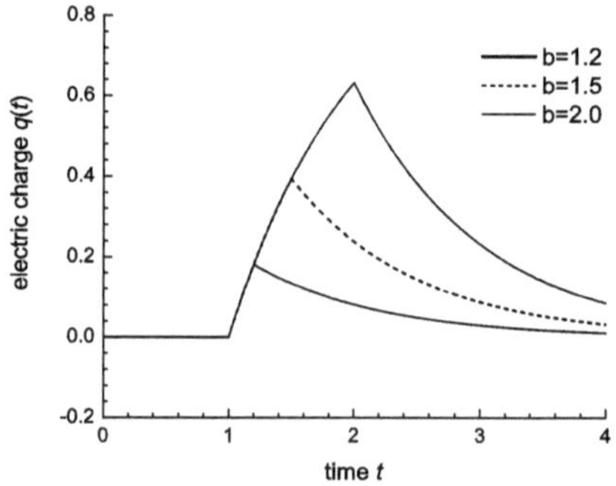

Fig. 13.14. Time dependence of the electric charge $q(t)$ described by (13.50), which is accumulated in the condenser in the RC circuit. The parameter a introduced in (13.49) is fixed at $a = 1.0$

13.5.2 Electric Circuits II

Next, in order to illustrate the use of **convolution integrals** in applications of Laplace transforms, we solve the previous equation (13.48) with respect to the current $i(t)$ instead of charge $q(t)$. We consider the differential equation

$$Ri(t) + \frac{1}{C}\int_0^t i(u)du = v(t), \tag{13.50}$$

with the rectangular voltage (13.49). The integral term on the left-hand side in (13.50) is rewritten as a convolution integral:

$$\int_0^t i(u)du = \int_0^t i(u)\theta(t-u)du = \theta(t) * i(t), \tag{13.51}$$

whose Laplace transform reads

$$L[\theta(t) * i(t)] = L[\theta(t)] \cdot L[i(t)] = \frac{1}{s}I(s).$$

Hence, applying the Laplace transformation of both sides of (13.50) yields

$$RI(s) + \frac{I(s)}{Cs} = \frac{v_0}{s}\left(e^{-as} - e^{-bs}\right), \tag{13.52}$$

which implies

$$I(s) = \frac{v_0}{R}\frac{e^{-as} - e^{-bs}}{s+\tau^{-1}} \quad (\tau = RC). \tag{13.53}$$

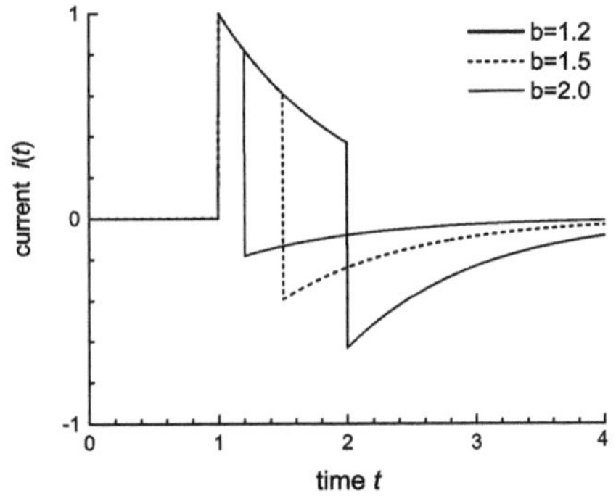

Fig. 13.15. Time dependence of the current $i(t)$ in the RC circuit described by (13.54). The parameter a introduced in (13.49) is fixed at $a = 1.0$

Using the inverse transformaion, we finally set

$$
\begin{aligned}
i(t) &= L^{-1}\left[I(s)\right] \\
&= \frac{v_0}{R}\left[e^{-(t-a)/\tau}\theta(t-a) - e^{-(t-b)/\tau}\theta(t-b)\right] \\
&= \begin{cases} 0 & t < a, \\ \dfrac{v_0}{R}e^{a/\tau}e^{-t/\tau} & a \le t < b, \\ \dfrac{v_0}{R}\left(e^{a/\tau} - e^{b/\tau}\right)e^{-t/\tau} & t \ge b. \end{cases}
\end{aligned}
$$

Figure 13.15 illustrates the time dependence of the current $i(t)$.

14

Wavelet Transformation

Abstract Similar to the Fourier and Laplace transforms, a wavelet transform is an integral transform of a function by using "wavelets." A wavelet is a mathematical mold with a finite-length and fast-decaying oscillating waveform, which is used to divide a given function into different scale components. Wavelet transforms have certain advantages over conventional Fourier transforms, as they can reveal the nature of a function in the time and frequency domains simultaneously.

14.1 Continuous Wavelet Analyses

14.1.1 Definition of Wavelet

This short chapter covers the minimum ground for understanding **wavelet analysis**. The concept of **wavelet** originates from the study of signal analysis, i.e., from the need in certain cases to analyze a signal in the time and frequency domains simultaneously. The crucial advantage of wavelet analyses is that they allow us to decompose complicated information contained in a signal into elementary functions associated with different time scales and different frequencies and to reconstruct it with high precision and efficiency. In the following discussions, we first determine what constitutes a wavelet and then describe how it is used in the transformation of a signal.

The primary question concerns the definition of a wavelet:

♠ **Wavelet:**

A **wavelet** is a real-valued function $\psi(t)$ having a localized waveform that satisfies the following criteria:

1. The integral of $\psi(t)$ is zero: $\int_{-\infty}^{\infty} \psi(t)dt = 0.$
2. The square of $\psi(t)$ integrates to unity: $\int_{-\infty}^{\infty} \psi(t)^2 dt = 1.$

3. The Fourier transform $\Psi(\omega)$ of $\psi(t)$ satisfies the **admissibility condition** expressed by

$$C_\Psi \equiv \int_0^\infty \frac{|\Psi(\omega)|^2}{\omega} d\omega < \infty. \tag{14.1}$$

Here, C_Ψ is called the **admissibility constant**, whose value depends on the chosen wavelet.

We restrict our attention to real-valued wavelets, although it is possible to define complex-valued wavelets as well. Observe that condition **2** above says that $\psi(t)$ has to deviate from zero at finite intervals of t. On the other hand, condition **1** tells us that any deviation above zero must be canceled out by a deviation below zero. Hence, $\psi(t)$ must oscillate across the t-axis like a wave. The following are the most important two examples of wavelets:

Examples **1.** The **Haar wavelet** (See Fig. 14.1a):

$$\psi(t) \equiv \begin{cases} -\frac{1}{\sqrt{2}}, & -1 < t \leq 0, \\ \frac{1}{\sqrt{2}}, & 0 < t \leq 1, \\ 0, & \text{otherwise}. \end{cases} \tag{14.2}$$

2. The **Mexican hat wavelet** (see Fig. 14.1b):

$$\psi(t) \equiv \frac{2\left(1 - \frac{t^2}{\sigma^2}\right) e^{-t^2/(2\sigma^2)}}{\sqrt{3}\sigma\pi^{1/4}}. \tag{14.3}$$

To form the Mexican hat wavelet (14.3), we start with the Gaussian function with mean zero and variance σ^2:

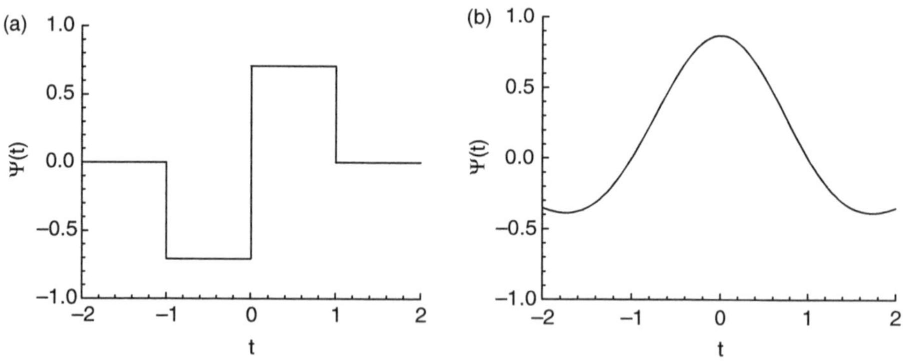

Fig. 14.1. **(a)** The Haar wavelet given by (14.2). **(b)** The Mexican hat wavelet given by (14.3) with $\sigma = 1$

$$f(t) \equiv \frac{e^{-t^2/(2\sigma^2)}}{\sqrt{2\pi\sigma^2}}.$$

If we take the negative of the second derivative of $f(t)$ with normalization for satisfying condition **2**, we obtain the Mexican hat wavelet (14.3). In the meantime, we proceed with our argument on the basis of that wavelet by setting $\sigma = 1$ and omitting the normalization constant for simplicity.

Remark. We know that all the derivatives of the Gaussian function may be used as wavelets. The most appropriate one many particular case depends on the application.

14.1.2 The Wavelet Transform

In mathematical terminology, the **wavelet transform** is known as a **convolution**; more precisely, it is a convolution of the wavelet function with a signal to be analyzed. In the convolution process, two parameters are involved that manipulate the function form of the wavelet. The first is the **dilatation parameter** denoted by a, which characterizes the dilation and contraction of the wavelet in the time domain (see Fig. 14.2a). For the Mexican hat wavelet, it is the distance between the center of the wavelet and its crossing of the time axis. The second is the **translation parameter** b, which governs the movement of the wavelet along the time axis (see Fig. 14.2b). With this notation, shifted and dilated versions of a Mexican hat wavelet are expressed by

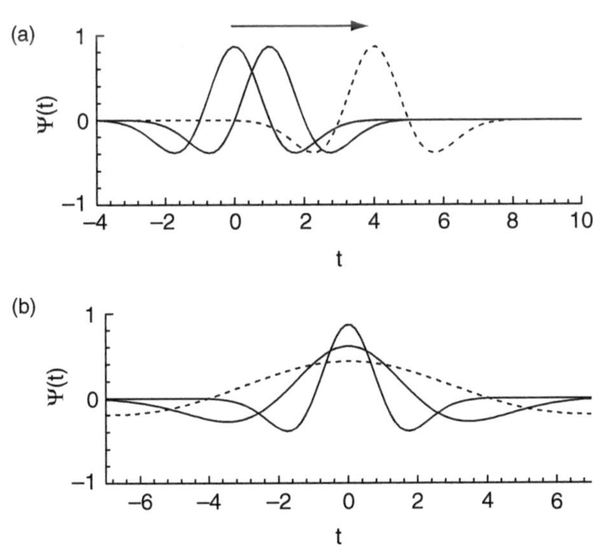

Fig. 14.2. Translation **(a)** and dilatation of a wavelet **(b)**

$$\psi\left(\frac{t-b}{a}\right) = \left[1 - \left(\frac{t-b}{a}\right)^2\right] e^{-[(t-b)/a]^2/2}, \tag{14.4}$$

where we have set $\sigma = 1$ in (14.3) and omitted the normalization factor for simplicity. We are now in a position to define the wavelet transform.

♠ **Wavelet transform:**

The **wavelet transform** $T(a,b)$ of a continuous signal $x(t)$ with respect to the wavelet $\psi(t)$ is defined by

$$T(a,b) = w(a) \int_{-\infty}^{\infty} x(t)\psi\left(\frac{t-b}{a}\right) dt, \tag{14.5}$$

where $w(a)$ is an appropriate **weight function.**

Typically, $w(a)$ is set to $1/\sqrt{a}$ because this choice yields

$$\int_{-\infty}^{\infty} \left[\frac{1}{\sqrt{a}}\psi\left(\frac{t-b}{a}\right)\right]^2 dt = \int_{-\infty}^{\infty} \psi(u)^2 du = 1 \quad \text{with } u = \frac{t-b}{a},$$

i.e., the normalization condition for the square integral of $\psi(t)$ remains invariant, which is why we use this value for the rest of this section.

The dilated and shifted wavelet is often written more compactly as

$$\psi_{a,b}(t) = \frac{1}{\sqrt{a}}\psi\left(\frac{t-b}{a}\right),$$

so that the transform integral may be written as

$$T(a,b) = \int_{-\infty}^{\infty} x(t)\psi_{a,b}(t)dt. \tag{14.6}$$

From here on, we use this notation and refer to $\psi_{a,b}(t)$ simply as the wavelet.

14.1.3 Correlation Between Wavelet and Signal

Having defined the wavelet and its transform, we are ready to see how the transform is used as a signal analysis tool. In plain words, the wavelet transform works as a mathematical microscope, where b is the location on the time series being viewed and a represents the magnification at location b.

Let us look at a simple example evaluating the wavelet transform $T(a,b)$. Figures 14.3 and 14.4 show the same sinusoidal waves together with Mexican hat wavelets of various locations and dilations. In Fig. 14.3a, the wavelet is located on a segment of the signal on which a positive part of the signal is fairly coincidental with that of the wavelet. This results in a large positive

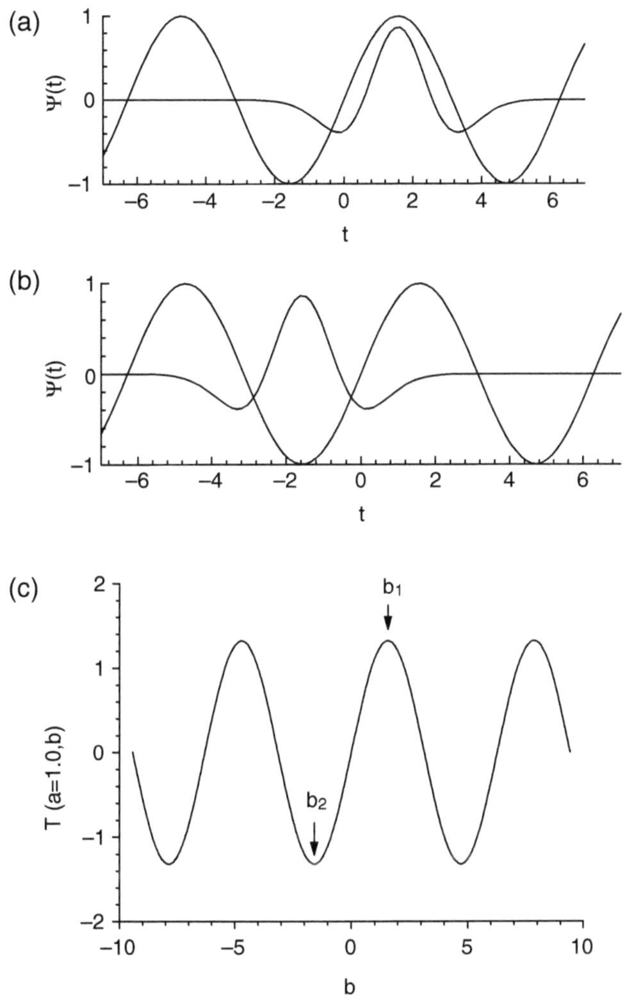

Fig. 14.3. **(a)**, **(b)** Positional relations between the wavelet (*thick*) and signal (*thin*). The wavelet in **(a)** located at $b_1 = \pi/2$ is in phase with the signal, which results in a large positive value of $T(a, b)$ at b_1. The wavelet in **(b)** located at $b_2 = -\pi/2$ is out of phase with the signal, which yields a large negative value of $T(b)$ at b_2. **(c)** The plot of $T(a = 1.0, b)$ as a function of b

value of $T(a, b)$ in (14.6). In Fig. 14.3b, the wavelet is moved to a new location where the wavelet and the signal are out of phase. In this case, the convolution expressed by (14.6) produces a large negative value of $T(a, b)$. In between these two extrema, the value of $T(a, b)$ decreases from a maximum to a minimum as shown in Fig. 14.3. The three figures thus clearly demonstrate how the wavelet transform $T(a, b)$ depends on the translation parameter b of the wavelet of interest.

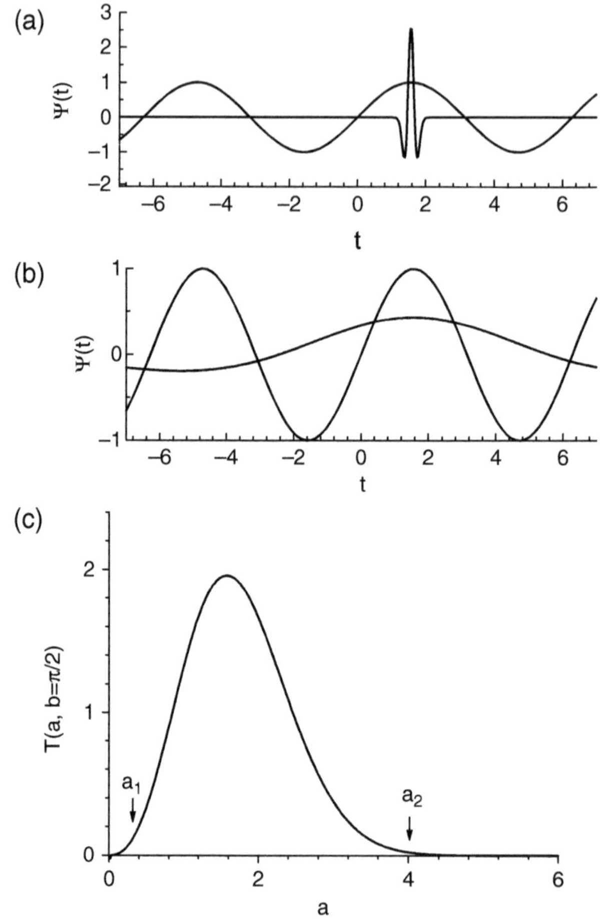

Fig. 14.4. Wavelets with $a = 0.33$ **(a)** and $a = 4.0$ **(b)**, in which $b = \pi/2$ is fixed. The resulting wavelet transform $T(a, b = \pi/2)$ as a function of a is given in **(c)**

In a similar way, Fig. 14.4 a–c shows the dependence of $T(a, b)$ on the dilatation parameter a. When a is quite small, the positive and negative parts of the wavelet are all convolved by roughly the same part of the signal $x(t)$, producing a value of $T(a, b)$ near zero (see Fig. 14.4a). Likewise, $T(a, b)$ tends to zero as a becomes very large (see Fig. 14.4b), since the wavelet covers many positive and negatively repeating parts of the signal. These latter two results indicate that when the dilatation parameter a is either very small or very large compared with the period of the signal, the wavelet transform $T(a, b)$ gives near-zero values.

Figure 14.5 shows a contour plot of $T(a, b)$ vs. a and b for a sinusoidal signal

$$x(t) = \sin t,$$

where the Mexican hat wavelet has been used. The light and shadowed regions indicate positive and negative magnitudes of $T(a,b)$, respectively. The near-zero values of $T(a,b)$ are evident in the plot at both large and small values of a. In addition, at intermediate values of a, we observe large undulations in $T(a,b)$ corresponding to the sinusoidal form of the signal. This wavelike behavior is accounted for by referring back to Figs. 14.3a–b and 14.4a–b, where wavelets move in and out of phase with the signal.

Therefore, when the wavelet matches the shape of the signal well at a specific scale and location, the transform value is high. On the other hand, if the wavelet and the signal do not correlated well, the transform value is low. Carrying out the process at various signal locations and for various wavelet scales, we can determine the correlation between the wavelet and the signal.

Remark. In Fig. 14.5, the maxima and minima of the transform occur at an a scale of one quarter of the period, $\pi/2$, of the sine wave $x(t) = \sin t$. This feature holds in general; correlation between the wavelet $\psi_{a,b}(t)$ and the signal $x(t)$ with a period p becomes a maximum at $a = p/4$.

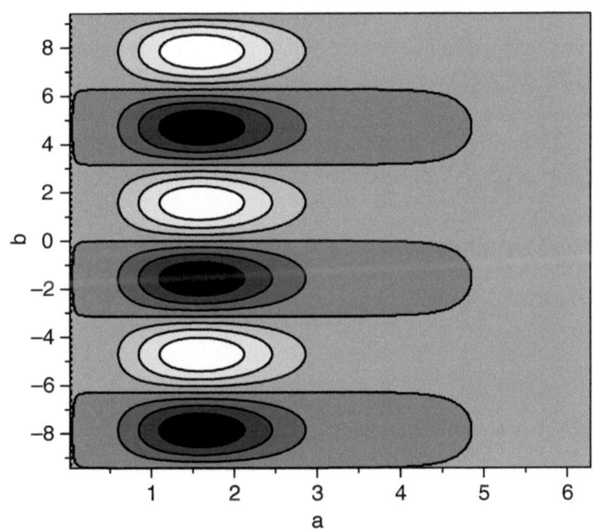

Fig. 14.5. Contour plot of the wavelet transform $T(a,b)$ of a sinusoidal wave $x(t) = \sin t$

14.1.4 Actual Application of the Wavelet Transform

The wavelet transformation procedure can be applied to signals that have a more complicated wave form than a simple sinusoidal wave. Figure 14.6 shows a signal

$$x(t) = \sin t + \sin 3t$$

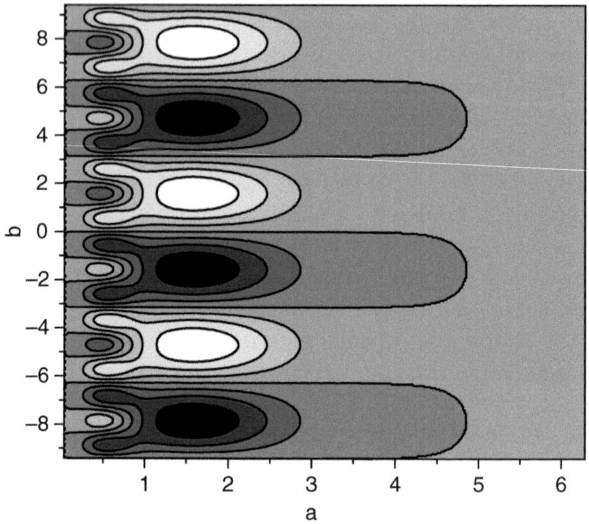

Fig. 14.6. Wavelet transform $T(a,b)$ of a complicated signal $x(t) = \sin t + \sin 3t$

composed of two sinusoidal waves with different frequencies. The wavelet transform $T(a,b)$ of $x(t)$ is plotted in Fig. 14.6. It is clear that the contribution from the wave with the higher-frequency oscillation appears at a smaller a scale. This clearly demonstrates the ability of the wavelet transform to decompose the original signal into its separate components.

14.1.5 Inverse Wavelet Transform

Similar to its Fourier counterpart, there is an **inverse wavelet transformation**, that enables us to reproduce the original signal $x(t)$ from its wavelet transform $T(a,b)$.

♠ **Inverse wavelet transform:**
If $x \in L^2(\boldsymbol{R})$, then f can be reconstructed by equation

$$x(t) = \frac{1}{C_\Psi} \int_{-\infty}^{\infty} db \int_0^{\infty} \frac{da}{a^2} T(a,b) \psi_{a,b}(t), \tag{14.7}$$

where the equality holds **almost everywhere**.

The proof of the equation is based on the lemma below.

♠ **Parseval identity for wavelet transform:**
Let $T_f(a,b), T_g(a,b)$ be the wavelet transform of $f(t), g(t) \in L^2(\boldsymbol{R})$, respectively, associated with the wavelet $\psi_{a,b}(t)$. Then we have

$$\int_0^\infty \frac{da}{a^2}\int_{-\infty}^{\infty} dbT_f(a,b)T_g^*(a,b) = C_\Psi \int_{-\infty}^{\infty} f(t)g(t)^* dt. \tag{14.8}$$

This identity is derived in Exercise **4**. We are now ready to prove the inverse transformation (14.7).

Proof **(of the inverse wavelet transformation):** Assume an arbitrary real function $g(t) \in L^2(\boldsymbol{R})$. It follows from the Parseval identity that

$$\begin{aligned} C_\Psi \int_{-\infty}^{\infty} f(t)g(t)dt &= \int_{-\infty}^{\infty} db \int_0^\infty \frac{da}{a^2} T_f(a,b)T_g(a,b) \\ &= \int_{-\infty}^{\infty} db \int_0^\infty \frac{da}{a^2} T_f(a,b) \int_{-\infty}^{\infty} g(t)\psi_{a,b}(t)dt \\ &= \int_{-\infty}^{\infty} dtg(t) \left[\int_{-\infty}^{\infty} db \int_0^\infty \frac{da}{a^2} T_f(a,b)\psi_{a,b}(t)\right]. \end{aligned}$$

Since $g(t)$ is arbitrary, the inverse equation (14.7) follows. ♣

14.1.6 Noise Reduction Technique

Suppose that the inverse transformation equation (14.7) is rewritten as

$$x^*(t) = \frac{1}{C_\Psi}\int_{-\infty}^{\infty} db \int_{a^*}^{\infty} \frac{da}{a^2} T(a,b)\psi_{a,b}(t),$$

the integration range with respect to a in an interval $[a^*, \infty)$ with $a^* > 0$. Then, the result $x^*(t)$ obtained on the left-hand side deviates from the original signal $x(t)$ owing to the lack of information for the scale from $a = 0$ to $a = a^*$. In applications, this deviation property is made use of as a noise reduction technique.

By way of a demonstration, Fig. 14.7a illustrates a segment of the signal

$$x(t) = \sin t + \sin 3t + R(t)$$

constructed from two sinusoidal waveforms plus a local burst of noise $R(t)$. The transform plot of the composite signal shows the two constituent waveforms at scales $a_1 = \pi/2$ and $a_2 = \pi/6$ in addition to a burst of noise around $b = 5.0$ in a high-frequency region (i.e., small a scale).

Now we try to remove the high-frequency noise component by means of the following reconstruction procedure. Figure 14.7b shows a reconstruction of the signal where we artificially set $T(a, b) = 0$ for $a < a^*$. In effect, we are reconstructing the signal using

$$x(t) = \frac{1}{C_\Psi}\int_{-\infty}^{\infty} db \int_{a^*}^{\infty} \frac{da}{a^2} T(a,b)\psi_{a,b}(t),$$

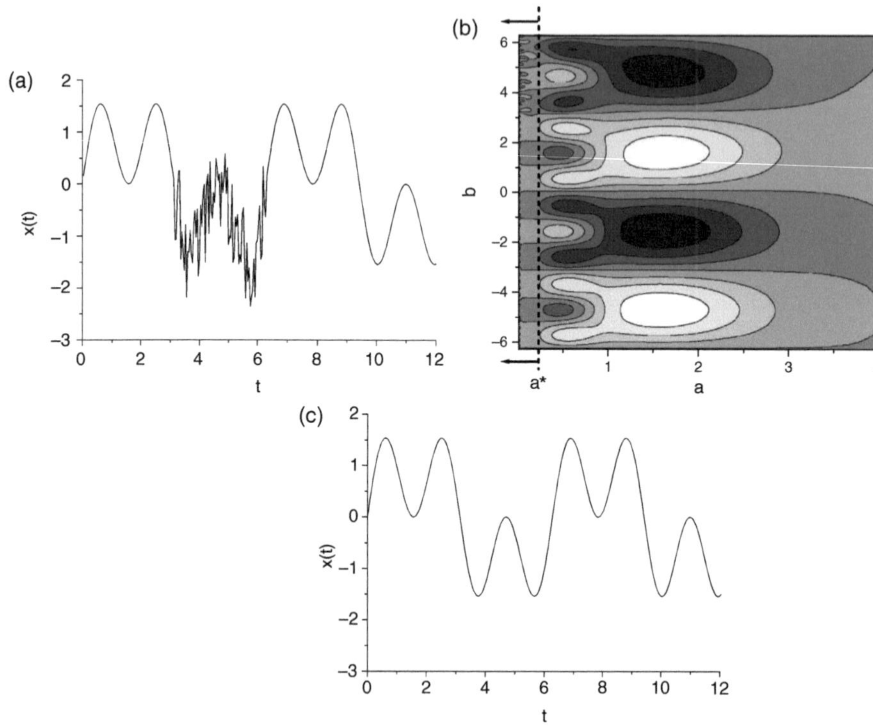

Fig. 14.7. Noise reduction procedure through wavelet transformation. **(a)** A signal $x(t) = \sin t + \sin 3t + R(t)$ with a local burst of noise $R(t)$. **(b)** The wavelet transform $T(a,b)$ of the $x(t)$. Noise reduction is accomplished through the inverse transformation of the $T(a,b)$ by applying an artifical condition of $T(a < a^*, b) = 0$. **(c)** The reconstructed signal $x^*(t)$ from the noise-reduction procedure

i.e., over a range of scales $[a^*, \infty)$. The lower integral limit, a^*, is the cut-off scale indicated by the dotted line in Fig. 14.7b. As a result, the high-frequency noise component evidently reduces in the reconstructed signal as shown in Fig. 14.7c. This simple noise reduction method is known as **scale-dependent thresholding**.

Exercises

1. Show that the Fourier transform of the Haar wavelet satisfies the admissible condition (14.1).

   Solution: The Fourier transform $\Psi(\omega)$ of the Haar wavelet $\psi(t)$ is given by

$$\Psi(\omega) = \int_0^{1/2} e^{-i\omega t} dt - \int_{1/2}^{1} e^{-i\omega t} dt = ie^{-i\omega/2} \frac{\sin^2(\omega/4)}{\omega/4}.$$

Hence, we have

$$C_\Psi = \int_0^\infty \frac{|\Psi(\omega)|^2}{\omega} d\omega = 16 \int_0^\infty \frac{\sin^4(\omega/4)}{\omega^3} d\omega < \infty. \quad ♣$$

2. Prove that the Fourier transform of $\psi_{a,b}(t)$ yields $\Psi_{a,b}(\omega) = \sqrt{a} e^{-ib\omega} \Psi(a\omega)$.

Solution: It readily follows that

$$\Psi_{a,b}(\omega) = \frac{1}{\sqrt{a}} \int_{-\infty}^{\infty} e^{-i\omega t} \psi_{a,b}(t) dt = \frac{1}{\sqrt{a}} \int_{-\infty}^{\infty} e^{-i\omega t} \psi\left(\frac{t-b}{a}\right) dt.$$

Set $u = (t-b)/a$ in the last integral to obtain

$$\Psi_{a,b}(\omega) = \frac{1}{\sqrt{a}} \int_{-\infty}^{\infty} e^{-i\omega \cdot (au+b)} \psi(u) a du = \sqrt{a} e^{-ib\omega} \Psi(a\omega). \quad ♣$$

3. Let $\psi(t)$ be a wavelet and $\phi(t)$ be a real, bounded, and integrable function. Show that the convolution $\psi * \phi$ is also a wavelet.

Solution: We first show that $\psi * \phi \in L^2(\boldsymbol{R})$. Observe that

$$\begin{aligned}
[\psi(t) * \phi(t)]^2 &= \left[\int_{-\infty}^{\infty} \psi(t-u)\phi(u) du\right]^2 \\
&= \left[\int_{-\infty}^{\infty} \psi(t-u)\, \phi(u)^{1/2}\, \phi(u)^{1/2} du\right]^2 \\
&\le \int_{-\infty}^{\infty} \psi(t-u)^2\, \phi(u) du \int_{-\infty}^{\infty} \phi(u') du'.
\end{aligned}$$

The integral $\int_{-\infty}^{\infty} \phi(u')du'$ is a constant, denoted by A. Integrate both sides with respect to t to obtain

$$\begin{aligned}
\int_{-\infty}^{\infty} [\psi(t) * \phi(t)]^2 dt &\le A \int_{-\infty}^{\infty} \phi(u) \left[\int_{-\infty}^{\infty} \psi(t-u)^2 dt\right] du \\
&= A \int_{-\infty}^{\infty} \phi(u) du \int_{-\infty}^{\infty} \psi(t)^2 dt = A^2 \int_{-\infty}^{\infty} \psi(t)^2 dt < \infty,
\end{aligned}$$

which clearly indicates that $\psi * \phi \in L^2(\boldsymbol{R})$. Next we show that the convolution $\psi * \phi$ satisfies the admissibility condition. In fact,

$$\begin{aligned}
\int_{-\infty}^{\infty} \frac{|\mathcal{F}[\psi * \phi]|^2}{\omega} d\omega &= \int_{-\infty}^{\infty} \frac{|\Psi(\omega)\Phi(\omega)|^2}{\omega} d\omega \\
&= \int_{-\infty}^{\infty} \frac{|\Psi(\omega)|^2}{\omega} \sup |\Phi(\omega)|^2 d\omega < \infty.
\end{aligned}$$

These two results implys that the convolution $\psi * \phi$ is a wavelet. ♣

4. Derive the Parseval identity for the wavelet transform (14.8).

Solution: The transform $T_f(a,b)$ reads

$$T_f(a,b) = \int_{-\infty}^{\infty} f(t)\psi_{a,b}^*(t)dt = \frac{1}{2\pi}\int_{-\infty}^{\infty} F(\omega)\sqrt{a}e^{-ib\omega}\Psi(a\omega)d\omega,$$

where we used the fact that $\Psi_{a,b}(\omega) = \sqrt{a}e^{-ib\omega}\Psi(a\omega)$. Similarly, we have $T_g(a,b) = \frac{1}{2\pi}\int_{-\infty}^{\infty} G(\omega)\sqrt{a}e^{-ib\omega}\Psi(a\omega)d\omega$. Hence, we have

$$\begin{aligned}
&\int_0^{\infty}\frac{da}{a^2}\int_{-\infty}^{\infty} dbT_f(a,b)T_g(a,b) \\
&= \int_0^{\infty}\frac{da}{a^2}\int_{-\infty}^{\infty} db\int_{-\infty}^{\infty} d\omega\int_{-\infty}^{\infty} d\omega'\frac{ae^{-ib(\omega+\omega')}}{(2\pi)^2}F(\omega)G(\omega')\Psi(a\omega)\Psi(a\omega') \\
&= \frac{1}{2\pi}\int_0^{\infty}\frac{da}{a}\int_{-\infty}^{\infty} d\omega\int_{-\infty}^{\infty} d\omega' F(\omega)G(\omega')\Psi(a\omega)\Psi(a\omega')\delta(\omega+\omega') \\
&= \frac{1}{2\pi}\int_0^{\infty}\frac{da}{a}\int_{-\infty}^{\infty} d\omega F(\omega)G(-\omega)\Psi(a\omega)\Psi(-a\omega).
\end{aligned}$$

Since $\psi(t)$ and $g(t)$ are both real, $\Psi(-a\omega) = \Psi(a\omega)^*$ and $G(-\omega) = G^*(\omega)$. Thus we have

$$\begin{aligned}
\int_0^{\infty}\frac{da}{a^2}\int_{-\infty}^{\infty} dbT_f(a,b)T_g(a,b) &= \frac{1}{2\pi}\int_{-\infty}^{\infty} d\omega F(\omega)G^*(\omega)\int_0^{\infty}\frac{|\Psi(x)|^2}{x}dx \\
&= C_{\Psi}\int_{-\infty}^{\infty} f(t)g(t)dt,
\end{aligned}$$

where $x \equiv a\omega$. This completes the proof. ♣

14.2 Discrete Wavelet Analysis

14.2.1 Discrete Wavelet Transforms

Having discussed the continuous wavelet transform, we move on to its discrete version, known as the **discrete wavelet transform**. In many applications, data are represented by a finite number of values, so it is important and often useful to consider the discrete version of a wavelet transform. We also can use an efficient numerical algorithm, called the **fast wavelet transform**, which allows us to compute the wavelet transform of the signal and its inverse quite efficiently.

We begin with the definition of a **discrete wavelet**. In the previous section, the wavelet function was defined at scale a and location b as

$$\psi_{a,b}(t) = \frac{1}{\sqrt{a}}\psi\left(\frac{t-b}{a}\right),$$

in which the values of parameters a and b can change continuously. We now want to discretize the values of a and b. One possible way to sample a and b is to use a logarithmic discretization of the a scale and link this to the size of the steps taken between b locations. This kind of discretization yields

$$\psi_{m,n}(t) = \frac{1}{\sqrt{a_0^m}}\psi\left(\frac{t-nb_0a_0^m}{a_0^m}\right), \tag{14.9}$$

where the integers m and n control the wavelet dilation and translation respectively; a_0 is a specified fixed dilation step parameter and b_0 is the location parameter. In the expression (14.9), the size of the translation steps, $\Delta b = b_0 a_0^m$, is directly proportional to the wavelet scale, a_0^m.

Common choices for discrete wavelet parameters a_0 and b_0 are $1/2$ and 1, respectively. This power-of-two logarithmic scaling of the dilation steps is known as the **dyadic grid arrangement**. Substituting $a_0 = 1/2$ and $b_0 = 1$ into (14.9), we obtain the **dyadic grid wavelet** represented by

$$\psi_{m,n}(t) = 2^{m/2}\psi\left(2^m t - n\right). \tag{14.10}$$

Using the dyadic grid wavelet of (14.10), we arrive at the **discrete wavelet transform** of a continuous signal $x(t)$:

♠ **Discrete wavelet transform:**

$$T_{m,n} = \int_{-\infty}^{\infty} x(t)\psi_{m,n}(t)dt = \int_{-\infty}^{\infty} x(t)2^{m/2}\psi\left(2^m t - n\right)dt. \tag{14.11}$$

Remark. Note that the discrete wavelet transform (14.11) differs from the discretized approximation of the continuous wavelet transform given by

$$T(a,b) = \int_{-\infty}^{\infty} x(t)\psi_{a,b}^*(t)dt \simeq \sum_{l=-\infty}^{\infty} x(l\Delta t)\psi_{a,b}^*(l\Delta t)\Delta t. \tag{14.12}$$

In (14.12), the integration variable t is discretized, and a and b are continuous whose values can be arbitrarily chosen. On the other hand, in the discrete wavelet transform (14.11), a and b are discretized and t remains continuous.

14.2.2 Complete Orthonormal Wavelets

The fundamental question is whether the original signal $x(t)$ can be constructed from the discrete wavelet transform $T_{m,n}$ through the relation

$$x(t) = \sum_{m=-\infty}^{\infty} \sum_{n=-\infty}^{\infty} T_{m,n} \psi_{m,n}(t). \tag{14.13}$$

As intuitively understood, the reconstruction equation (14.13) is justified if the discretized wavelets $\psi_{m,n}(t)$ are **orthonormal** and **complete**. The completeness of $\psi_{m,n}(t)$ implies that any function $x \in L^2(\boldsymbol{R})$ can be expanded by

$$x(t) = \sum_{m=-\infty}^{\infty} \sum_{n=-\infty}^{\infty} c_{m,n} \psi_{m,n}(t) \tag{14.14}$$

with appropriate expansion coefficients $c_{m,n}$. Hence, the orthonormality

$$\int_{-\infty}^{\infty} \psi_{m,n}(t) \psi_{m',n'}(t) dt = \delta_{m,n} \delta_{m',n'} \tag{14.15}$$

results in $c_{m,n} = T_{m,n}$ in (14.14) because

$$\begin{aligned}
T_{m,n} &= \int_{-\infty}^{\infty} x(t) \psi_{m,n}(t) dt = \int_{-\infty}^{\infty} \left[\sum_{m'=-\infty}^{\infty} \sum_{n'=-\infty}^{\infty} c_{m',n'} \psi_{m',n'}(t) \right] \psi_{m,n}(t) dt \\
&= \sum_{m'=-\infty}^{\infty} \sum_{n'=-\infty}^{\infty} c_{m',n'} \int_{-\infty}^{\infty} \psi_{m,n}(t) \psi_{m',n'}(t) dt \\
&= \sum_{m'=-\infty}^{\infty} \sum_{n'=-\infty}^{\infty} c_{m',n'} \delta_{m,n} \delta_{m',n'} \quad = c_{m,n}.
\end{aligned}$$

In general, however, the wavelets $\psi_{m,n}(t)$ given by (14.9) are neither orthonormal nor complete. We thus arrive at the following theorem:

♠ **Validity of the inverse transformation formula:**
The inverse transformation formula (14.13) is valid only for a limited class of sets of discrete wavelets $\{\psi_{m,n}(t)\}$ that is endowed with both orthonormality and completeness.

The simplest example of such desired wavelets is the **Haar discrete wavelet** presented below.

Examples The Haar discrete wavelet is defined by

$$\psi_{m,n}(t) = 2^{m/2} \psi(2^m t - n),$$

where

$$\psi(t) = \begin{cases} 1 & 0 \le t < 1/2, \\ -1 & 1/2 \le t < 1, \\ 0 & \text{otherwise.} \end{cases}$$

This wavelet is known to be orthonormal and complete; its orthonormality is verified in Exercise **1**.

14.2.3 Multiresolution Analysis

We know from Sect. 14.2.2 that in order to use equation (14.13), we must find an appropriate set of discrete wavelets $\{\psi_{m,n}\}$ that possess both orthonormality and completeness. In the remainder of this section, we describe a framework for constructing such discrete wavelets that is based on the concept of **multiresolution analysis**.

Multiresolution analysis involves a particular class of a set of function spaces. The greatest peculiarity is that it establishes a nesting structure of subspaces of $L^2(\boldsymbol{R})$ that allows us to construct a complete orthonormal set of functions (i.e., an **orthonormal basis**) for $L^2(\boldsymbol{R})$. The resulting orthonormal basis is simply the discrete wavelet $\psi_{m,n}(t)$ that yields the reconstruction equation (14.13).

♠ **Multiresolution analysis:** A multiresolution analysis involves a set of function spaces that consists of a sequence $\{\mathcal{V}_j \ : \ j \in \boldsymbol{Z}\}$ of closed subspaces of $L^2(\boldsymbol{R})$. Here the subspaces $\mathcal{V}_j$ satisfy the following conditions:

1. $\cdots \subset \mathcal{V}_{-2} \subset \mathcal{V}_{-1} \subset \mathcal{V}_0 \subset \mathcal{V}_1 \subset \mathcal{V}_2 \cdots \subset L^2(\boldsymbol{R})$.
2. $\bigcap_{j=-\infty}^{\infty} \mathcal{V}_j = \{0\}$.
3. $f(t) \in \mathcal{V}_j$ if and only if $f(2t) \in \mathcal{V}_{j+1}$ for all integers j.
4. There exists a function $\phi(t) \in \mathcal{V}_0$ such that the set $\{\phi(t-n),\ n \in \boldsymbol{Z}\}$ is an orthonormal basis for $\mathcal{V}_0$.

The function $\phi(t)$ introduced above is called the **scaling function** (or **father wavelet**). It should be emphasized that the above definition gives no information as to the existence of (or the way to construct) the function $\phi(t)$ satisfying condition **4**. However, once we find such a function $\phi(t)$, we can establish a multiresolution analysis $\{\mathcal{V}_j\}$ by defining the function space $\mathcal{V}_0$ spanned by the orthonormal basis $\{\phi(t-n),\ n \in \boldsymbol{Z}\}$ and then forming other subspaces $\mathcal{V}_j$ $(j \neq 0)$ successively by using the property denoted in condition **3**. If this is achieved, we say that *our scaling function* $\phi(t)$ *generates the multiresolution analysis* $\{\mathcal{V}_j\}$.

Remark. There is no straightforward way to construct a scaling function $\phi(t)$ or, equivalently, a multiresolution analysis $\{\mathcal{V}_j\}$. Nevertheless, many kinds of scaling functions have been discovered by means of sophisticated mathematical techniques. Here we omit the details of the derivations and just refer to the resulting scaling function at need.

Examples Consider the space $\mathcal{V}_m$ of all functions in $L^2(\boldsymbol{R})$ that are constant in each interval $[2^{-m}n, 2^{-m}(n+1)]$ for all $n \in \boldsymbol{Z}$. Obviously, the space $\mathcal{V}_m$ satisfies conditions **1**–**3** of a multiresolution analysis. Furthermore, it is easy to see that the set $\{\phi(t-n),\ n \in \boldsymbol{Z}\}$ depicted in Fig. 14.8, which is defined by

$$\phi(t) = \begin{cases} 1, & 0 \le t \le 1, \\ 0, & \text{otherwise}, \end{cases} \tag{14.16}$$

satisfies condition **4**. Hence, any function $f \in \mathcal{V}_0$ can be expressed by

$$f(t) = \sum_{n=-\infty}^{\infty} c_n \phi(t-n),$$

with appropriate constants c_n. Thus, the spaces $\mathcal{V}_m$ consist of the multiresolution analysis generated by the scaling function (14.16).

14.2.4 Orthogonal Decomposition

The importance of a multiresolution analysis lies in its ability to construct an orthonormal basis (i.e., a complete orthonormal set of functions) for $L^2(\boldsymbol{R})$.

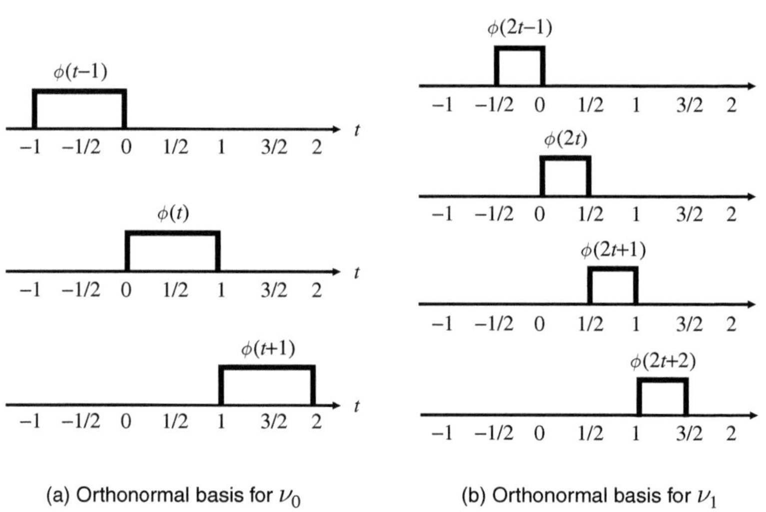

Fig. 14.8. Two different sets of functions: $\mathcal{V}_0$ and $\mathcal{V}_1$

In order to prove this statement, we first recall that a multiresolution analysis $\{\mathcal{V}_j\}$ satisfies the relation

$$\mathcal{V}_0 \subset \mathcal{V}_1 \subset \mathcal{V}_2 \subset \cdots \subset L^2.$$

We now define a space $\mathcal{W}_0$ as the **orthogonal complement** of $\mathcal{V}_0$ and $\mathcal{V}_1$, which yields

$$\mathcal{V}_1 = \mathcal{V}_0 \oplus \mathcal{W}_0. \tag{14.17}$$

The space $\mathcal{W}_0$ we have introduced is called the **wavelet space** of zero order: the reason for the name is clarified in Sect. 14.2.5. The relation (14.17) extends to

$$\mathcal{V}_2 = \mathcal{V}_1 \oplus \mathcal{W}_1 \quad = \mathcal{V}_0 \oplus \mathcal{W}_0 \oplus \mathcal{W}_1 \tag{14.18}$$

or, more generally, it gives

$$L^2 = \mathcal{V}_\infty \quad = \mathcal{V}_0 \oplus \mathcal{W}_0 \oplus \mathcal{W}_1 \oplus \mathcal{W}_2 \oplus \cdots , \tag{14.19}$$

where $\mathcal{V}_0$ is the initial space spanned by the set of functions $\{\phi(t-n),\ n \in \boldsymbol{Z}\}$. Figure 14.9 illustrates the nesting structure of the spaces $\mathcal{V}_j$ and $\mathcal{W}_j$ for different scales j.

Since the scale of the initial space is arbitrary, it can be chosen at a higher resolution such as

$$L^2 = \mathcal{V}_5 \oplus \mathcal{W}_5 \oplus \mathcal{W}_6 \oplus \cdots ,$$

or at a lower resolution such as

$$L^2 = \mathcal{V}_{-3} \oplus \mathcal{W}_{-3} \oplus \mathcal{W}_{-2} \oplus \cdots ,$$

or even at negative infinity, where (14.19) becomes

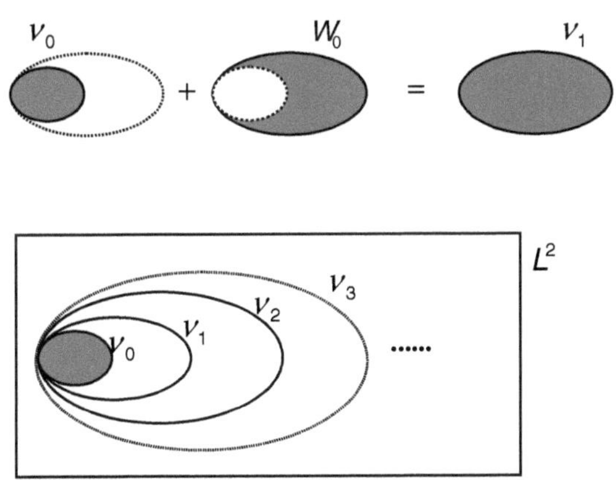

Fig. 14.9. Hierarchical structure of the spaces $\mathcal{V}_j$ and $\mathcal{W}_j$ as subspaces of L^2

$$L^2 = \cdots \oplus \mathcal{W}_{-1} \oplus \mathcal{W}_0 \oplus \mathcal{W}_1 \oplus \cdots . \tag{14.20}$$

The expression (14.20) is referred to as the **orthogonal decomposition** of the L^2 space and indicates that any function $x \in L^2(\boldsymbol{R})$ can be decomposed into the infinite sum of $g_j \in \mathcal{W}_j$:

$$x(t) = \cdots + g_{-1}(t) + g_0(t) + g_1(t) + \cdots . \tag{14.21}$$

14.2.5 Constructing an Orthonormal Basis

Let us further examine the orthogonal property of the wavelet spaces $\{\mathcal{W}_j\}$. From (14.17) and (14.18), we have

$$\mathcal{W}_0 \subset \mathcal{V}_1 \text{ and } \mathcal{W}_1 \subset \mathcal{V}_2.$$

In view of the definition of the multiresolution analysis $\{\mathcal{V}_j\}$, it follows that

$$f(t) \subset \mathcal{V}_1 \iff f(2t) \subset \mathcal{V}_2,$$

so

$$f(t) \in \mathcal{W}_0 \iff f(2t) \in \mathcal{W}_1. \tag{14.22}$$

Furtheremore, condition **4** in Sect. 14.2.3 results in

$$f(t) \in \mathcal{W}_0 \iff f(t-n) \in \mathcal{W}_0 \text{ for any } n \in \boldsymbol{Z}. \tag{14.23}$$

The two results (14.22) and (14.23) are ingredients for constructing the orthonormal basis of $L^2(R)$ that we are looking for, as demonstrated below.

We first assume that there exists a function $\psi(t)$ that leads to an orthonormal basis $\{\psi(t-n),\ n \in \boldsymbol{Z}\}$ for the space $\mathcal{W}_0$. Then, if we use the notation

$$\psi_{0,n}(t) \equiv \psi(t-n) \in \mathcal{W}_0,$$

it follows from (14.22) and (14.23) that its scaled version defined by

$$\psi_{1,n}(t) = \sqrt{2}\psi(2t-n)$$

serves as an orthonormal basis for $\mathcal{W}_1$. The term $\sqrt{2}$ was introduced to keep the normalization condition

$$\int_{-\infty}^{\infty} \psi_{0,n}(t)^2 dt = \int_{-\infty}^{\infty} \psi_{1,n}(t)^2 dt = 1.$$

By repeating the same procedure, we find that the function

$$\psi_{m,n}(t) = 2^{m/2}\psi(2^m t - n) \tag{14.24}$$

constitutes an orthonormal basis for the space $\mathcal{W}_m$. Applying these results to the expression (14.21), we have for any $x \in L^2(\boldsymbol{R})$,

$$\begin{aligned} x(t) &= \cdots + g_{-1}(t) + g_0(t) + g_1(t) + \cdots \\ &= \cdots + \sum_{n=-\infty}^{\infty} c_{-1,n}\psi_{-1,n}(t) + \sum_{n=-\infty}^{\infty} c_{0,n}\psi_{0,n}(t) + \sum_{n=-\infty}^{\infty} c_{1,n}\psi_{1,n}(t) + \cdots \\ &= \sum_{m=-\infty}^{\infty}\sum_{n=-\infty}^{\infty} c_{m,n}\psi_{m,n}(t). \end{aligned} \tag{14.25}$$

Hence, the family $\psi_{m,n}(t)$ represents an orthonormal basis for $L^2(\boldsymbol{R})$. The above arguments are summarized by the following theorem:

♠ **Theorem:**
Let $\{\mathcal{V}_j\}$ be a multiresolution analysis and define the space $\mathcal{W}_0$ by $\mathcal{W}_0 = \mathcal{V}_1 \backslash \mathcal{V}_0$. If a function $\psi(t)$ that leads to an orthonormal basis $\{\psi(t-n),\ n \in \boldsymbol{Z}\}$ for $\mathcal{W}_0$ is found, then the set of functions $\{\psi_{m,n},\ m,n \in \boldsymbol{Z}\}$ given by

$$\psi_{m,n}(t) = 2^{m/2}\psi(2^m t - n)$$

constitutes an orthonormal basis for $L^2(\boldsymbol{R})$.

Emphasis is placed on the fact that since $\psi_{m,n}(t)$ is the orthonormal basis for $L^2(\boldsymbol{R})$, the coefficients $c_{m,n}$ in (14.25) are identical to the discrete wavelet transform $T_{m,n}$ given by (14.11) (see Sect. 14.2.2). Therefore, the function $\psi(t)$ we introduce here is identified with the wavelet in the framework of continuous and discrete wavelet analysis, such as the Haar and the Mexican hat wavelets. In this sense, each $\mathcal{W}_m$ is referred to as the **wavelet space** and the function $\psi(t)$ is sometimes called the **mother wavelet**.

14.2.6 Two-Scale Relations

The preceding argument suggests that an orthonormal basis $\{\psi_{m,n}\}$ for $L^2(\boldsymbol{R})$ can be constructed by specifying the explicit function form of the mother wavelet $\psi(t)$. Thus the remaining task is to develop a systematic way of determining the mother wavelet $\psi(t)$ that leads to an orthonormal basis $\{\psi(t-n)\ n \in \boldsymbol{Z}\}$ for the space $\mathcal{W}_0 = \mathcal{V}_1 \backslash \mathcal{V}_0$ contained in a given multiresolution analysis. We shall see that the $\psi(t)$ can be found by examining the properties of the scaling function $\phi(t)$; we should recall that $\phi(t)$ yields an orthonormal basis $\{\phi(t-n)\ n \in \boldsymbol{Z}\}$ for the space $\mathcal{V}_0$. (In this context, the space $\mathcal{V}_j$ is sometimes referred to as the **scaling function space**.)

In this subsection, we make reference to an important feature of the scaling function $\phi(t)$ called the **two-scale relation**, which plays a key role in constructing the mother wavelet $\psi(t)$ of a given multiresolution analysis. We already know that all the functions in $\mathcal{V}_m$ are obtained from those in $\mathcal{V}_0$ through scaling by 2^m. Applying this result to the scaling function denoted by

$$\phi_{0,n}(t) \equiv \phi(t-n) \in \mathcal{V}_0,$$

we find that

$$\phi_{m,n}(t) = 2^{m/2}\phi(2^m t - n), \quad m \in \boldsymbol{Z} \tag{14.26}$$

is an orthonormal basis for $\mathcal{V}_m$. In particular, since $\phi \in \mathcal{V}_0 \subset \mathcal{V}_1$ and $\phi_{1,n}(t) = \sqrt{2}\phi(2t-n)$ is an orthonormal basis for $\mathcal{V}_1$, $\phi(t)$ can be expanded by $\phi_{1,n}(t)$. This is formally stated in the following theorem:

♠ **Two-scale relation:**

If the scaling function $\phi(t)$ generates a multiresolution analysis $\{\mathcal{V}_j\}$, it satisfies the recurrence relation:

$$\phi(t) = \sum_{n=-\infty}^{\infty} p_n \phi_{1,n}(t) = \sqrt{2} \sum_{n=-\infty}^{\infty} p_n \phi(2t-n), \tag{14.27}$$

where

$$p_n = \int_{-\infty}^{\infty} \phi(t)\phi_{1,n}(t)dt. \tag{14.28}$$

This recurrence equation is called the **two-scale relation** of $\phi(t)$ and the coefficients p_n are called the **scaling function coefficients.**

Remark. The two-scale relation is also referred to as the **multiresolution analysis equation**, the **refinement equation**, or the **dilation equation**, depending on the context.

Examples Consider again the space $\mathcal{V}_m$ of all functions in $L^2(\boldsymbol{R})$ that are constant on intervals $[2^{-m}n, 2^{-m}(n+1)]$ with $n \in \boldsymbol{Z}$. This multiresolution analysis is known to be generated by the scaling function $\phi(t)$ of (14.16). Substituting (14.16) into (14.28), we obtain

$$p_0 = p_1 = \frac{1}{\sqrt{2}} \quad \text{and} \quad p_n = 0 \quad \text{for } n \neq 0, 1.$$

Thus the two-scale relation reads

$$\phi(t) = \phi(2t) + \phi(2t-1).$$

This means that the scaling function $\phi(t)$ in this case is a linear combination of its contracted versions as depicted in Fig. 14.10.

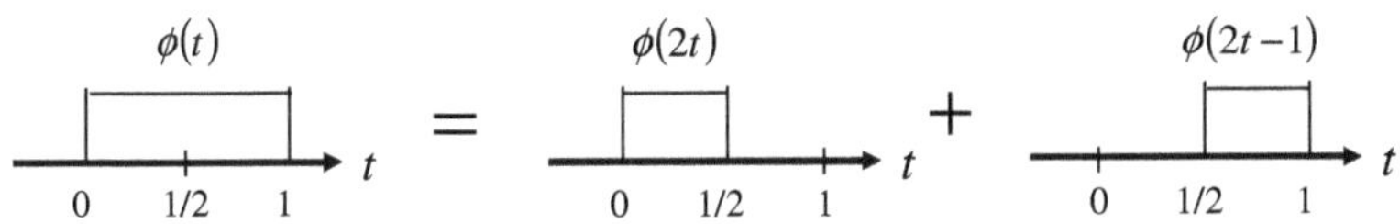

Fig. 14.10. Two-scale relation of $\phi(t)$

14.2.7 Constructing the Mother Wavelet

We are now in a position to determine the mother wavelet $\psi(t)$ that enables us to establish an orthonormal basis $\{\psi(t-n),\ n \in \boldsymbol{Z}\}$ for $L^2(\boldsymbol{R})$. Recall that a mother wavelet $\psi(t) = \psi_{0,0}(t)$ resides in a space $\mathcal{W}_0$ spanned by the next subspace of the scaling function $\mathcal{V}_1$, i.e., $\mathcal{W}_0 \subset \mathcal{V}_1$. Hence, in the same context as in the previous subsection, $\psi(t)$ can be represented by a weighted sum of the shifted scaling function $\phi(2t)$ by

$$\psi(t) = \sum_{n=-\infty}^{\infty} q_n \sqrt{2}\phi(2t-n), \quad n \in \boldsymbol{Z}. \tag{14.29}$$

The expansion coefficients q_n are called **wavelet coefficients** and are given by

$$q_n = (-1)^{n-1} p_{-n-1} \tag{14.30}$$

as stated below.

♠ **Theorem:**

If $\{\mathcal{V}_m\}$ is a multiresolution analysis with the scaling function $\phi(t)$, the mother wavelet $\psi(t)$ is given by

$$\psi(t) = \sqrt{2} \sum_{n=-\infty}^{\infty} (-1)^{n-1} p_{-n-1} \phi(2t-n), \quad n \in \boldsymbol{Z}, \tag{14.31}$$

where p_n is the scaling function coefficient of $\phi(t)$.

Remember that p_n in (14.31) is uniquely determined by the function form of the scaling function $\phi(t)$; See (14.28). Thus the above theorem states that the mother wavelet $\psi(t)$ is obtained once the scaling function $\phi(t)$ of a given multiresolution analysis is specified.

Remark. The relation $q_n = (-1)^n p_{1-n}$ employed in equation (14.31) is one possible choice for constructing the mother wavelet $\psi(t)$ from the father wavelet $\phi(t)$. In fact, there are alternative choices such as

$$q_n = (-1)^n p_{1-n}$$

or

$$q_n = (-1)^{n-1} p_{2N-1-n}$$

with certain $N \in \boldsymbol{Z}$. Hence, the mother wavelet $\psi(t)$ associated with a given multiresolution analysis is not unique. In practice, however, any preceding definition of q_n can be used to obtain a mother wavelet $\psi(t)$ because it leads to an orthonormal basis for the space $\mathcal{W}_0$.

The proof of equation (14.31) requires the following two lemmas:

♠ **Lemma 1:**
The Fourier transform $\Phi(\omega)$ of the scaling function $\phi(t)$ satisfies

$$\Phi(\omega) = M\left(\frac{\omega}{2}\right)\Phi\left(\frac{\omega}{2}\right),$$

where $M(\omega)$ is the **generating function of the multiresolution analysis** defined by

$$M(\omega) = \frac{1}{\sqrt{2}} \sum_{n=-\infty}^{\infty} p_n e^{-in\omega} \tag{14.32}$$

with the scaling function coefficient p_n of $\phi(t)$.

♠ **Lemma 2:**
The Fourier transform $F(\omega)$ of any function $f \in \mathcal{W}_0$ can be expressed by

$$F(\omega) = V(\omega) e^{i\omega/2} M^*\left(\frac{\omega}{2} + \pi\right) \Phi\left(\frac{\omega}{2}\right), \tag{14.33}$$

where $V(\omega)$ is a 2π-periodic function, i.e., $V(\omega) = V(\omega + 2\pi)$.

We should keep in mind that $V(\omega)$ is the only term on the right-hand side of (14.33) that depends on $f(t)$; the remainder term $e^{i\omega/2}M^*[(\omega/2)+\pi]\Phi(\omega/2)$ is independent of $f(t)$. The proofs of the two lemmas are outlined in Exercises **3** and **4**. Now we turn to a proof of equation (14.31) for the construction of the mother wavelet $\psi(t)$ from the scaling function $\phi(t)$.

Proof **(of Theorem):** Since the mother wavelet $\psi(t)$ gives an orthonormal basis $\{\psi(t-n),\ n \in \boldsymbol{Z}\}$ for the space $\mathcal{W}_0$, any function $f \in \mathcal{W}_0$ can be expressed by

$$f(t) = \sum_{n=-\infty}^{\infty} h_n \psi(t-n)$$

with appropriate coefficients h_n. Its Fourier transform $F(\omega)$ reads

$$F(\omega) = \left(\sum_{n=-\infty}^{\infty} h_n e^{-in\omega} \right) \Psi(\omega),$$

where the sum in parentheses is 2π-periodic. Comparing this with (14.33), we obtain

$$\Psi(\omega) = e^{i\omega/2} M^* \left(\frac{\omega}{2} + \pi \right) \Phi \left(\frac{\omega}{2} \right). \tag{14.34}$$

Substituting expression (14.32) into (14.34) yields

$$\begin{aligned}
\Psi(\omega) &= \frac{e^{i\omega/2}}{\sqrt{2}} \sum_{n=-\infty}^{\infty} p_n e^{in[(\omega/2)+\pi]} \Phi \left(\frac{\omega}{2} \right) \\
&= \frac{1}{\sqrt{2}} \sum_{n=-\infty}^{\infty} p_n e^{in\pi} e^{i(n+1)(\omega/2)} \Phi \left(\frac{\omega}{2} \right) \\
&= \frac{1}{\sqrt{2}} \sum_{k=-\infty}^{\infty} p_{-k-1} (-1)^{k-1} e^{-ik\omega/2} \Phi \left(\frac{\omega}{2} \right) \quad [k \equiv -n-1].
\end{aligned}$$

Take the inverse Fourier transform of the both sides to find

$$\begin{aligned}
\psi(t) &= \frac{1}{\sqrt{2}} \sum_{k=-\infty}^{\infty} p_{-k-1} (-1)^{k-1} \int_{-\infty}^{\infty} e^{-ik\omega/2} e^{i\omega t} \Phi \left(\frac{\omega}{2} \right) d\omega \\
&= \frac{2}{\sqrt{2}} \sum_{k=-\infty}^{\infty} p_{-k-1} (-1)^{k-1} \int_{-\infty}^{\infty} e^{i\omega'(2t-k)t} \Phi(\omega') d\omega' \quad [\omega' \equiv \omega/2] \\
&= \sqrt{2} \sum_{k=-\infty}^{\infty} p_{-k-1} (-1)^{k-1} \phi(2t-k).
\end{aligned}$$

This is our desired result (14.31). ♣

14.2.8 Multiresolution Representation

Through the discussions thus far, we have obtained an orthonormal basis consisting of scaling functions $\phi_{j,k}(t)$ and wavelets $\psi_{j,k}(t)$ that span all of $L^2(\boldsymbol{R})$. Since

$$L^2 = \mathcal{V}_{j_0} \oplus \mathcal{W}_{j_0} \oplus \mathcal{W}_{j_0+1} \oplus \cdots,$$

any function $x(t) \in L^2(\boldsymbol{R})$ can be expanded, e.g.,

$$x(t) = \sum_{k=-\infty}^{\infty} S_{j_0,k} \phi_{j_0,k}(t) + \sum_{k=-\infty}^{\infty} \sum_{j=j_0}^{\infty} T_{j,k} \psi_{j,k}(t). \tag{14.35}$$

Here, the initial scale j_0 could be zero or another integer or negative infinity as in (14.13), where no scaling functions are used. The coefficients $T_{j,k}$ are identified with the discrete wavelet transform given in (14.11). Often $T_{j,k}$ in (14.35) is called the **wavelet coefficient** and $S_{j,k}$ is called the **approximation coefficient**.

The representation (14.35) can be simplified by using the following notation. We denote the first summation on the right-hand side of (14.35) by

$$x_{j_0}(t) = \sum_{k=-\infty}^{\infty} S_{j_0,k}\phi_{j_0,k}(t). \tag{14.36}$$

Equation (14.36) is called the **continuous approximation** of the signal $x(t)$ at scale j_0. Observe that the continuous approximation approaches $x(t)$ in the limit of $j_0 \to \infty$, since in this case $L^2 = \mathcal{V}_\infty$. In addition, we introduce the notation

$$z_j(t) = \sum_{k=-\infty}^{\infty} T_{j,k}\psi_{j,k}(t), \tag{14.37}$$

where $z_j(t)$ is known as the **signal detail** at scale j. With these conventions, we can write (14.35) as

$$x(t) = x_{j_0}(t) + \sum_{j=j_0}^{\infty} z_j(t). \tag{14.38}$$

Equation (14.38) says that the original continuous signal $x(t)$ is expressed as a combination of its continuous approximation x_{j_0} at an arbitrary scale index j_0 added to a succession of signal details $z_j(t)$ from scales j_0 up to infinity.

Also noteworthy is the fact that due to the nested relation of $\mathcal{V}_{j+1} = \mathcal{V}_j \oplus W_j$, we can write

$$x_{j+1}(t) = x_j(t) + z_j(t). \tag{14.39}$$

This indicates that if we add the signal detail at an arbitrary scale (index j) to the continuous approximation at the same scale, we get the signal approximation at an increased resolution (i.e., at a smaller scale, index $j+1$). The important relation (14.39) between continuous approximations $x_j(t)$ and signal details $z_j(t)$ is called a **multiresolution representation**.

Exercises

1. Verify the orthonormality of the Haar discrete wavelet $\psi_{m,n}(t)$ defined by $\psi_{m,n}(t) = 2^{m/2}\psi(2^m t - n)$, where

$$\psi(t) = \begin{cases} 1 & 0 \le t < 1/2, \\ -1 & 1/2 \le t < 1, \\ 0 & \text{otherwise.} \end{cases}$$

Solution: First we note that the norm of $\psi_{m,n}(x)$ is unity:

$$\int_{-\infty}^{\infty} \psi_{m,n}(t)^2 dt = 2^{-m} \int_{-\infty}^{\infty} \left[\psi_{m,n}(2^{-m}t - n)\right]^2 dt$$
$$= 2^{-m} \cdot 2^m \int_{-\infty}^{\infty} \psi_{m,n}(u)^2 du = 1.$$

Thus, we obtain

$$I \equiv \int_{-\infty}^{\infty} \psi_{m,n}(t)\psi_{k,l}(t)dt = \int_{-\infty}^{\infty} 2^{-m/2}\psi(2^{-m}t - n)2^{-k/2}\psi(2^{-k}t - \ell)dt$$
$$= 2^{-m/2} \cdot 2^m \int_{-\infty}^{\infty} \psi(u)2^{-k/2}\psi[2^{m-k}(u+n) - \ell]dt. \qquad (14.40)$$

If $m = k$, the integral in the last line in (14.40) reads

$$\int_{-\infty}^{\infty} \psi(u)\psi(u + n - \ell)dt = \delta_{0,n-\ell} = \delta_{n,\ell},$$

since $\psi(u) \neq 0$ in $0 \leq u \leq 1$ and $\psi(u + n - \ell) \neq 0$ in $\ell - n \leq u \leq \ell - n + 1$, so that these intervals are disjoint unless $n = \ell$. Owing to symmetry if $m \neq k$, it suffices to look at the case of $m > k$. Set $r = m - k \neq 0$ in (14.40) to obtain

$$I = 2^{r/2} \int_{-\infty}^{\infty} \psi(u)\psi(2^r v + s)du$$
$$= 2^{r/2} \left[\int_0^{1/2} \psi(2^r v + s)du - \int_{1/2}^{1} \psi(2^r v + s)du\right],$$

which can be simplified as

$$I = \int_s^a \psi(x)dx - \int_a^b \psi(x)dx = 0, \qquad (14.41)$$

where $2^r u + s = x$, $a = s + 2^{r-1}$, $b = s + 2^r$. Observe that $[s, a]$ contains the interval $[0, 1]$ of the Haar wavelet $\psi(t)$, which implies that the first integral in (14.41) vanishes. Similarly, the second integral equals zero. We thus conclude that

$$I = \int_{-\infty}^{\infty} \psi_{m,n}(t)\psi_{k,\ell}dt = \delta_{m,k}\delta_{n,\ell},$$

which means that the Haar discrete wavelet $\psi_{m,n}(t)$ is orthonormal. ♣

2. Let $\phi \in L^2(\boldsymbol{R})$ and $\Phi(\omega)$ be the Fourier transform of $\phi(t)$. Prove that the system $\{\phi_{0,n} \equiv \phi(t-n), \quad n \in \boldsymbol{Z}\}$ is orthonormal if and only if $\sum_{k=-\infty}^{\infty} |\Phi(\omega + 2k\pi)|^2 = 1$ almost everywhere.

Solution: It is obvious that the Fourier transform of $\phi_{0,n}(t)$ reads $\Phi_{0,n}(\omega) = e^{-in\omega}\Phi(\omega)$. In view of the Parseval identity for the wavelet transform (14.8), we have

$$\begin{aligned}
&\int_{-\infty}^{\infty} \phi_{0,n}(t)\phi_{0,m}(t)dt = \int_{-\infty}^{\infty} \phi_{0,0}(t)\phi_{0,m-n}(t)dt \\
&= \frac{1}{2\pi}\int_{-\infty}^{\infty} \Phi_{0,0}(\omega)\Phi_{0,m-n}(\omega)d\omega = \frac{1}{2\pi}\int_{-\infty}^{\infty} e^{-i(m-n)\omega}\left[\Phi_{0,0}(\omega)\right]^2 d\omega \\
&= \frac{1}{2\pi}\sum_{k=-\infty}^{\infty}\int_{2\pi k}^{2\pi(k+1)} e^{-i(m-n)\omega}\left[\Phi_{0,0}(\omega)\right]^2 d\omega \\
&= \frac{1}{2\pi}\int_{0}^{2\pi} e^{-i(m-n)\omega}\sum_{k=-\infty}^{\infty}\left[\Phi_{0,0}(\omega)\right]^2 d\omega.
\end{aligned}$$

It thus follows from the completeness of $\{e^{-in\omega}, \quad n \in \boldsymbol{Z}\}$ in $L^2(0,2\pi)$ that $\int_{-\infty}^{\infty} \phi_{0,n}(t)\phi_{0,m}(t)dt = 0$ if and only if

$$\sum_{k=-\infty}^{\infty}\left[\Phi_{0,0}(\omega)\right]^2 = 1 \quad \text{almost everywhere.} \quad \clubsuit$$

3. Let $\Phi(\omega)$ be the Fourier transform of the scaling function $\phi(t)$ and let p_n be its scaling function coefficient. Prove that

$$\Phi(\omega) = M\left(\frac{\omega}{2}\right)\Phi\left(\frac{\omega}{2}\right) \quad \text{with} \quad M(\omega) = \frac{1}{\sqrt{2}}\sum_{n=-\infty}^{\infty} p_n e^{-in\omega}. \tag{14.42}$$

Solution: Since $\phi(t) = \sqrt{2}\sum_{n=-\infty}^{\infty} p_n\phi(2t-n)$, we have

$$\begin{aligned}
\Phi(\omega) &= \sqrt{2}\sum_{n=-\infty}^{\infty} p_n \int_{-\infty}^{\infty}\phi(2t-n)e^{-i\omega t}dt \\
&= \sqrt{2}\sum_{n=-\infty}^{\infty} p_n \int_{-\infty}^{\infty}\phi(t')e^{-i\omega(t'+n)/2}dt \quad (t' \equiv 2t-n) \\
&= \frac{1}{\sqrt{2}}\sum_{n=-\infty}^{\infty} p_n e^{-in\omega/2}\Phi\left(\frac{\omega}{2}\right) = M\left(\frac{\omega}{2}\right)\Phi\left(\frac{\omega}{2}\right). \quad \clubsuit
\end{aligned}$$

4. Let $f(t)$ be a function $f \in \mathcal{W}_0 = \mathcal{V}_1 \backslash \mathcal{V}_0$ for a given multiresolution analysis $\{\mathcal{V}_j\}$. Prove that its Fourier transform $F(\omega)$ necessarily takes the form

$$F(w) = V(\omega)e^{i\omega/2}M^*\left(\frac{\omega}{2}+\pi\right)\Phi\left(\frac{\omega}{2}\right), \tag{14.43}$$

where $V(\omega) = V(\omega + 2\pi)$.

Solution: Since $f \in \mathcal{W}_0$ and $\mathcal{V}_1 = \mathcal{V}_0 \oplus \mathcal{W}_0$, it follows that $f \in \mathcal{V}_1$ and is orthogonal to $\mathcal{V}_0$. Hence, we can write $f(t) = \sqrt{2}\sum_{n=-\infty}^{\infty} c_n\phi_{1,n}(t) = \sqrt{2}\sum_{n=-\infty}^{\infty} c_n\phi(2t-n)$, where $c_n = \int_{-\infty}^{\infty} f(t)\phi_{1,n}(t)dt$. Take the Fourier transform of both sides to obtain

$$F(w) = M_f\left(\frac{\omega}{2}\right)\Phi\left(\frac{\omega}{2}\right) \quad \text{with} \quad M_f(\omega) \equiv \frac{1}{\sqrt{2}}\sum_{n=-\infty}^{\infty} c_n e^{-in\omega}. \tag{14.44}$$

Evidently, $M_f(\omega)$ is a 2π-periodic function belonging to $L^2(0, 2\pi)$. Since f is orthogonal to $\mathcal{V}_0$, we have $\int_{-\infty}^{\infty} F(w)\Phi^*(\omega)e^{in\omega}d\omega = 0$, so

$$\int_{-\infty}^{\infty}\left[\sum_{k=-\infty}^{\infty} F(w+2k\pi)\Phi^*(\omega+2k\pi)\right]e^{in\omega}d\omega = 0.$$

Consequently, $\sum_{k=-\infty}^{\infty} F(w+2k\pi)\Phi^*(\omega+2k\pi) = 0$. Substituting (14.42) and (14.44) into this result, we obtain

$$\sum_{k=-\infty}^{\infty} M_f\left(\frac{\omega}{2}+k\pi\right)M^*\left(\frac{\omega}{2}+k\pi\right)\left|\Phi\left(\frac{\omega}{2}+k\pi\right)\right|^2 = 0.$$

Meanwhile we denote $M_f(\omega)M^*(\omega)$ and $|\Phi(\omega)|^2$ by $M_2(\omega)$ and $\Phi_2(\omega)$, respectively. By splitting the sum into even and odd integers k and then employing the 2π-periodicity of $M(\omega)$ and $M_f(\omega)$ [and thus $M_2(\omega)$], we have

$$\begin{aligned}
0 &= \sum_{k=-\infty}^{\infty} M_2\left(\frac{\omega}{2}+2k\pi\right)\Phi_2\left(\frac{\omega}{2}+2k\pi\right) \\
&\quad + \sum_{k=-\infty}^{\infty} M_2\left[\frac{\omega}{2}+(2k+1)\pi\right]\Phi_2\left[\frac{\omega}{2}+(2k+1)\pi\right] \\
&= M_2\left(\frac{\omega}{2}\right)\sum_{k=-\infty}^{\infty}\Phi_2\left(\frac{\omega}{2}+2k\pi\right) \\
&\quad + M_2\left(\frac{\omega}{2}+\pi\right)\sum_{k=-\infty}^{\infty}\Phi_2\left[\frac{\omega}{2}+(2k+1)\pi\right] \\
&= M_2\left(\frac{\omega}{2}\right) + M_2\left(\frac{\omega}{2}+\pi\right),
\end{aligned} \tag{14.45}$$

where we used the orthonormality condition with respect to the set of scaling functions $\{\phi_{0,k}(t)\}$. Finally, replacing ω in the last line in (14.45) by 2ω gives

$$\begin{vmatrix} M_f(\omega) & M^*(\omega+\pi) \\ -M_f(\omega+\pi) & M^*(\omega) \end{vmatrix} = 0, \tag{14.46}$$

which indicates the linear dependence of two vectors: $[M_f(\omega), -M_f(\omega+\pi)]$ and $[M^*(\omega+\pi), M^*(\omega)]$. Hence, there exists a function $\lambda(\omega)$ such that

$$M_f(\omega) = \lambda(\omega) M^*(\omega+\pi). \tag{14.47}$$

Since both M and M_f are 2π periodic, so is λ. Further, substituting (14.47) into (14.46) yields

$$\lambda(\omega) + \lambda(\omega+\pi) = 0, \tag{14.48}$$

which means that there exists a function $V(\omega)$ such that

$$\lambda(\omega) = e^{i\omega} V(\omega) \quad \text{and} \quad V(\omega) = V(\omega + 2\pi).$$

Eventually, the results (14.44), (14.47), and (14.48) lead to the desired representation (14.43). ♣

14.3 Fast Wavelet Transformation

14.3.1 Generalized Two-Scale Relations

We know that a signal $x(t) \in L^2(\boldsymbol{R})$ can be represented in terms of the continuous approximation $S_{m,n}$ and the discrete wavelet transform $T_{m,n}$ by

$$x(t) = \sum_{n=-\infty}^{\infty} S_{m_0,n}\phi_{m_0,n}(t) + \sum_{m=-\infty}^{\infty}\sum_{n=-\infty}^{\infty} T_{m,n}\psi_{m,n}(t),$$

where

$$\phi_{m,n}(t) = 2^{m/2}\phi(2^m t - n) \quad \text{and} \quad \psi_{m,n}(t) = 2^{m/2}\psi(2^m t - n). \tag{14.49}$$

[See (14.24) and (14.26).] In principle, both expansion coefficients $S_{m,n}$ and $T_{m,n}$ can be computed through the convolution integral defined by

$$S_{m,n} = \int_{-\infty}^{\infty} x(t)\phi_{m,n}(t)dt \ \text{and}\ T_{m,n} = \int_{-\infty}^{\infty} x(t)\psi_{m,n}(t)dt. \tag{14.50}$$

Actual computations of these integrals are very time-consuming. However, there is an efficient method for computing $S_{m,n}$ and $T_{m,n}$ at all m, known as the **fast wavelet transform**. This sophistuated method is based on recursive equations for $S_{m,n}$ and $T_{m,n}$ and thus is markedly suitable for numerical computations of wavelet analyses.

To proceed with the argument, we need some preliminary results. We know that the father wavelet $\phi(t)$ and the mother wavelet $\psi(t)$ can be described by a linear combination of contracted and shifted versions of $\phi(t)$ as follows:

$$\phi(t) = \sqrt{2} \sum_{k=-\infty}^{\infty} p_k \phi(2t-k) \text{ and } \psi(t) = \sum_{k=-\infty}^{\infty} (-1)^k p_{1-k} \phi(2t-k),$$

where p_n is the scaling function coefficient of $\phi(t)$. For convenience, we use an alternative definition $q_n = (-1)^n p_{1-n}$ of the wavelet coefficient q_n instead of the one used in (14.30). These facts immediately result in

$$\begin{aligned}
\phi_{m,n}(t) &= 2^{m/2} \phi\left(2^m t - n\right) = 2^{m/2} \sum_{k=-\infty}^{\infty} p_k \phi\left[2\left(2^m t - n\right) - k\right] \\
&= 2^{m/2} \sum_{k=-\infty}^{\infty} p_k 2^{-(m+1)/2} \phi_{m+1,2n+k}(t) \\
&= 2^{-1/2} \sum_{k=-\infty}^{\infty} p_k \phi_{m+1,2n+k}(t), \qquad (14.51)
\end{aligned}$$

and similarly,

$$\psi_{m,n}(t) = 2^{-1/2} \sum_{k=-\infty}^{\infty} q_k \phi_{m+1,2n+k}(t). \qquad (14.52)$$

The expressions (14.51) and (14.52) are generalizations of (14.27) and (14.31) applicable for $\phi(t)$ and $\psi(t)$.

♠ **Generalized two-scale relations:**

Given a multiresolution analysis, $\phi_{m,n}(t)$ and $\psi_{m,n}(t)$ are obtained from the set of functions $\{\phi_{m+1,2n+k}(t); \ -\infty < k < \infty\}$ by

$$\phi_{m,n}(t) = 2^{-1/2} \sum_{k=-\infty}^{\infty} p_k \phi_{m+1,2n+k}(t),$$

$$\psi_{m,n}(t) = 2^{-1/2} \sum_{k=-\infty}^{\infty} q_k \phi_{m+1,2n+k}(t).$$

14.3.2 Decomposition Algorithm

The fast wavelet transform consists of two main parts, called, respectively, the **decomposition algorithm** and the **reconstruction algorithm**, each of which gives a recursive relation between approximation coefficients $S_{m,n}$ and wavelet coefficients $T_{m,n}$ at neighboring scales. This subsection focuses on the former algorithm and in the next subsection deals with the latter.

Remark. In the literature about the fast wavelet transform, all of the terms below mean the same thing:

- discrete wavelet transform
- decomposition/reconstruction algorithm
- fast orthogonal wave transform
- multiresolution algorithm
- pyramid algorithm
- tree algorithm

The decomposition algorithm enables us to obtain $S_{m,n}$ and $T_{m,n}$ at all m smaller than a prescribed scale m_0, once $S_{m_0,n}$ is given. To attain our objective, we first derive a recursive formula for $S_{m,n}$ at two different scales, i.e., $S_{m,n}$ and $S_{m+1,n}$. From the expansion (14.49) and from the orthonormality of $\phi_{m,n}(t)$, it follows that

$$S_{m,n} = \int_{-\infty}^{\infty} x(t)\phi_{m,n}(t)dt.$$

Using the generalized two-scale relation (14.51), we can write

$$\begin{aligned} S_{m,n} &= \int_{-\infty}^{\infty} x(t) \left[\frac{1}{\sqrt{2}} \sum_{k=-\infty}^{\infty} p_k \phi_{m+1,2n+k}(t) \right] dt \\ &= \frac{1}{\sqrt{2}} \sum_{k=-\infty}^{\infty} p_k \left[\int_{-\infty}^{\infty} x(t)\phi_{m+1,2n+k}(t)dt \right] \\ &= \frac{1}{\sqrt{2}} \sum_{k=-\infty}^{\infty} p_k S_{m+1,2n+k}. \end{aligned}$$

Replacing the summation index k with $k - 2n$, we obtain

$$S_{m,n} = \frac{1}{\sqrt{2}} \sum_{k=-\infty}^{\infty} p_{k-2n} S_{m+1,k}, \tag{14.53}$$

which provides the approximation coefficients $S_{m,n}$ from $S_{m+1,n}$.

Similarly the wavelet coefficients $T_{m,n}$ can be found from the approximation coefficients at the previous scale:

$$T_{m,n} = \frac{1}{\sqrt{2}} \sum_{k=-\infty}^{\infty} q_k S_{m+1,2n+k} = \frac{1}{\sqrt{2}} \sum_{k=-\infty}^{\infty} q_{k-2n} S_{m+1,k}. \tag{14.54}$$

As a consequence, if we know the approximation coefficients $S_{m_0,k}$ at a specific scale m_0 then, through repeated application of (14.53) and (14.54), we can generate $S_{m,n}$ and $T_{m,n}$ at all $m < m_0$. This procedure, called the **decomposition algorithm**, which is based on (14.53) and (14.54) is the first half of the fast wavelet transform that allows us to compute the wavelet coefficients efficiently, rather than computing them laboriously from the convolution of (14.50).

14.3.3 Reconstruction Algorithm

We can go in the opposite direction and reconstruct $S_{m+1,n}$ from $S_{m,n}$ and $T_{m,n}$. We already know from (14.39) that $x_{m+1}(t) = x_m(t) + z_m(t)$, and we can expand this as

$$x_{m+1}(t) = \sum_{n=-\infty}^{\infty} S_{m,n}\phi_{m,n}(t) + \sum_{n=-\infty}^{\infty} T_{m,n}\psi_{m,n}(t).$$

Furthermore, using (14.51) and (14.52), we can expand this equation in terms of the scaling function at the previous scale:

$$\begin{aligned} x_{m+1}(t) = &\sum_{n=-\infty}^{\infty} S_{m,n} \frac{1}{\sqrt{2}} \sum_{k=-\infty}^{\infty} p_k \phi_{m+1,2n+k}(t) \\ &+ \sum_{n=-\infty}^{\infty} T_{m,n} \frac{1}{\sqrt{2}} \sum_{k=-\infty}^{\infty} q_k \phi_{m+1,2n+k}(t). \end{aligned}$$

Rearranging the summation indices, we get

$$\begin{aligned} x_{m+1}(t) = &\sum_{n=-\infty}^{\infty} S_{m,n} \frac{1}{\sqrt{2}} \sum_{k=-\infty}^{\infty} p_{k-2n} \phi_{m+1,k}(t) \\ &+ \sum_{n=-\infty}^{\infty} T_{m,n} \frac{1}{\sqrt{2}} \sum_{k=-\infty}^{\infty} q_{k-2n} \phi_{m+1,k}(t). \end{aligned} \tag{14.55}$$

We also know that we can expand $x_{m-1}(t)$ in terms of the approximation coefficients at scale $m-1$, i.e.,

$$x_{m+1}(t) = \sum_{k=-\infty}^{\infty} S_{m+1,k}\phi_{m+1,k}(t). \tag{14.56}$$

Equating the coefficients in (14.56) with (14.55) yields the reconstruction algorithm:

$$S_{m+1,n} = \frac{1}{\sqrt{2}} \sum_{k=-\infty}^{\infty} p_{n-2k} S_{m,k} + \frac{1}{\sqrt{2}} \sum_{k=-\infty}^{\infty} q_{n-2k} T_{m,k},$$

where we have swapped the indices k and n. Hence, at the scale $m+1$, the approximation coefficients $S_{m+1,n}$ can be found in terms of a combination of $S_{m,n}$ and $T_{m,n}$ at the next scale, m. The reconstruction algorithm is the second half of the fast wavelet transform.

Part V

Differential Equations

15

Ordinary Differential Equations

Abstract The main objective of this chapter is to ensure that the reader understands the "existence theorem" (Sect. 15.2.3) and the "unique theorem" (Sect. 15.2.4) for a first-order ordinary differential equation. These theorems prove the existence and uniqueness of a solution of the differential equation and delineate the conditions that should be satisfied by the functions that are to be differentiated.

15.1 Concepts of Solutions

15.1.1 Definition of Ordinary Differential Equations

Many physical laws are often formulated as **ordinary differential equations** (ODEs) whose unknowns are functions of a single variable. Below are basic notation and several important theorems that are used throughout this chapter. We start with the formal definition of ODEs.

♠ **Ordinary differential equations:**
An ordinary differential equation of order n is an equation

$$F\left[x, y(x), y'(x), \cdots, y^{(n)}(x)\right] = 0 \tag{15.1}$$

that is satisfied by the function $y(x)$ and its derivatives $y'(x), y''(x), \cdots, y^{(n)}(x)$ with respect to a single independent variable x.

Here, the **order** of a differential equation means the largest positive integer n for which an nth derivative appears in equation (15.1). For instance, a general form of the first-order differential equations is given by

$$F\left[x,\ y(x),\ y'(x)\right] = 0, \tag{15.2}$$

where F is a single-valued function on its arguments in some domain D. Hereafter we restrict our attention to the case where x is a real number.

Remark.

1. An ODE (15.1) is called a **linear ODE** if it is linear in the unknown function $y(x)$ and in all its derivatives; otherwise, it is **nonlinear**.
2. A linear ODE of order n is said to be **homogeneous** if it is of the form $a_n(x)y^{(n)} + a_(n-1)(x)y^{(n-1)} + ... + a_1(x)y' + a_0(x)y = 0$, where there is no term that contains a function of x alone.
3. The term **homogeneous** may have a totally different meaning specifically when a linear ODE is first order, which occurs if the ODE is written in the form
$$\frac{dy}{dx} = F\left(\frac{y}{x}\right). \tag{15.3}$$
Such equations can be solved in closed form by a change of variables $u = y/x$, which transforms the equation into the separable equation
$$\frac{dx}{x} = \frac{du}{F(u) - u}. \tag{15.4}$$

15.1.2 Explicit Solution

Let $y = \varphi(x)$ define y as a function of x on an interval $I = (a, b)$. We say that the function $\varphi(x)$ is an **explicit solution** or a simple **solution** of the ODE (15.1) if it satisfies the equation for *every* x in I. In mathematical symbols, this definition reads as follows:

♠ **Explicit solution of an ODE:**
A function $y = \varphi(x)$ defined on an interval I is a solution of the ODE (15.1) if
$$F\left[x, \varphi(x), \varphi'(x), \cdots, \varphi^{(n)}(x)\right] = 0$$
for *every* x in I.

Note that a real function should be a correspondence between two sets of real numbers. In this context, if an equation involving x and y does not define a real function, then it is not a solution of any ODE even if the equation formally satisfies the ODE. For example, the equation
$$y = \sqrt{-(1 + x^2)} \tag{15.5}$$
does not define a real function; therefore, it is not a solution of the ODE

$$x + yy' = 0 \tag{15.6}$$

even though the formal substitution of (15.5) into (15.6) yields an identity.

Examples **1.** The function

$$y = \log x + c, \ x > 0$$

is a solution of $y' = 1/x$ for all $x > 0$.

2. The function

$$y = \tan x - x, \quad x \neq \frac{2n+1}{2}\pi \ (n = 0, \pm 1, \pm 2, \cdots) \tag{15.7}$$

is a solution of

$$y' = (x+y)^2. \tag{15.8}$$

In fact, the substitution of y into (15.8) gives the identity $\tan^2 x = (x + \tan x - x)^2 = \tan^2 x$ in each of the intervals specified in (15.7).

Remark. Note that the ODE (15.8) is defined for all x, but its solution (15.7) is not defined for all x. Hence, the interval for which the function given by (15.7) may be a solution of (15.8) is a smaller set of the intervals in (15.7).

3. The function $y = |x|$ is a solution of

$$y' = 1 \quad \text{in the interval } x > 0,$$

and is also a solution of

$$y' = -1 \quad \text{in the interval } x < 0.$$

Remark. Observe that the function $y = |x|$ is defined for all x, whereas the corresponding ODEs are defined in only a restricted interval of x, in contrast to Example **2**.

15.1.3 Implicit Solution

It is sometimes not easy (or even impossible) to solve an equation of the form $g(x, y) = 0$ for y in terms of x. However, whenever it can be shown that an implicit function does satisfy a given ODE on an interval I, then the relation $g(x, y) = 0$ is called an **implicit solution** of the ODE. A formal definition is given below.

♠ **Implicit solution of an ODE:**
A relation $g(x,y)=0$ is an **implicit solution** of an ODE

$$F\left[x, y(x), y'(x), \cdots, y^{(n)}(x)\right] = 0$$

on an interval I if:

1. There exists a function $h(x)$ defined on I such that $g(x, h(x)) = 0$ for *every* x in I.
2. If $F\left[x, h(x), h'(x), \cdots, h^{(n)}(x)\right] = 0$ for *every* x in I.

Remark. It must be cautioned that $g(x,y)=0$ is merely an *equation*, and it is thus never a precise solution of an ODE, as only a *function* can be a solution of an ODE. What we mean in the above definition is that the function $h(x)$ defined by the relation $g(x,y)=0$ is the solution of the ODE.

Examples The equation

$$g(x,y) = x^2 + y^2 - 25 = 0$$

is an implicit solution of the ODE

$$F(x, y, y') = yy' + x = 0$$

on the interval $I : -5 < x < 5$. In fact, the function $h(x) = \sqrt{25 - x^2}$ defined on I yields

$$F[x, h(x), h'(x)] = \sqrt{25 - x^2}\left(-\frac{x}{\sqrt{25 - x^2}}\right) + x = 0$$

for every x on I.

15.1.4 General and Particular Solutions

We next observe that an ODE in general has many solutions. For example, the ODE

$$y' = e^x$$

can be solved as

$$y = e^x + c, \tag{15.9}$$

where c can take any numerical value. Similarly, if

$$y''' = e^x, \tag{15.10}$$

then its solution, obtained by integrating three times, is

$$y = e^x + c_1 x^2 + c_2 x + c_3, \tag{15.11}$$

where c_1, c_2, c_3 can take on any numerical values. Note that both (15.9) and (15.11) express infinitely many solutions since, which are constants, the c's can have infinitely many values. Figure 15.1 is a geometrical interpretation of this point. Each curve corresponds to a solution (15.11) for $c_2 = -5, 1, 4$ and $c_3 = -2, 1, 3$ while $c_1 = 1$ is fixed.

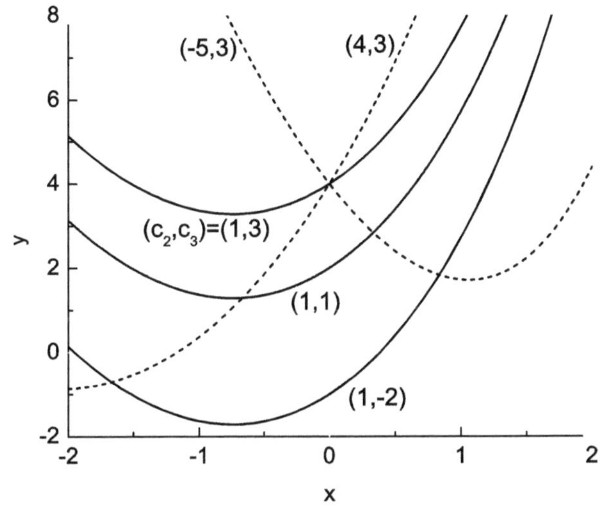

Fig. 15.1. Family of the infinitely many solutions (15.11) of the differential equation (15.10). *Solid* and *dotted curves* correspond to $c_2 = 1$ and $c_2 = 3$, respectively

The two examples above illustrate that solutions of an ODE may often be represented by a single equation involving an arbitrary constant c. Such a function involving an arbitary constant is called a **general solution** (or **complete integral** or **primitive integral**) of an ODE. Geometrically, these are infinitely many curves, one for each set of values of the c's. If we choose specific values of the c's, we obtain what is called a **particular solution** of that ODE.

Remark. From the examples above, the reader might assume that

(i) an ODE always has infinitely many solutions, or that
(ii) a solution of an nth order ODE always contains n arbitrary constants.

However, these two conjectures are false. For instance,

- The equation $(y'')^2 + y^2 = 0$ has only one solution $y = 0$ that possesses *no* arbitrary constant.
- The equation $|y'| + 1 = 0$ has *no* solution.
- The *first*-order equation $(y' - y)(y' - 2y) = 0$ has the solution $(y - c_1 e^x)(y - c_2 e^{2x}) = 0$ that has *two* (not *one*) arbitrary constants.

15.1.5 Singular Solution

Consider an ODE of the form

$$y - xy' = f(y'), \tag{15.12}$$

which is known as a **Clairaut equation**. We solve it by differentiating both sides to yield

$$y'' [f'(y') + x] = 0.$$

We thus have two possibilities. If we set $y'' = 0$, then $y = ax + b$ so that substitution back into the original equation (15.12) gives $b = f(a)$. Thus we have a general solution:

$$y = ax + f(a),$$

where a is an arbitrary constant. On the other hand, if we set

$$f'(y') + x = 0, \tag{15.13}$$

then eliminating y' between (15.13) and the original equation gives us a solution with *no* arbitrary constant, which is known as a **singular solution**. There are various other types of singular solutions, one of which is given below.

Examples Suppose the Clairaut equation to be of the form

$$y = xy' + (y')^2$$

and differentiate both sides to obtain

$$y''(x + 2y') = 0.$$

If we set $y'' = 0$, then the general solution reads

$$y = cx + c^2 \tag{15.14}$$

with an arbitrary constant c. However, if we choose the possibility that $2y' + x = 0$, then we have

$$x^2 + 4y = 0. \tag{15.15}$$

Remark. Geometrically, the singular solution (15.14) is an *envelope* of the family of integral curves defined by the general solution (15.15), as depicted in Fig. 15.2. The dotted parabola is the singular solution and the straight lines tangent to the parabola are the general solution.

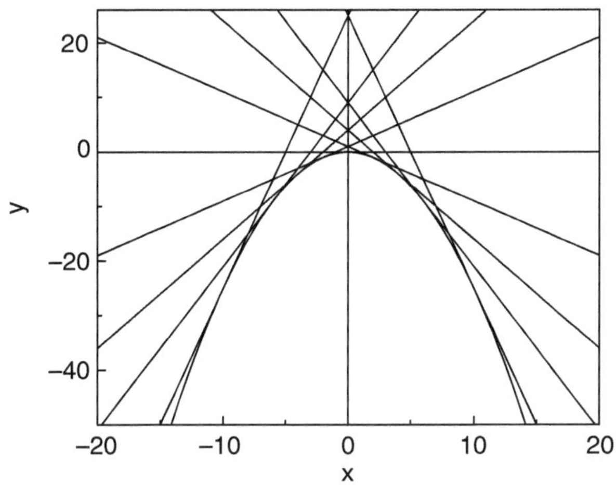

Fig. 15.2. The singular solution (15.14) is an envelope of the family of **integral curves** (see Sect. 15.1.6), which are defined by the general solution (15.15)

15.1.6 Integral Curve and Direction Field

Before closing this section, we must emphasize the geometric significance of a solution of a first-order ODE. In many practical problems, a rough geometrical approximation to a solution may be all that is needed rather then an evaluation of its explicit functional form. Let

$$y = f(x) \quad \text{or} \quad g(x, y) = 0$$

define a function of x whose derivative y' exists on an interval $I : a < x < b$. Then y' gives the direction of the tangent to the curve at each of these points. Therefore, finding a solution for

$$y' = F(x, y), \quad a < x < b \tag{15.16}$$

can be reduced to finding a curve on the $(x\text{-}y)$-plane whose slope at each of its points is given by (15.16). The relevant terminology is given below.

♠ **Integral curve:**
If a curve $y = f(x)$ [or $g(x, y) = 0$] satisfies a first-order ODE (15.16) on an interval I, then the graph of this function is called an **integral curve**.

Obviously, an integral curve is the graph of a function that is a solution of a first-order ODE (15.16). Therefore, even if we cannot find an elementary function that is a solution of (15.16), we can draw a small line element at any point on the $(x\text{-}y)$-plane for which x is in I to represent the slope of an integral curve. If this line is short enough, the curve itself over that length resembles the line. These lines are called **line elements** and an ensemble of such lines is called a **direction field**.

Exercises

1. Test whether the relation

$$xy^2 - e^{-y} - 1 = 0 \tag{15.17}$$

is an implicit solution of the ODE

$$\left(xy^2 + 2xy - 1\right) y' + y^2 = 0. \tag{15.18}$$

Solution: If we blindly differentiate both sides to yield

$$2xyy' + y^2 + e^{-y}y' = 0 \quad \left(2xy + e^{-y}\right) y' + y^2 = 0$$

and then eliminate e^{-y} from the final result by using (15.17), we obtain the ODE (15.18). This implies the possibility that (15.17) is an implicit solution of the ODE (15.18). The remaining task is, therefore, to determine the *interval* I on which we can define such a function $y = h(x)$ that satisfies the relation (15.17) for *every* x on I.

As a first step, we write (15.17) as

$$y = \pm\sqrt{\frac{1+e^{-y}}{x}},$$

which says that y is defined only for $x > 0$ since e^{-y} is always positive. Hence, the interval for which (15.17) may be a solution of (15.18) must exclude values of $x \leq 0$.

Next, we depict a graph of equation (15.17) on the $(x\text{-}y)$-plane (see Fig. 15.3). From the graph, we see that there are three choices for the function $y = h(x)$, each of which gives a one-to-one relation between x and y. If we choose the upper branch $(y > 0)$, then we can say that "(15.17) is an implicit solution of (15.18) *for all* $x >$ 0." If we choose either of the two lower branches (one is above the dashed line and the other is below), then we can say that "(15.17) is an implicit solution of (15.18) *only for* $x > x_0 \simeq 2.07$." ♣

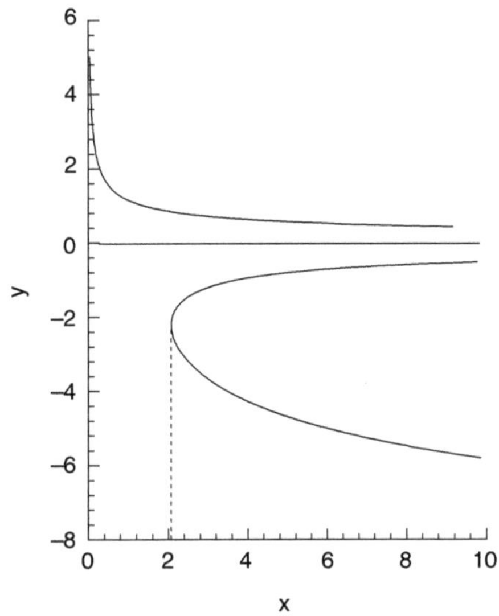

Fig. 15.3. The curve of the function (15.17)

15.2 Existence Theorem for the First-Order ODE

15.2.1 Picard Method

In this section, we consider a first-order ODE of the form

$$y'(x) = f(x, y(x)), \tag{15.19}$$

where f is some continuous function. Our main purpose is to prove that:

(i) a wide class of equations of the form (15.19) have solutions, and

(ii) solutions to **initial value problems**

$$y'(x) - f(x, y(x)), \quad y(x_0) - y_0$$

are unique. Statements **(i)** and **(ii)** are supported by the **existence theorem** and the **uniqueness theorem**, respectively, as is demonstrated in the subsequent subsections.

Our proof of the two theorems is based on that we call **Picard's method**, which gives solutions of an initial value problem

$$y'(x) = f\,(\,x, y(x)\,)\,, \quad y(x_0) = y_0, \tag{15.20}$$

where $f\,(\,x, y(x)\,)$ is assumed to be continuous and real-valued in a rectangle:

$$R: \ |x - x_0| < a, \ \ |y - y_0| < b \ \ (a, b > 0). \tag{15.21}$$

The key to Picard's method is to replace the differential equation in (15.20) by the equivalent integral form,

$$y(x) = y_0 + \int_{x_0}^{x} f\left(t, \ y(t)\right) dt, \tag{15.22}$$

which is an **integral equation** because the unknown function $y(x)$ appears in the integrand. That the integral equation (15.22) is equivalent to the original initial value problem can be checked by differentiating (15.22) on x.

Remark. Note that the initial condition $y(x_0) = y_0$ is automatically included in (15.22).

We now try to solve (15.22). As a crude approximation to a solution, we take the constant function $\varphi_0(x) = y_0$, which clearly satisfies the initial condition

$$\varphi_0(x_0) = y_0,$$

whereas it does not satisfy (15.22) in general. Nevertheless, if we substitute the constant function into $f(t, y(t))$ of (15.22), we have

$$\varphi_1(x) = y_0 + \int_{x_0}^{x} f(\ t, \varphi_0(t)\ )dt, \tag{15.23}$$

which is a closer approximation to a solution than $\varphi_0(x)$. By continuing the process, we have a sequence of functions $\{\varphi_n(x)\}$:

♠ **Successive approximation:**
Given an integral equation (15.22) with respect to $y(x)$, a set of functions defined by

$$\varphi_0(x) = y_0,$$
$$\varphi_n(x) = y_0 + \int_{x_0}^{x} f\left(t, \varphi_{n-1}(t)\right)dt. \ (n = 1, 2, \cdots) \tag{15.24}$$

is called a **successive approximation** to a solution of (15.22).

We understand intuitively that taking the limit $n \to \infty$ yields

$$\varphi_n(x) \to \varphi(x),$$

where $\varphi(x)$ is the exact solution of the integral equation (15.22). The convergence property of the sequence $\{\varphi_i(x)\}$ and the equivalence of the limit function $\varphi(x)$ to the solution of (15.22) are guaranteed if the integrand $f(x, y(x))$ satisfies several conditions as is demonstrated in Sect. 15.2.3.

In summary, we now know the following:

♠ **Picard method:**
The differential equation $y'(x) = f(x, y(x))$ for a given initial value $y(x_0) = y_0$ can be solved by starting with $\varphi_0(x) = y_0$ and then computing successive approximations (15.24). The process converges to a solution of the differential equation, where $f(x, y)$ satisfies several specific conditions given in Sect. 15.2.3.

15.2.2 Properties of Successive Approximations

We have previously assumed that $f(x, y(x))$ is continuous in the rectangle R defined in (15.21). Hereafter, we further assume that $f(x, y(x))$ is bounded on R, which means the existence of a constant $M > 0$ such that

$$|f(x, y(x))| \le M \quad \text{for all } (x, y) \in R.$$

In this case, the successive approximations $\{\varphi_n(x)\}$ show both the continuity and boundedness property stated below.

♠ **Continuity of successive approximations:**
Let $f(x, y)$ be continuous and bounded by $|f(x, y)| \le M$ in a rectangle

$$R: \ |x - x_0| < a, \quad |y - y_0| < b \ \ (a, b > 0).$$

Then, the successive approximations $\varphi_n(x)$ are continuous on the interval

$$I: |x - x_0| \le c \equiv \min\left[a, \frac{b}{M}\right].$$

♠ **Boundedness of successive approximations:**
Under the same conditions as above, the $\varphi_n(x)$ satisfy the inequality

$$|\varphi_n(x) - y_0| \le M|x - x_0|$$

for all x in I.

Remark. The condition $|f(x,y)| \leq M$ has an important geometric meaning in terms of the direction field. Since $y' = f(x,y)$, the direction field y' is bounded as $|y'| \leq M$, namely, $-M \leq y' \leq M$ for all points in R. Therefore, a solution curve $\varphi(x)$ that passes through $(x_0,\ y_0)$ must lie in the shadowed region in Fig. 15.4.

Proof (of the continuity). From (15.23), we have

$$|\varphi_1(x) - y_0| = \left| \int_{x_0}^{x} f(\, t, \varphi_0(t)\,)dt \right| \leq \int_{x_0}^{x} |f(\, t, y_0\,)|\, dt \leq M\,|x - x_0|\,, \quad (15.25)$$

since $\varphi_0(t) = y_0$ and $|f(x, y_0)| \leq M$. Now we tentatively assume that the theorem is true for a function φ_n with $n \geq 1$, and then prove inductively that it is also true for φ_n. By hypothesis, all points $(t, \varphi_{n-1}(t))$ for t in I are located within R. Hence, the function

$$F_{n-1}(t) = f(t, \varphi_{n-1}(t))$$

exists for t in I, which implies that

$$\varphi_n(x) = y_0 + \int_{x_0}^{x} F_{n-1}(t)dt$$

exists as a continuous function on I. ♣

Proof (of the boundedness). Since by hypothesis

$$|F_{n-1}(t)| = |f\,(t, \varphi_{n-1}(t))| \leq M,$$

we have

$$|\varphi_n(x) - y_0| \leq \left| \int_{x_0}^{x} F_{n-1}(t)dt \right| \leq \int_{x_0}^{x} |F_{n-1}(t)|\, dt \leq M|x - x_0|.$$

Therefore, the boundedness of $\varphi_n(x)$ has been proved by induction. ♣

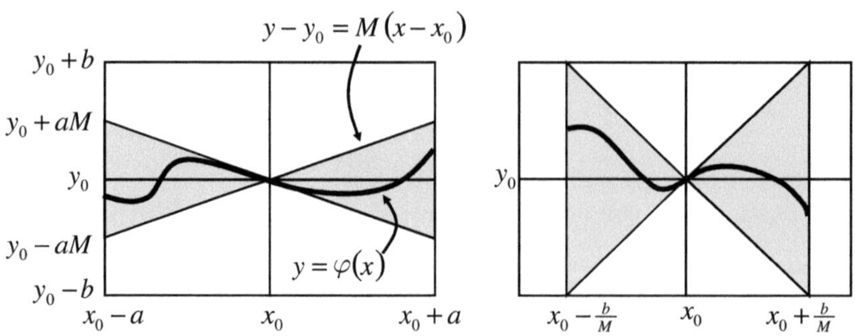

Fig. 15.4. Continuity and boundedness of a solution curve $\varphi(x)$ on the interval $I : |x - x_0| \leq c \equiv \min[a, \frac{M}{b}]$

15.2.3 Existence Theorem and Lipschitz Condition

Let $f(x, y(x))$ be a function defined for (x, y) in the rectangle R in the $(x$-$y)$-plane. We would like to verify the existence of solutions for the first-order ODEs expressed by

$$y'(x) = f(x, y(x))$$

by imposing a **Lipschitz condition**:

♠ **Lipschitz condition:**
We say that $f(x, y(x))$ satisfies a **Lipschitz condition** on a region R if there exists a constant $K > 0$ such that

$$|f(x, y(x)) - f(x, z(x))| \leq K\,|y(x) - z(x)| \tag{15.26}$$

for all $(x, y), (x, z) \in R$. Here the positive constant K is called the **Lipschitz constant.**

Our most important theorem is presented below.

♠ **Existence theorem:**
Suppose that

1. $f(x, y)$ is continuous and real-valued on the rectangle R.

2. $|f(x, y)| \leq M$ for all (x, y) in R.

3. f satisfies a Lipshitz condition with constant K in R.

Then the initial value problem

$$y'(x) = f(x, y(x)), \quad y(x_0) = y_0 \tag{15.27}$$

has at least one solution $y(x)$ in the interval

$$I : |x - x_0| \leq c \equiv \min\left[a, \frac{b}{M}\right].$$

Proof Consider the successive approximations $\{\varphi_n(x)\}$ to a solution of the initial value problem (15.27), wherein $f(x, y(x))$ is assumed to satisfy the Lipshitz condition (15.26). We would like to prove that **(i)** the limit function

$$\varphi(x) \equiv \lim_{n\to\infty} \varphi_n(x)$$

exists and **(ii)** that it is the solution of (15.27).

By definition of $\varphi_n(x)$, for $n \geq 1$, we obtain

$$\begin{aligned}
|\varphi_{n+1}(x) - \varphi_n(x)| &\leq \left| \int_{x_0}^{x} [\, f(t, \varphi_n(t)) - f(t, \varphi_{n-1}(t)) \,]\, dt \right| \\
&\leq \int_{x_0}^{x} |f(t, \varphi_n(t)) - f(t, \varphi_{n-1}(t))|\, dt \\
&\leq K \int_{x_0}^{x} |\varphi_n(t) - \varphi_{n-1}(t)|\, dt. \qquad (15.28)
\end{aligned}$$

Set $n = 1$ in (15.28) and substitute it in the result (15.25) to find

$$|\varphi_2(x) - \varphi_1(x)| \leq KM \frac{|x - x_0|^2}{2!}. \qquad (15.29)$$

Set $n = 2$ in (15.28) and use the result of (15.29) in the last term in (15.28). Continuing the process, we have

$$|\varphi_n(x) - \varphi_{n-1}(x)| \leq K^{n-1} M \frac{|x - x_0|^n}{n!}. \qquad (15.30)$$

Observe that the right-hand side of (15.30) is the nth term of the power series for $e^{K|x-x_0|}$ multiplied by M/K. This implies that the infinite series

$$\varphi_0(x) + \sum_{k=1}^{\infty} [\varphi_k(x) - \varphi_{k-1}(x)] \qquad (15.31)$$

is absolutely (and thus ordinary) convergent, ensuring the existence of the limit function $\varphi(x) \equiv \lim_{n\to\infty} \varphi_n(x)$. (See Sect. 3.2 for the convergence properties of Cauchy sequences.)

Next we prove statement **(ii)** above. Note that the nth partial sum of (15.31) is just $\varphi_n(x)$ and that the infinite series (15.31) equals the limit function $\varphi(x)$. Hence, we have from (15.30) and (15.31) that

$$\begin{aligned}
|\varphi(x) - \varphi_n(x)| &= \left| \sum_{k=n+1}^{\infty} [\varphi_k(x) - \varphi_{k-1}(x)] \right| \\
&\leq \sum_{k=n+1}^{\infty} |\varphi_k(x) - \varphi_{k-1}(x)| \leq \sum_{k=n+1}^{\infty} K^{k-1} M \frac{|x - x_0|^k}{k!} \\
&\leq \sum_{k=n+1}^{\infty} K^{k-1} M \frac{c^k}{k!} \leq \frac{M}{K} \alpha_n e^{Kc},
\end{aligned}$$

where

$$\alpha_n = \frac{(Kc)^{n+1}}{(n+1)!}.$$

Since α_n is the nth term of the power series of e^{Kc}, we have $\lim_{n\to\infty} \alpha_n = 0$. Therefore, the series of functions $\{\varphi_n(x)\}$ converges **uniformly** to $\varphi(x)$ in the interval $I: \ x \in [x_0 - c, \ x_0 + c]$, which means that

$$\lim_{n\to\infty} f(x, \varphi_n(x)) = f(x, \varphi(x)). \tag{15.32}$$

That being so, we can write

$$\begin{aligned}
\varphi(x) = \lim_{n\to\infty} \varphi_n(x) &= y_0 + \lim_{n\to\infty} \int_{x_0}^{x} f(t, \varphi_n(t))\, dt \\
&= y_0 + \int_{x_0}^{x} \lim_{x\to\infty} f(t, \varphi_n(t))\, dt \\
&= y_0 + \int_{x_0}^{x} f(t, \varphi(t))\, dt.
\end{aligned} \tag{15.33}$$

By differentiating on x, we have

$$\varphi(x)' = f(x, \varphi(x)), \quad \varphi(x_0) = y_0.$$

These ensure that $\varphi(x)$ is a solution of our initial value problem (15.27). ♣

15.2.4 Uniqueness Theorem

Next we examine the uniqueness of the solution $\varphi(x)$ that we found earlier using the Picard approximation method (see Sect. 15.2.1). This is described by the theorem below.

♠ **Uniqueness theorem:**
Let $f(x, y)$ be continuous and satisfy the Lipschitz condition (15.26) in the rectangle R. If φ and ψ are two solutions of the initial value problem (15.27) in an interval I containing x_0, then $\varphi(x) = \psi(x)$ for all x in I.

Proof We assume that both $\varphi(x)$and $\psi(x)$ are solutions of (15.27). For $x > x_0$, we have from (15.33) and the Lipschitz condition (15.26) that

$$\begin{aligned}
|\varphi(x) - \psi(x)| &\le \int_{x_0}^{x} |f(t, \varphi(t)) - f(t, \psi(t))|\, dt \\
&\le K \int_{x_0}^{x} |\varphi(t) - \psi(t)|\, dt.
\end{aligned} \tag{15.34}$$

This holds in the interval $I: \ x \in [x_0, \ x_0 + \delta]$ for arbitrary small $\delta > 0$. Since $|\varphi(x) - \psi(x)|$ is continuous in I, it has a maximum at some x on I, which we label μ. Equation (15.34) provides that

$$\mu \le K\mu |x - x_0| \le K\mu\delta \quad \text{for all } x \text{ in } I, \tag{15.35}$$

so we have

$$(1 - K\delta)\mu \le 0.$$

Note that by definition $\mu \ge 0$. Hence, if $K\delta < 1$, we have $\mu = 0$, which says that given any Lipshitz constant K, we can find a sufficiently small δ such that

$$\max |\varphi(x) - \psi(x)| = 0,$$

i.e.,

$$|\varphi(x) - \psi(x)| \equiv 0 \quad \text{for } x \in [x_0, x_0 + \delta].$$

Continuing this process yields the conclusion that $|\varphi(x) - \psi(x)| \equiv 0$ for all x in R. The same holds for the case $x < x_0$, completing the proof. ♣

15.2.5 Remarks on the Two Theorems

1. The existence and uniqueness theorems only ensure the existence and uniqueness of a solution. They do not tell us whether the solution can or cannot be expressed in terms of an elementary function form or help us to find the solution.
2. Arguments for real-valued functions given thus far are straightforwardly extended to the case that f is complex-valued. In this case we must admit complex-valued solutions and f must be defined for complex z. The set of points z satisfying $|z - z_0| \le b$ becomes a circle with a center z_0 and radius b, so domain R is no longer a rectangle.
3. The initial value problem

$$y'(x) = \sqrt{|y(x)|}, \quad y(0) = 0,$$

has two solutions,

$$y(x) \equiv 0 \quad \text{and} \quad y(x) = \begin{cases} x^2/4 & \text{if } x \ge 0, \\ -x^2/4 & \text{if } x < 0, \end{cases}$$

although $f(x, y) = \sqrt{|y|}$ is continuous for all y. The Lipshitz condition is violated in any region that includes the line $y = 0$ because for $y_1 = 0$ and positive y_2 we have

$$\frac{|f(x, y_2) - f(x, y_1)|}{|y_2 - y_1|} = \frac{\sqrt{y_2}}{y_2} = \frac{1}{\sqrt{y_2}} \quad (\sqrt{y_2} > 0) \tag{15.36}$$

and this can be made as large as we please by choosing y_2 sufficietly small, whereas the Lipshitz condition requires that the quotient on the left-hand side of (15.36) not exceed a fixed constant M.

Exercises

1. Using the Picard method, evaluate the successive approximation to the solution of the initial value problem

$$y'(x) = 1 + y(x)^2, \quad y(0) = 0.$$

Solution: Set $x_0 = 0$, $y_0 = 0$, $f(x,y) = 1 + y^2$ in (15.24) to find that

$$\varphi_n(x) = \int_0^x \left\{1 + [\varphi_{n-1}(t)]^2\right\} dt = x + \int_0^x [\varphi_{n-1}(t)]^2 \, dt.$$

Hence, we obtain

$$\varphi_1(x) = x + \int_0^x 0 \, dt = x, \quad \varphi_2(x) = x + \int_0^x t^2 dt = x + \frac{x^3}{3},$$

$$\varphi_3(x) = x + \int_0^x \left(t + \frac{t^3}{3}\right)^2 dt = x + \frac{x^3}{3} + \frac{2}{15}x^5 + \frac{1}{63}x^7, \text{ and so on.} \ \clubsuit$$

Remark. The exact solution of the above problem can be deduced by separating variables:

$$y(x) = \tan x = x + \frac{x^3}{3} + \frac{2}{15}x^5 + \frac{17}{315}x^7 + \cdots \quad \left(-\frac{\pi}{2} < x < \frac{\pi}{2}\right).$$

The first three terms of $\varphi_3(x)$ and of the series above are the same. The series converges only for $|x| < \pi/2$; therefore, all that we can expect is that our sequence $\varphi_1, \varphi_2, \cdots$ converges to a function that is the solution of our problem for $|x| < \pi/2$.

2. By applying the Picard method to

$$y'(x) = xy(x), \quad y(0) = 1, \tag{15.37}$$

show that the Picard series $\{\varphi_n(x)\}$ converges absolutely and uniformly.

Solution: The integral equation corresponding to (15.37) becomes

$$y(x) = 1 + \int_0^x ty(t)dt.$$

The iterative equation is written as $\varphi_0(x) = 0$ and

$$\varphi_{n+1}(x) = 1 + \int_0^x t\varphi_n(t)dt, (n = 1, 2 \cdots).$$

Thus, we easily find

$$\varphi_n(x) = 1 + \frac{x^2}{2} + \frac{1}{2!}\left(\frac{x^2}{2}\right)^2 + \cdots + \frac{1}{k!}\left(\frac{x^2}{2}\right)^n.$$

The nature of the convergence is obvious for all real x, since it is a partial sum for the Taylor series of the function $\varphi(x) = e^{x^2/2}$. This means that $\varphi_n(x) \to \varphi(x)$ as $n \to \infty$. ♣

3. For the equation given by

$$y'(x) = 2y(x)^{1/2}, \quad y(0) = 0,$$

check the uniqueness of the solution in connection with the Lipschitz condition.

Solution: This equation has the two solutions $y(x) = 0$, $y(x) = 16x^2$, although $f(x, y) = 2(y)^{1/2}$ is continuous for all y. The Lipschitz condition (15.26) is violated in any region that includes the line $y(x) = 0$ because for $y_1 = 0$ and $y_2 \neq 0$ we have

$$\frac{|f(x,y_2) - f(x,y_)|}{|y_2 - y_1|} = \frac{\sqrt{y_2}}{y_2} = \frac{1}{\sqrt{y_2}},$$

which diverges for $y_2 \to 0$, exceeding a fixed constant K. ♣

15.3 Sturm–Liouville Problems

15.3.1 Sturm–Liouville Equation

ODEs encountered in physics are often classified as **Sturm–Liouville equations**:

♠ **Sturm–Liouville equation:**
A Sturm–Liouville equation is a second-order homogeneous linear ODE of the form

$$-\frac{d}{dx}\left[p(x)\frac{dy}{dx}\right] + q(x)y + \lambda w(x)y = 0, \tag{15.38}$$

where λ is a parameter and p, q, w are real-valued continuous functions with $p(x) > 0$ and $w(x) > 0$. Here $w(x)$ is called a **weight function.**

Using the **Sturm–Liouville operator** L defined by

$$L = \frac{1}{w(x)}\left[-\frac{d}{dx}\left(p(x)\frac{d}{dx}\right) + q(x)\right] \tag{15.39}$$

we reduce the Sturm–Liouville equation (15.38) to the abbreviated form

$$Ly(x) = -\lambda y(x). \tag{15.40}$$

Examples The **Legendre equation**

$$\left(1-x^2\right)y'' - 2xy' + n(n+1)y = 0, \quad n \geq 1, \quad x \in [-1,\ 1]$$

is expressed as

$$\left[\left(1-x^2\right)y'\right]' + n(n+1)y = 0.$$

This is in the Sturm–Liouville form of $p = 1 - x^2$, $q = 0$, $w = 1$, and $\lambda = n(n+1)$.

Relevant terminology is given below.

♠ **Sturm–Liouville system:**

A **Sturm–Liouville system** consists of a Sturm–Liouville equation (15.38) on a finite closed interval $a \leq x \leq b$, together with two separated boundary conditions of the form

$$y(a) = \alpha y'(a) \text{ and } y(b) = \beta y'(b)$$

with α, β being real.

A nontrivial solution of a Sturm–Liouville system is called an **eigenfunction** and the corresponding λ is called an **eigenvalue**. The set of all eigenvalues of a Sturm–Liouville system is called the **spectrum** of the system.

Examples The Sturm–Liouville system consisting of the ODE

$$y'' + \lambda y = 0 \quad 0 \leq x \leq \pi$$

with the separated boundary conditions

$$y(0) = 0,\ y(\pi) = 0$$

has the eigenfunction

$$y_n(x) = \sin nx$$

and the eigenvalues

$$\lambda_n = n^2. \ (n = 1, 2, \cdots).$$

15.3.2 Conversion into a Sturm–Liouville Equation

Mathematically, Sturm–Liouville equations represent only a small fraction of the second-order differential equations. Nevertheless, any second-order equation of the form

$$a(x)y'' + b(x)y' + c(x)y + \lambda e(x)y = 0$$

can be transformed into a Sturm–Liouville equation by multiplying the factor

$$\xi(x) = \exp\left[\int^x \frac{b(s) - a'(s)}{a(s)} ds\right], \tag{15.41}$$

which yields a Sturm–Liouville form,

$$(\xi a y')' + \xi c y + \lambda \xi e y = 0,$$

with a nonnegative weight function $\xi(x)a(x)$.

Examples We show below that the **Hermite equation** of the form

$$y'' - 2xy' + 2\alpha y = 0 \tag{15.42}$$

can be transformed into a Sturm–Liouville equation. Substituting $a(x) = 1$ and $b(x) = -2x$ into (15.41) yields

$$\xi(x) = \exp\left[\int^x (-2s) ds\right] = e^{-x^2},$$

by which we multiplying both sides of (15.42), to obtain

$$e^{-x^2} y'' - 2xe^{-x^2} y' + 2\alpha e^{-x^2} y = \left(e^{-x^2} y'\right)' + 2\alpha e^{-x^2} y = 0,$$

This is the Sturm-Liouville form with

$$p(x) = e^{-x^2},\ q(x) = 0,\ w(x) = e^{-x^2},$$

and $\lambda = 2\alpha$.

15.3.3 Self-adjoint Operators

We know many facts about Sturm–Liouville problems. Below is an important concept regarding the nature of these problems.

♠ **Adjoint operator:**
The adjoint of an operator L, denoted by $L^\dagger$, is defined by

$$\int_a^b f^*(x)[Lg(x)]\rho(x)dx = \left\{\int_a^b g^*(x)[L^\dagger f(x)]\rho(x)dx\right\}^*. \tag{15.43}$$

Using **inner product notation**, we can write the definition of the adjoint operator (15.43) as

$$(f, Lg) = (g, L^\dagger f).$$

The most important terminology in this section is given below.

♠ **Self-adjoint operator:**
An operator L is called **self-adjoint** (or **Hermitian**) if

$$L = L^\dagger$$

or, in inner product notation,

$$(f, Lg) = (g, Lf)^*.$$

It should be noted that an operator is said to be self-adjoint only if certain boundary conditions are met by the functions f and g on which it acts. An illusrative example follows:

Examples Let us derive the required boundary conditions for the linear operator

$$L = \frac{d^2}{dx^2}$$

to be self-adjoint over the interval $[a, b]$. From the definition of self-adjoint operators, the operator L should satisfy the relation:

$$\int_a^b f^* \frac{d^2 g}{dx^2} dx = \left(\int_a^b g^* \frac{d^2 f}{dx^2} dx \right)^* . \tag{15.44}$$

Through integration by parts, the left-hand side gives

$$\int_a^b f^* \frac{d^2 g}{dx^2} dx = \left[f^* \frac{dg}{dx} \right]_a^b + \left[-g \frac{df^*}{dx} \right]_a^b + \int_a^b g \frac{d^2 f^*}{dx^2} dx. \tag{15.45}$$

From a comparison as (15.44) and (15.45), it follows that the operator L is Hermitian provided that

$$\left[f^* \frac{dg}{dx} \right]_a^b = \left[g \frac{df^*}{dx} \right]_a^b .$$

15.3.4 Required Boundary Condition

In the example in Sect. 15.3.3, we derived the required boundary condition for a specific Sturm–Liouville operator to be self-adjoint. For general Sturm–Liouville operators, such a required boundary condition is given by the following theorem.

♠ **Theorem:**
A Sturm–Liouville operator is self-adjoint on $[a, b]$ if any two eigenfunctions y_i and y_j of (15.38) satisfy the boundary condition

$$\left[p y_i^* y_j'\right]_a^b = 0. \tag{15.46}$$

Proof It follows from the explicit form of the Sturm–Liouville operator L that

$$(y_i, Ly_j) = -\frac{1}{w}\int_a^b y_i^* \left(p y_j'\right)' dx - \frac{1}{w}\int_a^b y_i^* q y_j dx. \tag{15.47}$$

The first integral is integrated by parts to give

$$-\frac{1}{w}\left[y_i^* p y_j'\right]_a^b + \frac{1}{w}\int_a^b (y_i^*)' p y_j' dx,$$

in which the first term vanishes because we have assumed the boundary condition (15.46). Integration by parts then yields

$$\frac{1}{w}\left[(y_i^*)' p y_j\right]_a^b - \frac{1}{w}\int_a^b \left[(y_i^*)' p\right]' y_j dx,$$

where the first term is again zero owing to our assumption. As a result, the sum of integrals I in (15.47) reads

$$(y_i, Ly_j) = \frac{1}{w}\int_a^b \left\{\left[-(y_i^*)' p\right]' y_j - y_i^* q y_j\right\} dx \tag{15.48}$$

$$= \frac{1}{w}\left\{-\int_a^b \left[y_j^* \left(p y_i'\right)' - y_i q y_j^*\right] dx\right\}^* = (y_j, Ly_i)^*, \tag{15.49}$$

which completes the proof. ♣

15.3.5 Reality of Eigenvalues

♠ **Theorem:** For a Sturm–Liouville system under the boundary condition (15.46), we have:

(a) All eigenvalues are real.
(b) Eigenfunctions corresponding to distinct eigenvalues are orthogonal.

Proof of (a). If an eigenfunction y_n belongs to the eigenvalue λ_n, then

$$\begin{aligned}\lambda_n^*(y_n, y_n) &= (\lambda_n y_n, y_n) = -(Ly_n, y_n) \\ &= -(y_n, Ly_n) = \lambda_n(y_n, y_n).\end{aligned}$$

This indicates that $\lambda_n^* = \lambda_n$ since $(y_n, y_n) > 0$. Therefore λ_n is real for all n. ♣

Proof of (b). According to the same argument as above,

$$\begin{aligned}\lambda_m(y_m, y_n) &= (\lambda_m y_m, y_n) = -(Ly_m, y_n) \\ &= -(y_m, Ly_n) = \lambda_n(y_m, y_n).\end{aligned}$$

Thus, for $\lambda_m \neq \lambda_n$, $(y_m, y_n) = 0$, which means that eigenfunctions corresponding to distinct eigenvalues are orthogonal. ♣

Remark. If eigenvalues are degenerate, say, $\lambda_m = \lambda_n \quad (m \neq n)$, an orthogonal set of eigenfunctions is constructed using the **Gram–Schmidt orthogonalization method**. Namely, we can choose the eigenfunctions to be orthogonal to each other with respect to the weight function w such that if $(y_m, y_n) \neq 0$, we replace y_n by $\tilde{y}_n = y_n - ay_m$ where a should be chosen to be $(y_m, \tilde{y}_n) = 0$.

Exercises

1. Show that the **Bessel equation** given by

$$x^2y'' + xy' + \left(x^2 - n^2\right)y = 0 \text{ with } n \geq 0 \text{ and } x \in (-\infty, \infty)$$

can be expressed in the form of a Sturm–Liouville equation.

Solution: After the transformation $x \to kx$, we have

$$[xy'(kx)]' + \left(-\frac{n^2}{x} + k^2x\right)y(kx) = 0, \quad n \geq 0,$$

where $p = x$, $q = -n^2x$, $w = x$, and the parameter $\lambda = k^2$ in (15.38). ♣

2. The **Bernoulli equation** is given as a nonlinear equation by

$$y' = a(x)y + b(x)y^k, \tag{15.50}$$

where $a(x)$, $b(x)$ are continuous functions in an interval I and k is an arbitrary constant.

(a) Show that the transformation $u = y^{1-k}$ provides an inhomogeneous linear equation for u.

(b) Find a solution for the transformed linear equation for u under the initial condition $u(x_0) = u_0$.

Solution:

(a) The transformed equation becomes $u' = (1-k)a(x)u + (1-k)b(x)$.

(b) The above equation can be reduced to an inhomogeneous linear equation of the form

$$u' = p(x)u + q(x),$$

where $p(x) = (1-k)a(x)$, $q(x) = (1-k)b(x)$ are continuous functions. Let $P(x)$ be a function whose derivative is $p(x)$ such that

$$P(x) = \int_{x_0}^{x} p(t)dt,$$

where x_0 is a fixed point in I. Multiplying both sides of (15.50) by $e^{P(x)}$ to, we have the relation

$$(e^P u)' = e^P(u' - pu) = e^P q.$$

Therefore, we obtain a solution such that

$$u(x) = u_0 e^{-P(x)} + e^{-P(x)} \int_{x_0}^{x} e^{P(s)} q(s) ds,$$

where u_0 comes from the initial condition. ♣

3. The **logistic equation** is a special type of Bernoulli equation given by

$$y' = ay - by^2, \tag{15.51}$$

where a, b are constants. Find a solution for the above by imposing the initial condition $y(x_0) = y_0$.

Solution: Using a solution for Exercise **2(b)** by setting $k = 2$, we have

$$y(x) = \frac{a}{b + (a/y_0 - b)e^{-a(x-x_0)}} .$$

Note that $y(x) = a/b$ as $x \to \infty$. ♣

4. The **Riccati equation** is a nonlinear equation given by

$$y' + p(x)y + q(x)y^2 = r(x). \tag{15.52}$$

(a) Assuming $u(x)$ to be a particular solution of the above, namely, a solution when we set $r(x) = 0$, show that $z(x)$ defined by $y(x) = u(x) + z(x)$ constitutes the Bernoulli equation.

(b) Show that the Riccati equation is reduced to a linear equation of the second order by the transformation $y = Qv'/v$.

Solution:

(a) Substituting $y = u + z$ into the equation, we have

$$\left[z' + p(z^2 + 2uz) + qz\right] + \left[u' + pu + qu^2 - r\right] = 0.$$

The second parenthesis vanishes and we have the Bernoulli equation such that $z' + (2up + q)z + pz^2 = 0$.

(b) The first order derivative gives

$$y' = Q\left(\frac{v''}{v} - \frac{v'^2}{v^2}\right) + Q'\frac{v'}{v}.$$

Thus, we have

$$Q\frac{v''}{v} + (pQ - 1)Q\frac{v'^2}{v^2} + (Q' + qQ)\frac{v'}{v} + r = 0.$$

Setting $Q = 1/p(x)$, we have $v'' + \left(q - \frac{p'}{p}\right)v' + prv = 0$. ♣

16

System of Ordinary Differential Equations

Abstract In this chapter we focus on an autonomous system (Sect. 16.3), which is a specific type of system of ordinary differential equations. Autonomous systems can be used to describe the dynamics of the physical objects that are encountered in physics and engineering problems, wherein the laws governing the motion of the objects are time-independent, namely, they hold true at all times. The stability of these dynamical systems is characterized by the critical point (Sect. 16.3.3), whose nature is revealed by the functional form of the autonomous systems.

16.1 Systems of ODEs

16.1.1 Systems of the First-Order ODEs

This section deals with n coupled ordinary differential equations (ODEs). The formal definition is stated below.

♠ **Systems of ODEs:**
A system of ODEs is given by

$$F_i\left[x; y_1, y_1', y_1'', \cdots, y_1^{(r_{i1})}; y_2, y_2', y_2'', \cdots, y_2^{(r_{i2})}; \cdots\right] \\ = 0 \quad (i = 1, 2, \cdots), \tag{16.1}$$

which involves a set of unknown functions $y_1(x), y_2(x), \cdots$ and their derivatives with respect to a single independent variable x.

For each ith equation of (16.1), we denote the highest order of the derivatives of y_j by r_{ij}. Hereafter, we consider the case of $r_{ij} \equiv 1$ for all i and j, i.e., a system of n ordinary differential equations (ODEs) of the first order expressed by

$$\begin{aligned} y_1'(x) &= f_1(x, y_1, y_2, \cdots, y_n), \\ y_2'(x) &= f_2(x, y_1, y_2, \cdots, y_n), \\ &\cdots \\ y_n'(x) &= f_2(x, y_1, y_2, \cdots, y_n). \end{aligned} \tag{16.2}$$

Here, $\{f_k\}$, $k = 1, 2, \cdots, n$ are single-valued continuous functions in a certain domain of their arguments and $\{y_k\}$, $k = 1, 2, \cdots, n$ are unknown complex functions of a real variable x.

16.1.2 Column-Vector Notation

For convenience we use **column-vector notation** for an ordered set of unknown functions $\{y_k(x)\}$ in which each $y_k(x)$ is called a component, which we denote by a bold-face letter:

$$\boldsymbol{y}(x) = [y_1(x), y_2(x), \cdots, y_n(x)]^{\mathrm{T}}, \tag{16.3}$$

where the norm of the vector is defined by

$$\|\boldsymbol{y}(x)\| = \left(|y_1|^2 + |y_2|^2 + \cdots + |y_n|^2\right)^{1/2}. \tag{16.4}$$

Using vector notation, we can express (16.2) in the concise form

$$\boldsymbol{y}'(x) = \boldsymbol{f}(x, \boldsymbol{y}(x)), \tag{16.5}$$

where the column vector $\boldsymbol{f}$ is defined by its components

$$\boldsymbol{f}(x, \boldsymbol{y}(x)) = [f_1, f_2, \cdots, f_n]^{\mathrm{T}}. \tag{16.6}$$

If there exists a set of functions $\boldsymbol{\varphi}(x) = (\varphi_1(x), \varphi_2(x), \cdots, \varphi_n(x))$ satisfying

$$\varphi_i(x)' = f_i\left(x, \varphi_1(x), \varphi_2(x), \cdots, \varphi_n(x)\right), \quad i = 1, 2, \cdots, n,$$

we say $\boldsymbol{\varphi}(x)$ is a solution of (16.2). The initial value problem consists of finding a solution $\boldsymbol{\varphi}(x)$ of (16.5) in I satisfying the initial condition $\boldsymbol{\varphi}(x_0) = \boldsymbol{y}_0 = (y_{10},\ y_{20}, \cdots, y_{n0})$.

16.1.3 Reducing the Order of ODEs

Let consider an nth order ODE of $u(x)$ given by

$$\frac{d^n u(x)}{dx^n} + p_1(x)\frac{d^{n-1}u(x)}{dx^{n-1}} + \cdots + p_n(x)u(x) = q(x). \tag{16.7}$$

We show that equation (16.7) can always be reduced to a system of n first-order differential equations, which is stated as follows:

♠ **Theorem:**
Given an nth-order ODE, it can always be reduced to a system of n first-order ODEs.

Proof We take $u(x)$ and its derivatives $u', u'', \cdots, u^{(n-1)}$ as new unknown functions defined by

$$y_k(x) \equiv \frac{d^{k-1}u(x)}{dx^{k-1}}, \quad k = 1, 2, \cdots, n. \tag{16.8}$$

It is evident that (16.7) is equivalent to the following set of equations:

$$y_1' = y_2, \ y_2' = y_3, \ \cdots, \ y_{n-1}' = y_n \tag{16.9}$$

and

$$y_n' = -p_1 y_n - p_2 y_{n-1} - \cdots - p_n y_1 + q. \tag{16.10}$$

Equations (16.9) and (16.10) can be written in a brief vector form as

$$\frac{d\boldsymbol{y}(x)}{dx} = \boldsymbol{f}(x, \boldsymbol{y}), \tag{16.11}$$

where the column vectors are defined as

$$\boldsymbol{y} = (y_1, y_2, \cdots, y_n)$$

and

$$\begin{aligned} \boldsymbol{f} &= \boldsymbol{f}(x, \boldsymbol{y}) \\ &= [y_2, \ y_3, \ \cdots, \ y_n, \ -p_1 y_n - p_2 y_{n-1} - \cdots - p_n y_1 + q]^{\mathrm{T}}. \end{aligned}$$ ♣

Example One of the most famous systems of the type (16.11) results from the equation of motion for a particle of mass m. For a mobile particle along the x-axis, the equation of motion is

$$m\frac{d^2x(t)}{dt^2} = F\left(t, x(t), \frac{dx(t)}{dt}\right), \tag{16.12}$$

where t is the time and F represents the force acting on the particle. To see how the second-order ODE (16.12) can be viewed as a system of the form (16.11), we make the following substitutions:

$$t \to x, \ \ x \to y_1, \ \ \frac{dx}{dt} \to y_2.$$

Then (16.12) is equivalent to a system of two equations:

$$\begin{aligned} y_1' &= y_2, \\ y_2' &= \frac{1}{m}F(x, y_1, y_2), \end{aligned}$$

which is of the form of $\boldsymbol{y}'(x) = \boldsymbol{f}(x, \boldsymbol{y})$.

16.1.4 Lipschitz Condition in Vector Spaces

The vector equation

$$\boldsymbol{y}'(x) = \boldsymbol{f}(x, \boldsymbol{y}) \tag{16.13}$$

is obviously analogous to the scalar equation

$$y'(x) = f(x, y).$$

This analogy implies the possibility that the definition of a Lipschitz condition can be extended to the vector equation. The extended Lipshitz condition provides a simple sufficient condition for the uniqueness and existence of solutions, which implies that all the theorems for the scalar equation can be generalized so as to hold for the vector equation.

♠ **Lipschitz condition for a vector function:**

A vector function $\boldsymbol{f}(x, \boldsymbol{y})$ in (16.13) is said to satisfy the Lipschitz condition on a region R if and only if

$$|\boldsymbol{f}(x, \boldsymbol{y}(x)) - \boldsymbol{f}(x, \boldsymbol{z}(x))| \leq K |\boldsymbol{y}(x) - \boldsymbol{z}(x)|,$$

$$(R: |x - x_0| \leq a, \ |\boldsymbol{y} - \boldsymbol{y}_0| \leq b, \ |\boldsymbol{z} - \boldsymbol{z}_0| \leq b). \tag{16.14}$$

for the Lipschitz constant K.

When $\boldsymbol{f}(x, \boldsymbol{y})$ satisfies the Lipschitz condition noted above, we see from (16.14) that

$$|f_k(x, y_1, y_2, \cdots, y_n) - f_k(x, z_1, z_2, \cdots, z_n)|$$
$$\leq K(|y_1 - z_1| + |y_2 - z_2| + \cdots + |y_n - z_n|) \quad (k = 1, 2, \cdots, n). \tag{16.15}$$

Using this, we can prove the theorem of the existence and uniqueness of solutions for the general vector equation (16.13). For instance, the uniqueness of the solution for (16.13) is straightforward as shown below. The right-hand side of (16.15) yields

$$K \sum_{k=1}^{n} |y_k(x) - z_k(x)| \leq K \sum_{k=1}^{n} \int_{x_0}^{x} |f_k(x, \boldsymbol{y}(x)) - f_k(x, \boldsymbol{z}(x))| \, dx$$
$$\leq nK^2 \int_{x_0}^{x} \sum_{k=1}^{n} |y_k(x) - z_k(x)| dx, \tag{16.16}$$

which holds for the interval I; $x \in [x_0, x_0 + \delta]$ for any small δ. Since the left-hand side of (16.16) is continuous on I, it has a maximum at some x, which we label μ. Then, the inequality (16.16) becomes

$$\mu \leq nK\mu(x - x_0) \leq nK\mu\delta,$$

which gives us $\mu(1 - nK\delta) \le 0$. For any small $\delta > 0$, we have $\mu = 0$, which indicates that $\sum |y_k - z_k| = 0$. The same holds true for the case $x < x_0$. Thus, the solution of (16.13) is unique.

Exercises

1. Consider a initial value problem given by

$$\boldsymbol{y}' = \boldsymbol{f}(x, \boldsymbol{y}), \quad \boldsymbol{y}(x_0) = \boldsymbol{y}_0,$$

defined on $\boldsymbol{R} : |x - x_0| \le a,\ |\boldsymbol{y} - \boldsymbol{y}_0| \le b,\ (a, b > 0)$. Assuming that $\boldsymbol{f}$ is continuous on $\boldsymbol{R}$, a sequence of successive approximations $\varphi_0, \varphi_1, \cdots$ is given by

$$\varphi_0(x) = \boldsymbol{y}_0$$

and

$$\varphi_{n+1}(x) = \boldsymbol{y}_0 + \int_{x_0}^{x} \boldsymbol{f}(t, \varphi_n(t))dt \quad \text{for } n = 1, 2, \cdots.$$

Using this procedure, find a sequence of successive approximations for

$$(y_1', y_2') = (y_2, -y_1), \quad \text{for } \boldsymbol{y}(0) = (0, 1).$$

Solution: Here $\boldsymbol{f}(x, \boldsymbol{y}) = (y_2, -y_1)$, so we have

$$\begin{aligned}
\varphi_0(x) &= (0, 1),\\
\varphi_1(x) &= (0, 1) + \int_0^x (1, 0)dt = (x, 1),\\
\varphi_2(x) &= (0, 1) + \int_0^x (1, -t)dt = (0, 1) + \left(x, -\frac{x^2}{2}\right) = \left(x, 1 - \frac{x^2}{2}\right).
\end{aligned}$$

Continuing with this process, we find the solution of the problem as $\varphi_k(x) \to \varphi(x) = (\sin x, \cos x)$. ♣

16.2 Linear System of ODEs

16.2.1 Basic Terminology

We now focus on a particular class of systems of ODEs called a **linear** system of first-order ODEs, described by

$$\frac{dy_1(x)}{dx} - \sum_{j=1}^{n} a_{1j}(x)y_j(x) = q_1(x),$$

$$\frac{dy_2(x)}{dx} - \sum_{j=1}^{n} a_{2j}(x)y_j(x) = q_2(x),$$
$$\cdots$$
$$\frac{dy_n(x)}{dx} - \sum_{j=1}^{n} a_{nj}(x)y_j(x) = q_n(x).$$

Here $a_{kj}(x)$ and $q_k(x)$ with $j, k = 1, 2, \cdots, n$ are continuous functions on x on some interval I. For convenience we use the vector representation given by

$$\frac{d\boldsymbol{y}(x)}{dx} - \boldsymbol{A}(x)\boldsymbol{y}(x) = \boldsymbol{q}(x), \tag{16.17}$$

where $\boldsymbol{A} = [a_{kj}]$ is an $n \times n$ matrix. Therefore, $\boldsymbol{Ay}$ stands for the matrix $\boldsymbol{A}$ applied to the column vector $\boldsymbol{y} = [y_1, y_2, \cdots, y_n]^{\mathrm{T}}$, namely, the **linear transform** of $\boldsymbol{y}$ by $\boldsymbol{A}$. The vector $\boldsymbol{q}$ is defined as $\boldsymbol{q} = [q_1, q_2, \cdots, q_n]^{\mathrm{T}}$. Given any $\boldsymbol{y}(x_0)$ for x_0 in I, there exists a unique solution $\boldsymbol{\varphi}(x)$ on I such that $\boldsymbol{\varphi}(x_0) = [y_1(x_0), y_2(x_0), \cdots, y_n(x_0)]^{\mathrm{T}}$.

The use of the linear operator L to (16.17) yields

$$L\left[\boldsymbol{y}(x)\right] = \boldsymbol{q}(x),$$

where the L is defined as

$$L = \frac{d}{dx} - \boldsymbol{A}. \tag{16.18}$$

If $\boldsymbol{q}(x) = 0$ for all x on I, (16.17) is said to be a **linear homogeneous system** of nth order, expressed by

$$\frac{d\boldsymbol{y}(x)}{dx} - \boldsymbol{A}(x)\boldsymbol{y}(x) = 0. \tag{16.19}$$

Otherwise, (16.17) is called **inhomogeneous**. A homogeneous system obtained from the inhomogeneous system (16.17) by setting $\boldsymbol{q}(x) \equiv \boldsymbol{0}$ is called the **reduced** or **complementary system**.

Remark. Note that every linear homogeneous system always has a trivial solution $\boldsymbol{\varphi}(x) = 0$, as can be immediately checked. From the uniqueness of the solution, therefore, there is no solution vanishing at *only* some point of x.

16.2.2 Vector Space of Solutions

Let $\boldsymbol{\varphi}_i(x)$ $(i = 1, 2, \cdots)$ be solutions for an n-dimensional linear homogeneous system

$$\boldsymbol{y}'(x) = \boldsymbol{A}(x)\boldsymbol{y}(x). \tag{16.20}$$

Referring to the axioms given in Sect. 4.2.1, it readily follows that the solutions $\{\boldsymbol{\varphi}_i(x)\}$ form a vector space V. Indeed, if $\boldsymbol{\varphi}_1(x)$ and $\boldsymbol{\varphi}_2(x)$ are solutions of (16.20), then $c_1\boldsymbol{\varphi}_1(x)+c_2\boldsymbol{\varphi}_2(x)$ with arbitrary constants c_1, c_2 is also a solution of (16.20), and so on.

Now we pose a question as to the dimension of the vector space V mentioned above. We have the answer in the following theorem:

♠ **Theorem:**
Solutions of the system (16.20) on an interval I form an n-dimensional vector space if the $n \times n$ matrix $\boldsymbol{A}(x)$ is continuous on I.

Proof The continuity of $\boldsymbol{A}(x)$ implies that all its components do not diverge. This allows us to set a constant K,

$$K = \max \sum_{i=1}^{n} |a_{ij}(x)|,$$

and it then follows that the vector $\boldsymbol{f}$ defined by $\boldsymbol{f}(x) = \boldsymbol{A}(x)\boldsymbol{y}(x)$ satisfies the Lipschitz condition:

$$|\boldsymbol{f}(x, \boldsymbol{y}) - \boldsymbol{f}(x, \boldsymbol{z})| \leq K |\boldsymbol{y} - \boldsymbol{z}| \quad \text{for } x \in I.$$

From the existence and uniqueness theorems we know that there are n solutions $\boldsymbol{\varphi}_i(x)$ of (16.20) such that each solution exists on the entire interval I and satisfies the initial condition

$$\boldsymbol{\varphi}_i(x_0) = \boldsymbol{e}_i \quad (i = 1, 2, \cdots, n) \quad \text{for } x_0 \in I, \tag{16.21}$$

where the $\boldsymbol{e}_i'$s are n linearly independent vectors.

We tentatively assume that the solutions $\boldsymbol{\varphi}_i$ are linearly *dependent* on I. Then there exist constants c_i, not all zero, such that

$$\sum_{i=1}^{n} c_i \boldsymbol{\varphi}_i(x) = 0 \quad \text{for every } x \text{ on } I.$$

In particular, setting $x = x_0$, and using the initial condition (16.21), we have

$$\sum_{i=1}^{n} c_i \boldsymbol{e}_i = 0,$$

which contradicts the assumed linear independence of $\boldsymbol{e}_i$. Hence, we conclude that the solutions $\boldsymbol{\varphi}_i$ are linearly *independent* on I.

Next we prove the completeness of $\{\boldsymbol{\varphi}_i(x)\}$; i.e., that every solution $\boldsymbol{\psi}(x)$ of (16.20) can be expanded as a linear combination of $\boldsymbol{\varphi}_i(x)$ satisfying the initial condition (16.21). Since the $\boldsymbol{e}_i$ are linearly independent in the n-dimensional Euclidean space E_n, they form a basis for E_n and there exist unique constants b_i such that the constant vector $\boldsymbol{\psi}(x_0)$ can be expressed as

$$\boldsymbol{\psi}(x_0) = \sum_{i=1}^{n} b_i \boldsymbol{e}_i. \tag{16.22}$$

Consider the vector

$$\boldsymbol{\varphi}(x) = \sum_{i=1}^{n} b_i \boldsymbol{\varphi}_i(x),$$

where the b_i are identical to those in (16.22). Clearly $\boldsymbol{\varphi}(x)$ is a solution of (16.20) on I. In addition, the initial value of $\boldsymbol{\varphi}$ reads

$$\boldsymbol{\varphi}(x_0) = \sum_{i=1}^{n} b_i \boldsymbol{e}_i,$$

so that $\boldsymbol{\varphi}(x_0) = \boldsymbol{\psi}(x_0)$. In view of the uniqueness theorem, we have

$$\boldsymbol{\varphi}(x) = \boldsymbol{\psi}(x) \quad \text{for every } x \text{ on } I.$$

This leads to the conclusion that every solution $\boldsymbol{\psi}(x)$ of an nth-order linear homogeneous system (16.20) is expressed by the unique linear combination

$$\boldsymbol{\psi}(x) = \sum_{i=1}^{n} b_i \boldsymbol{\varphi}_i(x) \quad \text{for every } x \text{ on } I,$$

where the b_i are uniquely determined once we have $\boldsymbol{\psi}(x)$. As a result, n solutions $\boldsymbol{\varphi}_i(x)$ of the system (16.20) form the basis for an n-dimensional vector space. ♣

16.2.3 Fundamental Systems of Solutions

Again let $\boldsymbol{\varphi}_i(x) = [\varphi_{1i}(x), \cdots, \varphi_{ni}(x)]^{\mathrm{T}}$ $(i = 1, 2, \cdots, n)$ be solutions of the linear homogeneous system (16.20) such that

$$\boldsymbol{\varphi}_i(x)' = \boldsymbol{A}(x)\boldsymbol{\varphi}_i(x) \quad \text{for all } i = 1, 2, \cdots, n.$$

Note here that $\{\boldsymbol{\varphi}_i(x)\}$ may or may not be linearly independent, since no initial condition is imposed (contrary to the case of (16.21)). Specifically, if the set $\{\boldsymbol{\varphi}_i(x)\}$ is endowed with the linear independence property, it is called the **fundamental system of solutions** of (16.20).

♠ **Fundamental system of solutions:**
A collection of n solutions $\{\boldsymbol{\varphi}_i(x)\}$ of an n-dimensional lienar homogeneous system is called a **fundamental system of solutions** of the system *if it is linearly independent.*

Remark. The significance of a fundamental system of solutions lies in the fact that it can describe any solution $\boldsymbol{\varphi}(x)$ of the corresponding linear homogeneous system. Consequently, the problem of finding a solution $\boldsymbol{\varphi}(x)$ becomes equivalent to that of finding n linearly independent solutions.

With this terminology, the theorem presented in Sect. 16.2.2 leads to the following result:

♠ **Theorem:**
A fundamental system of solutions exists for an arbitrary linear homogeneous system.

Example The second-order equation

$$y''(t) + y(t) = 0 \tag{16.23}$$

is equivalent to the two-dimensional linear system

$$\boldsymbol{u}'(t) = \boldsymbol{A}\boldsymbol{u}(t) \tag{16.24}$$

with

$$\boldsymbol{u}(t) = \begin{bmatrix} y'(t) \\ y(t) \end{bmatrix} \text{ and } \boldsymbol{A} = \begin{pmatrix} 0 & 1 \\ -1 & 0 \end{pmatrix}.$$

The fundamental system of solutions of (16.24) is given by

$$\varphi_1(t) = [\cos t, -\sin t]^{\mathrm{T}} \text{ and } \varphi_2(t) = [\sin t, \cos t]^{\mathrm{T}},$$

whose linear independence follows from the fact that $c_1 \sin t \pm c_2 \cos t \equiv 0$ implies $c_1 = c_2 = 0$. Furthermore, $\varphi_1(0) = (1,0)$ and $\varphi_2(0) = (0,1)$, so any solution $\varphi(t)$ is given by

$$\varphi(t) = a_0\varphi_1(t) + b_0\varphi_2(t) \quad \text{for } -\infty < t < \infty, \tag{16.25}$$

where $\varphi(0) = (a_0, b_0)$.

Remark. The solution $\varphi(t)$ in (16.25) corresponds to the solution of the second-order ODE (16.23) satisfying the initial conditions: $y(0) = a_0$ and $y'(0) = b_0$.

16.2.4 Wronskian for a System of ODEs

The theorems given in Sect. 16.2.2 and 16.2.3 ensure the existence of a fundamental system of solutions for any linear homogeneous system of the form

$$\boldsymbol{y}'(x) = \boldsymbol{A}(x)\boldsymbol{y}(x). \tag{16.26}$$

However, it provides no information as to whether a certain set of solutions is a fundamental system or not. In what follows, we consider the criteria concerning this issue. Following are preliminary concepts that we need in order to proceed.

♠ **Wronsky determinant:**

Let $\{\varphi_k(x)\}$ $(k = 1, 2, \cdots, n)$ be solutions of (16.26), where $\varphi_k(x) = [\varphi_{1k}(x), \cdots, \varphi_{nk}(x)]^{\mathrm{T}}$. Then the scalar function

$$W(x) = \det \begin{bmatrix} \varphi_{11} & \varphi_{12} & \cdots & \varphi_{1n} \\ \varphi_{21} & \varphi_{22} & \cdots & \varphi_{2n} \\ & \cdots & \cdots & \\ & \cdots & \cdots & \\ & \cdots & \cdots & \\ \varphi_{n1} & \varphi_{n2} & \cdots & \varphi_{nn} \end{bmatrix} \tag{16.27}$$

is called the **Wronsky determinant** (or the **Wronskian**) of the solutions $\{\varphi_k(x)\}$.

If $\{\varphi_k(x)\}$ is a fundamental system of solutions of (16.26), then the matrix corresponding to $W(x)$ is called a **fundamental matrix**. Hence, a fundamental matrix is a matrix whose columns form a fundamental system of solutions of (16.26).

Example For the two-dimensional system given in Sect. 16.2.3, the matrix

$$\Phi(t) = \begin{pmatrix} \cos t & \sin t \\ -\sin t & \cos t \end{pmatrix}, \quad -\infty < t < \infty$$

is a fundamental matrix and $W(t) \equiv 1$ for all t.

16.2.5 Liouville Formula for a Wronskian

The following theorem shows that given any n solutions of (16.26) and any t_0 in (r_1, r_2), we can completely determine the corresponding Wronskian without computing the $n \times n$ determinant.

♠ **Liouville formula:**

Let $\{\varphi_k(x)\}$ $(k = 1, 2, \cdots, n)$ be any n solution of (16.26) and let x_0 be in (r_1, r_2). Then the Wronskian of $\{\varphi_k(x)\}$ for $x \in (r_1, r_2)$ is given by

$$W(x) = W(x_0) \exp\left[\int_{x_0}^{x} \mathrm{tr}\boldsymbol{A}(s)ds\right].$$

See Exercise **2** for the proof. Since $\exp\left[\int_{x_0}^{x} \mathrm{tr}\boldsymbol{A}(s)ds\right]$ is never zero, the theorem implies that the Wronskian of any collection of n solutions of (16.26) is *identically zero* or *never zero* on (r_1, r_2). The latter case characterizes a fundamental system, as shown by the following theorem:

♠ **Theorem:**
A necessary and sufficient condition for $\{\varphi_k(x)\}$ $(k = 1, 2, \cdots, n)$ to be a fundamental system of solutions of (16.26) is that $W(x) \neq 0$ for $r_1 < x < r_2$.

Proof Let $\{\varphi_k(x)\}$ $(k = 1, 2, \cdots, n)$ be a fundamental system of solutions of (16.26) and let $\varphi(x)$ be any nontrivial solution. Then there exist $c_1, \cdots, c_n$ not all zero such that $\varphi(x) = \sum_{i=1}^{n} c_i \varphi_i(x)$, and by the uniqueness of the solutions the c_i are unique. If $\boldsymbol{c} = [c_1, \cdots, c_n]^{\mathrm{T}}$ and $\boldsymbol{\Phi}(x)$ is the fundamental matrix of $\{\varphi_k(x)\}$, then the previous relation can be written as

$$\varphi(x) = \boldsymbol{c}\boldsymbol{\Phi}(x).$$

For any x in (r_1, r_2), this is a system of n linear equations in the unknowns $c_1, \cdots, c_n$. Since this has a unique solution in $\boldsymbol{c}$, det$\boldsymbol{\Phi}$ cannot be zero, i.e.,

$$\det\boldsymbol{\Phi}(x) = W(x) \neq 0 \quad \text{for any } x \in (r_1, r_2).$$

Conversely, $W(x) \neq 0$ for $r_1 < x < r_2$, implies that the columns $\varphi_1(x), \cdots, \varphi_n(x)$ of $\boldsymbol{\Phi}(x)$ are linearly independent for $r_1 < x < r_2$. Since they are solutions of (16.26), they form a fundamental system ofsolutions. ♣

16.2.6 Wronskian for an nth-Order Linear ODE

The previous results for systems of ODEs can be applied to an nth-order linear equation

$$u^{(n)}(x) + a_1(x)u^{(n-1)}(x) + \cdots + a_n(x)u(x) = 0, \tag{16.28}$$

since (16.28) is transformed into a vector form as

$$\boldsymbol{y}' = \boldsymbol{A}\boldsymbol{y}, \tag{16.29}$$

where

$$\boldsymbol{y} = \begin{bmatrix} u \\ u' \\ \cdots \\ \cdots \\ u^{(n-1)} \end{bmatrix} \quad \text{and} \quad \boldsymbol{A} = \begin{bmatrix} 0 & 1 & 0 \cdots & & 0 \\ 0 & 0 & 1 \cdots & & 0 \\ \cdots & & & & \cdots \\ 0 & & & 0 & 1 \\ -a_n(x) & -a_{n-1}(x) & \cdots\cdots & -a_2(x) & -a_1(x) \end{bmatrix}.$$

Relevant terminology and theorems are given below.

♠ **Fundamental system of solutions:**
A collection

$$\xi_1(x), \cdots, \xi_n(x), \quad r_1 < x < r_2$$

of solutions of (16.28) is called a **fundamental system of solutions** of (16.28) if it is linearly independent.

♠ **Theorem:**
A fundamental system of solutions of equation (16.28) exists.

Proof We know that a fundamental system of solutions of (16.29) exists, and we express it by $\varphi_1(x), \cdots, \varphi_n(x)$, where $\varphi_k(x) = [\varphi_{1k}(x), \cdots, \varphi_{nk}(x)]^{\mathrm{T}}$. Furthermore, we may assume that given x_0 in (r_1, r_2),

$$\varphi_k(x_0) = [0, \cdots, 0, 1, 0, \cdots, 0]^{\mathrm{T}} \equiv \boldsymbol{e}_k, \quad k = 1, 2, \cdots, n,$$

where the single nonzero component 1 in $\boldsymbol{e}_k$ is assigned to the kth place in the square brackets. By the correspondence of solutions of (16.28) and (16.29), we have

$$\varphi_k(x) = \left[\xi_k(x), \xi_k'(x), \cdots, \xi_k^{(n-1)}(x)\right]^{\mathrm{T}}$$

for some solution $u(x) = \xi_k(x)$ of (16.28). The collection $\xi_1(x), \cdots, \xi_n(x)$ comprises distinct nontrivial solutions, since they satisfy distinct initial conditions and $\xi_k \equiv 0$ for $r_1 < x < r_2$ would imply that $\varphi_k(x) \equiv \mathbf{0}$, which is impossible.

Finally, if there existed constants $c_1, \cdots, c_n$ not all zero such that $\sum_{k=1}^n c_k \xi_k(x) \equiv 0$ for $r_1 < x < r_2$, then

$$\sum_{k=1}^n c_k \xi_k'(x) \equiv 0, \cdots, \sum_{k=1}^n c_k \xi_k^{(n-1)}(x) \equiv 0, \quad r_1 < x < r_2.$$

This implies that

$$\sum_{k=1}^n c_k \varphi_k(x) \equiv \mathbf{0}, \quad r_1 < x < r_2,$$

which contradicts the fact that $\{\varphi_k(x)\}$ is a fundamental system of (16.29). ♣

We now define the Wronskian of a collection of n solutions of (16.28).

♠ **Wronsky determinant:**
Given any collection $\xi_1(x), \cdots, \xi_n(x)$ of solutions of (16.28), then

$$W(x) = \det \begin{bmatrix} \xi_1 & \xi_2 & \cdots & \xi_n \\ \xi_1' & \xi_2' & \cdots & \xi_n' \\ & \cdots & \cdots & \\ & \cdots & \cdots & \\ & \cdots & \cdots & \\ \xi_1^{(n-1)} & \xi_2^{(n-1)} & \cdots & \xi_n^{(n-1)} \end{bmatrix} \tag{16.30}$$

is called the **Wronsky determinant** (or the **Wronskian**) of the solutions $\{\xi_k(x)\}$ $(k = 1, 2, \cdots, n)$.

As before, if $\xi_1(x), \cdots, \xi_n(x)$ make up a fundamental system of (16.28), then the matrix corresponding to $W(x)$ is called a **fundamental matrix**. In any case, note that the columns of the matrix corresponding to $W(x)$ are n solutions of the system (16.29). We may therefore immediately state a result analogous to the Liouville formula given in Sect. 16.2.4, noting that $\mathrm{tr}\boldsymbol{A}(x) = -a_1(x)$:

♠ **Theorem:**
The Wronskian $W(x)$ of any collection $\xi_1(x), \cdots, \xi_n(x)$ of solutions of (16.28) satisfies the relation

$$W(x) = W(x_0) \exp\left[-\int_{x_0}^{x} a_1(s) ds\right], \quad r_1 < x_0,\ x < r_2.$$

Finally, we have the result corresponding to the theorem in Sect. 16.2.4, for which the proof is virtually the same.

♠ **Theorem:**
A necessary and sufficient condition for $\xi_1(x), \cdots, \xi_n(x)$ to be a fundamental system of solutions of equation (16.28) is that

$$W(x) \neq 0 \quad \text{for } r_1 < x < r_2.$$

Example Assume a second-order equation

$$y''(x) + a(x)y(x) = 0.$$

For any two solutions $\xi_1(x)$ and $\xi_2(x)$, we have

$$W(x) = \det \begin{pmatrix} \xi_1(x) & \xi_2(x) \\ \xi_1'(x) & \xi_2'(x) \end{pmatrix} = \text{const.}$$

The constant is nonzero if and only if ξ_1 and ξ_2 are linearly independent.

Remark. The fact that linear independence implies a nonvanishing Wronskian is a property of solutions of *linear* equations; i.e., it does not hold for *nonlinear* equations. To see this, we consider the functions $\xi_1(x) = x^3$ and $\xi_2(x) = |x|^3$. They are linearly independent on $-\infty < x < \infty$, but

$$W(x) = \det \begin{pmatrix} x^3 & |x|^3 \\ 3x^2 & 3x|x| \end{pmatrix} = 0.$$

This results from the fact that $\xi_1(x)$ and $\xi_2(x)$ cannot both be solutions near $x = 0$ of a second-order linear equation. In fact, they both satisfy $\xi(0) = \xi'(0) = 0$ yet are distinct, which violates uniqueness.

16.2.7 Particular Solution of an Inhomogeneous System

We close this section by discussing an inhomogeneous linear equation

$$\frac{d\boldsymbol{y}(x)}{dx} - \boldsymbol{A}(x)\boldsymbol{y}(x) = \boldsymbol{q}(x). \tag{16.31}$$

Let $\boldsymbol{q}(x)$ be continuous on x on some interval I and let $\{\boldsymbol{\varphi}_k\}$ $(k = 1, 2, \cdots, n)$ be a fundamental system of solutions for the reduced equation of (16.31). A general solution of (16.31) can be written as the sum

$$\boldsymbol{\psi}(x) = \boldsymbol{\varphi}_p(x) + c_1\boldsymbol{\varphi}_1(x) + \cdots + c_n\boldsymbol{\varphi}_n(x), \tag{16.32}$$

where $\boldsymbol{\varphi}_p(x)$ is a **particular solution** of (16.31) with no adjustable parameter.

A particular solution can be obtained from a fundamental system $\{\boldsymbol{\varphi}_k\}$ $(k = 1, 2, \cdots n)$ of the reduced equation (16.19) by means of the **method of variation of constant parameters**. We assume a particular solution of the form

$$\boldsymbol{\varphi}_p(x) = C_1(x)\boldsymbol{\varphi}_1(x) + \cdots + C_n(x)\boldsymbol{\varphi}_n(x), \tag{16.33}$$

where the coefficients $\{C_k(x)\}$ $(k = 1, 2, \cdots, n)$ are not constants, but unknown functions of x. Differentiating (16.33) on x and substituting it into (16.31), we obtain

$$\sum_{k=1}^{n} \left[C_k(x)\boldsymbol{\varphi}_k'(x) + C_k'(x)\boldsymbol{\varphi}_k(x) - C_k(x)\boldsymbol{A}(x)\boldsymbol{\varphi}_k(x)\right] = \boldsymbol{q}(x). \tag{16.34}$$

Since $\{\boldsymbol{\varphi}_k\}$ $(k = 1, 2, \cdots, n)$ are solutions of the reduced equation (16.19), equation (16.34) yields

$$\sum_{k=1}^{n} \boldsymbol{\varphi}_k(x) C_k'(x) = \boldsymbol{q}(x). \tag{16.35}$$

If we express $\boldsymbol{\varphi}_k(x)$ by its components as

$$\boldsymbol{\varphi}_k(x) = [\varphi_{k1}(x), \varphi_{k2}(x), \cdots \varphi_{kn}(x)],$$

equation (16.35) becomes

$$\sum_{j=1}^{n} \varphi_{kj}(x) C_j'(x) = q_k(x), \tag{16.36}$$

or equivalently,

$$\begin{bmatrix} \varphi_{11} & \varphi_{12} & \cdots & \varphi_{1n} \\ \varphi_{21} & \varphi_{22} & \cdots & \varphi_{2n} \\ & & \cdots & \\ & & \cdots & \\ & & \cdots & \\ \varphi_{n1} & \varphi_{n2} & \cdots & \varphi_{nn} \end{bmatrix} \begin{bmatrix} C_1'(x) \\ C_2'(x) \\ \cdot \\ \cdot \\ \cdot \\ C_n'(x) \end{bmatrix} = \begin{bmatrix} q_1'(x) \\ q_2'(x) \\ \cdot \\ \cdot \\ \cdot \\ q_n'(x) \end{bmatrix}. \tag{16.37}$$

The matrix $[\varphi_{kj}]$ on the left-hand side of (16.37) satisfies $\det[\varphi_{kj}] \neq 0$ because of the linear independence of the fundamental system of solutions $\{\boldsymbol{\varphi}_k\}$. Hence, multiplying the **inverse matrix** (see Sect. 18.1.7) of $[\varphi_{kj}]$ by the both sides of (16.37), we have

$$C_k'(x) = p_k(x), \tag{16.38}$$

where $\{p_k(x)\}$, $k = 1, 2, \cdots, n$ are continuous functions obtained from (16.37). Thus once the differential equation (16.38) is solved with respect to $C_k(x)$, the solutions determine a particular solution of the form

$$\boldsymbol{\varphi}_p(x) = \sum_{k=1}^{n} C_k(x) \boldsymbol{\varphi}_k(x).$$

Exercises

1. Suppose $\varphi_1(x), \varphi_2(x)$ to be two solutions of the ODE $y'' + a_1 y' + a_2 y = 0$ on an interval I containing a point x_0. Show that

$$W(\varphi_1, \varphi_2)(x) = e^{-a_1(x-x_0)} W(\varphi_1, \varphi_2)(x_0).$$

Solution: We have $\varphi_1'' + a_1\varphi_1' + a_2\varphi_1 = 0$ and $\varphi_2'' + a_1\varphi_2' + a_2\varphi_2 = 0$. Multiplying the first equation by $-\varphi_2$, and the second by φ_1 and adding we obtain

$$(\varphi_1\varphi_2'' - \varphi_1''\varphi_2) + a_1(\varphi_1\varphi_2' - \varphi_1'\varphi_2) = 0.$$

Note that $W = \varphi_1\varphi_2{}' - \varphi_1{}'\varphi_2$ and $W' = \varphi_1\varphi_2{}'' - \varphi_1{}''\varphi_2$. Thus W satisfies the first-order equation:

$$W' + a_1 W = 0,$$

which implies $W(x) = ce^{-a_1 x}$ in which c is some constant. Setting $x = x_0$, we have $c = e^{-a_1 x_0} W(x_0)$, and thus

$$W(x) = e^{-a_1(x-x_0)} W(x_0). \quad ♣$$

2. Assume an n-dimensional linear homogeneous system $\boldsymbol{y}'(x) = \boldsymbol{A}(x)\boldsymbol{y}(x)$ on $I = (a, b)$, and let $\{\boldsymbol{g}_i(x)\}$ be any n solution. Show that the Wronskian of $\{\boldsymbol{g}_i(x)\}$ is given by

$$W(x) = W(x_0) \exp\left[\int_{x_0}^{x} \mathrm{tr}\boldsymbol{A}(s)ds\right], \quad \text{where } a < x_0 < b, \tag{16.39}$$

which is called the **Liouville formula.**

Solution: We show that $W(x)$ satisfies the differential equation $W'(x) = \mathrm{tr}\boldsymbol{A}(x)W(x)$ from which the conclusion (16.39) follows. The expansion by **cofactors** of $W(x)$ yields

$$W(x) = \sum_{j=1}^{n} \varphi_{ij}(x)\Delta_{ij}(x), \tag{16.40}$$

where $\varphi_{ij}(x)$ is the jth element of $\boldsymbol{\varphi}_i(x)$ and $\Delta_{ij}(x)$ is the cofactor of $W(x)$ (see Sect. 18.1.7 for the definition of the cofactor). Note that $\Delta_{ij}(x)$ does not contain the term $\varphi_{ij}(x)$. Hence, if $W(x)$ given in (16.40) is regarded as a function of the $\varphi_{ij}(x)$, we have $\partial W/\partial\varphi_{ij} = \Delta_{ij}(x)$ and, by the chain rule,

$$W(x)' = \sum_{i,j=1}^{n} \frac{\partial W}{\partial \varphi_{ij}} \varphi_{ij}(x)' = \sum_{i=1}^{n} \left[\sum_{j=1}^{n} \varphi_{ij}(x)'\Delta_{ij}(x)\right]. \tag{16.41}$$

We define $W_i(x)$ as

$$W_i(x) \equiv \det \begin{pmatrix} \varphi_{11}(x) & \cdots\cdots & \varphi_{1n}(x) \\ \cdots & & \cdots \\ \varphi_{i1}(x)' & \cdots\cdots & \varphi'_{in}(x) \\ \cdots & & \cdots \\ \varphi_{n1}(x) & \cdots\cdots & \varphi_{nn}(x) \end{pmatrix},$$

where all the elements in the ith row are differentiated. Then, the expression in the square brackets in (16.41) is the expansion of $W_i(x)$ by cofactors, so that

$$W(x)' = \sum_{i=1}^{n} W_i(x). \tag{16.42}$$

Furthermore, since $\varphi_{ij}(x)' = \sum_{k=1}^{n} a_{ik}(x)\varphi_{kj}(x)$, we have

$$W_i(x) \equiv \det \begin{pmatrix} \varphi_{11}(x) & \cdots\cdots & \varphi_{1n}(x) \\ \cdots & & \cdots \\ \sum_{k=1}^{n} a_{ik}\varphi_{k1}(x) & \cdots\cdots & \sum_{k=1}^{n} a_{ik}\varphi_{kn}(x) \\ \cdots & & \cdots \\ \varphi_{n1}(x) & \cdots\cdots & \varphi_{nn}(x) \end{pmatrix}.$$

Multiply the kth row ($k \neq i$) of the left matrix by $-a_{ik}(x)$ and then add it to the ith row. This process does not change the value of the determinant $W_i(x)$, but gives the relation

$$W_i(x) \equiv \det \begin{pmatrix} \varphi_{11}(x) & \cdots\cdots & \varphi_{1n}(x) \\ \cdots & & \cdots \\ a_{ii}\varphi_{i1}(x) & \cdots\cdots & a_{ii}\varphi_{in}(x) \\ \cdots & & \cdots \\ \varphi_{n1}(x) & \cdots\cdots & \varphi_{nn}(x) \end{pmatrix} = a_{ii}(x)W(x). \qquad (16.43)$$

From (16.42) and (16.43), we arrive at the desired result. ♣

16.3 Autonomous Systems of ODEs

16.3.1 Autonomous System

We noted earlier that an nth-order ODE reduces to the first-order form:

$$\boldsymbol{y}(x)' \equiv \frac{d}{dx}\begin{bmatrix} y_1(x) \\ y_2(x) \\ \cdots \\ y_n(x) \end{bmatrix} = \begin{bmatrix} F_1(x; y_1, y_2, \cdots, y_n) \\ F_2(x; y_1, y_2, \cdots, y_n) \\ \cdots \\ F_n(x; y_1, y_2, \cdots, y_n) \end{bmatrix} \equiv \boldsymbol{F}(x, \boldsymbol{y}), \qquad (16.44)$$

where $\boldsymbol{y}(x)$ and $\boldsymbol{F}(x, \boldsymbol{y})$ are n column vectors. Particularly important in many applications is the case where $\boldsymbol{F}(x, \boldsymbol{y})$ does not depend explicitly on x. Relevant terminology is given below.

♠ **Autonomous system of ODEs:**
A system of a first-order ODE of the form

$$\boldsymbol{y}'(x) = \boldsymbol{F}(\boldsymbol{y})$$

is called an **autonomous system**, wherein $\boldsymbol{F}$ does not depend explicitly on the independent variable x.

If $\boldsymbol{F}$ does depend explicitly on x, the system is said to be **nonautonomous**.

Example Consider a second-order ODE of the form

$$u''(x) = f\left(u(x), u'(x)\right).$$

Setting $y_1(x) = u(x)$ and $y_2(x) = u'(x)$, we have an autonomous system such as

$$\boldsymbol{y}'(x) = \frac{d}{dx}\begin{bmatrix} y_1(x) \\ y_2(x) \end{bmatrix} = \begin{bmatrix} y_2(x) \\ f(y_1(x), y_2(x)) \end{bmatrix} \equiv \boldsymbol{F}(\boldsymbol{y}).$$

16.3.2 Trajectory

As a prototype of autonomous systems of ODEs, we consider a two-dimensional system such that

$$\boldsymbol{y}'(t) = \frac{d}{dt}\begin{bmatrix} y_1(t) \\ y_2(t) \end{bmatrix} = \begin{bmatrix} f_1(y_1, y_2) \\ f_2(y_1, y_2) \end{bmatrix} = \boldsymbol{f}(\boldsymbol{y}), \tag{16.45}$$

where $y_1(t)$, $y_2(t)$ are unknown functions on t in some interval I. We assume that $f_1(y_1, y_2)$ and $f_2(y_1, y_2)$ are defined in some domain D and satisfy the Lipschitz condition on both $y_1(t)$ and $y_2(t)$. If t_0 is any real number and $(y_{10},\ y_{20}) \in D$ for any $y_{10} \equiv y_1(t_0)$ and $y_{20} \equiv y_2(t_0)$, the above hypotheses guarantee the existence and uniqueness of solutions for (16.45),

$$y_1(t) = \varphi_1(t), \quad y_2(t) = \varphi_2(t),$$

satisfying the initial conditions

$$\varphi_1(t_0) = y_{10}, \quad \varphi_2(t_0) = y_{20}.$$

We now consider a subdomain R of D in which $f_1(y_1, y_2)$ does not vanish. Then, we have in R the relation

$$\frac{dy_2}{dy_1} = \frac{dy_2}{dt}\frac{dt}{dy_1} = \frac{dy_2/dt}{dy_1/dt} = \frac{f_2(y_1, y_2)}{f_1(y_1, y_2)}, \tag{16.46}$$

which represents a direction field in (y_1-y_2)-plane as noted in Sect. 15.1.6. From the uniqueness theorem, there exists a unique integral curve of (16.46) in R satisfying the initial conditions. Such an integral curve on (y_1-y_2)-plane is called a **trajectory** of (16.45).

♠ **Theorem:** At most one trajectory passes through any point.

Proof This is obvious from the uniqueness of solutions. If not, two or more trajectories emerge from the crossing point chosen as an initial value point. ♣

Remark. When the vector field

$$\boldsymbol{v}(x, y) = \begin{bmatrix} x_1' \\ x_2' \end{bmatrix} = \begin{bmatrix} f_1(x_1, x_2) \\ f_2(x_1, x_2) \end{bmatrix}$$

describes the motion of a point in R, the domain R is called a **phase space** of the system (16.45).

16.3.3 Critical Point

Suppose that the autonomous system (16.45) has a time-independent solution expressed by

$$\varphi(t) = \boldsymbol{c} \in D,$$

where $\boldsymbol{c} = (c_1, c_2)$ is a constant vector. Then, no trajectory can pass through the point $\boldsymbol{c}$ (see the theorem in Sect. 16.3.3). In addition, we obviously have

$$\varphi'(t) = \boldsymbol{0} = \boldsymbol{f}(\boldsymbol{c}).$$

Conversely, if there exists a point $\boldsymbol{c}$ in R for which $\boldsymbol{f}(\boldsymbol{c}) = \boldsymbol{0}$, then the functions $\varphi(t) = \boldsymbol{c}$ are solutions of (16.45). The point $\boldsymbol{c}$ is said to be a **critical point** (or **singular point** or **point of equilibrium**).

♠ **Critical point:**
Assume an autonomous system

$$\boldsymbol{y}'(x) = \boldsymbol{F}(\boldsymbol{y}) \quad \text{for } \boldsymbol{y} \in D. \tag{16.47}$$

Then, any point $\boldsymbol{c} \in D$ that gives

$$\boldsymbol{F}(\boldsymbol{c}) = \boldsymbol{0}$$

is called a **critical point** of (16.47). Any other point in D is called a **regular point.**

16.3.4 Stability of a Critical Point

Let us discuss the **stability of a critical point** of an autonomous system (16.45) by analyzing trajectories of its solutions around the critical point.

We assume throughout that the function $\boldsymbol{F}(\boldsymbol{y})$ is differentiable of the first order on D, which guarantees the existence and uniqueness of solutions of the initial value problem (16.45). Then, the solutions of (16.45) can be conveniently pictured as curves in the phase space.

Now we consider a solution $\psi(x)$ of (16.45) that passes through the point η for x_0, where the distance between η and c is small. Let us now follow the trajectory that starts at a point η different from y_0, but near c. If the resulting motion ψ remains close to the critical point c for $x \geq x_0$, then the critical point is said to be **stable**, but if the solution ψ tends to return to the critical point c as x increases to infinity, then the critical point is said to be **asymptotically stable**. Finally, if the solution ψ leaves every small neighborhood of c, the critical point is said to be **unstable**. More precisely, we have the following definitions:

♠ **Stability of a critical point:**

Let c be a critical point of the autonomous system $\boldsymbol{y}'(x) = \boldsymbol{F}(\boldsymbol{y})$, so that $\boldsymbol{F}(\boldsymbol{c}) = \boldsymbol{0}$. The critical point c is called:

(i) **stable** when given a positive ε, there exists a δ so small that

$$|\boldsymbol{y}(0) - \mathbf{c}| < \delta \;\Rightarrow\; |\boldsymbol{y}(x) - \mathbf{c}| < \varepsilon \;\text{ for all } x > 0;$$

(ii) **asymptotically stable** when for some δ,

$$|\boldsymbol{y}(0) - \mathbf{c}| < \delta \;\Rightarrow\; \lim_{x\to\infty} |\boldsymbol{y}(x) - \mathbf{c}| = 0;$$

(iii) **strictly stable** when it is stable and asymptotically stable;
(iv) **neutrally stable** when it is stable but not asymptotically stable; and
(v) **unstable** when it is not stable.

16.3.5 Linear Autonomous System

An autonomous system $\boldsymbol{y}' = \boldsymbol{F}(\boldsymbol{y})$ is called **linear** if and only if all the elements F_i of $\boldsymbol{F}$ are linear homogeneous functions of the y_k, so that

$$\frac{dy_k}{dx} = a_{i1}y_1 + \cdots + a_{in}y_n \quad (i = 1, \cdots, n).$$

Hence, a linear autonomous system is just a (homogeneous) linear system of ODEs with *constant* coefficients. The analyses for linear systems are generally useful since we can always replace $F_i(\boldsymbol{y})$ by the linear terms of their Taylor expansions about a point $\boldsymbol{y} = \boldsymbol{y}_0$ for analyzing their local behavior.

We now discuss in detail the case $n = 2$ of linear plane autonomous systems of the form $\boldsymbol{y}' = \boldsymbol{A}\boldsymbol{y}$. Any such system is expressed by

$$\frac{dx}{dt} = ax + by, \quad \frac{dy}{dt} = cx + dy, \tag{16.48}$$

where $x = y_1$, $y = y_2$, and

$$A = \begin{pmatrix} a & b \\ c & d \end{pmatrix}$$

with a, b, c, d being constants. Observe that the simultaneous linear equations

$$ax + by = cx + dy = 0$$

have no solution except

$$x = y = 0$$

unless $\det \boldsymbol{A} = 0$. We thus see that the origin is the only critical point of the system (16.48) unless $ad = bc$.

Relevant terminology is given below.

♠ **Secular equation:**

If $(x(t), y(t))$ is a solution of (16.48), then $x(t)$ and $y(t)$ satisfy the equation:

$$u'' - (a+d)u' + (ad - bc)u = 0. \tag{16.49}$$

This equation is called the **secular equation** of the autonomous system (16.48).

Proof The first equation of (16.48) says that

$$by = x' - ax,$$

which implies that

$$x'' - ax' = by'.$$

We thus have

$$x'' - ax' = b(cx + dy) = bcx + d(x' - ax),$$

or equivalently,

$$x'' - (a+d)x' + (ad - bc)x = 0.$$

The proof for $y(t)$ is the same, replacing a with d and b with c. ♣

The secular equation (16.49) has an important property associated with the nature of the critical point. This is seen by introducing the concept of the **characteristic polynomial** P of (16.49) as

$$P \equiv \lambda^2 - (a+d)\lambda + (ad - bc) = \begin{vmatrix} a - \lambda & b \\ c & d - \lambda \end{vmatrix} = \det(\boldsymbol{A} - \lambda \boldsymbol{I}).$$

If λ_j $(j = 1, 2)$ are the roots of $P = 0$, then there exist nonzero **eigenvectors** (x_j, y_j) such that

$$\begin{pmatrix} a & b \\ c & d \end{pmatrix} \begin{pmatrix} x_j \\ y_j \end{pmatrix} = \begin{pmatrix} ax_j + by_j \\ cx_j + dy_j \end{pmatrix} = \lambda_j \begin{pmatrix} x_j \\ y_j \end{pmatrix}.$$

From this, it follows that the functions

$$e^{\lambda_1 t} \begin{pmatrix} x_1 \\ y_1 \end{pmatrix} \text{ and } e^{\lambda_2 t} \begin{pmatrix} x_2 \\ y_2 \end{pmatrix}$$

are a basis of vector-valued solutions of (16.48). We shall see later that the nature of a critical point of a system is completely determined by the values of the roots λ_1, λ_2.

16.4 Classification of Critical Points

The behavior of trajectories of a linear autonomous system

$$\frac{d}{dt} \begin{pmatrix} u \\ v \end{pmatrix} = \begin{pmatrix} a & b \\ c & d \end{pmatrix} \begin{pmatrix} u \\ v \end{pmatrix} = \boldsymbol{A} \begin{pmatrix} u \\ v \end{pmatrix} \tag{16.50}$$

near its critical point depends on the eigenvalues of the matrix $\boldsymbol{A}$, denoted by λ_1 and λ_2. There are five cases to consider and we discuss each in turn.

16.4.1 Improper Node

We first consider the case where λ_1 and λ_2 are *real, unequal,* and of *the same sign.* A critical point for this case is called an **improper node**. In this case, all the trajectories approach the critical point tangentially to the same straight line with increasing t.

Example An example of improper nodes is given by

$$\frac{d}{dt} \begin{pmatrix} u \\ v \end{pmatrix} = \begin{pmatrix} -2 & 0 \\ 0 & -3 \end{pmatrix} \begin{pmatrix} u \\ v \end{pmatrix}. \tag{16.51}$$

The eigenvalues are obviously $\lambda = -3, -2$ and the corresponding eigenvectors are $(1, 0)$ and $(0, 1)$. The general solution to (16.51) is

$$\begin{pmatrix} u \\ v \end{pmatrix} = c_1 \begin{pmatrix} 1 \\ 0 \end{pmatrix} e^{-2t} + c_2 \begin{pmatrix} 0 \\ 1 \end{pmatrix} e^{-3t}. \tag{16.52}$$

The trajectories given by (16.52) for several values of c_1 and c_2 are shown in Fig. 16.1.

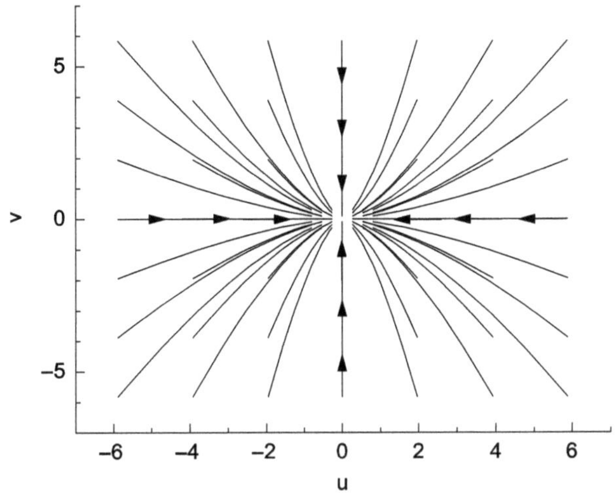

Fig. 16.1. Trajectories associated with the improper node of the system (16.51)

Remark. If the eigenvalues are real, unequal, and *positive* (contrary to the above example), then the trajectories are similar to those in Fig. 16.1 except that the directions of the arrows are reversed; in other words the trajectories recede from the critical point and go off toward infinity.

16.4.2 Saddle Point

We next consider the case where λ_1 and λ_2 are *real, unequal*, and of *the opposite sign*. In this case, the trajectories approach the critical point along one eigenvector direction and recede along the other eigenvector direction. The critical point in this case is called a **saddle point**.

Example Assume the system

$$\frac{d}{dt}\begin{pmatrix} u \\ v \end{pmatrix} = \begin{pmatrix} -1 & 1 \\ 0 & 2 \end{pmatrix}\begin{pmatrix} u \\ v \end{pmatrix}. \tag{16.53}$$

The eigenvalues are $\lambda_{\pm} = -1, 2$, and the corresponding eigenvectors are $(1, 0)$ and $(1, 3)$, respectively. The general solution to (16.53) is

$$\begin{pmatrix} u \\ v \end{pmatrix} = c_1 \begin{pmatrix} 1 \\ 0 \end{pmatrix} e^{-t} + c_2 \begin{pmatrix} 1 \\ 3 \end{pmatrix} e^{2t}. \tag{16.54}$$

The trajectories given by (16.54) are shown in Fig. 16.2. As (16.54) consists of an e^{-t} term and an e^{2t} term, the trajectories approach the origin along the eigenvector direction $(1, 0)$ and recede along the direction $(1, 3)$ as t increases.

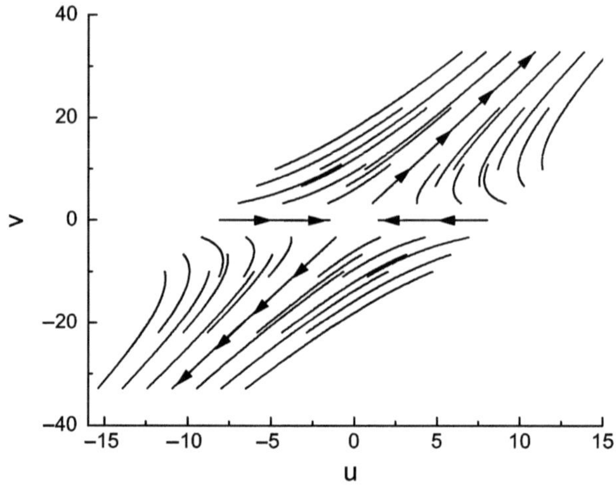

Fig. 16.2. Trajectories around the saddle point of the system (16.53)

16.4.3 Proper Node

We next consider the case of two roots of the characteristic equation being *real* and *equal*. This type of critical point is called a **proper node**.

Example We consider the critical point of the system

$$\frac{d}{dt}\begin{pmatrix} u \\ v \end{pmatrix} = \begin{pmatrix} \alpha & 0 \\ 0 & \alpha \end{pmatrix}\begin{pmatrix} u \\ v \end{pmatrix}. \tag{16.55}$$

The critical point occurs at the origin, with the degenerate eigenvalue being α. Generally when the eigenvalue λ of the characteristic equation is degenerate, the eigenvector is given by

$$u(t) = (c_1 + c_2 t)e^{\lambda t}, \quad v(t) = (c_3 + c_4 t)e^{\lambda t}. \tag{16.56}$$

Hence, we set $\lambda = \alpha$ in (16.56) and substitute the results into (16.55) to obtain $c_2 = c_4 = 0$. The solution to (16.55) is thus

$$u(t) = c_1 e^{\alpha t}, \quad v(t) = c_3 e^{\alpha t}. \tag{16.57}$$

Eliminating t from (16.57) yields the expression of the trajectories:

$$v = \frac{c_3}{c_1} u \text{ if } c_1 \neq 0$$

and

$$u = 0 \text{ if } c_1 = 0,$$

both of which are depicted in Fig. 16.3. The trajectories approach or recede from the origin, depending on the sign of α.

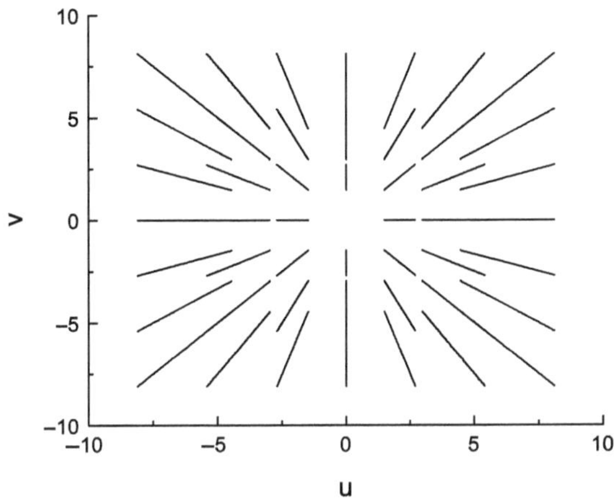

Fig. 16.3. Trajectories around the proper node of the system (16.55)

16.4.4 Spiral Point

So far, we have restricted our attention to cases where the two eigenvalues are real. Now we consider the case in which the two eigenvalues are **complex conjugates** of each other. The corresponding critical point is called a **spiral point** or a **focus**.

Example An example for this case is

$$\frac{d}{dt}\begin{pmatrix} u \\ v \end{pmatrix} = \begin{pmatrix} 0 & -1 \\ 2 & -2 \end{pmatrix}\begin{pmatrix} u \\ v \end{pmatrix}. \tag{16.58}$$

The critical point is at the origin, and the eigenvalues are $\lambda_{\pm} = -1 \pm i$ with coresponding eigenvectors $(1, 1 \mp i)$. The general solution to this system is

$$\begin{pmatrix} u \\ v \end{pmatrix} = c_1 \begin{pmatrix} 1 \\ 1-i \end{pmatrix} e^{(-1+i)t} + c_2 \begin{pmatrix} 1 \\ 1+i \end{pmatrix} e^{(-1-i)t}. \tag{16.59}$$

The result represents a family of curves that spiral into the critical point as t increases. Real components of the solutions $u(t)$ and $v(t)$ given by (16.59) are plotted in Fig. 16.4.

16.4.5 Center

The final class of critical points is called a **center**, for which the two eigenvalues are *pure imaginary*. In this case, trajectories consist of a family of closed loops centered about the critical point.

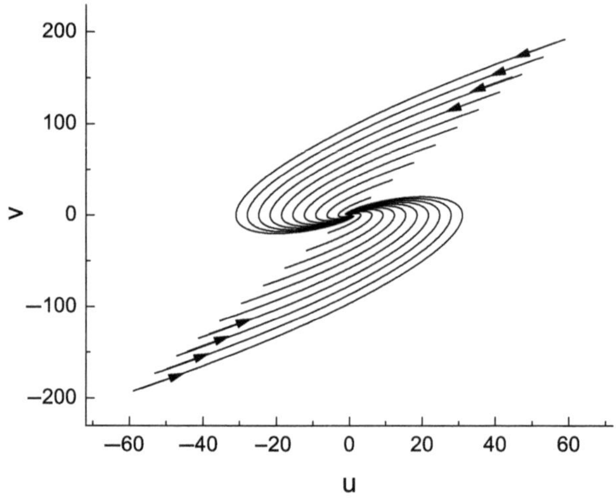

Fig. 16.4. Trajectories for the spiral point of the system (16.58)

Example Consider the system

$$\frac{d}{dt}\begin{pmatrix} u \\ v \end{pmatrix} = \begin{pmatrix} 1 & 2 \\ -1 & -1 \end{pmatrix}\begin{pmatrix} u \\ v \end{pmatrix}. \tag{16.60}$$

The eigenvalues and corresponding eigenvectors are $\lambda_{\pm} = \pm i$ and $(1 \pm i,\ -1)$, and the general solution reads

$$\begin{pmatrix} u \\ v \end{pmatrix} = c_1 \begin{pmatrix} 1+i \\ -1 \end{pmatrix} e^{it} + c_2 \begin{pmatrix} 1-i \\ -1 \end{pmatrix} e^{-it}. \tag{16.61}$$

Figure 16.5 shows several trajectories for different values of c_1 and c_2. All the trajectories represent periodic motion about the critical point.

16.4.6 Limit Cycle

Before closing this section, we have one more topic to discuss. Consider the system

$$\begin{aligned} x' &= x + y - x\left(x^2 + y^2\right), \\ y' &= -x + y - y\left(x^2 + y^2\right). \end{aligned} \tag{16.62}$$

The only critical point is at the origin. Letting $x = r\cos\theta$ and $y = r\sin\theta$, the system (16.62) becomes

$$r' = r\left(1 - r^2\right) \quad \text{and} \quad \theta' = -1. \tag{16.63}$$

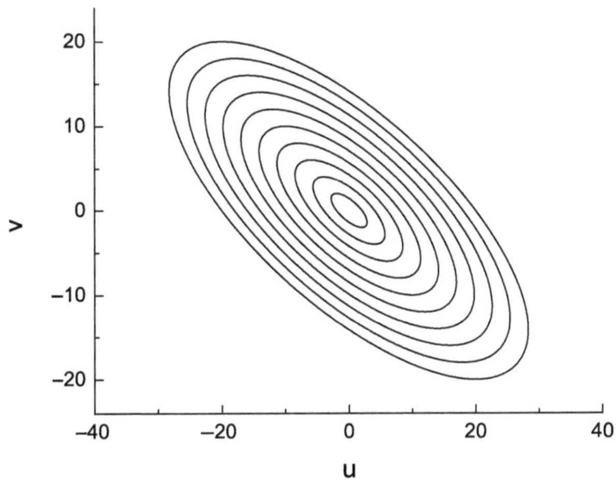

Fig. 16.5. Periodic trajectories for the **center** of the system (16.60)

The system (16.63) has a trivial solution of $r = 1$, $\theta = -t + \text{const}$, which represents periodic clockwise motion around the unit circle. We can find other solutions by solving (16.63). The equation for r is easy to solve, yielding

$$r(t) = \frac{1}{\sqrt{1 + \left(\frac{1 - r_0^2}{r_0^2}\right) e^{-2t}}},$$

where $r(0) = r_0$. Figure 16.6 shows $r(t)$ plotted against t for $r_0 > 1$ and $r_0 < 1$. The trajectories spiral in toward the unit circle as $t \to \infty$ if $r_0 > 1$

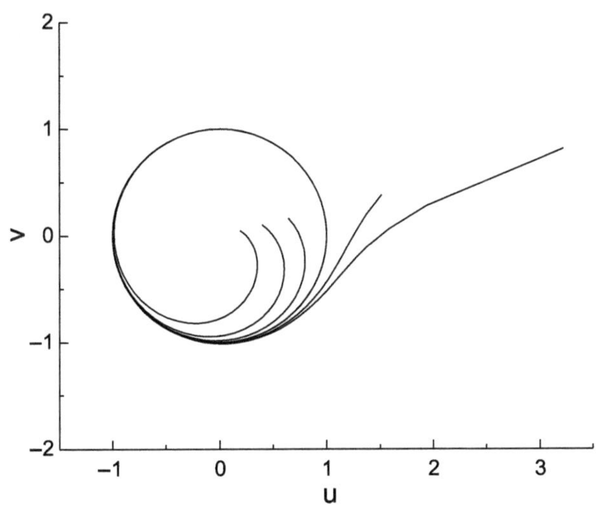

Fig. 16.6. Converging behavior of solutions of the system (16.62) to a limit cycle

and they spiral out toward the unit circle as $t \to \infty$ if $r_0 < 1$. Hence, all the trajectories spiral into the unit circle as $t \to \infty$.

The unit circle mentioned above is called a **limit cycle**. Limit cycles are important for determining the stability of the system, since the existence of a limit cycle ensures the existence of periodic solutions to a system.

Exercises

1. Consider the system given by

$$x' = e^x + \sin 5y - \cos 2y,$$
$$y' = x + 2\sin y.$$

Find the equilibrium point and describe the stability of the system around the point.

Solution: Expanding all functions on the right-hand side around $x = 0$, $y = 0$, we have

$$x' = x + 5y + g(x, y),$$
$$y' = x + 2y + h(x, y).$$

The functions g, h converge to zero faster than $\sqrt{x^2 + y^2}$, and the characteristic equation becomes

$$\begin{vmatrix} -\lambda + 1 & 5 \\ 1 & -\lambda + 2 \end{vmatrix} = 0,$$

whose roots are $(3 \pm \sqrt{21})/2$. Both of these are positive and the system is unstable. ♣

16.5 Applications in Physics and Engineering

16.5.1 Van der Pol Generator

As a physical example of a system in which a **limit cycle** may occur, we consider the following electric circuit consisting of a coil with inductance L and a condenser with capacitance C attached to a **tunnel diode** . A tunnel diode is a nonlinear element in the sense that it exhibits nonlinear current–voltage characteristics:

$$I(V) = I_0 - a(V - V_0) + b(V - V_0)^3. \tag{16.64}$$

It follows from Fig. 16.7 that a tunnel diode behaves like an ordinary resistor at low and high voltages, but like a *negative* resistor at intermediate voltages.

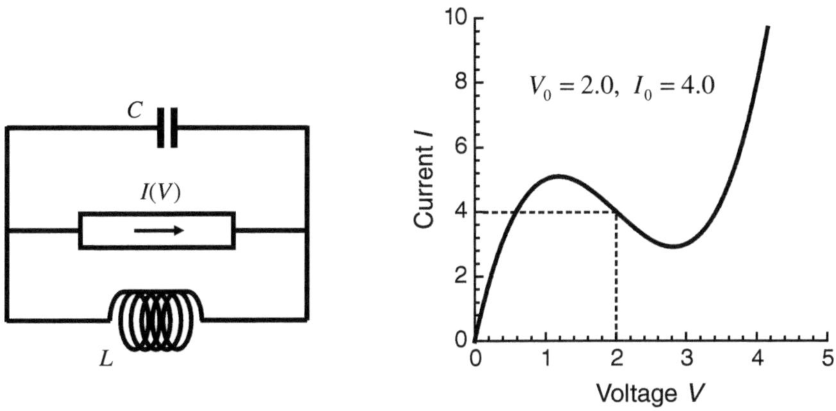

Fig. 16.7. *Left*: An electric circuit consisting of a coil with inductance L, a condenser with capacitance C, and a tunnel diode. *Right*: Plot of the nonlinear current–voltage characteristics $I(V)$ of the tunnel diode

Thus, a tunnel diode is expected to amplify small oscillations in the system, provided we choose the parameters in an appropriate manner.

The equation of motion for the LC circuit attached to a tunnel diode is obtained as described below. The **law of the conservation current flow** ensures that

$$I_L + I(V) + I_C = 0, \tag{16.65}$$

where

$$I_L = \frac{1}{L}\int V dt \quad \text{and} \quad I_C = C\frac{dV}{dt}. \tag{16.66}$$

Substituting (16.64) and (16.66) into (16.65) and then differentiating with respect to time, we get

$$\frac{d^2V}{dt^2} + \frac{1}{C}\left[-a + 3b(V - V_0)^2\right]\frac{dV}{dt} + \omega_0^2 V = 0,$$

where we introduce the resonant frequency ω_0 defined by the equation $\omega_0^2 = 1/(LC)$. For simplicity, we define a new variable

$$x = \frac{V - V_0}{V},$$

for which

$$\ddot{x} - \alpha(1 - \beta x^2)\dot{x} + \omega_0^2 x = 0 \quad \left(\dot{x} \equiv \frac{dx}{dt}\right) \tag{16.67}$$

with

$$\alpha = \frac{a}{C}, \quad \beta = \frac{3CV_0^2 b}{a}.$$

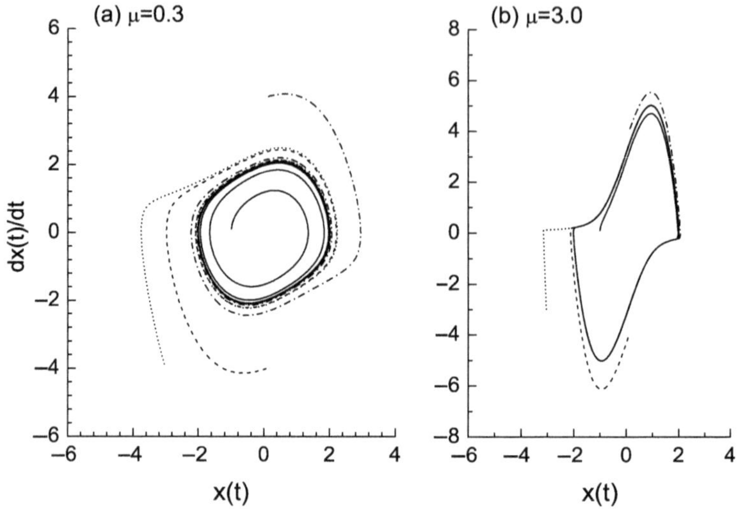

Fig. 16.8. Trajectories for the Van der Pol equation (16.68) with the parameter $\mu = 0.3$ for **(a)** and $\mu = 3.0$ for **(b)**

Equation (16.67) can be further simplified by replacing x by $x/\sqrt{\beta}$ and introducing a new time variable $\tilde{t} \equiv \omega_0 t$. Hence, we finally obtain

$$\ddot{x} - \mu(1 - x^2)\dot{x} + x = 0 \tag{16.68}$$

with the following key parameter:

$$\mu = \frac{\alpha}{\omega_0} = a\sqrt{\frac{L}{C}}.$$

The nonlinear differential equation (16.68) is known as the **Van der Pol equation**. As shown below, it describes self-sustaining oscillations in which energy is supplied to small oscillations and removed from large oscillations, which gives rise to the limit cycle in the phase space.

We can observe the self-exciting behavior of the system governed by (16.68) in the phase space plot in Fig. 16.8, where we set $\mu = 0.3$ and $\mu = 3.0$ for various initial points $\boldsymbol{x}_0 = (x(t = 0), \dot{x}(t = 0))$. We see that all the trajectories starting at a point $\boldsymbol{x}_0$ inside (or outside) a closed contour C move outward (or inward) as t increases and then converge to the contour C; such a characteristic closed contour is known as the limit cycle of the system. The shape of the limit cycle depends on the value of μ, as is evident from Fig. 16.8.

17

Partial Differential Equations

Abstract Broadly speaking, there are three classes of partial differential equations that are relevant to mathematical physics, as reflected in the section titles of this chapter. After examining the basic properties common to all the abovementioned classes of equations, we devote the balance of this chapter to a discussion of the mathematical essence of each class.

17.1 Basic Properties

17.1.1 Definitions

In this section we present the basic theory of **partial differential equations** (PDEs), an understanding of which is crucial is for describing or predicting the realm of nature. The formal definition is given below.

♠ **Partial differential equations:**

A partial differential equation of order r is a functional equation of the form

$$F\left(x_1, x_2, \cdots, x_n; \frac{\partial u}{\partial x_1}, \frac{\partial u}{\partial x_2}, \cdots, \frac{\partial u}{\partial x_n}; \frac{\partial^2 u}{\partial x_1^2}, \cdots\right) = 0, \tag{17.1}$$

which involves at least one rth-order partial derivative of the unknown function $u = u(x_1, x_2, \cdots, x_n)$ of independent variables $x_1, x_2, \cdots, x_n$.

In this chapter we often denote partial derivatives with subscripts such as

$$\frac{\partial u}{\partial x} \equiv u_x \equiv \partial_x u, \quad \frac{\partial^2 u}{\partial x \partial y} \equiv u_{xy} \equiv \partial_x \partial_y u,$$

We also use the shorthand

$$\partial_j \equiv \frac{\partial}{\partial x_j}, \quad \partial_i \partial_j = \frac{\partial^2}{\partial x_i \partial x_j},$$

Then, the general form (17.1) of a PDE is expressed as

$$F(x, y, \cdots, u, u_x, \cdots, u_{xx}, u_{xy}, \cdots) = 0, \tag{17.2}$$

where $u = u(x, y, \cdots)$ is the unknown function of independent variables $x, y, \cdots$. A **solution** (or **integral**) of a PDE is a function $\varphi(x, y, \cdots)$ satisfying equation (17.2) identically, at least in some region of the independent variables $x, y, \cdots$.

17.1.2 Subsidiary Conditions

The general solution of (17.1) depends on an arbitrary function. This statement is valid even for higher-order PDEs, indicating that a PDE has in general many solutions. Hence, in order to determine a unique solution, auxiliary conditions must be imposed. Such conditions are usually called **initial conditions** on time or **boundary conditions** for positions.

Initial condition:

In physics, an unknown function in a PDE usually involves independent variables of time t and position $x, y, \cdots$. Initial conditions for an unknown function are imposed on a particular (initial) time $t = t_0$ for an unknown function and/or its time derivatives.

Boundary condition:

Boundary conditions are imposed for an unknown function at the boundary or the infinity of a domain D in which the PDE is valid and are classified into two cases:

1. **Dirichlet condition** is the case in which an unknown function u is specified on the boundary of the domain D (often denoted by ∂D), where u is a function of time t and **position** $x, y, \cdots$.
2. **Neumann condition** is the case in which the normal derivative of an unknown function $\partial u / \partial n$ is specified.

17.1.3 Linear and Homogeneous PDEs

A PDE is called **linear** if and only if the F of (17.1) is a linear function of u and its derivatives. First we assume a first-order PDE with two independent variables x and y, whose general form reads

$$F(x, y;\ u, u_x, u_y) = 0. \tag{17.3}$$

Then, if it is linear, (17.3) can be expressed by

$$a(x, y)u_x + b(x, y)u_y + c(x, y)u = g(x, y), \tag{17.4}$$

where a, b, c, and g are given functions of x, y. Using the operator L, we express (17.4) by a simple form such that

$$Lu(x, y) = g(x, y), \tag{17.5}$$

where the operator L is defined by

$$L = a(x, y)\partial_x + b(x, y)\partial_y + c(x, y).$$

The linearity of PDEs guarantees that for any function u, v and any constant c the relations hold for

$$L(u + v) = Lu + Lv, \quad L(cu) = cL(u).$$

Examples

$$\begin{aligned}
&u_{xx} - e^{-x}u_{yy} = 0 && \text{(linear)}\\
&u_{xx} - e^{-x}u_{yy} = \sin x && \text{(linear)}\\
&uu_x + u_y = 0 && \text{(nonlinear)}\\
&xu_x + yu_y + u^2 = 0 && \text{(nonlinear)}
\end{aligned}$$

A linear equation is said to be **homogeneous** if the equation contains either the dependent variable u or its derivatives $u_x, u_y, \cdots$, not an independent variable such as $x, y, \cdots$. For instance, the PDE (17.5) is homogeneous if

$$g(x, y) \equiv 0,$$

since the equation

$$Lu(x, y) = 0 \tag{17.6}$$

involves only u, u_x, u_y and not x or y. On the other hand, if $g \neq 0$ in (17.5), it is called an **inhomogeneous** (or **nonhomogeneous**) linear equation. These statements are generally valid even for higher-order PDEs.

17.1.4 Characteristic Equation

We consider **a first-order homogeneous linear** PDE of the form

$$a(x, y)\partial_x u(x, y) + b(x, y)\partial_y u(x, y) = 0, \tag{17.7}$$

which is the most simple (and thus pedagogical) class of PDEs. In general, solutions of PDEs are described by *arbitrary* functions $f(p)$ of a particular independent variable p, wherein $p = p(x, y)$ is some combination of independent variables x and y. We verify this statement for the case of (17.7).

By the chain rule on the derivative, we have

$$\frac{\partial u}{\partial x} = \frac{\partial p}{\partial x}\frac{df}{dp}, \quad \frac{\partial u}{\partial y} = \frac{\partial p}{\partial y}\frac{df}{dp}. \tag{17.8}$$

Hence, the PDE (17.7) can be rewritten in the form

$$\left[a(x,y)\frac{\partial p}{\partial x} + b(x,y)\frac{\partial p}{\partial y}\right]\frac{df}{dp} = 0. \tag{17.9}$$

This implies that the function form of $f(p)$ may be arbitrary if $p = p(x,y)$ satisfies the equation

$$a(x,y)\frac{\partial p}{\partial x} + b(x,y)\frac{\partial p}{\partial y} = 0. \tag{17.10}$$

Therefore, an arbitrary function $f(p)$ such that the p satisfies (17.10) serves as the solution of the original PDE of (17.7). [The case of $\partial f/\partial p = 0$ gives a trivial solution of $f(p) = u(x,y) = \text{const}$, which we omit below.]

To obtain the solution $p = p(x,y)$ of the equation (17.10), we tentatively suppose that the function $p = p(x,y)$ takes a constant value along a curve $C: \ y = y(x)$ on the $(x$-$y)$-plane. Then, the total derivative of p on the curve C should vanish, so that

$$dp = \frac{\partial p}{\partial x}dx + \frac{\partial p}{\partial y}dy = 0. \tag{17.11}$$

From the correspondence between (17.10) and (17.11), we see that these are equivalent provided that

$$\frac{dy}{dx} = \frac{b(x,y)}{a(x,y)}, \quad a(x,y) \neq 0. \tag{17.12}$$

This is called the **characteristic equation** of the PDE (17.7) and its solution $y = y(x)$ is the **characteristic curve** of (17.7). From (17.11) and (17.12), therefore, we obtain the desired function form of $p = p(x,y)$ that makes an arbitrary function $f(p)$ the solution of the original PDE.

Examples We evaluate a general solution for (17.7) in the case that a, b are constant and nonzero coefficients. From (17.11) and (17.12), we obtain the characteristic curve (line) $p = bx - ay$. Then a general solution takes the form

$$u(x,y) = f(p) = f(bx - ay), \tag{17.13}$$

where f is an arbitrary function. The solution can be easily checked by taking derivatives using (17.8) and substituting those into (17.7). A less trivial example is given in Exercise **1**.

17.1.5 Second-Order PDEs

The general form of second-order linear PDEs is

$$\begin{aligned}\sum_{i,j=1}^{n} a_{ij}(x_1, x_2, \cdots, x_n)\partial_i\partial_j u &+ \sum_{j=1}^{n} a_i(x_1, x_2, \cdots, x_n)\partial_i u \\ +a_0(x_1, x_2, \cdots, x_n)u &= g(x_1, x_2, \cdots, x_n),\end{aligned} \tag{17.14}$$

where the unknown function u depends on n-independent variables denoted by $x_1, x_2, \cdots, x_n$. Note that $a_{ij} = a_{ji}$ since the mixed derivatives are equal. The form of (17.14) represents a very large class of PDEs. Among them, we restrict our attention to the case $g = 0$ with real constant coefficients, namely, second-order linear homogeneous PDEs. The general form of linear PDEs of second-order involving n independent variables with real constant coefficients is written as

$$\sum_{i,j=1}^{n} a_{ij}\partial_i\partial_j u + \sum_{j=1}^{n} a_i\partial_i u + a_0 u = 0. \tag{17.15}$$

The linear transformation of independent variables $\boldsymbol{x} = (x_1, x_2, \cdots, x_n)$ to $\boldsymbol{y} = (y_1, y_2, \cdots, y_n)$ is given by

$$\boldsymbol{y} = B\boldsymbol{x}, \tag{17.16}$$

or equivalently,

$$y_k = \sum_{m=1}^{n} b_{km}x_m,$$

where the b_{km} are elements of the $n \times n$ matrix B. Using the chain rule on the derivative, we have

$$\frac{\partial}{\partial x_j} = \sum_{k=1}^{n} \frac{\partial y_k}{\partial x_i}\frac{\partial}{\partial y_k}$$

and

$$\frac{\partial^2 u}{\partial x_i \partial x_j} = \left(\sum_{k=1}^{n} b_{ki}\frac{\partial}{\partial y_k}\right)\left(\sum_{m=1}^{n} b_{mj}\frac{\partial}{\partial y_m}\right)u. \tag{17.17}$$

Hence, the first term of (17.15) is converted to

$$\sum_{i,j=1}^{n} a_{ij}\partial_i\partial_j u = \sum_{k,m=1}^{n} (b_{ki}a_{ij}b_{mj})\,\partial_k\partial_m u,$$

which leads the relation

$$a_{ij} \to \sum_{k,m=1}^{n} b_{ki}a_{ij}b_{mj}. \tag{17.18}$$

Thus we obtain the PDE with new variables $y_1, y_2, \cdots, y_n$ by the transformation $A \to B^t AB$, where B^t is the transpose of B.

The appropriate choice of the matrix B makes it possible to diagonalize A such that

$$B^t AB = \begin{pmatrix} c_1 & & & \\ & c_2 & 0 & \\ & 0 & \ddots & \\ & & & c_n \end{pmatrix},$$

where $c_1, c_2, \cdots, c_n$ are the real eigenvalues of the matrix A. Thus, any PDE of the form (17.15) can be converted into a PDE with diagonal coefficients in terms of a linear transformation of a set of independent variables such as

$$\sum_{i=1}^{n} c_i \frac{\partial^2 u}{\partial y_i^2} + \sum_{i=1}^{n} d_i \frac{\partial u}{\partial y_i} = 0. \tag{17.19}$$

♠ **Theorem:**

By linear transformation of independent variables, the equation (17.15) can be reduced to the canonical form (17.19).

17.1.6 Classification of Second-Order PDEs

We can classify the types of PDEs depending on the positive or negative values of the coefficients $c_1, c_2, \cdots, c_n$ in (17.19) for the case $d_i \equiv 0$.

1. **Elliptic case:**
   If all the eigenvalues $c_1, c_2, \cdots, c_n$ are positive or negative, the PDE is called elliptic. A simple example is given by

$$\frac{\partial^2 u}{\partial y_1^2} + \frac{\partial^2 u}{\partial y_2^2} + \cdots = 0.$$

2. **Hyperbolic case:**
   In this case none of the $\{c_i\} : i = 1, 2, \cdots, n$ vanish and one of them has the opposite sign from $n-1$ than the others. For example,

$$\frac{\partial^2 u}{\partial y_1^2} - \frac{\partial^2 u}{\partial y_2^2} + \cdots = 0.$$

3. **Parabolic case:**
   If one of the $\{c_i\}$, $i = 1, 2, \cdots, n$ is zero and all the others have the same sign, the PDE is parabolic.

Below are basic PDEs in physics classified by the definition given above:

$$\textbf{Laplace equation:} \quad \Delta_n u = 0, \tag{17.20}$$

$$\textbf{Wave function:} \quad u_{tt} = \Delta_n u, \tag{17.21}$$

Here Δ_n means the **Laplacian** defined by $\Delta_n = \partial_1^2 + \partial_2^2 + \cdots + \partial_n^2$. The other important equation takes the form

$$u_t = \Delta_n u, \tag{17.22}$$

which we call the **diffusion equation**. The diffusion equation is different from the wave equation, where the time reversal symmetry $t \to -t$ holds. All of these equations (17.20)–(17.22) are **linear** since they are first degree in the dependent variable u.

Exercises

1. Find a general solution of the PDE of $u = u(x, y)$ given by

$$u_x + 2xy^2 u_y = 0. \tag{17.23}$$

Solution: The characteristic equation of (17.23) reads $dy/dx = 2xy^2$, which has the solution $y = 1/(p - x^2)$. Hence, we have $p = x^2 + (1/y)$, i.e., the general solution is given by

$$u(x, y) = f\left(x^2 + \frac{1}{y}\right).$$

In fact, $u(x, y)$ is a constant on the characteristic curve $y = 1/(p - x^2)$ whatever value p takes, as proved by

$$\frac{d}{dx} u\left(x, \frac{1}{p - x^2}\right) = \frac{\partial u}{\partial x} + \frac{2x}{(p - x^2)^2} \frac{\partial u}{\partial y} = \frac{\partial u}{\partial x} + 2xy^2 \frac{\partial u}{\partial y} = 0,$$

and similarly we have $du/dy = 0$. ♣

2. Classify second-order PDEs in two independent variables whose general form is given by

$$\partial_x^2 u + 2a_{12}\partial_x \partial_y u + a_{22}\partial_y^2 u = 0, \tag{17.24}$$

where a_{12}, a_{22} are real constants.

Solution: By completing the square, we can write (17.24) as

$$(\partial_x + a_{12}\partial_y)^2 u + (a_{22} - a_{12}^2)\partial_y^2 u = 0. \tag{17.25}$$

Here, let us introduce the new variables z and w by the linear transformation of the form $x = z$, $y = a_{12}z + (a_{22} - a_{12}^2)^{1/2}w$. We then have

$$\frac{\partial}{\partial z} = \frac{\partial}{\partial x} + a_{12}\frac{\partial}{\partial y}, \quad \frac{\partial}{\partial w} = (a_{22} - a_{12}^2)^{1/2}\frac{\partial}{\partial y},$$

so that for the case $a_{22} > a_{12}^2$ (17.25) gives

$$\frac{\partial^2 u}{\partial z^2} + \frac{\partial^2 u}{\partial w^2} = 0.$$

This is the elliptic case and is called the Laplace equation in the zw-plane. We easily see that for (17.25) the hyperbolic case is obtained for $a_{22} < a_{12}^2$. Thus, the second term of (17.25) determines the types of PDEs. ♣

17.2 The Laplacian Operator

17.2.1 Maximum and Minimum Theorem

We describe the fundamental properties of three operators, the Laplace, diffusion, and wave operators. These three operators are of great importance in the theory of PDEs. We begin with a description of the **Laplace operator** (or simply **Laplacian**) Δ_n on $\mathbf{R}^n$ defined by

$$\Delta_n = \sum_{i=1}^{n} \partial_i^2,$$

where n is a positive integer. The Laplacian is not only important in its own right, but also forms the spatial component of the **diffusion operator** $L_D = \partial_t - \Delta_n$ and the **wave operator** $L_W = \partial_t^2 - \Delta_n$, whose properties are discussed in Sect. 17.3 and 17.4.

First, we explain the **maximum principle** for the **Laplace equation** given by

$$\Delta_n u(\boldsymbol{x}) = 0,$$

whose solutions are called **harmonic functions**. Obviously, the one-dimensional case ($n = 1$) is trivial, so we consider the case where $n > 1$. Let D be a connected open set and u be an harmonic function in D with sup $u(\boldsymbol{x}) = A < \infty$ for $\boldsymbol{x} \in D$.

♠ **Maximum and minimum theorem:**

The maximum and minimum values of u are achieved on ∂D, say the boundary of D, and nowhere inside.

Before going to the proof, we examine certain properties of the solutions of **Poisson's equation** expressed by

$$\Delta_n u(\boldsymbol{x}) = -4\pi\rho(\boldsymbol{x}). \tag{17.26}$$

♠ **Lemma:**
If the function $\rho(\boldsymbol{x})$ in Poisson's equation (17.26) is positive (or negative) at a point $\boldsymbol{x}_0$, then the solution of (17.26) cannot attain its maximum (or minimum) value at the point $\boldsymbol{x}_0$.

Proof (of the lemma): If the function $u(\boldsymbol{x})$ satisfying (17.26) attains a minimum at a point $\boldsymbol{x}_0$, then it should attain a minimum with respect to each component $x_1, x_2, \cdots, x_n$ separately at that point. Then all the second-order derivatives of u would have to be nonnegative, which means that the left-hand side of (17.26), i.e., the sum of the second-order derivatives would have to be nonnegative. This result contradicts our hypothesis that $\rho(\boldsymbol{x})$ in (17.26) is positive. Hence, the first part of the lemma has been proved. The second part of the lemma is proved in a similar manner by assuming that $\rho(\boldsymbol{x})$ is negative. ♣

We are now ready to verify the maximum and minimum theorem.

Proof (of the maximum and minimum theorem): The proof is by contradiction. We first assume that

$$u(\boldsymbol{x}_0) > u_b + \varepsilon,$$

where u_b is the value of the function $u(\boldsymbol{x})$ at an arbitrary point on the boundary of the defining domain D. We further assume the function

$$v(\boldsymbol{x}) = u(\boldsymbol{x}) + \eta r(\boldsymbol{x})^2,$$

where

$$r(\boldsymbol{x})^2 = |\boldsymbol{x} - \boldsymbol{x}_0|^2$$

and η is some positive constant. It then follows that

$$\Delta_n v = \Delta_n u + 2n\eta = 2n\eta,$$

which says that the $v(\boldsymbol{x})$ is a solution of Poisson's equation (17.26) with negative $\rho(\boldsymbol{x})$. Note that $v(\boldsymbol{x}_0) = u(\boldsymbol{x}_0)$ and by hypothesis

$$u(\boldsymbol{x}_0) > u_b + \varepsilon = v_b + \varepsilon - \eta r^2.$$

Choosing η to be so small that throughout D

$$\varepsilon - \eta r^2 > \frac{\varepsilon}{2},$$

we obtain

$$v(\boldsymbol{x}_0) > v_b + \frac{\varepsilon}{2},$$

which implies that v attains its maximum somewhere within the domain D. This clearly contradicts the lemma above, so our assumption at the beginning of the proof was false. ♣

17.2.2 Uniqueness Theorem

The following theorem establishes the uniqueness of the solution of the Dirichlet problems for the Laplace equations.

♠ **Uniqueness theorem:**
If it exists, the solution of the Dirichlet problem for a Laplace equation is unique.

Proof Suppose that u_1 and u_2 are solutions on D for the Dirichlet problem such that

$$\Delta_n u = f(\boldsymbol{x}) \text{ in } D,$$

$$u = g(\boldsymbol{x}) \text{ on } \partial D.$$

Let $w = u_1 - u_2$, then $\Delta_n w = 0$ in D and $w = 0$ on ∂D. By the maximum (or minimum) principle, the point $\boldsymbol{x}_m$ (or $\boldsymbol{x}_M$) that minimizes (or maximizes) $w(\boldsymbol{x})$ should be located on the boundary of D. Hence, we have

$$0 = w(\boldsymbol{x}_m) \le w(\boldsymbol{x}) \le w(\boldsymbol{x}_M) = 0$$

for all $\boldsymbol{x} \in D$, which means that $w = 0$ and $u_1 = u_2$. ♣

17.2.3 Symmetric Properties of the Laplacian

The Laplacian is invariant under all rigid transformations such as translations and rotations. A translation from $\boldsymbol{x}$ to a new variable $\boldsymbol{x}'$ is given by

$$\boldsymbol{x}' = \boldsymbol{x} + \boldsymbol{a},$$

where $\boldsymbol{a}$ is a constant vector in n-dimensional space. The rotation is expressed by

$$\boldsymbol{x}' = B\boldsymbol{x}, \tag{17.27}$$

where B is an orthogonal matrix with the property $BB^t = B^tB = I$. Invariance under translations or rotations means simply that

$$\sum_{i=1}^{n} \frac{\partial^2}{\partial x_i^2} = \sum_{j=1}^{n} \frac{\partial^2}{\partial x'^2_j}.$$

The proof for translational invariance is simple, so we leave it to the reader. In physical systems, translational invariance is apparent because the physical laws are independent of the choice of coordinates.

A rotational invariance under (17.27) is proved by using the chain rule on the derivative such that

$$\sum_{k=1}^{n} \frac{\partial^2}{\partial x_k^2} = \sum_{ij}\sum b_{ik}b_{jk} \frac{\partial^2}{\partial x'_i \partial x'_j} = \sum_{i,j} \delta_{ij} \frac{\partial^2}{\partial x'_i \partial x'_j} = \sum_{i=1}^{n} \frac{\partial^2}{\partial x'^2_i},$$

where we have used the relation

$$\sum_{k}^{n} b_{ik}b_{jk} = (BB^t)_{ij} = \delta_{ij}.$$

Thus the proof has been completed.

Rotational invariance suggests that a two- or three-dimensional Laplacian should take a particularly simple form in polar or spherical coordinates.

Exercises

1. Find the harmonic function for a two-dimensional Laplace equation that is invariant under rotations.

 Solution: The two-dimensional Laplacian in polar coordinates is given by

$$\Delta_2 = \frac{\partial^2}{\partial r^2} + \frac{1}{r}\frac{\partial}{\partial r} + \frac{1}{r^2}\frac{\partial^2}{\partial \vartheta^2} \quad (r > 0),$$

 where we seek for solutions $u(r)$ depending only on r. Then we take the **radial part** of the Laplace equation, which gives $u_{rr} + \frac{1}{r}u_r = 0$ $(r > 0)$. This is the ODE whose solution is given by $u(r) = a \log r + b$ $(r > 0)$, where a, b are constants. Note that the form of the function $\log\ r$ is scale invariant under the dilatation transformation $r \to cr$ for a positive constant c. ♣

2. Find the harmonic function in three dimensions that is invariant under rotations.

Solution: The Laplacian in spherical coordinates takes the form

$$\Delta_3 = \frac{\partial^2}{\partial r^2} + \frac{2}{r}\frac{\partial}{\partial r} + \frac{1}{r^2 \sin\vartheta}\frac{\partial}{\partial \vartheta}\left(\sin\vartheta \frac{\partial}{\partial \vartheta}\right) + \frac{1}{r^2 \sin^2\vartheta}\frac{\partial^2}{\partial \phi^2} \quad (r > 0).$$

Since the solution depends only on r, we have the Laplace equation given by $u_{rr} + \frac{2}{r}u_r = 0$ $(r > 0)$. So we have $(r^2 u_r)_r = 0$ and the solution becomes $u = \frac{a}{r} + b$ $(r > 0)$, where a, b are constants. This is an important harmonic function that is not finite at the origin. ♣

3. Show that, for an arbitrary integer $n > 2$, the general form of solutions with rotational symmetry is given by

$$u(r) = ar^{2-n} + b \quad (n > 2, r > 0), \tag{17.28}$$

where a, b are constants.

Solution: This is shown by applying the chain rule to the derivative such that

$$\begin{aligned}\Delta_n u(r) &= \sum_{i=1}^{n} \partial_i \left[\frac{x_i}{r}u'(r)\right] = \sum_{i=1}^{n}\left[\frac{x_i^2}{r^2}u''(r) + \frac{1}{r}u'(r) - \frac{x_i^2}{r^3}u'(r)\right] \\ &= u''(r) + \frac{n-1}{r}u'(r),\end{aligned} \tag{17.29}$$

where the relation $\partial r/\partial x_i = x_i/r$ is used. If $\Delta_n u = 0$, (17.29) yields

$$\frac{u''(r)}{u'(r)} = \frac{1-n}{r}.$$

Integrating twice, we have (17.28). ♣

17.3 The Diffusion Operator

17.3.1 The Diffusion Equations in Bounded Domains

The diffusion equation describes physical phenomena such as **Brownian motion** of a particle or **heat flow**, whose general form is written as

$$L_D u(x_1, x_2, \cdots, t) = 0, \tag{17.30}$$

where L_D is the diffusion operator defined by

$$L_D = \partial_t - \sum_{i=1}^{n} \partial_i^2. \tag{17.31}$$

If the scale transformation $t \to Dt$ is used, we have the **diffusion equation**

$$\partial_t u - D\Delta u = 0,$$

where D is the **diffusion constant**. For heat flow u represents the temperature at position $\boldsymbol{x} = (x_1, x_2, \cdots)$ and time t, and for Brownian motion u is the probability of finding a particle at $\boldsymbol{x}$ and t. Hereafter, we treat the system of the unit diffusion constant $D = 1$. If we have to go back to the actual diffusion equation, we do the transformation $t \to Dt$ in the final solution.

17.3.2 Maximum and Minimum Theorem

We begin by describing the maximum principle for the diffusion equation defined in a bounded domain, from which we deduce the uniqueness of initial and boundary value problems.

♠ **Maximum and minimum theorem:**
Let D be a bounded domain in $\mathbf{R}^n$ and $0 < t < T < \infty$. If u is a real-valued continuous function, it takes its maximum either at the initial value ($t = 0$) or on the boundary ∂D.

Proof For any $\varepsilon > 0$, we set

$$v(\boldsymbol{x}, t) = u(\boldsymbol{x}, t) + \varepsilon \left|\boldsymbol{x}\right|^2,$$

for which we have

$$v_t - \Delta_n v = -2n\varepsilon < 0. \tag{17.32}$$

If the maximum for u occurs at an interior point $(\boldsymbol{x}_0, \ \mathbf{t}_0)$ in the domain $D \times [0, T]$, we know that the first derivatives $v_t, v_{x_1}, v_{x_2}, \cdots$ of v vanish there and that the second derivative $\Delta v \le 0$. This contradicts (17.32), so there is no interior maximum. Suppose now that the maximum occurs at $t = T$ on D; the time derivative v_t must be nonnegative there because

$$v(\boldsymbol{x}_0, T) \ge v(\boldsymbol{x}_0, T - \delta)$$

and

$$\Delta v \le 0,$$

which again contradicts (17.32). Therefore, the maximum must be at the initial time $t = 0$, namely, $D \times \{0\}$ or the boundary ∂D. Replacing u by $-u$, we see that the minimum is also achieved on either $D \times \{0\}$ or ∂D. ♣

17.3.3 Uniqueness Theorem

The maximum principle can be used to prove **uniqueness** for the **Dirichlet problem for the diffusion equation**. The conditions are given by

$$L_D u = f(\boldsymbol{x}, t) \quad \text{on } D \times (0, \infty)$$
$$u(\boldsymbol{x}, 0) = g(\boldsymbol{x}), \;\; u(\boldsymbol{x}, t) = h(t) \quad \text{on } \partial D$$

for given functions f, g, and h.

The following is an immediate corollary of the maximum and minimum theorem.

♠ **Uniqueness theorem:**
There is at most one solution of the Dirichlet problem for the diffusion equation.

Proof Let $u(\boldsymbol{x}, \ \mathbf{t})$ and $v(\boldsymbol{x}, t)$ be two solutions of (17.33) and $w = u - v$ be their difference. Hence, we have $L_D w = 0$, $w(\boldsymbol{x}, 0) = 0$, $w(0, t) = 0$, $w(\boldsymbol{x}, t) = 0$ on ∂D. By the maximum principle, $w(\boldsymbol{x}, t)$ has its maximum at the initial time or the boundary, exactly where w vanishes. Thus $w(\boldsymbol{x}, t) \le 0$. The same reasoning for the minimum shows that $w(\boldsymbol{x}, t) \ge 0$. Therefore $w(\boldsymbol{x}, t) = 0$, so that $u = v$ for all $t \ge 0$. ♣

17.4 The Wave Operator

17.4.1 The Cauchy Problem

The **wave operator** (or **d'Alemberian**) on $\mathbf{R}^n \times \mathbf{R}$ is expressed by

$$L = \partial_t^2 - \Delta_n = \partial_t^2 - \sum_{i=1}^{n} \partial_i^2, \tag{17.33}$$

from which we have the wave equation in the general form

$$Lu = \partial_t^2 u - \Delta_n u = 0. \tag{17.34}$$

The wave equation is the prototype of the hyperbolic PDEs and describes waves with unit velocity of propagation in homogeneous isotropic media. By making the transformation $t \to ct$, we have the standard form of the wave equation

$$\partial_t^2 u - c^2 \Delta_n u = 0, \tag{17.35}$$

where c is the wave velocity. The solution for (17.35) is obtained by transforming the time variable t into ct in the result of (17.34).

The initial value problem for the wave equation is called the **Cauchy problem** and is given by the inhomogeneous wave equation

$$\partial_t^2 u(\boldsymbol{x}, t) - \Delta_n u(\boldsymbol{x}, t) = f(\boldsymbol{x}, t) \tag{17.36}$$

under the two initial conditions

$$u(\boldsymbol{x},0) = \phi(\boldsymbol{x}),$$
$$\partial_t u(\boldsymbol{x},0) = \psi(\boldsymbol{x}),$$

where f, ϕ, ψ are continuous and differentiable given functions. For example, $f(\boldsymbol{x},t)$ provides an external force acting on the system described by (17.36).

The wave operator (17.33) is a linear operator, so the solution is the sum for the general solution of the homogeneous equation (17.34) and a particular solution for the inhomogeneous equation (17.36).

17.4.2 Homogeneous Wave Equations

First, we provide the solution for the one-dimensional homogeneous version ($f = 0$) of the Cauchy problem (17.36), in which the spatial part is defined on the whole region of one dimension $-\infty < x < \infty$. For example, consider the case of an infinitely long vibrating string. The wave equation is written as

$$u_{tt} - u_{xx} = 0, \tag{17.37}$$

which is a hyperbolic second-order PDE that we can express by

$$\left(\frac{\partial}{\partial t} - \frac{\partial}{\partial x}\right)\left(\frac{\partial}{\partial t} + \frac{\partial}{\partial x}\right) u = 0. \tag{17.38}$$

Let us set

$$u_t + u_x = v, \tag{17.39}$$

then the first-order PDE for $v(t,x)$ is obtained from (17.38) as

$$v_t - v_x = 0. \tag{17.40}$$

As shown earlier, (17.40) has a solution of the form

$$v(x,t) = g(x+t), \tag{17.41}$$

where g is any function. Thus we must solve (17.39) for u, which is given by

$$u_t + u_x = g(x+t). \tag{17.42}$$

One solution of (17.42) takes the following form:

$$u(x,t) = h(x+t), \tag{17.43}$$

which we can check through direct differentiation of (17.43) by setting $p = x+t$ such that

$$\frac{\partial u}{\partial x} = \frac{dh}{dp}\frac{\partial p}{\partial x} = h',$$

$$\frac{\partial u}{\partial t} = \frac{dh}{dp}\frac{\partial p}{\partial t} = h'.$$

Then we have

$$h(p) = \frac{1}{2}\int^{p} g(p)dp. \tag{17.44}$$

Another possibility is the general solution of the homogeneous equation obtained by setting $g = 0$ in (17.42), which takes the form

$$u = z(x - t). \tag{17.45}$$

Adding this to (17.44), we have the general expression of a solution,

$$u(x,t) = h(x+t) + z(x-t). \tag{17.46}$$

Now let us solve (17.46) under the initial conditions

$$u(x,0) = \phi(x),$$
$$u_t(x,0) = \psi(x),$$

where ϕ and ψ are given functions of x. From (17.46), we have the relations

$$\phi(x) = h(x) + z(x), \tag{17.47}$$
$$\psi(x) = h'(x) - z'(x). \tag{17.48}$$

By differentiating (17.47), we obtain $\phi' = h' + z'$. Combining this with (17.48), we have

$$h' = \frac{1}{2}(\phi' + \psi), \quad z' = \frac{1}{2}(\phi' - \psi).$$

Integrating on p yields

$$y(p) = \frac{1}{2}\phi(p) + \frac{1}{2}\int_0^p \psi dp + a,$$

$$z(p) = \frac{1}{2}\phi(p) + \frac{1}{2}\int_0^p \psi dp - a.$$

So, we get

$$u(x,t) = \frac{1}{2}\left[\phi(x+t) + \phi(x-t)\right] + \frac{1}{2}\int_{x-t}^{x+t} \psi(p)dp, \tag{17.49}$$

which is the solution for the initial value problem for the homogeneous equation (17.37).

17.4.3 Inhomogeneous Wave Equations

Next we solve the initial value problem for an inhomogeneous PDE ($f \neq 0$) by applying the **method of characteristic coordinates**. We transform the variables x, t into new variables $\xi = x + t,\ \eta = x - t$. The wave equation for the new variables yields

$$\partial_\eta \partial_\xi u = -\frac{1}{4} f\left(\frac{\xi+\eta}{2}, \frac{\xi-\eta}{2}\right).$$

This equation can be integrated with respect to η, leaving ξ as a constant. Thus we have

$$u_\xi = -\frac{1}{4}\int^\eta f d\eta, \tag{17.50}$$

where the lower limit of integration is arbitrary. Again we can integrate with respect to ξ:

$$u(\xi, \eta) = -\frac{1}{4}\int^\xi \int^\eta f\left(\frac{\xi+\mu}{2}, \frac{\xi-\eta}{2}\right) d\eta d\xi. \tag{17.51}$$

Here we consider the dependent variable u at a fixed point (ξ_0, η_0) defined by

$$\xi_0 = x_0 + t_0, \quad \eta_0 = x_0 - t_0. \tag{17.52}$$

We can evaluate (17.51) at the point (ξ_0, μ_0) and make a particular choice of the lower limits such that

$$u(\xi_0, \mu_0) = \frac{1}{4}\int_{-\infty}^{\xi} \int_{\eta_0}^{\infty} f d\eta d\xi.$$

Here we change the variables ξ, η into the original ones (x, t), and the Jacobian is the determinant of its coefficient matrix:

$$J = \det \begin{vmatrix} \dfrac{\partial \xi}{\partial x} & \dfrac{\partial \xi}{\partial t} \\[2ex] \dfrac{\partial \eta}{\partial x} & \dfrac{\partial \eta}{\partial t} \end{vmatrix} = 2.$$

Thus $d\xi d\eta = J dx dt = 2 dx dt$, so the double integral can be transformed as

$$u(x_0, t_0) = \frac{1}{2}\int_0^{t_0} \int_{x_0-(t_0-t)}^{x_0+(t_0-t)} f(x, t) dx dt$$

As a result, we have the following theorem:

♠ **Theorem:**
The unique solution of (17.36) on one spatial dimension is given by

$$u(x,t) = \frac{1}{2}\left[\phi(x+t) + \psi(x-t)\right] + \frac{1}{2}\int_{x-t}^{x+t} \psi(p)dp + \frac{1}{2}\int_0^t \int_{x-(t-t')}^{x+(t-t')} f(x',t')dx'dt',$$

where $\phi(x) = u(x,0)$ and $\psi(x) = u_t(x,0)$.

17.4.4 Wave Equations in Finite Domains

In this section we attempt to solve the wave equations defined in the region $D \times (0,\infty)$, where D is the bounded domain of $\boldsymbol{R}^n$. For this problem, we have to specify the initial conditions at $t=0$ as well as some boundary conditions on ∂D. As noted in Sect. 17.1.2, the commonly used boundary conditions are the **Dirichlet** and **Neumann** conditions. First we treat a homogeneous wave equation with no external term given by

$$\partial_t^2 u - \Delta_n u = 0, \tag{17.53}$$

where the initial condition is

$$u(\boldsymbol{x},0) = f(\boldsymbol{x}), \quad \partial_t u(\boldsymbol{x},0) = g(\boldsymbol{x}), \tag{17.54}$$

and the boundary conditions on ∂D are given by

$$u(\boldsymbol{x},t) = 0 \text{ or } \partial_n u(\boldsymbol{x},t) = 0. \tag{17.55}$$

Thus, when the boundary conditions are independent of t, the **method of separation of variables** is useful, i.e., we assume that the solution u takes the form

$$u(\boldsymbol{x},t) = X(\boldsymbol{x})T(t), \tag{17.56}$$

where X satisfies the boundary conditions (17.55) on ∂D. Substituting (17.56) into (17.53), we have

$$-\frac{\Delta X(\boldsymbol{x})}{X(\boldsymbol{x})} = -\frac{T''(t)}{T(t)} = \mu^2. \tag{17.57}$$

This defines a quantity μ^2 that must be constant since $\Delta X/X$ depends only on $\boldsymbol{x}$ and T''/T depends only on t. The reason for the positive constant $\mu^2 > 0$ will be seen later.

Equation (17.57) gives a pair of separate differential equations for $X(\boldsymbol{x})$ and $T(t)$: the one equation is

$$\Delta X(\boldsymbol{x}) = -\mu^2 X(\boldsymbol{x}) \tag{17.58}$$

that satisfies the given boundary conditions of (17.55), and the other is

$$T''(t) = -\mu^2 T(t) \tag{17.59}$$

in which $0 < t < \infty$. The solution for this ODE is obtained as

$$T(t) = a\cos\mu t + b\sin\mu t, \tag{17.60}$$

where a and b are constants that can be determined from the initial conditions.

Combining (17.60) with $X(\boldsymbol{x})$, the solution is expressed as

$$u(\boldsymbol{x}, t) = X_\mu(\boldsymbol{x})(a\cos\mu t + b\sin\mu t). \tag{17.61}$$

This is a **normal mode** of vibration with eigenfrequency μ and the general solution is obtained by the superposition of **normal modes.** Thus, we have the general solution of the form

$$u(\boldsymbol{x}, t) = \sum_n X_n(\boldsymbol{x})(a_n\cos\mu_n t + b_n\sin\mu_n t). \tag{17.62}$$

For example, from the initial conditions in (17.54), we have

$$\sum_n a_n X_n = f, \quad \sum_n \mu_n b_n X_n = g,$$

so the coefficients in (17.62) are given by

$$a_n = \langle f \mid X_n \rangle, \quad b_n = \langle g \mid F_n \rangle / \mu_n.$$

Exercises

1. Find the general solution for the wave equation defined on the one-dimensional bounded domain $(0, l) \times (0, \infty)$, which is given by

$$\partial_t^2 u - \Delta u = 0,$$

under the conditions: $u(x, 0) = f(x)$, $\partial_t u(x, 0) = g(x)$, $u(0, t) = u(l, t) = 0$.

Solution: The normalized eigenfunctions are $X_n = \sqrt{\frac{2}{l}}\sin\left(\frac{n\pi x}{l}\right)$, and the associated eigenfrequencies μ_n are the integer multiples of the fundamental frequency π/l. Thus, we obtain

$$u(x, t) = \sum \left(a_n\cos\frac{n\pi t}{l} + b_n\sin\frac{n\pi t}{l}\right)\sin\frac{n\pi x}{l},$$

where the coefficients are

$$a_n = \frac{2}{l}\int_0^l f(x)\sin\frac{n\pi x}{l}dx, b_n = \frac{1}{\pi n}\int_0^l g(x)\sin\frac{\pi x}{l}dx. \quad ♣$$

2. The differential equation under a point source $\boldsymbol{r} = 0$ at time $t_0 = 0$ in an infinite medium is given by

$$\partial_t G - \Delta_3 G = \delta(\boldsymbol{r})\delta(t). \tag{17.63}$$

Find the solution $G(\boldsymbol{r}, t)$ called **Green's function** by means of Fourier and Laplace transforms.

Solution: If we take a Fourier transform in space and a Laplace transform in time, (17.63) becomes $G(\boldsymbol{k}, \omega) = 1/[(2\pi)^3(\omega + k^2)]$. The inverse Laplace transform yields $G(\boldsymbol{k}, t) = e^{-k^2 t}/(2\pi)^3$. We then obtain Green's function $G(\boldsymbol{r}, t)$ by the spatial Fourier transform given by

$$G(\boldsymbol{r}, t) = \frac{1}{(2\pi)^3} \int e^{-k^2 t} e^{i\boldsymbol{k}\cdot\boldsymbol{r}} d^3\boldsymbol{k}.$$

Integrating over the angles of $\boldsymbol{r}$ yields

$$\begin{aligned} G(\boldsymbol{r}, t) &= \frac{1}{(2\pi)^2} \int_0^\infty e^{-k^2 t} \frac{\sin kr}{kr} k^2 dk = \frac{1}{4\pi r} \mathrm{Im} \int_{-\infty}^\infty e^{-k^2 t} e^{ikr} k dk \\ &= \frac{1}{4\pi r} \mathrm{Im}\, \frac{1}{i} \frac{\partial}{\partial r} \int_{-\infty}^\infty e^{-k^2 t} e^{ikr} dk, \end{aligned}$$

which gives Green's function in the form $G(\boldsymbol{r}, t) = e^{-r^2/4t}/(4\pi t)^{3/2}$, $t > 0$. For $t < 0$, $G(\boldsymbol{r}, t) = 0$. ♣

3. Find the half-space one-dimensional Green's function defined on $x > 0$ that satisfies the boundary condition of $G(x, t) = 0$ at $x_0 = 0$.

Solution: Using an image source of negative strength at $x = -x_0$, the solution is expressed by $G(x, t) = G_0(x - x_0, t) - G_0(x + x_0, t)$, $x > 0$, where $G_0(x, t) = e^{-x^2/4t}/(4\pi t)^{1/2}$. ♣

4. Consider the wave equation with a source term $h(\boldsymbol{r}, t)$ given by

$$\partial_t^2 f - \Delta_3 f = h(\boldsymbol{r}, t).$$

Show that the solution is expressed as

$$f(\boldsymbol{r}, t) = \frac{1}{4\pi} \int h(\boldsymbol{r}', t) \frac{1}{|\boldsymbol{r} - \boldsymbol{r}'|} d^3\boldsymbol{r}. \tag{17.64}$$

Solution: The Green function $G(\boldsymbol{r}, t)$ is defined as the solution satisfying the equation $\partial_t^2 G - \Delta_3 G = \delta(\boldsymbol{r})\delta(t)$. The spatial Fourier and temporal Laplace transform of the above is obtained by settings $\omega \to -\omega^2$ in Exercise **1** as $G(\boldsymbol{k}, \omega) = 1/[(2\pi)^3(-\omega^2 + k^2)]$. The inverse transform of the above gives $G(|\boldsymbol{r} - \boldsymbol{r}'|, t - t') = \delta(|\boldsymbol{r} - \boldsymbol{r}'|)\delta(t - t')/(4\pi|\boldsymbol{r} - \boldsymbol{r}'|)$. Since the physical system is invariant under the translations in space and time, the Green function depends only on relative space and time coordinates $|\boldsymbol{r} - \boldsymbol{r}'|$ and $t - t'$. The Green function has the property that the solution can be written as $f(\boldsymbol{r}, t) = \int G(\boldsymbol{r}, \boldsymbol{r}'; t, t')h(\boldsymbol{r}', t')d^3\boldsymbol{r}'dt$, so it is given by (17.64). ♣

6. Find the general solution for the wave equation

$$\partial_t^2 u - \Delta_3 u = 0,$$

where Δ_3 is the Laplacian defined by

$$\Delta_3 = \partial_x^2 + \partial_y^2 + \partial_z^2.$$

Solution: Since the system is isotropic and homogeneous, we can assume the solution in analogy with the one-dimensional case as $f(p) = f(\boldsymbol{n}\cdot x \pm t)$, where $\boldsymbol{n}$ is the unit vector that points along the direction of propagation of the wave and $\boldsymbol{n}\cdot\boldsymbol{x} = lx + my + nz$. From the chain rule on derivatives, we have $(l^2 + m^2 + n^2 - 1)f''(p) = 0$. Thus, f can be arbitrary since $l^2 + m^2 + n^2 = 1$. The general solution is given by $u(\boldsymbol{x}, t) = f(\boldsymbol{n}\cdot\boldsymbol{x} - t) + g(\mathbf{n}\cdot\boldsymbol{x} + t)$, where $\boldsymbol{n}$ is any unit vector. ♣

17.5 Applications in Physics and Engineering

17.5.1 Wave Equations for Vibrating Strings

In the previous sections, we rigorously studied the theories underlying the following three typical classes of partial differential equations: Laplace equations, diffusion equations, and wave equations. In this section, we attempt to formulate mathematical expressions for these classes on the basis of the associated physical phenomena; e.g., we will see that the mathematical form of the wave equation

$$\partial_t^2 u(x, t) = v^2 \partial_x^2 u(x, t) \tag{17.65}$$

is derived by considering the wavy motion of a that string. This will make clear why equations of the same form as (17.65) are called *wave* equations.

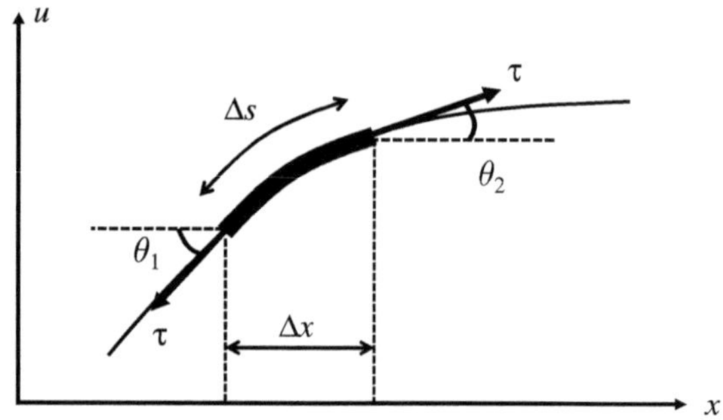

Fig. 17.1. Schematic of a thin stretched string and a line element Δs; the tensions exerted at both ends of the line element have equal magnitudes $\boldsymbol{\tau}$ but act in different directions

Suppose that a thin string is stretched between two fixed points with a tension $\boldsymbol{\tau}$ exerted at the two end points. We assume that the string is perfectly flexible, i.e., only tensile forces can be transmitted in the tangential direction. Then, as illustrated in Fig. 17.1, the magnitude of the tension exerted in the tangential direction is the same for every part of the string. It can be seen from the figure that the vertical components of the tension τ at the two ends of a line element with length Δs are $-\tau \sin\theta_1$ and $\tau \sin\theta_2$, where $\tau = |\boldsymbol{\tau}|$. Hence, the vertical component τ_u of the external force exerted on the line element is given by

$$\tau_u = \tau\left(\sin\theta_2 - \sin\theta_1\right). \tag{17.66}$$

The sine terms are rewritten in terms of the derivative $\partial_x u$ by using the following approximations:

$$\begin{aligned}\sin\theta_1 &= \frac{\partial u}{\partial s} = \partial_x u \frac{dx}{ds} = \partial_x u \frac{1}{\sqrt{1+(\partial_x u)^2}} \\ &\simeq \partial_x u\left[1 + \frac{1}{2}(\partial_x u)^2 + \cdots\right] \simeq \partial_x u\end{aligned}$$

where the relation $ds = \sqrt{(dx)^2 + (du)^2}$ was used, and

$$\begin{aligned}\sin\theta_2 &= \frac{\partial u(x+\Delta x)}{\partial s} = \frac{\partial u(x+\Delta x)}{\partial x}\frac{dx}{ds} \simeq \frac{\partial u(x+\Delta x)}{\partial x} \\ &= \partial_x u + \partial_x^2 u\big|_{x=\xi}\, \Delta x,\end{aligned}$$

where ξ is a constant satisfying the condition $x \le \xi \le x + \Delta x$. (The **mean value theorem** ensures the existence of such a constant ξ.) Substitution of the sine terms in (17.66) yields

$$\tau_u = \tau \, \partial_x^2 u\big|_{x=\xi} \, \Delta x.$$

Since the inertial force exerted by the line element is given by

$$\rho \Delta s \partial_t^2 u,$$

where, (ρ is the line density) we obtain the equation of motion for the string in the vertical direction:

$$\rho \frac{\Delta s}{\Delta x} \partial_t^2 u = \tau \, \partial_x^2 u\big|_{x=\xi} .$$

Taking the limit $\Delta x \to 0$ so that $\xi \to x$, and then approximating ds/dx as 1, we obtain the final result:

$$\partial_t^2 u = \frac{\tau}{\rho} \partial_t^2 u. \tag{17.67}$$

It is customary to denote the positive constant as τ/ρ, v^2, which allows us to write (17.67) in the familiar form (17.65).

17.5.2 Diffusion Equations for Heat Conduction

In this subsection, we attempt to derive mathematical expressions for diffusion equations by considering the physical phenomena that occur during heat conduction, i.e., the flow of heat in a certain medium from points at a high temperature to those at lower temperatures. This process takes place in such a manner that molecules in irregular motion exchange their kinetic energy by colliding with each other.

We aim to determine the amount of heat δQ penetrating an arbitrarily chosen surface element δS inside the medium per unit time (called the **heat flux**). In order to find δQ, we consider another surface element δS_1, which has the same magnitude as δS, parallel to δS and located at an orthogonal distance Δn from δS. We assume that δS is so small that the temperatures $u = u(x, y, z, t)$ on δS and $u_1 = u(x + \delta x, y + \delta y, z + \delta z, t)$ on δS_1 are constant over δS and δS_1, respectively.

From thermodynamics, we know that the magnitude of the flux of heat difference between u and u_1, denoted by δu, and the area of the surface element are related in the following manner:

$$\delta Q \cdot \delta n = \kappa \cdot \delta u \cdot \delta S, \tag{17.68}$$

where the value of the constant κ, called the **thermal conductivity**, depends on the medium. Dividing both sides of (17.68) by δn and taking the limit $\delta n \to 0$, we have

$$\delta Q = \kappa \frac{\partial u}{\partial n} \cdot \delta S.$$

Here, $\partial u/\partial n$ is the derivative in the direction normal to δS and is expressed as

$$\frac{\partial u}{\partial n} = \overline{\boldsymbol{n}} \cdot \nabla u,$$

where $\overline{\boldsymbol{n}}$ is the unit vector normal to δS. Thus, we obtain the flux passing through a volume element δv that is enclosed by a surface S:

$$\begin{aligned} Q &= \kappa \iint_S \frac{\partial u}{\partial n} dS = \kappa \iint_S \nabla u \cdot \overline{\boldsymbol{n}} dS \\ &= \kappa \iiint_V \nabla \cdot (\nabla u) dV. \end{aligned}$$

We now apply the mean value theorem to the volume integral over δv to obtain

$$Q = \kappa \nabla \cdot [\nabla u(x^*, y^*, z^*, t)] \, \delta v, \tag{17.69}$$

where (x^*, y^*, z^*) is a point in δv.

Apart from the above-mentioned discussion, we also see that Q is related to the temperature variation in δv with time. In fact, the temperature u in δv increases (or decreases) owing to the accumulation (or loss) of heat in δv at a rate of $\partial u / \partial t$. Therefore, the flow of heat into (or out of) δv can be written as

$$\rho \sigma \frac{\partial u}{\partial t} \delta v, \tag{17.70}$$

where ρ is the mass density and σ (the **specific heat**) is a characteristic of the medium. By setting (17.69) equal to (17.70), and allowing the volume element δv to shrink to a point, we have

$$\rho \sigma \partial_t u = \kappa \nabla \cdot (\nabla u) = \kappa \nabla^2 u.$$

Clearly, this result is of the same form as a diffusion equation that describes heat conduction in a medium with physical parameters ρ, σ, κ.

Part VI

Tensor Analyses

18

Cartesian Tensors

Abstract Tensors are geometric entities that provide concise mathematical frameworks for formulating problems in physics and engineering. The most important feature of tensors is their coordinate invariance: tensors are independent of the type of coordinate system chosen. This feature is similar to the condition that the length and direction of a geometric figure do not change, regardless of the coordinate system used for the algebraic expression. In contrast, the components of a tensor are coordinate-dependent in a structured routine. In this chapter, we discuss the ways in which the choice of a coordinate system affects the components of a tensor.

18.1 Rotation of Coordinate Axes

18.1.1 Tensors and Coordinate Transformations

A **tensor** is a natural generalization of a vector or a scalar encountered in elementary vector calculus. The latter two are, in fact, both special cases of tensors of order n, whose specification in a three-dimensional space requires 3^n numbers, called the **components** of the tensor. In this context, scalars are tensors of the zero order with a $3^0 = 1$ component and vectors are tensors of the first order with $3^1 = 3$ components.

Of importance is the fact that a tensor of order n is much more than just a set of 3^n numbers. The key property of tensors is adherence to the transformation law of its components under a change of coordinate system, say, from a rectangular to elliptic, polar, or other curvilinear coordinate system. If the coordinate system is changed to a new one, the components of a tensor change according to a characteristic transformation law. We shall see that this transformation law makes clear the physical (or geometrical) meaning of the tensor being invariant under a change of coordinate system. The coordinate-invariance-character of tensors answers the demand that the proper formulation of physical laws should be independent of the choice of coordinate systems.

It is obvious that physical processes must be independent of the coordinate system. However, what is not so trivial is what the coordinate-independence property of physical processes implies about the transformation law of mathematical objects (i.e., tensors). The study of these implications and the classification of physical quantities by means of the transformation laws constitute the primary content of this chapter. Emphasis is placed on the fact that all kinds of tensors are geometric objects whose representation (i.e., The values of its components) obey a characteristic transformation law under **coordinate transformation**.

18.1.2 Summation Convention

In order to simplify subsequent notation, we introduce the following two conventions:

♠ **Summation convention:**
When the same index appears repeatedly in one term, we carry out a summation with respect to the index. The range of summation is from 1 to n, where n is the number of dimensions of the space.

♠ **Range convention:**
All non repeated indices are understood to run from 1 to n.

These conventions are operative throughout this chapter unless specifically stated otherwise.

Example The summation convention yields the new notation as

$$a_i b_i \equiv \sum_{i=1}^{n} a_i b_i = a_1 b_1 + a_2 b_2 + \cdots + a_n b_n.$$

Similarly, if i and j have the range from 1 to 2, then

$$\begin{aligned} a_{ij} b_{ij} &= a_{1j} b_{1j} + a_{2j} b_{2j} \\ &= a_{11} b_{11} + a_{12} b_{12} + a_{21} b_{21} + a_{22} b_{22}, \end{aligned}$$

where it does not matter whether the first sum is carried out on i or j.

Remark. Repeated indices are referred to as **dummy indices** since, owing to the implied summation, any such pair may be replaced by any other pair of repeated indices without changing the meaning of the mathematical expression.

18.1.3 Cartesian Coordinate System

A tensor is a mathematical object composed of several components. The values of the components depend on the coordinate system to be employed, so are altered through a coordinate transformation even when the tensor itself remains unchanged. Among the many possible choices of coordinate transformations, a **rigid rotation** of a **rectangular Cartesian coordinate system** is the simplest. The remainder of this section is devoted to explaining the basic properties of the simplest coordinate transformations, as a preliminary for our subsequent study of tensors in terms of more general coordinate systems.
We begin with two formal definitions:

♠ **Cartesian coordinate system:**
A Cartesian coordinate system associates a unique ordered set of real numbers (**coordinates**) $(x_1, x_2, \cdots, x_n)$ with every point in a given n-dimensional space by reference to a set of directed straight lines (**coordinate axes**) $Ox_1, Ox_2, \cdots, Ox_n$ intersecting at the origin O.

♠ **Rectangular Cartesian coordinate system:**
If the three axes of a Cartesian coordinate system are mutually perpendicular, we have what is called a **rectangular Cartesian coordinate system**.

Figure 18.1 is a schematic illustration of a rectangular Cartesian coordinate system in three-dimensional space. Referring to this coordinate system, we denote the triples $(1,0,0)$, $(0,1,0)$, $(0,0,1)$ by $\boldsymbol{e}_1$, $\boldsymbol{e}_2$, $\boldsymbol{e}_3$, respectively. These triples are represented geometrically by mutually perpendicular unit arrows.

The set of Cartesian axes Ox_1, Ox_2, and Ox_3 is said to be **right-handed** if and only if the rotation needed to turn the x^1-axis into the direction of the x^2-axis through an angle $\angle x^1 O x^2 < \pi$ would propel a right-handed screw toward the positive direction of the x^3-axis. Conversely, if such a rotation

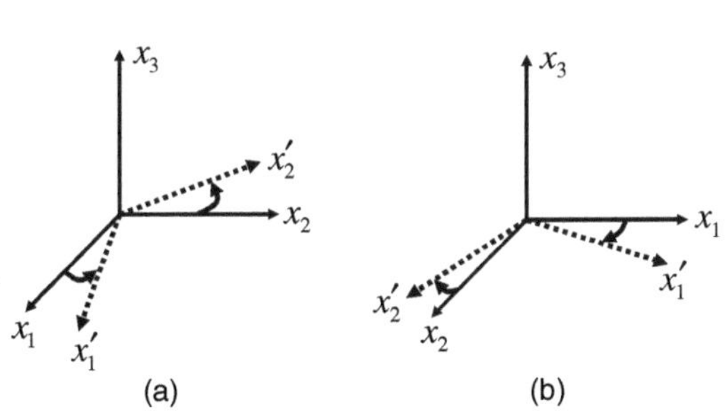

Fig. 18.1. **(a)** Right-handed and **(b)** left-handed Cartesian coordinate systems

propels a left-handed screw in the positive direction of the x^3-axis, the set of axes is said to be **left-handed.** In this section, we consider only Cartesian coordinate systems that are rectangular and right-handed.

18.1.4 Rotation of Coordinate Axes

We now formulate a rigid rotation of rectangular Cartesian axes. Assume a position arrow $\boldsymbol{r}$ whose components are given by (x_1, x_2, x_3) and (x'_1, x'_2, x'_3) in terms of two different rectangular coordinate systems having a common origin. We denote the set of unit orthogonal basis arrows associated with the unprimed and primed system by $\{\boldsymbol{e}_i\}$ and $\{\boldsymbol{e}'_i\}$, respectively. The transformation from one Cartesian coordinate system to another is called a **rigid rotation of Cartesian axes** and has the following property:

♠ **A rigid rotation of Cartesian axes:**
A rigid rotation of Cartesian axes is described by the transformation equations of coordinates x_k as

$$x'_j = R_{jk}\, x_k \quad \text{(summed over } k), \tag{18.1}$$

$$x_k = R_{jk}\, x'_j \quad \text{(summed over } j), \tag{18.2}$$

where $R_{jk} = \boldsymbol{e}'_j \cdot \boldsymbol{e}_k$ are **directed cosines** of $\boldsymbol{e}'_j$ associated with $\boldsymbol{e}_k$.

Remarks.

1. Each of the two indices j and k for R_{jk} refers to a different basis: the first index j refers to the *primed* set $\{\boldsymbol{e}'_j\}$, while the second index k refers to the *unprimed* one $\{\boldsymbol{e}_k\}$. This means that, in general, $R_{jk} \neq R_{kj}$.
2. The transformation coefficients R_{jk} do not constitute a tensor, but simply set of real numbers. (See the second remark in Sect. 18.2.3.)

Proof A geometric arrow $\boldsymbol{r}$ joining the origin O and the point P is expressed by

$$\boldsymbol{r} = x_k \boldsymbol{e}_k = x'_j \boldsymbol{e}'_j. \tag{18.3}$$

We expand $\boldsymbol{e}_k$ by the set of $\{\boldsymbol{e}'_j\}$ as

$$\boldsymbol{e}_k = (\boldsymbol{e}'_j \cdot \boldsymbol{e}_k)\, \boldsymbol{e}'_j = R_{jk} \boldsymbol{e}'_j, \tag{18.4}$$

where we use both the range and the summation conventions. Substituting (18.4) into (18.3), we obtain

$$x_k R_{jk} \boldsymbol{e}'_j = x'_j \boldsymbol{e}'_j,$$

and thus

$$(x_k R_{jk} - x'_j)\, \boldsymbol{e}'_j = \mathbf{0}.$$

Since the arrow set $\{\boldsymbol{e}'_j\}$ is linearly independent, the quantities in the parentheses equal zero, which results in the desired equation (18.1).

Similarly, expanding $\{\boldsymbol{e}'_j\}$ by $\{\boldsymbol{e}_k\}$ as

$$\boldsymbol{e}'_j = (\boldsymbol{e}_k \cdot \boldsymbol{e}'_j)\, \boldsymbol{e}_k = R_{jk}\boldsymbol{e}_k \tag{18.5}$$

and substituting it into (18.3), we arrive at equation (18.2). ♣

Remark. Observe that in the transformation law (18.1) and the expansion (18.5), R_{jk} acts on an *unprimed* entity (i.e., x_k or $\boldsymbol{e}_k$) to produce a primed one (i.e., x'_j or $\boldsymbol{e}'_j$). However, in (18.2) and (18.4), R_{jk} acts on a primed entity to produce an unprimed one. In all the cases above, we should make sure that the order of indices j and k attached to the coefficients R_{jk} remains unchanged: the first index, j, always refers to the primed entity and the second, k, to the unrpimed one.

18.1.5 Orthogonal Relations

The following theorem states an important property of the transformation coefficients R_{jk} that gives rise to a rigid rotation of Cartesian axes.

♠ **Orthogonal relations:**

The transformation coefficients R_{jk} for a rigid rotation of axes satisfy the conditions

$$R_{ik}R_{jk} = \delta_{ij}, \quad R_{ik}R_{i\ell} = \delta_{k\ell}. \tag{18.6}$$

Proof The first relation of (18.6) follows from a geometric formula for the angle θ between two basic arrows: $\boldsymbol{e}'_i$ and $\boldsymbol{e}'_j$. Taking the inner product of the two basic arrows $\boldsymbol{e}'_i = R_{ik}\boldsymbol{e}_k$ and $\boldsymbol{e}'_j = R_{j\ell}\boldsymbol{e}_\ell$, we have

$$\cos\theta = \boldsymbol{e}'_i \cdot \boldsymbol{e}'_j = R_{ik}R_{j\ell}(\boldsymbol{e}_k \cdot \boldsymbol{e}_\ell) = R_{ik}R_{j\ell}\delta_{k\ell} = R_{ik}R_{jk}. \tag{18.7}$$

If $i = j$, $\boldsymbol{e}'_i$ and $\boldsymbol{e}'_j$ coincide so that $\theta = 0$, whereas if $i \neq j$, $\boldsymbol{e}'_i$ and $\boldsymbol{e}'_j$ are orthogonal so that $\theta = \pi/2$. Hence, we have

$$R_{ik}R_{jk} = \begin{cases} 1 & \text{if } i = j, \\ 0 & \text{if } i \neq j. \end{cases}$$

The second equation of (18.6) can be verified in a similar manner by considering the angle between $\boldsymbol{e}_k$ and $\boldsymbol{e}_\ell$. ♣

The physical meaning of the relations (18.6) is rather obvious. They ensure that the axes of each set $\{\boldsymbol{e}'_i\}$ or $\{\boldsymbol{e}_k\}$ are mutually orthogonal, i.e.,

$$\boldsymbol{e}'_i \cdot \boldsymbol{e}'_j = \delta_{ij} \quad \text{and} \quad \boldsymbol{e}_k \cdot \boldsymbol{e}_\ell = \delta_{k\ell}.$$

18.1.6 Matrix Representations

Since the transformation coefficients R_{jk} have two subscripts, it is natural to display their values in matrix form. The notation $[R_{jk}]$ is used to denote the matrix having R_{jk} as the element in the jth row and kth column. In addition, when denoted by R, it represents a linear operator of a rotation of axes without reference to any of the values of its coefficients R_{jk}.

Example In two dimensions, a rigid rotation of rectangular axes for an angle θ is given by

$$\boldsymbol{e'}_i = R_{ij}\boldsymbol{e}_j, \quad [R_{ij}] = \begin{pmatrix} R_{11} & R_{12} \\ R_{21} & R_{22} \end{pmatrix} = \begin{pmatrix} \cos\theta & \sin\theta \\ -\sin\theta & \cos\theta \end{pmatrix}. \tag{18.8}$$

This rotation of axes gives rise to a coordinate transformation,

$$x'_i = R_{ij}x_j,$$

or equivalently,

$$\begin{pmatrix} x'_1 \\ x'_2 \end{pmatrix} = \begin{pmatrix} R_{11} & R_{12} \\ R_{21} & R_{22} \end{pmatrix} \begin{pmatrix} x_1 \\ x_2 \end{pmatrix} = \begin{pmatrix} \cos\theta & \sin\theta \\ -\sin\theta & \cos\theta \end{pmatrix} \begin{pmatrix} x_1 \\ x_2 \end{pmatrix}. \tag{18.9}$$

Remark. We comment briefly on the distinction between **active** and **passive tranformations**; since it often causes confusion. Throughout this chapter, we are concerned solely with passive transformations, for which physical entities of interest (e.g., the mass of a particle or a geometric arrow) remain unaltered and only the coordinate system is changed from $\{\boldsymbol{e}_i\}$ to $\{\boldsymbol{e'}_i\}$, as given by (18.8). In contrast, an active transformation alters the position and/or the direction of the physical entity itself, while the axes $\{\boldsymbol{e}_i\}$ remain fixed. In the latter case, a rotation of a geometric arrow $\boldsymbol{x}$ through an angle θ in two dimensions is described by

$$\boldsymbol{e'}_i = \boldsymbol{e}_i \quad \text{and} \quad \begin{pmatrix} x'_1 \\ x'_2 \end{pmatrix} = \begin{pmatrix} \cos\theta & -\sin\theta \\ \sin\theta & \cos\theta \end{pmatrix} \begin{pmatrix} x_1 \\ x_2 \end{pmatrix},$$

which obviously differs from those for a passive transformation. Figure 18.2 illustrates the difference between the two transformations.

It can be shown that the determinant of the matrix $[R_{k\ell}]$ reads

$$\det[R_{k\ell}] = \pm 1 \tag{18.10}$$

(see the proof in Exercise **5**). This means that there are two classes of rectangular Cartesian coordinate systems, corresponding to the positive and negative signs in (18.10). Throughout this section, we consider only cases of

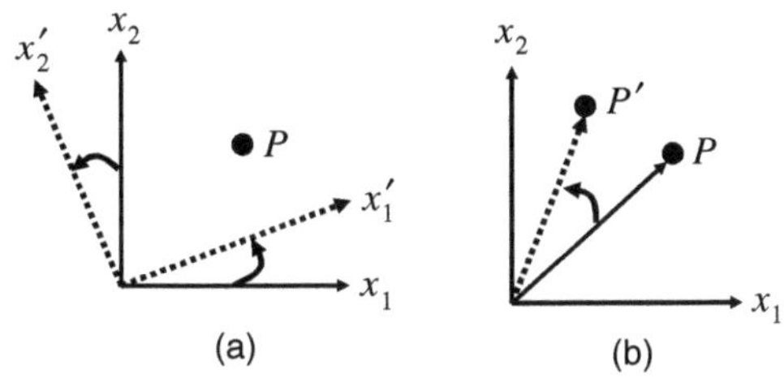

Fig. 18.2. Difference between a passive **(a)** and an active **(b)** transformation

$\det[R_{k\ell}] = +1$, which corresponds to our previous restriction to a single type of coordinate system (i.e., right-handed). We shall see in sect. 18.3.1 that the rotation of coordinate axes whose transformation coefficients give rise to $\det[R_{k\ell}] = -1$ yields the left-handed system, which for the moment is beyond our scope.

18.1.7 Determinant of a Matrix

We close this section by commenting on a formal definition of the determinant of a matrix and its relevant materials, as a preliminary for the exercises below.

1. The **determinant**

$$D = \det[a_{ij}] = \begin{vmatrix} a_{11} & a_{12} & \cdots & a_{1n} \\ a_{21} & a_{22} & \cdots & a_{2n} \\ \cdots & \cdots & & \cdots \\ a_{n1} & a_{n2} & \cdots & a_{nn} \end{vmatrix} \tag{18.11}$$

   of the square array of n^2 numbers (elements) a_{ij} is the sum of the $n!$ terms

$$(-1)^r a_{1k_1} a_{2k_2} \cdots a_{nk_n}, \tag{18.12}$$

   each corresponding to one of the $n!$ different ordered sets $k_1, k_2, \cdots, k_n$ obtained by r interchanges of elements from the ordered set $1, 2, \cdots, n$.
2. The **minor** (or **complementary minor**) M_{ij} of the elements a_{ij} in the nth-order determinant $D = \det[a_{ij}]$ is the $(n-1)$th-order determinant obtained from (18.11) on erasing the ith row and the jth column.

   Example Given a third-order determinant

$$D = \begin{vmatrix} a_{11} & a_{12} & a_{13} \\ a_{21} & a_{22} & a_{23} \\ a_{31} & a_{32} & a_{33} \end{vmatrix},$$

   its a_{12} minor is obtained by removing the first row and the second column in D, as expressed by

$$M_{12} = \begin{vmatrix} a_{21} & a_{23} \\ a_{31} & a_{33} \end{vmatrix}.$$

3. The **cofactor** C_{ij} of the element a_{ij} is defined by

$$C_{ij} = \frac{\partial D}{\partial a_{ij}},$$

or equivalently,

$$C_{ij} = (-1)^{i+j} M_{ij}.$$

4. A determinant D may be represented in terms of the elements and cofactors of any one row or column by

$$D = \det[a_{ij}] = \sum_{i=1}^{n} a_{ij} C_{ij} = \sum_{k=1}^{n} a_{jk} C_{jk} \quad \text{(with } j \text{ fixed).} \tag{18.13}$$

This is called **simple Laplace development** of a determinant D. (The proof is given in Exercise **2**). The expression (18.13) gives the same value for D regardless of the column or row [i.e., no matter what value of j in (18.13)] we choose in the expansion. Note also that for $j \neq h$,

$$\sum_{i=1}^{n} a_{ij} C_{ih} = \sum_{k=1}^{n} a_{jk} C_{hk} = 0.$$

Example The expansion by the first row gives

$$D = \begin{vmatrix} 1 & 3 & 0 \\ 2 & 6 & 4 \\ -1 & 0 & 2 \end{vmatrix} = 1 \begin{vmatrix} 6 & 4 \\ 0 & 2 \end{vmatrix} - 3 \begin{vmatrix} 2 & 4 \\ -1 & 2 \end{vmatrix} + 0 \begin{vmatrix} 2 & 6 \\ -1 & 0 \end{vmatrix} = -12.$$

Remark. In view of the expansion (18.13), an nth-order determinant D is represented by a linear combination of n numbers of $(n-1)$th-order determinants. Similarly, each of the latter $(n-1)$th-order determinants is in turn represented in terms of $n-1$ numbers of $(n-2)$th-order determinants, and so on. In a successive manner, we finally arrive at $n!$ numbers of first-order determinants (i.e., just $n!$ real numbers), each of which is expressed by (18.12).

Exercises

1. Check the validity of the orthogonal conditions (18.6) for the transformation coefficients R_{ij} in two dimensions.

   Solution: For instance, if we set $j = 1$ and $k = 2$, then

   $$R_{i1} R_{i2} = R_{11} R_{12} + R_{21} R_{22} = -\cos\theta \sin\theta + \cos\theta \sin\theta = 0,$$

   or if $j = k = 2$, we have

   $$R_{i2} R_{i2} = R_{21} R_{21} + R_{22} R_{22} = (-\sin\theta)^2 + \cos^2\theta = 1.$$

   Other equations can be proved in a similar way. ♣

2. Show that the expression (18.13) for a determinant

$$D = \det[a_{pq}] = \sum_{k=1}^{n} a_{ik} C_{ik} \text{ and } D = \sum_{i=1}^{n} a_{\ell i} C_{\ell i} \text{ for fixed } i$$

yields the same value of D no matter what value of i we choose.

Solution: We prove only the first formula, since the proof of the second is quite similar to that of the first.

It easily follows that the statement is true for a second-order determinant for which the expansions with fixed $i = 1$, $a_{11}a_{22} + a_{12}(-a_{21})$ and that with $i = 2$, $a_{21}(-a_{12}) + a_{22}a_{11}$, give the same value. By mathematical induction, we tentatively assume that the statement is true for an $(n-1)$th-order determinant and try to prove that it is also true for an nth-order determinant.

To do so, we expand D in terms of each of two arbitrary rows, say, the ith and the jth row with $i < j$, and compare the results.

(i) Let us first expand D by its ith row. A typical term in the expansion by the ith row reads

$$a_{ik} C_{ik} = a_{ik} \cdot (-1)^{i+k} M_{ik}, \tag{18.14}$$

where i is fixed and k runs from 1 to n. Since the minor M_{ik} of a_{ik} in D is an $(n-1)$th-order determinant, owing to the induction hypothesis it can be expanded by any row. Expand M_{ik} by its $(j-1)$th row. This row corresponds to the jth row of D, because M_{ik} does not contain elements of the ith row of D and $i < j$. Hence, the expansion of M_{ik} by its $(j-1)$th row consists of the linear combination of the elements $a_{j\ell}$ with $\ell = 1, 2, \cdots, k-1, k+1, \cdots, n$ (i.e., $\ell \neq k$). We distinguish between the two cases, $\ell < k$ and $\ell > k$, as follows.

For $\ell < k$, the element $a_{j\ell}$ belongs to the ℓth column of M_{ik}. Hence, the term involving $a_{j\ell}$ in the expansion of M_{ik} reads

$$a_{j\ell} \cdot (\text{cofactor of } a_{j\ell} \text{ in } M_{ik}) = a_{j\ell} \cdot (-1)^{(j-1)+\ell} M_{ikj\ell}. \tag{18.15}$$

Here $M_{ikj\ell}$ is the minor of $a_{j\ell}$ in M_{ik}, which is obtained from D by deleting the ith and jth rows and the kth and ℓth columns of D. Then it follows from (18.14) and (18.15) that the resulting terms in the expansion of D are of the form

$$a_{ik} a_{j\ell} \cdot (-1)^b M_{ikj\ell} \quad \text{with } b = i + k + j + \ell - 1. \tag{18.16}$$

If $\ell > k$, the only difference is that $a_{j\ell}$ belongs to the $(\ell-1)$th column of M_{ik} because M_{ik} does not contain elements of the kth column of D and $k < \ell$. This results in an additional minus sign in (18.15) and instead of (18.16) we obtain $-a_{ik}a_{j\ell}\cdot(-1)^{b+1}M_{ikj\ell}$ with the same value of b. In short, the expansion of D by the ith row yields

$$D = \sum_{k=1}^{n} a_{ik}C_{ik} = \sum_{k=1}^{n} a_{ik}\left[\sum_{\ell=1}^{k-1}(-1)^{b}a_{j\ell}M_{ikj\ell} + \sum_{\ell=k+1}^{n}(-1)^{b+1}a_{j\ell}M_{ikj\ell}\right] \tag{18.17}$$

with $b = i + k + j + \ell - 1$.

(ii) We next expand D by the jth row. A typical term in this expansion is

$$a_{j\ell}C_{j\ell} = a_{j\ell}\cdot(-1)^{j+\ell}M_{j\ell}. \tag{18.18}$$

By the induction hypothesis, we may expand the minor $M_{j\ell}$ of $a_{j\ell}$ in D by its ith row, which corresponds to the ith row of D since $j > i$.

For $k > \ell$, the element a_{ik} in that row belongs to the $(k-1)$th column of $M_{j\ell}$, because $M_{j\ell}$ does not contain elements of the ℓth column of D and $\ell < k$. Hence, the term involving a_{ik} in this expansion is

$$a_{ik}\cdot(\text{cofactor of } a_{ik} \text{ in } M_{j\ell}) = a_{ik}\cdot(-1)^{i+(k-1)}M_{ikj\ell}, \tag{18.19}$$

where the minor $M_{ikj\ell}$ of a_{ik} in $M_{j\ell}$ is obtained by deleting the ith and jth rows and the kth and ℓth columns of D, and is thus identical to $M_{ikj\ell}$ in (18.15). It follows from (18.18) and (18.19) that this yields a representation whose terms are identical to those given by (18.16) when $\ell < k$.

For $k < \ell$, the element a_{ik} belongs to the kth column of $M_{j\ell}$, so we get an additional minus sign and the result agrees with that characterized by (18.16). Hence, we conclude that the expansion of D by the $j(> i)$th row, $\sum_{k=1}^{n} a_{jk}C_{jk}$, is identical to the expansion (18.17).

The conclusions from the discussions in **(i)** and **(ii)** clearly show that the two expansions of D consist of the same terms, which completes our proof of the given statement. ♣

3. Let $b_{kj} = C_{jk}/D$, where C_{jk} is the cofactor of $[a_{jk}]$ in $D = \det[a_{jk}]$. Show that

$$b_{kj}a_{\ell k} = \delta_{j\ell} \quad \text{and} \quad b_{kj}a_{j\ell} = \delta_{k\ell}, \tag{18.20}$$

which means that the matrix $[b_{kj}]$ is the **inverse** of $[a_{jk}]$.

Solution: If follows that

$$b_{kj}a_{\ell k} = \frac{C_{jk}a_{\ell k}}{D} = \frac{a_{\ell 1}C_{j1} + a_{\ell 2}C_{j2} + \cdots + a_{\ell n}C_{jn}}{D}. \tag{18.21}$$

The discussion in Exercise **2** tells us that the sum in the numerator equals D when $\ell = j$ regardless of the value of j. Hence, we have

$$b_{kj}a_{\ell k} = 1 \quad \text{if } j = \ell.$$

We next consider the case of $j \neq \ell$. To do thus we replace the elements in the jth row of D by those in the $\ell(\neq j)$th row of D. The resulting determinant, denoted by $\tilde{D}$, reads

$$\tilde{D} = \begin{vmatrix} a_{11} & a_{12} & \cdots & a_{1n} \\ \cdots & \cdots & & \cdots \\ a_{j-1,1} & a_{j-1,2} & \cdots & a_{j-1,n} \\ a_{\ell,1} & a_{\ell,2} & \cdots & a_{\ell,n} \\ a_{j+1,1} & a_{j+1,2} & \cdots & a_{j+1,n} \\ \cdots & \cdots & & \cdots \\ a_{n,1} & a_{n,2} & \cdots & a_{n,n} \end{vmatrix} = 0,$$

since $\tilde{D}$ has two identical rows. Note that the expansion of $\tilde{D}$ by its ℓth row is $\tilde{D} = \sum_{p=1}^{n} a_{\ell p}C_{jp}$, which equals the sum in the numerator in (18.21). It thus follows that

$$b_{kj}a_{\ell k} = 0 \quad \text{if } j \neq \ell.$$

These arguments complete the proof of the first equation. The second can be verified in the same manner. ♣

4. Suppose a matrix $[Q_{kj}]$ defined by $Q_{kj} = C_{jk}/\det[R_{jk}]$, where the R_{jk} are the transformation coefficients of a rigid rotation of axes and the C_{jk} are the cofactors of $[R_{jk}]$ in $\det[R_{jk}]$. Prove that $Q_{kj} = R_{jk}$.

Solution: Apply the result of Exercise **3** to find that $Q_{kj}R_{\ell k} = \delta_{\ell j}$. Multiplying both sides by $R_{\ell m}$ and summing with respect to ℓ, we arrive at

$$Q_{kj}R_{\ell k}R_{\ell m} = \delta_{\ell j}R_{\ell m} = R_{jm}.$$

Then, the orthogonal relation $R_{\ell k}R_{\ell m} = \delta_{km}$ implies that $Q_{kj}\delta_{km} = Q_{mj} = R_{jm}$. ♣

5. Show that $\det[R_{jk}] = \pm 1$.

Solution: It follows from (18.20) that

$$\det[Q_{kj}R_{j\ell}] = \det[\delta_{k\ell}] = 1,$$

where the Q_{kj} are the same quantities as in Exercise **4**. From elementary linear algebra, we find that

$$\det[Q_{kj}R_{j\ell}] = \det[Q_{kj}]\det[R_{j\ell}] = (\det[R_{j\ell}])^2 ,$$

where the identity $Q_{kj} = R_{jk}$ was used to obtain the last term. Combining the two results above, we obtain $\det[R_{jk}] = \pm 1$. ♣

18.2 Cartesian Tensors

18.2.1 Cartesian Vectors

Having dealt with the rotation of coordinate axes, we are ready to introduce the concept of tensors and their transformation law in terms of Cartesian coordinate systems. Assume an ordered set of three quantities v_i $(i = 1, 2, 3)$ that are explicit or implicit functions of x_j. Let us see how the values of $v_i(x_j)$ change through a rigid rotation of the Cartesian axes. If they transform according to the law given below, the quantities v_i are called the **components** of a particular kind of tensor, i.e., of a **Cartesian vector** (or a **first-order Cartesian tensor**).

♠ **Cartesian vectors:**

A Cartesian vector $\boldsymbol{v}$ is an object represented by an ordered set of n functions $v_i(x_j)$ in terms of the x-coordinate system and by another set of n functions $v'_i(x'_j)$ in terms of the x'-coordinate system, where v'_i and v_i at each point are related by the transformation law:

$$v'_i = R_{ij} v_j \quad \text{and} \quad v_i = R_{ki} v'_k, \tag{18.22}$$

where $R_{ij} = \boldsymbol{e}'_i \cdot \boldsymbol{e}_j$.

Obviously, a vector $\boldsymbol{v}$ is a geometric object (like an arrow) so that it is uniquely determined independently of the coordinate system. On the other hand, the function form of the components $v_i(x_j)$ depend on our choice of coordinate system, even when we consider only the same vector $\boldsymbol{v}$. This is why the concepts of a *vector* and its *components* are inherently different from each other. (See also Sect. 18.2.2 on this point.)

We emphasize again that the index i of R_{ij} in (18.22) refers to the dashed (transformed) function v'_i, whereas the j refers to the undashed (original) one v_j. In the following, we consider several examples of ordered sets of functions v_i in two dimensions, which may or may not be a first-order Cartesian tensor.

Examples **1.** The ordered set of functions (v_i) $(i = 1, 2)$ with the components $v_1 = x_2$ and $v_2 = -x_1$.

Using the transformation law of coordinates $x'_i = R_{ij} x_j$, we set the following for each function:

$$\begin{aligned} v'_1 &= x'_2 = R_{21}x_1 + R_{22}x_2 = -x_1 \sin\theta + x_2 \cos\theta, \\ v'_2 &= -x'_1 = -R_{11}x_1 - R_{12}x_2 = -x_1 \cos\theta - x_2 \sin\theta. \end{aligned}$$

On the other hand, the functions v'_i should be obtained from v_i through the transformation law as

$$\begin{aligned} v'_1 &= R_{1k}v_k = v_1 \cos\theta + v_2 \sin\theta = x_2 \cos\theta - x_1 \sin\theta, \\ v'_2 &= R_{2k}v_k = -v_1 \sin\theta + v_2 \cos\theta = -x_2 \sin\theta - x_1 \cos\theta. \end{aligned}$$

The two expressions for v'_1 and v'_2 are identical to one another regardless of the values of θ. Therefore, the pair of functions $v_i(x_j)$ are components of a Cartesian vector.

2. The set v_i with $v_1 = x_2$ and $v_2 = x_1$.

Following the same procedure as above, we have

$$\begin{aligned} v'_1 &= x'_2 = -sx_1 + cx_2, \\ v'_2 &= x'_1 = cx_1 + sx_2 \end{aligned}$$

and

$$\begin{aligned} v'_1 &= cv_1 + sv_2 = cx_2 + sx_1, \\ v'_2 &= -sv_1 + cv_2 = -sx_2 + cx_1, \end{aligned}$$

where c and s represent $\cos\theta$ and $\sin\theta$, respectively. These two sets of expressions do not agree with each other. Hence, the pair (x_2, x_1) is not a first-order Cartesian tensor.

18.2.2 A Vector and a Geometric Arrow

The result of Example **2** in Sect. 18.2.1 might be confusing for some readers; the functions $v_i(x_1, x_2)$ given there are not components of a vector, although they *appear* to represent a geometric arrow in (x_1-x_2)-plane. To make this point clear, we have to comment on the difference between the formal definition of a *vector* as a first-order tensor and our familiar definition of a vector as a *geometric arrow.*

In elementary calculus, vectors are simply defined by a geometric arrow with certain length and direction, commonly denoted by a bold-face letter, say, $\boldsymbol{v}$. Owing to this definition, $\boldsymbol{v}$ is uniquely determined by specifying its length and direction, which are both independent of our choice of coordinate systems. However, the uniqueness disappears if it is defined algebraically by specifying its components v_k relative to given coordinate axes. The values of the components v_k depend on our choice of coordinate system even when the same arrow $\boldsymbol{v}$ is considered. Hence, when we apply a coordinate transformation, the values of v_k are altered in a way that preserves the length and direction of the arrow $\boldsymbol{v}$, according to (18.22), which is why we call the set of n functions v_k not a *vector*, but the *components* of a vector.

In short, we should always keep in mind that a vector is a geometric object independent of coordinate systems, whereas components of a vector are just mathematical representations of the vector with reference to a specific coordinate system. This caution applies to all the classes of tensors presented throughout this section.

Remark. Despite the above caution, we sometimes call a set of components of a tensor just a "tensor" to shorten our sentences. However, it is important to note an inherent difference between a tensor (=coordinate-*independent* object) and components of a tensor (=coordinate-*dependent* quantities).

18.2.3 Cartesian Tensors

We turn to a **second-order Cartesian tensor** that requires two subscripts to identify a particular element of the set.

♠ **Second-order Cartesian tensor:**

A second-order Cartesian tensor $\boldsymbol{T}$ is an object represented by an ordered set of two-index quantities T_{ij} in terms of the x-coordinate system and by another set of quantities $T'_{k\ell}$ in terms of the x'-coordinate system, where T_{ij} and $T'_{k\ell}$ at each point are related by

$$T'_{ij} = R_{ik}R_{j\ell}T_{kl}, \quad T_{k\ell} = R_{mk}R_{n\ell}T'_{mn}. \tag{18.23}$$

Here, the two-index quantities T_{ij} are called **components of the tensor** $\boldsymbol{T}$.

In a similar way, we may define a Cartesian tensor of general order as follows: The set of expressions $T_{ij\cdots k}$ are the components of a Cartesian tensor if, for all rotations of the axes, the expressions using the new coordinates $T'_{\ell m\cdots n}$ are given by

$$T'_{ij\cdots k} = R_{i\ell}R_{jm}\cdots R_{kn}T_{\ell m\cdots n}$$

and

$$T_{\ell m\cdots n} = R_{p\ell}R_{qm}\cdots R_{rn}T'_{pq\cdots r}.$$

It is apparent that an nth-order Cartesian tensor in three dimensions has 3^n components.

Example Assume two Cartesian vectors $\boldsymbol{a}$ and $\boldsymbol{b}$, each of which is represented by the components a_j and b_k associated with the same coordinate system. Then, it is possible to create nine products of the components expressed by

$$a_j b_k \quad (j,k = 1,2,3),$$

which is called an **outer product** (or **direct product**) of the vectors $\boldsymbol{a}$ and $\boldsymbol{b}$ (see also Sect. 18.4.3). The outer product consists of a second-order Cartesian tensor. In fact, since each a_i and b_j transforms as

$$a'_i = R_{ik}a_k \quad \text{and} \quad b'_j = R_{j\ell}b_\ell,$$

we have

$$T'_{ij} \equiv a'_i b'_j = R_{ik}R_{j\ell}a_k b_\ell = R_{ik}R_{j\ell}T_{k\ell}.$$

Remark. We emphasize that transformation coefficients, say, the R_{ij}, do not form a tensor and note the fact that the two indices i and j in the tensor T_{ij} refer to the *same* coordinate system, whereas those in the coefficients R_{ij} refer to *different* coordinate systems. Hence, T_{ij} and R_{ij} are inherently different from each other, though both require two indices.

18.2.4 Scalars

Contrary to the case of finite-order tensors, we now consider quantities that are unchanged by a rotation of axes, which are called **scalars** or **tensors of zero order** and contain only one component. The most obvious example is the square of the distance of a point from the origin, which must be invariant under any rotation of coordinate axes. Other examples of scalars are presented below.

Examples **1.** We show that the **scalar product** $\boldsymbol{u} \cdot \boldsymbol{v}$ is invariant under rotation. In the original (unprimed) system, the scalar product is given in terms of components by $u_i v_i$ and in the rotated (primed) system, it is given by

$$u'_i v'_i = (R_{ij}u_j)(R_{ik}v_k) = R_{ij}R_{ik}u_j v_k = u_j v_k \delta_{jk} = u_j v_j, \tag{18.24}$$

where the **orthogonal relation** $R_{ij}R_{ik} = \delta_{jk}$ in (18.6) was used. Since the expression in the rotated system is the same as that in the original system, the scalar product is indeed invariant under rotations.

2. If the v_i are the components of a vector, the divergence $\nabla \cdot \boldsymbol{v} = \partial v_i / \partial x_i$ becomes a scalar. This is proven as follows: In the rotated coordinate system, $\nabla \cdot \boldsymbol{v}$ is given by

$$\frac{\partial v'_i}{\partial x'_i} = \frac{\partial}{\partial x'_i}(R_{ik}v_k) = R_{ik}\frac{\partial v_k}{\partial x'_i},$$

where the elements $R_{ik} = \boldsymbol{e}_k \cdot \boldsymbol{e}'_i$ are not functions of position. Using the relation $\partial x_j / \partial x'_i = R_{ij}$ (see Exercise **2** below), we have

$$R_{ik}\frac{\partial v_k}{\partial x'_i} = R_{ik}\frac{\partial x_j}{\partial x'_i}\frac{\partial v_k}{\partial x_j} = R_{ik}R_{ij}\frac{\partial v_k}{\partial x_j} = \delta_{jk}\frac{\partial v_k}{\partial x_j} = \frac{\partial v_j}{\partial x_j}.$$

Finally, we obtain

$$\frac{\partial v'_i}{\partial x'_i} = \frac{\partial v_j}{\partial x_j}.$$

Exercises

1. Examine whether or not the ordered set of functions v_i defined by $v_1 = (x_1)^2$ and $v_2 = (x_2)^2$ constitute a vector. Here $(x_1)^2$ means the square of x_1.

   Solution: To examine the first function, v_1, alone is sufficient to show that this pair is not a vector. Evaluating v'_1 directly gives

   $$v'_1 = (x'_1)^2 = c^2(x_1)^2 + 2c(-s)x_1x_2 + (-s)^2(x_2)^2,$$

   whereas (18.8) requires that $v'_1 = cv_1 - sv_2 = c(x_1)^2 - s(x_2)^2$, which is different from the above. ♣

2. Show that $R_{ij} = \boldsymbol{e}'_i \cdot \boldsymbol{e}_j = \dfrac{\partial x'_i}{\partial x_j} = \dfrac{\partial x_j}{\partial x'_i}$ in Cartesian coordinate systems.

   Solution: Set $x'_i\boldsymbol{e}'_i = x_j\boldsymbol{e}_j$ and differentiate both sides with respect to x'_j to obtain $(\partial x'_i/\partial x_j)\boldsymbol{e}'_i = \boldsymbol{e}_j$. Taking the scalar product of both sides with $\boldsymbol{e}'_k$ yields

   $$\frac{\partial x'_i}{\partial x_j}\boldsymbol{e}'_i \cdot \boldsymbol{e}'_k = \frac{\partial x'_i}{\partial x_j}\delta_{ik} = \frac{\partial x'_k}{\partial x_j} \quad \text{and} \quad \boldsymbol{e}_j \cdot \boldsymbol{e}'_k = R_{kj}.$$

   Hence, we have $R_{kj} = \partial x'_k/\partial x_j$. Similarly, if we differentiate by x'_i (instead of x_j) the first identity yields $R_{kj} = \partial x_j/\partial x'_k$. ♣

3. Show that the **gradient of a vector** $\boldsymbol{v}$, denoted by $\nabla \boldsymbol{v}$, is a second-order tensor.

   Solution: Suppose that v_i represents the components of a vector $\boldsymbol{v}$ and consider the quantities generated by $T_{ij} = \partial v_i/\partial x_j$ $(i, j = 1, 2, 3)$. These nine quantities form the components of a second-order tensor, as can be seen from the fact that

   $$T'_{ij} = \frac{\partial v'_i}{\partial x'_j} = \frac{\partial(R_{ik}v_k)}{\partial x_\ell}\frac{\partial x_\ell}{\partial x'_j} = R_{ik}\frac{\partial v_k}{\partial x_\ell}R_{j\ell} = R_{ik}R_{j\ell}T_{k\ell}. \quad ♣$$

Remark. The concept (and its notation) $\nabla \boldsymbol{v}$ introduced above is not in the category of simple vector calculus. In fact, the quantity $\nabla \boldsymbol{v}$ is not a vector like $\nabla \times \boldsymbol{v}$ and $\nabla\phi$, but a second-order tensor.

18.3 Pseudotensors

18.3.1 Improper Rotations

So far our coordinate transformations have been restricted to rigid rotations described by an orthogonal matrix $[R_{ij}]$ with the property

$$|\mathsf{R}| \equiv \det[R_{ij}] = +1.$$

Such transformations are called **proper rotations**. We now broaden our discussion to include transformations that are described by an orthogonal matrix $[R_{ij}]$ for which

$$|\mathsf{R}| = -1.$$

The latter kind of transformations are called **improper rotations** (or **rotation with reflection**). Below are two examples of improper rotations.

(a) Inversion

The most obvious example of an improper rotation is an **inversion** of the coordinate axes through the origin represented by

$$\boldsymbol{e}'_i = -\boldsymbol{e}_i \quad \text{for all } i = 1, 2, 3.$$

In this case, a position arrow $\boldsymbol{x}$ is described in terms of the bases $\boldsymbol{e}'_i$ and $\boldsymbol{e}_i$, respectively, by

$$\boldsymbol{x} = x_i \boldsymbol{e}_i \quad \text{and} \quad \boldsymbol{x} = x'_i \boldsymbol{e}'_i = x'_i(-\boldsymbol{e}_i).$$

Equating them, we obtain

$$x_i = -x'_i = -\delta_{ij} x'_j,$$

which shows that an inversion of axes is expressed by $R_{ij} = -\delta_{ij}$. In fact, its determinant becomes

$$|\mathsf{R}| = \begin{vmatrix} -1 & 0 & 0 \\ 0 & -1 & 0 \\ 0 & 0 & -1 \end{vmatrix} = -1.$$

(b) Reflection

Another example is a **reflection** that reverses the direction of one basis:

$$\boldsymbol{e}'_i = -\boldsymbol{e}_i \quad \text{for a specified } i.$$

For the reflection of the x-axis, e.g., we have

$$|\mathsf{R}| = \begin{vmatrix} -1 & 0 & 0 \\ 0 & 1 & 0 \\ 0 & 0 & 1 \end{vmatrix} = -1.$$

Remark. Note that an inversion is different from a proper rotation only for an odd dimensionality. In a case of two or four dimensions, for instance, an inversion is the same as a proper rotation. In contrast, a reflection that changes the sign of only one coordinate is always different from a proper rotation regardless of the dimensionality.

Through an improper rotation, our initial **right-handed coordinate system** is changed into a **left-handed** one. This is illustrated schematically in Fig. 18.3. The reader should note that such a change cannot be accomplished by any kind of proper rotation.

18.3.2 Pseudovectors

Regardless of whether it is proper or improper, any rotation described by R_{ij} transforms the components v_i of a vector $\boldsymbol{v}$ as

$$v'_i = R_{ij} v_j.$$

This is because any real physical vector $\boldsymbol{v}$ may be considered as a geometrical object (i.e., an arrow in space), whose direction and magnitude cannot be altered merely by describing it in terms of a different coordinate system.

It is, however, possible to define another type of vector $\boldsymbol{w}$ whose components w_i transform as

$$\begin{aligned} w'_i &= R_{ij} w_j \quad \text{under proper rotations,} \\ w'_i &= -R_{ij} w_j \quad \text{under improper rotations,} \end{aligned}$$

or equivalently,

$$w'_i = |\mathsf{R}| R_{ij} w_j.$$

In this case, the w_i are no longer strictly the components of a true Cartesian vector. Rather, they are said to form the components of a **pseudovector** or a first-order Cartesian **pseudotensor**. A pseudovector may be alternatively referred to as an **axial vector**; correspondingly, a true vector may be called a **polar vector**.

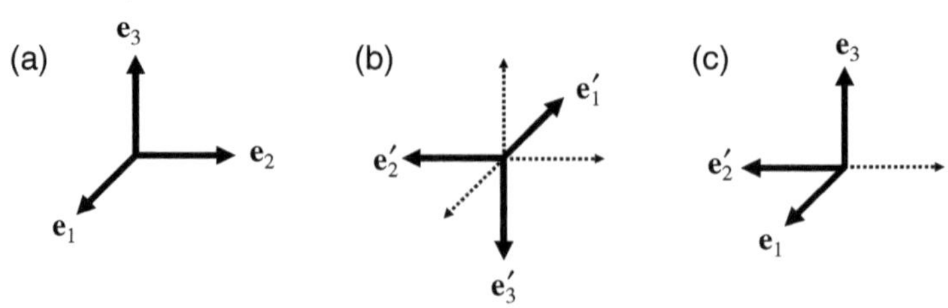

Fig. 18.3. Improper rotation: **(b)** inversion and **(c)** reflection of the right-handed Cartesian coordinate systsem depicted in **(a)**

Remark. A pseudovector should not be considered a real *geometric arrow* in space, since its direction is reversed by an improper transformation of the coordinate axes. This is illustrated in Fig. 18.4, where the pseudovector $\boldsymbol{w}$ is shown as a broken line to indicate that it is *not* a real physical vector.

Below is a summary of the discussion above.

♠ **Vectors and pseudovectors:**
Components v_i of a vector $\boldsymbol{v}$ transform as

$$v'_i = R_{ij} v_j$$

under a rigid rotation of Cartesian axes, whereas components w_i of a pseudovector $\boldsymbol{w}$ transform as

$$w'_i = |\mathsf{R}| R_{ij} w_j,$$

where $|\mathsf{R}|$ is the determinant of the transformation matrix $[R_{ij}]$.

Hence, the difference between a vector and a pseudovector manifests when applying an improper rotation that yields $|\mathsf{R}| = -1$.

Pseudovectors occur frequently in physics, although this fact is not usually pointed out explicitly. Following are physical examples of pseudovectors.

Examples The following three physical quantities are all pseudovectors.

1. Angular momentum of a moving particle, $\boldsymbol{L} = \boldsymbol{r} \times \boldsymbol{p}$, where $\boldsymbol{r}$ is the particle's position arrow and $\boldsymbol{p}$ its moment vector.
2. Torque on a particle, $\boldsymbol{N} = \boldsymbol{r} \times \boldsymbol{F}$, where $\boldsymbol{r}$ is the particle's position arrow and $\boldsymbol{F}$ the force acting on the particle.
3. Magnetic field, $\boldsymbol{B} = \boldsymbol{\nabla} \times \boldsymbol{A}$, defined by the rotation of the vector potential $\boldsymbol{A}$.

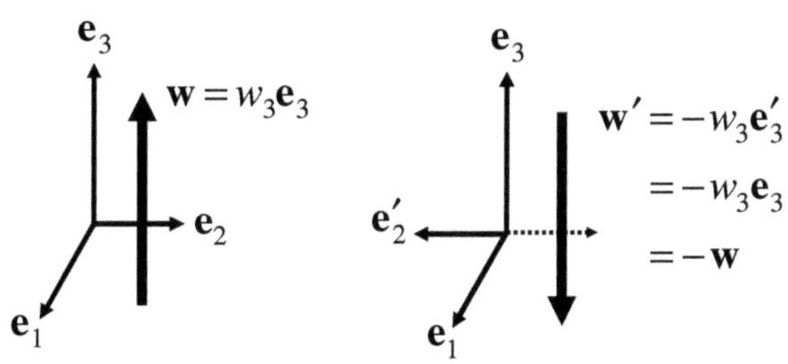

Fig. 18.4. Reversing behavior of the pseudovector $\boldsymbol{w}$ via the reflection of the $\boldsymbol{e}_2$-axis

It is noteworthy that each of these pseudovectors consists of a vector product of two vectors.

18.3.3 Pseudotensors

We can extend the notion of vectors and pseudovectors to objects with two or more subscripts. For instance, assume a quantity with components transforming as

$$T'_{ij} = R_{ik}R_{j\ell}T_{k\ell}$$

under proper rotations, but

$$T'_{ij} = -R_{ik}R_{j\ell}T_{k\ell}$$

under improper ones. Then, the T_{ij} are components of a **second-order Cartesian pseudotensor**. Similarly, Cartesian pseudotensors of arbitrary order are defined such that their components transform as

$$T'_{ij\cdots k} = |\mathsf{R}|R_{i\ell}R_{jm}\cdots R_{kn}T_{\ell m\cdots n}$$

where $|\mathsf{R}|$ is the determinant of the transformation matrix $[R_{ij}]$. Corresponding to these, zeroth-order objects may also be divided into scalars and pseudoscalars, the latter being invariant under rotation but changing sign on improper rotation.

18.3.4 Levi–Civita Symbols

A typical example of a third-order pseudotensor is the **Levi–Civita symbol** ε_{ijk}.

♠ **Levi–Civita symbol:**
The Levi–Civita symbol (or the **permutation symbol**), denoted by ε_{ijk}, takes the values $+1$ and -1 if the ordered set i, j, k is obtained by an even or odd permutation, respectively, of the set $1, 2, 3$.

Actually, ε_{ijk} takes the values

$$\varepsilon_{123} = \varepsilon_{231} = \varepsilon_{312} = +1,$$
$$\varepsilon_{213} = \varepsilon_{321} = \varepsilon_{132} = -1,$$

and $\varepsilon_{ijk} = 0$ if any two of the indices i, j, k are equal.

The pseudotensor property of ε_{ijk} follows from a convenient notation for the determinant $|\mathsf{A}|$ of a general 3×3 matrix $[A_{ij}]$ (see Exercise **2**):

$$|\mathsf{A}|\varepsilon_{lmn} = A_{i\ell}A_{jm}A_{kn}\varepsilon_{ijk}.$$

Certainly, this equation holds for the transformation matrix $[R_{ji}]$ for rigid rotation. Hence, we have

$$|\mathsf{R}|\varepsilon_{lmn} = R_{i\ell}R_{jm}R_{kn}\varepsilon_{ijk},$$

or equivalently,

$$\varepsilon_{ijk} = |\mathsf{R}|R_{i\ell}R_{jm}R_{kn}\varepsilon_{\ell mn}. \tag{18.25}$$

This shows that ε_{ijk} is a third-order Cartesian tensor.

The result (18.25) indicates more than the pseudotensorian character of ε_{ijk}. It clearly demonstrates that all of the components of ε_{ijk} are unaltered by any rotation of axes. Tensors endowed with this property are called **isotropic tensors** (**invariant tensors** or **fundamental tensors**). We know that there are no isotropic tensors of first order and that the only ones of second and third order are scalar multiples of δ_{ij} and ε_{ijk}, respectively. Additionally, the most general isotropic tensor of fourth order is given by

$$\lambda\delta_{ik}\delta_{mp} + \mu\delta_{im}\delta_{kp} + \nu\delta_{ip}\delta_{km},$$

with arbitrary constants λ, μ, ν. (Such a fourth-order isotropic tensor occurs in the **elasticity theory** of solids; see Sect. 18.5.4). All the isotropic tensors above are relevant to the description of the physical properties of an isotropic medium (i.e., a medium having the same properties regardless of the way in which it is orientated).

Exercises

1. Show that an angular momentum $\boldsymbol{L} = \boldsymbol{r} \times \boldsymbol{p}$ is a pseudovector.

Solution: Since the position vector $\boldsymbol{r}$ and the momentum vector $\boldsymbol{p}$ are vectors, they transform under certain rotations of the axes (proper and improper) as $r'_j = R_{jk}r_k$, $p'_m = R_{mn}p_n$. Hence, the components of $\boldsymbol{L}$ in a new coordinate system read

$$\begin{aligned}
L'_i &= \varepsilon'_{ijk}r'_jp'_k \\
&= (|\mathsf{R}|\; R_{i\ell}R_{jm}R_{kn}\;\varepsilon_{\ell mn})\,(R_{jq}r_q)\;(R_{ks}p_s) \\
&= |\mathsf{R}|\; R_{i\ell}\;(R_{jm}R_{jq})\;(R_{kn}R_{ks})\;\varepsilon_{\ell mn}r_qp_s \\
&= |\mathsf{R}|\; R_{i\ell}\;\delta_{mq}\delta_{ns}\;\varepsilon_{\ell mn}r_pp_q \\
&= |\mathsf{R}|\; R_{i\ell}\;\varepsilon_{\ell mn}r_mp_n \;\; = |\mathsf{R}|\; R_{i\ell}\;L_\ell,
\end{aligned}$$

which clearly indicates that the quantities L_i form the components of a first-order Cartesian pseudotensor (i.e., a pseudovector). ♣

2. Determine whether $|\mathsf{A}|\varepsilon_{\ell mn} = A_{i\ell}A_{jm}A_{kn}\varepsilon_{ijk}$ holds for a general matrix $[A_{ij}]$ in three dimensions.

Solution: Set $\ell = 1$, $m = 2$, $n = 3$, for instance, to find that the right-hand side reads

$$\begin{aligned} A_{i1}A_{j2}A_{k3}\varepsilon_{ijk} = \quad & A_{11}A_{22}A_{33} + A_{21}A_{32}A_{13} + A_{31}A_{12}A_{23} \\ & -A_{11}A_{32}A_{23} - A_{21}A_{12}A_{33} - A_{31}A_{22}A_{13} \quad = |\mathsf{A}|. \end{aligned}$$

Other cases can be proven in the same manner. ♣

3. Derive the identity: $\varepsilon_{ijk}\varepsilon_{klm} = \delta_{il}\delta_{jm} - \delta_{im}\delta_{jl}$.

Solution: We first note that the right-hand side of the above identity, $\delta_{il}\delta_{jm} - \delta_{im}\delta_{jl}$, reads

$$+1 \qquad \text{if } i = l \text{ and } j = m \neq i, \tag{18.26}$$

$$-1 \qquad \text{if } i = m \text{ and } j = l \neq i, \tag{18.27}$$

$$0 \qquad \text{otherwise.}$$

In the case of (18.26), the left-hand side of the desired identity is

$$\varepsilon_{ijk}\varepsilon_{klm} = \varepsilon_{ijk}\varepsilon_{kij} = (\varepsilon_{ijk})^2. \tag{18.28}$$

Since $i \neq j$, (18.28) takes the value $+1$ when $k \neq i$ and $k \neq j$. As a result, we successfully obtain the desired identity. A similar procedure reveals that $\varepsilon_{ijk}\varepsilon_{klm} = -1$ in the case of (18.27) and 0 otherwise. ♣

Remark. We should note that in (18.28), we have not summed with respect to i and j. This is because the second term in (18.28) was obtained by a substitution of particular values into the subscripts l and m, respectively.

18.4 Tensor Algebra

18.4.1 Addition and Subtraction

We demonstrate below the bases of tensor algebra that provide ways of constructing new tensors from old ones. For convenience, we may simply refer to T_{ij} as the tensor, but it should always be remembered that the T_{ij} are the components of $\boldsymbol{T}$ in a specific coordinate system.

The addition and subtraction of tensors are defined in an obvious fashion. If $A_{ij\cdots k}$ and $B_{ij\cdots k}$ are (the components of) tensors of the same order, then their sum and differences, $S_{ij\cdots k}$ and $D_{ij\cdots k}$ respectively, are given by

$$S_{ij\cdots k} = A_{ij\cdots k} + B_{ij\cdots k},$$

$$D_{ij\cdots k} = A_{ij\cdots k} - B_{ij\cdots k},$$

for each set of values $i, j, \cdots k$. Furthermore, the linearity of a rotation of coordinates immediately yields

$$\begin{aligned} R_{ip} S_{pq\cdots r} &= R_{ip} \left(A_{pq\cdots r} + B_{pq\cdots r} \right) = R_{ip} A_{pq\cdots r} + R_{pi} B_{pq\cdots r} \\ &= A'_{iq\cdots r} + B'_{iq\cdots r} = S'_{iq\cdots r}. \end{aligned}$$

18.4.2 Contraction

Next is an operation peculiar to tensor algebra that is of considerable importance in certain manipulations.

♠ **Contraction:**
Contraction is an operation that makes two of the indices equal and sums over all values of the equalized indices.

As an example, we consider a third-order tensor T_{ijk} whose transformation law is described by

$$T'_{ijk} = R_{i\ell} R_{jm} R_{kn} T_{\ell mn}. \tag{18.29}$$

Now we perform a contraction of this tensor with respect to j and k. Setting $j = k$ in (18.29) and summing over k, we get

$$T'_{ikk} = R_{i\ell} R_{km} R_{kn} T_{\ell mn} = R_{i\ell} \delta_{mn} T_{\ell mn} = R_{i\ell} T_{\ell nn},$$

where we used the orthogonality condition on the sum $R_{km} R_{kn}$. The result indicates that the quantity T_{ikk} forms the components of a tensor of order $1 = 3 - 2$. In general, contraction reduces the order of a tensor by two; contraction of an Nth-order tensor $T_{ij\cdots l\cdots m\cdots k}$ by making the subscripts l and m equal produces another tensor of order $N - 2$. In particular, if contraction is applied to a tensor of order 2, the result is a scalar.

18.4.3 Outer and Inner Products

Let us consider the multiplication of tensors. For example, we may take two tensors A_{ij} and $B_{k\ell m}$ of different order and simply write them in juxtaposition:

$$C_{ijk\ell m} \equiv A_{ij} B_{k\ell m}. \tag{18.30}$$

Then, the quantities are the components of a tensor of fifth-order, which follows immediately from the transformation law of tensors. Such a product of (18.30), in which all the indices are different from one another, is called an **outer product** of tensors.

Another kind of tensor, product of known as the **inner product** of tensors, is obtained from the outer product by contraction. For instance, putting $j = k$ in (18.30) results in

$$C_{ijj\ell m} \equiv A_{ij} B_{j\ell m}, \tag{18.31}$$

which consists of a third-order tensor as demonstrated in Sect. 18.4.2. Then, the right-hand side of (18.31) is called an inner product of the components of the tensors A_{ij} and $B_{k\ell m}$.

Examples The process of taking the scalar product of two vectors $\boldsymbol{u}$ and $\boldsymbol{v}$, expressed by $u_i v_i$, can be recast into tensor language as forming the outer product

$$T_{ij} \equiv u_i v_j$$

and then contracting it to give

$$T_{ii} = u_i v_i.$$

Using the concept of outer (and inner) product of tensors, we can write many familiar expressions of vector algebra as contracted tensors. For example, the vector product $\boldsymbol{a} = \boldsymbol{b} \times \boldsymbol{c}$ has

$$a_i = \varepsilon_{ijk} b_j c_k,$$

as its ith component, where ε_{ijk} is the Levi–Civita symbol introduced in Sect. 18.3.4. This notation clarifies the distinction between the pseudovector consisting of the components $\varepsilon_{ijk} b_j c_k$ and the second-order tensor composed of the outer product $b_i c_j$.

Remark. The outer product of two vectors is often denoted without reference to any coordinate system as

$$\boldsymbol{T} = \boldsymbol{u} \otimes \boldsymbol{v}. \tag{18.32}$$

This should not be confused with the vector product of two vectors, which is itself a pseudovector and is discussed in Sect. 18.3.2. The expression (18.32) gives the basis to which the components T_{ij} of the second-order tensor refer: since $\boldsymbol{u} = u_i \boldsymbol{e}_i$ and $\boldsymbol{v} = v_j \boldsymbol{e}_j$, we may write the tensor $\boldsymbol{T}$ as

$$\boldsymbol{T} = u_i \boldsymbol{e}_i \otimes v_j \boldsymbol{e}_j = u_i v_j \boldsymbol{u} \otimes \boldsymbol{v} = T_{ij} \boldsymbol{u} \otimes \boldsymbol{v}.$$

Furthermore, we have

$$\boldsymbol{T} = u_i \boldsymbol{e}_i \otimes v_j \boldsymbol{e}_j = u'_i \boldsymbol{e}'_i \otimes v'_j \boldsymbol{e}'_j,$$

which indicates that the quantities T'_{ij} are the components of the same tensor $\boldsymbol{T}$ but referred to a different coordinate system. These concepts can be extended to higher-order tensors.

We show below several expressions of vector algebra as contracted Cartesian tensors: the notation $[\boldsymbol{a}]_i$ indicates that one takes the ith component of the vector (or tensor) $\boldsymbol{a}$.

Examples

1. $\boldsymbol{a} \cdot \boldsymbol{b} = a_i b_i = \delta_{ij} a_i b_j$.

2. $[\boldsymbol{a} \cdot (\boldsymbol{b} \times \mathbf{c})]_i = \delta_{il} a_i [\boldsymbol{b} \times \mathbf{c}]_l = \delta_{il} a_i (\varepsilon_{ljk} b_j c_k) = \varepsilon_{ijk} a_i b_j c_k$.

3. $\nabla^2 \phi = \dfrac{\partial^2 \phi}{\partial x_i \partial x_i} = \delta_{ij} \dfrac{\partial^2 \phi}{\partial x_i \partial x_j}$.

4. $[\nabla \times \boldsymbol{v}]_i = \varepsilon_{ijk} \dfrac{\partial v_k}{\partial x_j}$.

5. $[\nabla(\nabla \cdot \boldsymbol{v})]_i = \dfrac{\partial}{\partial x_i} (\nabla \cdot \boldsymbol{v}) = \dfrac{\partial}{\partial x_i} \left(\dfrac{\partial v_j}{\partial x_j} \right) = \delta_{jk} \dfrac{\partial^2 v_j}{\partial x_i \partial x_k}$.

6. $[\nabla \times (\nabla \times \boldsymbol{v})]_i = \varepsilon_{ijk} \dfrac{\partial}{\partial x_j} [\nabla \times \boldsymbol{v}]_k = \varepsilon_{ijk} \varepsilon_{klm} \dfrac{\partial^2 v_m}{\partial x_j \partial x_l}$.

18.4.4 Symmetric and Antisymmetric Tensors

The order of subscripts attached to a tensor is important; in general, T_{ij} is not the same as T_{ji}. But there are some cases of interest as described below.

♠ **Symmetric and asymmetric tensor:**
If

$$T_{ij} = T_{ji},$$

holds for all i and j, the tensor composed of T_{ij} is called a **symmetric tensor**. Otherwise, if

$$T_{ij} = -T_{ji}, \tag{18.33}$$

the tensor is said to be **antisymmetric** (or **skew-symmetric**).

A tensor that is symmetric (or antisymmetric) in one coordinate system remains symmetric (or antisymmetric) in any other coordinate system. In fact, if T_{ij} is symmetric in a given system, i.e., $T_{ij} = T_{ji}$, then

$$T'_{ij} = R_{ik} R_{j\ell} T_{k\ell} = R_{j\ell} R_{ik} T_{\ell k} = T_{ji},$$

and similarly for antisymmetry and tensors of higher order.

Notably, every tensor can be resolved into symmetric and antisymmetric parts by the identity

$$T_{ij} = S_{ij} + A_{ij}, \tag{18.34}$$

where

$$S_{ij} = \frac{1}{2}(T_{ij} + T_{ji}) \text{ and } A_{ij} = \frac{1}{2}(T_{ij} - T_{ji}).$$

Evidently S_{ij} is a symmetric tensor since it is unaltered even if i and j are interchanged. In contrast, A_{ij} is an antisymmetric tensor since the signs of all the components are reversed by exchanging i and j. Then, S_{ij} and A_{ij} are called the **symmetric** and **antisymmetric parts** of T_{ij}, respectively.

18.4.5 Equivalence of an Antisymmetric Second-Order Tensor to a Pseudovector

It is noteworthy that in three dimensions, a second-order **antisymmetric tensor $\boldsymbol{W}$** is associated with a pseudovector $\boldsymbol{w}$. To see this, let the W_{ij} be components of an antisymmetric second-order tensor whose the transformation law reads

$$\begin{aligned} W'_{ij} &= R_{i\ell}R_{jm}W_{\ell m} \\ &= R_{i1}R_{j2}W_{12} + R_{i1}R_{j3}W_{13} + R_{i2}R_{j1}W_{21} + R_{i2}R_{j3}W_{23} \\ &\quad + R_{i3}R_{j1}W_{31} + R_{i3}R_{j2}W_{32}, \end{aligned} \tag{18.35}$$

since $W_{11} = W_{22} = W_{33} = 0$. Moreover, since $W_{\ell m} = -W_{m\ell}$, we can reduce (18.35) to the form

$$W'_{ij} = \sum_{(\ell,m)} (R_{i\ell}R_{jm} - R_{im}R_{j\ell})W_{\ell m}, \tag{18.36}$$

where the sum $\sum_{(\ell,m)}$ restricts the values of (ℓ, m) to $(1,2)$, $(2,3)$, or $(3,1)$.

Now we introduce the notation

$$w_1 \equiv W_{23} = W_{32}, \quad w_2 \equiv W_{31} = W_{13}, \quad w_3 \equiv W_{12} = W_{21},$$

or more concisely,

$$w_n \equiv W_{\ell m},$$

where ℓ, m, n is a cyclic permutation of the numbers $1, 2, 3$, i.e.,

$$(\ell, m, n) = (1,2,3), (2,3,1), (3,1,2).$$

Then (18.36) can be written as

$$w'_k = \sum_{(\ell,m,n)} (R_{i\ell}R_{jm} - R_{im}R_{j\ell})w_n, \tag{18.37}$$

in which i, j, k and ℓ, m, n are both cyclic permutations of $1, 2, 3$.

Noteworthy is the fact that (18.37) is equivalent to the transformation law of components w_k of a pseudovector $\boldsymbol{w}$. After some algebra, we see that equation (18.37) can be reduced to a more compact form as

$$w'_k = |\mathsf{R}| R_{kn} w_n, \tag{18.38}$$

which is nothing but the transformation law of a pseudovector. [See Exercise **2** for the proof of (18.38).]

We have now arrived at the following theorem:

♠ **Theorem:**

Assume a second-order antisymmetric tensor in three dimensions, whose components W_{ij} take the form

$$[W_{ij}] = \begin{pmatrix} 0 & W_{12} & -W_{31} \\ -W_{12} & 0 & W_{23} \\ W_{31} & -W_{23} & 0 \end{pmatrix}.$$

Then, the three components, W_{12}, W_{31}, and W_{23} can be associated with the pseudovector $\boldsymbol{w}$ whose components are given by

$$(w_1, w_2, w_3) = (W_{23}, W_{31}, W_{12}),$$

or more concisely,

$$w_i = \frac{1}{2}\varepsilon_{ijk} W_{jk}. \tag{18.39}$$

The right-hand side of (18.39) is a twice-contracted product of the third-order pseudotensor, ε^i_{jk}, and second-order tensor, W_{ij}; hence, it is a pseudovector.

Examples In physical applications, we often use the vector representation (18.39) of a second-order antisymmetric tensor. For instance, let us consider the equations of angular momentum of a moving particle with mass m. We assume that a force $\boldsymbol{F}$ acts on the particle located at $\boldsymbol{x}$. Then, with i and j each taking the values 1, 2, 3 we get

$$m(\ddot{x}_j x_k - \ddot{x}_k x_j) = F_j x_k - F_k x_j, \tag{18.40}$$

which gives us nine equations. Note that both sides of (18.40) are antisymmetric tensors. Among the nine equations, therefore, there are only three that are independent, $(j,k) = (1,2), (2,3), (3,1)$. So we can convert (18.40) into a more concise vector form as

$$mw_i = N_i,$$

where we have defined

$$w_i = \varepsilon_{ijk}(\ddot{x}_j x_k - \ddot{x}_k x_j) \quad \text{and} \quad N_i = \varepsilon_{ijk}(f_j x_k - f_k x_j).$$

18.4.6 Quotient Theorem

Sometimes it is necessary to clarify whether a set of functions, say, $\{a_i(x_j)\}$, forms the components of a vector or not. A direct method is to examine whether the functions satisfy a required transformation law under a rotation of axes, which is, however, troublesome in practice. In this subsection, we describe an alternative and more efficient method, called the **quotient law**, which is a simple indirect test for determining whether a given set of quantities forms the components of a tensor.

♠ **Quotient theorem:**
If $a_i v_i$ is a scalar for a vector $\boldsymbol{v}$ in any rotated coordinate system, then the a_i constitute the components of a vector $\boldsymbol{a}$.

Proof Suppose that we are given a set of n quantities a_i subject to the condition that $a_i v_i$ is a scalar for components v_i of arbitrary vector $\boldsymbol{v}$ in terms of an *arbitrarily rotated* coordinate system. We may then write

$$a_j v_j = \phi, \tag{18.41}$$

in which ϕ denotes a scalar. Denoting the (as yet unknown) transform of a_i by a'_j, we know that in the x'-coordinate system the condition (18.41) reads

$$a'_i v'_i = \phi'. \tag{18.42}$$

Since ϕ is a scalar, $\phi = \phi'$. Furthermore, since v_i are components of a vector, it follows that

$$v'_i = R_{ij} v_j.$$

Accordingly, subtracting (18.42) from (18.41) gives

$$(a_j - a'_i R_{ij})\, v_j = 0. \tag{18.43}$$

On the left-hand side, a summation over j is implied, so, we cannot assert directly that the coefficients of v_j vanish. However, since (18.43) should be valid for any coordinate system, we may specifically choose the coordinate system in which the components of $\boldsymbol{v}$ read $v_1 = 1$ and $v_{(i\neq 1)} = 0$. Equation (18.43) then reduces to

$$a_1 - R_{i1} a'_i = 0.$$

Similarly, choosing an appropriately rotated coordinate system that provides the components $v_2 = 1$ and $v_{(i\neq 2)} = 0$, we infer that

$$a_2 - R_{i2} a'_i = 0.$$

Continuing in this manner, we find that

$$a_j = R_{ij} a'_i \quad \text{for all } j.$$

Multiplying both sides by R_{kj} yields

$$R_{kj} a_j = R_{kj} R_{ij} a'_i = \delta_{ki} a'_i = a'_k,$$

i.e.,

$$a'_i = R_{ij} a_j,$$

which is the transformation law for components of a vector. We thus conclude that the a_i constitute the components of a vector, denoted by $\boldsymbol{a}$. ♣

Remark. In applications of the above theorem, one must be certain that the coordinate system employed is arbitrarily rotated, and this hypothesis represents a very strict condition that is not often satisfied.

18.4.7 Quotient Theorem for Two-Subscripted Quantities

As a second important case, assume a set of n^2 quantities a_{ij} such that $a_{ij} v_i v_j$ is a scalar ϕ for a vector $\boldsymbol{v}$ and for any rotated coordinate system. Our task is to examine whether such two-subscripted quantities a_{ij} constitute the components of a tensor of second order. We shall see, however, that the answer is negative. In fact, we can say nothing about the tensorian character of a_{ij} from the hypothesis noted above, which implies the need to modify the quotient theorem for two-subscripted quantities.

Developing the modified quotient theorem requires a discussion that parallels that given in Sect. 18.4.5. By hypothesis, we can set

$$a_{ij} v_i v_j = \phi$$

in the given x-coordinate system and similarly

$$a'_{k\ell} v'_k v'_\ell = \phi' \tag{18.44}$$

in the x'-coordinate system. In (18.44), we have denoted the as yet unknown transforms of a_{ij} by a'_{ij}. Using the transformation law of v_i as well as the fact that $\phi = \phi'$ gives us

$$(a_{ij} - R_{ki} R_{\ell j} a'_{k\ell})\, v_i v_j = 0. \tag{18.45}$$

As a summation is implied over i and j, we cannot infer directly that the coefficients of $v_i v_j$ vanish. Instead, we successively choose components $(v_1, v_2, v_3, \cdots)$ as $(1, 0, 0, \cdots)$ and $(0, 1, 0, \cdots)$, etc., to get

$$a_{11} - R_{k1} R_{\ell 1} a'_{k\ell} = 0, \quad a_{22} - R_{k2} R_{\ell 2} a'_{k\ell} = 0, \cdots . \tag{18.46}$$

These results imply that the terms a_{ij} with $i = j$ obey the transformation law of second-order tensors. Nevertheless, it tells us nothing about the terms involving a_{ij} with $i \neq j$. To further examine this point, we set components as $v_1 \neq 0$, $v_2 \neq 0$, and $v_i = 0$ for other i. Then, (18.45) becomes

$$(a_{11} - R_{k1}R_{\ell 1}a'_{k\ell})\, v_1 v_1 + (a_{12} - R_{k1}R_{\ell 2}a'_{k\ell})\, v_1 v_2$$
$$+ (a_{21} - R_{k2}R_{\ell 1}a'_{k\ell})\, v_2 v_1 + (a_{22} - R_{k2}R_{\ell 2}a'_{k\ell})\, v_2 v_2 = 0.$$

Owing to (18.46), we find that the coefficients of $v_1 v_1$ and $v_2 v_2$ vanish. Furthermore, since

$$R_{k1}R_{\ell 2}a'_{k\ell} = R_{k2}R_{\ell 1}a'_{\ell k}$$

is simply a relabeling of the indices k and ℓ, we see that

$$[(a_{12} + a_{21}) - (a'_{k\ell} + a'_{\ell k})\, R_{k2}R_{\ell 1}]\, v_1 v_2 = 0.$$

Thus, choosing $v_1 = 1$ and $v_2 = 1$ gives us

$$a_{12} + a_{21} = (a'_{k\ell} + a'_{\ell k})\, R_{k2}R_{\ell 1}.$$

Again, this process may be repeated to yield

$$a_{ij} + a_{ji} = (a'_{k\ell} + a'_{\ell k})\, R_{kj}R_{\ell i},$$

i.e.,

$$a'_{k\ell} + a'_{\ell k} = R_{kj}R_{\ell i}\, (a_{ij} + a_{ji})\,.$$

This is indeed the transformation law of a second-order tensor, but it refers to $a_{ij} + a_{ji}$, i.e., the symmetric part of $2a_{ij}$, and not to a_{ij} as such. Accordingly, the quotient theorem for this case must be stated as follows.

♠ **Quotient theorem for two-subscripted quantities:**

Suppose a set of n^2 quantities a_{ij} to be such that for a vector $\boldsymbol{v}$ and for any rotated system, the sum $a_{ij}v_i v_j$ is a scalar. Then the symmetric parts $(a_{ij} + a_{ji})/2$ of a_{ij} are the components of a second-order tensor.

Remark.

1. If in addition to the above hypothesis, we are given that the a_{ij} are symmetric, then the a_{ij} themselves are the components of a second-order tensor.
2. Nothing can be inferred about the tensorial character of the anti-symmetric part of a_{ij} from the above hypothesis that because part contributes nothing to the scalar ϕ, as seen from

$$(a_{ij} - a_{ji})v_i v_j = a_{ij}v_i v_j - a_{ji}v_i v_j = a_{ij}v_i v_j - a_{ij}v_j v_i = 0,$$

where in the last step the indices i and j are interchanged.

Example Using the quotient theorem, we show that the two-subscripted quantities a_{ij} given by

$$[a_{ij}] = \begin{bmatrix} (x_2)^2 & -x_1x_2 \\ -x_1x_2 & (x_1)^2 \end{bmatrix}$$

are the components of a second-order tensor. Note first that $a_{ij} = a_{ji}$ and that the outer product $x_k x_\ell$ is a second-order tensor. Contracting the quantities a_{ij} with the outer product $x_k x_\ell$, we obtain

$$a_{ij}x_ix_j = (x_2)^2(x_1)^2 - x_1x_2x_1x_2 - x_1x_2x_2x_1 + (x_1)^2(x_2)^2 = 0, \qquad (18.47)$$

in which the last term, 0, is a zeroth-order tensor. Since (18.47) holds for any rotated coordinate system, we conclude that a_{ij} is a second-order tensor.

Exercises

1. Derive the equation $\nabla \times (\nabla \times \boldsymbol{v}) = \nabla(\nabla \cdot \boldsymbol{v}) - \nabla^2 \boldsymbol{v}$.

   Solution: Straightforward calculations yield

$$\begin{aligned}[\nabla \times (\nabla \times \boldsymbol{v})]_i &= \varepsilon_{ijk}\varepsilon_{klm}\frac{\partial^2 v_m}{\partial x_j \partial x_l} = (\delta_{il}\delta_{jm} - \delta_{im}\delta_{jl})\frac{\partial^2 v_m}{\partial x_j \partial x_l} \\ &= \frac{\partial^2 v_j}{\partial x_j \partial x_i} - \frac{\partial^2 v_i}{\partial x_j \partial x_j} = \frac{\partial}{\partial x_i}\left(\frac{\partial v_j}{\partial x_j}\right) - \frac{\partial^2 v_i}{\partial x_j \partial x_j} \\ &= \frac{\partial}{\partial x_i}(\nabla \cdot \boldsymbol{v}) - \nabla^2 v_i = [\nabla(\nabla \cdot \boldsymbol{v})]_i - [\nabla^2 \boldsymbol{v}]_i \\ &= [\nabla(\nabla \cdot \boldsymbol{v}) - \nabla^2 \boldsymbol{v}]_i. \quad \clubsuit\end{aligned}$$

2. Derive the expression (18.38) using the result (18.37).

   Solution: We consider the vector products (in the sense of elementary vector calculus) of the transformed basis arrows $\boldsymbol{e}'_i$ given by $\boldsymbol{e}'_i \times \boldsymbol{e}'_j = (R_{i\ell}\boldsymbol{e}_\ell) \times (R_{jm}\boldsymbol{e}_m) = R_{i\ell}R_{jm}\boldsymbol{e}_\ell \times \boldsymbol{e}_m$. Forming the scalar product with $\boldsymbol{e}_n$ yields

$$(\boldsymbol{e}'_i \times \boldsymbol{e}'_j) \cdot \boldsymbol{e}_n = R_{i\ell}R_{jm}\,(\boldsymbol{e}_\ell \times \boldsymbol{e}_m) \cdot \boldsymbol{e}_n,$$

   where on the right-hand side only two terms survive for each fixed value of n since

$$(\boldsymbol{e}_\ell \times \boldsymbol{e}_m) \cdot \boldsymbol{e}_n = \begin{cases} +1 & \text{if } (\ell, m, n) = (1,2,3), (2,3,1), (3,1,2), \\ -1 & \text{if } (\ell, m, n) = (2,1,3), (3,2,1), (1,3,2), \\ 0 & \text{otherwise.} \end{cases}$$

   (Here we assume that the coordinate systems associated with $\{\boldsymbol{e}_i\}$ and $\{\boldsymbol{e}'_j\}$ are both right-handed.) Hence, we have

$$(\boldsymbol{e'}_i \times \boldsymbol{e'}_j) \cdot \boldsymbol{e}_n = R_{i\ell}R_{jm} - R_{im}R_{j\ell}, \tag{18.48}$$

where ℓ, m, n is a cyclic permutation of $1, 2, 3$. Moreover, since

$$\boldsymbol{e'}_i \times \boldsymbol{e'}_j = \boldsymbol{e'}_k = R_{kr}\boldsymbol{e}_r, \tag{18.49}$$

it follows from (18.48) and (18.49) that

$$R_{kr}\boldsymbol{e}_r \cdot \boldsymbol{e}_n = R_{kr}\delta_{rn} = R_{kn} = R_{i\ell}R_{jm} - R_{im}R_{j\ell},$$

where again i, j, k and ℓ, m, n are both cyclic permutations of $1, 2, 3$. If $\{\boldsymbol{e'}_\ell\}$ is left-handed, a similar procedure yields

$$R_{kn} = -\left(R_{i\ell}R_{jm} - R_{im}R_{j\ell}\right).$$

Substituting these results into (18.37), we finally arrive at the conclusion that

$$w'_k = |\mathsf{R}|R_{kn}w_n,$$

which is a transformation law for a pseudovector. ♣

3. Show that the process of contraction of an Nth-order tensor produces another tensor of order $N-2$.

Solution: Let $T_{ij\cdots l\cdots m\cdots k}$ be the components of an Nth-order tensor; then

$$T'_{ij\cdots l\cdots m\cdots k} = R_{ip}R_{jq}\cdots R_{lr}\cdots R_{ms}\cdots R_{kn}T_{pq\cdots r\cdots s\cdots n}.$$

Thus if, e.g., we make the two subscripts l and m equal and sum over all the values of these subscripts, we obtain

$$\begin{aligned} T'_{ij\cdots l\cdots m\cdots k} &= R_{ip}R_{jq}\cdots R_{lr}\cdots R_{ms}\cdots R_{kn}T_{pq\cdots r\cdots s\cdots n} \\ &= R_{ip}R_{jq}\cdots \delta_{rs}\cdots R_{kn}T_{pq\cdots r\cdots s\cdots n} \\ &= R_{ip}R_{jq}\cdots R_{kn}T_{pq\cdots r\cdots r\cdots n}, \end{aligned}$$

showing that $T_{ij\cdots l\cdots l\cdots k}$ are the components of a (different) Cartesian tensor of order $N-2$. ♣

18.5 Applications in Physics and Engineering

This section is devoted to illustrations of physical applications of second- and higher-order Cartesian tensors. We start with an example from mechanics and follow that by examples from electromagnetism and elasticity.

18.5.1 Inertia Tensor

Consider a collection of rigidly connected particles, wherein the αth particle has mass $m^{(\alpha)}$ and is positioned at $\boldsymbol{r}^{(\alpha)}$ with respect to the origin O. Suppose that the rigid assembly is rotating about an axis through O with **angular velocity** ω. The **angular momentum** $\boldsymbol{J}$ of the assembly is given by

$$J = \sum_\alpha \left(\boldsymbol{r}^{(\alpha)} \times \boldsymbol{p}^{(\alpha)} \right).$$

Here $\boldsymbol{p}^{(\alpha)} = m^{(\alpha)} \dot{\boldsymbol{r}}^{(\alpha)}$ and $\dot{\boldsymbol{r}}^{(\alpha)} = \boldsymbol{\omega} \times \boldsymbol{r}^{(\alpha)}$ for any α whose components are expressed in subscript form as

$$p_k^{(\alpha)} = m^{(\alpha)} \dot{x}_k^{(\alpha)} \quad \text{and} \quad \dot{x}_k^{(\alpha)} = \varepsilon_{klm} \omega_l x_m^{(\alpha)}.$$

Thus we obtain

$$\begin{aligned} J_i &= \sum_\alpha \sum_{j,k} \varepsilon_{ijk} x_j^{(\alpha)} \dot{p}_k^{(\alpha)} = \sum_\alpha \sum_{j,k,l,m} m^{(\alpha)} \varepsilon_{ijk} x_j^{(\alpha)} \varepsilon_{klm} \omega_l x_m^{(\alpha)} \\ &= \sum_\alpha \sum_{j,l} m^{(\alpha)} (\delta_{il}\delta_{jm} - \delta_{im}\delta_{jl}) x_j^{(\alpha)} x_m^{(\alpha)} \omega_l \\ &= \sum_\alpha \sum_l m^{(\alpha)} \left[\left(r^{(\alpha)} \right)^2 \delta_{il} - x_i^{(\alpha)} x_l^{(\alpha)} \right] \omega_l \equiv \sum_l I_{il} \omega_l, \end{aligned} \tag{18.50}$$

with the definition

$$I_{il} = \sum_\alpha m^{(\alpha)} \left[\left(r^{(\alpha)} \right)^2 \delta_{il} - x_i^{(\alpha)} x_l^{(\alpha)} \right]. \tag{18.51}$$

The set of quantities I_{il} forms a symmetric second-order Cartesian tensor; the symmetric property expressed by $I_{il} = I_{li}$ follows readily from (18.51). The fact that the I_{il} form tensors can be proved by applying the quotient rule (see Sect. 18.4.6) to equation (18.50), wherein J_i and ω_l are vectors. The tensor I_{il} is called the **inertia tensor** of the assembly with respect to O. As evident from (18.51), I_{il} depends only on the distribution of mass in the assembly and not on the direction or magnitude of the angular velocity of the assembly, $\boldsymbol{\omega}$.

If a continuous rigid body is considered, $m^{(\alpha)}$ is replaced by the mass distribution $\rho(\boldsymbol{r})$ and the summation $\sum_\alpha$ by the integral of $\int dV$ over the volume of the whole body. When expanded in Cartesian coordinates, the inertia tensor of a continuous body would have the form

$$\mathsf{I} = [I_{ij}] = \begin{pmatrix} \int (y^2+z^2)\rho dV & -\int xy\rho dV & -\int zx\rho dV \\ -\int xy\rho dV & \int (z^2+x^2)\rho dV & -\int yz\rho dV \\ -\int zx\rho dV & -\int yz\rho dV & -\int (x^2+y^2)\rho dV \end{pmatrix}.$$

The diagonal elements of this tensor are called the **moments of inertia** and the off-diagonal elements without the negative signs are known as the **products of inertia**.

It is possible to show that the **kinetic energy** K of the rotating system is given by $K = \frac{1}{2} I_{jl} \omega_j \omega_l$, which is a scalar obtained by twice contracting the vector ω_j with the inertia tensor I_{jl}. In fact, an argument parallel to that leading to (18.50) yields

$$K = \frac{1}{2}\sum_{\alpha} m^{(\alpha)}\dot{\boldsymbol{r}}^{(\alpha)} \cdot \dot{\boldsymbol{r}}^{(\alpha)} = \frac{1}{2}\sum_{\alpha} m^{(\alpha)} \sum_{j,k,l,m} \varepsilon_{ijk}\omega_j x_k^{(\alpha)} \varepsilon_{ilm}\omega_l x_m^{(\alpha)}$$
$$= \frac{1}{2}\sum_{j,l} I_{jl}\omega_j\omega_l.$$

This shows that the kinetic energy of the rotating body can be expressed as a scalar obtained by twice contracting the vector ω_j with the inertia tensor I_{jl}. Alternatively, since $J_j = I_{jl}\omega_l$, the kinetic energy may be written as $K = \frac{1}{2}J_j\omega_j$.

18.5.2 Tensors in Electromagnetism in Solids

Magnetic susceptibility and **electric conductivity** are also examples of physical quantities represented by second-order tensors. For the former, we have the standard expression

$$M_i = \sum_j \chi_{ij} H_j, \tag{18.52}$$

where $\boldsymbol{M}$ is the magnetic moment per unit volume and $\boldsymbol{H}$ is the magnetic field. Similarly, for the case of electric conductivity, we can write

$$j_i = \sum_j \sigma_{ij} E_j. \tag{18.53}$$

Here, the current density $\boldsymbol{j}$ (current per unit perpendicular area) is related to the electric field $\boldsymbol{E}$. In both cases, we have a vector on the left-hand side and the contraction of a second-order tensor with another vector on the right-hand side.

For isotropic media, the vector $\boldsymbol{M}$ is parallel to $\boldsymbol{H}$ and, similarly, the vector $\boldsymbol{j}$ is parallel to $\boldsymbol{E}$. Thus, the above tensors satisfy $\chi_{ij} = \chi\delta_{ij}$ and $\sigma_{ij} = \sigma\delta_{ij}$, respectively, resulting in $\boldsymbol{M} = \chi\boldsymbol{H}$ and $\boldsymbol{j} = \sigma\boldsymbol{E}$. However, for anistropic materials such as crystals, the magnetic susceptibility and electric conductivity may be different along different crystal axes, thus making χ_{ij} and σ_{ij} general second-order tensors (usually symmetric).

18.5.3 Electromagnetic Field Tensor

All the tensors that we have considered in this chapter so far relate to the three dimensions of space and they are defined as having a certain transformation property under spatial rotations. In this subsection, we shall have the occasion to use a tensor in the four dimensions of relativistic space-time; the tensor is the electromagnetic field tensor $F_{\mu\nu}$.

Recall that an **electromagnetic field** in free space is governed by the **Maxwell equations**, which take the form

$$\nabla \cdot \boldsymbol{B} = 0, \qquad \nabla \cdot \boldsymbol{E} = 4\pi k_1 \rho,$$
$$\nabla \times \boldsymbol{B} = 4\pi k_2 \boldsymbol{J} + \frac{k_2}{k_1}\frac{\partial \boldsymbol{E}}{\partial t}, \qquad \nabla \times \boldsymbol{E} = -k_3 \frac{\partial \boldsymbol{B}}{\partial t}.$$

Here $\boldsymbol{E}$ is the electric field intensity, $\boldsymbol{B}$ is the magnetic induction, ρ is the charge density, and $\boldsymbol{J}$ is the current density. There are several ways of defining the values of constants k_i ($i = 1, 2, 3$); indeed, their values depends on which system of unit we use. Typical examples are listed in Table 18.1.

The Maxwell equations take on a particularly simple and elegant form on introducing the **electromagnetic field tensor** $F_{\mu\nu}$ defined as

$$F_{\mu\nu} = \partial_\nu A_\mu - \partial_\mu A_\nu. \tag{18.54}$$

Here, $A_\mu = (\phi/c, -\boldsymbol{A})$ is called a **four potential**, determined by the scalar potential ϕ and the vector potential $\boldsymbol{A}$ that generate the fields $\boldsymbol{B} = \nabla \times \boldsymbol{A}$ and $\boldsymbol{E} = -\nabla\phi - \partial \boldsymbol{A}/\partial t$. The symbol ∂_μ in (18.54) denotes the partial derivatives with respect to the μth coordinate. Straightforward calculations yield

$$[F_{\mu\nu}] = \begin{bmatrix} 0 & E^1/c & E^2/c & E^3/c \\ -E^1/c & 0 & -B^3 & B^2 \\ -E^2/c & B^3 & 0 & -B^1 \\ -E^3/c & -B^2 & B^1 & 0 \end{bmatrix}, \tag{18.55}$$

where $\boldsymbol{E} = (E^1, E^2, E^3)$ and $\boldsymbol{B} = (B^1, B^2, B^3)$. We also introduce another relevant tensor defined by

$$[F^{\mu\nu}] = \begin{bmatrix} 0 & -E^1/c & -E^2/c & -E^3/c \\ E^1/c & 0 & -B^3 & B^2 \\ E^2/c & B^3 & 0 & -B^1 \\ E^3/c & -B^2 & B^1 & 0 \end{bmatrix}, \tag{18.56}$$

in which μ and ν are *superscripts* in opposed to (18.55), where they are *subscripts*. As a result, we can see that the Maxwell equations are equivalent

Table 18.1. Values of the constants $k_i (i = 1, 2, 3)$ in the Maxwell equations. μ_0, ε_0 and c are the permeability, permittivity, and speed of light in vacuum, respectively

System of Unit	k_1	k_2	k_3
MKSA	$1/(4\pi\varepsilon_0)$	$\mu_0/(4\pi)$	1
CGS-esu	1	$1/c^2$	1
CGS-emu	c^2	1	1
CGS-Gauss	1	$1/c$	$1/c$

to the following two field equations:

$$\sum_{\nu} \partial_\nu F^{\mu\nu} = \mu_0 j^\mu,$$

$$\partial_\sigma F_{\mu\nu} + \partial_\mu F_{\nu\sigma} + \partial_\nu F_{\sigma\mu} = 0,$$

where $j^\mu = (\rho c, \boldsymbol{J})$ is the **four-current density**.

Remark. The distinction between superscripts and subscripts on the symbol F shown in (18.55) and (18.56), respectively, is clarified in Chap. 19, which deals with non-Cartesian tensor calculus.

18.5.4 Elastic Tensor

Thus so far, we have focused on the physical applications of second-order tensors, which relate two vectors. Now, we extend this idea to a situation where a fourth-order tensor relates two physical second-order tensors. Such relationships commonly occur in **elasticity theory**. In the framework of this theory, the local deformation of an elastic body at any interior point P is described by a second-order symmetric tensor e_{ij} called the **strain tensor**, which is given by

$$e_{ij} = \frac{1}{2}\left(\frac{\partial u_i}{\partial x_j} + \frac{\partial u_j}{\partial x_i}\right),$$

where $\boldsymbol{u}$ is the **displacement vector** describing the strain of a small volume element. Similarly, we can describe the stress in the body at P by a second-order symmetric **stress tensor** p_{ij}; the quantity p_{ij} is the x_j-component of the stress vector acting across a plane through P, whose normal lies in the x_i-direction. A generalization of **Hooke's law** that relates the stress and strain tensors is

$$p_{ij} = \sum_{k,l} c_{ijkl} e_{kl}, \tag{18.57}$$

where c_{ijkl} is a fourth-order Cartesian tensor.

Specifically, for an isotropic medium, we must have an isotropic tensor for c_{ijkl}; the most general fourth-order isotropic tensor is

$$c_{ijkl} = \lambda \delta_{ij}\delta_{kl} + \eta \delta_{ik}\delta_{jl} + \nu \delta_{il}\delta_{jk}.$$

Substituting this into (18.57) yields

$$p_{ij} = \lambda \delta_{ij} \sum_k e_{kk} + \eta e_{ij} + \nu e_{ji}. \tag{18.58}$$

Note that e_{ij} is symmetric. Hence, if we write $\eta + \nu = 2\mu$, (18.58) takes the conventional form

$$p_{ij} = \lambda \sum_k e_{kk}\delta_{ij} + 2\mu e_{ij},$$

in which λ and μ are known as **Lamé constants**.

19

Non-Cartesian Tensors

Abstract Having discussed tensor theory based on Cartesian coordinates, we now move on to its counterpart, i.e., tensors described by curvilinear coordinate systems. The use of a curvilinear coordinate system endows the tensor calculus with the properties of "covariance" (Sect. 19.1.3) and "contravariance" (Sect. 19.1.4), both of which are new concepts originating from the nonorthogonality of the coordinate axes.

19.1 Curvilinear Coordinate Systems

19.1.1 Local Basis Vectors

We have thus far restricted our attention to the study of Cartesian tensors, where, from a practical stand point, only rigid rotations of axes (proper and/or improper) are taken into account as coordinate tranformations. However, we must free ourselves from this restriction and develop the tensor calculus in terms of **curvilinear coordinate systems**, In advanced mathematical physics, we often have to deal with tensor analysis on curved surfaces (or more abstract manifolds) on which orthonormal coordinate systems cannot be defined, and in such cases the theory developed thus far is entirely inadequate. This means that we have to formulate tensors and their transformations in terms of general curvilinear coordinate systems.

To begin with, we review some properties of general curvilinear coordinates. Suppose that the position of an arbitrary point P in a three-dimensional space has Cartesian coordinates x, y, z. In general, this position may be expressed in terms of three curvilinear coordinates u_1, u_2, u_3, which are functions of x, y, z as explicitly represented by

$$\begin{aligned} u_1 &= u_1(x, y, z), \\ u_2 &= u_2(x, y, z), \\ u_3 &= u_3(x, y, z). \end{aligned}$$

We denote by $\boldsymbol{r}$ the position arrow connecting the origin O and the point P. Obviously, the direction and magnitude of the arrow depend on the coordinates of P, which are symbolized by

$$\boldsymbol{r} = \boldsymbol{r}(u_1, u_2, u_3).$$

We now consider the partial derivative of $\boldsymbol{r}$ with respect to u_i, i.e.,

$$\boldsymbol{e}_i \equiv \frac{\partial \boldsymbol{r}}{\partial u_i}. \tag{19.1}$$

From the definition, the vectors $\boldsymbol{e}_i$ are directed along the corresponding coordinate lines at the point P. As a result, an infinitesimal vector displacement $d\boldsymbol{r}$ in curvilinear coordinates is given by

$$d\boldsymbol{r} = \frac{\partial \boldsymbol{r}}{\partial u_i} du_i = \boldsymbol{e}_i du_i,$$

where the summation convention is employed. The vectors $\boldsymbol{e}_i$ are referred to as **local basis vectors**. (In precise terminology, they are called **covariant local basis vectors**, as explained later.)

It is obvious from (19.1) that the vectors $\boldsymbol{e}_i$ are functions of the curvilinear coordinates u_i, namely, $\boldsymbol{e}_i = \boldsymbol{e}_i(u_1, u_2, u_3)$. This implies that the directions and magnitudes of the $\boldsymbol{e}_i$ vary from point to point in the space considered, which is in contrast to the case of a Cartesian coordinate system, where the basis vectors are spatially independent. Spatial dependence of basis vectors is actually one of the most important properties of curvilinear coordinate systems.

Another notable property of curvilinear coordinate systems is the fact that they allow us to define another useful set of three vectors at P as

$$\boldsymbol{\varepsilon}_i \equiv \nabla u_i.$$

Clearly the direction of $\boldsymbol{\varepsilon}_i$ is normal to the surface $u_i = \text{const}$; thus being different from the directions of any vectors $\boldsymbol{e}_i$ ($i = 1, 2, 3$) in general (see Fig. 19.1). Therefore, at each point P in a curvilinear coordinate system, there exist two sets of basis vectors defined by

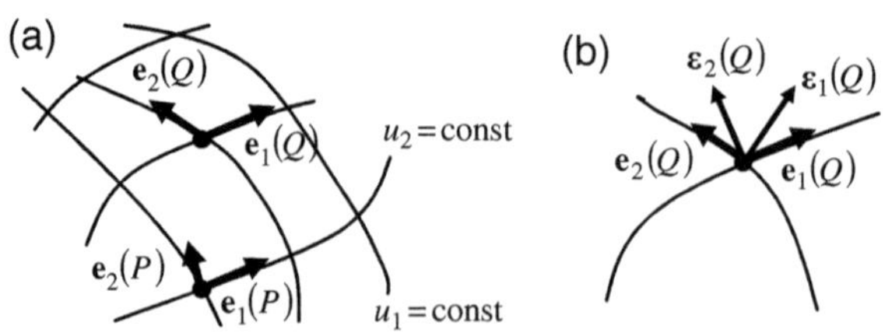

Fig. 19.1. (a) Spatial dependence of $\boldsymbol{e}_i$ in the curvilinear coordinate system. **(b)** Difference between $\boldsymbol{e}_i$ and $\boldsymbol{\varepsilon}_i$

$$e_i = \frac{\partial \boldsymbol{r}}{\partial u_i} \quad \text{and} \quad \boldsymbol{\varepsilon}_i = \nabla u_i. \tag{19.2}$$

In the tensor analysis, literature of the set of vectors $\boldsymbol{\varepsilon}_i$ introduced above is denoted by $\boldsymbol{e}^i$, the index being placed as a superscript to distinguish it from the first set of vectors $\boldsymbol{e}_i$. Relating to the notation above, we introduce a **modified summation convention** as follows: if we find a lower-case alphabetic index that appears twice, once as a subscript and once as a superscript, we sum over all the values that the index can take. In this convention, the curvilinear coordinates are denoted by u^1, u^2, u^3, with the index raised (see the remark in Sect. 19.1.3), to arrive at the following definition.

♠ **Local basis vectors:**
A curvilinear coordinate system is characterized by two sets of three vectors $\{\boldsymbol{e}_i\}$ and $\{\boldsymbol{e}^i\}$ defined by

$$\boldsymbol{e}_i = \frac{\partial \boldsymbol{r}}{\partial u^i} \quad \text{and} \quad \boldsymbol{e}^i = \nabla u^i.$$

Here, the $\boldsymbol{e}_i$ are referred to as the **covariant local basis vectors**, and the $\boldsymbol{e}^i$ as the **contravariant local basis vectors.**

The prefix "local" emphasizes the fact that the lengths and orientations of these basis vectors vary from point to point in the space; this fact is explicitly represented by

$$\boldsymbol{e}_i = \boldsymbol{e}_i(u^1, u^2, u^3) \quad \text{and} \quad \boldsymbol{e}^j = \boldsymbol{e}^j(u^1, u^2, u^3).$$

For the sake of conciseness, we omit the prefix in the subsequent discussions and use the terms **contravariant** (or **covariant**) **basis vectors**, bearing the locality in mind.

Remark.

1. In common practice indices that represent *contravariant* character are placed as superscripts and those indicating *covariant* character as subscripts.
2. For Cartesian coordinate systems, the two sets of basis vectors $\boldsymbol{e}_i$ and $\boldsymbol{e}^i$ are identical and, hence, there is no need to differentiate between contravariance and covariance.
3. In derivatives such as $\partial \boldsymbol{r}/\partial u^i$, the i is considered as a *sub*script.

19.1.2 Reciprocity Relations

Generally the covariant basis vectors $\boldsymbol{e}_1$, $\boldsymbol{e}_2$, and $\boldsymbol{e}_3$ are neither of unit length nor are they orthogonal to each other; this is also true for the contravariant basis vectors, $\boldsymbol{e}^1$, $\boldsymbol{e}^2$, and $\boldsymbol{e}^3$. Nevertheless, the sets $\{\boldsymbol{e}_i\}$ and $\{\boldsymbol{e}^j\}$ still have an important property as stated below.

♠ **Reciprocity relations:**

The sets of contravariant and covariant local basis vectors $\{\boldsymbol{e}_i\}$ and $\{\boldsymbol{e}_j\}$ satisfy the reciprocity relations such that

$$\boldsymbol{e}_i \cdot \boldsymbol{e}^j = \delta_i^j, \tag{19.3}$$

where the scalar product of the vectors is taken in the sense of elementary vector calculus.

Proof By using Cartesian representation, we have

$$\begin{aligned}
\boldsymbol{e}_i \cdot \boldsymbol{e}^j &= \frac{\partial \boldsymbol{r}}{\partial u^i} \cdot \nabla u^j = \left(\frac{\partial x}{\partial u^i}, \frac{\partial y}{\partial u^i}, \frac{\partial z}{\partial u^i} \right) \cdot \left(\frac{\partial u^j}{\partial x}, \frac{\partial u^j}{\partial y}, \frac{\partial u^j}{\partial z} \right) \\
&= \frac{\partial x}{\partial u^i}\frac{\partial u^j}{\partial x} + \frac{\partial y}{\partial u^i}\frac{\partial u^j}{\partial y} + \frac{\partial z}{\partial u^i}\frac{\partial u^j}{\partial z} \\
&= \frac{\partial u^j}{\partial u^i} = \delta_i^j. \quad ♣
\end{aligned}$$

Remark. The reciprocity relation (19.3) implies that each covariant (or contravariant) basis vector $\boldsymbol{e}_i$ (or $\boldsymbol{e}^i$) is perpendicular to all contravariant (or covariant) basis vectors $\boldsymbol{e}_k$ (or $\boldsymbol{e}^k$) except $k = i$. For instance, $\boldsymbol{e}_1$ is perpendicular to $\boldsymbol{e}^2$ and $\boldsymbol{e}^3$, but not to $\boldsymbol{e}^1$ in general. To be precise, the vectors $\boldsymbol{e}_1$ and $\boldsymbol{e}^1$ make an angle θ that satisfies

$$\boldsymbol{e}_1 \cdot \boldsymbol{e}^1 = |\boldsymbol{e}_1|\,|\boldsymbol{e}^1| \cos\theta = 1,$$

where $|\boldsymbol{e}_1| \neq 1$ and $|\boldsymbol{e}^1| \neq 1$.

19.1.3 Transformation Law of Covariant Basis Vectors

We are now in a position to discuss the concept of general transformations from one coordinate system, u^1, u^2, u^3, to another, u'^1, u'^2, u'^3. A coordinate transformation is described by using the three equations

$$u'^i = u'^i(u^1, u^2, u^3), \tag{19.4}$$

for $i = 1, 2, 3$, in which the new coordinates u'^i can be arbitrary functions of the old ones u^i. We assume that the transformation can be inverted, so that we can write the old coordinates in terms of the new ones as

$$u^i = u^i(u'^1, u'^2, u'^3).$$

We now formulate the transformation law of basis vectors. The two sets of basis vectors in the new coordinate system are given by

$$\boldsymbol{e}'_i = \frac{\partial \boldsymbol{r}}{\partial u'^i} \quad \text{and} \quad \boldsymbol{e}'^i = \nabla u'^i. \tag{19.5}$$

Using the chain rule, we find that the first set of basis vectors yields

$$\boldsymbol{e}'_i = \frac{\partial \boldsymbol{r}}{\partial u^j}\frac{\partial u^j}{\partial u'^i} = \frac{\partial u^j}{\partial u'^i}\boldsymbol{e}_j. \tag{19.6}$$

This describes the transformation behavior of the local covariant basis vectors from the unprimed one $\boldsymbol{e}_j$ to the primed one $\boldsymbol{e}'_i$ under the coordinate transformation (19.4). Note that the partial derivatives as well as the basis vectors in (19.6) vary from point to point. Hence, relation (19.6) is valid under the condition that all terms involved are evaluated at the same point P in the space being considered.

In the same manner, it follows that

$$\boldsymbol{e}_k = \frac{\partial \boldsymbol{r}}{\partial u^j} = \frac{\partial \boldsymbol{r}}{\partial u'^\ell}\frac{\partial u'^\ell}{\partial u^k} = \frac{\partial u'^\ell}{\partial u^k}\boldsymbol{e}'_\ell.$$

We thus have proved the following theorem:

♠ **Transformation law of covariant basis vectors:**

The sets of local covariant basis vectors $\{\boldsymbol{e}_i(u^j)\}$ and $\{\boldsymbol{e}'_k(u'^\ell)\}$ associated with two different curvilinear coordinate systems are related at a point P by

$$\boldsymbol{e}'_i = \frac{\partial u^j}{\partial u'^i}\boldsymbol{e}_j \quad \text{and} \quad \boldsymbol{e}_k = \frac{\partial u'^\ell}{\partial u^k}\boldsymbol{e}'_\ell, \tag{19.7}$$

in which the partial derivatives are to be evaluated at P.

Remark. Observe that in all the mathematical expressions above (and below), the summation convention is applied to the indices that are repeated in one term as both a subscript and a superscript. Indeed, it was to satisfy this summation convention that the coordinates were written as u^i rather than u_i.

19.1.4 Transformation Law of Contravariant Basis Vectors

Next we consider the transformation law of the contravariant basis vectors $\boldsymbol{e}^i = \nabla u^i$. Recall that in terms of a rectangular Cartesian coordinate system, the operator ∇ is expressed as

$$\nabla = \boldsymbol{i}\frac{\partial}{\partial x} + \boldsymbol{j}\frac{\partial}{\partial y} + \boldsymbol{k}\frac{\partial}{\partial z},$$

where $\boldsymbol{i}, \boldsymbol{j}, \boldsymbol{k}$ are mutually orthogonal basis vectors of unit length. It then, follows that

$$\boldsymbol{e}'^k = \nabla u'^k = \boldsymbol{i}\frac{\partial u'^k}{\partial x} + \boldsymbol{j}\frac{\partial u'^k}{\partial y} + \boldsymbol{k}\frac{\partial u'^k}{\partial z},$$

in which the first partial derivative reads

$$\frac{\partial u'^k}{\partial x} = \frac{\partial u^i}{\partial x}\frac{\partial u'^k}{\partial u^i}$$

and other derivatives are written in the same way. Hence, we have

$$\boldsymbol{e}'^k = \left(\boldsymbol{i}\frac{\partial u^i}{\partial x} + \boldsymbol{j}\frac{\partial u^i}{\partial y} + \boldsymbol{k}\frac{\partial u^i}{\partial z}\right)\frac{\partial u'^k}{\partial u^i} = \left(\nabla u^i\right)\frac{\partial u'^k}{\partial u^i} = \frac{\partial u'^k}{\partial u^i}\boldsymbol{e}^i.$$

Similarly, we have

$$\boldsymbol{e}^j = \nabla u^j = \left(\boldsymbol{i}\frac{\partial u'^\ell}{\partial x} + \boldsymbol{j}\frac{\partial u'^\ell}{\partial y} + \boldsymbol{k}\frac{\partial u'^\ell}{\partial z}\right) = \frac{\partial u^j}{\partial u'^\ell}\boldsymbol{e}'^\ell.$$

These results are summarized as follows:

♠ **Transformation law of contravariant basis vectors:**
The two sets of local contravariant basis vectors $\{\boldsymbol{e}^i(u^j)\}$ and $\{\boldsymbol{e}'^k(u'^\ell)\}$ are related at a point P by

$$\boldsymbol{e}'^k = \frac{\partial u'^k}{\partial u^i}\boldsymbol{e}^i \quad \text{and} \quad \boldsymbol{e}^\ell = \frac{\partial u^\ell}{\partial u'^j}\boldsymbol{e}'^j, \tag{19.8}$$

where the partial derivatives are again to be evaluated at P.

It should be emphasized again that, owing to the summation convention, the repeated indices in (19.8) appear once as a superscript and once as a subscript

19.1.5 Components of a Vector

Given the two bases $\boldsymbol{e}_i$ and $\boldsymbol{e}^i$, we may express a general geometric arrow $\boldsymbol{a}$ (i.e., a vector $\boldsymbol{a}$) equally well in terms of either basis as follows:

$$\boldsymbol{a} = a^1 \boldsymbol{e}_1 + a^2 \boldsymbol{e}_2 + a^3 \boldsymbol{e}_3 = a^i \boldsymbol{e}_i,$$

$$\boldsymbol{a} = a_1 \boldsymbol{e}^1 + a_2 \boldsymbol{e}^2 + a_3 \boldsymbol{e}^3 = a_i \boldsymbol{e}^i.$$

The a^i are called the **contravariant components** of the vector $\boldsymbol{a}$ and the a_i the **covariant components**. Both kinds of components a^i and a_j describe the same vector $\boldsymbol{a}$, but they are associated with different basis vectors $\boldsymbol{e}_i$ and $\boldsymbol{e}^j$, respectively. In plain words, a vector assigned at a point in a curvilinear coordinate system has two different expressions; say, (a^1, a^2, a^3) and (a_1, a_2, a_3) for the same vector $\boldsymbol{a}$. The tensorian characters of the two kinds of components are inherently different from each other, as we shall see in subsequent discussions.

For any vector $\boldsymbol{a}$, the two kinds of components a^i and a_i are readily obtained by forming the scalar products,

$$\boldsymbol{a} \cdot \boldsymbol{e}^i = a^j \boldsymbol{e}_j \cdot \boldsymbol{e}^i = a^j \delta^i_j = a^i$$

and

$$\boldsymbol{a} \cdot \boldsymbol{e}_i = a_j \boldsymbol{e}^j \cdot \boldsymbol{e}_i = a_j \delta^j_i = a_i,$$

where we have used the reciprocity relation (19.3). Furthermore, using the transformation law of $\boldsymbol{e}_i$ given in (19.7) gives us

$$\boldsymbol{a} = a'^i \boldsymbol{e}'_i = a^j \boldsymbol{e}_j = a^j \frac{\partial u'^i}{\partial u^j} \boldsymbol{e}'_i. \tag{19.9}$$

This provides the transformation law of the contravariant components of a vector such that

$$a'^i = \frac{\partial u'^i}{\partial u^j} a^j. \tag{19.10}$$

This relation is, in fact, the defining property for a set of quantities a^i to form the contravariant components of a vector. The formal statement is given below.

♠ **Contravariant component of a vector:**

Quantities a_i associated with a point P are said to be the contravariant components of a vector if these, quantities transform through the equation

$$a'^i = \frac{\partial u'^i}{\partial u^j} a^j, \tag{19.11}$$

where the partial derivatives are evaluated at P.

Remark. It might occur that a given ordered set of quantities a^k associated with a point P has nothing to do with a vector; only those sets satisfying the transformation law (19.11) serve as (contravariant) components of a vector.

Analogously to the case of (19.9), it follows from the identity for an arbitrary vector $\boldsymbol{a}$,

$$\boldsymbol{a} = a'_i \boldsymbol{e}'^i = a_j \boldsymbol{e}^j = a_j \frac{\partial u^j}{\partial u'^i} \boldsymbol{e}'^i,$$

that the transformation law of covariant components yields

$$a'_i = \frac{\partial u^j}{\partial u'^i} a_j. \tag{19.12}$$

Again we take this result as the defining property of the covariant components of a vector.

♠ **Covariant components of a vector:**
Quantities a_i associated with a point P are said to be the covariant components of a vector if those quantities transform through the equation

$$a'_k = \frac{\partial u^\ell}{\partial u'^k} a_\ell, \tag{19.13}$$

where the partial derivatives are evaluated at P.

Remark. Other textbooks may use the expression "**contravariant** (or **covariant**) **vector**," which is a distinctly different concept from a vector $\boldsymbol{a}$ or its components a^i (or a_i) that we have just defined. Say, rather, that a contravariant vector is a *collection* of ordered triples,

$$\left\{ \left(a^1, a^2, a^3\right), \left(a'^1, a'^2, a'^3\right), \left(a''^1, a''^2, a''^3\right), \cdots \right\},$$

in which all the ordered triples consist of contravariant components of the *same* vector $\boldsymbol{a}$ associated with different coordinate systems. We should make sure that a contravariant (or covariant) vector is not expressed by a geometric arrow as is done for a vector.

19.1.6 Components of a Tensor

We now define geometric objects of the contravariant class, which are more complicated in character than vectors and begin with the following:

♠ **Contravariant component of a tensor:**
Index quantities T_{jk} associated with a point P are said to be **contravariant components** of a tensor if these quantities transform according to the equation

$$T'^{jk} = T^{\ell m} \frac{\partial u'^j}{\partial u^\ell} \frac{\partial u'^k}{\partial u^m}. \tag{19.14}$$

There is no difficulty in defining covariant tensors of higher orders. For a tensor of second order, e.g., we have the definition below.

♠ **Covariant components of a tensor:**
Index quantities T_{jk} are said to be covariant components of a second-order tensor if these quantities transform according to the equation

$$T'_{jk} = T_{\ell m} \frac{\partial u^\ell}{\partial u'^j} \frac{\partial u^m}{\partial u'^k}. \tag{19.15}$$

We shall see later that there are many examples of tensors of this kind in physics and engineering. The moment of inertia, the stress of elasticity, and the electromagnetic field are cases in point; if their components in terms of certain coordinate systems are evaluated, they all turn out to obey the transformation law (19.14).

In terminology, all quantities satisfying (19.11), (19.13) and (19.14), (19.15) are called **components of a first-order tensor** and **components of a second-order tensor**, respectively; the order goes as the number of indices attached. The definitions of tensors of higher orders are given through a straightforward generalization of the above. Conversely, we can define a **tensor of zero order**, called a **scalar**, that involves no index so that its single component (i.e., the scalar itself) is constant under any coordinate transformation; namely,

$$T' = T.$$

Such a quantity is called an **invariant**.

Remark. For any components of tensors, the number of indices is independent of the number of dimensions of the space considered. The definitions above for vectors, tensors, and scalars are all valid for an arbitrary n-dimensional space.

19.1.7 Mixed Components of a Tensor

Having defined contravariant and covariant components of a tensor, we can now define another class of components, called **mixed components of a tensor**, that involve the two character simulteneously.

♠ **Mixed components of a tensor:**
Index quantities ${T^i}_{jk}$ are said to be the mixed components of a tensor of the third order if these quantities transform according to the equation

$$T'^i{}_{jk} = T^\ell{}_{mn} \frac{\partial u'^i}{\partial u^\ell} \frac{\partial u^m}{\partial u'^j} \frac{\partial u^n}{\partial u'^k}.$$

Clearly, ${T^i}_{jk}$ transforms contravariantly with respect to the first index i but covariantly with respect to the other indices j and k.

If we consider the components of higher-order tensors in non-Cartesian coordinates, there are even more possibilities. As an example, let us consider a second-order tensor $\boldsymbol{T}$. Using the outer product notation, we may write $\boldsymbol{T}$ in three different ways:

$$\boldsymbol{T} = T^{ij} \boldsymbol{e}_i \otimes \boldsymbol{e}_j = T^i_j \boldsymbol{e}^i \otimes \boldsymbol{e}_j = T_{ij} \boldsymbol{e}^i \otimes \boldsymbol{e}^j,$$

where T^{ij}, T^i_j, and T_{ij} are called the **contravariant**, **mixed**, and **covariant components** of $\boldsymbol{T}$, respectively. It is important to remember that these three sets of quantities form the components of the same tensor $\boldsymbol{T}$ but refer to different tensor bases made up from the basis vectors of the coordinate system. Again, if we use Cartesian coordinates, the three sets of components are identical.

We may generalize the above equation to higher-order components. An object $T^{\alpha\cdots\beta}{}_{\gamma\cdots\delta}$ is called a component of type (n, m) in which the integers n and m represent the numbers of superscripts and subscripts, respectively. By definition, components carrying only superscripts (i.e., $m = 0$) or those carrying only subscripts (i.e., $n = 0$) are referred to as the contravariant and covariant components, respectively; all others are called mixed components.

Remark. The order of indices needs caution. For instance, we shall see later that in general

$$T^i{}_j \neq T_j{}^i.$$

Nevertheless, we can write T^i_j with no clarification of the order of i and j if no ambiguity occurs or the order of indices is irrelevant.

19.1.8 Kronecker Delta

The **Kronecker delta** is a special kind of a second-order tensor that has mixed components given by δ^i_j, and is defined as follows:

$$\delta^i_j = \begin{cases} 1 & (i = j), \\ 0 & (i \neq j). \end{cases}$$

As these are mixed components of a tensor, they transform as

$$\delta'^i_j = \frac{\partial u'^i}{\partial u^\ell}\frac{\partial u^m}{\partial u'^j}\delta^\ell_m = \frac{\partial u'^i}{\partial u^\ell}\frac{\partial u^\ell}{\partial u'^j} = \frac{\partial u'^i}{\partial u'^j} = \begin{cases} 1, & \text{if } i = j, \\ 0, & \text{if } i \neq j, \end{cases}$$

since in the last partial derivative, u'^i and u'^j are independent coordinates. Thus, we obtain the result

$$\delta'^i_j = \delta^i_j, \tag{19.16}$$

which means that the tensor consisting of δ^i_j has the same components in all coordinate systems. This is why the tensor consisting of δ^i_j is called the **fundamental mixed tensor**.

Remark. The components δ^{ij} (or δ_{ij}) are of no special importance, since they do not satisfy the invariance condition (19.16), which means that their values change when we use other coordinate systems. A exception is the case of rectangular coordinate systems, where the contravariant and covariant tensors become identical, so that we have $\delta^{ij} = \delta^i_j = \delta_{ij}$.

19.2 Metric Tensor

19.2.1 Definition

We now introduce important quantities that describe the geometric character of the space arithmetized by a certain curvilinear coordinate system. We know that the scalar product of a vector $\boldsymbol{a}$ and local basis vectors $\boldsymbol{e}_i$ and $\boldsymbol{e}^j$ yields

$$a_j = \boldsymbol{a} \cdot \boldsymbol{e}_j = a^i (\boldsymbol{e}_i \cdot \boldsymbol{e}_j) \quad \text{and} \quad a^j = \boldsymbol{a} \cdot \boldsymbol{e}^j = a_i \left(\boldsymbol{e}^i \cdot \boldsymbol{e}^j\right). \tag{19.17}$$

Now we introduce the following notation:

$$\boldsymbol{e}_i \cdot \boldsymbol{e}_j = g_{ij} = g_{ji}$$

and

$$\boldsymbol{e}^i \cdot \boldsymbol{e}^j = g^{ij} = g^{ji}.$$

We can then write (19.17) in the form

$$a_j = g_{jk} a^k \quad \text{and} \quad a^j = g^{jk} a_k.$$

These equations express the covariant components of the vector $\boldsymbol{a}$ in terms of its contravariant components, and vice versa. We shall see that the nine quantities g_{ik} form a second-order tensor called a **metric tensor**.

♠ **Metric tensor:**
Two-index quantities defined by

$$g_{ij} = \boldsymbol{e}_i \cdot \boldsymbol{e}_j \quad \text{and} \quad g^{k\ell} = \boldsymbol{e}^k \cdot \boldsymbol{e}^\ell \tag{19.18}$$

serve as covariant and contravariant components of a second-order tensor called a **metric tensor.**

The proof of the tensor character for the above is given in Exercise **1**.

Remark.

1. Since both $\boldsymbol{e}_i$ and $\boldsymbol{e}^j$ are functions of the coordinates, so are the quantities g_{ij} and g^{ij}.

2. The mixed components g^i_j of the metric tensor are identical to those of δ^i_j since, by definition, we have

$$g^i_j = \boldsymbol{e}^i \cdot \boldsymbol{e}_j = \delta^i_j.$$

Examples We calculate the elements g_{ij} for cylindrical coordinates, where $(u^1, u^2, u^3) = (\rho, \phi, z)$ and ρ and ϕ are related to Cartesian coordinates x and y as $x = \rho \cos\phi$ and $y = \rho \sin\phi$. Hence, the position vector $\boldsymbol{r}$ of any point may be written as

$$\boldsymbol{r} = \rho\cos\phi \boldsymbol{i} + \rho\sin\phi \boldsymbol{j} + z\boldsymbol{k},$$

where $\boldsymbol{i}, \boldsymbol{j}, \boldsymbol{k}$ are orthogonal basis vectors. By definition, we have

$$\boldsymbol{e}_i = \frac{\partial \boldsymbol{r}}{\partial \rho} = \cos\phi\boldsymbol{i} + \sin\phi\boldsymbol{j}, \tag{19.19}$$

$$\boldsymbol{e}_j = \frac{\partial \boldsymbol{r}}{\partial \phi} = -\rho\sin\phi\boldsymbol{i} + \rho\cos\phi\boldsymbol{j}, \tag{19.20}$$

$$\boldsymbol{e}_k = \frac{\partial \boldsymbol{r}}{\partial z} = \boldsymbol{k}. \tag{19.21}$$

Thus the components of the metric tensor $[g_{ij}] = [\boldsymbol{e}_i \cdot \boldsymbol{e}_j]$ are found to be

$$[g_{ij}] = \begin{pmatrix} 1 & 0 & 0 \\ 0 & \rho^2 & 0 \\ 0 & 0 & 1 \end{pmatrix}.$$

19.2.2 Geometric Role of Metric Tensors

The quantities g_{ik} (or g^{ik}) describe the fundamental geometric character of a space arithmetrized by a certain u^i-coordinate system with a **basis** $\{\boldsymbol{e}_i\}$.

A geometric role for g_{ij} was implied in the definition (19.18), where g_{ij} equals the scalar product of the two covariant local basis vectors $\boldsymbol{e}_i$ and $\boldsymbol{e}_j$. Hence, g_{ij} determines the angles of local basis vectors $\boldsymbol{e}_i$ and $\boldsymbol{e}_j$ at each point and thus describes the coordinate(u^k)-dependence of the vectors $\boldsymbol{e}_i = \boldsymbol{e}_i(u^k)$ and $\boldsymbol{e}_j = \boldsymbol{e}_j(u^k)$ that span the space being considered. This implies the possibility that the metric tensor $\boldsymbol{g}$ rather than the basis vectors can be regarded as a more fundamental object determining the geometric nature of the space in question. Indeed, we can establish the framework of tensor calculus based on a knowledge of the spatial dependence of the metric tensor $\boldsymbol{g}$ without any information about the local basis vectors. This point is dealt with in §20.3.5 Sect. 20.3.5.

The role of g_{ij} in determining the geometric nature of the space also follows from another stand point as shown below. Let ds be the arc length between two infinitely close points. We denote by $d\boldsymbol{r}$ the vector joining the two points, whose covariant components are du_i and contravariant components du^i. Then, since $d\boldsymbol{r} = \boldsymbol{e}_i du^i = \boldsymbol{e}^k du_k$, we have

$$
\begin{aligned}
(ds)^2 &= |d\boldsymbol{r}|^2 = d\boldsymbol{r} \cdot d\boldsymbol{r} \\
&= \boldsymbol{e}_i du^i \cdot \boldsymbol{e}_k du^k = \boldsymbol{e}_i du^i \cdot \boldsymbol{e}^k du_k = \boldsymbol{e}^i du_i \cdot \boldsymbol{e}^k du_k,
\end{aligned}
$$

or

$$(ds)^2 = g_{ik} du^i du^k, \tag{19.22}$$

$$(ds)^2 = g^{ik} du_i du_k, \tag{19.23}$$

$$(ds)^2 = du_i du^i. \tag{19.24}$$

Since $(ds)^2$ is a scalar, all of the quantities on the right-hand sides are also scalars. It should also be noted that in (19.22) and (19.23), the du^i (or du_k) are contravariant (or covariant) components of a vector. Hence, in view of the quotient theorem regarding two-index quantities (see Sect 18.4.7), it turns out that the symmetric quantities g_{ik} (or g^{ik}) form covariant (or contravariant) components of a second-order tensor.

19.2.3 Riemann Space and Metric Tensor

We have seen that in terms of tensor calculus, the metric tensor $\boldsymbol{g}$ rather than the local basis vectors $\boldsymbol{e}_i$ and $\boldsymbol{e}^j$ is a more fundamental object in determining geometric properties of the space being considered. In fact, an abstract space of points to which we assign a certain class of a second-order tensor $\boldsymbol{g}$ at each point is referred to by a special name as stated below, which gives a formal definition of the metric tensor $\boldsymbol{g}$ in the language of tensor calculus.

♠ **Riemann space:**

A finite-dimensional space of points labeled by an ordered set of real coordinates $u^1, u^2, \cdots, u^n$ is called a **Riemann space** if it is possible to define two-index quantities g_{ij} that possess the following properties:

1. Each entity $g_{ij}(u^1, u^2, \cdots, u^n)$ is a real single-valued function of the coordinates and has continuous partial derivatives.

2. $g_{ik}(u^1, u^2, \cdots, u^n) = g_{ki}(u^1, u^2, \cdots, u^n)$.

3. $g = \det[g_{ik}] \neq 0$.

The tensor $\boldsymbol{g}$ formed by the two-index quantities g_{ij} noted above is called a **metric tensor** of the space.

Remark. Note that the above definition of a metric tensor is free of the concept of local basis vectors.

In this context, the superscripted components g^{ij} are defined by

$$g^{ik} g_{kj} = \delta^i_j \quad \text{or} \quad g^{ik} = \frac{C^{ik}}{g},$$

where $C^{ik}(= C^{ki})$ is the cofactor of g_{ik} in the determinant $g = \det[g_{ik}]$. (See Exercise **2** for the proof of the above.)

Our familiar Euclidean space is a particular class of Riemann space as stated below.

♠ **Flat Riemann space:**

A Riemann space is **flat** if and only if it admits a system of rectangular Cartesian coordinates $x^1, x^2, \cdots, x^n$ such that at every point of the space,

$$(ds)^2 = \varepsilon_1 \left(dx^1\right)^2 + \varepsilon_2 \left(dx^2\right)^2 + \cdots + \varepsilon_n \left(dx^n\right)^2, \tag{19.25}$$

where each ε_i equals either $+1$ or -1.

♠ **Euclidean space:**

A **Euclidean space** is a flat Riemann space for which all ε_i in (19.25) are equal to $+1$.

19.2.4 Elements of Arc, Area, and Volume

Below we describe several useful relations in connection with the elements of arc length, areas, and volumes in terms of metric tensors.

1. **Element of arc length:**
The element of arc length ds_i along a particular coordinate curve u^i with fixed i is

$$ds_i = |d\boldsymbol{r}| = |\boldsymbol{e}_i|du^i = \sqrt{\boldsymbol{e}_i \cdot \boldsymbol{e}_i}du^i = \sqrt{g_{ii}}du^i \quad \text{(no summation over } i\text{)}.$$

2. **Element of area**
The element of area $d\sigma_1$ in the coordinate surface $u^1 = \text{const}$; for instance, reads

$$\begin{aligned} d\sigma_1 &= |d\boldsymbol{r}_2 \times d\boldsymbol{r}_3| = |\boldsymbol{e}_2 \times \boldsymbol{e}_3|\, du^2 du^3 \\ &= \sqrt{(\boldsymbol{e}_2 \times \boldsymbol{e}_3) \cdot (\boldsymbol{e}_2 \times \boldsymbol{e}_3)}\, du^2 du^3 \\ &= \sqrt{(\boldsymbol{e}_2 \cdot \boldsymbol{e}_2)(\boldsymbol{e}_3 \cdot \boldsymbol{e}_3) - (\boldsymbol{e}_2 \cdot \boldsymbol{e}_3)(\boldsymbol{e}_2 \cdot \boldsymbol{e}_3)}\, du^2 du^3 \\ &= \sqrt{g_{22}g_{33} \quad (g_{23})^2}\, du^2 du^3. \end{aligned}$$

Similarly, we have

$$d\sigma_2 = \sqrt{g_{33}g_{11} - (g_{13})^2}\, du^3 du^1,$$

$$d\sigma_3 = \sqrt{g_{11}g_{22} - (g_{12})^2}\, du^1 du^2,$$

which are summarized by

$$d\sigma_i = \sqrt{g_{jj}g_{kk} - (g_{jk})^2}\, du^j du^k \quad \text{(no summation over } j \text{ and } k\text{)},$$

where i, j, k is a cyclic permutation of the numbers $1, 2, 3$.

3. **Element of volume**
Finally, we can derive the equation for the element of volume as

$$\begin{aligned} dV &= |(d\boldsymbol{r}_1 \times d\boldsymbol{r}_2) \cdot d\boldsymbol{r}_3| = |(\boldsymbol{e}_1 \times \boldsymbol{e}_2) \cdot \boldsymbol{e}_3|\, du^1 du^2 du^3 \\ &= \sqrt{g} du^1 du^2 du^3, \end{aligned}$$

where $g = \det[g_{ik}]$. [Proof of the identity $(\boldsymbol{e}_1 \times \boldsymbol{e}_2) \cdot \boldsymbol{e}_3 = g$ is given in Exercise **2**.]

Our results are summarized as:

♠ **Theorem:**
Elements of arc length ds_i, area $d\sigma_i$, and volume dV, respectively, are represented in terms of curvilinear coordinate systems by

$$ds_i = \sqrt{g_{ii}}du^i \quad \text{(no sum over } i\text{)},$$

$$d\sigma_i = \sqrt{g_{jj}g_{kk} - (g_{jk})^2}\, du^j du^k \quad \text{(no sum over } j \text{ and } k\text{), and}$$

$$dV = \sqrt{g}\, du^1 du^2 du^3,$$

where i, j, k is a cyclic permutation of $1, 2, 3$.

19.2.5 Scale Factors

In this subsection, we consider the case of orthogonal coordinate systems, for which the basic descriptive quantities are the **scale factors** (or the **metric coefficients**) h_1, h_2, h_3, defined by

$$h_1 = \sqrt{g_{11}}, \quad h_2 = \sqrt{g_{22}}, \quad h_3 = \sqrt{g_{33}}.$$

Obviously, they satisfy the equation

$$(ds)^2 = \left(h_1 du^1\right)^2 + \left(h_2 du^2\right)^2 + \left(h_3 du^3\right)^2.$$

Furthermore, since $g_{ij} = 0$ for $i \neq j$, we have

$$ds_i = h_i du^i \quad \text{(no sum over } i\text{)},$$

$$d\sigma_i = h_j h_k du^j du^k \quad \text{(no sum over } j \text{ and } k\text{)},$$

$$dV = h_1 h_2 h_3 du^1 du^2 du^3,$$

where i, j, k is a cyclic permutation of $1, 2, 3$.

Examples **1.** In rectangular Cartesian coordinates,

$$(ds)^2 = (dx)^2 + (dy)^2 + (dz)^2,$$

so

$$h_1 = h_2 = h_3 = 1.$$

2. In cylindrical coordinates,

$$(ds)^2 = (dR)^2 + (Rd\theta)^2 + (dz)^2,$$

so

$$h_1 = 1, \quad h_2 = R, \quad h_3 = 1.$$

3. In spherical coordinates,

$$(ds)^2 = (dR)^2 + (Rd\theta)^2 + (R\sin\theta d\phi)^2,$$

so

$$h_1 = 1, \quad h_2 = R, \quad h_3 = R\sin\theta.$$

19.2.6 Representation of Basis Vectors in Derivatives

It is often desirable to represent local covariant basis vectors $\boldsymbol{e}_i$ as well as components of metric tensors $g_{ij} = \boldsymbol{e}_i \cdot \boldsymbol{e}_j$ at a point $\boldsymbol{r}$ in terms of derivatives of $\boldsymbol{r}$ with respect to coordinates the u^i.

Suppose the relation between a system of curvilinear coordinates u^1, u^2, u^3 and an underlying system of rectangular coordinates $x_1, x_2, x_3 (= x, y, z)$ is given by

$$u^i = u^i(x_k) \quad \text{and} \quad x_k = x_k(u^i), \tag{19.26}$$

where the Jacobian

$$J = \left| \frac{\partial u^i}{\partial x_k} \right|$$

is neither zero nor infinite. Writing the latter equation in (19.26) more concisely as

$$\boldsymbol{r} = \boldsymbol{r}(u^i),$$

where $\boldsymbol{r} = x_k \boldsymbol{i}_k$ is the position arrow of an arbitrary point, we find

$$d\boldsymbol{r} = \frac{\partial \boldsymbol{r}}{\partial u^i} du^i.$$

It then follows that

$$(ds)^2 = d\boldsymbol{r} \cdot d\boldsymbol{r} = \frac{\partial \boldsymbol{r}}{\partial u^i} \cdot \frac{\partial \boldsymbol{r}}{\partial u^j} du^i du^j,$$

which implies that the vectors of the local basis are

$$\boldsymbol{e}_i = \frac{\partial \boldsymbol{r}}{\partial u^i}$$

and the metric tensor is

$$g_{ij} = \frac{\partial \boldsymbol{r}}{\partial u^i} \cdot \frac{\partial \boldsymbol{r}}{\partial u^j} = \frac{\partial x_k}{\partial u^i} \boldsymbol{i}_k \cdot \frac{\partial x_\ell}{\partial u^j} \boldsymbol{i}_\ell = \frac{\partial x_k}{\partial u^i} \frac{\partial x_\ell}{\partial u^j} (\boldsymbol{i}_k \cdot \boldsymbol{i}_\ell) = \frac{\partial x_k}{\partial u^i} \frac{\partial x_k}{\partial u^j}.$$

This leads to the following expression for the scale factors (for the case of orthogonal coordinate systems):

$$h_i = \sqrt{\left(\frac{\partial x_1}{\partial u^i} \right)^2 + \left(\frac{\partial x_2}{\partial u^i} \right)^2 + \left(\frac{\partial x_3}{\partial u^i} \right)^2}.$$

19.2.7 Index Lowering and Raising

In curvilinear coordinate systems, it is possible to express a scalar product of two vectors via several different subscript forms. For instance, the scalar product of two vectors $\boldsymbol{a}$ and $\boldsymbol{b}$ may be written using their contravariant or covariant components:

$$\boldsymbol{a} \cdot \boldsymbol{b} = a^i \boldsymbol{e}_i \cdot b^j \boldsymbol{e}_j = g_{ij} a^i b^j \tag{19.27}$$

and

$$\boldsymbol{a} \cdot \boldsymbol{b} = a_i \boldsymbol{e}^i \cdot b_j \boldsymbol{e}^j = g^{ij} a_i b_j. \tag{19.28}$$

Furthermore, we may express the scalar product in terms of the contravariant components of one vector and the covariant components of the other:

$$\boldsymbol{a} \cdot \boldsymbol{b} = a_i \boldsymbol{e}^i \cdot b^j \boldsymbol{e}_j = a_i b^j \delta^i_j = a_i b^i \tag{19.29}$$

and

$$\boldsymbol{a} \cdot \boldsymbol{b} = a^i \boldsymbol{e}_i \cdot b_j \boldsymbol{e}^j = a^i b_j \delta^j_i = a^i b_i. \tag{19.30}$$

By comparing the four alternative expressions (19.27)–(19.30) for $\boldsymbol{a} \cdot \boldsymbol{b}$, we can deduce the following useful property of g_{ij} and g^{ij}. From (19.27) and (19.30) we see that the identity

$$g_{ij} a^i b^j = a^i b_i$$

holds for any arbitrary a^i. Hence, we have

$$g_{ij} b^j = b_i. \tag{19.31}$$

which illustrates the fact that the covariant components g_{ij} can be used to lower an index b^j. In other words, it provides a means of obtaining the covariant components b_i of a vector from its contravariant components b^j. By a similar argument, we have

$$g^{ij} b_j = b^i,$$

where the contravariant components g^{ij} are used to raise the index j attached to b_j.

♠ **Index lowering and raising (I):**

For any vector $\boldsymbol{a}$, its components a^i and a_i are related via the components of the metric tensor as

$$a_i = g_{ik} a^k \quad \text{and} \quad a^j = g^{j\ell} a_\ell.$$

The above discussion regarding vectors can be extended to tensors of arbitrary rank. For example, the contraction with g_{ij} results in a lowering of the corresponding index:

$$T_{ij} = g_{ik} T^k_{\ \bullet j} = g_{jk} T_i^{\ \bullet k}. \tag{19.32}$$

Here the dots ($\bullet$) in the mixed components emphasize the order of occurrence of the indices; in fact, in general, $T_i^{\ \bullet l} \neq T^l_{\ \bullet i}$. Repeated contraction with g_{ij} yields

$$T_{ij} = g_{ik} g_{jl} T^{kl}.$$

Similarly, contraction with g^{ij} raises an index, i.e.,

$$T^{ij} = g^{ik} T_k^{\bullet\, j} = g^{ik} g^{jl} T_{kl}. \tag{19.33}$$

Comparable arguments are applicable to local basis vectors $\boldsymbol{e}_i$ and $\boldsymbol{e}^j$ as stated below.

♠ **Index lowering and raising (II):**
Local basis vectors $\boldsymbol{e}_i$ and $\boldsymbol{e}^k$ are related as

$$\boldsymbol{e}_i = g_{ik}\boldsymbol{e}^k \quad \text{and} \quad \boldsymbol{e}^j = g^{j\ell}\boldsymbol{e}_\ell.$$

Proof Since $\boldsymbol{a} = a^i \boldsymbol{e}_i = a_j \boldsymbol{e}^j = a^k g_{kj} \boldsymbol{e}^j$, we have

$$a^1 \left(\boldsymbol{e}_1 - g_{1j}\boldsymbol{e}^j\right) + a^2 \left(\boldsymbol{e}_2 - g_{2j}\boldsymbol{e}^j\right) + a^3 \left(\boldsymbol{e}_3 - g_{3j}\boldsymbol{e}^j\right) = \boldsymbol{0},$$

which holds for any vector $\boldsymbol{a}$. Hence, $\boldsymbol{e}_k - g_{kj}\boldsymbol{e}^k = 0$ for all k, i.e.,

$$\boldsymbol{e}_k = g_{kj}\boldsymbol{e}^j.$$

Similarly, we have

$$\boldsymbol{e}^k = g^{kj}\boldsymbol{e}_j. \quad ♣$$

Exercises

1. Show that the quantities $g_{ij} = \boldsymbol{e}_i \cdot \boldsymbol{e}_j$ form the covariant components of a second-order tensor.

   Solution: In the new (primed) coordinate system we have $g'_{ij} = \boldsymbol{e'}_i \cdot \boldsymbol{e'}_j$. Using the transformation law (19.7) of covariant basis vectors, we have

$$g'_{ij} = \left(\frac{\partial u^k}{\partial u'^i}\boldsymbol{e}_k\right) \cdot \left(\frac{\partial u^l}{\partial u'^j}\boldsymbol{e}_l\right) = \frac{\partial u^k}{\partial u'^i}\frac{\partial u^l}{\partial u'^j}\left(\boldsymbol{e}_k \cdot \boldsymbol{e}_l\right) = \frac{\partial u^k}{\partial u'^i}\frac{\partial u^l}{\partial u'^j} g_{kl}.$$

   This clearly indicates that the g_{ij} are covariant components of a second-order tensor (i.e., the metric tensor $\boldsymbol{g}$). A similar argument shows that the quantities g^{ij} form the contravariant components of $\boldsymbol{g}$, which transform as follows:

$$g'^{ij} = \frac{\partial u'^i}{\partial u^k}\frac{\partial u'^j}{\partial u^l} g^{kl}. \quad ♣$$

2. Show that the matrix $[g^{ij}]$ is the inverse of the matrix $[g_{ij}]$.

For an arbitrary vector $\boldsymbol{a}$, we find $a^i = g^{ij}a_j = g^{ij}g_{jk}a^k$. Since $\boldsymbol{a}$ is arbitrary, we must have

$$g^{ij}g_{jk} = \delta^i_k = \begin{cases} 1 & i = k, \\ 0 & i \neq k. \end{cases} \tag{19.34}$$

This clearly indicates that the matrices $[g_{ij}]$ and $[g^{ij}]$ are inverse to each other. ♣

3. Show that $\sqrt{g} = (\boldsymbol{e}_i \times \boldsymbol{e}_j) \cdot \boldsymbol{e}_k$ and $1/\sqrt{g} = (\boldsymbol{e}^i \times \boldsymbol{e}^j) \cdot \boldsymbol{e}^k$, where i, j, k is a cyclic permutation of the numbers $1, 2, 3$.

Solution: By direct calculations, we obtain

$$g^{i\ell} = \boldsymbol{e}^i \cdot \boldsymbol{e}^\ell = \frac{(\boldsymbol{e}_j \times \boldsymbol{e}_k) \cdot (\boldsymbol{e}_m \times \boldsymbol{e}_n)}{[\boldsymbol{e}_i \cdot (\boldsymbol{e}_j \times \boldsymbol{e}_k)]\,[\boldsymbol{e}_\ell \cdot (\boldsymbol{e}_m \times \boldsymbol{e}_n)]}, \tag{19.35}$$

where i, j, k and ℓ, m, n are cyclic permutations of the ordered set of numbers $1, 2, 3$. The numerator in (19.35) reads

$$\begin{aligned}
(\boldsymbol{e}_j \times \boldsymbol{e}_k) \cdot (\boldsymbol{e}_m \times \boldsymbol{e}_n) &= [(\boldsymbol{e}_j \times \boldsymbol{e}_k) \times \boldsymbol{e}_m] \cdot \boldsymbol{e}_n \\
&= [(\boldsymbol{e}_j \cdot \boldsymbol{e}_m)\,\boldsymbol{e}_k - (\boldsymbol{e}_k \cdot \boldsymbol{e}_m)\,\boldsymbol{e}_j] \cdot \boldsymbol{e}_n \\
&= (\boldsymbol{e}_j \cdot \boldsymbol{e}_m)\,(\boldsymbol{e}_k \cdot \boldsymbol{e}_n) - (\boldsymbol{e}_k \cdot \boldsymbol{e}_m)\,(\boldsymbol{e}_j \cdot \boldsymbol{e}_n) \\
&= \begin{vmatrix} \boldsymbol{e}_j \cdot \boldsymbol{e}_m & \boldsymbol{e}_k \cdot \boldsymbol{e}_m \\ \boldsymbol{e}_j \cdot \boldsymbol{e}_n & \boldsymbol{e}_k \cdot \boldsymbol{e}_n \end{vmatrix} = \begin{vmatrix} g_{jm} & g_{km} \\ g_{jn} & g_{kn} \end{vmatrix} = C^{i\ell}.
\end{aligned}$$

Here, $C^{i\ell}$ is the cofactor of $g_{i\ell}$ in the determinant $g = \det[g_{i\ell}]$. Comparing the results with the definition $g^{i\ell} = C^{i\ell}/g$, we find that

$$g = [\boldsymbol{e}_i \cdot (\boldsymbol{e}_j \times \boldsymbol{e}_k)]\,[\boldsymbol{e}_\ell \cdot (\boldsymbol{e}_m \times \boldsymbol{e}_n)],$$

which is equivalent to

$$g = [\boldsymbol{e}_i \cdot (\boldsymbol{e}_j \times \boldsymbol{e}_k)]^2, \quad \text{i.e.,} \quad \sqrt{g} = \pm \boldsymbol{e}_i \cdot (\boldsymbol{e}_j \times \boldsymbol{e}_k),$$

where the plus sign is chosen if the given basis is right-handed. In a similar manner, via the relations $g_{i\ell} = C_{i\ell}$ and $\det[\delta^k_i] = \det[g_{ij}g^{jk}] = \det[g_{ij}]\det[g^{jk}] = 1$, we obtain

$$\frac{1}{\sqrt{g}} = \pm \boldsymbol{e}^i \cdot \left(\boldsymbol{e}^j \times \boldsymbol{e}^k\right). \quad ♣$$

19.3 Christoffel Symbols

19.3.1 Derivatives of Basis Vectors

Several new concepts are required for the differentiation of vectors or tensors with respect to curvilinear coordinates. Recall that in a general curvilinear coordinate system, the basis vectors $\boldsymbol{e}_i$ and $\boldsymbol{e}^i$ are functions of the coordinates. This implies that differentiation of vectors (say, $\boldsymbol{v} = v^i \boldsymbol{e}_i$) or tensors (say, $\boldsymbol{T} = T_i^j \boldsymbol{e}^i \otimes \boldsymbol{e}_j$) involves their derivatives, such as $\partial \boldsymbol{e}_i / \partial u^j$.

Suppose that the derivative $\partial \boldsymbol{e}_i / \partial u^j$ can be written as a linear combination of the basis vectors $\boldsymbol{e}_k$ as denoted by

$$\frac{\partial \boldsymbol{e}_i}{\partial u^j} = \Gamma_{ij}^k \boldsymbol{e}_k, \tag{19.36}$$

the symbol Γ_{ij}^k being the coefficients associated with the kth component of the linear combination. Using the reciprocity relation $\boldsymbol{e}^i \cdot \boldsymbol{e}_j = \delta_j^i$, we write this as

$$\Gamma_{ij}^k = \boldsymbol{e}^k \cdot \frac{\partial \boldsymbol{e}_i}{\partial u^j}. \tag{19.37}$$

This three-index symbol is called a **Christoffel symbol**. In a similar manner as above, we can show that the derivative of the contravariant basis vectors reads

$$\frac{\partial \boldsymbol{e}^i}{\partial u^j} = -\Gamma_{kj}^i \boldsymbol{e}^k. \tag{19.38}$$

Details of the derivation are given in Exercise **1**.

We shall see that Christoffel symbols play a key role in defining the derivatives of vectors and tensors in terms of general coordinate systems. A more formal definition of Christoffel symbols in terms of metric tensors is given in Sect. 19.3.4.

Remark. It is clear from (19.37) that in Cartesian coordinate systems, $\Gamma_{ij}^k = 0$ for all values of the indices i, j, and k, owing to the identity: $\partial \boldsymbol{e}_i / \partial u^j \equiv 0$.

Example 4. Let us calculate the Christoffel symbols Γ_{ij}^m for cylindrical coordinates, where $(u^1, u^2, u^3) = (\rho, \phi, z)$, and the position vector $\boldsymbol{r}$ of any point may be written

$$\boldsymbol{r} = \rho \cos\phi \boldsymbol{i} + \rho \sin\phi \boldsymbol{j} + z\boldsymbol{k}.$$

From this we find that the covariant basis vectors are given by

$$\boldsymbol{e}_\rho = \frac{\partial \boldsymbol{r}}{\partial \rho} = \cos\phi \boldsymbol{i} + \sin\phi \boldsymbol{j}, \tag{19.39}$$

$$\boldsymbol{e}_\phi = \frac{\partial \boldsymbol{r}}{\partial \phi} = -\rho \sin\phi \boldsymbol{i} + \rho \cos\phi \boldsymbol{j}, \tag{19.40}$$

$$e_z = \frac{\partial \boldsymbol{r}}{\partial z} = \boldsymbol{k}. \tag{19.41}$$

It is a straightforward mother to show that the only derivatives of these vectors that are nonzero with respect to the coordinates are

$$\frac{\partial \boldsymbol{e}_\rho}{\partial \phi} = \frac{1}{\rho}\boldsymbol{e}_\phi, \quad \frac{\partial \boldsymbol{e}_\phi}{\partial \rho} = \frac{1}{\rho}\boldsymbol{e}_\phi, \quad \frac{\partial \boldsymbol{e}_\phi}{\partial \phi} = -\rho \boldsymbol{e}_\rho.$$

Thus, from (19.37), we immediately have

$$\Gamma^2_{12} = \Gamma^2_{21} = \frac{1}{\rho} \quad \text{and} \quad \Gamma^1_{22} = -\rho. \tag{19.42}$$

19.3.2 Nontensor Character

Despite their appearance, the Christoffel symbols Γ^k_{ij} do not form the components of a third-order tensor.

♠ **Theorem:**
Christoffel symbols Γ^k_{ij} do not form any kind of tensor.

Proof This is verified by considering their transformation behavior under a general coordinate transformation. In a transformed coordinate system, we have

$$\Gamma'^k_{ij} = \boldsymbol{e}'^k \cdot \frac{\partial \boldsymbol{e}'_i}{\partial u'^j}. \tag{19.43}$$

Applying the transformation law of local basis vectors, we obtain

$$\begin{aligned}
\Gamma'^k_{ij} &= \left(\frac{\partial u'^k}{\partial u^n}\boldsymbol{e}^n\right) \cdot \frac{\partial}{\partial u'^j}\left(\frac{\partial u^l}{\partial u'^i}\boldsymbol{e}_l\right) \\
&= \left(\frac{\partial u'^k}{\partial u^n}\boldsymbol{e}^n\right) \cdot \left[\frac{\partial^2 u^l}{\partial u'^i \partial u'^j}\boldsymbol{e}_l + \left(\frac{\partial u^l}{\partial u'^i}\frac{\partial \boldsymbol{e}_l}{\partial u'^j}\right)\right] \\
&= \frac{\partial u'^k}{\partial u^n}\frac{\partial^2 u^l}{\partial u'^i \partial u'^j}(\boldsymbol{e}^n \cdot \boldsymbol{e}_l) + \frac{\partial u'^k}{\partial u^n}\frac{\partial u^l}{\partial u'^i}\left(\boldsymbol{e}^n \cdot \frac{\partial \boldsymbol{e}_l}{\partial u'^j}\right) \\
&= \frac{\partial u'^k}{\partial u^n}\frac{\partial^2 u^l}{\partial u'^i \partial u'^j}\delta^n_l + \frac{\partial u'^k}{\partial u^n}\frac{\partial u^l}{\partial u'^i}\frac{\partial u^m}{\partial u'^j}\left(\boldsymbol{e}^n \cdot \frac{\partial \boldsymbol{e}_l}{\partial u^m}\right) \\
&= \frac{\partial u'^k}{\partial u^l}\frac{\partial^2 u^l}{\partial u'^i \partial u'^j} + \frac{\partial u'^k}{\partial u^n}\frac{\partial u^l}{\partial u'^i}\frac{\partial u^m}{\partial u'^j}\Gamma^n_{lm}.
\end{aligned} \tag{19.44}$$

Hence, the presence of the first term in the last line in (19.44) prevents the Γ^k_{ij} from forming a third-order tensor. ♣

19.3.3 Properties of Christoffel Symbols

Christoffel symbols Γ_{ij}^k satisfy the following relations:

1. $\Gamma_{ij}^k = \Gamma_{ji}^k$.

2. $\dfrac{\partial g_{ij}}{\partial u^k} = g_{\ell j}\Gamma_{ik}^{\ell} + g_{i\ell}\Gamma_{jk}^{\ell}$.

3. $\dfrac{\partial g^{ij}}{\partial u^k} = -g^{i\ell}\Gamma_{\ell k}^{j} - g_{j\ell}\Gamma_{\ell k}^{i}$.

4. $\Gamma_{ik}^k = \dfrac{\partial}{\partial u^i}\log\sqrt{|g|} = \dfrac{1}{\sqrt{|g|}}\dfrac{\partial\sqrt{|g|}}{\partial u^i}$.

Proofs of these relations are given in Exercises **2**–**4**.

Remark. Some textbooks refer to our three-index symbol Γ_{ij}^k defined by (19.37) as the **Christoffel symbol of the second kind** and use the following notation:

$$\Gamma_{ij}^k = \left\{ \begin{matrix} k \\ i\ j \end{matrix} \right\} = \boldsymbol{e}_k \cdot \frac{\partial \boldsymbol{e}_i}{\partial u^j}. \tag{19.45}$$

As a counterpart, we may define the **Christoffel symbol of the first kind** $[k, ij]$ by

$$[k,\ ij] = \boldsymbol{e}^k \cdot \frac{\partial \boldsymbol{e}_i}{\partial u^j}. \tag{19.46}$$

Note that the index k on the right-hand side of (19.46) is a *superscript*, whereas that of (19.45) is a *subscript*. These two kinds of Christoffel symbols are related to each other as

$$\left\{ \begin{matrix} k \\ i\ j \end{matrix} \right\} = g^{k\ell}[\ell,\ ij].$$

19.3.4 Alternative Expression

In principle, we can calculate the Γ_{ij}^k in a given coordinate system using the expression (19.37) based on $\boldsymbol{e}_i$. However, it is simple to use an alternative expression in terms of the metric tensor g_{ij} and its derivatives as stated below.

♠ **Theorem:**
Christoffel symbols are expressed as

$$\Gamma_{ij}^m = \frac{1}{2}g^{mk}\left(\frac{\partial g_{jk}}{\partial u^i} + \frac{\partial g_{ki}}{\partial u^j} - \frac{\partial g_{ij}}{\partial u^k}\right).$$

Proof Recall the relation

$$\frac{\partial g_{ij}}{\partial u^k} = g_{\ell j}\Gamma^{\ell}_{ik} + g_{i\ell}\Gamma^{\ell}_{jk} \tag{19.47}$$

given in Sect. 19.3.3. By cyclically permuting the free indices i, j, k, we obtain two further equivalent relations:

$$\frac{\partial g_{jk}}{\partial u^i} = \Gamma^{\ell}_{ji} g_{\ell k} + \Gamma^{\ell}_{ki} g_{j\ell} \tag{19.48}$$

and

$$\frac{\partial g_{ki}}{\partial u^j} = \Gamma^{\ell}_{kj} g_{\ell i} + \Gamma^{\ell}_{ij} g_{k\ell}. \tag{19.49}$$

Then, subtracting (19.47) from the sum of (19.48) and (19.49), we find

$$\begin{aligned}
&\frac{\partial g_{jk}}{\partial u^i} + \frac{\partial g_{ki}}{\partial u^j} - \frac{\partial g_{ij}}{\partial u^k} \\
&= \Gamma^{\ell}_{ji} g_{\ell k} + \Gamma^{\ell}_{ki} g_{j\ell} + \Gamma^{\ell}_{kj} g_{\ell i} + \Gamma^{\ell}_{ij} g_{k\ell} - \Gamma^{\ell}_{ik} g_{\ell j} - \Gamma^{\ell}_{jk} g_{i\ell} \\
&= \left(\Gamma^{\ell}_{ji} g_{\ell k} + \Gamma^{\ell}_{ij} g_{k\ell}\right) + \left(\Gamma^{\ell}_{ki} g_{j\ell} - \Gamma^{\ell}_{ik} g_{\ell j}\right) + \left(\Gamma^{\ell}_{kj} g_{\ell i} - \Gamma^{\ell}_{jk} g_{il}\right) \\
&= 2\Gamma^{\ell}_{ij} g_{k\ell} + 0 + 0 \quad = 2\Gamma^{\ell}_{ij} g_{k\ell},
\end{aligned} \tag{19.50}$$

where we have used the symmetry properties: $g_{ij} = g_{ji}$ and $\Gamma^{\ell}_{ij} = \Gamma^{\ell}_{ji}$. Contracting both sides with g^{mk} yields

$$g^{mk}\left(\frac{\partial g_{jk}}{\partial u^i} + \frac{\partial g_{ki}}{\partial u^j} - \frac{\partial g_{ij}}{\partial u^k}\right) = 2g^{mk}\Gamma^{\ell}_{ij} g_{k\ell} = 2\Gamma^{\ell}_{ij}\delta^m_{\ell} = 2\Gamma^m_{ij},$$

i.e.,

$$\Gamma^m_{ij} = \frac{1}{2} g^{mk}\left(\frac{\partial g_{jk}}{\partial u^i} + \frac{\partial g_{ki}}{\partial u^j} - \frac{\partial g_{ij}}{\partial u^k}\right). \quad \clubsuit \tag{19.51}$$

This result enables us to compute the Christoffel symbol of a given coordinate system from information about the metric tensor.

Examples We again evaluate the Christoffel symbols Γ^m_{ij} for cylindrical coordinates. Using (19.51) and the fact that $g_{11} = 1$, $g_{22} = \rho^2$, $g_{33} = 1$ and the other components are zero, we see that the only three nonzero Christoffel symbols are indeed $\Gamma^2_{12} = \Gamma^2_{21}$ and Γ^1_{22}. Given by

$$\Gamma^2_{12} = \Gamma^2_{21} = \frac{1}{2g_{22}}\frac{\partial g_{22}}{\partial u^1} = \frac{1}{2\rho^2}\frac{\partial}{\partial \rho}\rho^2 = \frac{1}{\rho}, \tag{19.52}$$

$$\Gamma^1_{22} = -\frac{1}{2g_{11}}\frac{\partial g_{22}}{\partial u^1} = -\frac{1}{2}\frac{\partial}{\partial \rho}\rho^2 = -\rho, \tag{19.53}$$

they agree with the expressions in (19.42).

Remark. The result (19.51) implies that the Christoffel symbol of the first kind $[k,\ ij]$ mentioned in (19.46) is written as

$$[k,\ ij] = \frac{1}{2}\left(\frac{\partial g_{jk}}{\partial u^i} + \frac{\partial g_{ki}}{\partial u^j} - \frac{\partial g_{ij}}{\partial u^k}\right).$$

Exercises

1. Derive equation (19.38).

Solution: By differentiating the reciprocity relation $\boldsymbol{e}^i \cdot \boldsymbol{e}_j = \delta^i_j$ with respect to the coordinates, we have

$$\frac{\partial \boldsymbol{e}^i}{\partial u^k} \cdot \boldsymbol{e}_j + \boldsymbol{e}^i \cdot \frac{\partial \boldsymbol{e}_i}{\partial u^k} = \frac{\partial \delta^i_j}{\partial u^k}. \tag{19.54}$$

The right-hand side of (19.54) vanishes since the element δ^i_j consists of the contants +1 and 0, which are independent of the coordinates u^k. Hence, using (19.37), we obtain

$$\frac{\partial \boldsymbol{e}^i}{\partial u^k} \cdot \boldsymbol{e}_j + \Gamma^i_{jk} = 0. \tag{19.55}$$

Similar to the case of (19.36), for the moment we write the derivative $\partial \boldsymbol{e}^i / \partial u^j$ as a linear combination of the basis vector $\boldsymbol{e}^\ell$ as

$$\frac{\partial \boldsymbol{e}^i}{\partial u^k} = B^\ell_{ik} \boldsymbol{e}^\ell. \tag{19.56}$$

Substituting (19.56) into (19.55), we obtain $B^j_{ik} = -\Gamma^i_{jk}$. Consequently, we have $\dfrac{\partial \boldsymbol{e}^i}{\partial u^k} = -\Gamma^i_{\ell k} \boldsymbol{e}^\ell$, or equivalently (by interchanging the subscripts),

$$\frac{\partial \boldsymbol{e}^i}{\partial u^j} = -\Gamma^i_{kj} \boldsymbol{e}^k. \quad \clubsuit$$

2. Show that $\Gamma^k_{ij} = \Gamma^k_{ji}$.

Solution: It follows that $\dfrac{\partial \boldsymbol{e}_i}{\partial u^j} = \dfrac{\partial}{\partial u^j}\dfrac{\partial \boldsymbol{r}}{\partial u^i} = \dfrac{\partial}{\partial u^i}\dfrac{\partial \boldsymbol{r}}{\partial u^j} = \dfrac{\partial \boldsymbol{e}_j}{\partial u^i}$, which yields $\Gamma^k_{ij} = \boldsymbol{e}^k \cdot \dfrac{\partial \boldsymbol{e}_i}{\partial u^j} = \boldsymbol{e}^k \cdot \dfrac{\partial \boldsymbol{e}_j}{\partial u^i} = \Gamma^k_{ji}$. ♣

3. Show that $\dfrac{\partial g_{ij}}{\partial u^k} = g_{\ell j}\Gamma^\ell_{ik} + g_{i\ell}\Gamma^\ell_{jk}$.

Solution: Derivatives of the metric tensor $g_{ij} = \boldsymbol{e}_i \cdot \boldsymbol{e}_j$ with respect to u^k read

$$\frac{\partial g_{ij}}{\partial u^k} = \frac{\partial \boldsymbol{e}_i}{\partial u^k} \cdot \boldsymbol{e}_j + \boldsymbol{e}_i \cdot \frac{\partial \boldsymbol{e}_j}{\partial u^k} = \Gamma^\ell_{ik} \boldsymbol{e}_\ell \cdot \boldsymbol{e}_j + \boldsymbol{e}_i \cdot \Gamma^\ell_{jk} \boldsymbol{e}_\ell$$
$$= \Gamma^\ell_{ik} g_{\ell j} + \Gamma^\ell_{jk} g_{i\ell}. \quad \clubsuit$$

4. Show that $\dfrac{\partial g}{\partial u^k} = g g^{ji} \dfrac{\partial g_{ij}}{\partial u^k}$, where $g = \det[g_{ij}]$.

Solution: We know that the determinant g is given by (see Sect. 18.1.7)

$$g = \sum_{j=1}^{n} g_{ij} C^{ij} \quad \text{(with } i \text{ fixed)},$$

where C^{ij} is the cofactor of the element g_{ij} in g. Partially differentiating both sides with respect to g_{ij} gives

$$\frac{\partial g}{\partial g_{ij}} = C^{ij}. \tag{19.57}$$

Since $g^{ij} = C^{ij}/g$ (see Sect. 19.2.4), it follows from (19.57) that

$$\frac{\partial g}{\partial u^k} = \frac{\partial g}{\partial g_{ij}} \frac{\partial g_{ij}}{\partial u^k} = C^{ij} \frac{\partial g_{ij}}{\partial u^k} = g g^{ji} \frac{\partial g_{ij}}{\partial u^k}. \quad \clubsuit$$

5. Show that $\Gamma^k_{ik} = \dfrac{\partial}{\partial u^i} \log \sqrt{g}$.

Solution: According to the expression (19.51), we have

$$\Gamma^i_{ki} = \frac{1}{2} g^{i\ell} \left(\frac{\partial g_{i\ell}}{\partial u^k} + \frac{\partial g_{k\ell}}{\partial u^i} - \frac{\partial g_{ki}}{\partial u^\ell} \right).$$

The last two terms in the parentheses cancel out because

$$g^{i\ell} \frac{\partial g_{k\ell}}{\partial u^i} = g^{\ell i} \frac{\partial g_{ki}}{\partial u^\ell} = g^{i\ell} \frac{\partial g_{ki}}{\partial u^\ell},$$

where we have interchanged the dummy indices i and l in the first equality, and have used the symmetry of the metric tensor in the second. Hence, we set

$$\Gamma^i_{ki} = \frac{g^{i\ell}}{2} \frac{\partial g_{i\ell}}{\partial u^k}. \tag{19.58}$$

This can be further simplified by using the result of Exercise **4** as

$$\Gamma^i_{ki} = \frac{1}{2g} \frac{\partial g}{\partial u^k} = \frac{1}{2g} \frac{\partial g}{\partial \sqrt{g}} \frac{\partial \sqrt{g}}{\partial u^k} = \frac{1}{\sqrt{g}} \frac{\partial \sqrt{g}}{\partial u^k} = \frac{\partial}{\partial u^k} \log \sqrt{g}. \quad \clubsuit$$

19.4 Covariant Derivatives

19.4.1 Covariant Derivatives of Vectors

The derivatives of a scalar in terms of Cartesian coordinates work as covariant components of a vector. This is also true for the case of general coordinate systems, as can be shown by considering the differential of a scalar

$$d\phi = \frac{\partial \phi}{\partial u^i} du^i.$$

Since the du^i are contravariant components of a vector and $d\phi$ is a scalar, we see from the quotient law that the quantities $\partial \phi / \partial u^i$ must form covariant components of a vector.

Except for a scalar, however, the derivatives of a general tensor do not necessarily form the component of another tensor. To see this, we consider the derivative of the covariant components v^i of a vector $\boldsymbol{v}$ with respect to a general coordinate u^j. In a new (primed) coordinate, it reads

$$\begin{aligned}\frac{\partial v'^i}{\partial u'^j} &= \frac{\partial u^k}{\partial u'^j}\frac{\partial v'^i}{\partial u^k} = \frac{\partial u^k}{\partial u'^j}\frac{\partial}{\partial u^k}\left(\frac{\partial u'^i}{\partial u^l} v^l\right) \\ &= \frac{\partial u^k}{\partial u'^j}\frac{\partial u'^i}{\partial u^l}\frac{\partial v^l}{\partial u^k} + \frac{\partial u^k}{\partial u'^j}\frac{\partial^2 u'^i}{\partial u^k \partial u^l} v^l. \end{aligned} \tag{19.59}$$

The presence of the second term in the last line of (19.59) prevents the derivative $\partial v^i / \partial x^j$ from obeying the transformation law of the components of a second-order tensor. The nontensor character stems from the fact that the second-order derivative,

$$\frac{\partial^2 u'^i}{\partial u^k \partial u^l}, \tag{19.60}$$

involved in the last line of (19.59) does not vanish. In fact, the first-order derivative $\partial u'^i / \partial u^l$ is not constant in non-Cartesian coordinates, whereas it is constant in Cartesian coordinates [so that the term (19.60) vanishes in the latter case].

In the context above, it is natural to introduce a new class of differentiation that turns the derivatives of components of a tensor into components of another tensor. This is achieved with the help of the Christoffel symbols discussed in Sect. 19.3. Let us consider the derivative of a vector $\boldsymbol{v}$ with respect to the coordinates u^j. We find

$$\frac{\partial \boldsymbol{v}}{\partial u^j} = \frac{\partial v^i}{\partial u^j}\boldsymbol{e}_i + v^i \frac{\partial \boldsymbol{e}_i}{\partial u^j}, \tag{19.61}$$

where the second term arises because, in general, the basis vectors $\boldsymbol{e}_i$ are not constant. Using (19.36), we write

$$\frac{\partial \boldsymbol{v}}{\partial u^j} = \frac{\partial v^i}{\partial u^j}\boldsymbol{e}_i + v^i \Gamma_{ij}^k \boldsymbol{e}_k.$$

Since i and k are dummy indices, we may interchange them to obtain

$$\frac{\partial \boldsymbol{v}}{\partial u^j} = \frac{\partial v^i}{\partial u^j}\boldsymbol{e}_i + v^k \Gamma_{kj}^i \boldsymbol{e}_i$$

$$= \left(\frac{\partial v^i}{\partial u^j} + v^k \Gamma_{kj}^i \right) \boldsymbol{e}_i. \tag{19.62}$$

The quantity in parentheses is referred to specificcally as the **covariant derivative of a vector**:

♠ **Covariant derivative of a vector:**
The quantities defined by

$$v^i_{\ ;j} \equiv \frac{\partial v^i}{\partial u^j} + \Gamma_{kj}^i v^k. \tag{19.63}$$

are called **covariant derivatives** of contravariant components v^i of a vector $\boldsymbol{v}$ with respect to u^j. Here, the semicolon subscript on the left-hand side denotes covariant differentiation.

Using this notation, we may write the derivative of a vector in the very compact form

$$\frac{\partial \boldsymbol{v}}{\partial u^j} = v^i_{\ ;j}\boldsymbol{e}_i.$$

The corresponding result for the *covariant* components v_i can be found in a similar way by considering the derivative of $\boldsymbol{v} = v_i \boldsymbol{e}^i$ and using (19.38) to obtain

$$v_{i\,;j} = \frac{\partial v_i}{\partial u^j} - \Gamma_{ij}^k v_k. \tag{19.64}$$

19.4.2 Remarks on Covariant Derivatives

1. The arrangement of indices i, j, k in the Christoffel symbols in (19.63) and (19.64) can be determined systematically in the following manner. First, the index to which the derivative is taken (i.e., j in this case) is the last subscript on the Christoffel symbol. Secondly, the other index appearing on the left-hand side (i.e., i in this case) also appears in the Christoffel symbol on the right-hand side without raising or lowering. The remaining index can then be arranged in only one.

2. Similar to $v^i{}_{;j}$, a comparable short-hand notation for partial derivatives is obtained by replacing the semicolon by a comma such as

$$v^i{}_{,j} \equiv \frac{\partial v^i}{\partial u^j} \quad \text{and} \quad v_{i\,,j} \equiv \frac{\partial v_i}{\partial u^j}.$$

3. In Cartesian coordinates, all the Γ^i_{kj} are zero, so the covariant derivative reduces to the simple partial derivative, say, $v^i{}_{;j} \equiv v^i{}_{,j}$.

19.4.3 Covariant Derivatives of Tensors

Covariant derivatives of higher-order tensors can be defined by a procedure similar to the one for vectors. As an example, let us consider the derivative of the second-order tensor $\boldsymbol{T}$ with respect to the coordinate u^k. Expressing $\boldsymbol{T}$ in terms of its contravariant components, we have

$$\begin{aligned}\frac{\partial \boldsymbol{T}}{\partial u^k} &= \frac{\partial}{\partial u^k}\left(T^{ij}\boldsymbol{e}_i \otimes \boldsymbol{e}_j\right) \\ &= \frac{\partial T^{ij}}{\partial u^k}\boldsymbol{e}_i \otimes \boldsymbol{e}_j + T^{ij}\frac{\partial \boldsymbol{e}_i}{\partial u^k} \otimes \boldsymbol{e}_j + T^{ij}\boldsymbol{e}_i \otimes \frac{\partial \boldsymbol{e}_j}{\partial u^k}. \end{aligned} \tag{19.65}$$

Using Christoffel symbols, we obtain

$$\frac{\partial \boldsymbol{T}}{\partial u^k} = \frac{\partial T^{ij}}{\partial u^k}\boldsymbol{e}_i \otimes \boldsymbol{e}_j + T^{ij}\Gamma^{\ell}_{ik}\boldsymbol{e}_l \otimes \boldsymbol{e}_j + T^{ij}\boldsymbol{e}_i \otimes \Gamma^{\ell}_{jk}\boldsymbol{e}_l.$$

Interchanging the dummy indices i and l in the second term and j and l in the third term on the right-hand side, we set

$$\frac{\partial \boldsymbol{T}}{\partial u^k} = \left(\frac{\partial T^{ij}}{\partial u^k} + \Gamma^i_{lk}T^{lj} + \Gamma^j_{lk}T^{il}\right)\boldsymbol{e}_i \otimes \boldsymbol{e}_j,$$

where the expression in parentheses is the required covariant derivative defined by

$$T^{ij}_{\ ;k} = \frac{\partial T^{ij}}{\partial u^k} + \Gamma^i_{lk}T^{lj} + \Gamma^j_{lk}T^{il}. \tag{19.66}$$

Using the notation (19.66), we can write the derivative of the tensor $\boldsymbol{T}$ with respect to u^k as

$$\frac{\partial \boldsymbol{T}}{\partial u^k} = T^{ij}_{\ ;k}\boldsymbol{e}_i \otimes \boldsymbol{e}_j.$$

Results similar to (19.66) can be obtained for the covariant derivatives of the mixed and covariant components of a second-order tensor. Collecting all of these results leads to the following:

♠ **Covariant derivative of a tensor:**

Covariant derivatives of components of a second-order tensor $\boldsymbol{T}$ are given by

$$T^{ij}_{\ \ ;k} = T^{ij}_{\ \ ,k} + \Gamma^i_{lk} T^{lj} + \Gamma^j_{lk} T^{il},$$

$$T^i_{j\ ;k} = T^i_{j\ ,k} + \Gamma^i_{lk} T^l_j - \Gamma^\ell_{jk} T^i_l,$$

$$T_{ij\ ;k} = T_{ij\ ,k} - \Gamma^\ell_{ik} T_{lj} - \Gamma^\ell_{jk} T_{il},$$

where the comma notation means the taking of partial derivatives.

The position of the indices in the expressions is very systematic. We focus on the index i or j on the left-hand side. First, the index k to which the derivative is taken should be the last subscript on the Christoffel symbol. Next, if the index (i or j) on the left-hand side is a superscript, then the corresponding term on the right-hand side containing a Christoffel symbol is attached to a plus sign. In contrast, when the index on the left-hand side is a subscript, the corresponding term on the right is attached to a minus sign. We can extend this in a straightforward manner to tensors with an arbitrary number of contravariant and covariant indices.

Remark.

1. All of the quantities $T^{ij}_{\ \ ;k}$, $T^i_{j\ ;k}$, and $T_{ij\ ;k}$ are the components of the same third-order tensor $\nabla \boldsymbol{T}$ with respect to different tensor bases, i.e.,

$$\nabla \boldsymbol{T} = T^{ij}_{\ \ ;k} \boldsymbol{e}_i \otimes \boldsymbol{e}_j \otimes \boldsymbol{e}^k = T^i_{j\ ;k} \boldsymbol{e}_i \otimes \boldsymbol{e}^j \otimes \boldsymbol{e}^k = T_{ij\ ;k} \boldsymbol{e}^i \otimes \boldsymbol{e}^j \otimes \boldsymbol{e}^k.$$

2. In general, we may call the $v^i_{\ ;j}$ the covariant derivative of $\boldsymbol{v}$ and denote it by $\nabla \boldsymbol{v}$. In Cartesian coordinates, its components are just $\partial v^i / \partial x^j$.
3. Given a metric tensor $\boldsymbol{g}$, the covariant derivatives of its components, $g_{ij;\,k}$ and $g^{ij}_{\ \ ;k}$, are identically zero in terms of arbitrary coordinates. This is called **Ricci's theorem**, for which we give the proof in Exercises **2** and **3**).

19.4.4 Vector Operators in Tensor Form

This subsection is devoted to finding expressions for vector differential operators such as grad, div, rot, and the Laplacian in tensor form that are valid in general coordinate systems. In principle, they are obtained in a straightforward manner by replacing the partial derivative given in Cartesian coordinates with covariant derivatives. These tensor forms, however, can be simplified by using the metric tensor g_{ij} as shown below.

1. **Gradient:** The gradient of a scalar ϕ in a general coordinate system is given by

$$\nabla\phi = \phi_{;i}\boldsymbol{e}^i = \frac{\partial\phi}{\partial u^i}\boldsymbol{e}^i, \tag{19.67}$$

since the covariant derivative of a scalar is the same as its partial derivative.

2. **Divergence:** The tensor form of the divergence of a vector $\boldsymbol{v}$ is given by

$$\nabla\cdot\boldsymbol{v} = v^i_{;i} = \frac{\partial v^i}{\partial u^i} + \Gamma^i_{ki}v^k. \tag{19.68}$$

Observe that the index i appears twice in the Christoffel symbol. Using the expression (see Sect. 19.3.3)

$$\Gamma^k_{ik} = \frac{\partial}{\partial u^i}\log\sqrt{g} = \frac{1}{\sqrt{g}}\frac{\partial\sqrt{g}}{\partial u^i},$$

we obtain a more compact form:

$$\begin{aligned} v^i_{;i} &= \frac{\partial v^i}{\partial u^i} + \Gamma^i_{ki}v^k = \frac{\partial v^i}{\partial u^i} + \left(\frac{1}{\sqrt{g}}\frac{\partial\sqrt{g}}{\partial u^k}\right)v^k \\ &= \frac{1}{\sqrt{g}}\sqrt{g}\left(\frac{\partial v^k}{\partial u^k}\right) + \frac{1}{\sqrt{g}}\left(\frac{\partial\sqrt{g}}{\partial u^k}\right)v^k = \frac{1}{\sqrt{g}}\frac{\partial}{\partial u^k}\left(\sqrt{g}v^k\right). \end{aligned} \tag{19.69}$$

3. **Laplacian:** The tensor form of the Laplacian $\nabla^2\phi$ is obtained by making use of the following relation:

$$v^i_{;i} = \nabla\cdot\boldsymbol{v} = \nabla\cdot(\nabla\phi) = \nabla^2\phi,$$

where we assume that $\boldsymbol{v} = \nabla\phi$. From (19.67), we have

$$v_i\boldsymbol{e}^i = \boldsymbol{v} = \nabla\phi = \frac{\partial\phi}{\partial u^i}\boldsymbol{e}^i.$$

Thus the covariant components of $\boldsymbol{v}$ are given by

$$v_i = \frac{\partial\phi}{\partial u^i},$$

and its contravariant components v^i can be obtained by raising the index using the metric tensor:

$$v^j = g^{jk}v_k = g^{jk}\frac{\partial\phi}{\partial u^k}.$$

Substituting this into (19.69), we finally arrive at

$$\nabla^2\phi = v^j_{;j} = \frac{1}{\sqrt{g}}\frac{\partial}{\partial u^j}\left(\sqrt{g}g^{jk}\frac{\partial\phi}{\partial u^k}\right). \tag{19.70}$$

4. **Rotation:** In general curvilinear coordinates, the operation $\nabla \times \boldsymbol{v}$ is defined by

$$[\nabla \times \boldsymbol{v}]_{ij} = v_{i\,;j} - v_{j\,;i}, \tag{19.71}$$

which forms covariant components of an antisymmetric tensor. The right-hand side of (19.71) can be simplified as

$$\begin{aligned} v_{i\,;j} - v_{j\,;i} &= \frac{\partial v_i}{\partial u^j} - \Gamma^{\ell}_{ij} v_\ell - \frac{\partial v_j}{\partial u^i} + \Gamma^{\ell}_{ji} v_\ell \\ &= \frac{\partial v_i}{\partial u^j} - \frac{\partial v_j}{\partial u^i}, \end{aligned} \tag{19.72}$$

where the Christoffel symbols cancel out owing to their symmetric properties. Therefore, components of the tensor $\nabla \times \boldsymbol{v}$ can be written in terms of partial derivatives as

$$[\nabla \times \boldsymbol{v}]_{ij} = \frac{\partial v_i}{\partial u^j} - \frac{\partial v_j}{\partial u^i}. \tag{19.73}$$

Our results are summarized as follows:

♠ **Vector operators in tensor forms:**

1. $\nabla \phi = \dfrac{\partial \phi}{\partial u^i} \boldsymbol{e}^i.$ **2.** $\nabla \cdot \boldsymbol{v} = \dfrac{1}{\sqrt{g}} \dfrac{\partial}{\partial u^i} \left(\sqrt{g} v^i \right).$

3. $\nabla^2 \phi = \dfrac{1}{\sqrt{g}} \dfrac{\partial}{\partial u^i} \left(\sqrt{g} g^{ik} \dfrac{\partial \phi}{\partial u^k} \right).$ **4.** $[\nabla \times \boldsymbol{v}]_{ij} = \dfrac{\partial v_i}{\partial u^j} - \dfrac{\partial v_j}{\partial u^i}.$

Exercises

1. Prove that the covariant derivatives $v^k_{\ ;j}$ form a second-order tensor of type $(1,1)$.

Solution: Employ the transformation laws of v^k and Γ^k_{pj} [see (19.44)] to obtain

$$\begin{aligned} v^k_{\ ;j} &= \frac{\partial v^k}{\partial u^j} + \Gamma^k_{pj} v^p \\ &= \frac{\partial}{\partial u^j} \left(\frac{\partial u^k}{\partial u'^q} v'^q \right) + \left(\frac{\partial u^k}{\partial u'^q} \frac{\partial u'^r}{\partial u^p} \frac{\partial u'^s}{\partial u^j} \Gamma'^q_{\ rs} + \frac{\partial u^k}{\partial u'^q} \frac{\partial^2 u'^q}{\partial u^p \partial u^j} \right) \left(\frac{\partial u^p}{\partial u'^t} v'^t \right) \\ &= \frac{\partial^2 u^k}{\partial u'^a \partial u'^q} \frac{\partial u'^a}{\partial u^j} v'^q + \frac{\partial u^k}{\partial u'^q} \frac{\partial u'^t}{\partial u^j} \frac{\partial v'^q}{\partial u'^t} \\ &\quad + \frac{\partial u^k}{\partial u'^q} \left(\frac{\partial u'^r}{\partial u^p} \frac{\partial u'^s}{\partial u^j} \Gamma'^q_{\ rs} + \frac{\partial^2 u'^q}{\partial u^p \partial u^j} \right) \frac{\partial u^p}{\partial u'^t} v'^t. \end{aligned} \tag{19.74}$$

The sum of the terms involving second derivatives is zero; this is seen by taking a partial derivative with respect to u'^q in the expression

$$\frac{\partial u^k}{\partial u'^a}\frac{\partial u'^a}{\partial u^j} = \delta^k_j,$$

which yields

$$\begin{aligned}\frac{\partial}{\partial u'^q}\left(\frac{\partial u^k}{\partial u'^a}\frac{\partial u'^a}{\partial u^j}\right) &= \frac{\partial^2 u^k}{\partial u'^q \partial u'^a}\frac{\partial u'^a}{\partial u^j} + \frac{\partial u^k}{\partial u'^a}\frac{\partial^2 u'^a}{\partial u'^q \partial u^j} \\ &= \frac{\partial^2 u^k}{\partial u'^q \partial u'^a}\frac{\partial u'^a}{\partial u^j} + \frac{\partial u^k}{\partial u'^a}\frac{\partial u^m}{\partial u'^q}\frac{\partial^2 u'^a}{\partial u^m \partial u^j} \\ &= 0. \end{aligned} \tag{19.75}$$

From (19.74) and (19.75), it follows that

$$\begin{aligned} v^k_{\ ;j} &= \frac{\partial u^k}{\partial u'^q}\frac{\partial u'^t}{\partial u^j}\frac{\partial v'^q}{\partial u'^t} + \frac{\partial u^k}{\partial u'^q}\delta'^r_{\ t}\frac{\partial u'^s}{\partial u^j}\Gamma'^q_{\ rs}v^t \\ &= \frac{\partial u^k}{\partial u'^q}\frac{\partial u'^s}{\partial u^j}\left(\frac{\partial v'^q}{\partial u'^s} + \Gamma'^q_{\ ts}v'^t\right) \\ &= \frac{\partial u^k}{\partial u'^q}\frac{\partial u'^s}{\partial u^j}v'^q_{\ ;s}, \end{aligned}$$

in which the last term in the last line, $v'^q_{\ ;s}$, represents the covariant derivative of v'^q with respect to the *primed* coordinates u'^s. Hence, we see that the $v^k_{\ ;j}$ form a second-order tensor of type $(1,1)$. ♣

2. Show that the metric tensor is a **covariant constant**, i.e., the covariant derivative of any component is identically zero: $g_{kp;\,j} = 0$. This result is known as **Ricci's theorem.**

Solution: It follows that

$$\begin{aligned} g_{kp;\,j} &= g_{kp,\,j} - \Gamma^r_{kj}g_{rp} - \Gamma^r_{pj}g_{kr} \\ &= g_{kp,\,j} - \frac{1}{2}\delta^s_p\left(g_{js,\,k} + g_{sk,\,j} - g_{kj,\,s}\right) - \frac{1}{2}\delta^s_k\left(g_{js,\,p} + g_{sp,\,j} - g_{pj,\,s}\right) \\ &= 0. \end{aligned}$$ ♣

3. Show that δ^p_k and g^{kp} are also covariant constants.

Solution: We have $\delta^p_{k\,;\,j} = \delta^p_{k,\,j} - \Gamma^p_{qj}\delta^q_k - \Gamma^q_{kj}\delta^p_q = 0$, which completes our first proof. Next, observe that $\delta^k_j = g_{jp}g^{pk}$ to find the identity

$$0 = \delta^k_{j\,;\,q} = \left(g_{jp}g^{pk}\right)_{;\,q} = g_{jp\,;\,q}g^{pk} + g_{jp}g^{pk}_{\ ;\,q}.$$

Since g_{jp} is a covariant constant, the first term in the last expression has the value zero. Multiplication by g^{jr} produces the desired result. ♣

Remark. Owing to Ricci's theorem and its two corollaries noted above, the components of the metric tensor can be regarded as constants under covariant differentiation. Thus, e.g.,

$$g_{i\ell}A^{\ell}{}_{;\,k} = \left(g_{i\ell}A^{\ell}\right)_{;\,k} = A_{i;\,k},$$
$$g_{i\ell}T^{\ell m}{}_{;\,k} = \left(g_{i\ell}T^{\ell m}\right)_{;\,k} = T_i{}^m{}_{;\,k},$$
$$T_{ik;\,\ell}g^{im}g^{kn} = \left(T_{ik}g^{im}g^{kn}\right)_{;\,\ell} = T^{mn}{}_{;\,\ell},$$

and so on.

3. Use (19.70) to find expressions for $\nabla^2\phi$ and $\nabla\cdot\boldsymbol{v}$ in an orthogonal coordinate system with scale factors h_i $(i = 1, 2, 3)$.

Solution: For an orthogonal coordinate system $\sqrt{g} = h_1h_2h_3$; further, $g^{ii} = 1/h_i^2$ for fixed i and $g^{ij} = 0$ for $i \neq j$. Therefore, from (19.70), we set

$$\nabla^2\phi = \frac{1}{h_1h_2h_3}\frac{\partial}{\partial u^j}\left(\frac{h_1h_2h_3}{h_j^2}\frac{\partial\phi}{\partial u^j}\right).$$

In a similar manner, we have

$$\nabla\cdot\boldsymbol{v} = \frac{1}{h_1h_2h_3}\frac{\partial}{\partial u^j}\left(h_1h_2h_3v^i\right). \quad \clubsuit$$

.

19.5 Applications in Physics and Engineering

19.5.1 General Relativity Theory

It cannot be denied that the **general relativity theory** is one of the most famous and beautiful applications of non-Cartesian tensor calculus in physics. This section outlines the concepts one needs in order to understand the general theory of relativity, which is necessary for obtaining the gravitational field equation and relevant tensorial quantities that are involved with the equation.

Before proceeding to the argument, let us point out that the notion of **geometric curvature** is central to general relativity, which quantifies the curvature of space at any given point in the space considered. In Sect. 19.2.3, we learned that a space is a **flat** locally (or entirety), if there exist coordinates x^i such that the line element through a limited region (or the whole) can be written as

$$(ds)^2 = \varepsilon_i(dx^i)^2,$$

where $\varepsilon = \pm 1$. However, if we employ a different coordinate system x'^i, the line element $(ds)^2$, in general, is not of the above form, but reads as

$$(ds)^2 = g_{ij} dx^i dx^j$$

with the appropriate metric tensor g_{ij}. Hence, we require a means of identifying a flat space directly from the metric g_{ij}, *independent* of our choice of coordinate system. Such a coordinate-independent way of defining the curvature of a space leads to the field equation of gravity, i.e., **Einstein's field equation**, described in Sect. 19.5.4.

19.5.2 Riemann Tensor

The curvature of space can be quantified in a manner independent of the coordinate system by changing the order of **covariant differentiation**. Covariant differentiation is a generalization of partial differentiation, in which interchanging the order of differentiation changes the result. To illustrate this, let us consider an arbitrary vector field with covariant components v_i. The covariant derivative of v_i is given by [see (19.64)]

$$v_{i\,;j} = \frac{\partial v_i}{\partial u^j} - \Gamma^{\ell}_{ij} v_{\ell}.$$

A second covariant differentiation then yields

$$\begin{aligned}(v_{i\,;j})_{;k} &= \frac{\partial v_{i\,;j}}{\partial u^k} - \Gamma^m_{ik} v_{m\,;j} - \Gamma^m_{jk} v_{i\,;m} \\ &= \frac{\partial^2 v_i}{\partial u^j \partial u^k} - \left(\frac{\partial \Gamma^{\ell}_{ij}}{\partial u^k}\right) v_{\ell} - \Gamma^{\ell}_{ij}\left(\frac{\partial v_{\ell}}{\partial u^k}\right) \\ &\quad - \Gamma^m_{ik}\left(\frac{\partial v_m}{\partial u^j} - \Gamma^{\ell}_{mj} v_{\ell}\right) - \Gamma^m_{jk}\left(\frac{\partial v_i}{\partial u^m} - \Gamma^{\ell}_{im} v_{\ell}\right).\end{aligned}$$

By interchanging the indices j and k to obtain the expression corresponding to $(v_{i\,;k})_{;j}$ and then subtracting the expression we set from the above relation gives us

$$(v_{i\,;j})_{;k} - (v_{i\,;k})_{;j} = R^{\ell}_{ijk} v_{\ell},$$

where

$$R^{\ell}_{ijk} = \frac{\partial \Gamma^{\ell}_{ik}}{\partial u^j} - \frac{\partial \Gamma^{\ell}_{ij}}{\partial u^k} + \Gamma^m_{ik}\Gamma^{\ell}_{mj} - \Gamma^m_{ij}\Gamma^{\ell}_{mk}. \tag{19.76}$$

The quantity R^{ℓ}_{ijk} shown on the left-hand side is called the **Riemann tensor** (or **curvature tensor**). Since Christoffel sumbols Γ^k_{ij} are functions of the metric tensor g_{ij}, (19.76) indicates that the Riemann tensor is defined in terms of the metric tensor and its first and second derivatives.

Recall that if the space being considered is flat, we may choose coordinates such that Γ_{ij}^k and its derivatives vanish. Therefore, we have

$$R_{ijk}^{\ell} = 0 \tag{19.77}$$

at every point in the flat region. In fact, it is possible to show that (19.77) is a necessary and sufficient condition for the region of a space to be flat. Consequently, we conclude the following: when the Riemann tensor satisfies (19.77), it indicates that the region of a space is flat and when it does not satisfy (19.77), the region is curved.

Two relevant quantities are obtained by contracting the Riemann tensor. One is the **Ricci tensor** defined by

$$R_{ij} \equiv R_{ijk}^k$$

and the other is the **scalar curvature** (or **Ricci scalar**) given by

$$R \equiv g^{ij} R_{ij} = R_i^i.$$

These two quantities are important for introducing the **Einstein tensor**

$$G^{ij} = R^{ij} - \frac{1}{2} g^{ij} R,$$

which describes the space-time curvature in the field equation of general relativity.

19.5.3 Energy–Momentum Tensor

We now wish to determine the form of the gravitational field equation that, in the weak limit of a static gravitational field, reduces, to the classical **Newtonian field of gravity** described by

$$\nabla^2 \Phi = 4\pi G \rho. \tag{19.78}$$

Here, Φ is the **potential field** that corresponds to the space-time curvature in relativistic theory, G is the **universal gravitational constant**, and ρ is the mass-density distribution of matter. Note that (19.78) is a form of **Poisson's equation** with $4\pi G\rho$ as the source term. This implies the presence of a corresponding source term associated with the space-time curvature in Einstein's field equation. This source term is given by the **energy-momentum tensor** T^{ij} defined by

$$T^{ij} = \rho u^i u^j.$$

Here, ρ is the density of matter, u^i is the **four-velocity** represented by $u^i = (u^0, u^1, u^2, u^3) = (\gamma c, \gamma \boldsymbol{v})$, where c is the velocity of light, $\boldsymbol{v}$ is the three-dimensional velocity (nonrelativistic) of a particle, and $\gamma = (1 - v^2/c^2)^{-1/2}$.

The physical interpretations of the components of the energy-momentum tensor are:

T^{00} : the energy density of the particles.
T^{0i} : the energy flux (the heat conduction) in the ith direction.
T^{i0} : the momentum density in the ith direction.
T^{ij} : the flow of the ith-component momentum in the jth direction (i.e., the random thermal motions giving rise to viscous stress).

19.5.4 Einstein Field Equation

The parameters necessary to obtain Einsteinfs field equation, which relates the geometric space-time curvature to the density of mass-energy, are already on hand. One side of the equation should comprise the measure of the density of mass-energy, i.e, the stress-energy tensor T_{ij}, and the other side should consist of a measure of the curvature involving the Ricci curvature R_{ij} and scalar curvature R. By making this equation consistent with Newtonfs equation of motion in the limit of a weak gravitational force as well as with several postulates from a physical standpoint, Einstein's field equation is obtained in the following form:

$$R_{ij} - \frac{1}{2} g_{ij} R = \frac{8\pi G}{c^4} T_{ij}. \tag{19.79}$$

Given the matter source T_{ij}, this tensor equation is composed of ten partial differential equations for the metric tensor $g_{ij}(x)$. Apparently, the tensor equation is analogous to the **Maxwell equations** that determine the electromagnetic field given the charge and current densities (see Sect. 18.5.3). Unlike the Maxwell equations, however, the differential equations of gravitational theory are nonlinear, which make them very difficult to solve. Surprisingly, despite the nonlinearity, a number of exact solutions have been obtained owing to the presence of symmetries in space-time, which restrict the possible forms of the metric.

Remark. Einstein's field equation in (19.79) is the most fundamental equation in classical physics. The explicit form of the equation can be derived from a few arguments. However, it cannot be derived from other physical principles since there is no theory that is more fundamental.

20

Tensor as Mapping

Abstract In this chapter, we show that tensors can be identified with mathematical operators that transform elements from one abstract vector space to another. This viewpoint on tensors is apparently different from those presented in Chaps. 18 and 19, where tensors have been identified as sets of index quantities subject to a transformation law under changes of coordinate systems. However, the viewpoint presented here turns out to be consistent with those presented in the previous two chapters when we introduce the concept of inner product into the abstract vector space (Sect. 20.3.4).

20.1 Vector as a Linear Function

20.1.1 Overview

In Chaps. 19 and 20 tensors are defined as collections of index quantities that obey characteristic transformation laws under a change of coordinate systems. In this chapter we present an alternative definition of tensors; that does not require specification of a coordinate system, so that it is suitable for more general tensor analyses describing geometric properties of abstract vector spaces other than our familiar three-dimensional Euclidean space.

The crucial point is that in this alternative definition, a tensor is considered not as a set of index quantities but as an operator (linear function or mapping) acting on vector spaces. For instance, a second-order tensor $\boldsymbol{T}$ is identified with a linear function that associates two vectors $\boldsymbol{v}$ and $\boldsymbol{w}$ with a real number $c \in \boldsymbol{R}$, which is symbolized by

$$\boldsymbol{T}(\boldsymbol{v}, \boldsymbol{w}) = c.$$

Emphasis should be placed on the fact that such a generalized definition of tensors applies to all kinds of general vector spaces (finite-dimensional), regardless of whether or not they possess geometric properties such as the **distance**, **norm**, or **inner product** of their elements (see Sect. 4.2.1). In fact, the tensors we discussed earlier belong to a specific class of more general tensors, in the sense that they were defined solely on the threedimensional Euclidean space, a particular class of vector spaces endowed with the inner product property. However, we shall see that, the concept of tensor can be extended beyond inner product spaces by introducing the more general definition referred to above.

Throughout the following discussion, we restrict our arguments to finite-dimensional vector spaces over $\boldsymbol{R}$ in order to provide a minimal course for general tensor calculus.

20.1.2 Vector Spaces Revisited

To begin with, we briefly review the definition of abstract vector spaces. A **vector space** (or **linear space**) V over $\boldsymbol{R}$ is a set of elements called **vectors** that have two operations, **addition** and **scalar multiplication**, and a distinguishing element $\mathbf{0} \in V$. Here, addition (denoted by $+$) assigns to each pair of elements $\boldsymbol{v}, \boldsymbol{w} \in V$ a third element $\boldsymbol{v}+\boldsymbol{w} \in V$ and the scalar multiplication assigns an element $c\boldsymbol{v} \in V$ to each $\boldsymbol{v} \in V$ and $c \in \boldsymbol{R}$. By definition, all of the elements $\boldsymbol{v}, \boldsymbol{w}, \boldsymbol{x} \in V$ and all $a, b \in \boldsymbol{R}$ must satisfy the following axioms.

1. The commutative law for $+$, i.e., $\boldsymbol{v}+\boldsymbol{w} = \boldsymbol{w}+\boldsymbol{v}$.
2. The associative law for $+$, i.e., $(\boldsymbol{v}+\boldsymbol{w})+\boldsymbol{x} = \boldsymbol{v}+(\boldsymbol{w}+\boldsymbol{x})$.
3. Existence of identity for $+$, i.e., $\boldsymbol{v}+\mathbf{0} = \boldsymbol{v}$.
4. Existence of negatives, i.e., there is $-\boldsymbol{v}$ such that $\boldsymbol{v}+(-\boldsymbol{v}) = \mathbf{0}$.
5. $a(\boldsymbol{v}+\boldsymbol{w}) = a\boldsymbol{v}+a\boldsymbol{w}$.
6. $(a+b)\boldsymbol{v} = a\boldsymbol{v}+b\boldsymbol{v}$.
7. $(ab)\boldsymbol{v} = a(b\boldsymbol{v})$.
8. $1\boldsymbol{v} = \boldsymbol{v}$.

Given two vector spaces V and W, it is possible to set a function f so that

$$f: V \to W.$$

The function f is called a **linear function** (or **linear mapping**) of V into W if for all $\boldsymbol{v}_1, \boldsymbol{v}_2 \in V$ and $c \in \boldsymbol{R}$ it yields

$$\begin{aligned} f(\boldsymbol{v}_1+\boldsymbol{v}_2) &= f(\boldsymbol{v}_1)+f(\boldsymbol{v}_2), \\ f(c\boldsymbol{v}_1) &= cf(\boldsymbol{v}_1). \end{aligned}$$

20.1.3 Vector Spaces of Linear Functions

In elementary calculus, the concepts of vectors and linear functions are distinguished from one another: vectors are elements of a vector space and linear functions provide a correspondence between them. However, in view of the axioms **1** to **8** above, we observe that the set of linear functions $f, g, \cdots$ of V into W also forms a vector space in which addition and scalar multiplication, respectively, are defined by

$$(f + g)\boldsymbol{v} = f(\boldsymbol{v}) + g(\boldsymbol{v}), \tag{20.1}$$

and

$$(cf)\boldsymbol{v} = cf(\boldsymbol{v}), \tag{20.2}$$

where $\boldsymbol{v} \in V$ and $f(\boldsymbol{v}), g(\boldsymbol{v}) \in W$. We denote by $\mathcal{L}(V, W)$ a vector space spanned by a set of linear functions f as

$$f : V \to W.$$

It is a trivial matter to verify that $f+g$ and cf are also linear functions and so belong to the same vector space $\mathcal{L}(V, W)$. These arguments are summarized by the follwing important theorem:

♠ **Vector space of linear functions:**
Let V and W be vector spaces. A set of linear functions $f : V \to W$ forms a vector space denoted by $\mathcal{L}(V, W)$.

This theorem states that the linear functions $f_1, f_2, \cdots$ of V into W are elements of a vector space $\mathcal{L}(V, W)$, analogous to vectors $\boldsymbol{v}_1, \boldsymbol{v}_2, \cdots$ being elements of a vector space V. This analogy implies that a linear function $f \in \mathcal{L}(V, W)$ can be regarded as a *vector* and, conversely, that a vector $\boldsymbol{v} \in V$ can be regarded as a *linear function*. Such identifying vectors and linear functions is crucially important for obtaining a generalized definition of tensors that is free of the concept of inner product and the specification of a coordinate system.

20.1.4 Dual Spaces

Let V^* denote a set of all linear functions such as

$$f : V \to \boldsymbol{R}.$$

(Note that the asterisk (*) in V^* does not mean complex conjugate.) Then, since

$$V^* = \mathcal{L}(V, \boldsymbol{R}),$$

it follows that V^* is a vector space. The vector space V^* is called the **dual space** (or **conjugate space**) of V, whose elements $f \in V^*$ associate a vector $\boldsymbol{v} \in V$ with a real number $c \in \boldsymbol{R}$, symbolized as

$$f(\boldsymbol{v}) = c.$$

Particularly important elements of V^* are linear functions

$$\varepsilon^i : V \to \boldsymbol{R} \quad (i = 1, 2, \cdots, n)$$

that associate a **basis** vector $\boldsymbol{e}_i \in V$ with the unit number 1. In fact, a set of such linear functions $\{\varepsilon^j\}$ serves as a basis of the dual space V^* as stated below.

♠ **Dual basis:**
For each basis $\{\boldsymbol{e}_i\}$ for V, there is a unique basis $\{\varepsilon^j\}$ for V^* such that

$$\varepsilon^j(\boldsymbol{e}_i) = \delta_i^j. \tag{20.3}$$

The linear functions $\varepsilon^j : V \to \boldsymbol{R}$ defined by (20.3) make up the **dual basis** to the basis $\{\boldsymbol{e}_i\}$ of V.

Proof Let us verify that the set of $\{\varepsilon^j\}$ defined by (20.3) serves as a basis of V^*. Recall that in finite dimensions, a basis of a vector space V is defined as a set of linearly independent vectors that spans all of V. To show linear independence, we assume that $a_j \varepsilon^j = 0$. Then we have

$$a_j \varepsilon^j(\boldsymbol{e}_i) = a_j \delta_i^j = a_i = 0 \quad \text{for all } i,$$

which implies that $\{\varepsilon^j\}$ is linearly independent. ♣

Remark. Raising of the index j attached to ε^j is intentional, as this convention is necessary to provide a consistent notation of components of generalized tensors, demonstrated in Sect. 20.3.

Examples Expand a vector $\boldsymbol{v} \in V$ as

$$\boldsymbol{v} = v^i \boldsymbol{e}_i,$$

to find that

$$\varepsilon^j(\boldsymbol{v}) = \varepsilon^j\left(v^i \boldsymbol{e}_i\right) = v^i \varepsilon^j\left(\boldsymbol{e}_i\right) = v^i \delta_i^j = v^j.$$

This indicates that ε^j is the linear function that scans the jth component of $\boldsymbol{v}$ with respect to the basis $\{\boldsymbol{e}_i\}$.

20.1.5 Equivalence Between Vectors and Linear Functions

If V is a vector space and $\tau \in V^*$, then τ is a function of the variable $\boldsymbol{v} \in V$ that generates a real number denoted by $\tau(\boldsymbol{v})$. Owing to the identification of vectors and linear functions, however, it is possible to reverse our reasoning and consider $\boldsymbol{v}$ as a *function* of the *variable* τ, again with the real value $\boldsymbol{v}(\tau) = \tau(\boldsymbol{v})$. When we take this approach, $\boldsymbol{v}$ is a linear function on V^*.

Remark. The two views contrasted above are both asymmetric, but this asymmetry can be eliminated by introducing the notation

$$\langle\ ,\ \rangle : V \times V^* \to \boldsymbol{R},$$

which gives

$$\langle \boldsymbol{v}, \tau \rangle = \tau(\boldsymbol{v}) = \boldsymbol{v}(\tau) \in \boldsymbol{R}.$$

Here $\langle\ ,\ \rangle$ is a function of two variables $\boldsymbol{v}$ and τ, called the **natural pairing** of V and V^* into $\boldsymbol{R}$. It is easy to verify that $\langle\ ,\ \rangle$ is bilinear.

The concepts and notation introduced in Sect. 20.1.3 and 20.1.4 serve as preliminaries for the discussions in the following sections.

20.2 Tensor as Multilinear Function

20.2.1 Direct Product of Vector Spaces

To arrive at the new definition of tensors we are seeking requires three more concepts, demonstrated in Sect. 20.2.1–20.2.2.

The first is the direct product of vector spaces; if V and W are vector spaces, then we can establish a new vector space by forming the **direct product** (or **Cartesian product**) $V \times W$ of the two spaces. The direct product $V \times W$ consists of *ordered* pairs $(\boldsymbol{v}, \boldsymbol{w})$ with $\boldsymbol{v} \in V$ and $\boldsymbol{w} \in W$, as symbolized by

$$V \times W = \{(\boldsymbol{v}, \boldsymbol{w}) \mid \boldsymbol{v} \in V, \boldsymbol{w} \in W\}.$$

The addition and scalar multiplication of the elements are defined by

$$\begin{aligned}
(\boldsymbol{v}, \boldsymbol{w}_1) + (\boldsymbol{v}, \boldsymbol{w}_2) &= (\boldsymbol{v}, \boldsymbol{w}_1 + \boldsymbol{w}_2), \\
(\boldsymbol{v}_1, \boldsymbol{w}) + (\boldsymbol{v}_2, \boldsymbol{w}) &= (\boldsymbol{v}_1 + \boldsymbol{v}_2, \boldsymbol{w}), \\
c(\boldsymbol{v}, \boldsymbol{w}) &= (c\boldsymbol{v}, \boldsymbol{w}) \ = (\boldsymbol{v}, c\boldsymbol{w}).
\end{aligned}$$

The linear dimension of the resulting vector spaces $V \times W$ equals the product of the linear dimensions of V and W. The elements $(\boldsymbol{v}, \boldsymbol{w})$ of the direct product $V \times W$ is sometimes noted by $\boldsymbol{vw}$.

Remark. The reader should note a distinction between the direct product $V \times W$ and the **direct sum** $V + W$ of the two vector spaces. A direct sum $V + W$ consists of all pairs $(\boldsymbol{v}, \boldsymbol{w}) \equiv (\boldsymbol{w}, \boldsymbol{v})$ with $\boldsymbol{v} \in V$ and $\boldsymbol{w} \in W$ for which addition and scalar multiplication are defined by

$$(\boldsymbol{v}_1, \boldsymbol{w}_1) + (\boldsymbol{v}_2, \boldsymbol{w}_2) = (\boldsymbol{v}_1 + \boldsymbol{v}_2, \boldsymbol{w}_1 + \boldsymbol{w}_2), \quad c(\boldsymbol{v}, \boldsymbol{w}) = (c\boldsymbol{v}, c\boldsymbol{w}).$$

The linear dimension is thus equal to the sum of the dimensions of V and W. Every linear vector space of dimension greater than one can be represented by a direct sum of nonintersecting subspaces.

20.2.2 Multilinear Functions

Let V_1, V_2 and W be vector spaces. A function

$$f : V_1 \times V_2 \to W$$

is called **bilinear** if it is linear in each variable, i.e., if,

$$\begin{aligned} f(a\boldsymbol{v}_1 + b\boldsymbol{v}'_1,\ \boldsymbol{v}_2) &= af(\boldsymbol{v}_1, \boldsymbol{v}_2) + bf(\boldsymbol{v}'_1, \boldsymbol{v}_2), \\ f(\boldsymbol{v}_1, a\boldsymbol{v}_2 + b\boldsymbol{v}'_2) &= af(\boldsymbol{v}_1, \boldsymbol{v}_2) + bf(\boldsymbol{v}'_1, \boldsymbol{v}_2). \end{aligned}$$

The extension of this definition to functions of more than two variables is simple. Indeed, functions such as

$$f : V_1 \times V_2 \times \cdots \times V_n \to W \tag{20.4}$$

are called **multilinear functions**, more specifically n**-linear functions**, for which the defining relation is

$$\begin{aligned} f(\boldsymbol{v}_1, \cdots, a\boldsymbol{v}_i + b\boldsymbol{v}'_i, \cdots, \boldsymbol{v}_n) =& af(\boldsymbol{v}_1, \cdots, \boldsymbol{v}_i, \cdots, \boldsymbol{v}_n) \\ &+ bf(\boldsymbol{v}_1, \cdots, \boldsymbol{v}'_i, \cdots, \boldsymbol{v}_n). \end{aligned}$$

An n-linear function can be multiplied by a scalar and two n-linear functions can be added; in each case the result is an n-linear function. Thus, the set of n-linear functions given in (20.4) forms a vector space denoted by $\mathcal{L}(V_1 \times \cdots \times V_n,\ W)$.

20.2.3 Tensor Product

Suppose that $\tau^1 \in V_1^*$ and $\tau^2 \in V_2^*$, i.e., τ^1 and τ^2 are linear real-valued functions on V_1 and V_2, respectively. We can then form a bilinear real-valued function such as

$$\tau^1 \otimes \tau^2 : V_1 \times V_2 \to \boldsymbol{R},$$

which is represented by

$$\tau^1 \otimes \tau^2(\boldsymbol{v}_1, \boldsymbol{v}_2) = \tau^1(\boldsymbol{v}_1)\tau^2(\boldsymbol{v}_2). \tag{20.5}$$

Note that the right-hand side of (20.5) is just the product of two real numbers: $\tau^1(\boldsymbol{v}_1)$ and $\tau^2(\boldsymbol{v}_2)$. The bilinear function $\tau^1 \otimes \tau^2$ is called the **tensor product** of τ^1 and τ^2. Clearly, since τ^1 and τ^2 are separately linear, so is $\tau^1 \otimes \tau^2$. Hence, the set of the tensor product $\tau^1 \otimes \tau^2$ forms a vector space $\mathcal{L}(V_1 \times V_2,\ \boldsymbol{R})$.

Recall that the vectors $\boldsymbol{v} \in V$ can be regarded as linear functions acting on V^*. In this context, we can also construct tensor products of two vectors. For example, let $\boldsymbol{v}_1 \in V_1$ and $\boldsymbol{v}_2 \in V_2$ and define the tensor product

$$\boldsymbol{v}_1 \otimes \boldsymbol{v}_2 : V_1^* \times V_2^* \to \boldsymbol{R}$$

by

$$\boldsymbol{v}_1 \otimes \boldsymbol{v}_2(\tau_1, \tau_2) = \boldsymbol{v}_1(\tau_1)\boldsymbol{v}_2(\tau_2) = \tau_1(\boldsymbol{v}_1)\tau_2(\boldsymbol{v}_2). \tag{20.6}$$

This shows that the tensor product $\boldsymbol{v}_1 \otimes \boldsymbol{v}_2$ can be considered a bilinear function acting on $V_1^* \times V_2^*$, similar to $\tau_1 \otimes \tau_2$ being a bilinear function on $V_1 \times V_2$, which indicates that the set of $\boldsymbol{v}_1 \otimes \boldsymbol{v}_2$ form a vector space $\mathcal{L}(V_1^* \times V_2^*, \boldsymbol{R})$.

Furthermore, given a vector space V, we can construct mixed types of tensor products such as

$$\boldsymbol{v} \otimes \tau : \ V^* \times V \to \boldsymbol{R}$$

given by

$$\boldsymbol{v} \otimes \tau(\phi, \boldsymbol{u}) = \boldsymbol{v}(\phi)\tau(\boldsymbol{u}) = \phi(\boldsymbol{v})\boldsymbol{u}(\tau), \tag{20.7}$$

where $\boldsymbol{u}, \boldsymbol{v} \in V$ and $\phi, \tau \in V^*$. In a straightforward extrapolution, it is possible to develop tensor products of more than two linear functions or vectors such as

$$\boldsymbol{v}_1 \otimes \boldsymbol{v}_2 \otimes \cdots \boldsymbol{v}_r \otimes \tau^1 \otimes \tau^2 \otimes \cdots \tau^s, \tag{20.8}$$

which act on the vector space

$$V^* \times V^* \times \cdots \times V^* \times V \times V \times \cdots \times V,$$

where V^* appears r times and V s times. Similar to the previous cases, the set of tensor products (20.8) forms a vector space denoted by

$$\mathcal{L}\left[(V^*)^r \times V^s, \ \boldsymbol{R}\right],$$

where $(V^*)^r$ and V^s are direct products of V^* with r factors and those of V with s factors, respectively.

20.2.4 General Definition of Tensors

We finally arrive at the following generalized definition of a tensor.

♠ **Tensor:**
Let V be a vector space with a dual space V^*. Then a **tensor of type** (r, s), denoted by $\boldsymbol{T}_s^r$, is a multilinear function

$$\boldsymbol{T}_s^r : (V^*)^r \times (V)^s \to \boldsymbol{R}.$$

The number r is called the **contravariant degree** of the tensor, and s is called the **covariant degree** of the tensor.

♠ **Tensor space:**

The set of all tensors $\boldsymbol{T}_s^{\,r}$ for fixed r and s forms a vector space, called a **tensor space**, denoted by

$$\mathcal{T}_s^r(V) \equiv \mathcal{L}[(V^*)^r \times V^s,\ \boldsymbol{R}].$$

As an example, let $\boldsymbol{v}_1, \cdots, \boldsymbol{v}_r \in V$ and $\tau^1, \cdots, \tau^s \in V^*$ and define the tensor product (i.e., multilinear function)

$$\boldsymbol{T}_s^{\,r} \equiv \boldsymbol{v}_1 \otimes \cdots \otimes \boldsymbol{v}_r \otimes \tau^1 \otimes \cdots \otimes \tau^s, \tag{20.9}$$

which yields for $\theta^1, \cdots, \theta^r \in V^*$ and $\boldsymbol{u}_1, \cdots, \boldsymbol{u}_s \in V$,

$$\begin{aligned} &\boldsymbol{v}_1 \otimes \cdots \otimes \boldsymbol{v}_r \otimes \tau^1 \otimes \cdots \otimes \tau^s(\theta^1, \cdots, \theta^r, \boldsymbol{u}_1, \cdots, \boldsymbol{u}_s) \\ &= \boldsymbol{v}_1(\theta^1) \cdots \boldsymbol{v}_r(\theta^r)\tau^1(\boldsymbol{u}_1) \cdots \tau^s(\boldsymbol{u}_s) \\ &= \prod_{i=1}^{r}\prod_{j=1}^{s} \boldsymbol{v}_i(\theta^i)\tau^j(\boldsymbol{u}_j). \end{aligned} \tag{20.10}$$

Observe that each $\boldsymbol{v}$ in the tensor product (20.10) requires an element $\theta \in V^*$ to produce a real number $\boldsymbol{\theta}$, which is why the number of factors of V^* in the direct product (20.9) equals the number of $\boldsymbol{v}$'s in the tensor product (20.10).

In particular, a tensor of type $(0,0)$ is defined as a **scalar**, so $\mathcal{T}_0^0(V) = \boldsymbol{R}$; a tensor of type $(1,0)$, an ordinary vector, is called a **contravariant vector**; and one of type $(0,1)$, a linear function, is called a **covariant vector**. More generally, a tensor of type $(r,0)$ is called a **contravariant tensor** of rank (or degree) r and one of type $(0,s)$ is called a **covariant tensor** of rank (or degree) s.

Remark. We can form a tensor product of two tensors $\boldsymbol{T}_s^{\,r}$ and $\boldsymbol{U}_\ell^k$ such as

$$\boldsymbol{T}_s^{\,r} \otimes \boldsymbol{U}_\ell^k :\ (V^*)^{r+k} \times V^{s+\ell} \to \boldsymbol{R},$$

which is a natural generalization of tensor products given in (20.5), (20.6), and (20.7). It is easy to prove that the tensor product is associative and distributive over tensor addition, but not commutative.

20.3 Components of Tensors

20.3.1 Basis of a Tensor Space

In physical applications of tensor calculus, it is necessary to choose a basis for the vector space V and one for its dual space V^* to represent the tensors by a set of real numbers (i.e., components). The need for this process is analogous to the cases of elementary vector calculus, in which linear operators are often represented by arrays of numbers, i.e., by matrices referring to a chosen basis of the space. A basis of our tensor space $\mathcal{T}_s^r(V) \equiv \mathcal{L}[(V^*)^r \times V^s,\ \boldsymbol{R}]$ is defined as follows.

♠ **Basis for the tensor space:**
Let $\{\boldsymbol{e}_i\}$ and $\{\varepsilon^j\}$ be a basis in V and V^*, respectively. Then, a **basis for the tensor space** $\mathcal{T}_s^r(V)$ is a set of all tensor products:

$$\boldsymbol{e}_{i_1} \otimes \cdots \otimes \boldsymbol{e}_{i_r} \otimes \varepsilon^{j_1} \otimes \cdots \otimes \varepsilon^{j_s}. \tag{20.11}$$

♠ **Components of a tensor:**
The components of any tensor $\boldsymbol{A} \in \mathcal{T}_s^r(V)$ are the real numbers given by

$$A_{j_1 \cdots j_s}^{i_1 \cdots i_r} = \boldsymbol{A}\left(\varepsilon^{i_1}, \cdots, \varepsilon^{i_r}, \boldsymbol{e}_{j_1}, \cdots, \boldsymbol{e}_{j_s}\right).$$

Remark. **1.** A useful result of the theorem is the relation

$$\boldsymbol{A} = A_{j_1 \cdots j_s}^{i_1 \cdots i_r} \boldsymbol{e}_{i_1} \otimes \cdots \otimes \boldsymbol{e}_{i_r} \otimes \varepsilon^{j_1} \otimes \cdots \otimes \varepsilon^{j_s}.$$

2. Note that for every factor in the basis of $\mathcal{T}_s^r(V)$, there are N possibilities. (For instance, we have N choices for $\boldsymbol{e}_{i_1}$ in which $i_1 = 1, 2 \cdots, N$.) Thus, the number of possible tensor products represented by (20.11) is N^{r+s}.

Examples **1.** A tensor space $\mathcal{T}_0^1(V)$ has a basis $\{\boldsymbol{e}_i\}$ so that an element (i.e., a contravariant vector) $\boldsymbol{v} \in \mathcal{T}_0^1(V)$ can be expanded by

$$\boldsymbol{v} = v^i \boldsymbol{e}_i,$$

where the real numbers $v^i = \boldsymbol{v}(\varepsilon^i)$ are called the components of $\boldsymbol{v} : V \to \boldsymbol{R}$.

2. A tensor space $\mathcal{T}_1^0(V)$ has a basis $\{\varepsilon^j\}$ so that an element (i.e., a covariant vector) $\tau \in \mathcal{T}_1^0(V)$ can be expanded by

$$\tau = \tau_j \varepsilon^j,$$

where the real numbers $\tau_j = \tau(\boldsymbol{e}_j)$ are called the components of $\tau : V^* \to \boldsymbol{R}$.

3. A tensor space $\mathcal{T}_1^2(V)$ has a basis $\{\boldsymbol{e}_i \otimes \boldsymbol{e}_j \otimes \varepsilon^k\}$ so that for any $\boldsymbol{A} \in \mathcal{T}_1^2(V)$ we have

$$\boldsymbol{A} = A_k^{ij} \boldsymbol{e}_i \otimes \boldsymbol{e}_j \otimes \varepsilon^k,$$

where the real numbers

$$A_k^{ij} = \boldsymbol{A}\left(\varepsilon^i, \varepsilon^j, \boldsymbol{e}_k\right)$$

are the components of $\boldsymbol{A}: V \times V \times V^* \to \boldsymbol{R}$.

20.3.2 Transformation Laws of Tensors

The components of a tensor depend on the basis in which they are described. If the basis is changed, the components change. The relation between components of a tensor in different bases is called the **transformation law** for that particular tensor. Let us investigate this concept.

Assume two different bases of V, denoted by $\{\boldsymbol{e}_i\}$ and $\{\boldsymbol{e}'_i\}$. Similarly, we denote by $\{\varepsilon^j\}$ and $\{\varepsilon'^j\}$ two different bases of V^*. We can find appropriate transformation matrices $[R_i^j]$ and $[S_k^\ell]$ that satisfy

$$\boldsymbol{e}'_i = R_i^j \boldsymbol{e}_j \quad \text{and} \quad \varepsilon'^k = S_\ell^k \varepsilon^\ell .$$

Then, for a tensor $\boldsymbol{T}$ of type $(1,2)$, we have

$$\begin{aligned} T'^i_{jk} &= \boldsymbol{T}\left(\varepsilon'^i, \boldsymbol{e}'_j, \boldsymbol{e}'_k\right) = \boldsymbol{T}\left(S_\ell^i \varepsilon^\ell, R_j^m \boldsymbol{e}_m, R_k^n \boldsymbol{e}_n\right) \\ &= S_\ell^i R_j^m R_k^n \boldsymbol{T}\left(\varepsilon^\ell, \boldsymbol{e}_m, \boldsymbol{e}_n\right) \\ &= S_\ell^i R_j^m R_k^n T_{mn}^\ell , \end{aligned} \tag{20.12}$$

which is the transformation law of the components of the tensor $\boldsymbol{T}$ of type $(1,2)$.

Remember that in the coordinate-*dependent* treatment of tensors (as shown in Chaps. 18 and 19), the result (20.12) was considered to be the defining relation for a tensor of type $(1,2)$. In other words, a tensor of type $(1,2)$ was defined as a collection of numbers T_{np}^m that transform to another collection of numbers T'^i_{jk} according to the rule in (20.12) when the basis is changed. In our current (i.e., coordinate-*free*) treatment of tensors, it is not necessary to introduce a basis to define a tensor; a basis must be introduced only when the components of a tensor are needed. The advantage of this approach is obvious, since a $(1,2)$–type tensor has 27 components in three dimensions and 64 components in four dimensions, and all of these can be represented by the single symbol $\boldsymbol{T}$.

Remark. Note that the above arguments do not downplay the role of components. In fact, when it comes to actual calculations, we are forced to choose a basis and manipulate the components.

20.3.3 Natural Isomorphism

We comment below on an important property that is specific to components of tensors $\boldsymbol{A} \in \mathcal{T}_1^1(V)$. We know that tensors $\boldsymbol{A} \in \mathcal{T}_1^1(V)$ are bilinear functions such as

$$\boldsymbol{A} : V^* \times V \to \boldsymbol{R}$$

and that their components A^i_j are defined by

$$A^i_j = \boldsymbol{A}(\varepsilon^i, \boldsymbol{e}_j), \tag{20.13}$$

where each ε^i and $\boldsymbol{e}_j$ is a basis of V^* and V, respectively. Now we consider the matrix

$$[A^i_j] = \begin{bmatrix} A^1_1 & A^1_2 & \cdots & A^1_n \\ A^2_1 & & \cdots & \cdots \\ \cdots & & & \cdots \\ A^n_1 & \cdots & & A^n_n \end{bmatrix}, \tag{20.14}$$

whose elements A^i_j are the same as those given in (20.13). We shall see that (20.14) is the matrix representation of a linear operator A in terms of the basis $\{\boldsymbol{e}_i\}$ of V, which associates a vector $\boldsymbol{v} \in V$ with another $\boldsymbol{u} \in V$, i.e.,

$$\mathsf{A} : V \to V.$$

A formal statement on concerning this point is given below.

♠ **Natural isomorphism:**

For any vector space V, there is a one-to-one correspondence (called a **natural isomorphism**) between a tensor $\boldsymbol{A} \in \mathcal{T}^1_1(V)$ and a linear operator $\mathsf{A} \in \mathcal{L}(V, V)$.

Proof We write the tensor $\boldsymbol{A} \in \mathcal{T}^1_1(V)$ as

$$\boldsymbol{A} = A^i_j \boldsymbol{e}_i \otimes \varepsilon^j.$$

Given any $\boldsymbol{v} \in V$, we obtain

$$\begin{aligned} \boldsymbol{A}(\boldsymbol{v}) &= \left(A^i_j \boldsymbol{e}_i \otimes \varepsilon^j\right)(\boldsymbol{v}) = A^i_j \boldsymbol{e}_i \left[\varepsilon^j(\boldsymbol{v})\right] \\ &= A^i_j \boldsymbol{e}_i \left[\varepsilon^j(v^k \boldsymbol{e}_k)\right] = A^i_j \boldsymbol{e}_i \left(v^k \delta^j_k\right) \\ &= A^i_j v^j \boldsymbol{e}_i. \end{aligned} \tag{20.15}$$

Observe that the $A^i_j v^j$ in the last term are real numbers and that $\boldsymbol{e}_i \in V$. Hence, the object $\boldsymbol{A}(\boldsymbol{v})$ is a linear combination of bases $\boldsymbol{e}_i$ for V, i.e.,

$$\boldsymbol{A}(\boldsymbol{v}) \in V.$$

Denoting $\boldsymbol{A}(\boldsymbol{v})$ in (20.15) by $\boldsymbol{u} = u^i \boldsymbol{e}_i$, we find that

$$u^i = A^i_j v^j, \tag{20.16}$$

in which u^i, v^j are contravariant components of the vectors $\boldsymbol{u}, \boldsymbol{v} \in V$, respectively, in terms of the basis $\{\boldsymbol{e}_i\}$. The result (20.16) is identified with a matrix equation:

$$\begin{bmatrix} u^1 \\ u^2 \\ \cdots \\ u^n \end{bmatrix} = \begin{bmatrix} A_1^1 \; A_2^1 \cdots A_n^1 \\ A_1^2 \qquad\quad \cdots \\ \cdots \\ A_1^n \; A_2^n \cdots A_n^n \end{bmatrix} \begin{bmatrix} v^1 \\ v^2 \\ \cdots \\ v^n \end{bmatrix}. \tag{20.17}$$

Thus we can see that given $\boldsymbol{A} \in \mathcal{T}_1^1(V)$, its components form the matrix representation $[A_j^i]$ of a linear operator A that transforms a vector $\boldsymbol{v} \in V$ into another vector $\boldsymbol{u} \in V$.

Conversely, for any given linear operator on V with a matrix representation $[A_j^i]$ in terms of a basis of V, there exists a tensor $\boldsymbol{A} \in \mathcal{T}_1^1(V)$. This suggests a one-to-ome correspondence between the tensor space $\mathcal{T}_1^1(V)$ and the vector space $\mathcal{L}(V, V)$ comprising the linear mapping $f : V \to V$. ♣

A parallel discussion serves for a linear operator on V^*. In fact, for any $\tau \in V^*$, we have

$$\begin{aligned} \boldsymbol{A}(\tau) &= \left(A_j^i \boldsymbol{e}_i \otimes \varepsilon^j\right)(\tau) = A_j^i \left[\boldsymbol{e}_i(\tau)\right] \varepsilon^j \\ &= A_j^i \left[\boldsymbol{e}_i\left(\tau_k \varepsilon^k\right)\right] \varepsilon^j = A_j^i \left(\tau_k \delta_i^k\right) \varepsilon^j \\ &= A_j^i \tau_i \varepsilon^j, \end{aligned}$$

which means that $\boldsymbol{A}(\tau)$ is a linear combination of bases ε^j for V^*, i.e.,

$$\boldsymbol{A}(\tau) \in V^*.$$

Denoting $\boldsymbol{A}(\tau)$ by $\theta = \theta_j \varepsilon^j$, we obtain

$$\theta_j = A_j^i \tau_i, \tag{20.18}$$

where θ_j and τ_i are (covariant) components of the *vectors* $\theta, \tau \in V^*$ in terms of the basis $\{\varepsilon^i\}$. Using the same matrix representation of $[A_j^i]$ as in (20.17), we can rewrite (20.18) as

$$[\theta_1, \cdots, \theta_n] = [\tau_1, \cdots, \tau_n] \begin{bmatrix} A_1^1 \; A_2^1 \cdots A_n^1 \\ A_1^2 \qquad\quad \cdots \\ \cdots \\ A_1^n \; A_2^n \cdots A_n^n \end{bmatrix},$$

which describes a linear mapping from a vector $\tau \in V^*$ to another vector $\theta \in V^*$ through the linear operator with the matrix representation $[A_j^i]$.

We thus conclude that there is a natural isomorphism among the three vecror spaces:

$$\mathcal{T}_1^1(V) = \mathcal{L}(V^* \times V,\ \boldsymbol{R}), \quad \mathcal{L}(V, V), \quad \text{and } \mathcal{L}(V^*, V^*).$$

Owing to this isomorphism, these three vector spaces can be treated as though they are the same.

20.3.4 Inner Product in Tensor Language

As noted at the beginning of this chapter, our current discussion is applicable to any kind of vector space regardless of whether or not it is endowed with inner product properties. If the spaces we are dealing are inner product spaces, then all the results of Chaps. 18 and 19 are reproduced, owing the assumption that only a Euclidean space (i.e., a real inner product space; see Sect. 19.2.3) is considered there. In this subsection, we shall see that this is true, but we first have to introduce the concept of inner product in connection with our current vector spaces. Such a discussion enables us to make the correspondence between the two views of tensors clear:

tensors are *sets of index-quantities* (in Chaps. 18 and 19),

and

tensors are *linear mappings* (in Chap. 20).

Below is the definition of the inner product in the language of tensor calculus.

♠ **Inner product:**
An inner product, denoted by (,), is a real-valued function such as

$$(\, , \,) : V \times V \to \boldsymbol{R},$$

which has the following properties:

(i) it is **nondegenerate**, i.e.,

$$(\boldsymbol{u}, \boldsymbol{v}) = 0 \quad \text{for all } \boldsymbol{v} \quad \Longleftrightarrow \quad \boldsymbol{u} \equiv \boldsymbol{0},$$

(ii) it is **symmetric**, i.e., $(\boldsymbol{u}, \boldsymbol{v}) = (\boldsymbol{v}, \boldsymbol{u})$,
(iii) it is **positive definite**, i.e., $(\boldsymbol{u}, \boldsymbol{v}) > 0$ whenever $\boldsymbol{u} \neq \boldsymbol{0}$, and
(iv) it is **bilinear**, i.e., $(a\boldsymbol{u} + b\boldsymbol{v}, \boldsymbol{w}) = a(\boldsymbol{u}, \boldsymbol{v}) + b(\boldsymbol{v}, \boldsymbol{w})$.

Remark. The set of four axioms above is a restatement of those presented in Sect. 4.1.3 for real vector spaces.

By definition, the inner product of $\boldsymbol{v}$ and $\boldsymbol{w}$ reads

$$(\boldsymbol{v}, \boldsymbol{w}) = (v^i \boldsymbol{e}_i, w^j \boldsymbol{e}_j) = v^i w^j (\boldsymbol{e}_i, \boldsymbol{e}_j),$$

where $(\boldsymbol{e}_i, \boldsymbol{e}_j)$ as well as $(\boldsymbol{v}, \boldsymbol{w})$ are certain real numbers. Then, if we establish a matrix $[g_{ij}]$ with the entities

$$g_{ij} \equiv (\boldsymbol{e}_i, \boldsymbol{e}_j), \tag{20.19}$$

we have

$$(\boldsymbol{v}, \boldsymbol{w}) = g_{ij} v^i w^j,$$

which reproduces the previous notation (18.27) obtained via the coordinate-*dependent* treatment of tensors.

Remark. The notation in (20.19) seems to imply that the function (,) is written in terms of the dual basis $\varepsilon^i \in V^*$ by

$$(\ ,\ ) = g_{ij} \varepsilon^i \otimes \varepsilon^j. \tag{20.20}$$

However, this is not the case because (20.20) does not satisfy the symmetric property required by **(ii)** in the above definition of an inner product.

20.3.5 Index Lowering and Raising in Tensor Language

We demonstrate below another important consequence of the notation in (20.19). Since the inner product $(\boldsymbol{v}, \boldsymbol{w})$ is a bilinear function with variables $\boldsymbol{v}$ and $\boldsymbol{w}$, it is a linear function of $\boldsymbol{w}$ if we fix $\boldsymbol{v}$. Assume a function $\nu : V \to \boldsymbol{R}$ defined by

$$\nu(\boldsymbol{w}) \equiv (\boldsymbol{v}, \boldsymbol{w}).$$

Clearly, ν is a linear function of $\boldsymbol{w}$ and $\nu \in V^*$. Hence, ν can be expanded by the dual basis $\varepsilon^j \in V^*$, which results in

$$\begin{aligned} \nu = \nu_j \varepsilon^j &= \nu(\boldsymbol{e}_j)\varepsilon^j \\ = (\boldsymbol{v}, \boldsymbol{e}_j)\varepsilon^j &= (v^i \boldsymbol{e}_i, \boldsymbol{e}_j)\varepsilon^j \\ = v^i g_{ij} \varepsilon^j. & \end{aligned}$$

This indicates that the components ν_j of the linear function ν are given by

$$\nu_j = g_{ij} v^i.$$

Denoting ν_j by v_j, we obtain

$$v_j = g_{ij} v^i, \tag{20.21}$$

which is identified with the index lowering of v^i by the use of g_{ij}. Emphasis is placed on the fact that the result (20.21) gives a one-to-one relation between $\boldsymbol{v} \in V$ and $\nu \in V^*$ via the entities $g_{ij} = (\boldsymbol{e}_i, \boldsymbol{e}_j)$. That is, going from a vector $\boldsymbol{v} \in V$ to its unique image $\nu \in V^*$ is achieved by simply lowering the index of the contravariant component of $\boldsymbol{v}$ through relation (20.21).

The counterpart of (20.21), index raising, is obtained by noting the fact that by hypothesis the matrix $[g_{ij}]$ is nondegenerate. This implies the existence of the inverse matrix denoted by $[g^{ij}]$. Multiplying the elements g^{kj} by both sides of (20.21) yields

$$g^{kj} v_j = g^{kj} g_{ij} v^i = g^{kj} g_{ji} v^i = \delta^k_i v^i = v^k,$$

i.e.,

$$v^i = g^{ij} v_j.$$

We have thus shown that the introduction of the matrix $[g_{ij}]$ composed of the real numbers g_{ij} defined by (20.19) provides a bridge between the two apparently different viewpoints (those in Chaps. 18, 19 and in Chap. 20) regarding tensors.

Part VII

Appendixes

A

Proof of the Bolzano–Weierstrass Theorem

A.1 Limit Points

In this appendix we prove the **Bolzano–Weierstrass theorem**, first introduced in Sect. 2.2.2, which guarantees the existence of a **limit point** in some sets of real numbers. For a better understanding, we begin with a brief review of the basic properties of limit points.

Below is we repeat the definition of a limit point from Sect. 1.1.5.

♠ **Limit point:**

A point $x \in \boldsymbol{R}$ is called a **limit point** of a set $S \subseteq \boldsymbol{R}$ if every neighborhood V of x contains an element different from x.

We denote by $\hat{S}$ the set of limit points of S. A point in S that is not a limit point of S is called an **isolated point** of S. A limit point is often referred to as a **cluster point** or **accumulation point**.

Observe that $x \in \hat{S}$ if and only if every neighborhood of x contains an infinite number of points of S. This is so because if a neighborhood

$$V = (x - \delta, x + \delta)$$

of a limit point x contains only a finite number of points, say $a_1, a_2, \cdots, a_n$, where $a_i \neq x$, then there is a positive number ε such that

$$\varepsilon = \min_{1 \le i \le n} |a_i - x|.$$

Since x is a limit point of S, there is a point $a \in S$ such that $a \neq x$ and $|x - a| < \varepsilon$. This means that $a \in V$ but $a \neq a_i$ for any i, which contradicts the assumption that V contains only n points of S. The implication in the other direction is obvious.

A **finite set** cannot have a limit point, since any neighborhood of a limit point must contain an infinite number of elements of the set. On the other hand, an **infinite set** may or may not have a limit point.

A.2 Cantor Theorem

The previous discussion raises the question: When does a set possess a limit point? The following theorem serves as a lemma to answer this question. Meanwhile we denote by $\ell(I) = b-a$ the length of any closed interval $I = [a, b]$ with $a \leq b$.

♠ **Cantor theorem:**

Let (I_n) be a sequence of nonempty, closed, and bounded intervals. If $I_{n+1} \subseteq I_n$ for every $n \in \boldsymbol{N}$, then the intersection $\bigcap_{n=1}^{\infty} I_n$ is not empty. Furthermore, if

$$\inf\{\ell(I_n) \ : \ n \in \boldsymbol{N}\} = 0,$$

then $\bigcap_{n=1}^{\infty} I_n$ is a single point.

Proof Suppose $I_n = [a_n, b_n]$ and $I = \bigcap_{n=1}^{\infty} I_n$. Using the nested property of the intervals I_n, we have

$$m \geq p \ \Rightarrow \ I_m \subseteq I_p \ \Rightarrow \ a_p \leq a_m < b_m \leq b_p. \tag{A.1}$$

Clearly, the set $S = \{a_n \ : \ n \in \boldsymbol{N}\} \subset \boldsymbol{R}$ is not empty and is bounded above by b_1. Hence, the set S has a least upper bound, which we call x. If we can prove that $x \in I$, we can conclude that I is not empty. In fact, this can be proven by observing that

$$x \in I_k \quad \text{for all } k \in \boldsymbol{N},$$

i.e.,

$$a_k \leq x \leq b_k \quad \text{for all } k \in \boldsymbol{N}. \tag{A.2}$$

First, it is obvious from the definition of x that $a_k \leq x$ for all k. Second, b_k for arbitrary k satisfies $a_n \leq b_k$ for all $n \in \boldsymbol{N}$, i.e., b_k is an upper bound of S. In fact, if $n \leq k$ then, by (A.1), $a_n \leq a_k \leq b_k$; and if $n > k$ then, again by (A.1), $a_n \leq b_n \leq b_k$. Finally, it follows that $x \leq b_k$ for all $k \in \boldsymbol{N}$, since x is a *least* upper bound of S, whereas b_k is just an upper bound of S. Thus we can conclude that (A.2) is true.

Now we consider the second statement in the above theorem. Suppose that

$$\inf\{\ell(I_n) \ : \ n \in \boldsymbol{N}\} = 0,$$

and let $x, y \in I$. It then follows that $x, y \in I_n$ for every n, which implies that

$$|x - y| \leq \ell(I_n) \quad \text{for all } n \in \boldsymbol{N},$$

so

$$|x - y| \leq \inf\{\ell(I_n) \ : \ n \in \boldsymbol{N}\} = 0.$$

This means that $x = y$, i.e., the interval

$$I = \bigcap_{n=1}^{\infty} I_n$$

is a single point. ♣

A.3 Bolzano–Weierstrass Theorem

We are now ready to prove the Bolzano–Weierstrass theorem, which gives us sufficient conditions for the existence of a limit point in a set.

♠ **Bolzano–Weierstrass theorem:**
Every infinite and bounded subset of $\boldsymbol{R}$ has at least one limit point in $\boldsymbol{R}$.

Proof Let S be an infinite and bounded set of real numbers. Being bounded, S is contained in some bounded closed interval $I_0 = [a_0, b_0]$. First we bisect I_0 into the two subintervals

$$I_0' = \left[a_0, \frac{a_0 + b_0}{2}\right], \; I_0'' = \left[\frac{a_0 + b_0}{2}, b_0\right].$$

Since $S \subseteq (I_0' \cup I_0'')$ is infinite, at least one of the two sets $S \cap I_0'$ and $S \cap I_0''$ is infinite. Let $I_1 = [a_1, b_1] = I_0'$ if $S \cap I_0'$ is infinite; otherwise let $I_1 = I_0''$. We then have

$$I_1 \subseteq I_0, \quad \ell(I_1) = \frac{b_0 - a_0}{2}.$$

Now we bisect I_1 into the subintervals

$$I_1' = \left[a_1, \frac{a_1 + b_1}{2}\right], \; I_1'' = \left[\frac{a_1 + b_1}{2}, b_1\right],$$

one of which necessarily intersects S in an infinite set. Let $I_2 = [a_2, b_2] = I_1'$ if $S \cap I_1'$ is infinite; otherwise let $I_2 = I_1''$. Continuing in this fashion, we obtain the intervals $I_i = [a_i, b_i]$, $0 \leq i \leq n$, which satisfy

$$I_i \subseteq I_{i-1}, \quad \ell(I_i) = \frac{b_0 - a_0}{2^i},$$

and the fact that $S \cap I_i$ is infinite. We bisect I_n again to obtain

$$I_n' = \left[a_n, \frac{a_n + b_n}{2}\right], \; I_n'' = \left[\frac{a_n + b_n}{2}, b_n\right].$$

Since $I_n = I_n' \cup I_n''$ and $S \cap I_n$ is infinite, either $S \cap I_n'$ is infinite wherein we set $I_{n+1} = [a_{n+1}, b_{n+1}] = I'_n$, or $S \cap I''_n$ is infinite where $I_{n+1} = I''_n$ is chosen. Now we see that

$$I_{n+1} \subseteq I_n, \quad \ell(I_{n+1}) = \frac{b_0 - a_0}{2^{n+1}} = \frac{\ell(I_n)}{2}$$

and that $S \cap I_{n+1}$ is infinite. By induction, therefore, it is show that there exists a sequence (I_n) of nonempty, closed, and bounded intervals. In view of Cantor's theorem, we see that the intersection $\bigcap_{n=1}^{\infty} I_n$ contains at least one single point x. We now complete our proof by showing that $x \in \hat{S}$.

Suppose that $x \in \bigcap_{n=1}^{\infty} I_n$. Given any $\varepsilon > 0$, we can choose $n \in \boldsymbol{N}$ so that

$$\frac{b_0 - a_0}{2^n} < \varepsilon,$$

or equivalently,

$$|I_n| < \varepsilon.$$

This, together with the fact that $x \in I_n$ for all n, implies that

$$I_n \subset (x - \varepsilon, x + \varepsilon).$$

Since I_n contains an infinite number of elements of S, so does the neighborhood $(x - \varepsilon, x + \varepsilon)$ of x; hence $x \in \hat{S}$. ♣

B

Dirac δ Function

B.1 Basic Properties

In this appendix, we review the properties and various expressions of **Dirac's δ function**. The first thing to be noted is that the δ function is not a function at all. A function is a rule that assigns another number to each number in a set of numbers. However, the δ function as used in physics is rather a shorthand notation for a complicated limiting process whose use greatly simplifies calculations. It takes on a meaning only when it appears under an integral sign, in which case it behaves as

$$\int_{-\infty}^{\infty} f(x)\delta(x)dx = f(0), \tag{B.1}$$

For the special case of $f(x) \equiv 1$. We have

$$\int_{-\infty}^{\infty} \delta(x)dx = 1. \tag{B.2}$$

If the singular point is located at an arbitrary point x_0, then

$$\int_{-\infty}^{\infty} f(x)\delta(x - x_0)dx = f(x_0). \tag{B.3}$$

Except at the singular point $x_0 = 0$,

$$\delta(x) = 0. \tag{B.4}$$

Thus $\delta(x)$ vanishes at all points where its argument is not zero, but at that one point it is undefined. Nevertheless its behavior near this point is all that matters.

Let $\delta_a(x)$ be a set of functions parametrized by the index a that has the properties

$$\lim_{a\to 0} \delta_a(x) = 0 \qquad \text{for all } x \neq 0,$$

$$\lim_{a\to 0} \int_{-\infty}^{+\infty} f(x)\delta_a(x)dx = f(0). \tag{B.5}$$

In precise terms the original equations defining the δ function must be interpreted as standing for the limiting processes of (B.5).

B.2 Representation as a Limit of Function

In what follows, we look at several sets of functions that are endowed the properties described in (B.5).

1. The limit of a box function

The simplest example is the function $\delta_c(x)$ defined (for $c > 0$) by

$$\delta_c(x) \equiv \begin{cases} 1/c & \text{for } |x| \leq c/2, \\ 0 & \text{for } |x| > c/2. \end{cases} \tag{B.6}$$

Clearly, $\lim_{c\to 0} \delta_c(x) = 0$ at all $x \neq 0$ and $\int_{-\infty}^{+\infty} \delta_c(x)dx = 1$ independent of c. In addition, we have

$$\lim_{c\to 0} \int_{-\infty}^{\infty} f(x)\delta_c(x)dx = f(0), \tag{B.7}$$

which can be shown formally for continuous functions $f(x)$ as

$$\begin{aligned} \lim_{c\to 0} \int_{-\infty}^{\infty} f(x)\delta_c(x)dx &= \lim_{c\to 0} \int_{-c/2}^{c/2} f(x)\delta_c(x)dx = \lim_{c\to 0} \frac{1}{c} \int_{-c/2}^{c/2} f(x)dx \\ &= \lim_{c\to 0} \frac{f(\xi c)}{c} \int_{-c/2}^{c/2} dx = \lim_{c\to 0} f(\xi c). \end{aligned}$$

In the last line, the mean value theorem of integral calculus was employed with the definition $-1/2 < \xi < 1/2$. Letting $c \to 0$, we obtain (B.7).

2. The limit of a Gaussian function

The sequence of the Gaussian distribution function

$$\delta_a(x) \equiv \frac{1}{a\sqrt{\pi}} e^{-x^2/a^2}$$

provides another representation of the δ function. Note that $\lim_{a\to 0}\delta_a(x) = 0$ at all $x \neq 0$, $\int_{-\infty}^{+\infty}\delta_a(x)dx = 1$ independent of a, and $\lim_{a\to 0}\int_{-\infty}^{\infty} f(x)\delta_a(x)\, dx = f(0)$. The entire contribution to the integral, as $a \to 0$, comes from the neighborhood of $x = 0$. Therefore, we may write symbolically,

$$\delta(x) = \lim_{a\to 0}\delta_a(x) = \lim_{a\to 0}\frac{1}{a\sqrt{\pi}}e^{-x^2/a^2}.$$

3. The limit of a Lorentzian function

Another useful representation for the δ function is

$$\delta(x) = \lim_{\epsilon\to 0}\delta_\epsilon(x) \equiv \lim_{\epsilon\to 0}\frac{1}{\pi}\frac{\epsilon}{x^2+\epsilon^2},$$

which the reader can work out as in the above example.

4. The $n \to \infty$ limit of a sequence of functions

The final representation of the δ function is slightly different from the preceding three and plays a central role in the proof of the Weierstrass theorem, as demonstrated in Appendix C. It is defined as

$$\delta_n(x) = \begin{cases} c_n(1-x^2)^n & \text{for } 0 \le |x| \le 1 \quad (n = 1,2,3\cdots), \\ 0 & \text{for } |x| > 1, \end{cases} \tag{B.8}$$

where the constant c_n must be determined so that

$$\int_{-1}^{1}\delta_n(x)dx = 1. \tag{B.9}$$

The functions $\delta_n(x)$ form a sequence whose limit is a δ function. This representation of the δ function differs from the others in that the defining parameter n increases to infinity, rather than decreasing continuously to zero.

B.3 Remarks on Representation 4

We show below that representation **4** above has the conditions (B.5) required for identification with Dirac's δ function. At first, we determine the normalization constant c_n. From the hypothesis (B.9), we have

$$\frac{1}{c_n} = \int_{-1}^{1}(1-x^2)^n dx = 2\int_0^1 (1-x^2)^n dx. \tag{B.10}$$

Making the change of variable $x = \sin\theta$, we obtain

$$\frac{1}{c_n} = 2\int_{\pi/2}^{0} \cos^{2n+1}\theta d\theta = \frac{2^{n+1}n!}{1\cdot 3\cdot 5\cdots(2n+1)}, \tag{B.11}$$

which becomes

$$c_n = \frac{(2n+1)!}{2^{2n+1}(n!)^2}. \tag{B.12}$$

Next we consider the asymptotic behavior of c_n as $n \to \infty$. It follows from (B.10) that

$$\frac{1}{c_n} = 2\int_0^1 (1-x^2)^n dx \geq 2\int_0^{1/\sqrt{n}} (1-x^2)^n dx, \tag{B.13}$$

since $1/\sqrt{n}$ for all $n = 1, 2, \cdots$ and the integrand is positive throughout $[0, 1]$. Now we consider the function

$$g(x) \equiv (1-x^2)^n - (1-nx^2).$$

Since $g(0) = 0$ and

$$g'(x) = 2nx\left[1-(1-x^2)^{n-1}\right] > 0 \quad \text{for all } 0 < x \leq 1,$$

$g(x)$ must be monotonically increasing in the interval $[0, 1]$. Therefore, $g(x) \geq 0$, or equivalently,

$$(1-x^2)^n \geq 1 - nx^2$$

for all x in $[0, 1]$. Using this inequality in (B.13), we have

$$\frac{1}{c_n} \geq 2\int_0^{1/\sqrt{n}} (1-nx^2)^n dx = \frac{4}{3\sqrt{n}} > \frac{1}{\sqrt{n}},$$

i.e.,

$$c_n < \sqrt{n}. \tag{B.14}$$

This result implies that the limit $n \to \infty$ of the function $\delta_n(x)$ given in (B.8) equals zero for all $x \neq 0$.

Finally, we examine the validity of the relation:

$$\lim_{n\to\infty}\int_{-\infty}^{\infty} f(x)\delta_n(x)dx = f(0). \tag{B.15}$$

To prove this, it suffices to verify that the contribution to the integral $\int_{-1}^{1}\delta_n(x)dx$ comes increasingly from the neighborhood surrounding the origin as $n \to \infty$. Note that for $0 < \varepsilon < 1$,

$$\int_{-1}^{-\varepsilon} \delta_n(x)dx = \int_{\varepsilon}^{1} \delta_n(x)dx, \tag{B.16}$$

since $\delta_n(x)$ is an even function of x. Now

$$\int_\varepsilon^1 \delta_n(x)dx < \sqrt{n}\int_\varepsilon^1 (1-x^2)^n dx < \sqrt{n}(1-\varepsilon^2)^n(1-\varepsilon) < \sqrt{n}(1-\varepsilon^2)^n, \quad \text{(B.17)}$$

where we employ the fact that $(1-x^2)^n$ in the interval $[\varepsilon, 1]$ takes its maximum at $x = \varepsilon$. It is obvious that the decreasing behavior of the term $(1-\varepsilon^2)^n$ with n dominates increasing behavior of the term $\sqrt{n}$, so that

$$\lim_{n\to\infty}\int_\varepsilon^1 \delta_n(x)dx = 0. \quad \text{(B.18)}$$

The results (B.16) and (B.18) justify the desired relation (B.15).

C

Proof of Weierstrass Approximation Theorem

♠ **Weierstrass approximation theorem:**
If a function $f(x)$ is continuous on the closed interval $[a, b]$, there exists a polynomial

$$G_n(x) = \sum_{k=0}^{n} c_k x^k \tag{C.1}$$

that converges uniformly to $f(x)$ on $[a, b]$.

To prove this, we may assume without loss of generality that $f(x)$ is defined on $[0, 1]$ and that $f(0) = f(1) = 0$. Outside the interval $[0, 1]$, we may define $f(x)$ to be identically zero. Then, the relevant polynomial (C.1) is given by the integral form as

$$G_n(x) = \int_{-1}^{1} f(x+t)\delta_n(t)dt, \quad 0 \le x \le 1. \tag{C.2}$$

Here $\delta_n(t)$ is the sequence of the functions represented by

$$\delta_n(t) = \begin{cases} c_n(1-t^2)^n & \text{for } -1 \le t \le 1, \\ 0 & \text{for } |t| > 1, \end{cases}$$

where c_n is

$$c_n = \frac{(2n+1)!}{2^{2n+1}(n!)^2}, \quad \text{so that} \quad \int_{-1}^{1} \delta_n(t)dt = 1.$$

(In fact, the sequence $\delta_n(t)$ as $n \to \infty$ does have the properties characterizing a δ **function**; see Appendix B). Since $f(x)$ is assumed to vanish outside the interval $[0, 1]$, (C.2) can be rewritten as

$$G_n(x) = \int_{-x}^{1-x} f(x+t)\delta_n(t)dt.$$

By a change of variable $t \to t - x$, we obtain

$$G_n(x) = \int_0^1 f(t)\delta_n(t-x)dt = \int_0^1 f(t)c_n\left[1-(t-x)^2\right]^n dt.$$

This last integral shows that $G_n(x)$ is a polynomial of degree $2n$ in x. In what follows, we prove that the sequence of polynomials given by $\{G_1(x), G_2(x), \cdots\}$ converges uniformly to $f(x)$.

Since $f(x)$ is continuous on $[0,1]$, there exists an appropriate infinitesimal δ such that for a given $\varepsilon > 0$,

$$|f(x+\delta) - f(x)| < \varepsilon$$

for all x in $[0,1]$. Now, we use (C.2) for $G_n(x)$ to obtain the quantity

$$\begin{aligned}|G_n(x) - f(x)| &= \left|\int_{-1}^1 [f(x+t) - f(x)]\,\delta_n(t)dt\right| \\ &\le \int_{-1}^1 |f(x+t) - f(x)|\,\delta_n(t)dt, \end{aligned} \tag{C.3}$$

where $\delta_n(t) \ge 0$ on $t \in [0,1]$. If the last term in (C.3) vanishes as $n \to \infty$, the proof of the theorem is complete.

To show this, we break up the range of integration into three parts,

$$\int_{-1}^1 = \int_{-1}^{-\gamma} + \int_{-\gamma}^{\gamma} + \int_{\gamma}^1,$$

where γ is a certain infinitesimal number. Since $f(x)$ is continuous on a closed interval, it is bounded there. Let the maximum value of $|f(x)| = M$. Then the last term of integrals becomes

$$\begin{aligned}\int_{\gamma}^1 |f(x+t) - f(x)|\,\delta_n(t)dt &\le \int_{\gamma}^1 |f(x+t)|\,\delta_n(t)dt + \int_{\gamma}^1 |f(x)|\,\delta_n(t)dt \\ &\le 2M\int_{\gamma}^1 \delta_n(t)dt \ < 2M\sqrt{n}(1-\gamma^2)^n, \end{aligned} \tag{C.4}$$

where we have used the inequality (B.17). Similar arguments yield

$$\int_{-1}^{-\gamma} |f(x+t) - f(x)|\,\delta_n(t)dt < 2M\sqrt{n}(1-\gamma^2)^n. \tag{C.5}$$

Finally, the remaining integral, $\int_{-\gamma}^{\gamma}$, is estimated by using the continuity of $f(x)$, which guarantees that for any ε we can find an infinitesimal γ that satisfies

$$|t| < \gamma \ \Rightarrow \ |f(x+t) - f(x)| < \varepsilon.$$

This yields

$$\int_{-\gamma}^{\gamma} |f(x+t) - f(x)| \, \delta_n(t) dt < \varepsilon \int_{-\gamma}^{\gamma} \delta_n(t) dt < \varepsilon, \tag{C.6}$$

since $\int_{-\gamma}^{\gamma} \delta_n(t) dt < 1$.

Collecting the results of (C.3)–(C.6), we have

$$|G_n(x) - f(x)| < 4M\sqrt{n}(1-\gamma^2)^n + \varepsilon.$$

The value of $\sqrt{n}(1-\gamma^2)^n$ for $0 < \gamma < 1$ can be set arbitrarily small for large enough n and, in particular, smaller than ε. Therefore, there exists an N such that

$$n > N \;\; \Rightarrow \;\; |G_n(x) - f(x)| < \varepsilon$$

for any arbitrarily small preassigned ε where N does not depend on x. This means that the sequence of polynomials $G_n(x)$ converges uniformly to the continuous function $f(x)$ on $[0,1]$, which completes the proof. We emphasize that the above discussion holds for an arbitrary continuous function on an arbitrary finite closed interval $[a, b]$, as was indicated at the outset.

Remark. It should be emphasized that our initial hypothesis that $f(0) = f(1) = 0$ imposes *no* limitation on the validity of the proof. To see this, we now suppose that $f(x)$ is defined on $[a, b]$. Then, the function $g(x)$ defined by

$$g\left(\frac{x-a}{b-a}\right) \equiv f(x)$$

yields $f(a) = g(0)$ and $f(b) = g(1)$, where any x in the interval $[a, b]$ corresponds to z in $[0,1]$. Furthermore, by introducing the function

$$h(z) = g(z) - g(0) - z\,[g(1) - g(0)]$$

for z in $[0,1]$, we have $h(0) = 0$ and $h(1) = 0$. We can show that the polynomial $G_n(x)$ that approximates the original function $f(x)$ also approximates the modified function $h(z)$ by replacing the variable x in $G_n(x)$ by z.

D

Tabulated List of Orthonormal Polynomial Functions

Hermite Polynomials $H_n(x)$

Orthogonality:

$$\int_{-\infty}^{\infty} e^{-x^2} H_m(x) H_n(x) dx = 2^n n! \sqrt{\pi} \delta_{mn}.$$

Rodrigues formula:

$$H_n(x) = (-1)^n e^{x^2} \frac{d^n}{dx^n} \left(e^{-x^2} \right).$$

Differential equation:

$$\frac{d^2}{dx^2} H_n(x) - 2x \frac{d}{dx} H_n(x) + 2n H_n(x) = 0.$$

Recurrence formula:

$$H_{n+1}(x) - 2x H_n(x) + 2n H_{n-1}(x) = 0.$$

Generating functions:

$$g(t,x) = e^{2xt-t^2} = \sum_{n=0}^{\infty} \frac{H_n(x)}{n!} t^n.$$

Laguerre Polynomials $L_n^{\nu}(x)$

Orthogonality:

$$\int_0^{\infty} x^{\nu} e^{-x} L_m^{\nu}(x) L_n^{\nu}(x) dx = \frac{\Gamma(n+\nu+1)}{\Gamma(n+1)} \delta_{mn}.$$

Rodrigues formula:

$$L_n^\nu(x) = \frac{x^{-\nu}}{n!} e^x \frac{d^n}{dx^n} \left(e^{-x} x^{\nu+n}\right).$$

Differential equation:

$$x \frac{d^2}{dx^2} L_n^\nu(x) + (\nu + 1 - x) \frac{d}{dx} L_n^\nu(x) + n L_n^\nu(x) = 0.$$

Recurrence formula:

$$(n+1) L_{n+1}^\nu(x) - (2n + \nu + 1) L_n^\nu(x) - (n + \nu) L_{n-1}^\nu(x) = 0.$$

Generating functions:

$$g(t, x) = \frac{e^{-xt/(1-t)}}{(1-t)^{\nu+1}} = \sum_{n=0}^{\infty} L_n^\nu(x) t^n.$$

Jacobi Polynomials $G_n^{(\nu,\mu)}(x)$

Orthogonality:

$$\int_{-1}^{1} (1+x)^\mu (1-x)^\nu G_m^{(\nu,\mu)}(x) G_n^{(\nu,\mu)}(x) dx = \frac{2^{\mu+\nu+1} \Gamma(n+\mu+1) \Gamma(n+\nu+1)}{n!(2n+\mu+\nu+1)\Gamma(n+\nu+\mu+1)} \delta_{mn}.$$

Rodrigues formula:

$$G_n^{(\nu,\mu)}(x) = \frac{(-1)^n}{2^n n!} (1-x)^{-\nu} (1+x)^{-\mu} \frac{d^n}{dx^n} \left[(1-x)^{\nu+n} (1+x)^{\mu+n}\right].$$

Differential equation:

$$\begin{aligned}(1-x^2) \frac{d^2}{dx^2} G_n^{(\nu,\mu)}(x) + [\mu - \nu - (\nu+\mu+2)x] \frac{d}{dx} G_n^{(\nu,\mu)}(x) \\ + n(n+\nu+\mu+1) G_n^{(\nu,\mu)}(x) &= 0.\end{aligned}$$

Recurrence formula:

$$\begin{aligned}&G_0^{(\nu,\mu)}(x) = 1, \quad G_1^{(\nu,\mu)}(x) = \frac{1}{2}\{(\nu+\mu+2)x + (\nu^2 - \mu^2)\}, \\ &2(n+1)(n+\nu+\mu+1)(2n+\nu+\mu) G_{n+1}^{(\nu,\mu)}(x) \\ &-(2n+\nu+\mu+1)\left[(2n+\nu+\mu)(2n+\nu+\mu+2)x + \nu^2 - \mu^2\right] G_n^{(\nu,\mu)}(x) \\ &-2(n+\nu)(n+\mu)(2n+\nu+\mu+2) G_{n-1}^{(\nu,\mu)}(x) \quad = 0.\end{aligned}$$

Generating functions:

$$g(t,x) = \frac{2^{\nu+\mu}}{(1-2xt+t^2)^{1/2}\left\{1-t+(1-2xt+t^2)^{1/2}\right\}^{\nu}\left\{1+t+(1-2xt+t^2)^{1/2}\right\}^{\mu}}$$
$$= \sum_{n=0}^{\infty} G_n^{(\nu,\mu)}(x)t^n.$$

Gegenbauer Polynomials $C_n^\lambda(x)$

Orthogonality:

$$\int_{-1}^{1}\left(1-x^2\right)^{\lambda-\frac{1}{2}} C_m^\lambda(x)C_n^\lambda(x)dx = \frac{\sqrt{\pi}\Gamma(n+2\lambda)\Gamma\left(\lambda-\frac{1}{2}\right)}{n!(n+\lambda)\Gamma(2\lambda)\Gamma(\lambda)}\delta_{mn}.$$

Rodrigues formula:

$$C_n^\lambda(x) = \frac{(-1)^n\Gamma(n+2\lambda)\Gamma[\lambda+\frac{1}{2}]}{2^n n!\Gamma[n+\lambda+\frac{1}{2}]\Gamma(2\lambda)}(1-x^2)^{-\lambda+\frac{1}{2}}\frac{d^n}{dx^n}\left[(1-x^2)^{n+\lambda-\frac{1}{2}}\right].$$

Differential equation:

$$(1-x^2)\frac{d^2}{dx^2}C_n^\lambda(x) - (2\lambda+1)x\frac{d}{dx}C_n^\lambda(x) + n(n+2\lambda)C_n^\lambda(x) = 0.$$

Recurrence formula:

$$(n+1)C_{n+1}^\lambda(x) - 2(n+\lambda)xC_n^\lambda(x) - (n+2\lambda-1)C_{n-1}^\lambda(x) = 0.$$

Generating functions:

$$g(t,x) = \frac{1}{(1-2xt+t^2)^\lambda} = \sum_{n=0}^{\infty} C_n^\lambda(x)t^n.$$

Legendre Polynomials $P_n(x)$

Orthogonality:

$$\int_{-1}^{1} P_m(x)P_n(x)dx = \frac{2}{2n+1}\delta_{mn}.$$

Rodrigues formula:

$$P_n(x) = \frac{(-1)^n}{2^n n!}\frac{d^n}{dx^n}\left[(1-x^2)^n\right].$$

Index